Green Chemistry in Nanotechnology for Sustainable Synthesis, Characterization, and Applications of Nanoparticles

Nanotechnology has become one of the most revolutionary scientific domains of the 21st century, providing answers for healthcare, renewable energy, environmental cleanup, and industrial advancements. Nonetheless, the swift expansion of this technology necessitates a comparable dedication to environmental responsibility and safety. *Green Chemistry in Nanotechnology for Sustainable Synthesis, Characterization, and Applications of Nanoparticles* meets this requirement by combining green chemistry principles with cutting-edge nanoscience.

This edited collection offers an extensive summary of sustainable nanomaterial creation, starting with fundamental principles and synthesis methods based on renewable resources. It explores eco-friendly and bioinspired methods for producing gold, silver, calcium, and metal oxide nanoparticles, backed by in-depth case studies. Extensive coverage is provided on advanced characterization techniques, delivering insights into structural, morphological, and functional assessment. The book examines various applications, such as food packaging, antimicrobial surfaces, catalytic processes, energy devices, and biomedical technologies.

A specific part on nanotoxicology showcases contemporary safety evaluation techniques, including organs-on-chips and 3D cell culture systems, while conversations regarding ethics and sustainability stress responsible innovation. The combination of scientific, safety, and social factors sets this work apart in its breadth.

By connecting core concepts with real-world uses, the book acts as an essential tool for researchers, scholars, and industry experts aiming to create nanotechnology solutions that are not only efficient but also safe, sustainable, and ethically responsible. Policymakers and regulatory organizations will also appreciate it for developing frameworks that guarantee nanotechnology aids in advancing global sustainability objectives.

This book merges green chemistry with nanotechnology, enabling synthesis from sustainable pathways and characterization through advanced techniques jointly with applications and ethical considerations all under one roof. Works by world top-rated scientists propose renewable-resource-based, bioinspired, and eco-friendly routes toward synthesis of widely varying nanoparticles that make sense in real life. Case studies with applications in health care, food packaging, catalysis, and industrial plus environmental remediation work supported by the most advanced tools of nanotoxicological assessment, such as organs-on-chips and 3D cell cultures, are also presented here. Scientific novelty will be integrated with safety and sustainability so that learners can not only create new forms of nanomaterials but also "create safely." By combining breadth, depth, and practical insight, this work stands as an indispensable guide for researchers, industry professionals, policymakers, and educators shaping the future of sustainable nanotechnology.

Green Chemistry and Sustainability

This series covers books in the field of chemistry that aim to provide sustainable solutions and environment-friendly options. With real-life applications, books in this series cover range of topics including materials chemistry, industrial chemistry, and most importantly, green chemistry. Ethical use of technology is the central idea of these books where scientific advancement should be accompanied by sustainability.

Green Chemistry in Nanotechnology for Sustainable Synthesis, Characterization, and Applications of Nanoparticles
Hadagali Ashoka, B.E. Rangaswamy, Usha Desai, Rakesh Sengupta, and B. S. Maddodi

Green Chemistry in Nanotechnology for Sustainable Synthesis, Characterization, and Applications of Nanoparticles

Edited by Hadagali Ashoka, B. E. Rangaswamy,
Usha Desai, Rakesh Sengupta, and B. S. Maddodi

CRC Press
Taylor & Francis Group
Boca Raton London New York

CRC Press is an imprint of the
Taylor & Francis Group, an **informa** business

Designed cover image: Shutterstock

First edition published 2026
by CRC Press
2385 NW Executive Center Drive, Suite 320, Boca Raton FL 33431

and by CRC Press
4 Park Square, Milton Park, Abingdon, Oxon, OX14 4RN

CRC Press is an imprint of Taylor & Francis Group, LLC

ISBN: 978-1-041-08236-1 (hbk)
ISBN: 978-1-041-08237-8 (pbk)
ISBN: 978-1-003-64443-9 (ebk)

DOI: 10.1201/9781003644439

Typeset in Times
by Apex CoVantage, LLC

Contents

About the Authors

Ashoka Hadagali, a faculty member at B.M.S. College of Engineering since 2009, specializes in biotechnology and diabetes modeling. With a Ph.D. from IIT Bombay, his research bridges natural therapeutics and pharmacology. He is a passionate educator, prolific researcher, and active contributor to scientific journals and professional communities.

B. E. Rangaswamy is Registrar at Visvesvaraya Technological University, Belagavi, Karnataka. He is Professor of Biotechnology and a renowned expert in aerobiology, microbial diversity, and bio-pollution. With over 80 research papers, numerous national and international presentations, and patents, he has led impactful projects funded by premier agencies like DST, DBT, AICTE, VGST, and VTU other agencies and is the recipient of multiple prestigious awards, including the PH Gregory Award, Rajaram Bapu National Award, and Karnataka Rajyothsava Award by Davangere District. Dr. Rangaswamy has significantly contributed to sustainable technologies, public health research, and rural development through vermicomposting, beekeeping, and biodegradable product innovations, inspiring future scientists and fostering community welfare.

Usha Desai, Professor and Dean Research and Development at S.E.A College of Engineering & Technology, Bengaluru, is an accomplished researcher in biomedical signal processing, AI, and brain–computer interfaces (BCIs), with 21 years' academic experience. Dr. Desai has contributed over 70 publications in impactful journals, books, and global conferences.

Rakesh Sengupta has been a researcher and assistant professor in SIAS, Krea University, Sri City, India, since 2025. His interdisciplinary background spans cognitive and computational neuroscience and AI. He worked before at the University of Trento, Italy, and York University, Canada, as visiting scholar and postdoc, respectively. He has a growing scholarly record with 28 indexed publications, over 300 citations, and an H index of 9.

Balakrishna S. Maddodi, Academician and NSS Program Officer at Manipal Institute of Technology, India, is an expert in engineering geology, environmental management, GIS, remote sensing, solid waste management, and green chemistry, with over 27 years of influential teaching and research experience in diverse areas of education and innovation.

Contributors

Agbaogun, Kolawole I.
Institute of Physics, University of Munster
Munster, Germany

Ajjappala Brijesh
Department of Molecular Biology and Biotechnology, Keladi Shivappa Nayaka University of Agricultural and Horticultural Sciences, College of Agriculture
Shivamogga, Karnataka, India

Ali Ebrahim
Department of Civil Engineering, Vel Tech Rangarajan Dr. Sagunthala R&D Institute of Science and Technology
Chennai, Tamil Nadu, India

Athreya, Akshata G.
Department of Biotechnology, Acharya Institute of Technology
Bangalore, India

Gurudatta B V
Assistant Professor, Department of Human Genetics, National Institute of Mental Health and Neurosciences
Bangalore, India

Bandaru Manideep
Department of Biomedical Sciences, School of Health Sciences, The Apollo University
Chittoor, Andhra Pradesh, India

Benyeogor, Mbadiwe S.
Advanced Aerospace Engines Laboratory, National Space Research and Development Agency (NASRDA)
Oka Akoko, Ondo State, Nigeria

Bhat, Savithri
Department of Biotechnology, B.M.S. College of Engineering
Bangalore, Karnataka, India

Bhimavaram Keerthi
Department of Biomedical Sciences, School of Health Sciences, The Apollo University
Chittoor, Andhra Pradesh, India

Biswas, Joydeep
Department of Chemistry, Sikkim Manipal Institute of Technology, Sikkim Manipal University
Majitar, Sikkim, India

Chaitanya Lakshmi S R
Department of Biomedical Sciences, School of Health Sciences, The Apollo University
Chittoor, Andhra Pradesh, India

Jayashree D R
Department of Biotechnology, M. S. Ramaiah College of Arts, Science and Commerce
Bangalore, India

Desai, Usha
S.E.A College of Engineering and Technology
Bengaluru, India

Dheeraj Kumar
Department of Mechanical Engineering, Sikkim Manipal Institute of Technology, Sikkim Manipal University
Majitar, Sikkim, India

Dhruva Kumar
Department of Mechanical Engineering, Sikkim Manipal Institute of Technology, Sikkim Manipal University
Majitar, Sikkim, India

Dileep Kumar M G
Department of Biotechnology, M. S. Ramaiah College of Arts, Science and Commerce
Bangalore, India

Dixit, Mukta
Department of Biotechnology, B.M.S. College of Engineering
Bangalore, Karnataka, India

Dutta, Sharmistha
Department of Biotechnology, M.S. Ramaiah Institute of Technology
Bangalore, Karnataka, India

G., Lakshmi Priya
Centre for Advanced Materials and Innovative Technologies, School of Electronics Engineering, Vellore Institute of Technology
Chennai, India

G.K. Prashanth
Research & Development Centre, Department of Chemistry, Sir M. Visvesvaraya Institute of Technology
Bengaluru, India

G., Satheesh Babu
R.V. College of Engineering
Bengaluru, Karnataka, India

Gaurav Rajasekhar
Department of Biomedical Sciences, School of Health Sciences, The Apollo University
Chittoor, Andhra Pradesh, India

Ghosh, Mithun Kumar
Department of Chemistry, Medi-Caps University
Indore, Madhya Pradesh, India

Shamantha Gowda T S
Department of Biotechnology, R.V. College of Engineering
Bengaluru, Karnataka, India

Aishwarya H S
Department of Biotechnology, B.M.S. College of Engineering
Bangalore, Karnataka, India

H.S. Lalithamba
Department of Chemistry, Siddaganga Institute of Technology
Tumkur, India

H., Muktha
Department of Biotechnology, M. S. Ramaiah College of Arts, Science and Commerce
Bangalore, India

Hadagali, Ashoka
Department of Biotechnology, B.M.S. College of Engineering
Bangalore, Karnataka, India

Hemavathi Brijesh
Department of Biotechnology, School of Applied Sciences, REVA University
Bengaluru, Karnataka, India

Intisar Albander
Department of Biology, College of Sciences, University of Basrah
Basrah, Iraq

Jain, Ishika
Department of Biotechnology, M. S. Ramaiah College of Arts, Science and Commerce
Bangalore, India

Jeevan Kumar M.
Department of Biomedical Sciences, School of Health Sciences, The Apollo University
Chittoor, Andhra Pradesh, India

Johnson, Asha P
Centre for Pharmaceutical Nanotechnology, Apollo Institute of Pharmaceutical Sciences, The Apollo University
Chittoor, Andhra Pradesh, India

Jubu, Peverga Rex
Department of Industrial Physics, College of Physical Sciences, Joseph Sarwuan Tarka University
Makurdi, Benue State, Nigeria

K., Gurushantha
Department of Chemistry, Ramaiah Institute of Technology
Bengaluru, Karnataka, India

K.V. Rashmi
Department of Chemistry and Biochemistry, Jain (Deemed-to-be University), School of Science
Bengaluru, India

Ravi-Kumar Kadeppagari
Centre for Incubation, Innovation, Research and Consultancy (CIIRC), Department of Food Technology, Jyothy Institute of Technology
Tataguni, Off Kanakapura Main Road, Bengaluru, India

Kandasamy, S.
Department of Civil Engineering, Vel Tech Rangarajan Dr. Sagunthala R&D Institute of Science and Technology
Chennai, Tamil Nadu, India

Kevin M.
Department of Biotechnology, Sri Venkateswara College of Engineering
Pennalur, Sriperumbudur, Tamilnadu, India

Kumar, Anjan P
Department of Biotechnology, B.M.S. College of Engineering
Bangalore, Karnataka, India

Kumari, Neeraj
School of Biological and Environmental Sciences, Shoolini University of Biotechnology and Management Sciences
Solan, India

Kuthambakam Baby Rani
Department of Biomedical Sciences, School of Health Sciences, The Apollo University
Chittoor, Andhra Pradesh, India

Lawal, Olusegun I.
Advanced Aerospace Engines Laboratory, National Space Research and Development Agency (NASRDA)
Oka Akoko, Ondo State, Nigeria

Lighittha, P. R.
Department of Electronics and Computer Engineering, School of Electronics Engineering, Vellore Institute of Technology
Chennai, India

Liny, P.
Department of Biotechnology, Acharya Institute of Technology
VTU, Bangalore, Karnataka, India

Lokesh Ravi
Department of Biomedical Sciences, School of Health Sciences, The Apollo University
Chittoor, Andhra Pradesh, India

M., Ismail Shareef
Department of Biotechnology, Acharya Institute of Technology
VTU, Bangalore, Karnataka, India

M., Sundara Srivathsan
Department of Electronics and Computer Engineering, School of Electronics Engineering, Vellore Institute of Technology
Chennai, India

Mahapatra, Susanta Sinha
Department of Chemistry, BIT Mesra
Ranchi, India

Mahesh, R.
Department of Chemistry, R.V. College of Engineering, Bengaluru
Karnataka, India

Majhi, Shyam Sundar
Department of Chemistry, BIT Mesra, Off Campus Deoghar
Jharkhand, India

Mbachu, Nnaemeka M.
Institute of Physics, University of Munster
Munster, Germany

Mbalaha, Zendesha Stephen
Department of Industrial Physics, College of Physical Sciences, Joseph Sarwuan Tarka University
Makurdi, Benue State, Nigeria

N.P. Bhagya
Department of Chemistry, Sai Vidya Institute of Technology
Bengaluru, India

N., Pannirselvam
Department of Civil Engineering, SRM Institute of Science and Technology
Chennai, Tamil Nadu, India

Roopashree B N
Department of Chemistry, Ramaiah Institute of Technology
Bengaluru, Karnataka, India

Nair, Raji V.
Department of Chemistry, Sri Sathya Sai Institute of Higher Learning (Deemed to be University),
Anantapur Campus, Anantapur, Andhra Pradesh, India

Olaiya, Kabiru Alani
Department of Mechanical Engineering, Lagos State University of Science and Technology
Lagos, Nigeria

Prashanth Kumar H P
Department of Biotechnology and Genetics, M.S. Ramaiah College of Arts, Science and Commerce-Autonomous, Bengaluru
Karnataka, India

P., Praveen Kumar K.
Department of Biotechnology, Sri Venkateswara College of Engineering
Pennalur, Sriperumbudur, Tamilnadu, India

Sushma P S
Department of VLSI Design and Technology, NMAM Institute of Technology, Nitte (DU)
Udupi, Karnataka, India

Patil, Jagadish H.
Department of Chemical Engineering, RV College of Engineering
VTU, Bangalore, Karnataka, India

Prakash, Suraj
School of Biological and Environmental Sciences, Shoolini University
Solan, Himachal Pradesh, India

Preethi J.
Department of Biotechnology, Sri Venkateswara College of Engineering
Pennalur, Sriperumbudur, Tamil Nadu, India

Puri, Sunil
School of Biological and Environmental Sciences, Shoolini University
Solan, Himachal Pradesh, India

R., Prithvika
Department of Biotechnology, School of Applied Sciences, REVA University
Bengaluru, Karnataka, India

R., Sowbhagya
Department of Biotechnology, M. S. Ramaiah College of Arts, Science and Commerce
Bangalore, India

R. Venkatesh
Department of Physics, PSNA College of Engineering and Technology (Autonomous)
Dindigul, Tamil Nadu, India

Radha
School of Biological and Environmental Sciences, Shoolini University
Solan, Himachal Pradesh, India

Rajasekaran Subbarayan
Centre for Advanced Biotherapeutics and Regenerative Medicine, Faculty of Research, Chettinad Hospital and Research Institute, Chettinad Academy of Research and Education
Kelambakkam, India

Rao, Srilatha
Department of Chemistry, Nitte Meenakshi Institute of Technology
Bengaluru, India

Ravindra H J
Department of Physics, Shri Madhwa Vadiraja Institute of Technology and Management, Bantakal (Affiliated to VTU, Belagavi)
Karnataka, India

Roy, Snigdha
Department of Chemistry, Banwarilal Bhalotia College
Asansol, West Burdwan, West Bengal, India

Rupendra Shrestha
Department of Natural and Applied Sciences, Nexus Institute of Research and Innovation (NIRI)
Lalitpur, Nepal

S., Gopinath M.
Department of Studies in Biotechnology, Davangere University, Shivagangothri, Tholahunase, Davangere
Karnataka, India

S., Narendra Kumar
Department of Biotechnology, R.V. College of Engineering
Bengaluru, Karnataka, India

S., Pavithra
Department of Biotechnology, School of Applied Sciences, REVA University
Bengaluru, Karnataka, India

S., Pavithra Poornima
Department of Electronics and Communication, Shri Madhwa Vadiraja Institute of Technology and Management, Bantakal, (Affiliated to VTU, Belagavi)
Karnataka, India

Poornima DS
Department of Biomedical Sciences, School of Health Sciences, The Apollo University
Chittoor, Andhra Pradesh, India

S., Shashidhar
Department of Chemistry, New Horizon College of Engineering
Bengaluru, Karnataka, India

Sahu, Chandan Kumar
Centre for Incubation, Innovation, Research and Consultancy (CIIRC), Department of Food Technology, Jyothy Institute of Technology, Tataguni, Off Kanakapura Main Road
Bengaluru, India

Sampathkumar Natarajan
Department of Chemistry, SSM College of Arts and Science
Dindigul, Tamil Nadu, India

Shashank Vasanthkumar
Centre for Incubation, Innovation, Research and Consultancy (CIIRC), Department of Food Technology, Jyothy Institute of Technology, Tataguni
Off Kanakapura Main Road, Bengaluru, India

Shrividya G.
Department of Electronics and Communication Engineering, NMAM Institute of Technology, Nitte (DU)
Udupi, Karnataka, India

Srinidhi Varadharajan
Department of Biotechnology, Sri Venkateswara College of Engineering
Pennalur, Sriperumbudur, Tamilnadu, India

Srinivasarao Gopisetti
Department of Mechanical Engineering BIT Mesra, Off Campus Deoghar
Jharkhand, India

Subramani K.
Department of Physics, PSNA College of Engineering and Technology (Autonomous)
Dindigul, Tamil Nadu, India

Upadhyay, Ritesh Kumar
Department of Mechanical Engineering BIT Mesra, Off Campus Deoghar
Jharkhand, India

Varkar, Ansh
Department of Biotechnology, Acharya Institute of Technology
VTU, Bangalore, Karnataka, India

Venkatesh Sivalingam
Department of Biomedical Sciences, School of Health Sciences, The Apollo University
Chittoor, Andhra Pradesh, India

Manjunath Yadav S M
Department of Biotechnology, R.V. College of Engineering
Bengaluru, Karnataka, India

Preface

Nanotechnology is one of the most promising scientific disciplines of the twenty first century. It has the potential to revolutionize several scientific fields, including materials science, medicine, energy, food, and environmental cleanup. However, the traditional approaches to the synthesis of nanomaterials encounter problems of hazardous chemicals, toxic byproducts, and weight problems of ecological and human safety. In this scenario, the adoption of the principles of green chemistry is the most appropriate and constructive.

This book, *Green Chemistry in Nanotechnology for Sustainable Synthesis, Characterization, and Applications of Nanoparticles*, integrates several pioneering research works from various branches of disciplines to demonstrate the potential of responsible advancement of sustainable principles to nanotechnology. This book also presents the new methods of green synthesis of nanomaterials; the state-of-the-art techniques to characterize them; and their diverse applications, including but not limited to catalysis, medicine, food, electronics, energy, and environmental protection. It also attempts to solve the challenging problems of nanotoxicity and bioethics and the relevant laws that enable legal sustainable use.

The chapters in this book are organized to give a complete view:

Sustainable Synthesis Approaches: Chapters 1 to 8 focus on green chemistry principles, bioinspired strategies, renewable resource use, and eco-friendly synthetic methods. They include innovative ways to create nanoparticles like gold, calcium, and silver.

Catalysis and Advanced Materials: Chapters 9 to 12 explore processes driven by nanocatalysts, electrochemical applications, and the latest tools needed to evaluate green nanomaterials.

Industrial and Food Applications: Chapters 13 and 14 highlight how nanoparticles are changing the food industry and industrial processes while ensuring eco-safety.

Nanotoxicity and Ethical Issues: Chapters 15 to 20 discuss the molecular mechanisms of nanotoxicity, organ-on-chip platforms, safety assessments, and ethical concerns in advancing nanotechnology responsibly.

Biomedical and Healthcare Applications: Chapters 18, 19, 24, and 26 showcase innovations in therapy and medical devices powered by sustainable nanomaterials.

Emerging Nanocomposites and Engineering Applications: Chapters 21 to 23 examine hybrid nanostructures, biomedical implants, and reinforced concrete systems. These chapters show how green nanotechnology connects with engineering.

Biogenic and Resource-Based Nanomaterials: Chapters 5, 6, 7, and 25 provide examples of plant-based, microbial, and bioinspired synthesis methods with proven environmental and biomedical uses.

Together, these contributions offer a useful reference for students, researchers, educators, and industry professionals involved in green chemistry and nanotechnology. The book seeks to inspire future innovations that balance technology with environmental care and human welfare. We sincerely thank all contributing authors for their knowledge and commitment and the editorial and publishing teams for their ongoing support. We hope this volume will serve as both a resource of knowledge and a spark for more sustainable innovations in the exciting field of nanotechnology.

1 Green Chemistry Approaches for Sustainable Nanomaterial Synthesis and Catalysis

Raji V Nair

1.1 INTRODUCTION

1.1.1 Background

In the past decades, several nanostructures have revolutionized diverse fields like material science, engineering, medicine, and environmental applications. Nanomaterials, with dimensions typically in the range of 1–100 nm, possess unique properties, such as high surface area-to-volume ratios, quantum effects, and enhanced reactivity (Cao, 2004). These properties make them suitable for applications ranging from drug delivery systems and biosensors to water purification and catalysis. Primarily, two approaches have been implemented for the preparation of nanomaterials: the top-down approach and bottom-up approach (Rao et al., 2006). However, traditional methods of nanomaterial synthesis often involve hazardous chemicals and high energy inputs and generate toxic byproducts, raising significant concerns about environmental sustainability and human health risks (Jain et al., 2024). The meager knowledge of the risks associated with nanotechnology and the absence of regulatory guidelines about nanotechnology are the key concerns. On the other hand, the potential of nanotechnology is far reaching, and hence researchers are exploring nanomaterials for several applications with urgency. This calls for a paradigm shift in the synthetic strategies adopted for nanomaterials (Baig et al., 2021).

1.1.2 Green Chemistry and Nanotechnology

The concept of green chemistry provides a set of principles to reduce or eliminate the use and generation of hazardous substances in chemical processes (Anastas & Warner, 1998). When these principles are integrated into nanotechnology, they encourage the development of nanomaterials through environmentally benign synthesis routes, utilizing renewable raw materials, non-toxic solvents, and energy-efficient processes. Green nanotechnology is expected to pave the way for sustainable development by reducing health and environmental hazards (Hamers, 2017).

1.1.3 The Need for Sustainable Nanomaterials

The wide-ranging application potential of nanomaterials should not be limited on regard of environmental and health concerns. Hence it is significant to follow sustainable synthetic approaches causing less harm to the environment and living things (Sheldon et al., 2007). Sustainability needs to be maintained throughout its lifecycle, starting from raw material sourcing and synthesis to disposal or degradation. Along with the global efforts to reduce carbon footprints, decrease reliance on non-renewable resources, and create circular economies, the development of sustainable nanomaterials is crucial (Wiederrecht et al., 2023).

1.1.4 Objectives and the Scope of the Chapter

The current chapter aims to explore the application of green chemistry principles in nanotechnology, recent advancements in the eco-friendly synthesis of nanoparticles, and the use of renewable resources for nanomaterial synthesis. The chapter will also discuss the importance of green chemistry–integrated sustainable catalysis and reaction engineering strategies in reducing environmental impact while enhancing efficiency. Additionally, shortfalls in the field will be reviewed, and the future direction for sustainable green nanotechnology will be proposed.

The objectives of the chapter are

1. To review eco-friendly synthesis methods for nanomaterials, including plant-mediated, microbial, renewable resource-utilized, solvent-free, and low-energy methods.
2. To examine the advantages and challenges of employing green synthesis routes compared to conventional chemical methods.
3. To evaluate the role of sustainable nanomaterials in catalysis and reaction engineering, including applications in environmental remediation, energy conversion, and chemical transformations.
4. To identify the challenges and concerns associated with the application of green nanomaterials for catalysis.

DOI: 10.1201/9781003644439-1

1.2 PRINCIPLES OF GREEN CHEMISTRY IN NANOTECHNOLOGY

1.2.1 Overview of Green Chemistry

Green chemistry is defined as "the design of chemical products and processes that reduce or eliminate the use and generation of hazardous substances" (Anastas & Warner, 1998). The idea of "green chemistry" presents the need for developing novel reaction conditions favoring sustainability in terms of resource utilization, energy efficiency, product toxicity, operational mode of reaction, and health and environmental impacts. It will indeed provide a sustainable framework to guide innovation in nanotechnology, minimizing environmental and health hazards while maximizing material efficiency.

The 12 principles of green chemistry are discussed in Section 1.2.2.

1.2.2 Principle-Wise Application in Nanotechnology

1.2.2.1 Prevention

Avoiding waste generation is the first of the 12 principles of green chemistry. Waste prevention is better than treating it after formation. A material realizing no value and with no energy remaining unutilized can be considered waste. The waste generated can affect the environment differently depending on its nature, property, amount, or method of release (Sheldon, 2008). When using toxic raw materials for synthesis in large amounts, there exists a chance that waste will be generated into the environment. In view of nanoparticle synthesis, phyto-assisted synthesis of nanomaterials reduced the release of toxic byproducts compared to the chemical route of synthesis (Figure 1.1). For example, using aloe vera plant extracts as reducing agents for the synthesis of gold and silver nanoparticle avoids the use of reducing agents like sodium borohydride and trisodium citrate (Chandran et al., 2006).

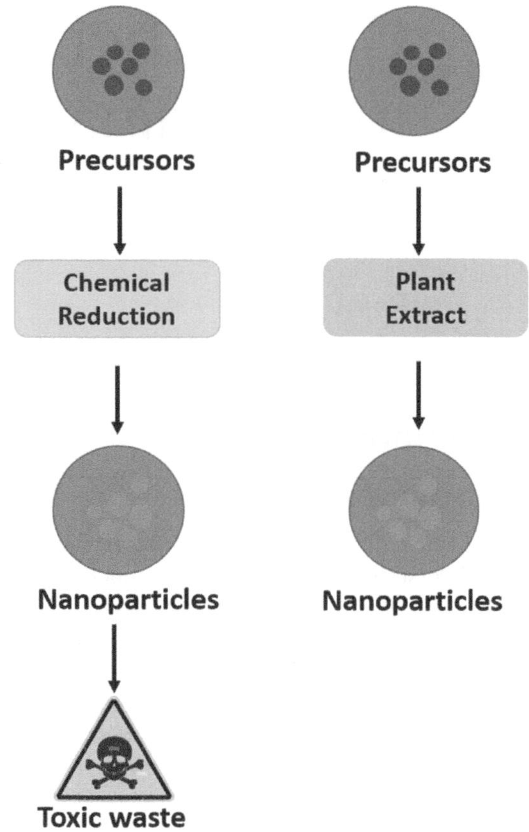

FIGURE 1.1 Conventional vs. green synthesis of nanoparticles.

1.2.2.2 Atom Economy

The concept of atom economy refers to the maximum use of raw materials so that the final product contains the maximum number of atoms from the reactants (Trost, 1991). Atomic economy is measured as the ratio of the molecular weight of the reaction product over that of all the reactants used in the reaction. In the synthesis of nanomaterials, methods like the sol-gel technique and bio-reduction processes are notable for their high atom economy. Both methods align with green chemistry principles by maximizing the incorporation of reactant atoms into the final product, reducing hazardous substances, and promoting energy-efficient processes.

1.2.2.3 Less Hazardous Chemical Synthesis

The principle of less hazardous chemical synthesis refers to design synthetic routes that are non-toxic to human health and the environment. The toxic chemicals employed in conventional reaction methods are replaced with green alternatives. For the synthesis of nanomaterials, a commonly used reducing agent is hydrazine, which is hazardous. In green nanotechnology, it can be replaced with least toxic reducing agents like plant polyphenols, amino acids, and sugars (Awan et al., 2023). Boruah et al. synthesized gold nanoparticles (AuNPs) utilizing polyphenols from *Camellia sinensis* (green tea) (2012).

1.2.2.4 Designing Safer Chemicals

Researchers have largely focused on designing chemicals for specific functions, often giving limited consideration to the toxicity of the chemicals used in these reactions. For sustainability, it is crucial to understand the safety of each chemical and its impact on the environment and human health while designing reaction pathways (DeVito & Garrett, 1996). This understanding will enable the development of environmentally benign and biocompatible products. Biocompatible nanoparticles (like chitosan-coated NPs) are examples of safer nanomaterials (Frank et al., 2020).

1.2.2.5 Safer Solvents and Auxiliaries

Solvents used in reaction systems are a major concern in green chemistry research, as they make up the lion's share of industrial waste. Use safer solvents wherever possible.

The volatility and solubility of solvents contribute to air, water, and land pollution and also affect the health of the workers, sometimes leading to serious accidents. Commonly suggested green solutions include recovery and reuse and use of solvent-less systems (Tanaka & Toda, 2000). In green nanotechnology, water and supercritical fluids are preferred over toxic organic solvents (Kerton & Marriott, 2013). Microwave- and ultrasound-assisted aqueous synthesis of ZnO, TiO_2, CeO_2, and Fe_3O_4 nanoparticles are greener alternatives (Amiri et al., 2025; Dey et al., 2025; Golestan et al., 2025; Nandhini et al., 2025).

1.2.2.6 Design for Energy Efficiency

Green chemistry supports employing chemical reactions with reduced energy barriers or with increased energy efficiency. Rising concerns over the depletion of non-renewable resources and the increase in energy consumption have urged the need for developing more energy-efficient processes and searching for renewable energy or non-depleting resources in a time frame applicable to human scale. Synthetic approaches adopted for nanomaterials like low-temperature synthesis, microwave-assisted techniques, and biological methods are consuming reduced energy. Many green nanoparticle syntheses occur at room temperature using plant extract. Room-temperature biosynthesis of silver NPs (AgNPs) using *Ocimum sanctum* extract is an example of low energy–consuming synthesis (Ramteke et al., 2013).

1.2.2.7 Use of Renewable Resources

Most reaction products are manufactured at the expense of non-renewable resources, especially petroleum products. Thus there is a concern about depletion of the resources. It is hence significant to design chemical reactions utilizing renewable resources for the synthesis of nanomaterials. Starch, cellulose, chitosan, lignin, plant polyphenols, and algae are sustainable feedstocks in nanomaterial synthesis. Cellulose-based carbon nanodots were synthesized by Ana and Camacho via the hydrothermal method (2019). Polyphenols can act as reducing and capping agents in the green synthesis of metal nanoparticles. Algae, being rich in polysaccharides and other bioactive compounds, offer a renewable source for the biosynthesis of various nanomaterials with applications in biomedicine and environmental remediation. Incorporating these renewable resources into nanomaterial synthesis not only addresses the depletion of non-renewable resources but also aligns with the principles of green chemistry by promoting atom economy, reducing hazardous substances, and minimizing environmental impact.

1.2.2.8 Reducing Derivatives

Avoiding unnecessary derivatization that requires additional reagents and generating added waste is another approach for green chemistry. Green synthesis of nanoparticles typically avoids protective groups or complex processing. Direct one-pot green synthesis of metal oxide NPs using plant extracts is an example of reduced derivatization (Lakshminarayanan et al., 2021; Sivasankarapillai et al., 2023). Plant extracts serve as both reducing and capping agents in the synthesis of metal nanoparticles, facilitating reactions under mild conditions without the need for toxic chemicals or complex procedures. For instance, the aqueous extract of *Rosa canina* L. (rosehip) has been utilized to synthesize gold and silver nanoparticles in a one-pot process at room temperature. The natural antioxidants present in the extract effectively reduce metal ions, leading to the formation of stable nanoparticles without additional reagents.

1.2.2.9 Catalysis

Catalysis is a cornerstone of green chemistry, enhancing reaction efficiency by lowering energy requirements and minimizing waste. The integration of nanotechnology and biocatalysis has further advanced sustainable practices in chemical synthesis. Catalysts are used for augmenting reaction efficiency by lowering energy inputs, so the use of catalytic reagents implies the usage of reduced raw materials and generation of less waste. Nanocatalysts are now widely employed in chemical reactions which are reusable and highly active (Somwanshi et al., 2020). Due to their high surface area-to-volume ratio, nanocatalysts exhibit superior reactivity and selectivity compared to their bulk counterparts. This heightened activity allows for reactions to proceed under milder conditions, reducing energy consumption and the need for excess reagents. At present, enzymes are used as biocatalysts for the green synthesis of nanomaterials (Buller et al., 2023). Enzymes, renowned for their specificity and efficiency, have been effectively harnessed in nanotechnology for the green synthesis of nanoparticles. By facilitating reactions under mild conditions and eliminating the need for hazardous chemicals, enzymatic methods offer an eco-friendly alternative to conventional nanoparticle synthesis techniques. This approach not only minimizes environmental impact but also allows precise control over nanoparticle characteristics, aligning seamlessly with the principles of green chemistry.

1.2.2.10 Design for Degradation

Designing biodegradable materials and chemicals breaking down into innocuous substances after use is an essential measure for meeting green chemistry targets. Biodegradable products can reduce waste accumulation and hence environmental pollution. Biodegradable polymers and natural product–based NPs degrade more safely than traditional materials (Cao et al., 2022; Zhang et al., 2014). Biodegradable polymers, including starch, gelatin, chitosan, and cellulose, have been extensively utilized in the green synthesis of nanoparticles. These natural polymers not only serve as reducing and stabilizing agents during nanoparticle formation but also ensure that the resulting nanomaterials are environmentally benign. Starch- or gelatin-stabilized NPs exhibit high adsorption capacities and

TABLE 1.1
Green Chemistry Principles in Nanotechnology

Green Chemistry Principle	Nanotechnology Application	Example
Prevention	Avoid toxic waste in NP synthesis	Plant-based AgNPs (Jain et al., 2021)
Atom Economy	Efficient conversion of precursors	Glucose-mediated AuNPs (Suvarna et al., 2017)
Less Hazardous Synthesis	Non-toxic reducing agents	Tulsi leaf extract (Gautam et al., 2017)
Safer Chemicals	Biocompatible NPs	Chitosan-coated NPs (Unsoy et al., 2012)
Safer Solvents	Water-based synthesis	ZnO NP in aqueous media (Behnajady et al., 2011)
Energy Efficiency	Room-temperature synthesis	Neem AgNPs (Verma & Mehata, 2016)
Renewable Feedstocks	Plant, algae, polysaccharides	Lignin-stabilized AgNPs (Aadil et al., 2016)
Reducing Derivatives	One-pot synthesis	Direct AuNP formation (Kim et al., 2006)
Catalysis	Reusable nanocatalysts	PdNPs in Suzuki coupling ((Pérez-Lorenzo, 2012))
Degradation	Eco-friendly breakdown	Starch-based FeNPs (Dasgupta et al., 2022)
Real-Time Analysis	UV-vis, fluorescence	*In-situ* NP monitoring (Ferreira et al., 2022)
Accident Prevention	Safer procedures	Low-temperature synthesis (Khorrami et al., 2019)

degrade safely in soil and water, minimizing ecological impact (Sooch & Mann, 2021).

1.2.2.11 Real-Time Analysis for Pollution Prevention

In green chemistry, the analytical procedures involved in a study should generate less waste and not cause environmental or health hazards (Rocha et al., 2001). This approach emphasizes the use of safer solvents, reduction of sample sizes, and elimination of hazardous reagents. Flow analysis strategies, such as sequential injection and multicommutation, have been developed to achieve these goals by enabling reagent-less procedures and online waste treatment or recycling. The monitoring of reactions to prevent formation of hazardous substances is significant for maintaining green principles. Spectrophotometric analysis like *in-situ* UV-vis spectroscopy and fluorescence spectroscopy helps in tracking nanoparticle formation. Recent studies have highlighted the use of fluorescent graphene quantum dots in fluorescence sensing, discussing various detection mechanisms and the challenges associated with their synthesis and application. The real-time analysis in the progress of formation of biosynthesized quantum dots (QDs) is generally effected through fluorescence (Patra et al., 2023).

1.2.2.12 Inherently Safer Chemistry for Accident Prevention

Chemistry working environments are often dangerous, involving highly volatile and flammable solvents and also pressurized systems. While designing chemical reactions, steps should be taken to avoid the potential for accidents such as the case of the Bhopal tragedy (Mishra et al., 2009). The use of hydrothermal plant-based synthesis of nanomaterials improves safety (Singh et al., 2025). This method involves conducting reactions in sealed vessels at elevated temperatures and pressures utilizing water as the solvent. Such conditions not only reduce the need for hazardous organic solvents but also enhance reaction efficiency and product purity. Moreover, the use of plant extracts as reducing and stabilizing agents eliminates the necessity for toxic chemicals, further enhancing the safety profile of the synthesis process.

1.2.3 Summary Table

A summary of this section is provided in Table 1.1.

1.3 RENEWABLE RESOURCES FOR NANOMATERIAL SYNTHESIS

Nanomaterial synthesis traditionally relies on synthetic chemicals and solvents derived from fossil fuels. However, sustainable nanosynthesis includes the synthesis of nanomaterials using natural processes and principles to limit the usage of toxic substances and reduce the generation of waste. Green synthesis aims at lowering the adverse environmental impacts of conventional synthesis methods. At the same time, beneficial natural processes like self-assembly and bio-mineralization provide important insights into the nanocircular economy, providing more advantages. The concept of recycling and reusability encourages the optimum economic use of resources. Also, in the green chemistry context, focus is shifting to renewable biological resources. These resources are biodegradable, abundant, cost effective, and environmentally benign—making them ideal for sustainable nanoparticle synthesis.

Common renewable sources used for the synthesis of nanomaterials are:

- **Plant extracts** (rich in phytochemicals)
- **Agricultural residues** (rice husk, banana peel, coconut shell)
- **Biopolymers** (chitosan, cellulose, starch)
- **Algae and microbes**

1.3.1 Plant-Based Synthesis of Nanomaterials

Researchers are interested in biological synthesis, including the use of natural reducing, capping and stabilizing agents, without using toxic, expensive chemicals and high energy consumption. Nowadays, nanomaterials are potential candidates for biological applications, and hence biocompatibility of nanoparticles is a major factor of concern. Phytomediated synthesis of nanoparticles is considered the most reliable environmentally sustainable synthesis method and is expected to produce biocompatible nanoparticles. Plants are rich in biomolecules like flavonoids, alkaloids, terpenoids, phenolics, and sugars that are excellent reducing agents (Aslam et al., 2021; Kandar, 2020; Ovais et al., 2018). These phytoconstituents can act as a scaffold, thereby playing the role of reducing as well as capping agents in the biosynthesis of nanoparticles (Singh et al., 2023). Figure 1.2 illustrates a schematic representation of green synthesis using plant extracts.

The advantages associated with phyto-assisted green synthesis are biocompatibility, ease of availability, scalability, and energy efficiency (Bao et al., 2021). Now material scientists are focusing on green nanoparticles to be utilized for various applications. Table 1.2 lists examples of nanoparticles synthesized using plant extracts.

1.3.2 Agricultural Waste as Raw Material

As part of the utilization of waste generated, agro-wastes are being explored for the synthesis of nanomaterials. Agro-wastes are abundant and rich in carbonaceous and functional compounds that can act as reducing, capping, or carbonizing agents. For example, as listed in Table 1.2, areca nut husks, lemon peels and litchi peels, the waste dumped after farming or consumption, are employed for the synthesis of palladium, titanium dioxide, and copper nanoparticles. Similarly, banana peels are also used for the synthesis of nanoparticles. Banana peels are composed of biopolymers such as cellulose, hemicelluloses, pectin, lignin, and proteins and could be efficiently used for the synthesis of NPs. Pectin present in the peel plays a major role in the surface functionalization of NPs. More examples are:

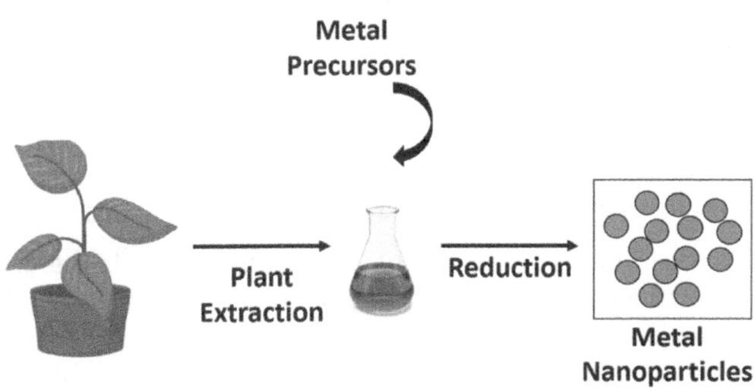

FIGURE 1.2 Green synthesis of nanoparticles using plant extracts.

Created in https://BioRender.com

TABLE 1.2
Nanoparticles Synthesized Using Plant Extracts

No.	Plant	Plant Part	Nanoparticle Synthesized	Reference
1	*Achillea biebersteinii*	Aerial parts	Silver	(Chahrdoli et al., 2020)A
2	*Aloe vera*	Flower	Copper	(Karimi & Mohsenzadeh, 2015)
3	*Areca catechu*	Husk	Palladium	(Hegde et al., 2021)
4	*Citrus limon*	Peel	Titanium dioxide	(Nabi et al., 2022)
5	*Litchi chenesis*	Peel	Carbon	(Chaudhary & Bhowmick, 2015)
6	*Ocimum sanctum*	Leaves	Platinum	(Prabhu & Gajendran, 2017)
7	*Prosopis juliflora*	Leaves	Zinc oxide	(Mydeen et al., 2020)
8	*Stevia rebaudiana*	Leaves	Gold	(Sadeghi et al., 2015)

- **Rice husk** → SiO_2 nanoparticles (Agi et al., 2020)
- **Banana peels** → Ag, ZnO nanoparticles (Al-Khaial et al., 2024; Goh et al., 2023)
- **Coconut shells** → Activated carbon for nanoparticle support (Jamaludin et al., 2023)
- **Sugarcane bagasse** → Carbon quantum dots (Luo et al., 2024)
- **Date seeds** →ZnO nanoparticles (Hussien, 2023)

While using agricultural waste as raw material, it uses a variety of green chemistry values, such as valorization of waste, circular economy, and zero-waste synthesis strategies (Flores-Contreras et al., 2024).

1.3.3 Biopolymers in Nanoparticle Synthesis

Biopolymers like chitosan, chitin, starch, gelatin, cellulose, pectin, xanthan gum, collagen, hyaluronic acid, silk, fibrin, casein, and alginate are now used as stabilizers and templates during nanoparticle formation. Biopolymer-capped nanoparticles find applications in drug delivery and bio-imaging, as they improve biocompatibility and aid in targeted delivery (George et al., 2020). Chitosan, an exoskeleton of shellfish, is a widely used biopolymer with cationic properties with comparable antimicrobial and antioxidant properties. Chitosan–nanoparticle composites act as potential candidates for various biomedical, environmental, and food industry applications (Mohammad et al., 2017). Chitosan used as a stabilizing agent in the synthesis of iron oxide nanoparticles has been demonstrated to possess increased antimicrobial activity. Increased antimicrobial activity was due to the increased stability of the iron oxide nanoparticles owing to the coating with chitosan (Nehra et al., 2018).

1.3.4 Algae and Microorganisms as Bio-Factories

Microalgae, bacteria, and fungi can biosynthesize nanoparticles intracellularly or extracellularly by secreting enzymes or metabolites (Shah et al., 2025). Among the various biological sources for the green synthesis of nanoparticles, green synthesis mediated by microorganisms has acquired a distinct place due to their high growth rate, ease of cultivation, and ability to grow in ambient conditions of temperature, pH, and pressure (Ali et al., 2020). Self-replicating systems, non-toxic byproducts, eco-friendly growth conditions, and versatility are the benefits associated with microbial synthesis of nanoparticles (Singh et al., 2016). A few examples of nanoparticles synthesized using microorganisms as reducing agents are listed in Table 1.3.

1.3.5 Challenges and Considerations

While the synthesis methods are promising, a few challenges are included in biosynthesis based on renewable resources. The standardization of reaction conditions and optimization of pH, concentration, and temperature is crucial, as a small change in these parameters can affect the size and shape of the nanoparticles. Slow production rates and the risk factors associated with the purification of nanoparticles can remain as an obstacle in the scaling up of products from lab scale to industry level (Altammar, 2023).

1.3.6 Future Prospects

The integration of bioinformatics tools into the selection of optimal bio-resources can reduce the wastage of resources and time. Additional techno-economic assessments are needed for the scaling-up procedure for the synthesis of nanoparticles. A consolidated regulatory framework needs to be put forward to support the green synthesis of nanomaterials.

TABLE 1.3
Microorganisms Used for the Synthesis of Nanoparticles

No.	Organism	Nanoparticle Synthesized	References
	Bacteria		
1	*Escherichia coli*	Pt, Pd	(Lin et al., 2014)
2	*Pseudomonas auriginosa*	Au	(Husseiny et al., 2007)
3	*Lactobacillus acidophilus*	Ag	(Rajesh et al., 2015)
	Fungi		
4	*Aspergillus fumigatus*	Ag	(Shahzad et al., 2019)
5	*Fusarium oxysporum*	CdS	(Ahmad et al., 2002)
6	*Neurospora crassa*	Pt	(Castro-Longoria et al., 2012)
7	*Penicillium chrysogenum*	Au	(Magdi & Bhushan, 2015)
	Algae		
8	*Spirulina platensis*	Au	(Kalabegishvili et al., 2012)
9	*Bifurcaria bifurcate*	CuO	(Abboud et al., 2014)
10	*Ulva fasciata*	ZnO	(Alsaggaf et al., 2021)

1.4 ECO-FRIENDLY SYNTHESIS OF NANOPARTICLES

Traditional synthesis of nanoparticles often involves hazardous chemicals, high energy input, and complex waste management. Green synthesis aims to overcome these limitations by using non-toxic, energy-efficient, and biodegradable methods. Eco-friendly synthesis broadly includes:

- **Biological synthesis** (plants, microbes, algae)
- **Low-energy physical methods**
- **Hybrid approaches** using biopolymers or waste materials

1.4.1 Biological Synthesis Methods

Plant extracts contain bioactive molecules (flavonoids, terpenoids, phenols, alkaloids) that serve as reducing and stabilizing agents. In plant-mediated biosynthesis, prepared aqueous plant extracts were mixed with metal salt precursors and subjected to reduction at room temperature. The nanoparticles formed were collected via ultracentrifugation. For instance, *Azadirachta indica* leaf extract reduces Ag^+ to Ag^0 nanoparticles at room temperature (Shakeel Ahmed et al., 2016).

Microorganisms like bacteria, fungi, and yeast are capable of enzymatic or metabolic nanoparticle reduction. Two modes of synthesis occur in microorganisms, intracellular synthesis and extracellular synthesis. In intracellular synthesis, the metal ions in precursors will enter the cell, and the reduction occurs intracellularly. On the other hand, secreted enzymes from the microorganisms reduce the metal ions outside the cell in extracellular synthesis. In *Bacillus subtilis,* extracellular gold nanoparticles are synthesized by reductase enzymes secreted by them (El-Shanshoury et al., 2011). Algae offer a unique platform due to their rapid growth, photosynthetic efficiency, and biochemical diversity. *Chlorella vulgaris* is used to synthesize gold and silver nanoparticles via extracellular routes (Rajkumar et al., 2021).

1.4.2 Low-Energy Physical Methods

Physical methods are often energy intensive, but green adaptations are still possible. Microwave- and ultrasound-assisted synthesis are environmentally friendly low-energy methods. Monodispersed nanoparticle synthesis can be achieved in these methods with a short reaction time and at the expense of a reduced amount of solvent (Ahmed et al., 2023). In ultrasound-assisted synthesis, the ultrasound induces acoustic cavitation, generating localized high temperature and pressure without bulk heating. Zou et al. achieved ultrasound-assisted synthesis of CuO nanoparticles using cotton fibers as a template agent (Zou et al., 2012).

1.4.3 Hybrid Green Synthesis Techniques

Another eco-friendly approach is combining biological materials with green physical techniques for improved efficiency and particle control. For example, a sonochemical-assisted synthesis strategy has been adopted for copper oxide nanoparticles by using the aqueous extract of the *Ficus johannis* plant (Rajabi et al., 2023). As pure biological methods fail in synthesizing monodispersed nanoparticles, hybrid approaches can improve control over size and shape of the nanoparticles. Reaction rate, product yield, stability, scalability, and biocompatibility can be improved by hybrid green synthesis techniques.

1.4.4 Characterization of Green Synthesized Nanoparticles

In nanomaterial synthesis, precise characterization is required to confirm their formation (Kumar & Dixit, 2017). The morphology of the nanoparticles is visualized by electron microscopy techniques like transmission electron microscopy (TEM) and scanning electron microscopy (SEM). Composition analysis is generally employed by Fourier-transform infrared spectroscopy (FTIR), UV-vis spectroscopy, X-ray photoelectron spectroscopy (XPS), and energy dispersive X-ray analysis (EDX). Crystal structure is studied by X-ray diffraction (XRD). The dynamic light scattering technique is adopted for understanding hydrodynamic diameter and surface charge. Noble metal nanoparticles have a unique surface plasmon resonance (SPR) peak. For example, in the case of silver nanoparticle formation, the SPR peak obtained from UV-vis spectra will confirm the formation of AgNPs. An SPR peak around 420 nm is highly specific for AgNPs (Patil & Chougale, 2021).

1.4.5 Limitations and Challenges

- Difficulty in **standardizing extract concentrations**
- Variability in **particle shape and yield**
- **Scale-up** issues in microbial synthesis
- Lack of **automated process control**

Thus, eco-friendly synthesis of nanoparticles is a rapidly growing domain at the intersection of green chemistry and nanotechnology. It offers a low-cost, sustainable, and environmentally benign route to nanomaterials, ideal for biomedical, agricultural, and environmental applications.

1.5 SUSTAINABLE CATALYSIS AND REACTION ENGINEERING

Catalysis a green chemistry principle, and sustainable catalysis refers to the design of catalysts and catalytic processes that minimize environmental impact and use renewable non-toxic feedstock. The catalyst of a reaction should be efficient, selective, durable, and reusable. They can maximize atom economy, save energy, and produce less waste (Gómez-López et al., 2020).

1.5.1 Green Catalysts: Design and Characteristics

A green nanocatalyst is a catalyst at nanoscale designed for implementing green chemistry principles. They are meant for utilizing renewable resources and possess high efficiency, selectivity, and turnover frequency. Green nanocatalysts will be more stable under mild reaction conditions and generally minimize byproduct formation (Shi et al., 2017). Heterogeneous gold, palladium, and copper nanoparticles using hydrotalcite as the support material have been widely used for environmentally benign organic transformations such as aerobic oxidation of alcohols, lactonization of diols, and deoxygenation of epoxides and nitro aromatic compounds (Abad et al., 2005; Mitsudome et al., 2008; Mori et al., 2004). They are widely accepted due to their high catalytic activities and selectivities, wide applicability, and reusability. Replacing these chemically stabilized metal nanoparticles with nanoparticles stabilized by phytoconstituents can improve the sustainability of the reactions. A few examples of different types of green catalysts are listed in Table 1.4.

TABLE 1.4
Different Types of Green Catalysts

Type	Description	Examples	References
Metal nanoparticles	Noble/transition metals, often stabilized with plant-based ligands	Pd, Ag, Au, Cu	(Bej et al., 2016), (Salehi-Khojin et al., 2013), (Sharma et al., 2007), (Hang et al., 2016)
Metal oxide nanoparticles	Common in photocatalysis, redox catalysis	TiO_2, ZnO, Fe_3O_4	(Bagheri et al., 2014)
Carbon-based nanomaterials	Derived from biomass; support or active site	Graphene, CNTs, biochar	(Mohapatra et al., 2023)
Biopolymer-stabilized	Embedded in chitosan, alginate, gelatin	Chitosan–Ag, Cellulose–Pd	(Murugadoss & Chattopadhyay, 2007), (Baran et al., 2018)
Enzyme-nano hybrids	Bio-catalysts immobilized on nanostructures	Lipase@ZnO, HRP@Au NPs	(Shah et al., 2015), (Grosu et al., 2019)

1.5.2 Nanocatalysts in Green Reactions

1.5.2.1 Photocatalysis

Photocatalysts use the light energy, sunlight or visible light, to accelerate a chemical reaction without being consumed during the process. Photocatalysts are generally used in water splitting, pollutant degradation, and CO_2 reduction. When exposed to sunlight, they absorb photons and get excited. Upon excitation, electron–hole pairs will be generated, which drive redox reactions. Nanophotocatalysts are chemically stable and active over many cycles. Some examples of green nanophotocatalysts are TiO_2–NPs green-synthesized using lemon peel extract (Narh et al., 2024), ZnO synthesized using *Padina pavonica* extract (Alprol et al., 2024), and Fe_2O_3@biochar systems (Herath et al., 2022).

1.5.2.2 Hydrogenation and Oxidation

Nanocatalysts are extremely efficient for hydrogenation and oxidation, as they have high surface area, increased active sites, and unique electronic properties. Nanocatalysts are superior to other catalysts as they enable hydrogenation (the addition of hydrogen to unsaturated organic compounds) at lower temperature and pressure with improved selectivity. Noble metal NPs (Au, Pt, Pd) supported on green carriers and bimetallic nanoparticles are the generally used green nanocatalysts (Imran et al., 2017; Zhao et al., 2021). Similarly, nanocatalysts promote oxidation using mild oxidants (like air or H_2O_2) under green conditions (Sadri et al., 2014).

1.5.2.3 Coupling Reactions

Coupling reactions are vital in organic chemistry, forming carbon–carbon (C-C) and carbon–heteroatom (C-X) bonds. Common coupling reactions are Suzuki, Heck, Sonogashira, and Buchwald–Hartwig reactions. Pd-loaded nanocellulose-based materials are used in Suzuki coupling (Riva et al., 2022).

1.5.3 Reactor Design for Sustainable Nanocatalysts

Modern green nanocatalytic processes go beyond the material, and they incorporate reaction engineering innovations to reduce waste and increase efficiency. Hence, reactor design plays a critical role in achieving efficient, scalable, and eco-friendly production and use of nanocatalysts. A well-designed reactor ensures minimal energy input, maximum catalyst performance, waste reduction, and safe operation, all aligning with green chemistry principles (Feng et al., 2022; Kim et al., 2018; Santana & Skrabalak, 2020).

1.5.3.1 Continuous Flow Systems

Continuous flow reactors (CFRs) work by constantly adding reactants to the system and taking out the products. This method has many benefits compared to traditional batch processes. In nanocatalysis, CFRs are especially useful because they help keep the right mix of chemicals, which lowers the need for solvents and reduces extra reactants. Additionally, these reactors require less energy because they transfer heat and matter more efficiently. The method is inherently safer when reactive or hazardous compounds are used in small amounts. CFRs are ideal for mass production of nanocatalysts due to scalability and can efficiently scale up from laboratory scale to industrial plant without any quality loss in the final product. Using tools like FTIR and UV-vis spectroscopy during the process lets scientists monitor and adjust the reaction in real time. Moreover, CFRs are perfect for making nanoparticles continuously, as they allow precise control over how particles form and grow, which is important for getting the right size and shape.

1.5.3.2 Microreactors

Microreactors are reaction systems on a small scale, with channel sizes in the micrometer range, and provide a very large surface-to-volume ratio, thus remarkably promoting heat and mass transfer. This provides unprecedented control over reaction system parameters (e.g., temperature, pressure, and residence time) and thus makes microreactors suited for application in nanocatalysis where uniform particle size and shape are required. These advantages are particularly useful for photocatalytic and enzymatic reactions, which need continual light illumination or a precise environment for the reaction. The use of microreactors also allows the fast screening of optimal catalytic conditions and formulations, enabling faster development and optimization of nanocatalysts. Moreover, their modularity allows simple incorporation with automated and continuous flow setups, which is ideal for basic research and scale-up applications. Microreactors can thus make a valuable contribution to the realization of green and sustainable catalysis by reduced reagent consumption and allowing safe and efficient processing.

1.5.3.3 Immobilized Catalyst Reactors

Nanocatalysts can be firmly immobilized within solid supports such as silica, alumina, polymers, or biocompatible materials such as alginate beads, enabling reuse in the catalyst operation while preventing leaching of the catalyst. Immobilization leads to increased stability, durability, and dispersion of the catalyst, leading to enhanced catalytic efficiency. These reactors are often designed as fixed or packed-bed systems in which a flow of the reactant stream passes through a matrix of stationary catalyst, thus providing continuous operation while avoiding a series of recovery steps for catalyst removal. The use of immobilized catalysts is especially valuable in heterogeneous catalysis and biocatalysis, where they guarantee operational robustness and minimum contamination. Cleaning the product streams with an immobilized catalyst would, therefore, enhance the principles of process intensification and waste minimization. Furthermore, immobilized systems can work under batch or continuous flow processes, enhancing versatility and sustainability in nanocatalytic applications.

1.5.4 Recyclability and Lifecycle of Nanocatalysts

Recyclability and lifecycle are critical factors while considering the green viability of nanocatalysts. Recyclability refers to the potential of a catalyst to be recovered and reused for several cycles without losing its catalytic activity, sensitivity and selectivity. Green metrics are the quantitative tools for evaluating the efficiency and sustainability of nanocatalysts (Kowtharapu et al., 2023). For examples, bio-synthesized AgNPs demonstrated recyclability for about seven cycles toward 4-nitrophenol reduction without any significant reactivity loss (Nanaei et al., 2019). Common methods employed for recovery are centrifugation and membrane filtration. Magnetic separation is adopted for magnetic nanoparticles like iron oxide. Lifecycle of a catalysts refers to the fate of a nanomaterial from synthesis to disposal. Lifecycle analysis of a nanoparticle helps to quantify energy use, environmental impact, and overall sustainability (Nizam et al., 2021).

Several risk factors associated with nanocatalysts, like leaching and environmental toxicity, can be prevented by using green synthesized nanoparticles stabilized with biocompatible functionalities. Furthermore, proper design for biodegradation or recyclability can contribute to sustainability of catalysts.

1.6 APPLICATIONS OF GREEN NANOMATERIALS IN INDUSTRY AND ENVIRONMENT

The present global scenario raises concerns regarding environmental pollution, resource conservation, and demands for sustainability and environmental accountability. Industries are actively exploring green nanotechnology methods to replace conventional materials that are energy-intensive, non-biodegradable, or hazardous. Green nanomaterials synthesized via eco-friendly methods offer several advantages over conventionally synthesized nanomaterials. Reduced ecotoxicity, lower carbon footprint, enhanced biocompatibility, biodegradability, and recyclability are a few among them.

1.6.1 Industrial Applications of Green Nanomaterials

1.6.1.1 Pharmaceutical Applications

The use of nanoparticles in pharmaceutical applications offers numerous advantages over traditional drug delivery systems. The targeted delivery of therapeutic agents to specific tissues is possible with nanopharmaceuticals with reduced side effects. Nanoformulations enhance the stability and bioavailability of active pharmaceutical ingredients after administration and possess superior safety and efficacy profiles over non-targeted modes of drug delivery. Controlled drug release over desired timeframes, passive targeting, and drug accumulation in malignant tumors and other pathological sites via the enhanced permeability and retention (EPR) effect are the major advantages of nanoparticle-mediated targeted drug delivery. Moreover, nanopharmaceutical products are more cost effective compared to their traditional counterparts (Petrovic et al., 2024). The development of green nanotechnology seems to be a potential alternative for diseases diagnosis and treatment by preparation of various sized and shaped nanomaterials. Green-synthesized nanoparticles are widely studied for antimicrobial, anticancer, and drug delivery potential. Furthermore, green-synthesized metal and metal oxide nanoparticles have been shown to possess prominent physical, chemical, and biological properties that can be efficiently utilized for pharmaceutical and biomedical applications. For instance, AuNPs and AgNPs synthesized using plant extracts proved to be biocompatible, being stabilized by phytoconstituents. For example, silver nanoparticles from *Scutellaria barbata* aqueous extract were used in wound healing (Veeraraghavan et al., 2021). Gold nanoparticles synthesized using the halophilic microalga *Dunaliella salina* exhibited selective cytotoxicity toward the MCF-7 breast cancer cell line while remaining nontoxic to the normal MCF-10A cell line (Singh et al., 2019). Biopolymers such as alginate and gelatin are also used in the delivery of probiotic bacteria to the intestine. *Faecalibacterium prausnitzii* encapsulated layer-by-layer with glycol chitosan and riboflavin-conjugated sodium alginate was used as a probiotic carrier (Qiu et al., 2024).

1.6.1.2 Agricultural Applications

Recently nanomaterials have gained attention in agricultural applications, with several studies showing significantly better crop growth and yield as compared to cultivation using traditional fertilizers and pesticides. Also, nanomaterials have presented promising results in solving problems associated with conventional fertilizers, such as low nutrient uptake efficiency and unregulated nutrient release with no precise control of nutrient delivery. Moreover, nanomaterials have positively impacted plant growth, with higher chlorophyll content in leaves, increased root/shoot lengths, and better stress tolerance. Furthermore, nanomaterials are extremely capable of disease detection in plants and soil remediation (Singh et al., 2022). Green nanomaterials also have the potential to transform agriculture by improving crop yield, combating diseases, and aiding in pest management. Thus, green nanotechnology is revolutionizing agriculture and horticulture through precision farming, controlled release fertilizers, nanopesticides, and soil remediation. Zn-encapsulated chitosan nanoparticles with antifungal and seed growth–promoting activity have been tested for maize crop cultivation to avert the Zn deficiency problem in soil (Choudhary et al., 2019). Similarly, smart green nanofertilizers have been applied in agriculture to improve crop production. Banana peel nanofertilizer extract has been substantiated as a biostimulant in the tomato and fenugreek germination process (Hussein et al., 2019).

1.6.1.3 Food Industry

Nanotechnology is used for developing eco-friendly nanoparticles and is a competent method to purify wastewater in the food industry, thereby improving food quality. Nanoparticles are widely employed in nanopackaging, nanonutraceuticals, and nanocarriers (Osman et al., 2024). In the food industry, the main objective is to protect the safety and freshness of food, and maintenance of the freshness of fruits and vegetables for extended periods presents a critical challenge due to potential microbial contamination and inherent natural deterioration. Consequently, there is a persistent need to improve the food packaging process to extend the shelf life of these products (Ezhilarasi et al., 2013). Green nanomaterials have been extensively applied to the food sector, including anticaking agents, nanoadditives, and antimicrobial agents. They are valuable in food packaging due to their potential to act as quality sensors, capable of monitoring and recording food quality. Green nanomaterials offer new opportunities for food-industry applications, such as nutritional additives, stronger flavorings, and antibacterial ingredients for food packaging (Edo et al., 2025; Sani et al., 2022; Shafiq et al., 2020). Green nanomaterials can extend shelf life and provide real-time spoilage indicators (Edo et al., 2025). A green, safe food packaging paper with enhanced barrier and antibacterial activity has been developed using a new natural coating formulation of carboxy methyl cellulose and cellulose nanocrystal immobilized AgNPs (He et al., 2021).

1.6.1.4 Textile Industry

In the textile sector, green nanomaterials provide antimicrobial, UV-protective, and self-cleaning properties without toxic chemical treatments (Kumari et al., 2024). Silver-coated fabrics have been evaluated for a number of properties such as electrical conductivity, physiological comfort, antibacterial, and durability (Ali et al., 2018). These properties, achieved using green-synthesized nanoparticles, can ensure sustainable development. Nanoparticles synthesized using plant extracts are used in the development of bioactive textiles, with emphasis on the utilization of citrus fruits and plant extracts (R. Ali et al., 2023). Durable protection against ultraviolet radiation has also been achieved by means of metal oxide nanoparticles synthesized using the low-temperature sonochemical method and its deposition on porous textiles for a broad range of functional applications. The ultraviolet-resistant property of TiO_2 has been exploited for modifying cotton fibers to develop durable ultraviolet resistance properties (Morshed et al., 2020).

1.6.1.5 Cosmetics and Personal Care

Nanomaterials are used in various cosmetic applications like body care, hair care, decorative cosmetics, skin care, sun care, and oral care (Fytianos et al., 2020). Nanomaterial-based cosmetics show unique advantages compared to micro-scale cosmetics. The use of nanomaterials by the cosmetic industry aims for long-lasting effects and increased stability. The high surface area of nanomaterials allows for more efficient transport of the ingredients to the skin. Some of the main targets of using nanomaterials in cosmetics could be efficient penetration into skin for improved delivery of the ingredients of the product, new color elements (e.g., in lipsticks and nail polishes), transparency (e.g., in sunscreens), and long-lasting effects (e.g., in makeup). The ultimate goal of the cosmetic industries when using nanomaterials is to deliver the right amount of ingredients to the desired parts of the body and to attain long-term stability. Currently, the most common use of nanomaterials in cosmetics is in skincare products, particularly in sunscreens to act as UV filters (Ahmad et al., 2018). But the long-term use of nanoparticles is a great worry for consumers. Replacing nanomaterials with green nanoparticles can enhance biocompatibility and sustainability. Eco-friendly nanomaterials such as green-synthesized ZnO are used in sunscreens as UV blockers (Elbrolesy et al., 2023), face masks as antibacterial agents (Qadir et al., 2024), and anti-aging creams as nanoemulsions for delivery (Koppolu, 2017).

1.6.2 Environmental Application

1.6.2.1 Wastewater Treatment

The world's population is affected by a shortage of safe water for drinking purposes, and clean water is the primary requirement due to population growth and prolonged drought, so wastewater is treated and reused. Wastewater treatments are done by implementing traditional methods, including physical, chemical, and biological treatment units, but there is still an urgent need to employ advanced treatment processes for treating wastewater to meet the demands for clean water. Nanotechnology is developing as an alternative to traditional wastewater treatment methods with large surface area-to-mass ratios, and nanoparticle-based materials have high adsorption capacity, but the toxicity of nanoparticles is a great concern. Hence, green nanoparticles are providing an alternative method for elimination of pollutants from water bodies (Neeti et al., 2023). The application of green nanoparticles provides an economical, convenient, and eco-friendly way of wastewater treatment. Different types of biosynthesized nanomaterials are used in the controlling of water pollution, such as zinc oxide nanoparticles, phytogenic magnetic nanoparticles, biopolymer-coated metal nanoparticles, silver-impregnated cyclodextrin nanocomposites, silver nanoparticles in aloe vera plant extract, and *Photocatalytic titania* nanoparticles (Singh et al., 2020). Green nanomaterials like iron oxide, biochar-supported metal NPs, and graphene oxide show remarkable efficiency in removing heavy metals (Pb, Cd, As), organic pollutants (dyes, pesticides), microbes, and pathogens (Devatha et al., 2016; Obayomi et al., 2023; Taha & Daffalla, 2023).

For example, activated carbon magnetic composite from banana peel (*Musa acuminata*) has been employed for recovering iron metal ions (Yanti et al., 2023). Green-synthesized zinc oxide nanoparticles using *Phoenix dactylifera* waste were also utilized as a bioreductant for effective dye degradation and antibacterial performance in wastewater treatment (Rambabu et al., 2021).

1.6.2.2 Air Purification

Green nanomaterials can serve as nanocatalysts to treat toxic organic pollutants in a safer way, including photodegradation to less or nontoxic products. Photocatalytic nanomaterials synthesized using green methods are effective in breaking down volatile organic compounds (VOCs) and NOx and SOx gases (Saika Ahmed et al., 2023; Sharma et al., 2022). Nitrous oxide (N_2O) and VOCs are a group of air contaminants that reduce air quality. Benzene, toluene, ethylbenzene, and xylene (BTEX) are important components of VOCs and NOx. They can cause respiratory diseases, eye and skin irritation, and memory loss (Park et al., 2014). Green nanomaterials are used as excellent adsorbents, catalysts, and sensing materials due to their large specific surface areas and high reactivity (Sharma et al., 2022). A chitosan/poly(vinyl alcohol) air filtration membrane fabricated via green approach of electrospinning and decorated with silica nanoparticle and silver nanoparticles (synthesized *in-situ* by UV curing) was proved to possess excellent filtration performance and efficient antimicrobial properties (Zhu et al., 2019).

1.6.2.3 Soil Remediation and Detoxification

Heavy metal contamination of agricultural soil is a major problem, as it is rapidly increasing. Heavy metal contamination can have an effect on the growth and development of plants. With increased awareness through international conventions, there has been a growing need to eliminate persistent contaminants from soil through effective bioremediation and proper management of contaminated sites that have been impacted by the deposition of toxic chemical pollutants. The biomagnification and bioaccumulation of these toxic compounds pose a greater risk for the cycle of the food chain and are more persistent in tertiary consumers, so there is a need to remove these pollutants at the initial stage. Green synthesis can be viewed as an effective tool to reduce the detrimental effects of traditional nanoparticle synthesis methods commonly used in laboratories and industries. Biosynthesis of NPs and the use of nanobiotechnology help to remediate heavy metal–contaminated soil. Green-synthesized nanoparticles are used to immobilize heavy metals, break down persistent organic pollutants, and support microbial biodegradation (Nakum & Bhattacharya, 2022). Biochar-supported Fe NPs reduce hexavalent chromium (Cr^{6+}) in contaminated soils (Su et al., 2016). Biochar can be prepared via the thermal treatment of various biological wastes, including crop residues. For instance, nano-zerovalent iron (nZVI) derived from waste tea could be applied for Cr(VI) reduction in the soil (Chrysochoou et al., 2012). Iron oxide nanoparticles prepared using leaf extract could stabilize Cd and As in the soil through co-precipitation (Lin et al., 2019; Su et al., 2020). *Prunus cerasifera* leaf extract–mediated zinc oxide nanoparticles (ZnO NPs) were used as bio-cleaning photocatalysts against organic pollutants, that is, bromocresol green, bromophenol blue, methyl red, and methylene blue. They were also tested against resistant microbes *Aspergillus niger, Aspergillus flavus, Aspergillus fumigatus, Aspergillus terreus, Penicillium chrysogenum, Fusarium solani, Lasiodiplodia theobromae, Xanthomonas axonopodis pv. citri*, and *Psuedomonas syringae* for the development of a new generation of antimicrobial agents (Jaffri & and Ahmad, 2019).

The application of green nanomaterials in various sectors promises cleaner technologies, enhanced efficiency, and significant environmental benefits. Continued progress in process standardization, regulatory evolution, and interdisciplinary research will ensure these materials play a defining role in a sustainable future. However, a concrete regulatory framework needs to be created for green nanomaterials. Currently very few regulatory guidelines are formulated on the testing, evaluation, and disposal of nanoparticulate materials. Although developed countries and many developing countries have invested in research in nanoscience and nanotechnology, the studies are not adequate on social justice, ethics, civil liberties, and good laboratory practice, including proper nanomaterial disposal issues. Nanotechnology practitioners are also not updated with the latest critical issues associated with nanomaterials.

1.7 CHALLENGES, RISKS, AND REGULATORY FRAMEWORKS

While green nanomaterials promise a more sustainable future, their widespread application raises important questions about human and environmental safety, long-term toxicity, and biodegradability. There is an urgent requirement for robust regulations for the ethical production and usage of green nanomaterials. If the fate of nanomaterials is not well understood or they are not properly managed, even green nanomaterials may cause unforeseen biological and environmental impacts (Ahmed et al., 2022). The challenges faced in terms of the development of green nanomaterials are the inconsistency of the bio source, lack of standardization of synthetic parameters and characterization, difficulty in scaling up, and lack of studies regarding toxicity of the nanoparticles (Azamal Husen & Siddiqi, 2014). The variation in microbial content in a culture and phytoconstituents in plants can lead to batch-to-batch variation in the morphology of the nanoparticles synthesized. Hence, the reaction parameters cannot be standard for a particular procedure. Similarly, the availability of each biosource may vary year-round, and hence the scaling-up of procedures remain a bottleneck. Whether they are synthesized by the chemical or green route, a toxicity study of each nanoparticle is crucial. Currently, limited data are available regarding the toxicity of nanomaterials. Bioaccumulation of nanoparticles and enzyme alterations caused by nanoparticles are concerns regarding them (Mariano et al., 2020). Bioaccumulation of nanoparticles leads to their accumulation in terrestrial and land food chains, raising environmental and health concerns. Magnetic nanoparticles tend to accumulate in sediments and remain as a threat to living things, even though they are recyclable (Pradhan et al., 2021). Furthermore, nanoparticles undergo transformations like redox, aggregation, or ligand exchange reactions in environments, leading to the formation of toxic species (Abbas et al., 2020). Toxicity studies of green-synthesized ZnO NPs using papaya fruit extract on zebrafish pointed out that the mortality rate was high at higher concentrations of NPs (Zavitri et al., 2023). Moreover, it was reported that workers handling nanoparticles in open systems also face risk. Thus, to minimize the risks associated with health, safety, and environmental hazards in research and occupational settings, guidelines for safe handling, wise use, and disposal of nanoparticles have to be provided (Amoabediny et al., 2009).

1.8 FUTURE PROSPECTS AND CONCLUSIONS

As the global demand for sustainable materials continues to rise, green nanotechnology offers a promising pathway toward environmental, economic, and societal harmony. However, realizing this vision requires not only technological innovation but also systemic integration across sectors, disciplines, and global borders. The development of smart nanomaterials with the integration of artificial intelligence (AI) can be considered an emerging trend in nanotechnology (Husen & Siddiqi, 2023). Smart nanomaterials can be applied across a range of areas, such as biomedicine, pharmaceuticals, food science and packaging, sensing, cosmetics and dermatology, gas, oil, energy, wastewater and environment, textiles, agriculture, and forestry sectors. The integration of AI in precision agriculture, where farmers respond in real time to changes in crop growth with nanotechnology and artificial intelligence, offers exciting opportunities for sustainable food production (Zhang et al., 2021). But the convergence of AI approaches and nano-enabled agriculture is in its infancy and needs to be explored further. However, both smart nanomaterials and smart materials share concerns associated with their functionality. In addition to that, smart nanomaterials bring issues and challenges that have been identified for materials at the nanoscale. Accordingly, any new material or product should not only be functional and cost efficient but also safe and sustainable so as to ensure compliance with regulations, acceptance by consumers and users, and consequently fast and successful access to the market (Gottardo et al., 2021). Thus, for developing green nanotechnology, it is crucial to enable a circular economy. There is a need to move from linear "make-use-dispose" models to closed-loop systems that make use of waste biomass as feedstocks and recover and reuse nanomaterials for the next cycle of reactions. In a sustainable approach, product design should consider end-of-life management, where materials are responsibly recovered, recycled, or disposed of after their useful life in order to minimize environmental impact (Asmatulu et al., 2012). It is important to educate and train the workforce by implementing an integrated curriculum of green chemistry, nanotechnology, and environmental laws. As there is a lack of a proper regulatory framework, policy making with global collaboration to provide a toxicity database and technology transfer is needed. Also, research laboratories should be equipped with infrastructure for sustainable synthesis. As green nanotechnology expands, addressing its risks, inconsistencies, and regulatory gaps is critical to unlocking its full potential responsibly. Transparent governance, validated risk assessments, and inclusive ethics are the cornerstones of a truly sustainable nano-future.

Thus, green nanomaterials represent a transformative approach to nanotechnology, aligning innovation with sustainability and environmental responsibility. By integrating green chemistry principles into nanomaterial synthesis, the ecological footprints can be significantly reduced and high-performance materials can be developed for diverse applications. While research in green nanotechnology progresses, with constant highlighting of eco-friendly methods and lifecycle analysis, the full potential of green nanotechnology can be realized in a sustainable future. The development of green nanomaterials not only addresses the emerging environmental challenges but also opens new frontlines in safe, scalable, and accountable nanoscience.

REFERENCES

Aadil, K. R., Barapatre, A., Meena, A. S., & Jha, H. (2016). Hydrogen peroxide sensing and cytotoxicity activity of Acacia lignin stabilized silver nanoparticles. *International Journal of Biological Macromolecules*, *82*, 39–47. https://doi.org/10.1016/j.ijbiomac.2015.09.072

Abad, A., Concepción, P., Corma, A., & García, H. (2005). A collaborative effect between gold and a support induces the selective oxidation of alcohols. *Angewandte Chemie International Edition*, *44*(26), 4066–4069. https://doi.org/10.1002/anie.200500382

Abbas, Q., Balal, Y., Habib, U., Ubaid, A. M., Sik, O. Y., & Rinklebe, J. (2020). Environmental transformation and nano-toxicity of engineered nano-particles (ENPs) in aquatic and terrestrial organisms. *Critical Reviews in Environmental Science and Technology*, *50*(23), 2523–2581. https://doi.org/10.1080/10643389.2019.1705721

Abboud, Y., Saffaj, T., Chagraoui, A., El Bouari, A., Brouzi, K., Tanane, O., & Ihssane, B. (2014). Biosynthesis, characterization and antimicrobial activity of copper oxide nanoparticles (CONPs) produced using brown alga extract (Bifurcaria bifurcata). *Applied Nanoscience*, *4*, 571–576. https://doi.org/10.1007/s13204-013-0233-x

Agi, A., Junin, R., Jaafar, M. Z., Mohsin, R., Arsad, A., Gbadamosi, A., . . . Gbonhinbor, J. (2020). Synthesis and application of rice husk silica nanoparticles for chemical enhanced oil recovery. *Journal of Materials Research and*

Technology, *9*(6), 13054–13066. https://doi.org/10.1016/j.jmrt.2020.08.112

Ahmad, A., Mukherjee, P., Mandal, D., Senapati, S., Khan, M. I., Kumar, R., & Sastry, M. (2002). Enzyme mediated extracellular synthesis of CdS nanoparticles by the fungus, Fusarium oxysporum. *Journal of the American Chemical Society*, *124*(41), 12108–12109. https://doi.org/10.1021/ja027296o

Ahmad, U., Ahmad, Z., Khan, A. A., Akhtar, J., Singh, S. P., & Ahmad, F. J. (2018). Strategies in development and delivery of nanotechnology based cosmetic products. *Drug Research (Stuttg)*, *68*(10), 545–552. https://doi.org/10.1055/a-0582-9372

Ahmed, A., Shahadat, M., ul Islam, S., Adnan, R., Mohamad Ibrahim, M. N., & Ullah, Q. (2023). Synthesis, characterization, and properties of green carbon nanodots. In *Green Carbon Materials for Environmental Analysis: Emerging Research and Future Opportunities* (pp. 25–39). ACS Publications. https://doi.org/10.1021/bk-2023-1441.ch002

Ahmed, S., Ahmad, M., Swami, B. L., & Ikram, S. (2016). Green synthesis of silver nanoparticles using Azadirachta indica aqueous leaf extract. *Journal of Radiation Research and Applied Sciences*, *9*(1), 1–7. https://doi.org/10.1016/j.jrras.2015.06.006

Ahmed, S., Ara, G., & Susan, M. A. B. H. (2023). Green nanomaterials for photocatalytic degradation of toxic organic compounds. *Current Pharmaceutical Biotechnology*, *24*(1), 118–144. https://doi.org/10.2174/1389201023666211231100843

Ahmed, S. F., Mofijur, M., Rafa, N., Chowdhury, A. T., Chowdhury, S., Nahrin, M., . . . Ong, H. C. (2022). Green approaches in synthesising nanomaterials for environmental nanobioremediation: Technological advancements, applications, benefits and challenges. *Environmental Research*, *204*, 111967. https://doi.org/10.1016/j.envres.2021.111967

Ali, A., Baheti, V., Militky, J., & Khan, Z. (2018). Utility of silver-coated fabrics as electrodes in electrotherapy applications. *Journal of Applied Polymer Science*, *135*(23), 46357. https://doi.org/10.1002/app.46357

Ali, M. A., Ahmed, T., Wu, W., Hossain, A., Hafeez, R., Islam Masum, M. M., . . . Li, B. (2020). Advancements in plant and microbe-based synthesis of metallic nanoparticles and their antimicrobial activity against plant pathogens. *Nanomaterials*, *10*(6), 1146. https://www.mdpi.com/2079-4991/10/6/1146

Ali, R., Karamat, U., Nazir, H. S., Baig, M. M., & Farrukh, M. (2023). Antimicrobial activity of cotton fibres treated with particles extracted from citrus plants: A review. *Textile & Leather Review*, 6(2), Article 8. https://doi.org/10.15240/tul/008/2023-2-008

Al-Khaial, M. Q., Chan, S. Y., Abu-Zurayk, R. A., & Alnairat, N. (2024). Biosynthesis and characterization of zinc oxide nanoparticles (ZnO-NPs) utilizing banana peel extract. *Inorganics*, *12*(4), 121. https://doi.org/10.3390/inorganics12040121

Alprol, A. E., Eleryan, A., Abouelwafa, A., Gad, A. M., & Hamad, T. M. (2024). Green synthesis of zinc oxide nanoparticles using Padina pavonica extract for efficient photocatalytic removal of methylene blue. *Scientific Reports*, *14*(1), 32160. https://doi.org/10.1038/s41598-024-80757-9

Alsaggaf, M. S., Diab, A. M., ElSaied, B. E., Tayel, A. A., & Moussa, S. H. (2021). Application of ZnO nanoparticles phycosynthesized with Ulva fasciata extract for preserving peeled shrimp quality. *Nanomaterials*, *11*(2), 385. https://doi.org/10.3390/nano11020385

Altammar, K. A. (2023). A review on nanoparticles: Characteristics, synthesis, applications, and challenges. *Frontiers in Microbiology*, *14*, 1155622. https://doi.org/10.3389/fmicb.2023.1155622

Amiri, M., Zinatloo-Ajabshir, S., Ahmadi-Zeidabadi, M., & Sharifianjazi, F. (2025). Innovative ultrasound assisted synthesis of sponge like cerium dioxide nanostructure using Rosa Damascena extract and its efficient performance for cancer therapy. *Scientific Reports*, *15*(1), 933. https://doi.org/10.1038/s41598-025-85137-5

Amoabediny, G. H., Naderi, A., Malakootikhah, J., Koohi, M. K., Mortazavi, S. A., Naderi, M., & Rashedi, H. (2009). Guidelines for safe handling, use and disposal of nanoparticles. *Journal of Physics: Conference Series*, *170*(1), 012037. https://doi.org/10.1088/1742-6596/170/1/012037

Ana, J. C. S., & Camacho, D. H. (2019). Influence of precursor size in the hydrothermal synthesis of cellulose-based carbon nanodots and its application towards solar cell sensitization. *Materials Chemistry and Physics*, *228*, 187–193. https://doi.org/10.1016/j.matchemphys.2019.02.073

Anastas, P. T., & Warner, J. C. (1998). Principles of green chemistry. *Green Chemistry: Theory and Practice*, *29*, 14821–14842.

Aslam, M., Abdullah, A. Z., & Rafatullah, M. (2021). Recent development in the green synthesis of titanium dioxide nanoparticles using plant-based biomolecules for environmental and antimicrobial applications. *Journal of Industrial and Engineering Chemistry*, *98*, 1–16. https://doi.org/10.1016/j.jiec.2021.04.010

Asmatulu, E., Twomey, J., & Overcash, M. (2012). Life cycle and nano-products: End-of-life assessment. *Journal of Nanoparticle Research*, *14*(3), 720. https://doi.org/10.1007/s11051-012-0720-0

Awan, M. N. S., Razzaq, H., & Qaisar, S. (2023). Hydrazine by conducting polymers nanocomposites: A review. *Journal of Chemical Reviews*, *5*(3).

Bagheri, S., Muhd Julkapli, N., & Bee Abd Hamid, S. (2014). Titanium dioxide as a catalyst support in heterogeneous catalysis. *The Scientific World Journal*, *2014*(1), 727496. https://doi.org/10.1155/2014/727496

Baig, N., Kammakakam, I., & Falath, W. (2021). Nanomaterials: A review of synthesis methods, properties, recent progress, and challenges. *Materials Advances*, *2*(6), 1821–1871. https://doi.org/10.1039/D0MA00807A

Bao, Y., He, J., Song, K., Guo, J., Zhou, X., & Liu, S. (2021). Plant-extract-mediated synthesis of metal nanoparticles. *Journal of Chemistry*, *2021*(1), 6562687. https://doi.org/10.1155/2021/6562687

Baran, N. Y., Baran, T., & Menteş, A. (2018). Production of novel palladium nanocatalyst stabilized with sustainable chitosan/cellulose composite and its catalytic performance in Suzuki-Miyaura coupling reactions. *Carbohydrate Polymers*, *181*, 596–604. https://doi.org/10.1016/j.carbpol.2017.11.107

Behnajady, M. A., Modirshahla, N., & Ghazalian, E. (2011). Synthesis of ZnO nanoparticles at different conditions: A comparison of photocatalytic activity. *Digest Journal of Nanomaterials and Biostructures*, *6*(1), 467–474.

Bej, A., Ghosh, K., Sarkar, A., & Knight, D. W. (2016). Palladium nanoparticles in the catalysis of coupling reactions.

RSC Advances, *6*(14), 11446–11453. https://doi.org/10.1039/C5RA26304B

Boruah, S. K., Boruah, P. K., Sarma, P., Medhi, C., & Medhi, O. K. (2012). Green synthesis of gold nanoparticles using Camellia sinensis and kinetics of the reaction. *Advanced Materials Letters*, *3*(6), 481–486. https://doi.org/10.5185/amlett.2012.icnano.103

Buller, R., Lutz, S., Kazlauskas, R., Snajdrova, R., Moore, J., & Bornscheuer, U. (2023). From nature to industry: Harnessing enzymes for biocatalysis. *Science*, *382*(6673), eadh8615. https://doi.org/10.1126/science.adh8615

Cao, F., Gui, S.-Y., Gao, X., Zhang, W., Fu, Z.-Y., Tao, L.-M., . . . Wang, X. (2022). Research progress of natural product-based nanomaterials for the treatment of inflammation-related diseases. *Materials & Design*, *218*, 110686. https://doi.org/10.1016/j.matdes.2022.110686

Cao, G. (2004). *Nanostructures & Nanomaterials: Synthesis, Properties & Applications*. Imperial College Press.

Castro-Longoria, E., Moreno-Velázquez, S. D., Vilchis-Nestor, A. R., Arenas-Berumen, E., & Avalos-Borja, M. (2012). Production of platinum nanoparticles and nanoaggregates using Neurospora crassa. *Journal of Microbiology and Biotechnology*, *22*(7), 1000–1004. https://doi.org/10.4014/jmb.1110.10085

Chahrdoli, A., Qalekhani, F., Ghowsi, M., Nemati, H., Shokoohinia, Y., & Fattahi, A. (2020). Achillea wilhelmsii C. Koch mediated blood compatible silver nanoparticles. *Materials Today Communications*, *25*, 101577. https://doi.org/10.1016/j.mtcomm.2020.101577

Chandran, S. P., Chaudhary, M., Pasricha, R., Ahmad, A., & Sastry, M. (2006). Synthesis of gold nanotriangles and silver nanoparticles using Aloevera plant extract. *Biotechnology Progress*, *22*(2), 577–583. https://doi.org/10.1021/bp0501423

Chaudhary, V., & Bhowmick, A. K. (2015). Green synthesis of fluorescent carbon nanoparticles from lychee (Litchi chinensis) plant. *Korean Journal of Chemical Engineering*, *32*, 1707–1711. https://doi.org/10.1007/s11814-014-0381-z

Choudhary, R. C., Kumaraswamy, R. V., Kumari, S., Sharma, S. S., Pal, A., Raliya, R., . . . Saharan, V. (2019). Zinc encapsulated chitosan nanoparticle to promote maize crop yield. *International Journal of Biological Macromolecules*, *127*, 126–135. https://doi.org/10.1016/j.ijbiomac.2018.12.274

Chrysochoou, M., Johnston, C. P., & Dahal, G. (2012). A comparative evaluation of hexavalent chromium treatment in contaminated soil by calcium polysulfide and green-tea nanoscale zero-valent iron. *Journal of Hazardous Materials*, *201–202*, 33–42. https://doi.org/10.1016/j.jhazmat.2011.11.003

Dasgupta, N., Nayak, M. A., & Gauthier, M. (2022). Starch-stabilized iron oxide nanoparticles for the photocatalytic degradation of methylene blue. *Polysaccharides*, *3*(3), 655–670. https://doi.org/10.3390/polysaccharides3030038

Devatha, C., Thalla, A. K., & Katte, S. Y. (2016). Green synthesis of iron nanoparticles using different leaf extracts for treatment of domestic waste water. *Journal of Cleaner Production*, *139*, 1425–1435. https://doi.org/10.1016/j.jclepro.2016.09.019

DeVito, S., & Garrett, R. (1996). *Designing Safer Chemicals: Green Chemistry for Pollution Prevention*. ACS Publications.

Dey, A., Vashishtha, L., & Gogate, P. R. (2025). Ultrasound-assisted sol–gel synthesis of N-TiO2, Fe-TiO2, and TiO2 with application in treatment of a commercial effluent. *Journal of Chemical Technology & Biotechnology*. https://doi.org/10.1002/jctb.7808

Edo, G. I., Mafe, A. N., Ali, A., Akpoghelie, P. O., Yousif, E., Isoje, E. F., . . . Essaghah, A. E. A. (2025). Advancing sustainable food packaging: The role of green nanomaterials in enhancing barrier properties. *Food Engineering Reviews*, 1–35. https://doi.org/10.1007/s12393-025-09407-8

Elbrolesy, A., Abdou, Y., Elhussiny, F., & Morsy, R. (2023). Novel green synthesis of UV-sunscreen ZnO nanoparticles using solanum lycopersicum fruit extract and evaluation of their antibacterial and anticancer activity. *Journal of Inorganic and Organometallic Polymers and Materials*, *33*(12), 3750–3759. https://doi.org/10.1007/s10904-023-02744-3

El-Shanshoury, A. E.-R. R., ElSilk, S. E., & Ebeid, M. E. (2011). Extracellular biosynthesis of silver nanoparticles using Escherichia coli ATCC 8739, Bacillus subtilis ATCC 6633, and Streptococcus thermophilus ESh1 and their antimicrobial activities. *International Scholarly Research Notices*, *2011*(1), 385480. https://doi.org/10.5402/2011/385480

Ezhilarasi, P. N., Karthik, P., Chhanwal, N., & Anandharamakrishnan, C. (2013). Nanoencapsulation techniques for food bioactive components: A review. *Food and Bioprocess Technology*, *6*(3), 628–647. https://doi.org/10.1007/s11947-012-0944-0

Feng, H., Zhang, Y., Liu, J., & Liu, D. (2022). Towards heterogeneous catalysis: A review on recent advances of depositing nanocatalysts in continuous–flow microreactors. *Molecules*, *27*(22), 8052. https://www.mdpi.com/1420-3049/27/22/8052

Ferreira, L. F., Giordano, G. F., Gobbi, A. L., Piazzetta, M. H., Schleder, G. R., & Lima, R. S. (2022). Real-time and in situ monitoring of the synthesis of silica nanoparticles. *ACS Sensors*, *7*(4), 1045–1057. https://doi.org/10.1021/acssensors.1c02697

Flores-Contreras, E. A., González-González, R. B., Pablo Pizaña-Aranda, J. J., Parra-Arroyo, L., Rodríguez-Aguayo, A. A., Iñiguez-Moreno, M., . . . Parra-Saldívar, R. (2024). Agricultural waste as a sustainable source for nanoparticle synthesis and their antimicrobial properties for food preservation. *Frontiers in Nanotechnology*, *6*, 1346069. https://doi.org/10.3389/fnano.2024.1346069

Frank, L. A., Onzi, G. R., Morawski, A. S., Pohlmann, A. R., Guterres, S. S., & Contri, R. V. (2020). Chitosan as a coating material for nanoparticles intended for biomedical applications. *Reactive and Functional Polymers*, *147*, 104459. https://doi.org/10.1016/j.reactfunctpolym.2019.104459

Fytianos, G., Rahdar, A., & Kyzas, G. Z. (2020). Nanomaterials in cosmetics: Recent updates. *Nanomaterials*, *10*(5), 979. https://doi.org/10.3390/nano10050979

Gautam, P., Kumar, S., Tomar, M., Singh, R., & Acharya, A. (2017). Biologically synthesized gold nanoparticles using Ocimum sanctum (Tulsi leaf extract) induced anti-tumor response in a T cell daltons lymphoma. *Journal of Cell Science & Therapy*, *8*(2). https://doi.org/10.4172/2157-7013.1000278

George, A., Sanjay, M., Srisuk, R., Parameswaranpillai, J., & Siengchin, S. (2020). A comprehensive review on chemical properties and applications of biopolymers and their composites. *International Journal of Biological Macromolecules*, *154*, 329–338. https://doi.org/10.1016/j.ijbiomac.2020.03.120

Goh, H. T., Cheok, C. Y., & Yeap, S. P. (2023). Green synthesis of silver nanoparticles using banana peel extract and application on banana preservation. *Food Frontiers*, *4*(1), 283–288. https://doi.org/10.1002/fft2.206

Golestan, N. Z., Javanbakht, S., Nazeri, M. T., & Shaabani, A. (2025). Ultrasound-assisted synthesis of MWCNT/Fe3O4/Cu (BDC) nanocomposite: An efficient fast-release antibacterial carrier for potential suppository administration. *Materials Chemistry and Physics*, *332*, 130283. https://doi.org/10.1016/j.matchemphys.2024.130283

Gómez-López, P., Puente-Santiago, A., Castro-Beltrán, A., do Nascimento, L. A. S., Balu, A. M., Luque, R., & Alvarado-Beltrán, C. G. (2020). Nanomaterials and catalysis for green chemistry. *Current Opinion in Green and Sustainable Chemistry*, *24*, 48–55. https://doi.org/10.1016/j.cogsc.2020.03.001

Gottardo, S., Mech, A., Drbohlavová, J., Małyska, A., Bøwadt, S., Riego Sintes, J., & Rauscher, H. (2021). Towards safe and sustainable innovation in nanotechnology: State-of-play for smart nanomaterials. *NanoImpact*, *21*, 100297. https://doi.org/10.1016/j.impact.2021.100297

Grosu, E.-F., Froidevaux, R., & Carja, G. (2019). Horseradish peroxidase-AuNP/LDH heterostructures: Influence on nanogold release and enzyme activity. *Gold Bulletin*, *52*, 87–97. https://doi.org/10.1007/s13404-019-00256-y

Hamers, R. J. (2017). Nanomaterials and global sustainability. *Accounts of Chemical Research*, *50*(3), 633–637. https://doi.org/10.1021/acs.accounts.6b00634

Hang, L., Zhao, Y., Zhang, H., Liu, G., Cai, W., Li, Y., & Qu, L. (2016). Copper nanoparticle@ graphene composite arrays and their enhanced catalytic performance. *Acta Materialia*, *105*, 59–67. https://doi.org/10.1016/j.actamat.2015.12.029

He, Y., Li, H., Fei, X., & Peng, L. (2021). Carboxymethyl cellulose/cellulose nanocrystals immobilized silver nanoparticles as an effective coating to improve barrier and antibacterial properties of paper for food packaging applications. *Carbohydrate Polymers*, *252*, 117156. https://doi.org/10.1016/j.carbpol.2020.117156

Hegde, R. V., Ghosh, A., Jadhav, A. H., Nizam, A., Patil, S. A., Peter, F., & Dateer, R. B. (2021). Biogenic synthesis of Pd-nanoparticles using Areca Nut Husk Extract: A greener approach to access α-keto imides and stilbenes. *New Journal of Chemistry*, *45*(35), 16213–16222. https://doi.org/10.1039/D1NJ02858H

Herath, A., Navarathna, C., Warren, S., Perez, F., Pittman Jr, C. U., & Mlsna, T. E. (2022). Iron/titanium oxide-biochar (Fe2TiO5/BC): A versatile adsorbent/photocatalyst for aqueous Cr (VI), Pb2+, F-and methylene blue. *Journal of Colloid and Interface Science*, *614*, 603–616. https://doi.org/10.1016/j.jcis.2022.01.067

Husen, A., & Siddiqi, K. S. (2014). Phytosynthesis of nanoparticles: Concept, controversy and application. *Nanoscale Research Letters*, *9*(1), 229. https://doi.org/10.1186/1556-276X-9-229

Husen, A., & Siddiqi, K. S. (2023). *Advances in Smart Nanomaterials and Their Applications*. Elsevier.

Hussein, H., Shaarawy, H., Hussien, N. H., & Hawash, S. (2019). Preparation of nano-fertilizer blend from banana peels. *Bulletin of the National Research Centre*, *43*(1), 1–9. https://doi.org/10.1186/s42269-019-0058-1

Husseiny, M., Abd El-Aziz, M., Badr, Y., & Mahmoud, M. (2007). Biosynthesis of gold nanoparticles using Pseudomonas aeruginosa. *Spectrochimica Acta Part A: Molecular and Biomolecular Spectroscopy*, *67*(3–4), 1003–1006. https://doi.org/10.1016/j.saa.2006.09.028

Hussien, N. A. (2023). Antimicrobial potential of biosynthesized zinc oxide nanoparticles using banana peel and date seeds extracts. *Sustainability*, *15*(11), 9048. https://doi.org/10.3390/su15119048

Imran, M., Yousaf, A. B., Zhou, X., Jiang, Y.-F., Yuan, C.-Z., Zeb, A., . . . Xu, A.-W. (2017). Pd/TiO nanocatalyst with strong metal–support interaction for highly efficient durable heterogeneous hydrogenation. *The Journal of Physical Chemistry C*, *121*(2), 1162–1170. https://doi.org/10.1021/acs.jpcc.6b10274

Jaffri, S. B., & Ahmad, K. S. (2019). Neoteric environmental detoxification of organic pollutants and pathogenic microbes via green synthesized ZnO nanoparticles. *Environmental Technology*, *40*(28), 3745–3761. https://doi.org/10.1080/09593330.2018.1488888

Jain, K., Takuli, A., Gupta, T. K., & Gupta, D. (2024). Rethinking nanoparticle synthesis: A sustainable approach vs. traditional methods. *Chemistry—An Asian Journal*, *19*(21), e202400701. https://doi.org/10.1002/asia.202400701

Jain, N., Jain, P., Rajput, D., & Patil, U. K. (2021). Green synthesized plant-based silver nanoparticles: Therapeutic prospective for anticancer and antiviral activity. *Micro and Nano Systems Letters*, *9*(1), 5. https://doi.org/10.1186/s40486-021-00131-6

Jamaludin, N. F. M., Jamali, N. S., Abdullah, L. C., Idrus, S., Engliman, N. S., & Abdul, P. M. (2023). Biohydrogen production with utilisation of magnetite nanoparticles embedded in granular activated carbon from coconut shell. *International Journal of Hydrogen Energy*, *48*(31), 11695–11708. https://doi.org/10.1016/j.ijhydene.2022.12.073

Kalabegishvili, T., Kirkesali, E., & Rcheulishvili, A. (2012). *Synthesis of Gold Nanoparticles by Blue-Green Algae Spirulina Platensis*. Frank Laboratory of Neutron Physics.

Kandar, C. C. (2020). Secondary metabolites from plant sources. In *Bioactive Natural Products for Pharmaceutical Applications* (pp. 329–377). Springer. https://doi.org/10.1007/978-3-030-54027-2_10

Karimi, J., & Mohsenzadeh, S. (2015). Rapid, green, and eco-friendly biosynthesis of copper nanoparticles using flower extract of Aloe vera. *Synthesis and Reactivity in Inorganic, Metal-Organic, and Nano-Metal Chemistry*, *45*(6), 895–898. https://doi.org/10.1080/15533174.2013.862644

Kerton, F. M., & Marriott, R. (2013). *Alternative Solvents for Green Chemistry*. Royal Society of Chemistry.

Khorrami, S., Zarepour, A., & Zarrabi, A. (2019). Green synthesis of silver nanoparticles at low temperature in a fast pace with unique DPPH radical scavenging and selective cytotoxicity against MCF-7 and BT-20 tumor cell lines. *Biotechnology Reports*, *24*, e00393. https://doi.org/10.1016/j.btre.2019.e00393

Kim, K. H., Lee, O. K., & Lee, E. Y. (2018). Nano-immobilized biocatalysts for biodiesel production from renewable and sustainable resources. *Catalysts*, *8*(2), 68. https://www.mdpi.com/2073-4344/8/2/68

Kim, K.-S., Choi, S., Cha, J.-H., Yeon, S.-H., & Lee, H. (2006). Facile one-pot synthesis of gold nanoparticles using alcohol ionic liquids. *Journal of Materials Chemistry*, *16*(14), 1315–1317. https://doi.org/10.1039/B601478J

Koppolu, R. (2017). *Green Synthesis of Formulated Zinc Oxide Nanoparticles for Chemical Protection of Skin Care and Related Applications*. Texas A&M University-Kingsville.

Kowtharapu, L. P., Katari, N. K., Muchakayala, S. K., & Marisetti, V. M. (2023). Green metric tools for analytical methods assessment critical review, case studies and crucify.

TrAC Trends in Analytical Chemistry, *166*, 117196. https://doi.org/10.1016/j.trac.2023.117196

Kumar, A., & Dixit, C. K. (2017). Methods for characterization of nanoparticles. In *Advances in Nanomedicine for the Delivery of Therapeutic Nucleic Acids* (pp. 43–58). Elsevier. https://doi.org/10.1016/B978-0-08-100557-6.00003-1

Kumari, I., Kaur, S., & Das, R. (2024). Green nanomaterials in textile industry. In *Green Synthesis of Nanomaterials: Biological and Environmental Applications* (pp. 114–130). https://doi.org/10.1002/9781119900931.ch6

Lakshminarayanan, S., Shereen, M. F., Niraimathi, K., Brindha, P., & Arumugam, A. (2021). One-pot green synthesis of iron oxide nanoparticles from Bauhinia tomentosa: Characterization and application towards synthesis of 1, 3 diolein. *Scientific Reports*, *11*(1), 8643. https://doi.org/10.1038/s41598-021-87960-y

Lin, I. W.-S., Lok, C.-N., & Che, C.-M. (2014). Biosynthesis of silver nanoparticles from silver (i) reduction by the periplasmic nitrate reductase c-type cytochrome subunit NapC in a silver-resistant *E. coli*. *Chemical Science*, *5*(8), 3144–3150. https://doi.org/10.1039/C4SC00138A

Lin, J., He, F., Su, B., Sun, M., Owens, G., & Chen, Z. (2019). The stabilizing mechanism of cadmium in contaminated soil using green synthesized iron oxide nanoparticles under long-term incubation. *Journal of Hazardous Materials*, *379*, 120832. https://doi.org/10.1016/j.jhazmat.2019.120832

Luo, Z., Wu, C., Yan, M., Yu, X., Yu, X., Qian, Q., . . . Long, G. (2024). Enhancing detectivity of organic photodetectors through the biomass carbon quantum dots from sugarcane bagasse. *Materials Chemistry and Physics*, *316*, 129056. https://doi.org/10.1016/j.matchemphys.2024.129056

Magdi, H. M., & Bhushan, B. (2015). Extracellular biosynthesis and characterization of gold nanoparticles using the fungus Penicillium chrysogenum. *Microsystem Technologies*, *21*, 2279–2285. https://doi.org/10.1007/s00542-015-2666-5

Mariano, S., Panzarini, E., Inverno, M. D., Voulvoulis, N., & Dini, L. (2020). Toxicity, bioaccumulation and biotransformation of glucose-capped silver nanoparticles in green microalgae chlorella vulgaris. *Nanomaterials*, *10*(7), 1377. https://www.mdpi.com/2079-4991/10/7/1377

Mishra, P. K., Samarth, R. M., Pathak, N., Jain, S. K., Banerjee, S., & Maudar, K. K. (2009). Bhopal gas tragedy: Review of clinical and experimental findings after 25 years. *International Journal of Occupational Medicine and Environmental Health*, *22*(3), 193. https://doi.org/10.2478/v10001-009-0028-1

Mitsudome, T., Mikami, Y., Ebata, K., Mizugaki, T., Jitsukawa, K., & Kaneda, K. (2008). Copper nanoparticles on hydrotalcite as a heterogeneous catalyst for oxidant-free dehydrogenation of alcohols. *Chemical Communications*, *39*, 4804–4806. https://doi.org/10.1039/B809012B

Mohammad, F., Al-Lohedan, H. A., & Al-Haque, H. N. (2017). Chitosan-mediated fabrication of metal nanocomposites for enhanced biomedical applications. *Advanced Materials Letters*, *8*(2), 89–100. https://doi.org/10.5185/amlett.2017.6925

Mohapatra, L., Cheon, D., & Yoo, S. H. (2023). Carbon-based nanomaterials for catalytic wastewater treatment: A review. *Molecules*, *28*(4), 1805. https://doi.org/10.3390/molecules28041805

Mori, K., Hara, T., Mizugaki, T., Ebitani, K., & Kaneda, K. (2004). Hydroxyapatite-supported palladium nanoclusters: A highly active heterogeneous catalyst for selective oxidation of alcohols by use of molecular oxygen. *Journal of the American Chemical Society*, *126*(34), 10657–10666. https://doi.org/10.1021/ja0488683

Morshed, M. N., Shen, X., Deb, H., Al Azad, S., Zhang, X., & Li, R. (2020). Sonochemical fabrication of nanocryatalline titanium dioxide (TiO2) in cotton fiber for durable ultraviolet resistance. *Journal of Natural Fibers*. https://doi.org/10.1080/15440478.2018.1465506

Murugadoss, A., & Chattopadhyay, A. (2007). A 'green'chitosan–silver nanoparticle composite as a heterogeneous as well asmicro-heterogeneous catalyst. *Nanotechnology*, *19*(1), 015603. https://doi.org/10.1088/0957-4484/19/01/015603

Mydeen, S. S., Kumar, R. R., Kottaisamy, M., & Vasantha, V. (2020). Biosynthesis of ZnO nanoparticles through extract from Prosopis juliflora plant leaf: Antibacterial activities and a new approach by rust-induced photocatalysis. *Journal of Saudi Chemical Society*, *24*(5), 393–406. https://doi.org/10.1016/j.jscs.2020.03.003

Nabi, G., Ain, Q.-U.-., Tahir, M. B., Nadeem Riaz, K., Iqbal, T., Rafique, M., . . . Rizwan, M. (2022). Green synthesis of TiO2 nanoparticles using lemon peel extract: Their optical and photocatalytic properties. *International Journal of Environmental Analytical Chemistry*, *102*(2), 434–442. https://doi.org/10.1080/03067319.2020.1722816

Nakum, J., & Bhattacharya, D. (2022). Various green nanomaterials used for wastewater and soil treatment: A mini-review. *Frontiers in Environmental Science*, *9*, 724814. https://doi.org/10.3389/fenvs.2021.724814

Nanaei, M., Nasseri, M. A., Allahresani, A., & Kazemnejadi, M. (2019). Phoenix dactylifera L. extract: Antioxidant activity and its application for green biosynthesis of Ag nanoparticles as a recyclable nanocatalyst for 4-nitrophenol reduction. *SN Applied Sciences*, *1*(8), 853. https://doi.org/10.1007/s42452-019-0895-4

Nandhini, J., Karthikeyan, E., Sheela, M., Bellarmin, M., Kannan, B. G., Pavithra, A., . . . Kumar, S. R. (2025). Optimization of microwave-assisted green synthesis of zinc oxide nanoparticles using Ocimum americanum and Euphorbia hirta extracts: In vitro evaluation of antioxidant, anti-inflammatory, antibacterial, cytotoxicity, and wound healing properties. *Intelligent Pharmacy*, *3*(1), 90–109. https://doi.org/10.1016/j.ipha.2024.09.003

Narh, D., Sampson, B., Ocrah Junior, S., Pokuaa Manu, G., Agyei-Tuffour, B., Nyankson, E., & Kwame Efavi, J. (2024). Green synthesis of citrus sinensis peel extract-mediated Ag-TiO2 and its application as a photocatalyst for organic molecules and antimicrobial agent. *Journal of Nanotechnology*, *2024*(1), 9169241. https://doi.org/10.1155/2024/9169241

Neeti, K., Singh, R., & Ahmad, S. (2023). The role of green nanomaterials as effective adsorbents and applications in wastewater treatment. *Materials Today: Proceedings*, *77*, 269–276. https://doi.org/10.1016/j.matpr.2022.11.300

Nehra, P., Chauhan, R., Garg, N., & Verma, K. (2018). Antibacterial and antifungal activity of chitosan coated iron oxide nanoparticles. *British Journal of Biomedical Science*, *75*(1), 13–18. https://doi.org/10.1080/09674845.2017.1347362

Nizam, N. U. M., Hanafiah, M. M., & Woon, K. S. (2021). A content review of life cycle assessment of nanomaterials: Current practices, challenges, and future prospects. *Nanomaterials*, *11*(12), 3324. https://doi.org/10.3390/nano11123324

Obayomi, K. S., Lau, S. Y., Danquah, M. K., Zhang, J., Chiong, T., Meunier, L., . . . Rahman, M. M. (2023). Green synthesis of graphene-oxide based nanocomposites for efficient removal of methylene blue dye from wastewater. *Desalination*, *564*, 116749. https://doi.org/10.1016/j.desal.2023.116749

Osman, A. I., Zhang, Y., Farghali, M., Rashwan, A. K., Eltaweil, A. S., Abd El-Monaem, E. M., . . . Yap, P.-S. (2024). Synthesis of green nanoparticles for energy, biomedical, environmental, agricultural, and food applications: A review. *Environmental Chemistry Letters*, *22*(2), 841–887. https://doi.org/10.1007/s10311-023-01682-3

Ovais, M., Khalil, A. T., Islam, N. U., Ahmad, I., Ayaz, M., Saravanan, M., . . . Mukherjee, S. (2018). Role of plant phytochemicals and microbial enzymes in biosynthesis of metallic nanoparticles. *Applied Microbiology and Biotechnology*, *102*, 6799–6814. https://doi.org/10.1007/s00253-018-9146-7

Park, J., Lee, L., Byun, H., Ham, S., Lee, I., Park, J., . . . Tsai, P. (2014). A study of the volatile organic compound emissions at the stacks of laboratory fume hoods in a university campus. *Journal of Cleaner Production*, *66*, 10–18. https://doi.org/10.1016/j.jclepro.2013.11.024

Patil, R. B., & Chougale, A. D. (2021). Analytical methods for the identification and characterization of silver nanoparticles: A brief review. *Materials Today: Proceedings*, *47*, 5520–5532. https://doi.org/10.1016/j.matpr.2021.03.384

Patra, S., Singh, M., Subudhi, S., Mandal, M., Nayak, A. K., Sahu, B. B., & Mahanandia, P. (2023). One-step green synthesis of in–situ functionalized carbon quantum dots from Tagetes patula flowers: Applications as a fluorescent probe for detecting Fe3+ ions and as an antifungal agent. *Journal of Photochemistry and Photobiology A: Chemistry*, *442*, 114779. https://doi.org/10.1016/j.jphotochem.2023.114779

Pérez-Lorenzo, M. (2012). Palladium nanoparticles as efficient catalysts for Suzuki cross-coupling reactions. *The Journal of Physical Chemistry Letters*, *3*(2), 167–174. https://doi.org/10.1021/jz2013984

Petrovic, S., Bita, B., & Barbinta-Patrascu, M.-E. (2024). Nanoformulations in pharmaceutical and biomedical applications: Green perspectives. *International Journal of Molecular Sciences*, *25*(11), 5842. https://doi.org/10.3390/ijms25115842

Prabhu, N., & Gajendran, T. (2017). Green synthesis of noble metal of platinum nanoparticles from Ocimum sanctum (Tulsi) Plant-extracts. *IOSR Journal of Biotechnology and Biochemistry*, *3*(1), 107–112. https://doi.org/10.9790/264X-0301107112

Pradhan, A., Fernandes, M., Martins, P. M., Pascoal, C., Lanceros-Méndez, S., & Cássio, F. (2021). Can photocatalytic and magnetic nanoparticles be a threat to aquatic detrital food webs? *Science of the Total Environment*, *769*, 144576. https://doi.org/10.1016/j.scitotenv.2020.144576

Qadir, M. B., Muhammad Awais, R., Ali, Z., Khaliq, Z., Arshad, S. N., Alkorbi, A. S., . . . Harraz, F. A. (2024). A novel antibacterial surgical face mask with green synthesized Zinc Oxide nanoparticles and Azadirachta Indica extract. *Journal of Engineered Fibers and Fabrics*, *19*, 15589250241262322. https://doi.org/10.1177/15589250241262322

Qiu, B., Jing, H., Han, S., Zhang, S., Ba, F., Xu, L., . . . Yao, M. (2024). Enhanced gut microbiota delivery of a model probiotic (Faecalibacterium prausnitzii): Layer-by-layer encapsulation using riboflavin-conjugated sodium alginate and glycol chitosan. *Food Hydrocolloids*, *154*, 110055. https://doi.org/10.1016/j.foodhyd.2024.110055

Rajabi, H. R., Alvand, Z. M., & Mirzaei, A. (2023). Sonochemical-assisted synthesis of copper oxide nanoparticles with the plant-mediated approach and comparative evaluation of some biological activities. *Environmental Science and Pollution Research*, *30*(57), 120236–120249. https://doi.org/10.1007/s11356-023-30684-5

Rajesh, S., Dharanishanthi, V., & Kanna, A. V. (2015). Antibacterial mechanism of biogenic silver nanoparticles of Lactobacillus acidophilus. *Journal of Experimental Nanoscience*, *10*(15), 1143–1152. https://doi.org/10.1080/17458080.2014.985750

Rajkumar, R., Ezhumalai, G., & Gnanadesigan, M. (2021). A green approach for the synthesis of silver nanoparticles by Chlorella vulgaris and its application in photocatalytic dye degradation activity. *Environmental Technology & Innovation*, *21*, 101282. https://doi.org/10.1016/j.eti.2020.101282

Rambabu, K., Bharath, G., Banat, F., & Show, P. L. (2021). Green synthesis of zinc oxide nanoparticles using Phoenix dactylifera waste as bioreductant for effective dye degradation and antibacterial performance in wastewater treatment. *Journal of Hazardous Materials*, *402*, 123560. https://doi.org/10.1016/j.jhazmat.2020.123560

Ramteke, C., Chakrabarti, T., Sarangi, B. K., & Pandey, R.-A. (2013). Synthesis of silver nanoparticles from the aqueous extract of leaves of Ocimum sanctum for enhanced antibacterial activity. *Journal of Chemistry*, *2013*(1), 278925. https://doi.org/10.1155/2013/278925

Rao, C. N. R., Müller, A., & Cheetham, A. K. (2006). *The Chemistry of Nanomaterials: Synthesis, Properties and Applications*. John Wiley & Sons.

Riva, L., Nicastro, G., Liu, M., Battocchio, C., Punta, C., & Sacchetti, A. (2022). Pd-loaded cellulose nanosponge as a heterogeneous catalyst for Suzuki–Miyaura coupling reactions. *Gels*, *8*(12), 789. https://doi.org/10.3390/gels8120789

Rocha, F. R., Nóbrega, J. A., & Fatibello Filho, O. (2001). Flow analysis strategies to greener analytical chemistry. An overview. *Green Chemistry*, *3*(5), 216–220. https://doi.org/10.1039/B103187M

Sadeghi, B., Mohammadzadeh, M., & Babakhani, B. (2015). Green synthesis of gold nanoparticles using Stevia rebaudiana leaf extracts: Characterization and their stability. *Journal of Photochemistry and Photobiology B: Biology*, *148*, 101–106. https://doi.org/10.1016/j.jphotobiol.2015.03.025

Sadri, F., Ramazani, A., Massoudi, A., Khoobi, M., Tarasi, R., Shafiee, A., . . . Joo, S. W. (2014). Green oxidation of alcohols by using hydrogen peroxide in water in the presence of magnetic Fe3O4 nanoparticles as recoverable catalyst. *Green Chemistry Letters and Reviews*, *7*(3), 257–264. https://doi.org/10.1080/17518253.2014.939721

Salehi-Khojin, A., Jhong, H.-R. M., Rosen, B. A., Zhu, W., Ma, S., Kenis, P. J., & Masel, R. I. (2013). Nanoparticle silver catalysts that show enhanced activity for carbon dioxide electrolysis. *The Journal of Physical Chemistry C*, *117*(4), 1627–1632. https://doi.org/10.1021/jp310509z

Sani, M. A., Tavassoli, M., Azizi-Lalabadi, M., Mohammadi, K., & McClements, D. J. (2022). Nano-enabled plant-based colloidal delivery systems for bioactive agents in foods: Design, formulation, and application. *Advances in Colloid and Interface Science*, *305*, 102709. https://doi.org/10.1016/j.cis.2022.102709

Santana, J. S., & Skrabalak, S. E. (2020). Continuous flow routes toward designer metal nanocatalysts. *Advanced Energy Materials*, *10*(11), 1902051. https://doi.org/10.1002/aenm.201902051

Shafiq, M., Anjum, S., Hano, C., Anjum, I., & Abbasi, B. H. (2020). An overview of the applications of nanomaterials and nanodevices in the food industry. *Foods*, *9*(2), 148. https://doi.org/10.1021/jp310509z

Shah, E., Mahapatra, P., Bedekar, A. V., & Soni, H. P. (2015). Immobilization of Thermomyces lanuginosus lipase on ZnO nanoparticles: Mimicking the interfacial environment. *RSC Advances*, *5*(33), 26291–26300. https://doi.org/10.1039/C5RA02249E

Shah, S. T., Sari, I. P., Yanto, D. H. Y., Chowdhury, Z. Z., Bashir, M. N., Badruddin, I. A., . . . Lee, J. S. (2025). Nature's nanofactories: Biogenic synthesis of metal nanoparticles for sustainable technologies. *Green Chemistry Letters and Reviews*, *18*(1), 2448171. https://doi.org/10.1080/17518253.2024.2448171

Shahzad, A., Saeed, H., Iqtedar, M., Hussain, S. Z., Kaleem, A., Abdullah, R., . . . Aihetasham, A. (2019). Size-controlled production of silver nanoparticles by Aspergillus fumigatus BTCB10: Likely antibacterial and cytotoxic effects. *Journal of Nanomaterials*, *2019*(1), 5168698. https://doi.org/10.1155/2019/5168698

Sharma, K., Tyagi, S., Vikal, S., Devi, A., Gautam, Y. K., & Singh, B. P. (2022). Green nanomaterials for remediation of environmental air pollution. In *Handbook of Green and Sustainable Nanotechnology: Fundamentals, Developments and Applications* (pp. 1–26). Springer. https://doi.org/10.1007/978-3-030-69023-6_66-1

Sharma, N. C., Sahi, S. V., Nath, S., Parsons, J. G., Gardea-Torresde, J. L., & Pal, T. (2007). Synthesis of plant-mediated gold nanoparticles and catalytic role of biomatrix-embedded nanomaterials. *Environmental Science & Technology*, *41*(14), 5137–5142. https://doi.org/10.1021/es062929a

Sheldon, R. A. (2008). E factors, green chemistry and catalysis: An odyssey. *Chemical Communications*, *29*, 3352–3365. https://doi.org/10.1039/B803584A

Sheldon, R. A., Arends, I., & Hanefeld, U. (2007). *Green Chemistry and Catalysis*. John Wiley & Sons.

Shi, G., Li, Y., Xi, G., Xu, Q., He, Z., Liu, Y., . . . Cai, J. (2017). Rapid green synthesis of gold nanocatalyst for high-efficiency degradation of quinclorac. *Journal of Hazardous Materials*, *335*, 170–177. https://doi.org/10.1016/j.jhazmat.2017.04.042

Singh, A. K., Tiwari, R., Singh, V. K., Singh, P., Khadim, S. R., Singh, U., . . . Asthana, R. K. (2019). Green synthesis of gold nanoparticles from Dunaliella salina, its characterization and in vitro anticancer activity on breast cancer cell line. *Journal of Drug Delivery Science and Technology*, *51*, 164–176. https://doi.org/10.1016/j.jddst.2019.02.023

Singh, H., Desimone, M. F., Pandya, S., Jasani, S., George, N., Adnan, M., . . . Alderhami, S. A. (2023). Revisiting the green synthesis of nanoparticles: Uncovering influences of plant extracts as reducing agents for enhanced synthesis efficiency and its biomedical applications. *International Journal of Nanomedicine*, 4727–4750. https://doi.org/10.2147/IJN.S419369

Singh, K., Kumar, A., Kumar, D., Kumar, A., Kumar, A., Mahatha, S. K., & Praveenkumar, S. (2025). Unveiling the role of temperature on structural, compositional, morphological, thermal and optical properties of hydrothermally synthesized SnS2 nanostructures. *Inorganic Chemistry Communications*, *171*, 113548. https://doi.org/10.1016/j.inoche.2024.113548

Singh, P., Kim, Y.-J., Zhang, D., & Yang, D.-C. (2016). Biological synthesis of nanoparticles from plants and microorganisms. *Trends in Biotechnology*, *34*(7), 588–599.

Singh, P., Yadav, S. K., & Kuddus, M. (2020). Green nanomaterials for wastewater treatment. In S. Ahmed & W. Ali (Eds.), *Green Nanomaterials: Processing, Properties, and Applications* (pp. 227–242). Singapore: Springer. https://doi.org/10.1007/978-981-15-3560-4_9

Singh, P. M., Tiwari, A., Maity, D., & Saha, S. (2022). Recent progress of nanomaterials in sustainable agricultural applications. *Journal of Materials Science*, *57*(24), 10836–10862. https://doi.org/10.1007/s10853-022-07259-9

Sivasankarapillai, V. S., Krishnamoorthy, N., Eldesoky, G. E., Wabaidur, S. M., Islam, M. A., Dhanusuraman, R., & Ponnusamy, V. K. (2023). One-pot green synthesis of ZnO nanoparticles using Scoparia Dulcis plant extract for antimicrobial and antioxidant activities. *Applied Nanoscience*, *13*(9), 6093–6103. https://doi.org/10.1007/s13204-022-02610-7

Somwanshi, S. B., Somvanshi, S. B., & Kharat, P. B. (2020). Nanocatalyst: A brief review on synthesis to applications. Paper presented at the *Journal of Physics: Conference Series*. https://doi.org/10.1088/1742-6596/1644/1/012046

Sooch, B. S., & Mann, M. K. (2021). Nanoreinforced biodegradable gelatin based active food packaging film for the enhancement of shelf life of tomatoes (Solanum lycopersicum L.). *Food Control*, *130*, 108322. https://doi.org/10.1016/j.foodcont.2021.108322

Su, B., Lin, J., Owens, G., & Chen, Z. (2020). Impact of green synthesized iron oxide nanoparticles on the distribution and transformation of As species in contaminated soil. *Environmental Pollution*, *258*, 113668. https://doi.org/10.1016/j.envpol.2019.113668

Su, H., Fang, Z., Tsang, P. E., Zheng, L., Cheng, W., Fang, J., & Zhao, D. (2016). Remediation of hexavalent chromium contaminated soil by biochar-supported zero-valent iron nanoparticles. *Journal of Hazardous Materials*, *318*, 533–540. https://doi.org/10.1016/j.jhazmat.2016.07.039

Suvarna, S., Das, U., Kc, S., Mishra, S., Sudarshan, M., Saha, K. D., . . . Narayana, Y. (2017). Synthesis of a novel glucose capped gold nanoparticle as a better theranostic candidate. *PLoS ONE*, *12*(6), e0178202. https://doi.org/10.1371/journal.pone.0178202

Taha, A., & Daffalla, S. (2023). Biochar derived from palm waste supported greenly synthesized MnO2 nanoparticles as a novel adsorbent for wastewater treatment. *Catalysts*, *13*(2), 451. https://doi.org/10.3390/catal13020451

Tanaka, K., & Toda, F. (2000). Solvent-free organic synthesis. *Chemical Reviews*, *100*(3), 1025–1074. https://doi.org/10.1021/cr940089p

Trost, B. M. (1991). The atom economy—a search for synthetic efficiency. *Science*, *254*(5037), 1471–1477. https://doi.org/10.1126/science.1962206

Unsoy, G., Yalcin, S., Khodadust, R., Gunduz, G., & Gunduz, U. (2012). Synthesis optimization and characterization of chitosan-coated iron oxide nanoparticles produced for biomedical applications. *Journal of Nanoparticle Research*, *14*, 1–13. https://doi.org/10.1007/s11051-012-0964-8

Veeraraghavan, V. P., Periadurai, N. D., Karunakaran, T., Hussain, S., Surapaneni, K. M., & Jiao, X. (2021). Green synthesis of silver nanoparticles from aqueous extract of Scutellaria barbata and coating on the cotton fabric for antimicrobial applications and wound healing activity in fibroblast cells (L929). *Saudi Journal of Biological Sciences*, *28*(7), 3633–3640. https://doi.org/10.1016/j.sjbs.2021.05.007

Verma, A., & Mehata, M. S. (2016). Controllable synthesis of silver nanoparticles using Neem leaves and their antimicrobial activity. *Journal of Radiation Research and Applied Sciences*, *9*(1), 109–115. https://doi.org/10.1016/j.jrras.2015.11.001

Wiederrecht, G. P., Bachelot, R., Xiong, H., Termentzidis, K., Nominé, A., Huang, J., . . . Ostraat, M. (2023). *Nanomaterials and Sustainability*. ACS Publications. https://doi.org/10.1021/acsenergylett.3c01303

Yanti, I., Sationo, P. P., Winata, W. F., Anugrahwati, M., Anas, A. K., & Swasono, Y. A. (2023). Effectiveness of activated carbon magnetic composite from banana peel (Musa acuminata) for recovering iron metal ions. *Case Studies in Chemical and Environmental Engineering*, *8*, 100378. https://doi.org/10.1016/j.cscee.2023.100378

Zavitri, N. G., Syahbaniati, A. P., Primastuti, R. K., Putri, R. M., Damayanti, S., & Wibowo, I. (2023). Toxicity evaluation of zinc oxide nanoparticles green synthesized using papaya extract in zebrafish. *Biomedical Reports*, *19*(6), 96. https://doi.org/10.3892/br.2023.1678

Zhang, P., Guo, Z., Ullah, S., Melagraki, G., Afantitis, A., & Lynch, I. (2021). Nanotechnology and artificial intelligence to enable sustainable and precision agriculture. *Nature Plants*, *7*(7), 864–876. https://doi.org/10.1038/s41477-021-00946-6

Zhang, Z., Ortiz, O., Goyal, R., & Kohn, J. (2014). Biodegradable polymers. In *Handbook of Polymer Applications in Medicine and Medical Devices* (pp. 303–335). https://doi.org/10.1016/B978-0-323-22805-3.00013-X

Zhao, X., Chang, Y., Chen, W.-J., Wu, Q., Pan, X., Chen, K., & Weng, B. (2021). Recent progress in Pd-based nanocatalysts for selective hydrogenation. *ACS Omega*, *7*(1), 17–31. https://doi.org/10.1021/acsomega.1c06244

Zhu, M., Xiong, R., & Huang, C. (2019). Bio-based and photocrosslinked electrospun antibacterial nanofibrous membranes for air filtration. *Carbohydrate Polymers*, *205*, 55–62. https://doi.org/10.1016/j.carbpol.2018.09.075

Zou, Y., Li, Y., Guo, Y., Zhou, Q., & An, D. (2012). Ultrasound-assisted synthesis of CuO nanostructures templated by cotton fibers. *Materials Research Bulletin*, *47*(11), 3135–3140. https://doi.org/10.1016/j.materresbull.2012.08.020

2 Green Chemistry and Bioinspiration in Nanotechnology

Pavithra Poornima S and Ravindra H J

2.1 INTRODUCTION

Nanomaterials and the associated technological evolution have attracted researchers worldwide due to widespread application in the fields of agriculture, health care, automobiles, electronics, and the environment. The unique materials, along with their tunability in size and shape, allow the production of desired properties. Basically, nanomaterials are those in which at least one dimension is in the nm range. When the particle sizes are less than 100 nm most of these materials exhibit properties which are significantly different from those of their bulk counterparts. This feature makes nanomaterials unique and suitable for various applications.

2.2 GENERAL APPROACHES TO SYNTHESIS OF NANOMATERIALS

Nanomaterials of different dimensions with controllable particle sizes and morphologies can be precisely produced by optimizing synthesis methods. There are generally two routes of synthesis, as depicted in Figure 2.1.

In the top-down (TD) approach, a bulk material will be broken down to fabricate nanoparticles. Common methods used in TD approaches are mechanical milling, etching, laser ablation, sputtering, and electro explosion [1]. However, TD approaches face significant limitations when scaling below 100 nm, particularly in the fabrication of electronic integrated circuits [2].

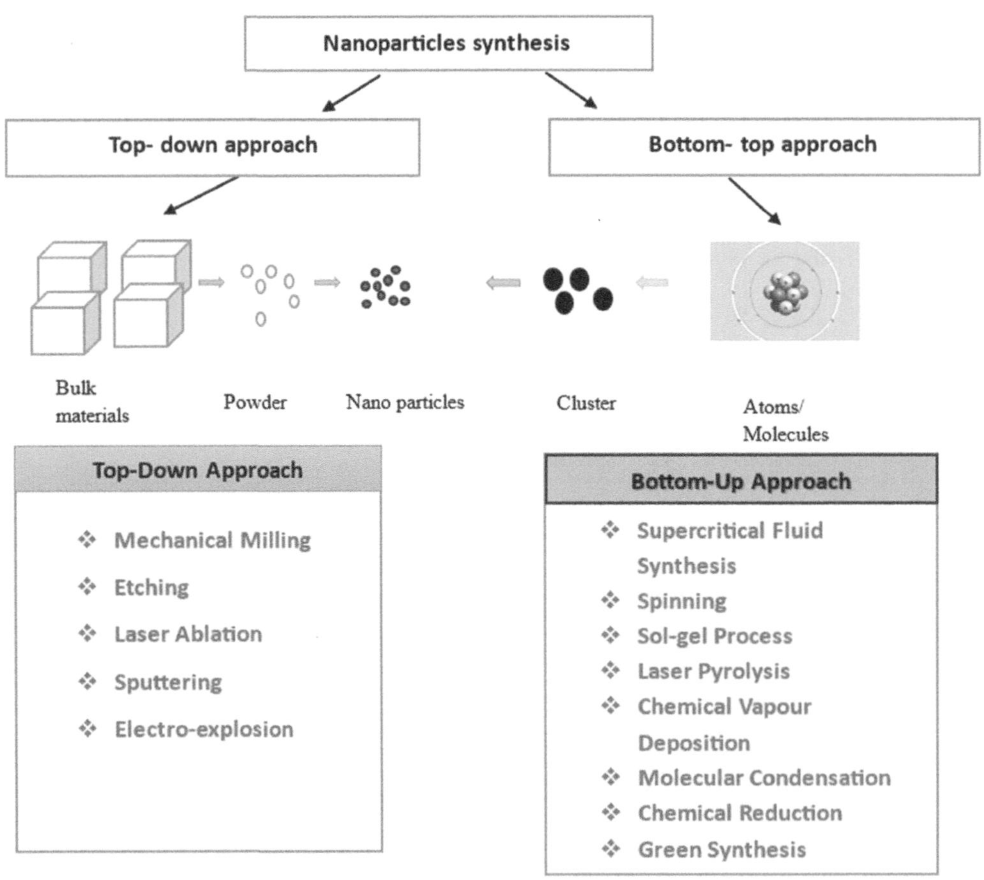

FIGURE 2.1 General approaches to synthesis of nanomaterials.

DOI: 10.1201/9781003644439-2

In bottom-up (BU) approaches, nanomaterials are synthesized by combining sub- or nanoscale objects such as atoms or molecules to build up nanostructures. The BU approach allows a more controlled system design compared to the TD approach and can be seen as the natural progression of supramolecular chemistry [2]. BU methods such as supercritical fluid synthesis, spinning, sol-gel process, laser pyrolysis, chemical vapor deposition, molecular condensation, chemical reduction, and green synthesis are novel methods.

2.3 WHY GREEN SYNTHESIS

Green synthesis represents a sustainable and eco-friendly approach to producing nanomaterials, leveraging biological systems such as plants, bacteria, fungi, algae, and other natural resources. This method adheres to green chemistry principles by avoiding hazardous chemicals, reducing energy consumption, and minimizing environmental impact. Unlike traditional physical and chemical methods, which often involve high temperatures, toxic reagents, and non-renewable resources, green synthesis utilizes phytochemicals—such as flavonoids, alkaloids, terpenoids, and phenolics—as natural reducing and stabilizing agents that facilitate the conversion of metal ions into nanoparticles.

This strategy offers multiple advantages: it reduces pollution by employing renewable resources, enhances the biocompatibility of nanoparticles for use in biomedical fields like drug delivery and imaging, lowers production costs through the use of readily available materials and mild reaction conditions, and allows for the synthesis of a wide variety of metallic and metal oxide nanoparticles with controlled shapes and sizes. Recent studies have demonstrated that nanoparticles synthesized via green methods exhibit significant antimicrobial, anticancer, and catalytic activities. For example, silver nanoparticles synthesized using *Curcuma longa* (turmeric) extract have shown potent antibacterial properties, further highlighting the potential of green synthesis in nanotechnology and medicine [3].

2.4 SYNTHESIS OF NANOPARTICLES VIA BIO/GREEN SYNTHESIS

Bio/green synthesis of nanoparticles represents a sustainable method that employs biological materials like plant extracts, microorganisms, or enzymes to fabricate nanoparticles in an eco-friendly manner. Among these, plant-mediated synthesis is widely favored due to its simplicity, rapidity, and cost effectiveness. Without the use of hazardous chemicals, phytochemicals found in seeds, leaves, or fruit extracts stabilize nanoparticles by acting as natural reducing and capping agents. This approach supports the principles of green chemistry, offering a cleaner and safer alternative for nanoparticle production with potential applications in medicine, agriculture, and environmental remediation [4].

2.5 NATURE-INSPIRED NANOMATERIAL SYNTHESIS

Nature-inspired methods for synthesizing nanomaterials get inspiration from biological systems and processes that occur naturally. These approaches are often energy efficient and environmentally friendly. The following are some nature-inspired methods.

2.5.1 Biomimetic Nanotechnology

The term "biomimetics" originates from the Greek word *biomimesis*, which means "imitation of life". The term was first used in 1957 by Otto Schmitt, a polymath who created a physical device that mimicked the electrical behavior of a nerve while pursuing his doctoral studies. Related ideas include biomimicry, biognosis, and bionics, which was first presented by Jack Steele at Wright-Patterson Air Force Base in Dayton, Ohio, in 1960 [5].

Biomimetic synthesis involves methodologies that closely emulate natural processes, often employing living organisms or using biologically derived materials such as proteins, enzymes, or cellular structures [6]. This synthesizes inorganic products under mild, environmentally friendly conditions. Bioinspired material synthesis draws on the fundamental ideas of natural biomineralization and applies them to the creation of inorganic materials using entirely artificial systems. Material bionics refers to the replication of nature's structural designs, reproducing functional features found in biological systems through specialized processing technique [7].

Some graft methods like autografts and allografts have been applied to bone reconstruction and repair [8]. However, although these grafts are highly biocompatible, though they have some drawbacks, including limited availability, risk of immune rejection, and suboptimal performance. To overcome these limitations, researchers shifted their focus to biomimetic materials to mimic the structure and properties of natural tissue for enhanced performance and biocompatibility [8].

Replicating the mechanical characteristics of cortical bone in synthetic biomaterials is quite challenging due to the complexity of its microstructure. Reproducing it requires an accurate composition and spatial arrangement of hydroxyapatite (HA) and collagen across multiple length scales.

2.5.1.1 Nacre/Mother-of-Pearl–Inspired Nanotechnology

The inner layer of abalone shells contains nacre (mother of pearl), a naturally occurring substance made of calcium carbonate and chitin [9]. It has mechanical toughness and strength arise from the brick-and-mortar arrangement of aragonite platelets, while organic biopolymers act as a lubricant, as seen in Figure 2.2. This structure can absorb energy and resist cracking due to controlled deformation, similar to bone.

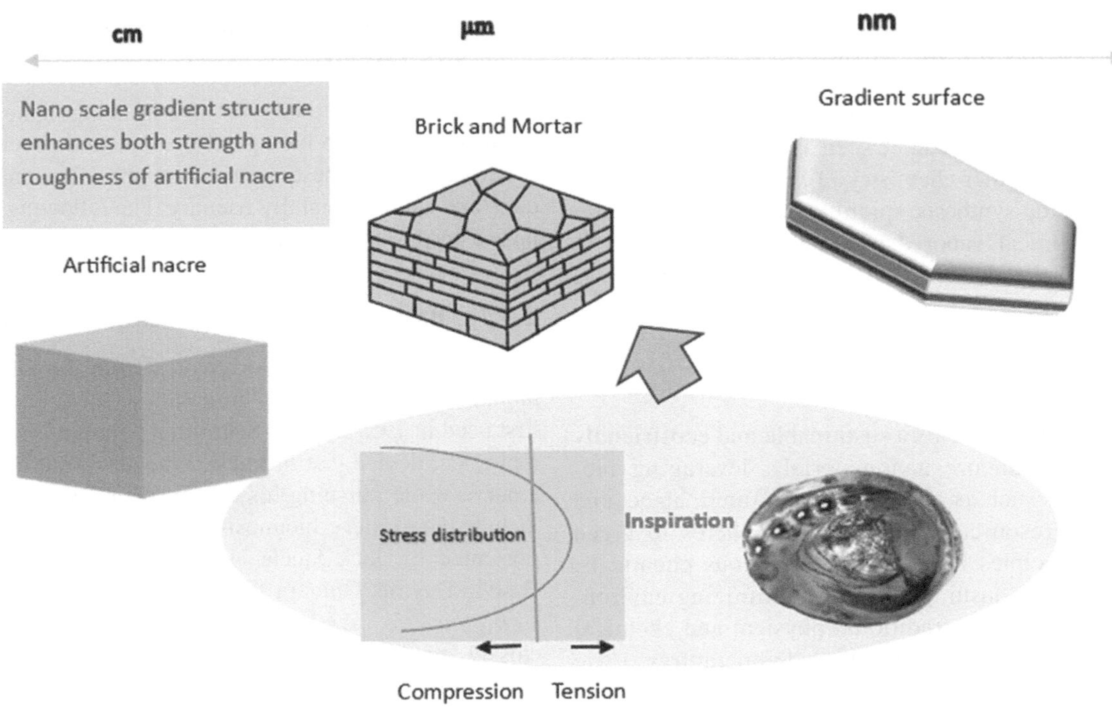

FIGURE 2.2 Nacre/mother-of-pearl–inspired nanotechnology.

Researchers found that nacre-based implants exhibit high fracture toughness (~9 MPa·m$^{1/2}$), a Young's modulus of 64–73 GPa, and a flexural strength of 210 MPa, making them durable and biocompatible orthopedic implants [8]. Libouban et al. observed surface erosion on large orthopedic screws (3.5 mm diameter) made from *Pinctada maxima* nacre, as indicated by a decrease in the fractal dimension of screw boundaries affected by the erosive process.

Green-synthesized nacre-based nanoparticles (NNPs), as illustrated in Figure 2.3, have a wide range of practical applications. When incorporated into hydrogels of hydrogels and biopolymers, they facilitate wound healing and support bone regeneration by promoting mineralization and cell growth. NNPs can also be used in drug delivery systems to enable safe and controlled release of medications. In addition, they function effectively as antimicrobial coatings to stop the growth of bacteria. Furthermore, nacre-based materials can be used to remove dyes and heavy metals from the environment. These applications highlight the safe and environmentally responsible potential of nacre-based nanoparticles.

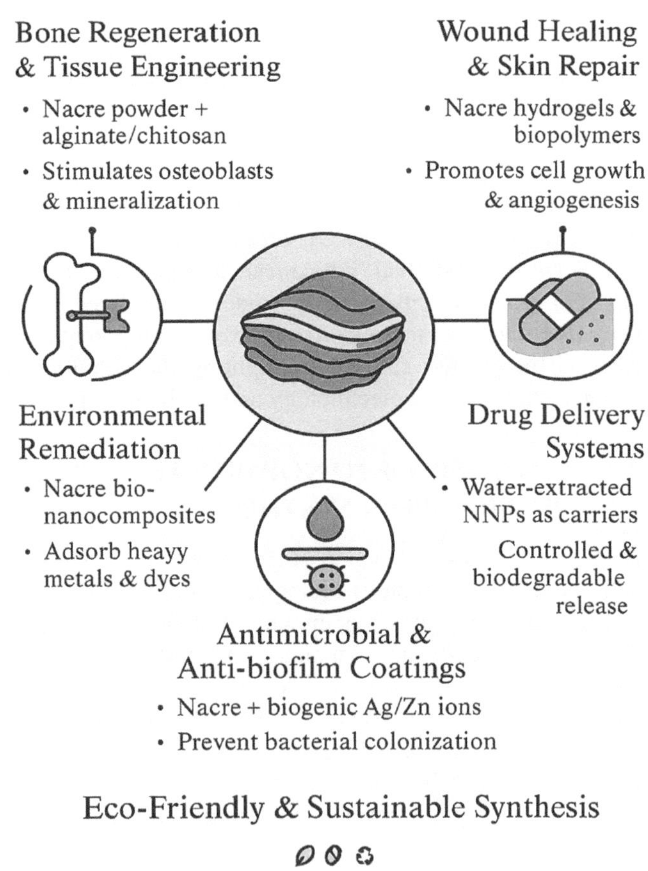

FIGURE 2.3 Applications of green-synthesized nacre-based nanoparticles.

2.5.1.2 Spider Silk–Inspired Nanomaterials

For millions of years, nature has served as an excellent source of inspiration for advanced material design. Spider silk stands out due to its extraordinary combination of toughness and strength, surpassing many synthetic materials.

Researchers Jin and Kaplan studied high-performance biomimetic fibers inspired by the biological mechanism behind silk production in insects and spiders [10]. Their studies revealed that silk proteins such as fibroin self-assemble into micellar structures approximately 100–200 nm in diameter. As the protein concentration increases, these micelles gradually aggregate into globules or gel-like states, maintaining solubility through a careful distribution of hydrophobic and hydrophilic domains.

When subjected to mechanical forces such as shearing or stretching, these assemblies transform into organized fibers, imitating the natural silk-spinning process of spiders. This has paved the way for innovation in synthetic silk materials for textiles, biomedical scaffolds, and lightweight composites.

Another example of bioinspiration is the vibrant and attractive colors seen in butterfly wings, particularly in *Morpho* species. These butterfly species exhibits a brilliant blue color not due to pigments, but due to interactions with light arising from highly ordered nanoscale architectures. These structures within the wing scales contains layered arrangements of chitin and air, which interact with light through interference and diffraction, producing vivid, angle-dependent structural colors.

In 2003, Vukusic and Sambles emphasized that the remarkable optical effects produced by naturally occurring nanostructures have inspired the development of advanced optical materials such as sensors, anti-counterfeiting measures, and color display systems by mimicking these biological designs [11].

2.5.1.3 Gecko-Inspired Nanotechnology

Another remarkable example of nanotechnology in nature is the gecko's ability to seemingly defy gravity. They can climb quickly, run along smooth vertical surfaces at up to 20 body lengths per second, and even land upside down on the ceiling without the need for liquids or suction [12].

The microscopic studies have shown that, the gecko's foot contains almost 500,000 keratinous hairs, known as setae. Each setae is approximately 30–130-μm-long and has a diameter nearly one-tenth that of human. Researchers Autumn and Peattie first performed direct measurements of the adhesive force produced by a single seta by using a 2D micro-electromechanical system (MEMS) force sensor along with wire as a force gauge. Their finding reveals that a single seta could produce an adhesive force ten times larger than previously estimated whole animal estimates. Moreover, they observed that macroscopic orientation and preloading of seta were found to enhance attachment, increasing adhesive forces by up to 600-fold compared to the internal frictional force of the material. These discoveries have driven the development of synthetic adhesives that mimic the gecko inspired adhesion mechanics. Such biomimetic adhesives show promising potential applications in diverse fields such as robotics, medical devices, and materials science, where reversible and strong adhesion is critical.

Because of their strong yet reversible adherence, gecko-inspired dry adhesives, which imitate the micro/nanoscale structure of gecko feet, offer a wide range of creative applications, as shown in Figure 2.4. These adhesives are employed in microgrippers, climbing robots, and self-cleaning surfaces to allow mobility on a variety of surfaces. They are also integrated into stimuli-responsive systems that activate under specific external condition as well as in biomedical adhesives designed for non-invasive, reusable attachment. Moreover, their ability to provide flexible and transient adhesion is critical in development of energy and sensor devices, where gecko inspired dry adhesives have proven to be indispensable [13].

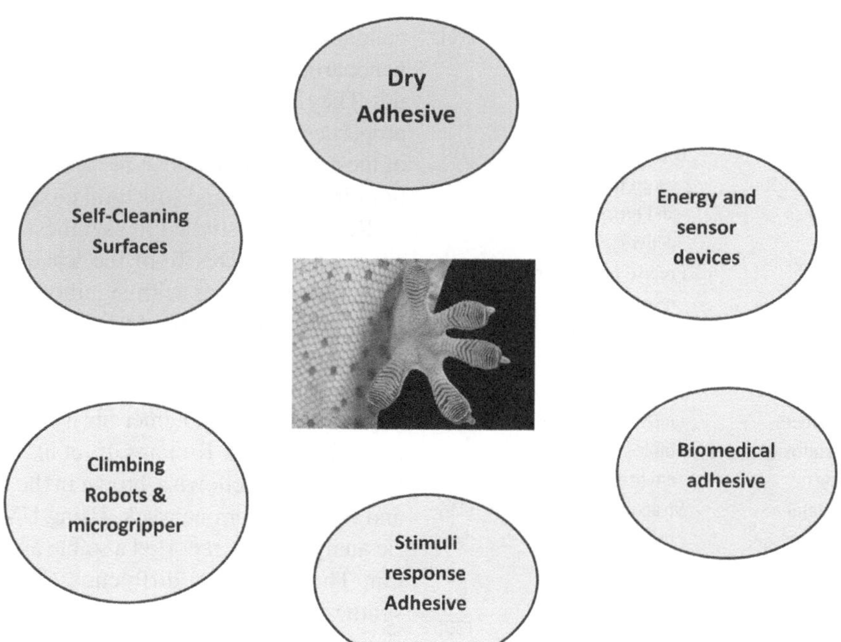

FIGURE 2.4 Applications of gecko-inspired nanoparticles.

2.5.1.4 Biomimetic Drug Delivery Systems

A drug delivery system refers to the methods and technologies used to transport a therapeutic compound to its specific within the body to achieve a desired optimal therapeutic effect. An effective system ensure that the drug reaches the targeted location at the right time in the correct dosage, while minimizing organ damage and undesirable effects.

Biomimetic drug delivery systems draw inspiration from nature, particularly from the biological structure and mechanical process guide substances in the body efficiently. Numerous studies in the field of biomimetic drug delivery systems [14] and approaches offer promising advancements in the development of safer, more targeted, and more effective drug delivery technologies, some of these approaches are listed in Table 2.1.

2.5.2 Green Synthesis (Biogenic Synthesis)

In recent years, the development of green chemistry methods for metal nanoparticle synthesis has become a major area of research. Scientists are actively exploring eco-friendly methods for producing well-characterized nanoparticles. Among the various biological methods, the plant-based synthesis has emerged as an highly promising strategy for large-scale biosynthesis. Plant-mediated synthesis of nanomaterials offers several advantages, including environmental friendliness, cost effectiveness, and scalability, making it superior to the other biological organisms such as microorganisms [21, 22].

TABLE 2.1
Biomimetic Drug Delivery Systems

Biomimetic System	Inspired by	Biological Function Mimicked	Ref.
Hydrogels	Human tissues (e.g., cartilage, extracellular matrix)	Moisture retention and controlled substance diffusion	[15]
Micelles	Cell membranes/ bile salts	Encapsulation of hydrophobic substances in aqueous systems	[16]
Liposomes	Phospholipid bilayer of cell membranes	Biocompatible cargo transport and targeted delivery	[16]
Dendrimers	Enzymes/ proteins with branching structures	Precise molecular recognition and cargo release	[17]
Polymeric carriers	Natural biopolymers (e.g., chitosan, alginate)	Degradability and interaction with biological environments	[18]
Nanotubes/ nanofibers	Cytoskeletal elements (e.g., microtubules, collagen)	Structural integrity and transport within cells	[19]
Nanoparticles	Viruses, exosomes	Targeted delivery and immune evasion	[20]

Nanoparticles produced through plant-mediated methods are generally more stable, and their synthesis is faster. Moreover, these nanoparticles exhibit greater diversity in shape and size compared to those generated by other biological systems. The advantages of plants and plant-derived materials, such as cost-effectiveness, eco-friendliness, and the ability to produce nanoparticles with defined sizes and shapes, have encouraged researchers to investigate the mechanisms of metal ion uptake, bio reduction processes, and the overall pathways of nanoparticle formation in plants. This section explores different plant species that have been employed in the biosynthesis of metal nanoparticles, shedding light on their role in advancing sustainable nanotechnology [23].

2.5.2.1 Plant-Mediated Synthesis of Nanoparticles

Plant-mediated nanoparticle synthesis is a sustainable and eco-friendly alternative to traditional chemical and physical processes. This environmentally friendly method enables the production of a variety of metal and metal oxide nanoparticles by utilizing the natural reducing, capping, and stabilizing agents present in plant extracts, including flavonoids, phenolics, and alkaloids. The properties of nanoparticles are greatly influenced by variables like temperature, pH, reaction time, and extract concentration. Because of their biocompatibility, affordability, and low environmental impact, these biosynthesized nanoparticles have found applications in biomedicine, environmental remediation, catalysis, and agriculture [24].

Researchers have extensively studied the development of nanomaterials using metallic nanoparticles, particularly silver nanoparticles (AgNPs), due to their notable biological properties. They have observed that AgNPs exhibit strong antibacterial properties against a wide range of bacteria. The effectiveness of these nanoparticles can be significantly influenced by variations in their size and structural morphology [24].

Researchers studied the synthesis and characterization of silver nanoparticles from the leaves of *Eugenia roxburghii DC,* a plant species with significant pharmaceutical importance. This species produces a tasty edible wild fruit with high mineral and vitamin content. Researchers also evaluated its effectiveness in inhibiting production of biofilm. Biofilm is an extracellular polymer fibril which helps bacteria adhere to surfaces. Alok Kumar Giri et al. observed a color change from green to yellowish-brown in the mixture of plant extract and AgNPs after one week. Using UV–Vis spectrophotometric analysis, they reported a stable and prominent peak at 417 nm. Through X-ray diffraction (XRD), the crystallinity of synthesized nanoparticles was confirmed and the particle size was found to be approximately 35 nm [25].

The variety of plant species utilized as biological factories for the synthesis of metal and metal oxide nanoparticles is demonstrated in an extensive tabular summary of plant-mediated nanoparticle synthesis (Table 2.2). The table includes details like the type of nanoparticle synthesized (such as silver, gold, or zinc oxide), the synthesis conditions, the applications of the nanoparticles, and the plant source (leaves, roots, bark, etc.) as illustrated in Figure 2.5. This summary sheds light on how phytochemicals affect the size, shape, and stability of nanoparticles and offers insight into their potential for use in industrial, environmental, and biomedical settings.

High-resolution transmission electron microscopy (TEM) was used to analyze the shape and size characteristics of colloidal nanoparticles with the help of a carbon-coated copper grid. TEM images revealed that the nanoparticles had an average estimated size of 24 nm [25].

In this study, a distinct difference was observed in the Congo red agar (CRA) plate assay. Bacteria on the control plate (without AgNPs) exhibited a characteristic color change, indicating biofilm production. In contrast, no color change was seen in the AgNP-treated plate, confirming that the biosynthesized silver nanoparticles effectively inhibited bacterial biofilm formation.

2.5.2.2 Microbial Synthesis Using Bacteria and Fungi

Microorganisms can synthesize nanoparticles either intra- or extracellularly, and this process can occur in various plant parts such as roots, leaves, stems, flowers, and seeds, as shown in Table 2.3. This section discusses endophytic microorganisms including bacteria and fungi. An endophyte is an organism (usually a bacterium or fungus) that lives within plant tissues. The steps involved in the mechanism of nanoparticle synthesis are bio-reduction, growth, and stabilization.

2.5.2.3 Algae-Mediated Nanoparticle Synthesis

Algae offer a flexible and sustainable resource for the environmentally friendly synthesis of metal nanoparticles. They are abundant in bioactive metabolites such as polysaccharides, proteins, phenolics, pigments, and amino acids, which can act as reducing and capping agents during nanoparticle biosynthesis.

Reducing agents convert metal ions (e.g., Ag^+ or Au^{3+}) into their elemental form (Ag^0 or Au^0), facilitating the formation of nanoparticles. Capping agents, on the other hand, bind to the nanoparticle surface, stabilizing the particles by preventing agglomeration and controlling their growth and morphology. These capping biomolecules play an important role in determining the size, shape, stability, and biological activity of the synthesized nanoparticles [43].

Table 2.4 summarizes various algal species utilized for the green synthesis of silver and gold NPs (AuNPs),

TABLE 2.2
Applications and Types of Nanoparticles with Plant-Based Inspiration

Plant Origin	NPs	Size (nm)	Morphology	Applications	Ref
Azadirachta indica (neem)	Gold, silver, and silver-gold alloys	50–100	Spherical, triangular, hexagonal	Remediation of toxic metals	[26]
Ocimum sanctum (tulsi; leaf extract)	Gold and silver	30 and 10–20	Crystalline, hexagonal, triangular, and spherical	Biolabeling, biosensor	[27]
Banana Peel	Silver	23.7	Crystalline, spherical	Biosensing, imaging, drug delivery, nanodevice fabrication, and in medicine	[28]
Hibiscus subdariffa leaf	ZnO	12–46	Crystalline, spherical	Antibacterial and anti diabetic	[29]
Camellia sinensis (green tea)	Fe_3O_4 (iron oxide)	10–40	Spherical, quasi-spherical	MRI contrast agent, drug delivery, magnetic separation	[30]
Syzygium aromaticum (clove)	CuO	~ 44	Crystalline, spherical, rod-shaped	Antibacterial, catalysis, wastewater treatment	[31]
Azadirachta indica (neem)	TiO_2	124	Rods, spheres, flower-like	Photocatalyst, self-cleaning	[32]
Cinnamomum spp.	Pd	12–26	Spherical, triangular, rectangular, cubic, and decahedron	Catalysis, cancer therapy	[33]
Camellia spp.	Pt	16–18	Irregular, spherical	Catalysis, drug delivery	[34]
Allium sativum (garlic)	Se	7–45	Spherical, amorphous	Nutritional, anticancer	[35]
Tamarindus indica	Ni	20–40	Spherical, polycrystalline	Catalysis, sensor application	[36]
Aloe vera	CeO_2	8–25	Spherical, cubic	Antioxidants, neuroprotection	[37]

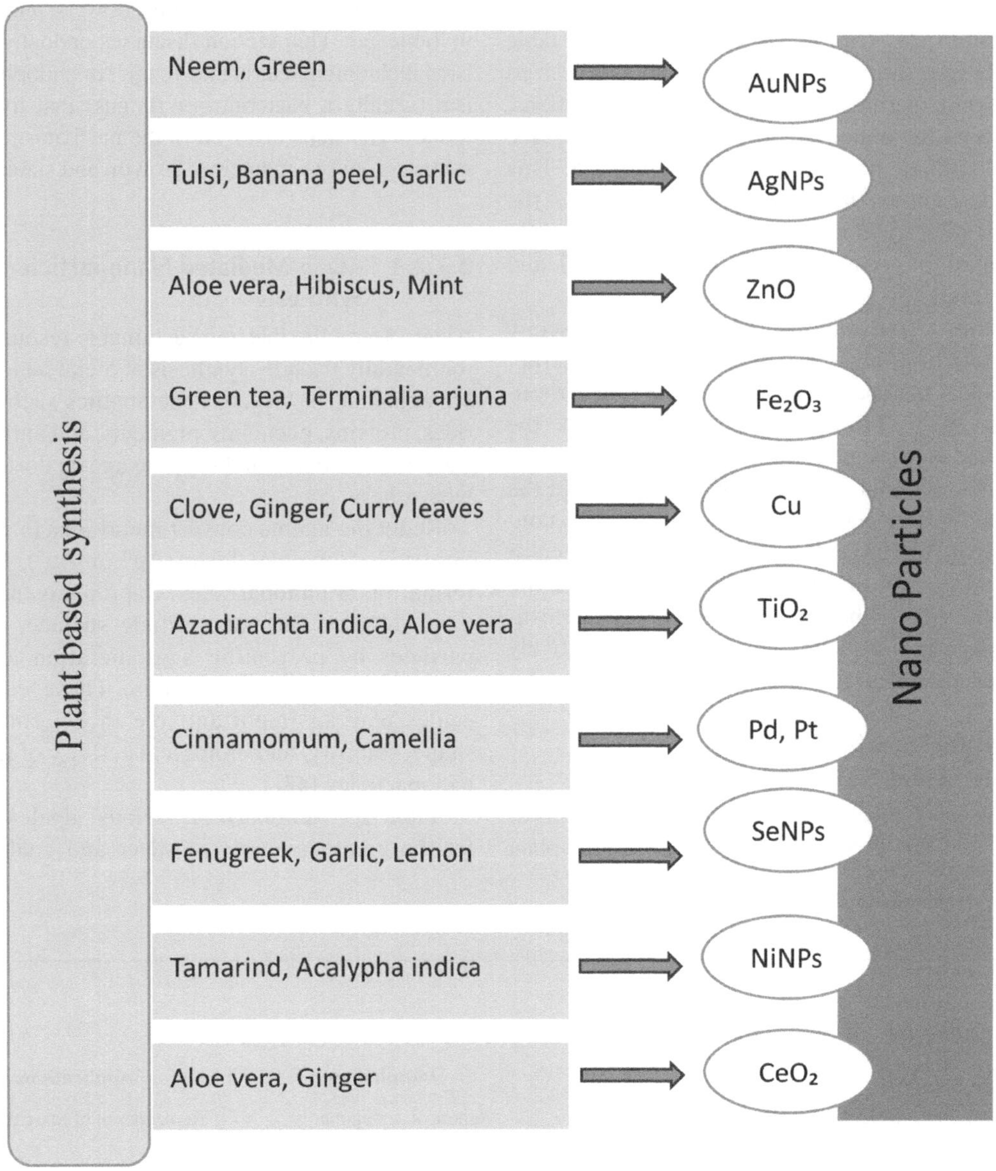

FIGURE 2.5 Plant-based synthesis of nanoparticles.

TABLE 2.3
Microbial Synthesis Using Bacteria and Fungi

Plants	Endophytes	UV-Vis	Shapes	Sizes of NP	NPs	Activity	Reference
Coriandrum sativum	*Bacillus siamensis* strain C1	200–800 nm	Spherical (TEM and SEM)	25–50nm	Silver (SEM and EDS)	Antibacterial	[38]
Coffea arabica L	*Pseudomonas fluorescens*	450–650 nm	Spherical to triangle to hexagonal	5–50nm	Gold	Antibacterial	[39]
Cinnamomum zeylanicum	*Lasiodiplodia theobromae*	300–700 nm	Spherical, oval	76nm	Silver	Antibacterial	[40]
Oxalis corniculate	*Streptomyces zaomyceticus Oc-5*	400 nm	Spherical	78–80nm	Copper oxide	Antimicrobial, antioxidant and anticancer	[41]
Ocimum sanctum	*Streptomyces coelicolor*	440 nm	Spherical and ellipsoidal	25 nm	MgO	Antimicrobial	[42]

TABLE 2.4
Algae-Mediated Nanoparticle Synthesis

Algal Species	Type	NP Type	Bioactive Components (Reducing and Capping Agents)	UV-Vis λmax (nm)	Shape	Size (nm)	Activity	Ref
Sargassum wightii	Brown algae	Silver (AgNPs)	Polysaccharides, phenolics	~420–430	Spherical	10–50	Antibacterial against *E. coli, S. aureus*	[44]
Chlorella vulgaris	Micro algae	Silver (AgNPs)	Proteins, chlorophyll, flavonoids	~410–425	Mostly spherical	20–80	Antifungal, anticancer	[45]
Ulva lactuca	Green macro algae	Silver (AgNPs)	Sulfated polysaccharides	~415	Spherical	15–70	Broad-spectrum antimicrobial	[46]
Spirulina platensis	Cyanobacteria	Silver (AgNPs)	Phycocyanin, proteins, amino acids	~420	Variable (sphere, rod)	30–50	Anti-inflammatory, antioxidant	[47]
Sargassum muticum	Brown algae	Gold (AuNPs)	Phenolic compounds, polysaccharides	~550	Spherical	15–40	Potential anticancer and catalytic activity	[48]
Gelidium amansii	Red algae	Gold (AuNPs)	Proteins, polysaccharides	~525	Spherical	27–54 nm	Antimicrobial, antioxidant	[49]

highlighting their source bioactive compounds, particle characteristics, and associated biological activities:

2.5.2.4 Honey-Mediated Green Synthesis of Nanoparticles

Natural honey is valued as one of the oldest known foods and is an excellent source of energy, with notable medicinal properties. It is produced by the honey bee, *Apis mellifera* (*A. mellifera*), from plant nectar, secretions, and excretions. Form ancient times, natural honey has been an excellent medicinal source, and a Sumerian tablet recorded honey as a drug and ointment. Honey is a sticky and viscous mixture composed of 0–85% carbohydrate (mainly glucose and fructose); 15–17% water; 0.1–0.4% protein, 0.2% ash; and minor quantities of amino acids, enzymes, and vitamins as well as other substances such as phenolic antioxidants. The color of honey can range from colorless to dark amber or black depending on botanical origin, age, and storage condition. However, its transparency depends on the number of particles, such as pollen [50].

Researchers have observed that honey-based synthesis offers several advantages over microorganism-based methods. The separation of nanoparticles from microorganisms is challenging, as it requires careful handling and involves a significant time lag in the nanoparticle conversion process [51].

Daizy Philip et al. published a study related to gold nanoparticles. Philip demonstrated that by adjusting the concentrations of chloroauric acid ($HAuCl_4$) and honey in aqueous solutions, it is possible to synthesize gold nanoparticles with different shapes and sizes at room temperature. The study suggested that, in addition to facilitating nanoparticle synthesis, honey can also act as a capping agent and help stabilize the nanoparticles [52]. Fructose acts as primary reducing agent of honey for the reaction, while the presence of protein in honey is responsible for the stability of the nanoparticles. Many researchers have been inspired by honey and synthesized nanoparticles like silver, gold, and platinum, as listed in Table 2.5.

TABLE 2.5
Honey-Based Nanoparticle Synthesis

Nanoparticle Type	Reducing Agent	Stabilizing Agent	Key Observations/ Applications	Ref
Gold	Fructose, vitamin C, H_2O_2, gluconic acid	Proteins in honey	Reduction speed increased with honey volume; nanoparticle stabilization	[52]
Gold	Honey	Honey	9.9 ± 3.7 nm size; moderate antibacterial activity against Gram+ and Gram– strains	[50]
Silver	Honey	Honey	Stable for 5 months; particle size decreased with increased honey concentration and pH	[53]
Silver	Fructose	Proteins in honey	Effective corrosion inhibitors; antimicrobial agent	[54]
Silver	Honey	Honey	Size ~4 nm; high crystallinity; protein capping via carboxylate group	[55]
Silver	Honey	Honey	Size 11.9 ± 5.25 nm; strong antimicrobial and antifungal activity	[50]
Palladium	Honey	Honey	5–40 nm size; used as catalyst in Suzuki coupling and hydrogenation	[56]

(*Continued*)

TABLE 2.5 ***(Continued)***
Honey-Based Nanoparticle Synthesis

Nano-particle Type	Reducing Agent	Stabilizing Agent	Key Observations/ Applications	Ref
Carbon	Honey (precursor)	Polysorbate, PEG	Size ~7 nm; used for real-time photoacoustic imaging	[57]
Carbon Dots	Honey	Honey	Size ~2 nm; QY ~19.8%; used in biosensing, bioimaging, fluorescent staining	[58]
Platinum	Honey	Proteins in honey	Size ~2.2 nm; catalytic properties for dye formation; nanowire formation	[59]

2.5.2.5 Fruit Extract–Inspired Green Synthesis

Green synthesis of NPs using biological materials offers a sustainable and environmentally friendly alternative to conventional chemical and physical methods. Among biological sources, fruits have emerged as effective reducing and stabilizing agents due to their rich phytochemical compounds such as vitamin C, flavonoids, phenolic acids, tannins, sugars, and enzymes. Utilizing fruit extracts ranging from edible part to peels and wastes supports waste recycling methods. These compounds facilitate the reduction of metal ions and stabilize nanoparticles such as gold, silver, and ZnO, enabling their use in healthcare, catalysis, environmental applications, and related fields. Table 2.6 presents various fruit extract used in the green synthesis of nanoparticles, highlighting the type, morphology, and size of the nanoparticles produced.

2.5.2.6 Seed-Mediated Nanoparticle Synthesis

Seed extracts from plants such as *Moringa oleifera*, sesame, citrus, litchi, and grape serve as efficient bio-reducing agents for green synthesis of nanoparticles. These seeds are rich in antioxidants, phenolic compounds, and proteins that help reduce metal ions into stable nanoparticles. *Moringa oleifera* seeds, in particular, are noted for their strong coagulant and antimicrobial properties, while sesame and grape seeds exhibit strong antioxidant activity. Citrus and litchi seeds also provide bioactive molecules that support the formation of nanoparticles with uniform morphology. This green synthesis strategy is both eco-friendly and cost-effective, offering significant potential for applications in medicine, agriculture, and environmental remediation. Table 2.7 represents the summary of the seed-mediated green synthesis of nanoparticles, detailing their type, morphology, and size, derived from various plant seeds.

TABLE 2.6
Fruit extract inspired green synthesis

Common Name	Scientific Name	Biological Extract	Type of NP	Morphology	NP Size	Ref
Strawberry	*Fragaria ananassa*	Whole fruit	Copper oxide	Sphere	10–30 nm	[60]
Guava	*Psidium guajava*	Whole fruit	Copper oxide	Flakes	15–20 nm	[61]
Pomegranate	*Punica granatum*	Juice	Gold	Triangular, pentagonal, hexagonal, and sphere	1–100 nm	[62]
Watermelon	*Citrullus lanatus*	Rind	Gold	Sphere	20–140 nm	[63]
Papaya	*Carica papaya*	Juice	Silve	Sphere	75 nm	[64]
Grape	*Vitis vinifera*	Whole fruit	Silver	Sphere	30–40 nm	[65]
Pineapple	*Ananas comosus*	Juice	Zinc oxide	No dimension	30–57 nm	[66]
Pomegranate	*Punica granatum*	Peel	Zinc oxide	Sphere	32.98–81.84 nm	[67]

TABLE 2.7
Seed-Based Nanoparticle Synthesis

Common Name	Scientific Name	Biological Extract	Type of NPs	Morphology	NP Size (nm)	Ref
Moringa	*Moringa oleifera*	Seed extract	Silver (AgNPs)	Spherical	10–50	[68]
Sesame	*Sesamum indicum*	Seed extract	Iron (FeNPs)	Spherical	~18	[69]
Citrus	*Citrus* spp.	Seed extract	Zinc oxide (ZnO NPs)	Spherical	20–60	[70]
Litchi	*Litchi chinensis*	Seed extract	Silver (AgNPs)	Hexagonal	80-120	[71]
Grape	*Vitis vinifera*	Seed extract	Silver/gold (Ag/Au NPs)	Mostly spherical	10–45	[72]

2.5.3 Sensors and Energy Devices Inspired by Nature

Living organisms generally possess a special ability to sense and react to environmental changes by collecting, processing and reacting to external stimuli [73, 74]. For instance, butterflies [75], chameleons [76], and even polar bears experiencing thermal fluctuation [77] all exhibit structural color changes. In particular, butterfly wings exhibit highly organized micro- or nanostructures, eye-catching wing coloration, a light-trapping mechanism, and stimuli responsive behavior [78]. Researchers have replicated these natural features through bioinspired approaches, translating biological structures and intelligent systems into synthetic or man-made materials for applications such as sensors, photovoltaics [79], robotics [80], and biomedicine [81].

The biological characteristics of butterfly wings were first investigated by Hooke in 1665. Later, H. Ghiradella and colleagues [82] examined chitin—a polysaccharide [$C_8 H_{13} O_5 N]_n$) and the primary structural component of butterfly wing scales—and reported that its nanoscale architecture, with features only a few hundred nanometers in size, creates variations in refractive indices that results in structural color [83, 84].

Sensors are advanced intelligent systems that detect and respond to changes in the environment [85] such as variations in humidity, temperature, vapors, and pH level. Even though sensors are used commercially in monitoring systems, they often face challenges such as low selectivity and sensitivity and complex fabrication processes [74]. To overcome these limitations, researchers turned to nature for inspiration. Butterfly wings are known for their micro- and nanostructures which enable selective and precise responses to external stimuli. These structures are not only lightweight and environmentally friendly but also relatively to fabricate. Such bioinspired sensors have found applications in biomedical diagnostics, security devices, and environmental monitoring technologies [86].

2.5.3.1 Thermal Sensors Inspired by Nature

Thermal or temperature sensors are designed to monitor the temperature to ensure safety and efficiency in and around the system. This type of sensor works through thermosensitive materials which absorb heat and convert temperature variation into measurable signals by altering its structure and material properties [87].

Various types of thermal sensors such as thermistors, thermocouples, and resistant temperature detectors (RTDs) are increasing in demand due to their stability, accuracy, and reliability although they may have limitations in complexity, and low spatial resolution. Nature offers valuable insights: fire beetles detect heat via infrared receptors located in their metathoracic pit organs [88].

Similarly, butterfly wing–inspired sensors [89] have a nanostructure capable of translating external stimuli into visible optical signals. These structures reacts to change in temperature through thermal expansion and show a corresponding shift in the refractive index. Studies shown that Morpho butterflies display different color changes when exposed to infrared radiation, inspiring researchers to develop a model for visually responsive thermal sensors [90].

To enhance the thermal detection range and functionality, researchers have integrated butterfly wing–inspired nanostructures with smart polymers and semiconductors, resulting in more sensitive and adaptive thermal sensing devices [91].

2.5.3.1.1 Chlorophyll-Inspired Thermal Sensing

In nature, chlorophyll is responsible for the green color, but it undergoes a visible transformation to brown or black when exposed to extreme temperatures. Inspired by this, a precursor molecular sensor (PMS) has been developed to exhibit a change of color at 180°C. When integrating PMS with a flame warning system, it is capable of producing a colorful alarm at 270°C within 20 secs [92, 93].

2.5.3.2 Bioinspired Medical Sensors for Real-Time Health Monitoring

The human sensory system is highly complex for sight, sound, taste, smell, and touch [94] and neurodegenerative diseases like Alzheimer's and Parkinson's can significantly affect these sensory functions. Similarly, glaucoma, a progressive eye disease [95], damages the optical nerves, resulting in irreversible blindness if left untreated. To support diagnosis and monitoring, advanced medical sensors are being developed to detect key physiological signals. Traditional devices such as electrograms typically rely on bulky equipment or invasive attachment methods, limiting patient comfort and practical usage [96].

Bioinspired medical sensors offer an innovative solution by mimicking natural systems. These sensors are flexible, lightweight, cost-effective, and designed to conform naturally to the contours of human tissue. They integrate functional materials with microfluidic and electronic systems for accurate and continuous monitoring of physiological parameters [97].

For instance, a wearable biosensor inspired by *Morpho menelaus* [98] butterfly wings has been developed to assist in monitoring neurodegenerative diseases. It incorporates a microfluidic layer, which enhances fluorescence signal for biomarker detection, along with a conductive ink-based electronic network to track physiological signals such as tremors. This hybrid sensor, operable via smartphone, facilitates simultaneous biochemical and physical monitoring, offering a valuable tool for early diagnosis and personalized healthcare applications [99].

Another example is, biophotonic intraocular pressure (IOP) sensors inspired by *Chorinea faunus* butterfly wings, with nanostructured membranes, have been developed for long-term glaucoma monitoring, showing excellent optical performance and biocompatibility in animal models [95]. Similarly, the design of gyroid-shaped poly(lactic acid) (PLA) scaffolds demonstrated superior porosity, mechanical strength, and longevity suitable for tissue engineering [100].

Additionally, a wearable sensor inspired by peacock feathers [101], designed to detect tear-based biomarkers for diabetes, offers a non-invasive method for glucose monitoring. These examples highlight how nature-inspired designs enable flexible, lightweight, and efficient sensors for real-time health monitoring and disease diagnosis.

2.5.3.3 Bioinspired Vapor Sensors

Toxic and volatile organic compounds (VOCs) like hazardous gases and vapors are harmful pollutants if leakage occurs [102] and require fast and selective detection. Traditional sensors composed of thick films result in a low diffusion rate and struggle from low sensitivity due to limited surface areas [103, 104].

Nature-inspired designs, particularly from butterfly wings, offer a solution by enabling sensor systems with high surface area, rapid response time, high sensitivity, and porous structure. These wings feature porous, hierarchical nanostructures that allow rapid vapor diffusion and produce visible color changes when exposed to specific vapors. This makes them useful for real-time, selective sensing characteristics [105].

Recent developments include using *Papilio paris* wing templates to fabricate a biomass carbon-(C/TiO_2) TiO_2 sensor with high surface area and strong vapor responsiveness when exposed to benzene and dimethylbenzene vapors at 300°C [106].

2.5.3.3.1 Chemical Warfare Agent Detection Using Bioinspired Sensors

Chemical warfare agents (CWAs) pose a serious threat to public health and security when released into the environment. Conventional detection techniques using chemiluminescence and porous silicon sensors are unsuitable for long-term use because of their poor sensitivity and selectivity, sensor degradation, high cost, and lack of portability [107].

The distinctive nanostructures of *Morpho didius* butterfly wings have served as an inspiration for recent developments in sensor technology. These nanostructures have been used to detect chemical warfare simulants like dichloropentane (a simulated mustard gas) and dimethyl methyl phosphonate (a simulated nerve agent) vapors [108]. These bioinspired sensors combine speed, sensitivity, and selectivity, which are essential traits for advanced environmental and security monitoring systems. Each analyte can be measured at parts per million thanks to the periodic structures inside the wings that reflect visible light in a measurable way.

2.5.3.4 Bioinspired Anti-Counterfeit Security Systems

Counterfeiting of important documents and currencies poses significant security risks [109]. Multi-layered security features, including overt markers that are visible to the unaided eye, semi-covert markers detectable by machines, and covert markers identifiable using forensic tools and specialized equipment, are commonly found in anti-counterfeit systems.

The primary objective is to develop an extremely intricate, one-of-a-kind, and personalized system that is virtually impossible to replicate [108]. Current anti-counterfeit solutions make extensive use of holograms, a printing technology that has made it feasible to replicate even the most complex holographic patterns [110]. Butterfly wings in particular have remarkable optical qualities that are not only resistant to fading but also highly responsive to a variety of stimuli, including mechanical, chemical, and electrical forces, as well as changes in pH and temperature.

These bioinspired nano- and microarchitectures of wings are resistant against counterfeiting because of their optical clarity, robustness, and individuality [107, 108]. Another innovation was the creation of templates based on polyvinylidene fluoride (PVDF) films, inspired by morphology and used for secure authentication. Through electronic devices, this bioinspired PVDF demonstrated exceptional mechanical strength (2.97 MPa) and flexibility (1000 bending cycles), as well as distinctiveness, durability, and perception [111].

Additionally, femtosecond laser processing was utilized to bleach the wing surface and imprint a digital signature of authentication on *Issoria lathonia* wings, which were used to create write-only memory security features. This method creates long-lasting, non-replicable security tags and is inexpensive, scalable, straightforward, and easily integrated with current systems [112].

2.5.3.5 Energy Applications Inspired by Butterfly Wings

Focus has shifted to unconventional energy sources like solar, wind, and tidal energy due to the growing demand for energy production. The energy conversion efficiency of the technologies is often limited. Because of their distinct porous structures, large surface area, and ability to manipulate light, butterfly wings offers inspiration for bioinspired techniques.

Butterfly wings improve charge transfer kinetics; facilitate simple semiconductor integration for easy replication; and are suitable for low-cost, simple-to-fabricate, environmentally friendly energy materials. These bioinspired systems contribute to energy harvesting, energy storage, and environmental remediation.

2.5.3.5.1 Photocatalysis

Photocatalysis uses light to activate catalysts for redox reactions by absorbing photons and generating energized electrons and holes. It offers eco-friendly pollutant degradation even at low concentrations. Researchers turned to butterfly-inspired photocatalysts to overcome issues like large band gaps and electron-hole recombination by utilizing their surface area and structural precision, which enhance charge separation and light absorption.

For instance, *Morpho didius* wings were used to fabricate 3D CdS/Au photocatalysts, which achieved hydrogen gas production (221.8 μmol/h). Similarly, polylactic acid gyroid structures implanted with TiO_2 showed effective methylene blue dye degradation under UV light.

Cicada wing–inspired fabrication of Ag–TiO_2 photocatalysts demonstrated effective methylene blue dye degradation within 15 minutes due to enhancement of surface plasmon resonance and improved light absorption [113].

Energy-harvesting technologies have the potential to convert renewable energy from the environment. Three energy harvesting methods are nanogenerators, biofuel cells, and photovoltaic cells. Conventional thin film solar systems have limited efficiency due to their limited quantum conversional outputs and absorption coefficient. Recently, bio-photosynthetic architectures have inspired the development of more efficient designs with enhanced solar absorption and increased surface area.

For example, the development of highly absorbing photovoltaic active layers inspired by the wings of the black butterfly (*Pachliopta aristolochiae*) has resulted in nanopatterned photovoltaic layers consisting of a thin a-Si:H layer deposited on a glass substrate. These layers are composed of a mixture of poly(methyl methacrylate) and polystyrene mixed in methyl ethyl ketone.

The nanopattern significantly boosts light absorption from 90% at normal incident angles as high as 200% at specific angles. Although output power and storage still requires improvement, nanogenerators transform mechanical energy into electrical energy. The drawbacks of conventional power generation systems include their high cost, complex design, and poor energy conversion.

Triboelectric nanogenerators (TENGs), particularly butterfly-inspired versions (B-TENGs), offer several benefits, including low cost, high reliability, and high efficiency. They can also generate sufficient power to light approximately 180 LEDs using energy harvested from water waves. Biofuels convert chemical energy to electrical energy, but their efficiency is limited by type and availability biomolecule. Efficiency, flexibility, surface area, and energy harvesting capacity are all increased by incorporating bioinspired architectures. Dielectric nanocomposites inspired by spider silk, for example, dielectric nanocomposites inspired by spider silk have demonstrated a 200% improvement in energy density at 1500°C.

2.5.3.6 Lotus Effect–Based Green Synthesis of Nanoparticles

The lotus (Nelumbo nucifera), a superhydrophobic plant often regarded as the "king plant" among semi-aquatic species, develops peltate leaves up to 30 cm in diameter [110], as shown in Figure 2.6. It is also considered a symbol of purity, as the plant is able to unfold perfectly clean leaves even when emerging from mud and dirty water [111]. Lotus leaves exhibit excellent self-cleaning behavior due to their non-wettability property.

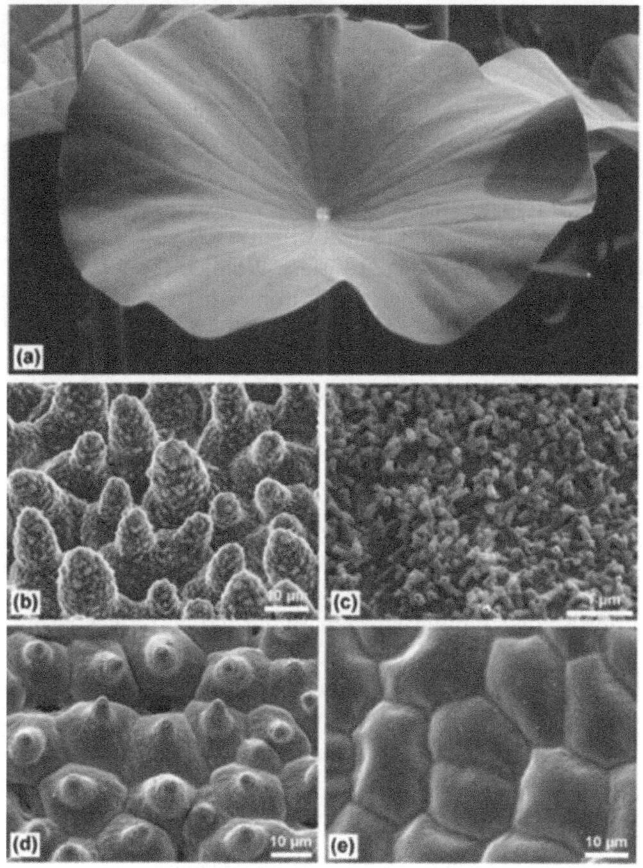

FIGURE 2.6 Water repellence and structure of lotus leaves.

Reproduced from (https://doi.org/10.3762/bjnano.2.19 © 2011 H. J. Ensikat et al., published by Beilstein-Institute, distributed under the terms of the Creative Commons Attribution 4.0 International License, https://creativecommons.org/licenses/by/4.0).

As shown in Figure 2.6, lotus leaves exhibit remarkable water repellence due to a hierarchical surface structure composed of papillae and wax tubules. SEM imaging reveals these features, while critical-point drying dissolves the wax, making stomata more visible. The underside of the leaf shows convex cells without the stomata.

2.5.3.6.1 *Lotus Effect*

In the late 1970s, Prof. Wilhelm Barthlott, a German botanist, started his research on lotus leaf surface roughness, adhesion, and water repellence [112] and found that these are the keystones for the self-cleaning mechanism. He found that adhesion between water droplets and dust particles is higher than the droplet and the leaf surface, which makes the spherical droplet carry dust while rolling off the lotus leaf. This phenomenon is called the "lotus effect" [112].

2.5.3.6.2 *Contact and Tilt Angles*

Numerous studies focused on the micro/nanostructure, wetting properties, and application of lotus leaf.

Lotus leaves can preserve superhydrophobicity for a long time due to their self-healing property [114]. The angle between the droplet and surface is called the contact angle, which describes how a surface interacts with droplets or how well it attracts or repels water. Superhydrophobic surfaces are defined as those with a contact angle greater than 150° [110].

The minimum tilt angle at which a water droplet begins to roll off or slide down a surface is called the tilt angle. It is measured by placing a water droplet on the surface that is tilted gradually. Typically, the sliding angle of the lotus leaf is very low, less than 10°. The lower the angle, the more easily water rolls off, carrying away dirt particles and enabling the self-cleaning mechanism.

2.5.3.6.3 *Morphology of Lotus Leaves*

The outermost cell layer of the lotus leaf is made of epidermal cells, which are irregular polygons contributing microscale roughness. Each epidermal cell gives rise to a domelike extension called a papilla, which is important for water repellence. Papillae are densely packed to form a hierarchical microstructure and reduce the contact area between water droplets and the surface of the leaf [110].

2.5.3.6.4 *Mechanical Protection of Wax Crystals by Papillae*

Water repellence happens when water droplets contact the wax crystal. The waxes are soft materials deposited on the epidermal cells and are considered one of the most important protective properties found on leaves. Neinhuis et al. reported that papillae protect wax crystals located between them. Even when the wax on the papillae tips is damaged, the wax between the papillae remains intact. Consequently, lotus leaves maintain the water repellence property for their lifetime. Researchers have also observed a reduction in water-repellent efficiency by wiping the leaf, however, it destroyed only the wax on the papiland the leaf remained superhydrophobic. In contrast, on non-papillose surfaces, the waxes are completely destroyed, causing the contact angle to decrease from 160° to 130° [110].

2.5.3.6.5 *Wax Tubules*

Wax tubules are tiny tubular structures made of wax crystals that are commonly found on the surfaces of some insect bodies, plant leaves, and stems. These features are part of the natural wax coating known as cuticular wax and play an essential role in protection, self-cleaning, and water resistance. Table 2.8 presents a comparison of the upper and lower sides of a lotus leaf. The wax tubules on the upper and lower sides of leaves can differ significantly in structure, density, and function, depending on the plant species and environmental adaptations.

TABLE 2.8
Comparison of Wax Tubules on Lotus Leaf Surfaces [110]

Parameter	Upper Side of Lotus Leaf	Lower Side of Lotus Leaf
Wax Type	Short, thin wax tubules	Typical nonacosanol tubules
Length of Wax Tubules	0.3–1 µm	1–2 µm
Thickness (Diameter)	80–120 nm	~150 nm
Density of Tubules	~200 tubules per 10 µm²	~63 tubules per 10 µm²
Spacing between Tubules	Very small (closer packed)	Larger spacing compared to upper side
Effect on Capillary Pressure	Higher capillary pressure due to smaller spacing	Lower capillary pressure due to larger spacing

2.5.3.6.6 *Double Roughness Structure*

In nature, many insects and plants have superhydrophobic surfaces [115] with double-roughness structure (hierarchical surface structure). For example, lotus leaves exhibit a double-roughness surface which combines micro- and nanoscale textures, which are useful for the self-cleaning mechanism and liquid repellent process. This phenomenon was studied by Lu et al., who developed durable, fluorine-free films by pretreatment a glass substrate with epoxy silane (ES) and employing aerosol-assisted chemical vapor deposition of polydimethylsiloxane (PDMS) and SiO_2 nanoparticles [116]. It was also observed that ES used as a coating binder, helped increase both surface roughness and transparency.

2.5.3.6.7 *Stimuli-Responsive Organic Crystals and Photoresponsive Superhydrophobic Surfaces*

Researchers studied stimuli-responsive organic materials that change their structure based on external stimuli such as heat and light. Naumov et al. demonstrated the mechanical response of these materials to light or temperature changes using photo- and thermo-salient crystals. Researchers have acheived controlled surface topography on microcrystalline films of a photochromatic diarylethene compound [115, 117].

2.5.3.6.8 *Applications of Lotus Effect*

Due to their water-repellent and self-cleaning properties, these materials are widely used in textiles, coatings, electronics, and solar panels. They are also applied in oil–water separation, corrosion protection, and the development of antibacterial surfaces, employing materials such as TiO_2 and SiO_2, as shown in Figure 2.7 [118–123].

FIGURE 2.7 Applications inspired by the lotus effect.

2.6 CONCLUSION

The integration of green chemistry and bioinspired approaches in nanotechnology represents a transformative step toward sustainable scientific development. By harnessing natural processes and biological systems—ranging from plant extracts to microbial and animal-inspired mechanisms—researchers can synthesize nanoparticles in more eco-friendly, cost-effective, and biocompatible ways. These methods significantly reduce the use of toxic chemicals and energy-intensive procedures, aligning with global efforts to minimize environmental impact.

However, while green and bioinspired methods offer numerous advantages, such as scalability, safety, and versatility, they also present certain limitations. Challenges include the variability in biological materials, difficulty in controlling nanoparticle size and shape consistently, and need for standardization in synthesis protocols. Despite these hurdles, continued research and technological advancements hold promise for overcoming these drawbacks, ultimately making green nanotechnology a viable and impactful field across biomedical, environmental, and industrial domains.

REFERENCES

[1] N. Baig, I. Kammakakam, W. Falath, and I. Kammakakam, "Nanomaterials: A review of synthesis methods, properties, recent progress, and challenges," Mar. 21, 2021, *R Soc Chem*. doi: 10.1039/d0ma00807a.

[2] A. Escudero *et al.*, "Molecular bottom-up approaches for the synthesis of inorganic and hybrid nanostructures," Jul. 01, 2021, *MDPI AG*. doi: 10.3390/inorganics9070058.

[3] S. Aththanayaka, G. Thiripuranathar, and S. Ekanayake, "Emerging advances in biomimetic synthesis of nanocomposites and potential applications," Dec. 2022, *Elsevier Ltd*. doi: 10.1016/j.mtsust.2022.100206.

[4] M. S. Samuel *et al.*, "A review on green synthesis of nanoparticles and their diverse biomedical and environmental applications," May 01, 2022, *MDPI*. doi: 10.3390/catal12050459.

[5] B. Bhushan and Y. C. Jung, "Natural and biomimetic artificial surfaces for superhydrophobicity, self-cleaning, low adhesion, and drag reduction," Jan. 01, 2011, *Elsevier Ltd*. doi: 10.1016/j.pmatsci.2010.04.003.

[6] S. Tabassum-Abbasi, A. Abbasi, R. Rajalakshmi, P. Patnaik, and T. Abbasi, "Biomimetic synthesis of nanoparticles: State-of-the-art," Jun. 01, 2023, *Technoscience Publications*. doi: 10.46488/NEPT.2023.v22i02.005.

[7] G. M. Luz and J. F. Mano, "Biomimetic design of materials and biomaterials inspired by the structure of nacre," Apr. 28, 2009, *R Soc*. doi: 10.1098/rsta.2009.0007.

[8] P. Tabrizian, S. A. Davis, and B. Su, "From bone to nacre - development of biomimetic materials for bone implants: A review," *Biomater Sci*, 2024, doi: 10.1039/d4bm00903g.

[9] H. Libouban, F. Pascaretti-Grizon, G. Camprasse, S. Camprasse, and D. Chappard, "In vivo erosion of orthopedic screws prepared from nacre (mother of pearl)," *Orthopaedics and Traumatology: Surgery and Research*, vol. 102, no. 7, pp. 913–918, Nov. 2016, doi: 10.1016/j.otsr.2016.06.012.

[10] H. J. Jin and D. L. Kaplan, "Mechanism of silk processing in insects and spiders," *Nature*, vol. 424, pp. 1057–1061, Aug. 2003, doi: 10.1038/nature01809.

[11] P. Vukusic and J. R. Sambles, "Photonic structures in biology," 2003. doi: 10.1038/nature01941.

[12] K. Autumn and A. M. Peattie, "Mechanisms of Adhesion in Geckos 1," 2002. [Online]. Available: http://icb.oxfordjournals.org/

[13] T. G. Chen, A. Cauligi, S. A. Suresh, M. Pavone, and M. R. Cutkosky, "Testing Gecko-Inspired Adhesives with Astrobee Aboard the International Space Station: Readying the Technology for Space," *IEEE Robot Autom Mag*, vol. 29, no. 3, pp. 24–33, Sep. 2022, doi: 10.1109/MRA.2022.3175597.

[14] S. Venkatesh, M. E. Byrne, N. A. Peppas, and J. Z. Hilt, "Applications of biomimetic systems in drug delivery," Nov. 2005. doi: 10.1517/17425247.2.6.1085.

[15] Z. Yao, J. Xu, J. Shen, L. Qin, and W. Yuan, "Biomimetic hierarchical nanocomposite hydrogels: From design to biomedical applications," Nov. 2022, *MDPI*. doi: 10.3390/jcs6110340.

[16] B. Ma, W. Zhuang, G. Liu, and Y. Wang, "A biomimetic and pH-sensitive polymeric micelle as carrier for paclitaxel

delivery," *Regen Biomater*, vol. 5, no. 1, pp. 15–24, Feb. 2018, doi: 10.1093/rb/rbx023.

[17] J. Wang, B. Li, L. Qiu, X. Qiao, and H. Yang, "Dendrimer-based drug delivery systems: History, challenges, and latest developments," Dec. 2022, *BioMed Central Ltd.* doi: 10.1186/s13036-022-00298-5.

[18] H. Omidian, R. L. Wilson, and S. Dey Chowdhury, "Injectable biomimetic gels for biomedical applications," Jul. 2024, *Multidisciplinary Digital Publishing Institute (MDPI).* doi: 10.3390/biomimetics9070418.

[19] Q. He, H. Möhwald, and J. Li, "Layer-by-layer assembled nanotubes as biomimetic nanoreactors for calcium carbonate deposition," *Macromol Rapid Commun*, vol. 30, pp. 1538–1542, Sep. 2009, doi: 10.1002/marc.200900261.

[20] K. G. Gareev *et al.*, "Biomimetic nanomaterials: Diversity, technology, and biomedical applications," Jul. 2022, *MDPI.* doi: 10.3390/nano12142485.

[21] S. Iravani, "Green synthesis of metal nanoparticles using plants," *Green Chem*, vol. 13, no. 10, pp. 2638–2650, Jan. 2011, doi: 10.1039/c1gc15386b.

[22] N. S. Alsaiari *et al.*, "Plant and microbial approaches as green methods for the synthesis of nanomaterials: Synthesis, applications, and future perspectives," Jan. 2023, *MDPI.* doi: 10.3390/molecules28010463.

[23] O. Messaoudi and M. Bendahou, "Biological synthesis of nanoparticles using endophytic microorganisms: Current development," in *Nanotechnology and the Environment*, IntechOpen, 2020. doi: 10.5772/intechopen.93734.

[24] N. Rani, P. Singh, S. Kumar, P. Kumar, V. Bhankar, and K. Kumar, "Plant-mediated synthesis of nanoparticles and their applications: A review," Jul. 2023, *Elsevier Ltd.* doi: 10.1016/j.materresbull.2023.112233.

[25] A. K. Giri *et al.*, "Green synthesis and characterization of silver nanoparticles using Eugenia roxburghii DC. extract and activity against biofilm-producing bacteria," *Sci Rep*, vol. 12, no. 1, Dec. 2022, doi: 10.1038/s41598-022-12484-y.

[26] S. S. Shankar, A. Rai, A. Ahmad, and M. Sastry, "Rapid synthesis of Au, Ag, and bimetallic Au core-Ag shell nanoparticles using Neem (Azadirachta indica) leaf broth," *J Colloid Interface Sci*, vol. 275, pp. 496–502, Jul. 2004, doi: 10.1016/j.jcis.2004.03.003.

[27] D. Philip and C. Unni, "Extracellular biosynthesis of gold and silver nanoparticles using Krishna tulsi (Ocimum sanctum) leaf," *Physica E Low Dimens Syst Nanostruct*, vol. 43, pp. 1318–1322, May 2011, doi: 10.1016/j.physe.2010.10.006.

[28] H. M. M. Ibrahim, "Green synthesis and characterization of silver nanoparticles using banana peel extract and their antimicrobial activity against representative microorganisms," *J Radiat Res Appl Sci*, vol. 8, pp. 265–275, Jul. 2015, doi: 10.1016/j.jrras.2015.01.007.

[29] N. Bala *et al.*, "Green synthesis of zinc oxide nanoparticles using Hibiscus subdariffa leaf extract: Effect of temperature on synthesis, anti-bacterial activity and anti-diabetic activity," *RSC Adv*, vol. 5, pp. 4993–5003, 2015, doi: 10.1039/c4ra12784f.

[30] V. C. Karade *et al.*, "Greener synthesis of magnetite nanoparticles using green tea extract and their magnetic properties," *Mater Res Express*, vol. 4, Sep. 2017, doi: 10.1088/2053-1591/aa892f.

[31] P. Kaur, S. Kumar, H. Kaur, J. Gaur, and D. Rani, "A green approach: Syzygium aromaticum mediated synthesis of CuO nanoparticles with potential for wastewater treatment," *Asian J Chem*, vol. 36, no. 7, pp. 1559–1570, 2024, doi: 10.14233/ajchem.2024.31377.

[32] R. Sankar, K. Rizwana, K. S. Shivashangari, and V. Ravikumar, "Ultra-rapid photocatalytic activity of Azadirachta indica engineered colloidal titanium dioxide nanoparticles," *Applied Nanoscience (Switzerland)*, vol. 5, pp. 731–736, Aug. 2015, doi: 10.1007/s13204-014-0369-3.

[33] S. A. Fahmy, E. Preis, U. Bakowsky, and H. M. E. S. Azzazy, "Palladium nanoparticles fabricated by green chemistry: Promising chemotherapeutic, antioxidant and antimicrobial agents," Sep. 2020, *MDPI AG.* doi: 10.3390/MA13173661.

[34] S. Azizi, M. M. Shahri, H. S. Rahman, R. A. Rahim, A. Rasedee, and R. Mohamad, "Green synthesis palladium nanoparticles mediated by white tea (Camellia sinensis) extract with antioxidant, antibacterial, and antiproliferative activities toward the human leukemia (MOLT-4) cell line," Dec. 2017, *Dove Medical Press Ltd.* doi: 10.2147/IJN.S149371.

[35] J. Vyas and S. Rana, "Antioxidant activity and green synthesis of selenium nanoparticles using allium sativum extract," *Int J Phytomedicine*, vol. 9, p. 634, Dec. 2017, doi: 10.5138/09750185.2185.

[36] M. M. Naik *et al.*, "A facile green synthesis of Nickel Ferrite nanoparticles using tamarindus indica seeds for magnetic and photocatalytic studies," Jun. 2021. doi: 10.21203/rs.3.rs-568580/v1.

[37] S. Sebastiammal, S. Sonia, J. Henry, and A. Lesly Fathima, "Green synthesis of cerium oxide nanoparticles using aloevera leaf extract and its optical properties," *Songklanakarin J Sci Technol*, vol. 43, pp. 582–587, Mar. 2021, doi: 10.14456/sjst-psu.2021.78.

[38] E. Ibrahim *et al.*, "Biosynthesis of silver nanoparticles using endophytic bacteria and their role in inhibition of rice pathogenic bacteria and plant growth promotion," *RSC Adv*, vol. 9, pp. 29293–29299, 2019, doi: 10.1039/c9ra04246f.

[39] B. Syed, M. Nagendra, N. Prasad, and S. Satisha, "Endogenic mediated synthesis of gold nanoparticles bearing bactericidal activity," *J Microsc Ultrastruct*, vol. 4, p. 162, 2016, doi: 10.1016/j.jmau.2016.01.004.

[40] R. S, S. A. M, M. Adnan, S. K. N, R. K, and H. S, "Synthesis, characterization and applications of endophytic fungal nanoparticles," *Inorg Nano-Metal Chem*, vol. 51, pp. 280–287, 2021, doi: 10.1080/24701556.2020.1784231.

[41] S. E. D. Hassan *et al.*, "Endophytic actinomycetes Streptomyces spp mediated biosynthesis of copper oxide nanoparticles as a promising tool for biotechnological applications," *J Biol Inorg Chem*, vol. 24, pp. 377–393, May 2019, doi: 10.1007/s00775-019-01654-5.

[42] S. H. El-Moslamy, "Bioprocessing strategies for cost-effective large-scale biogenic synthesis of nano-MgO from endophytic Streptomyces coelicolor strain E72 as an anti-multidrug-resistant pathogens agent," *Sci Rep*, vol. 8, Dec. 2018, doi: 10.1038/s41598-018-22134-x.

[43] R. Javed, M. Zia, S. Naz, S. O. Aisida, N. ul Ain, and Q. Ao, "Role of capping agents in the application of nanoparticles in biomedicine and environmental remediation: recent trends and future prospects," Dec. 01, 2020, *BioMed Central Ltd.* doi: 10.1186/s12951-020-00704-4.

[44] K. Govindaraju, V. Kiruthiga, V. G. Kumar, and G. Singaravelu, "Extracellular synthesis of silver nanoparticles by a marine alga, Sargassum wightii grevilli and their

Antibacterial effects," *J Nanosci Nanotechnol*, Sep. 2009, pp. 5497–5501. doi: 10.1166/jnn.2009.1199.

[45] N. E. A. El-Naggar, M. H. Hussein, S. A. Shaaban-Dessuuki, and S. R. Dalal, "Production, extraction and characterization of Chlorella vulgaris soluble polysaccharides and their applications in AgNPs biosynthesis and biostimulation of plant growth," *Sci Rep*, vol. 10, Dec. 2020, doi: 10.1038/s41598-020-59945-w.

[46] H. E. Lashgarian, M. Karkhane, A. K. Alhameedawi, and A. Marzban, "Phyco-mediated synthesis of Ag/AgCl nanoparticles using ethanol extract of a marine green algae, Ulva Fasciata delile with biological activity," *Biointerface Res Appl Chem*, vol. 11, no. 6, pp. 14545–14554, 2021, doi: 10.33263/BRIAC116.1454514554.

[47] G. Sharma, N. D. Jasuja, M. Kumar, and M. I. Ali, "Biological synthesis of silver nanoparticles by cell-free extract of spirulina platensis," *J Nanotechnol*, vol. 2015, 2015, doi: 10.1155/2015/132675.

[48] F. Namvar *et al.*, "Green synthesis and characterization of gold nanoparticles using the marine macroalgae Sargassum muticum," *Res Chem Intermed*, vol. 41, pp. 5723–5730, Aug. 2015, doi: 10.1007/s11164-014-1696-4.

[49] A. Pugazhendhi, D. Prabakar, J. M. Jacob, I. Karuppusamy, and R. G. Saratale, "Synthesis and characterization of silver nanoparticles using Gelidium amansii and its antimicrobial property against various pathogenic bacteria," *Microb Pathog*, vol. 114, pp. 41–45, Jan. 2018, doi: 10.1016/j.micpath.2017.11.013.

[50] E. R. Balasooriya, C. D. Jayasinghe, U. A. Jayawardena, R. W. D. Ruwanthika, R. M. De Silva, and P. V. Udagama, "Honey mediated green synthesis of nanoparticles: New era of safe nanotechnology," 2017, *Hindawi Ltd.* doi: 10.1155/2017/5919836.

[51] P. B. Olaitan, O. E. Adeleke, and I. O. Ola, "Honey: A reservoir for microorganisms and an inhibitory agent for microbes," *Afr Health Sci*, vol. 7, pp. 159–165, Sep. 2007, doi: 10.5555/afhs.2007.7.3.159.

[52] D. Philip, "Honey mediated green synthesis of gold nanoparticles," *Spectrochim Acta A Mol Biomol Spectrosc*, vol. 73, pp. 650–653, Aug. 2009, doi: 10.1016/j.saa.2009.03.007.

[53] M. Rafique, I. Sadaf, M. S. Rafique, and M. B. Tahir, "A review on green synthesis of silver nanoparticles and their applications," Oct. 2017, *Taylor and Francis Ltd.* doi: 10.1080/21691401.2016.1241792.

[54] G. Madhu, A. S. Kumar, and S. K. Nair, "Sunlight-induced honey-mediated green synthesis of silver nanoparticles," in *AIP Conference Proceedings*, American Institute of Physics Inc., Oct. 2019. doi: 10.1063/1.5130311.

[55] D. Philip, "Honey mediated green synthesis of silver nanoparticles," *Spectrochim Acta A Mol Biomol Spectrosc*, vol. 75, pp. 1078–1081, Mar. 2010, doi: 10.1016/j.saa.2009.12.058.

[56] S. M. Reddy, K. K. R. Datta, C. Sreelakshmi, M. Eswaramoorthy, and B. V. S. Reddy, "Honey mediated green synthesis of Pd nanoparticles for suzuki coupling and hydrogenation of conjugated olefins," *Nanosci Nanotech Let*, vol. 4, pp. 420–425, Apr. 2012, doi: 10.1166/nnl.2012.1331.

[57] L. Wu *et al.*, "A green synthesis of carbon nanoparticles from honey and their use in real-time photoacoustic imaging," *Nano Res*, vol. 6, pp. 312–325, 2013, doi: 10.1007/s12274-013-0308-8.

[58] X. Yang, Y. Zhuo, S. Zhu, Y. Luo, Y. Feng, and Y. Dou, "Novel and green synthesis of high-fluorescent carbon dots originated from honey for sensing and imaging," *Biosens Bioelectron*, vol. 60, pp. 292–298, Oct. 2014, doi: 10.1016/j.bios.2014.04.046.

[59] R. Venu, T. S. Ramulu, S. Anandakumar, V. S. Rani, and C. G. Kim, "Bio-directed synthesis of platinum nanoparticles using aqueous honey solutions and their catalytic applications," *Colloids Surf A Physicochem Eng Asp*, vol. 384, pp. 733–738, Jul. 2011, doi: 10.1016/j.colsurfa.2011.05.045.

[60] S. Hemmati, A. Ahmeda, Y. Salehabadi, A. Zangeneh, and M. M. Zangeneh, "Synthesis, characterization, and evaluation of cytotoxicity, antioxidant, antifungal, antibacterial, and cutaneous wound healing effects of copper nanoparticles using the aqueous extract of Strawberry fruit and L-Ascorbic acid," *Polyhedron*, vol. 180, Apr. 2020, doi: 10.1016/j.poly.2020.114425.

[61] A. Relhan, S. Guleria, A. Bhasin, A. Mirza, and J. L. Zhou, "Biosynthesized copper oxide nanoparticles by Psidium guajava plants with antibacterial, antidiabetic, antioxidant, and photocatalytic capacity," *Biomass Convers Biorefin*, 2024, doi: 10.1007/s13399-024-05544-y.

[62] S. S. Dash and B. G. Bag, "Synthesis of gold nanoparticles using renewable Punica granatum juice and study of its catalytic activity," *Appl Nanosci (Switzerland)*, vol. 4, pp. 55–59, Jan. 2014, doi: 10.1007/s13204-012-0179-4.

[63] J. K. Patra and K. H. Baek, "Novel green synthesis of gold nanoparticles using Citrullus lanatus rind and investigation of proteasome inhibitory activity, antibacterial, and antioxidant potential," *Int J Nanomedicine*, vol. 10, pp. 7253–7264, Dec. 2015, doi: 10.2147/IJN.S95483.

[64] N. Mude, A. Ingle, A. Gade, and M. Rai, "Synthesis of silver nanoparticles using callus extract of Carica papaya: A first report," *J Plant Biochem Biotechnol*, vol. 18, pp. 83–86, 2009, doi: 10.1007/BF03263300.

[65] G. Gnanajobitha *et al.*, "Fruit-mediated synthesis of silver nanoparticles using Vitis vinifera and evaluation of their antimicrobial efficacy," *J Nanostructure Chem*, vol. 3, Dec. 2013, doi: 10.1186/2193-8865-3-67.

[66] R. A. R. Ahmad *et al.*, "Biosynthesis of zinc oxide nanoparticles by using fruits extracts of Ananas comosus and its antibacterial activity," *Malays J Fund Appl Sci*, vol. 15, pp. 268–273, Mar. 2019, doi: 10.11113/mjfas.v15n2.1217.

[67] S. N. A. Mohamad Sukri, K. Shameli, M. Mei-Theng Wong, S. Y. Teow, J. Chew, and N. A. Ismail, "Cytotoxicity and antibacterial activities of plant-mediated synthesized zinc oxide (ZnO) nanoparticles using Punica granatum (pomegranate) fruit peels extract," *J Mol Struct*, vol. 1189, pp. 57–65, Aug. 2019, doi: 10.1016/j.molstruc.2019.04.026.

[68] N. Ghosh, S. Paul, and P. Basak, "Silver nanoparticles of Moringa Oleifera – green synthesis, characterisation and its antimicrobial efficacy," *J Drug Deliv Ther*, vol. 0, Jun. 2014, doi: 10.22270/jddt.v0i0.906.

[69] F. Bano, M. Baber, A. Ali, Z. Shah, and S. A. Muhammad, "Biosynthesis, characterization, and biological activities of iron nanoparticles using Sesamum indicum seeds extract," *Pharmacogn Mag*, vol. 13, no. 49, pp. S33–S36, Jan. 2017, doi: 10.4103/0973-1296.203985.

[70] A. Molla *et al.*, "Biosynthesis and characterization of ZnO nanoparticles using citrus reticulata peel followed by photocatalytic, antibacterial, and antioxidative nanotherapeutic

attributes assessment supported by computer simulation," *Int J Nanomedicine*, vol. 20, pp. 4803–4829, 2025, doi: 10.2147/IJN.S493905.

[71] L. G. R, E. Jenitta Emima Packiyam, and S. D. Shetty, "International journal of current research in chemistry and pharmaceutical sciences Green synthesis of Silver Nano particles from Litchi seeds and its Antimicrobial activity," *Int J Curr Res Chem Pharm Sci*, vol. 6, no. 2, pp. 30–36, 2019, doi: 10.22192/ijcrcps.

[72] J. Michailidu, O. Maťátková, I. Kolouchová, J. Masák, and A. Čejková, "Silver nanoparticle production mediated by vitis vinifera cane extract: Characterization and antibacterial activity evaluation," *Plants*, vol. 11, Feb. 2022, doi: 10.3390/plants11030443.

[73] X. Zhang *et al.*, "The pathway to intelligence: Using stimuli-responsive materials as building blocks for constructing smart and functional systems," Mar. 2019, *Wiley-VCH Verlag*. doi: 10.1002/adma.201804540.

[74] Y. Zhu, W. Zhang, and D. Zhang, "Fabrication of sensor materials inspired by butterfly wings," Jul. 01, 2017, *Wiley-Blackwell*. doi: 10.1002/admt.201600209.

[75] Z. Wang and Z. Guo, "Biomimetic superwettable materials with structural colours," *Chem Commun*, vol. 53, pp. 12990–13011, 2017, doi: 10.1039/c7cc07436k.

[76] G. Isapour and M. Lattuada, "Bioinspired stimuli-responsive color-changing systems," May 2018, *Wiley-VCH Verlag*. doi: 10.1002/adma.201707069.

[77] A. Du *et al.*, "Multifunctional silica nanotube aerogels inspired by polar bear hair for light management and thermal insulation," *Chem Mater*, vol. 30, pp. 6849–6857, Oct. 2018, doi: 10.1021/acs.chemmater.8b02926.

[78] Y. Wu, K. Zhang, and B. Yang, "Ordered hybrid micro/nanostructures and their optical applications," Apr. 2019, *Wiley-VCH Verlag*. doi: 10.1002/adom.201800980.

[79] H. Zhou *et al.*, "Bio-inspired photonic materials: Prototypes and structural effect designs for applications in solar energy manipulation," *Adv Funct Mater*, vol. 28, Jun. 2018, doi: 10.1002/adfm.201705309.

[80] D. Kong *et al.*, "Bioinspired co-design of tactile sensor and deep learning algorithm for human–robot interaction," *Adv Intell Syst*, vol. 4, Jun. 2022, doi: 10.1002/aisy.202200050.

[81] T. R. Prianka *et al.*, "Recent exploration of bio-mimetic nanomaterial for potential biomedical applications," Dec. 2018, *Elsevier Ltd*. doi: 10.1016/j.msec.2018.09.012.

[82] H. Ghiradella and W. Radigan, "Development of butterfly scales. II. Struts, lattices and surface tension," *J Morphol*, vol. 150, pp. 279–297, 1976, doi: 10.1002/jmor.1051500202.

[83] J. E. Shea, T. D. Crawford, M. T. Zanni, G. V. Hartland, and W. Aumiller, "50 and 100 years ago in the journal of physical chemistry 2023," Mar. 2023, *Am Chem Soc*. doi: 10.1021/acs.jpcb.3c01129.

[84] N. Dushkina and A. Lakhtakia, "Structural colors," in *Engineered Biomimicry*, Elsevier Inc., 2013, pp. 267–303. doi: 10.1016/B978-0-12-415995-2.00011-8.

[85] C. Qu *et al.*, "Bioinspired flexible volatile organic compounds sensor based on dynamic surface wrinkling with dual-signal response," *Small*, vol. 15, Apr. 2019, doi: 10.1002/smll.201900216.

[86] Q. Li, Q. Zeng, L. Shi, X. Zhang, and K. Q. Zhang, "Bioinspired sensors based on photonic structures of Morpho butterfly wings: A review," Mar. 2016, *R Soc Chem*. doi: 10.1039/c5tc04029a.

[87] Q. Shen *et al.*, "Bioinspired infrared sensing materials and systems," Jul. 2018, *Wiley-VCH Verlag*. doi: 10.1002/adma.201707632.

[88] D. Klocke, A. Schmitz, H. Soltner, H. Bousack, and H. Schmitz, "Infrared receptors in pyrophilous ('fire loving') insects as model for new un-cooled infrared sensors," *Beilstein J Nanotechnol*, vol. 2, pp. 186–197, 2011, doi: 10.3762/bjnano.2.22.

[89] F. Lu, Q. Tan, Y. Ji, Q. Guo, Y. Guo, and J. Xiong, "A novel metamaterial inspired high-temperature microwave sensor in harsh environments," *Sensors (Switzerland)*, vol. 18, Sep. 2018, doi: 10.3390/s18092879.

[90] A. D. Pris *et al.*, "Towards high-speed imaging of infrared photons with bio-inspired nanoarchitectures," *Nat Photonics*, vol. 6, pp. 195–200, Mar. 2012, doi: 10.1038/nphoton.2011.355.

[91] H. Xue *et al.*, "Toward the burgeoning optical sensors with ultra-precision hierarchical structures inspired by butterflies," Aug. 2021, *John Wiley and Sons Inc*. doi: 10.1002/admi.202100142.

[92] X. Li, J. Sánchez del Río Saez, X. Ao, A. Yusuf, and D. Y. Wang, "Highly-sensitive fire alarm system based on cellulose paper with low-temperature response and wireless signal conversion," *Chem Eng J*, vol. 431, Mar. 2022, doi: 10.1016/j.cej.2021.134108.

[93] T. Fu *et al.*, "Bioinspired color changing molecular sensor toward early fire detection based on transformation of phthalonitrile to phthalocyanine," *Adv Funct Mater*, vol. 29, Feb. 2019, doi: 10.1002/adfm.201806586.

[94] Y. H. Jung, B. Park, J. U. Kim, and T. il Kim, "Bioinspired electronics for artificial sensory systems," *Adv Mater*, vol. 31, Aug. 2019, doi: 10.1002/adma.201803637.

[95] W. Fink, "Nature-inspired sensors news-and-views," *Nat Nanotechnol*, vol. 13, pp. 437–438, Jun. 2018, doi: 10.1038/s41565-018-0164-5.

[96] T. Q. Trung and N. E. Lee, "Flexible and stretchable physical sensor integrated platforms for wearable human-activity monitoring and personal healthcare," *Adv Mater*, vol. 28, pp. 4338–4372, Jun. 2016, doi: 10.1002/adma.201504244.

[97] R. Dong, Y. Liu, L. Mou, J. Deng, and X. Jiang, "Microfluidics-based biomaterials and biodevices," *Adv Mater*, vol. 31, Nov. 2019, doi: 10.1002/adma.201805033.

[98] Z. He, A. Elbaz, B. Gao, J. Zhang, E. Su, and Z. Gu, "Disposable morpho menelaus based flexible microfluidic and electronic sensor for the diagnosis of neurodegenerative disease," *Adv Healthc Mater*, vol. 7, Mar. 2018, doi: 10.1002/adhm.201701306.

[99] B. Gao, X. Wang, T. Li, Z. Feng, C. Wang, and Z. Gu, "Gecko-inspired paper artificial skin for intimate skin contact and multisensing," *Adv Mater Technol*, vol. 4, Jan. 2019, doi: 10.1002/admt.201800392.

[100] V. Narasimhan *et al.*, "Multifunctional biophotonic nanostructures inspired by the longtail glasswing butterfly for medical devices," *Nat Nanotechnol*, vol. 13, pp. 512–519, Jun. 2018, doi: 10.1038/s41565-018-0111-5.

[101] B. Gao, Z. He, B. He, and Z. Gu, "Wearable eye health monitoring sensors based on peacock tail-inspired inverse opal carbon," *Sens Actuators B Chem*, vol. 288, pp. 734–741, Jun. 2019, doi: 10.1016/j.snb.2019.03.029.

[102] T. Jiang, Z. Peng, W. Wu, T. Shi, and G. Liao, "Gas sensing using hierarchical micro/nanostructures of Morpho

butterfly scales," *Sens Actuators A Phys*, vol. 213, pp. 63–69, Jul. 2014, doi: 10.1016/j.sna.2014.04.002.
[103] Y. Zhu, H. Su, Y. Chen, Z. Jin, J. Xu, and D. Zhang, "A facile synthesis of PdO-decorated SnO2 nanocomposites with open porous hierarchical architectures for gas sensors," *J Am Ceram Soc*, vol. 99, pp. 3770–3774, 2016, doi: 10.1111/jace.14378.
[104] F. Song, H. Su, J. Han, J. Xu, and D. Zhang, "Controllable synthesis and gas response of biomorphic SnO2 with architecture hierarchy of butterfly wings," *Sens Actuators B Chem*, vol. 145, pp. 39–45, Mar. 2010, doi: 10.1016/j.snb.2009.11.011.
[105] R. A. Potyrailo, N. Karker, M. A. Carpenter, and A. Minnick, "Multivariable bio-inspired photonic sensors for non-condensable gases," *J Opt (United Kingdom)*, vol. 20, Feb. 2018, doi: 10.1088/2040-8986/aa9ecf.
[106] L. Gao *et al.*, "Biomorphic carbon-doped TiO2 for photocatalytic gas sensing with continuous detection of persistent volatile organic compounds," *ACS Appl Nano Mater*, vol. 1, pp. 1766–1775, Apr. 2018, doi: 10.1021/acsanm.8b00209.
[107] Y. Meng, F. Liu, M. M. Umair, B. Ju, S. Zhang, and B. Tang, "Patterned and iridescent plastics with 3D inverse opal structure for anticounterfeiting of the banknotes," *Adv Opt Mater*, vol. 6, Apr. 2018, doi: 10.1002/adom.201701351.
[108] S. Berthier, J. Boulenguez, and Z. Bálint, "Multiscaled polarization effects in Suneve coronata (Lepidoptera) and other insects: Application to anti-counterfeiting of banknotes," *Appl Phys A Mater Sci Process*, vol. 86, pp. 123–130, Jan. 2007, doi: 10.1007/s00339-006-3723-9.
[109] R. A. Lee and P. W. Leech, "Development of optically variable devices for document security in CSIRO," *Journal of the Coatings Association of Australia*, vol. 42, no. 8, pp. 22–27, Aug. 2005.
[110] H. J. Ensikat, P. Ditsche-Kuru, C. Neinhuis, and W. Barthlott, "Superhydrophobicity in perfection: The outstanding properties of the lotus leaf," *Beilstein J Nanotechnol*, vol. 2, pp. 152–161, 2011, doi: 10.3762/bjnano.2.19.
[111] F. Müller, W. Michel, V. Schlicht, A. Tietze, and P. Winter, "Self-cleaning surfaces using the lotus effect," in *Handbook for Cleaning/Decontamination of Surfaces*, vol. 2, Elsevier, 2007, pp. 791–811. doi: 10.1016/B978-044451664-0/50024-3.
[112] W. Barthlott and C. Neinhuis, "Purity of the sacred lotus, or escape from contamination in biological surfaces," *Planta*, vol. 202, pp. 1–8, 1997, doi: 10.1007/s004250050096.
[113] I. Zada *et al.*, "The highly efficient photocatalytic and light harvesting property of Ag-TiO2 with negative nano-holes structure inspired from cicada wings," *Sci Rep*, vol. 7, no. 1, Dec. 2017, doi: 10.1038/s41598-017-17479-8.
[114] L. Wang *et al.*, "Mechanism of self-recovery of hydrophobicity after surface damage of lotus leaf," *Plant Methods*, vol. 20, Dec. 2024, doi: 10.1186/s13007-024-01174-7.
[115] R. Nishimura *et al.*, "Fractal surfaces of molecular crystals mimicking lotus leaf with phototunable double roughness structures," *J Am Chem Soc*, vol. 138, pp. 10299–10303, Aug. 2016, doi: 10.1021/jacs.6b05562.
[116] F. Chen, J. J. Kalmoni, S. Li, and C. J. Carmalt, "Robust, fluorine-free superhydrophobic films on glass via epoxysilane pretreatment," *Langmuir*, Jan. 2025, doi: 10.1021/acs.langmuir.4c02630.
[117] K. Uchida, "Photoinduced reversible topographical changes on diarylethene microcrystalline surface with wetting property," in *New Frontiers in Photochromism*, Springer Japan, 2013, pp. 41–60. doi: 10.1007/978-4-431-54291-9_3.
[118] S. Toth *et al.*, "Soiling and cleaning: Initial observations from 5-year photovoltaic glass coating durability study," *Sol Energy Mater Sol Cells*, vol. 185, pp. 375–384, Oct. 2018, doi: 10.1016/j.solmat.2018.05.039.
[119] B. K. Dejene and T. M. Geletaw, "A review of plant-mediated synthesis of zinc oxide nanoparticles for self-cleaning textiles," Nov. 2023, *Emerald Publishing*. doi: 10.1108/RJTA-12-2022-0154.
[120] M. Liu, Y. Hou, J. Li, and Z. Guo, "Stable superwetting meshes for on-demand separation of immiscible oil/water mixtures and emulsions," *Langmuir*, vol. 33, pp. 3702–3710, Apr. 2017, doi: 10.1021/acs.langmuir.7b00658.
[121] W. Zhao and H. Lu, "Self-cleaning performance of superhydrophilic coatings for dust deposition reduction on solar photovoltaic cells," *Coatings*, vol. 11, Sep. 2021, doi: 10.3390/coatings11091059.
[122] H. S. El-Beltagi *et al.*, "Biosynthesis of zinc oxide nanoparticles via neem extract and their anticancer and antibacterial activities," *PeerJ*, vol. 12, 2024, doi: 10.7717/peerj.17588.
[123] A. Anna Thomas, R. M. Varghese, and S. Rajeshkumar, "Antimicrobial effects of copper nanoparticles with green tea and neem formulation.," *Bioinformation*, vol. 18, pp. 284–288, 2022, doi: 10.6026/97320630018284.

3 Harnessing Renewable Resources for Sustainable Nanomaterial Synthesis

Principles, Challenges, and Future Trends

Snigdha Roy

3.1 INTRODUCTION

With the increasing prospects of nanomaterial applications, the global demand for the production of nanosubstances at large scale has grown many times over. Scientists and researchers have started to focus more on the use of various renewable resources for the preparation of different nanoparticles (NPs), considering their eco-friendly nature to aid human and environmental health. Traditional applications of harmful chemicals and toxic reagents have been widely replaced by their green alternatives such as plant extracts, microorganisms, agricultural waste, and other forms of biomass.

This paradigm shift follows the principles of green chemistry and promotes a circular economy by utilizing low-cost, abundant, and biodegradable natural precursors. Several research outcomes have already shown that these renewable starting materials lead to the formation of high-quality NPs with controlled size, shapes, and tailor-made functionalities, offering better performance either as adsorbents, catalysts, sensors, electronics, or biomedical devices. However, this transition brings about some synthetic challenges, too, mainly because of inconsistencies in biological sources and a lack of standardized methodologies to fabricate similar morphology in NPs. This chapter outlines the utilization of various natural resources for the preparation of nanoparticles, with their key applications and synthetic challenges, and explores the possible solutions that could lead towards the development of a new generation of sustainable nanotechnology.

3.2 TYPES OF RENEWABLE RESOURCES FOR NANOMATERIAL SYNTHESIS

Research into renewable resource–based nanomaterial synthesis has expanded rapidly because these materials have sustainable characteristics. Three main categories of renewable sources serve as starting materials for nanomaterial synthesis: plant-based extracts, microbial entities, and agricultural and industrial waste products (Bhardwaj et al., 2020). Renewable resources present different strengths for nanomaterial synthesis, including reduced costs and universal access, together with low environmental impact. This section discusses the synthesis of NPs from various sustainable resources and highlights their functions that help prepare nanosized particles of different types.

3.2.1 Plant-Based Resources

Plants have become a common choice in green nanoparticle synthesis due to their economic value; readily available nature; and biochemical elements like flavonoids, alkaloids, and phenolics that function as reducing agents to produce metal NPs from various metals. Apart from metal and metal oxide–based NPs, plant-based biomass has also been utilized for the synthesis of carbon NPs such as carbon nanotubes and graphenes.

3.2.1.1 Synthesis of Metal and Metal Oxide–Based Nanoparticles

Numerous studies (Radulescu et al., 2023) have demonstrated that extracts from different plant components like leaves, stems, seeds, and roots possess an outstanding ability to synthesize silver, gold, and copper oxide NPs. Such resources provide simple synthetic procedures where the natural molecules serve both as the reducing and stabilizing agents, thereby eliminating the requirement for extra chemicals. Plant-based synthetic methods are a safe and environmentally friendly way to make NPs because they don't need as many harmful chemicals. Silver NPs can be made by distilling *Azadirachta indica* (neem) water to make antibacterial compounds. Gold NPs synthesized using *Cinnamomum verum* (cinnamon) leaf extract show antioxidant properties (Table 3.1).

Plant polysaccharides (Makarov et al., 2014) can be alternative natural resources for the synthesis of NPs with better control over their sizes and shapes. Owing to the presence of versatile functional groups such as aminos, hydroxyls, aldehydes, and carboxyls, plant polysaccharides can also act as effective stabilizing, reducing, and scaffold-providing molecular entities. With the help of these multiple functional groups, polysaccharides can interact with metal ions and thereby create a protective layer around the ions that prevent them from aggregating together. Polysaccharides like chitosan and cellulose can also regulate the growth of NPs in a more controlled manner because of their

DOI: 10.1201/9781003644439-3

TABLE 3.1
Fabrication of Metal-Based NPs from Plant Resources

Resources	Metal Salts Used for Nanoparticle Synthesis	Functions of Plant Resources	Examples
Plant extracts rich in secondary metabolites, such as flavonoids, polyphenols, terpenoids, alkaloids	$CuSO_4$, $AgNO_3$, $AuCl_3$	Reducing and stabilizing agents	1. Neem, cinnamon, coriander, and aloe vera (AgNPs) 2. Drumstick tree, cinnamon (AuNPs)
Plant polysaccharides such as starch, cellulose, pectin, chitosan, alginate	$AgNO_3$, $AuCl_3$, $CuSO_4$, Fe_3O_4, $Zn(NO_3)_2$, $PtCl_2$	Reducing, stabilizing, and scaffold-generating agents	1. Corn, potatoes, rice starch (AgNPs) 2. Cellulose from wood, cotton, and other fibrous plants (AuNPs) 3. Chitosan from exoskeletons of fungal cell walls and crustaceans (CuNPs) 4. Brown algae (Fe_3O_4 NPs) 5. Cell wall pectin from citrus fruits and apples (AgNPs) 6. Bamboo cellulose, cotton cellulose, gum arabic, orange peel (ZnO NPs)
Plant proteins	$HAuCl_4$, AgCl, $ZnCl_2$, $FeSO_4$, $Fe(NO_3)_3$, $NiCl_2$	Reducing and stabilizing agents	1. Soybean and wheat gluten proteins (Ag and AuNPs) 2. Peanut and chickpea proteins (Fe_3O_4 NPs) 3. Moringa and rice proteins (ZnO NPs) 4. Sesame and sunflower seed proteins (CuO NPs) 5. Peanut and soybean proteins (NiO NPs)
Plant essential oils like terpenes, phenolic compounds	$Cu(NO_3)_2$, $FeSO_4$, $CuSO_4$, $AgNO_3$	Reducing and capping agents	1. Eugenol from clove, cinnamaldehyde from cinnamon (AgNPs) 2. Cineole from eucalyptus, linalool from lavender (AuNPs) 3. Gingerol from ginger, eugenol from basil (CuO NPs)
Plant-based secondary metabolites	$AgNO_3$, $HAuCl_4$, $CuCl_2$, $FeCl_3$, $Zn(NO_3)_2$	Reducing, stabilizing and shape-controlling agents	1. Terpenoida and tannins from neem (AuNPs) 2. Catechins and polyphenols from tea (ZnO NPs) 3. Curcumin and alkaloids from turmeric (CuO NPs) 4. Phenolics and flavonoids from aloe vera (AgNPs) 5. Tannins from acacia and chestnut (Fe_3O_4 NPs)

highly organized rigid structural properties. The cellulose of plant cell walls can produce cellulose nanocrystals and nanofibers that act as the templates for the growth of various metal NPs. Some polysaccharides like pectin and alginate can also lead to the production of biodegradable NPs, which have been widely used in different biomedical applications, including targeted drug delivery, tissue engineering, and wound healing. The reducing sugar and hydroxyl groups present in starch obtained from various plants like potatoes, corn, and rice have been used to prepare silver, gold, and copper NPs (CuNPs) (Table 3.1). Cationic polysaccharide chitosan-coated NPs are used as antimicrobial agents, while brown algae–derived alginate NPs are applied in controlled drug release therapy.

Plant-based protein molecules have also been exploited for the successful green synthesis of NPs. The presence of various functionalities like amino, thiol, carboxyl, and hydroxyl help these proteins act as potential reducing and capping agents to stabilize different metal NPs. A wide range of cereal, legume, seed, fruit, leaf, and tuber-based proteins are used to synthesize diverse metal nanomaterials.

The next eco-friendly method for nanoparticle synthesis utilizes various plant-based essential oils. These oils are rich in different bioactive compounds like phenolic, terpenes, flavonoids, and alkaloids which can participate in the metal reduction process by electron donation, prevent the agglomeration of NPs by effective capping of the metal surface with the help of suitable functional groups, and regulate the applications of NPs in various fields by modulating their size and shape. The majority of these NPs are used for various biomedical purposes like biosensors, imaging, drug delivery, and catalysis. The monoterpenoid alcohols linalool and eugenol extracted from basil plants are used for CuO NP synthesis with antimicrobial and catalysis properties. Curcumin from turmeric and carnosic acid from rosemary are applied for the preparation of iron oxide (Fe_3O_4) NP salts (Table 3.1). Tea tree–derived terpinen-4-ol and cinnamaldehyde from cinnamon are used for AgNP synthesis. Other terpenoids like citral from lemongrass and menthol from peppermint are utilized for ZnO nanoparticle synthesis, which has great applications in various sunscreen and antimicrobial coating agents.

Different secondary metabolites of plants, including alkaloids, flavonoids, phenolics, and tannins, can also be utilized as effective reducing and stabilizing agents in metal nanoparticle synthesis. Varying concentrations of the metabolites can cause a distinct change in the morphology of the synthesized NPs. Applications of various metabolites lead to the generation of rod, spherical, or triangular-shaped NPs. Flavonoid- and polyphenol-based NPs can also exhibit antioxidant and antimicrobial properties.

3.2.1.2 Synthesis of Carbon-Based Nanoparticles

The natural carbon content of various plant parts such as coconut shells, sugarcane, rice husks, banana peels, and coffee grounds is used for the preparation of hollow, cylindrical, nanosized carbon tubes and the single-layered honeycomb-like planar structure of the graphene sheet (Tripathi, Pavelyev, & Islam, 2017). Most of these plant masses are first decomposed and carbonized under high temperature and pressure. The following step involves chemical vapor deposition (CVD) in the presence of suitable transition metal catalysts like Fe, Ni, and Co. As an alternative process of pyrolysis under inert atmosphere, hydrothermal carbonization can also be conducted in an aqueous medium to transform the organic materials of the plant bodies into carbon nanostructures. The hydrothermal treatment requires the heating of the water-dispersed biomass at 120–250°C in an autoclave, which is much lower than the temperature of 300–900°C used in the case of pyrolytic processes. This relatively low energy-consuming hydrothermal carbonization process leads to carbon NPs (CNPs) with more oxygen-rich functional groups and thereby makes them more appropriate for biomedical applications. Sometimes other solvents like ethanol and ethylene glycol are also used to dissolve the biomass under pressure and heating conditions, followed by the microwave-assisted carbonization process. The microwave radiation makes the process faster and can control the growth of the NPs' size more efficiently. The role of the transition metal catalysts in the nucleation steps of the carbon nanomaterials is also crucial. They help to break down the plant biomass into carbon radicals first, which will later diffuse and get absorbed on the bed of the catalytic surface. After the metal reaches its carbon saturation point, the excess carbon vapor starts to precipitate out as carbon self-assembles into different nanostructural forms. Depending on the various substrates (SiO_2, Al_2O_3, quartz, etc.) and catalysts used, the sizes and shapes of the synthesized carbon NPs can also vary (Table 3.2).

Reducing the particle size even more, quantum dots of various kinds can also be produced using plant biomass. Three types of quantum dots, carbon quantum dots (CQDs), metallic quantum dots (MQDs), and graphene quantum dots (GQDs), are synthesized from different plant resources following a similar methodology of pyrolysis and hydrothermal carbonization of either plant extracts or metal ion solution–mixed plant extracts (S et al., 2023). Carbon-rich plant parts like coconut shells, rice husks, neem leaves, wood chips, and aloe vera are utilized for synthesizing CQDs and GQDs, while metallic salts like $Zn(NO_3)_2$, $CdCl_2$, and $AgNO_3$ are used to prepare MQDs (Table 3.3). QDs prepared by such methods exhibit unique electronic and size-dependent optical properties, mainly because of the quantum confinement effect. In optoelectronics, bioimaging, sensors, drug delivery, photocatalysis, and solar cells, these QDs find wide applications.

TABLE 3.2
Metal-Catalyzed Preparation of Carbon NPs

Substrate	Metal Catalyst	Method of Preparation	Effect on Carbon Nanoparticle Growth
Silica	Fe, Ni, Co	CVD, plasma-enhanced CVD	Vertically aligned carbon nanotubes
Alumina	Fe, Ni, Co	Thermal CVD, catalyst impregnation	Uniformly distributed carbon nanotubes (CNTs), highly pure CNPs
Quartz	Fe, Ni	CVD, plasma-enhanced CVD	Graphene layers, high yield of CNPs
Iron foil	Self-catalyzed	Thermal CVD	Multiwalled CNTs
Magnesium oxide (MgO)	Fe, Co	CVD	Highly crystalline CNPs

TABLE 3.3
Carbon Dots from Plant Resources

Type of QD	Plant Resource	Method of Preparation	Properties and Application
CQDs	Fruit peels, lignocellulosic biomass, leaves	Pyrolysis and hydrothermal carbonization	Biocompatible, fluorescent markers for bioimaging, drug delivery and pH sensors
GQDs	Coconut shells, wood chips, coffee grounds, sugarcane, rice husks	Pyrolysis and hydrothermal carbonization	Highly conductive; application in electronics, LEDs, and transistors
MQDs	Neem leaves, tea extract, lemon extract	Green synthesis in presence of plant based reducing and stabilizing agents	Tunable fluorescence, semiconductor, photocatalysis

3.2.2 Microorganisms in Nanomaterial Synthesis

Scientific studies have indicated that different microorganisms such as bacteria, fungi, yeast, and algae can also be exploited as renewable resources to synthesize NPs through diverse enzymatic and non-enzymatic mechanisms (Singh

TABLE 3.4
Utilization of Microorganisms for Metal NP Synthesis

Type of Micro-organism	Specific Microorganism Used in Biogenic Synthesis	Metal NP	Application
Bacteria	*Escherichia coli*, *Pseudomonas aeruginosa*	Au	Biosensors, cancer therapy
	Pseudomonas aeruginosa, *Bacillus subtilis*	Ag	Antibacterial agents
	Shewanella oneidensis, *Geobacter sulfurreducens*	Fe_3O_4	Drug delivery, magnetic resonance imaging, cancer treatment
	Bacillus mycoides, *Pseudomonas putida*	TiO_2	Photocatalysis, UV-blocking agents, chemical stability
Fungus	*Aspergillus flavus*, *Penicillium* sp.	ZnO	Antimicrobial agents, sunscreen
	Verticillium sp., *Fusarium oxysporum*	Au	Cancer therapy
Algae	*Spirulina platensis*, *Ulva lactuca*	ZnO	Antimicrobial coatings
	Chlorella vulgaris, *Sargassum muticum*	Fe_3O_4	MRI agent, cancer therapy
Yeast	*Schizosaccharomyces pombe*	CdS QDs	Fluorescent imaging, sensors

et al., 2023). These microbes can follow both the extracellular and intracellular route to furnish nanoparticle synthesis from metal ion solutions. In the intracellular method, bacteria and algae first absorb metal ions from an external source, and through passive diffusion or active transport methods, metal ions enter into the fungi and yeast bodies. Various intracellular enzymes like nitrate reductase and hydrogenase and metabolites such as proteins, peptides, and polysaccharides present inside these microbes help reduce the metal ions to form NPs, which then accumulate inside the cells and can be extracted later. In the case of extracellular synthesis, the metal ions get converted into NPs in the outside medium with the help of the secreted enzymes and other biomolecules from these same microbes. NPs thus synthesized through bioreduction processes are applied for multiple purposes, including medical, pharmaceutical, cosmetics, sensors, bioimaging, and photocatalysis (Table 3.4).

3.2.3 Utilizing Agricultural and Industrial Waste

Science has found promise in using agricultural as well as industrial waste materials to produce renewable nanomaterials (Sanoja-López et al., 2025). Several waste materials that originate from plant residues and animal manure, together with food processing byproducts, exist in high abundance while frequently ending up as pollutants in the environment. Scientists use these waste materials as eco-friendly precursors for nanomaterial creation and to decrease the burden of environmental waste. In this sustainable and cost-effective nanotechnology, various agricultural wastes such as leaves, fruit peels, and seeds are used as natural reducing and stabilizing agents for metal-based nanoparticle synthesis (Table 3.5). Such plant residues contain flavonoids, tannins, and phenolic-based different secondary metabolites that show antioxidant properties and play crucial roles in plant defense mechanisms.

Animal wastes like bones, eggshells, blood, feathers, and dung can also act as natural precursors for the synthesis of NPs of various substances (Table 3.6). Different vital

TABLE 3.5
Metal-Based NP Synthesis from Agricultural Waste

Metal-Based NP	Plant Waste	Key Metabolite	Applications
Ag	Banana peel	Polyphenols, flavonoids, alkaloids, terpenoids, and polysaccharides	Antimicrobial coatings
Au	Tea waste	Flavonoids, tannins, proteins, and amino acids	Biosensors and cancer therapy
Fe_3O_4	Corn husk	Lignin; cellulose; phenolic compounds such as ferulic acid, caffeic acid, and flavonoids (apigenin, quercetin)	Wastewater treatment, magnetic NPs
ZnO	Coconut husk waste	Phenolic compounds such as gallic acid, catechins, flavonoids, lignin, and polysaccharides	Antibacterial agents
SiO_2	Rice husk waste	Silica in ash, lignin, cellulose, and hemicellulose	Drug delivery and cosmetic products

TABLE 3.6
NP Synthesis from Animal Waste

Animal Waste	Key Ingredient	NPs	Application
Animal dung	Microorganisms for metal NPs and carbon sources for CNPs	Ag, Au, carbon NPs	Antibacterial coatings, nanofertilizers, supercapacitors, heavy metal adsorptions
Eggshells	Calcium carbonate ($CaCO_3$)	CaO and hydroxyapatite NPs	Drug delivery and bone regeneration
Animal bones and teeth	Hydroxyapatite, collagen	Hydroxyapatite	Dental and orthopedic and applications
Feathers and wool	Keratin, proteins	Keratin-based	Cosmetics and wound repairing
Animal blood	Hemoglobin, protein, and iron	Fe_3O_4	MRI imaging and drug delivery

biomaterials present in such waste products are cost-effectively and sustainably utilized for the fabrication of NPs.

In addition to these agricultural and animal wastes, industrial waste has also been repurposed to produce useful NPs of various origins. Waste materials collected from metallic and electronic industries often contain metal residues that can be recycled to obtain various metal-based NPs. For example, Fe_3O_4 NPs can be synthesized using iron slag and other steel industry–produced waste products. Different electronic wastes can provide Ag- and Cu-based NPs. Likewise, various chemical industry–derived pollutants can be converted into useful nanomaterials for future applications. Hydrocarbon waste from petroleum industries is a precursors for carbon dots and graphene NPs. Silica and alumina-rich fly ash collected from coal power plants can be utilized for the production of ceramic and catalyst NPs. Textile and paper industry waste products contain cellulose and lignin in high amounts that can be fruitfully used as the stabilizing agents in nanoparticle synthesis.

3.3 CHALLENGES IN THE SYNTHESIS OF NANOMATERIALS FROM RENEWABLE RESOURCES

There are multiple obstacles that stand in the way of increasing the application and commercial readiness of nanomaterials derived from renewable resources. Professional research in this field continues to make important progress, yet scalability issues, together with lacking reproducibility and narrow renewable resource selection, remain persistent obstacles. The current synthesis difficulties of nanomaterials derived from renewable resources will receive analysis, followed by proposed solutions and future research pathways for development.

3.3.1 Problems in Scaling Up the Synthesis Process

The main barrier to achieving large-scale manufacturing of green nanomaterials is in making their production processes more scalable (Paliwal, Babu, & Palakurthi, 2014). Mass production for industries presents an overwhelming challenge to scaling up existing laboratory-scale methods that synthesize NPs through plant extracts, microorganisms, and agricultural waste. While there is a lot of information on how to take green synthetic methods to industrial manufacturing scales, not much of it talks about how to get consistent nanomaterial results from various types of renewable resources along with process scalability. A significant obstacle exists because of the lack of consistent composition of the resource materials and availability in sufficient quantities and concentrations needed for large-scale industrial production. It's hard to get plant-based extracts for mass production, and the chemicals that make NPs change depending on the type of plant, its growth conditions, and when it is harvested. The extraction procedure demands extended operation time and expensive costs to run, which creates additional hurdles for expanding the scale of green synthesis production methods. Impurities present in the resources often interfere with the sizes of the newly synthesized NPs, their yield, and their stability. Researchers are working on three ways to make green synthesis processes better: finding a lot of sustainable resources, making extraction techniques better for quality control, and making nanomaterials in large quantities. Advancements in bioreactor technology and continuous flow system support could potentially increase the synthesis process scale.

3.3.2 Reproducibility of Nanomaterial Synthesis

The green synthesis of nanomaterials faces difficulties because the production method struggles to achieve consistent results. Numerous research studies have demonstrated the successful production of NPs via renewable materials, yet researchers face challenges to achieve batch-to-batch consistency and scale up success when using these resources. When the composition of renewable resources changes, as do reaction conditions like temperature, pH, and reaction time, NPs lose their shape and their ability to stay stable. The insufficient reproduction abilities of uniform nanomaterials create significant issues for real-world implementation due to mandatory requirements of quality consistency. The standardization of synthesis and the understanding of nanoparticle reduction/stabilization mechanisms serve as the primary solutions to this challenge. Successful commercialization of green-synthesized nanomaterials depends greatly on developing stable and reliable production techniques.

3.3.3 Limited Range of Renewable Resources

Plant extracts, alongside agricultural waste and microorganisms, form common resources to synthesize nanomaterials, yet the existing renewable supplies available for this synthesis operation are insufficient. Current nanoparticle research focuses on only a few plant species and microorganisms, while scientists require more investigation into multiple renewable resources that can generate unusual materials. Agricultural and industrial waste offers an untapped potential for the sustainable production of nanomaterials. New renewable material sources need systematic exploration together with advanced methods for their efficient use as future research areas.

3.3.4 Storage Challenge and Limited Commercial Adoption

Biogenic NPs synthesized from sustainable resources often aggregate or degrade over time in the absence of suitable stabilizing agents. The proper maintenance of their stability, shelf life, and reactivity pose great barriers in the commercialization process of these NPs. Moreover, due to the high production cost, lack of scalability, and reproducibility issues, a majority of industries hesitate to take up these green synthetic methods and prefer to follow conventional chemical processes.

3.3.5 Toxicity and Environmental Impact

Despite its reputation as an environmentally friendly process, the lifetime effects of nanomaterial production on ecosystems warrant further investigation. Green method synthesis does not eliminate all toxic properties in NPs, since their toxicity levels depend on their dimensions together with their electrical properties and substance composition. Additionally, the byproducts and the waste produced during the synthesis of these nanomaterials can also put the environment at danger. Hence, the environmental assessment of green nanomaterials should span production through their entire life cycle until disposal. Rigorous research should be conducted to determine the toxic potential of various green nanomaterials (Thu et al., 2023), especially those utilized for biomedical and food applications, along with developing techniques that can reduce their possible harmful effects.

3.4 POSSIBLE MEASURES TO SOLVE THE CHALLENGES

The attainment of maximum benefits from the use of renewable green nanomaterials depends primarily on resolving the challenging issues and hurdles which still exist in the developmental pathways of sustainable nanotechnology.

Critical focus should be given to alleviate the scalability and high production cost issues in the preparation of nanomaterials. Budget-friendly, abundant feedstocks should be used more and more. The application of various advanced synthetic instruments like bioreactors, spray dryers, continuous flow reactors, high-pressure autoclaves, and ultrasonic homogenizers can result in better yield, especially in industrial-level production.

High performance tools like TEM, SEM, DLS, FTIR, and XRD should be utilized for rapid and in-depth characterization of biogenic NPs. To overcome the challenges of nanoparticle agglomeration and degradation, innovative encapsulation techniques using lipids and biopolymers are currently being explored, with the aim of enhancing their shelf life and stability (Shishir et al., 2018).

Clear knowledge on the interaction mechanisms of NPs with biological and environmental objects can greatly help to standardize and develop new-generation nanomaterials with better functionality and efficiency. Scientists and researchers have a pivotal role to play here. Their innovations, inventions, and expertise can bring about a remarkable change, particularly in standardizing the synthetic protocols for the production of nanomaterials with consistent properties from natural variations. Structure–properties database analysis and state-of-the-art AI techniques can be utilized extensively to accomplish this work (Olawade et al., 2024). Safety issues in the bioapplications of NPs can be rigorously checked by developing different eco-toxicity testing models and integrating in silico toxicity prediction tools into nanomaterial research methodologies. Both in vitro and in vivo toxicity studies should be mandatory to carry out to declare the biocompatibility of any NP.

Today's research should also attempt to harness sustainable nanomaterials with energy storage technology and environmental cleaning solutions to obtain better functionality alongside environmental advantages.

Finally, better commercialization of green NPs by forming suitable regulatory guidelines can help future generations technologically progress towards sustainability.

3.5 FUTURE PROSPECTS

The future prospects of nanomaterials obtained from renewable resources are highly encouraging and follow the global challenge towards green chemistry and a sustainable economy. Green nanotechnology possesses outstanding potential to guide multiple industries toward sustainable operations because the market seeks more environmentally friendly materials.

Coconut shells and other biomass derived NPs have been effectively utilized for the development of supercapacitors and bio-batteries. To improve the light absorbance capacity and overall efficiency of solar cells, cellulose and lignin-based CQDs are currently applied widely in photovoltaic devices.

Various biogenic metallic NPs, such as iron, copper oxide (CuO), and magnesium oxide (MgO), have shown significant potential in absorbing heavy metals like lead (Pb^{2+}), cadmium (Cd^{2+}), and chromium (Cr^{6+}); organic dyes; and pathogens from polluted water. Methylene blue and methyl orange dyes have been degraded by almost 90% with the help of CuO NPs (Muhammad et al., 2024b). The photocatalytic activity promoted by bioderived zinc oxide (ZnO) and titanium dioxide (TiO_2) NPs has been successfully employed to disinfect contaminated water (Habib et al., 2023) by producing reactive oxygen species under light exposure.

To satisfy the growing demand of eco-certified personal care products, plant and natural waste–derived NPs are now in broad application. Research on various natural colorants like beetroot, curcumin, and anthocyanins is in continuous progress to make toxin-free, bio-based UV protectors, sunscreen, and other beauty products. Fully edible cellulose nanocrystals are another example being explored for use in anti-aging products.

In the case of biomedical applications, green nanotechnology has already exhibited several promising functions. Some notable examples include the use of lignin-based NPs produced from agricultural waste as pH-sensitive drug-releasing carriers for cancer treatment; CQDs synthesized from agricultural waste as sensors for glucose and cancer biomarker detection; and potato, corn, and rice-derived starch NPs as vaccine adjuvants for better immune response. The fabrication of multifunctional green NPs that can be used for diagnostic, therapeutic, and tracking purposes represents a next-generation target in the field of nanobiotechnology. Future research also aims at the incorporation of different natural peptides and ligands into existing nanodrug delivery systems to improve their targeting efficiency and ability to evade the immune system. Like bioapplications, in agriculture, too, the controlled release of nano-fertilizers and pesticides is a major focus of today's research and innovation to reduce the overuse of harmful chemicals and the impacts of environmental pollution (Haydar, Ghosh, & Roy, 2024).

3.6 CONCLUSION

The utilization of renewable resources in the synthesis of nanomaterials enhances the importance of nanotechnology to a great extent. All the recent advancements attained in this field have clearly demonstrated the feasibility and beneficial impacts of using such natural feedstock for the preparation of nanosized metal, metal oxide, and carbon particles. In comparison to traditional chemical-based methods, these approaches definitely reduce the environmental footprint of nanomaterial production and also provide a cost-effective, sustainable manufacturing process. However, synthetic process and stabilization related few challenges still persist in the production pathway. These factors need to be critically addressed to pave the progress of this fascinating technology towards its highest potential. Interdisciplinary collaborations, adequate investment in research infrastructure, and the development of a strong regulatory system can be implemented to attain better results both in terms of production and commercialization. Not only that, the modern AI-driven optimization methodologies are now planned to be explored along with the green chemistry principles for the development of efficient nanotechnological applications. With all these effective and targeted efforts, green nanotechnology is expected to build a sustainable and resilient future soon.

REFERENCES

Bhardwaj, B., Singh, P., Kumar, A., Kumar, S., & Budhwar, V. (2020). Eco-friendly greener synthesis of nanoparticles. *Advanced Pharmaceutical Bulletin*, *10*(4), 566–576. https://doi.org/10.34172/apb.2020.067

Habib, S., Rashid, F., Tahir, H., Liaqat, I., Latif, A. A., Naseem, S., Khalid, A., Haider, N., Hani, U., Dawoud, R. A., Modafer, Y., Bibi, A., & Jefri, O. A. (2023). Antibacterial and cytotoxic effects of biosynthesized zinc oxide and titanium dioxide nanoparticles. *Microorganisms*, *11*(6), 1363. https://doi.org/10.3390/microorganisms11061363

Haydar, M. S., Ghosh, D., & Roy, S. (2024). Slow and controlled release nanofertilizers as an efficient tool for sustainable agriculture: Recent understanding and concerns. *Plant Nano Biology*, *7*, 100058. https://doi.org/10.1016/j.plana.2024.100058

S, A. K., M, D. K., Saikia, M., N, R. D., & A, S. (2023). A review on plant derived carbon quantum dots for bio-imaging. *Materials Advances*, *4*(18), 3951–3966. https://doi.org/10.1039/d3ma00254c

Makarov, V. V., Love, A. J., Sinitsyna, O. V., Makarova, S. S., Yaminsky, I. V., Taliansky, M. E., & Kalinina, N. O. (2014). "Green" nanotechnologies: Synthesis of metal nanoparticles using plants. *Acta Naturae*, *6*(1), 35–44. https://doi.org/10.32607/20758251-2014-6-1-35-44

Olawade, D. B., Ige, A. O., Olaremu, A. G., Ijiwade, J., & Adeola, A. O. (2024). The synergy of artificial intelligence and nanotechnology towards advancing innovation and sustainability—A mini-review. *Nano Trends*, 100052. https://doi.org/10.1016/j.nwnano.2024.100052

Paliwal, R., Babu, R. J., & Palakurthi, S. (2014). Nanomedicine scale-up technologies: Feasibilities and challenges. *AAPS PharmSciTech*, *15*(6), 1527–1534. https://doi.org/10.1208/s12249-014-0177-9

Radulescu, D., Surdu, V., Ficai, A., Ficai, D., Grumezescu, A., & Andronescu, E. (2023). Green Synthesis of metal and metal oxide nanoparticles: A review of the principles and biomedical applications. *International Journal of Molecular Sciences*, *24*(20), 15397. https://doi.org/10.3390/ijms242015397

Sanoja-López, K. A., Cedeño-Muñoz, J. S., Rivadeneira-Mendoza, B. F., Vergara-Romero, A., Luque, R., & Rodríguez-Díaz, J. M. (2025). Agricultural residues to high-value nanomaterials: Pathways to sustainability. *Resources Conservation & Recycling Advances*, 200243. https://doi.org/10.1016/j.rcradv.2025.200243

Shishir, M. R. I., Xie, L., Sun, C., Zheng, X., & Chen, W. (2018). Advances in micro and nano-encapsulation of bioactive compounds using biopolymer and lipid-based transporters. *Trends in Food Science & Technology*, *78*, 34–60. https://doi.org/10.1016/j.tifs.2018.05.018

Singh, N. A., Narang, J., Garg, D., Jain, V., Payasi, D., Suleman, S., & Swami, R. K. (2023). Nanoparticles synthesis via microorganisms and their prospective applications in agriculture. *Plant Nano Biology*, *5*, 100047. https://doi.org/10.1016/j.plana.2023.100047

Thu, H. E., Haider, M., Khan, S., Sohail, M., & Hussain, Z. (2023). Nanotoxicity induced by nanomaterials: A review of factors affecting nanotoxicity and possible adaptations. *OpenNano*, *14*, 100190. https://doi.org/10.1016/j.onano.2023.100190

Tripathi, N., Pavelyev, V., & Islam, S. S. (2017). Synthesis of carbon nanotubes using green plant extract as catalyst: Unconventional concept and its realization. *Applied Nanoscience*, *7*, 557–566. https://doi.org/10.1007/s13204-017-0598-3

4 Evaluating Sustainable Nanomaterial Synthesis from Renewable Resources

Methods, Challenges, and Pathways for Green Innovation

R. Venkatesh, Natarajan Sampathkumar, and K. Subramani

4.1 INTRODUCTION

Nanomaterials have attracted considerable attention because of their distinctive properties, including high surface area, mechanical strength, and optical properties, which render them promising candidates for applications in multiple industries, like electronics, medicine, energy, and environmental technologies. With the continuous development of nanotechnology, more demand is arising for novel and sustainable approaches for synthesizing nanomaterials that have fewer environmental effects and less dependence on non-renewable resources [1].

4.1.1 Overview of Nanomaterial Synthesis

Nanomaterials may be synthesized by a range of methods, commonly divided into two broad categories: top-down and bottom-up approaches. Top-down strategies involve breaking up bulk material to nanoscale structures, while bottom-up techniques build up nanomaterial from molecular or atomic building units such as chemical vapor deposition, sol-gel operations, and wet chemical procedures [2]. Traditional nanomaterial synthesis is often based on unsafe chemicals, heavy energy use, and expensive raw materials, all of which work against environmental appeal. Therefore, researchers need greener and sustainable methods to replace traditional nanomaterial synthesis.

4.1.2 Importance of Sustainability in Nanotechnology

The fast expansion of nanotechnology has focused attention on the possibility of its impact on the surrounding environment and health, both in positive and negative ways. Many of the methods of synthesis for the most common nanomaterials in traditional approaches are based on the use of toxic chemicals, high energy requirements, and hazardous byproducts, contributing to sustainability issues. The nanotechnology world is working diligently to include green chemistry. Nanotechnology for sustainability involves energy use, renewable resources, and waste minimization [3]. All or some of these actions are necessary to ensure the future of nanotechnology and to enable global efforts to address environmental problems such as climate change and resource depletion.

4.1.3 Role of Renewable Resources in Green Nanomaterial Production

Renewable feedstocks, including plant-based materials, agricultural waste, and biodegradable polymers, provide sustainable and waste-free solutions for nanomaterial synthesis. Renewable feedstocks are quickly available, biodegradable, and produced by utilizing less impact environmentally compared to fossil-based feedstocks. The use of renewable feedstocks offers the potential to develop bio-based nanomaterials with reduced toxicity, reduced energy usage, and improved environmental compatibility [4]. Besides, these sources are also obtained from waste materials, such as food processing waste or wood biomass, further sustainable in a circular economy. Bio-based synthesis routes can involve plant extracts, microbes, and other natural templates that facilitate nanostructure formation without the necessity of toxic chemicals or solvents [5]. Table 4.1 emphasizes how renewable resources help in the production of sustainable and green nanomaterials and also fit well with a circular economy strategy.

4.1.4 Objectives of the Chapter

This chapter intends to examine the ability of renewable resources to contribute to sustainable nanomaterial synthesis, highlighting challenges, opportunities, and future directions for applying green chemistry principles to nanomaterial synthesis.

More specifically, the goals are to:

1. Examine the existing bio-based nanomaterial synthesis methods and materials.
2. Assess the environmental advantages, in terms of reduced energy usage and carbon emissions, of utilizing renewable resources for nanomaterial production.

DOI: 10.1201/9781003644439-4

TABLE 4.1
Impact of Renewable Resources on Sustainable Nanomaterial Production

Category	Description	Examples	Benefits	Contribution to Circular Economy
Plant-Based Materials	Use of plant-derived substances for nanomaterial synthesis.	Plant extracts, cellulose, lignin	Biodegradable, low toxicity, environmentally compatible, abundant, and renewable.	Utilizes agricultural waste, reducing the need for new resources.
Agricultural Byproducts	Agricultural waste products used as feedstocks for nanomaterial production.	Fruit peels, spent grains, nutshells	Reduces waste, cost-effective, promotes sustainable agriculture practices.	Recycles waste from food production, contributing to resource efficiency.
Biodegradable Polymers	Polymers derived from renewable sources that decompose naturally.	Polylactic acid (PLA), chitosan	Low environmental impact, non-toxic, reduces dependency on petroleum-based plastics.	Reduces plastic waste and promotes biodegradable alternatives.
Microorganisms	Utilizing microbes for the synthesis of nanomaterials without harmful chemicals.	Fungi, bacteria, algae	Eco-friendly, low-energy processes, natural templating for nanomaterials.	Converts organic waste into nanomaterials, supporting waste valorization.
Natural Templates	Using natural structures, such as plant templates, for shaping nanomaterials.	Tree bark, leaves, seeds	Minimal processing required, biodegradable, reduces the use of synthetic templates.	Supports sustainable sourcing of raw materials from nature.

3. Identify main challenges in the scale-up of bio-based synthesis of nanomaterials for industrial purposes.
4. Discuss innovative technologies, such as green catalysts and solventless processes, which would further maximize sustainability.
5. Suggest future research directions and innovations which would propel the commercialization of green nanomaterials.

Through the accomplishment of these goals, this chapter will help in the development of green nanotechnology and offer useful information on how renewable resources can be utilized to produce more sustainable and environmentally friendly nanomaterials for various applications.

4.2 METHODOLOGY

The research methodology was applied to investigate the synthesis of sustainable nanomaterials using renewable feedstocks. These involve material selection, synthesis conditions, experimental methods, and analytical techniques employed to assess the bio-based nanomaterials' properties. Further, this section outlines the approach used to gather and analyze the data to assess the environmental and performance parameters of these nanomaterials.

4.2.1 Material Selection: Renewable Feedstocks

Choosing renewable feedstocks is imperative for attaining the synthesis of sustainable nanomaterials (Figure 4.1). The renewable feedstocks used for the current study come from biological, agricultural, and forestry waste materials, along with plant compounds. The feedstocks used are:

- **Agricultural Waste**: Agronomy byproducts, such as corn stalk, rice husks, and bagasse of sugarcane containing cellulose, hemicellulose, and lignin.
- **Plant Extracts**: Organic substances extracted from plants like leaf extracts, fruit peels, and algae, which are high in polyphenols, flavonoids, and other bioactive molecules.
- **Biodegradable Polymers**: Nature-based polymers like chitosan (derived from shrimp shell), starch, and cellulose, which can be easily fabricated at the nanoscale.
- **Food Industry Byproducts**: Waste organic material generated during food production, like citrus peel, tea waste, and fruit seeds, which contain natural antioxidants and metals amenable to nanoparticle synthesis.

These feedstocks were chosen due to their availability, biodegradability, and processability for nanomaterial synthesis. They offer a sustainable and renewable alternative to traditional feedstocks that are based on petrochemical-derived chemicals.

4.2.2 Synthesis Routes for Bio-Based Nanomaterials

A number of synthesis pathways are used for bio-based nanomaterial production. All of these techniques focus on being environmentally friendly, as they do not use harmful chemicals, minimize the energy required, and follow principles of green chemistry.

1. **Green Synthesis via Plant Extracts**: Utilizing aqueous plant extracts (e.g., tea, neem, or aloe vera) as reducing agents and stabilizers in order to minimize metal salts in nanoparticles.

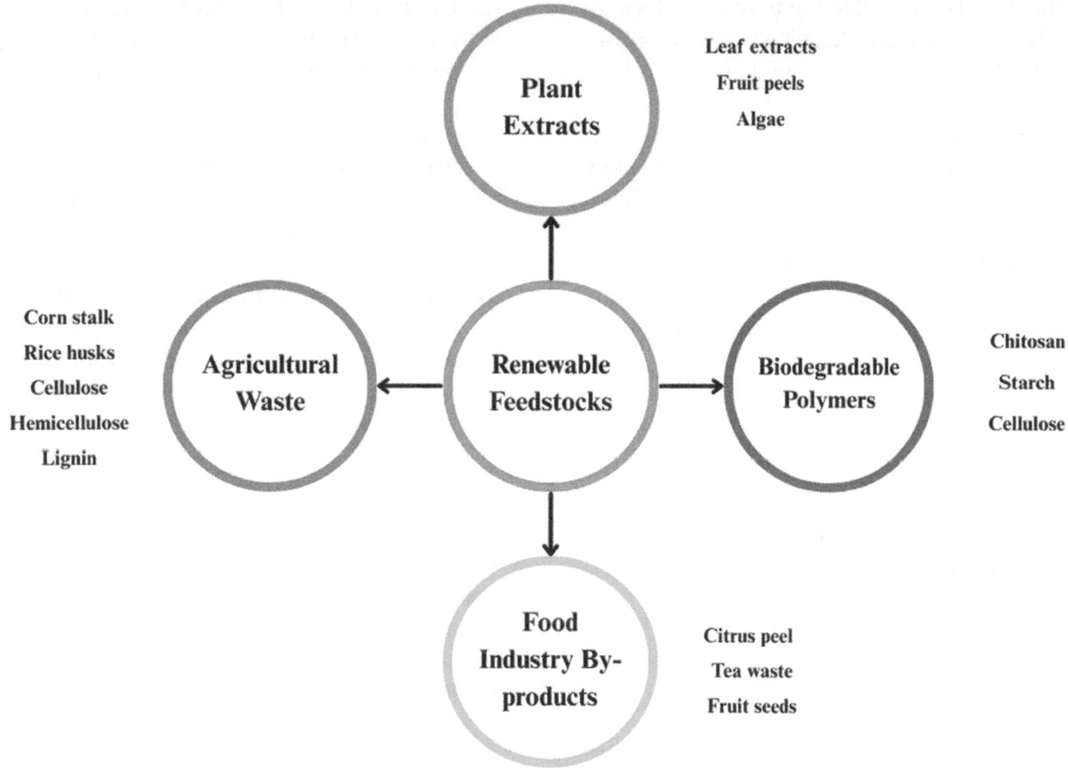

FIGURE 4.1 Sources of renewable raw materials for sustainable production.

2. **Sol-Gel Process Using Biopolymers**: Employing the use of degradable polymers like chitosan or starch to form nanoparticles and nanocomposites. Metal oxide nanoparticles are fabricated using the sol-gel route and can be used for making different materials that have more mechanical or chemical properties.
3. **Microbial Synthesis**: Using microbes such as fungi, bacteria, or algae in the biosynthesis of nanomaterials has been documented. The microbes can catalyze metal ions in particle synthesis, which allows the nanoparticles to be prepared in a cost-effective and green manner.
4. **Hydrothermal and Solvothermal Methods**: These methods use renewable solvents, like water or ethanol, to prepare metal oxide nanomaterials, which greatly reduces the organic solvent in comparison to previous methods.

These methods were selected for their ability to minimize the detrimental effects on the environment, eliminate energy costs, and create functional nanomaterials from renewable sources [6].

4.2.3 Experimental Setup and Techniques

The experimental protocol for the production of the bio-based nanomaterials consisted of the following technical steps (Figure 4.2).

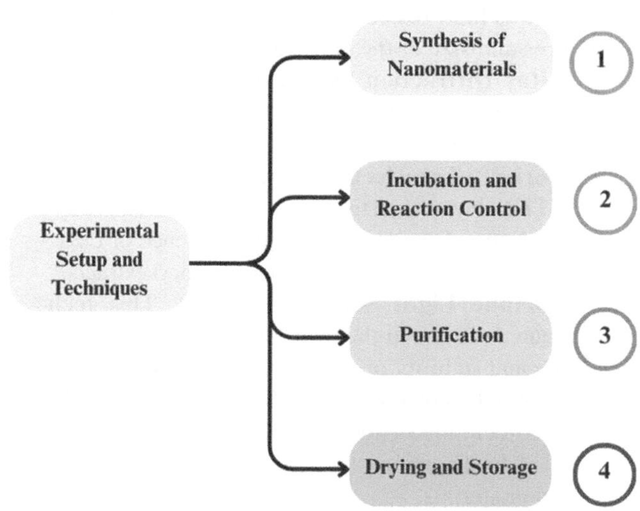

FIGURE 4.2 Key steps in the synthesis of bio-based nanomaterials.

1. **Synthesis of Nanomaterials**: Green feedstocks were used according to the synthesis routes selected. For example, plant extracts were made by macerating fresh plant materials, extracting them in distilled water, filtering the extract, and mixing it in a metal salt solution (i.e., silver nitrate or copper sulfate) to reduce the nanoparticles.
2. **Incubation and Reaction Control**: Microbial cultures were incubated under optimal conditions,

while metal salts were added to induce the formation of the nanoparticles. Reaction times, temperature, and pH ranges were measured and controlled to enhance the shape and size of the nanoparticles.

3. **Purification**: The nanomaterials were purified by centrifugation, filtration, and removing unreacted precursors or byproducts once the nanoparticles were synthesized.
4. **Drying and Storage**: The synthesized nanomaterials were dried in a vacuum oven and stored in an inert atmosphere to prevent oxidation or degradation.

The laboratory environment was kept at a consistent temperature and humidity level to allow reproducibility of the results.

4.2.4 Characterization Methods for Nanomaterials

To understand the physical, chemical, and morphological characteristics of the nanomaterials synthesized [7], the following characterization techniques were used:

1. **Scanning Electron Microscopy (SEM)**: This technique was used to study the surface morphology and particle size distribution of the nanoparticles.
2. **Transmission Electron Microscopy (TEM)**: It allowed high-resolution study of the internal structures and size of the particles.
3. **X-Ray Diffraction (XRD)**: It was used to study the crystalline structure of the nanomaterials and assist with understanding the composition.
4. **Fourier Transform Infrared Spectroscopy (FTIR)**: It was used to identify the functional groups, which may help demonstrate the presence of bioactive molecules from plant (or biopolymer) extracts.
5. **Dynamic Light Scattering (DLS)**: This technique determined the hydrodynamic size distribution and stability of nanoparticles in a solution.
6. **Energy Dispersive X-Ray Spectroscopy (EDX)**: This technique can be paired with SEM and/or TEM to determine the elemental composition of nanomaterials.

These techniques were used to confirm that the properties of the nanomaterials synthesized from the renewable materials met the properties needed for the stated end use.

4.2.5 Data Collection and Analysis Approach

Data acquisition focused on several of the elemental performance metrics for the synthesized bio-based nanomaterials:

1. **Yield and Efficiency**: The number of nanomaterials produced per unit of renewable feedstock was quantified to provide a measure of the efficiency of the synthetic process.
2. **Energy Use**: The energy used in each synthesis procedure was quantified and compared to conventional nanomaterial synthesis procedures to assess the sustainability of the processes.
3. **Environmental Impact**: The carbon footprint and other impacts were estimated based on the energy use and chemical interactions in the synthesis.
4. **Material Properties**: Characterization information such as size distributions, shapes, and stability measurements was used to determine if the bio-based nanomaterials were viable nanomaterials for target applications.

Statistical analysis was carried out to assess the reproducibility of results and patterns or correlations across feedstock type, synthesis method, and material properties. Comparisons of performance to traditional nanomaterial synthesis approaches were made in order to showcase the environmental and economic benefits of utilizing renewable materials.

4.3 PARAMETER ANALYSIS

The most important parameters that determine the sustainability of renewable resource–based nanomaterial synthesis. These parameters are crucial in assessing the environmental and economic viability of bio-based nanomaterials against traditional synthesis techniques. The subsequent analyses discuss energy usage, carbon emissions, material efficiency, yield, and comparison of renewable feedstocks for nanomaterial synthesis.

4.3.1 Energy Consumption in Bio-Based vs Conventional Synthesis

It is energy that is most important to the sustainability of nanomaterial production. Conventional techniques, such as chemical vapor deposition or high-temperature processes, require significant energy from high temperatures, vacuum, and poisonous chemicals. On the other hand, bio-based approaches have milder conditions, green solvents such as water or ethanol, and typically do not require high temperatures or energy-intensive apparatus.

A comparative study of energy consumption revealed that bio-based processes for the production of silver nanoparticles (AgNPs) with plant extracts consumed 30–40% less energy compared to traditional chemical processes [8]. For gold nanoparticles, reduction with plant extracts consumed just half the energy of traditional processes such as thermal deposition. This reduced energy consumption makes bio-based synthesis eco-friendlier and more sustainable.

4.3.2 Carbon Footprint Assessment of Nanomaterial Production

The carbon footprint for nanomaterial synthesis quantifies the overall greenhouse gas (GHG) emission during the course of synthesis from raw material, to synthesis, to purification. Classical methods involving the use of metal salts, organic solvents, and high-temperature reactions generally result in high carbon emissions by virtue of dependency on fossil energy and energy-draining processes [9].

Bio-based synthesis, however, has the potential to minimize carbon emissions. For instance, a lifecycle study of the production of carbon nanotubes (CNTs) from bio-derived glucose resulted in a 55% drop in carbon emissions compared to conventional processes that utilize petrochemical feedstocks. Likewise, synthesizing silver nanoparticles using green tea extract resulted in a 40% drop in emissions compared to traditional chemical routes [10]. This is due to the fact that bio-based processes utilize renewable feedstocks, consume less energy, and use environmentally friendly solvents.

Through the utilization of renewable resources, bio-based synthesis mitigates GHG emissions, aiding in efforts to achieve global carbon neutrality and sustainable production.

4.3.3 Material Efficiency and Yield

Material efficiency is a significant consideration in determining the sustainability of nanomaterial manufacturing. Bio-based synthesis tends to be more material efficient since plant-based and microbial systems naturally function better at synthesizing nanoparticles. The degree of nanoparticle formation based on renewable feedstocks relies on parameters such as precursor concentration, reaction time, and temperature.

For example, in the bio-synthesis of silver nanoparticles from green tea extract, as much as 90% of the metal precursor was reduced to nanoparticles, compared to the 60–70% yields obtained in conventional chemical reduction processes [11]. Plant extracts also yield nanoparticles with more uniform sizes and improved stability, resulting in less material loss and increased yields.

But bio-based synthesis processes are variable in yield, based on the feedstock and method applied. To date, scaling up these operations remains a challenge, and improvements continue to be made toward a consistent yield for large-scale production.

4.3.4 Comparison of Renewable Feedstocks for Nanomaterial Fabrication

The selection of renewable feedstocks is a major factor in determining the overall effectiveness and sustainability of the nanomaterial synthesis process. Several renewable resources, including agricultural waste, plant extracts, and biodegradable polymers, have been investigated for their potential to synthesize nanomaterials.

Agricultural Waste: Agricultural residues like corn husks, rice husks, and sugarcane bagasse are readily available and low in cost. They are rich in cellulose, hemicellulose, and lignin, which may be utilized to form cellulose-based nanomaterials or even as reducing agents for the formation of metal nanoparticles. Several studies have reported that these feedstocks have superior material availability but may require additional pretreatment to optimize their use in nanomaterial synthesis.

Plant Extracts: Vegetation-based feedstocks such as tea, neem, and aloe vera have emerged in the spotlight because they consist of a large percentage of bioactive molecules such as polyphenols, flavonoids, and alkaloids, which can be used as reducing agents during nanoparticle synthesis without any addition of chemicals. Plant-based feedstocks are inexpensive and can be easily found, and plant extract bioactive molecules can provide nanoparticles with certain desired properties like antimicrobial properties. However, the heterogeneity of extract composition can influence nanoparticle production consistency and scalability.

Biodegradable Polymers: Biopolymers such as chitosan, starch, and cellulose are common in the preparation of nanocomposites and nanocoating. They are biodegradable, non-toxic, and environmentally friendly. Chitosan, which is derived from crustacean shells, has proved especially efficient in the preparation of metal and metal oxide nanoparticles, providing biocompatibility and stability advantages. Their availability and price can depend on location and supply chains.

Food Industry Byproducts: Food industry wastes, including citrus peels, tea waste, and fruit seeds, are potential renewable feedstocks for nanomaterial synthesis. These byproducts are rich in polyphenolic compounds and act as reducing agents during nanoparticle synthesis due to their antioxidant content. Their utilization not only minimizes waste but also brings an aspect of circularity to nanomaterial synthesis.

Comparative analyses of these feedstocks show that the best compromise among material availability, price, and nanoparticle quality is usually delivered by plant extracts, but due to its high availability and cheap price, agricultural waste can become more appealing to be used for large-scale industrial production [12]. Table 4.2 points out the most distinctive differences in terms of energy usage between bio-based and traditional nanomaterial fabrication, highlighting

TABLE 4.2
Comparative Energy Consumption in Bio-Based and Conventional Nanomaterial Synthesis

Factor	Bio-Based Synthesis	Conventional Synthesis	Key Differences
Energy Consumption	Lower energy requirement, operates at milder conditions (e.g., room temperature, water/ethanol solvents)	High energy demand, requires elevated temperatures and vacuum conditions	Bio-based methods use less energy (30–40% lower for silver nanoparticles)
Energy Efficiency	30–40% lower than conventional methods for silver nanoparticles, 50% of the energy for gold nanoparticles	High energy use for chemical vapor deposition (CVD) or thermal deposition	Bio-based methods show higher energy efficiency in nanomaterial synthesis
Temperature and Equipment	Milder conditions, no need for high-temperature or high-energy equipment	Requires specialized equipment for high temperatures and vacuum conditions	Bio-based synthesis uses simpler, less energy-intensive equipment
Solvents	Green solvents (e.g., water, ethanol)	Toxic solvents (e.g., organic solvents, acids)	Bio-based synthesis avoids the use of toxic solvents, enhancing sustainability
Environmental Impact	Reduced environmental impact due to lower energy use and green solvents	Higher environmental impact from energy-intensive processes and toxic byproducts	Bio-based methods are more environmentally friendly due to reduced energy needs and toxic waste

advantages of bio-based approaches in both energy efficiency and environmental sustainability [13].

4.4 RESULTS AND DISCUSSION

The findings of the experimental research on nanomaterial synthesis with the aid of renewable materials are revealed and discussed here. This includes an examination of the environmental consequences, the scalability challenges in using bio-based methods for synthesis, and a comparison with conventional synthesis. Additionally, case studies on the successful green production of nanomaterials are reviewed in order to identify practical applications and potential commercialization [14].

4.4.1 Summary of Experimental Findings

The experimental findings revealed that renewable feedstocks such as plant extracts, agricultural byproducts, and biodegradable polymers can be successfully utilized for the synthesis of nanomaterials.

Nanoparticle Formation: Bio-based synthesis techniques yielded nanoparticles of varying sizes, ranging from 10 to 100 nm, based on the applied feedstock and synthesis technique. For instance, silver nanoparticles synthesized via green tea extract showed uniform spherical particles with a mean size of 25 nm, while nanoparticles synthesized via neem extract showed a mean size of 35 nm.

Material Efficiency: The conversion efficiencies for bio-based synthesis were high, with the majority of processes being 80–90% material efficient, indicating the ability of these processes to minimize material waste relative to traditional processes, which have lower conversion efficiencies (60–70%).

Stability: The nanomaterials derived from bio-based sources were more stable, especially those that were prepared using plant extracts rich in bioactive molecules, which stabilized the nanoparticles and hindered agglomeration with time.

Surface Area and Functionalization: The as-synthesized nanoparticles exhibited larger surface areas compared to their counterparts obtained by conventional means, which is useful for examples like catalysis and drug delivery. Furthermore, plant-based nanoparticles frequently exhibited functional groups (e.g., hydroxyl, carbonyl) on the surface, which can be utilized for further functionalization.

4.4.2 Environmental Impact: Reduction in Energy and Carbon Emissions

The environmental footprint of bio-based synthesis of nanomaterials was assessed based on energy consumption and carbon emissions.

Energy Reduction: The energy used for bio-based synthesis processes was always lower than that of traditional synthesis processes. For example, the energy used in synthesizing silver nanoparticles using green tea extract was 40% less than traditional chemical reduction processes involving high temperatures and the use of toxic solvents. Likewise, microbial synthesis processes utilized much less energy, particularly when ambient temperatures were used.

Carbon Emissions: Bio-based approaches showed between a 30 and 50% reduction in carbon emissions when compared with traditional methods. The carbon footprint of production for silver nanoparticles was lowered from about 120 g CO_2 per gram nanoparticle in traditional methods to only 60 g CO_2 per gram when a bio-based process was used [15]. This decrease largely stemmed from less energy being demanded and the implementation of renewable feedstocks, whose use has fewer environmental costs, particularly in processing and extraction.

The findings indicate that bio-based nanomaterial synthesis routes hold great promise to minimize the environmental impact of nanotechnology manufacture, and as such, they represent a very appealing alternative to traditional, energy-hungry processes.

4.4.3 Challenges in Scaling Bio-Based Nanomaterial Synthesis

Although bio-based nanomaterial synthesis holds the promise of a sustainable future, some challenges have to be overcome before widespread industrial application. One of these challenges is maintaining reproducibility in the size, shape, and characteristics of the nanoparticles, as plant extracts and agricultural wastes may differ in composition. To achieve consistent results, standardization of feedstocks and optimal reaction conditions are necessary. Scalability is also an issue, since bio-based processes tend to take longer reaction times and more feedstock, and therefore are less efficient at scale [16]. Also, optimizing separation and purification processes for larger quantities is needed in order to preserve quality. Although renewable feedstocks are cheap, the cost of bio-based manufacture can still be more expensive than conventional methods because more feedstock is required and additional purification processes are needed. Maximizing production yields and lowering costs is central to making bio-based synthesis more economically attractive.

4.4.4 Comparative Analysis of Bio-Based and Conventional Methods

A side-by-side comparison of bio-based and traditional nanomaterial synthesis processes shows numerous significant differences based on environmental effect, material use, and general sustainability.

The comparison shows that bio-based synthesis processes have great environmental advantages, especially with respect to energy efficiency and carbon emissions, but have scalability and cost-effectiveness challenges [17]. Table 4.3 lists the main factors in bio-based and traditional nanomaterial synthesis processes and identifies the sustainability benefits of bio-based processes as having lower energy requirements, less carbon emissions, and greater material efficiency.

TABLE 4.3
Comparison of Bio-Based vs. Conventional Nanomaterial Synthesis Methods

Parameter	Bio-Based Synthesis	Conventional Synthesis
Energy Consumption	30–40% lower than conventional methods	High energy consumption, especially in thermal and chemical processes
Carbon Emissions	30–50% reduction in emissions	Higher carbon footprint due to energy-intensive processes and chemical use
Material Efficiency	80–90% yield	60–70% yield
Toxicity of Byproducts	Minimal toxicity (bio-degradable)	Hazardous chemical byproducts (e.g., solvents)
Cost	Higher due to feedstock preparation	Typically lower for mass production
Reproducibility	Can vary with feedstock composition	More consistent due to controlled industrial processes

4.4.5 Case Studies of Successful Green Nanomaterial Production

A number of successful case studies indicate the viability and promise of bio-based nanomaterial production:

Green Tea Extract-Mediated Silver Nanoparticles: A study revealed that silver nanoparticles biosynthesized from green tea extract had superior antimicrobial activity, comparable to nanoparticles synthesized through traditional chemical routes. The biosynthesis involved much less energy and resulted in a greener process with lower waste production [18].

Gold Nanoparticles via Plant Biomass: Researchers proved the feasibility of gold nanoparticle synthesis through use of biomass extracted from coffee wastes. Coffee acted both as the reducing agent as well as a stabilizer of the resulting particles, forming colloidal gold with sizes between 20 and 30 nm [19]. The reaction was scalable to large scales with little energy intake and no residues.

Chitosan-Coated Nanoparticles for Drug Delivery: A research study involving the use of chitosan-coated nanoparticles for drug delivery proved that chitosan, a biopolymer extracted from shrimp shells, was utilized to create biodegradable and biocompatible nanocarriers. The application of chitosan not only imparted efficient drug encapsulation but also minimized environmental footprint through the use of seafood industrial wastes [20].

TABLE 4.4
Case Studies on Bio-Based Nanomaterial Production and Their Applications

Nanomaterial	Feedstock	Method Used	Size Range	Energy Consumption	Advantages	Applications
Silver Nanoparticles	Green tea extract	Bio-synthesis using plant extract as reducing agent	NA	Lower energy compared to conventional methods	Excellent antimicrobial properties, environmentally friendly, reduced waste	Antimicrobial coatings, medical devices
Gold Nanoparticles	Coffee grounds (plant biomass)	Coffee grounds used as reducing and stabilizing agents	20–30 nm	Minimal energy consumption, scalable	Stable nanoparticles, environmentally benign, cost-effective	Catalysis, sensors, electronics
Chitosan-Coated Nanoparticles	Shrimp shells (biopolymer)	Chitosan-coated nanoparticles for drug delivery	NA	Reduced environmental impact, utilizes waste products	Biocompatible, biodegradable, effective drug encapsulation	Drug delivery systems, biomedicine

These case studies demonstrate that bio-based synthesis of nanomaterials can not only offer sustainable alternatives to traditional approaches but also present certain benefits in specific applications like antimicrobial coatings, drug delivery systems, and catalysis. Table 4.4 summarizes the main features of the successful case studies, indicating the feedstocks, approaches, size ranges, energy requirements, benefits, and applications of bio-based nanomaterial synthesis. It reflects how bio-based processes provide environmental and practical advantages over traditional approaches.

4.5 FUTURE WORK

The future of synthesis of bio-based nanomaterials is highly promising, yet a number of avenues need exploration and development in order to support large-scale industrial applications.

4.5.1 Scaling Bio-Based Nanomaterial Synthesis for Industrial Applications

One of the biggest problems in the synthesis of bio-based nanomaterials is upscaling from lab scale to industrial manufacture. To overcome this, some key steps are essential: initial optimization of reaction conditions like temperature, pH, and precursor concentration to produce identical, high-quality nanoparticles at a bigger scale. Second, the application of process intensification methods such as continuous flow reactors may improve efficiency by minimizing reaction time and energy usage. Third, standardization and automation will play a central role in managing large quantities of feedstock, ensuring consistency, and living up to industrial quality standards [21]. These developments will close the gap between laboratory-scale research and large-scale, affordable industrial production.

4.5.2 Advancements in Green Catalysts and Solvent-Free Processes

The creation of green catalysts and solventless synthesis techniques represents a significant challenge to further decrease the environmental footprint of nanomaterial manufacturing.

Biocatalysts: Enzymes and other bio-based catalysts may be used more extensively in nanomaterial production. Biocatalysts possess high selectivity and activity, minimizing the use of strong chemicals and solvents. Enzymatic reactions can also occur under gentler conditions, with less energy needed and less byproduct formation.

Solvent-Free Synthesis: Most traditional nanomaterial synthesis processes utilize toxic organic solvents that add to pollution and toxic waste. Development of solvent-free synthesis methods, such as solid-phase synthesis or only using water as the solvent, could have a very significant impact on minimizing the environmental impact. Green solvents, such as ionic liquids or supercritical fluids, can also be investigated for their ability to substitute existing solvents without diminishing the efficiency or quality of nanoparticle generation.

These developments would lead to further reduction of environmental damage and enhance the sustainability of the nanomaterial synthesis process [22].

4.5.3 Exploration of Novel Renewable Feedstocks

Though renewable feedstocks such as agricultural wastes have come a long way, there remains great untapped potential in new and unusual resources. Future studies can investigate waste agricultural and food-processing biomass, like fruit peels, spent grains, and nutshells, which are so readily available yet so frequently underutilized. Examining their chemical makeup can reveal new opportunities for creating more types of nanomaterials. Marine products, such as seaweeds and algae, also show potential because of their high polysaccharide, protein, and polyphenolic compound content, which may be employed in the synthesis of nanoparticles [23]. Forestry waste, including wood chips and sawdust, also contains lignocellulose, which may be employed in the preparation of nanocellulose or metal oxide nanoparticles. Through the investigation of these varied and new feedstocks, scientists can increase the sustainability, diversity, and efficiency of bio-based nanomaterial synthesis while reducing waste.

4.5.4 Integration of Nanomaterial Synthesis with Circular Economy Models

One of the important future directions in sustainable synthesis of nanomaterials is incorporating it into circular economy models, which aim at minimizing waste, enhancing resource efficiency, and lengthening the lifetime of materials. Some of the strategies involve recycling nanomaterials from use in catalysis, water treatment, and electronics, with research on recovering and reusing nanoparticles. Another strategy is the development of bio-based nanomaterials as biodegradable and recyclable, such that they can be degraded into non-toxic fragments or reused simply, minimizing the demand for fresh resources and lowering environmental impact [24]. Waste valorization, in which byproducts from synthesis (e.g., plant extracts) are converted into feedstocks or value-added products such as biofuels or bioplastics, can also enhance sustainability. By embracing the principles of circular economy, nanomaterial synthesis is more sustainable, with additional business opportunities created around resource efficiency and waste minimization.

4.5.5 Research on Commercial Viability and Cost-Effectiveness

To transition from lab success to broad commercial use, additional research must be conducted to determine the cost-effectiveness and feasibility of bio-based nanomaterials. Priority areas for attention are lowering production costs by maximizing yields, reducing energy consumption, and enhancing feedstock extraction. Determining market opportunities in sustainable electronics, biodegradable packaging, and pharmaceuticals is also essential to increasing production. Furthermore, life cycle assessments (LCAs) will facilitate an understanding of the environmental, economic, and social effects of bio-based and traditional nanomaterials to inform decisions on the most sustainable means of production [25]. Commercial viability is crucial to embedding bio-based nanomaterials in the mainstream industries.

4.6 CONCLUSION

In summary, producing nanomaterials through bio-synthetic approaches represents a paradigm shift in nanomaterials, provides an environmentally sustainable alternative to conventional approaches, and is also one of the most efficient ways to use our energy supplies and materials because of the significant energy consumption and environmental implications of various chemical and metallurgical processes. It allows the use of renewable feedstocks, such as agricultural waste, plant extracts, and biodegradable polymers, which enables the employment of more environmentally sound processes but allows for additional capability, value, and economic value, such as antimicrobial properties and modularity for many applications. Despite the challenges that remain in bringing bio-based nanomaterials to large scale production, consistency, and optimization, ongoing research into green catalysts, solvent-free processes, and novel feedstocks like marine biomass and food waste are promising avenues for increased commercial engagement. In addition, by embedding bio-based nanomaterials into circular economy concepts, it is evident that the value can be increased further by reducing waste and resource consumption. The growing interest and pursuit of sustainable innovation across many global industries is fortuitous timing for bio-based nanomaterials, creating opportunities for those who are committed to building a greener, more robust future for nanotechnology.

REFERENCES

[1] Chavda, V. P., Acharya, D., Hala, V., Daware, S., & Vora, L. K. (2023). Sunscreens: A comprehensive review with the application of nanotechnology. *Journal of Drug Delivery Science and Technology*, 86, 104720. https://doi.org/10.1016/j.jddst.2023.104720

[2] Mekuye, B., & Abera, B. (2023). Nanomaterials: An overview of synthesis, classification, characterization, and applications. *Nano Select*, 4(8), 486–501. https://doi.org/10.1002/nano.202300038

[3] Serrano, E., Rus, G., & García-Martínez, J. (2009). Nanotechnology for sustainable energy. *Renewable and Sustainable Energy Reviews*, 13(9), 2373–2384. https://doi.org/10.1016/j.rser.2009.06.003

[4] Chakrabartty, I., Hakeem, K. R., Mohanta, Y. K., & Varma, R. S. (2022). Greener nanomaterials and their diverse applications in the energy sector. *Clean Technologies and Environmental Policy*, 24(10), 3237–3252. https://doi.org/10.1007/s10098-022-02368-0

[5] Aristri, M. A., Lubis, M. A. R., Iswanto, A. H., Fatriasari, W., Sari, R. K., Antov, P., Gajtanska, M., Papadopoulos, A. N., & Pizzi, A. (2021). Bio-based polyurethane resins derived from tannin: Source, synthesis, characterization, and application. *Forests*, 12(11), 1516. https://doi.org/10.3390/f12111516

[6] Stubbs, S., Yousaf, S., & Khan, I. (2022). A review on the synthesis of bio-based surfactants using green chemistry principles. *DARU Journal of Pharmaceutical Sciences*, 30(2), 407–426. https://doi.org/10.1007/s40199-022-00450-y

[7] Munaweera, I., & Madhusha, M. L. C. (2023). Chemical characterization techniques in detail. *Characterization Techniques for Nanomaterials*, 28–54. https://doi.org/10.1201/9781003354185-4

[8] Kumar, M., Saini, R. V., Gupta, M., & Singh, R. (2024). Green synthesis of silver nanoparticle (Cha-AgNPs) using Chenopodium album extract and evaluation of their antifungal potential against pathogenic fungi. *Biomass Conversion and Biorefinery*. https://doi.org/10.1007/s13399-024-05721-z

[9] Kartal, M. T. (2021). The role of consumption of energy, fossil sources, nuclear energy, and renewable energy on environmental degradation in top-five carbon producing countries. *Renewable Energy*, 184, 871–880. https://doi.org/10.1016/j.renene.2021.12.022

[10] Niesiobędzka, J., & Datta, J. (2023). Challenges and recent advances in bio-based isocyanate production. *Green Chemistry*, 25(7), 2482–2504. https://doi.org/10.1039/d2gc04644j

[11] Kaliammal, R., Parvathy, G., Maheshwaran, G., Velsankar, K., Devi, V. K., Krishnakumar, M., & Sudhahar, S. (2021). Zephyranthes candida flower extract mediated green synthesis of silver nanoparticles for biological applications. *Advanced Powder Technology*, 32(11), 4408–4419. https://doi.org/10.1016/j.apt.2021.09.045

[12] Shahbaz, A., Hussain, N., Saleem, M. Z., Saeed, M. U., Bilal, M., & Iqbal, H. M. (2022). Nanoparticles as stimulants for efficient generation of biofuels and renewables. *Fuel*, 319, 123724. https://doi.org/10.1016/j.fuel.2022.123724

[13] Sezali, N. A. A., Ong, H. L., Villagracia, A. R., & Hoang, T. (2024). Bio-based nanomaterials for energy application: A review. *Vietnam Journal of Chemistry*, 62(1), 1–12. Portico. https://doi.org/10.1002/vjch.202300158

[14] Daryoush, B., & Darvish, A. (2013). A case study and review of nanotechnology and nanomaterials in green architecture. *Research Journal of Environmental and Earth Sciences*, 5(2), 78–84. https://doi.org/10.19026/rjees.5.5641

[15] Rahuman, H. B. H., Dhandapani, R., Narayanan, S., Palanivel, V., Paramasivam, R., Subbarayalu, R., Thangavelu, S., & Muthupandian, S. (2022). Medicinal plants mediated the green synthesis of silver nanoparticles and their biomedical applications. *IET Nanobiotechnology*, 16(4), 115–144. https://doi.org/10.1049/nbt2.12078

[16] Debroy, A., Joshi, S., Yadav, M., & George, N. (2023). *Green Synthesis of Nanoparticles from Bio-Waste for Potential Applications: Current Trends, Challenges, and Prospects*. Elsevier eBooks (pp. 431–466). https://doi.org/10.1016/b978-0-323-91149-8.00009-0

[17] Selvaraj, V. K., & Subramanian, J. (2022). A comparative study on bio-based PU foam reinforced with nanoparticles for EMI-shielding applications. *Polymers*, 14(16), 3344. https://doi.org/10.3390/polym14163344

[18] Rolim, W. R., Pelegrino, M. T., de Araújo Lima, B., Ferraz, L. S., Costa, F. N., Bernardes, J. S., Rodigues, T., Brocchi, M., & Seabra, A. B. (2019). Green tea extracts mediated biogenic synthesis of silver nanoparticles: Characterization, cytotoxicity evaluation and antibacterial activity. *Applied Surface Science*, 463, 66–74. https://doi.org/10.1016/j.apsusc.2018.08.203

[19] El-Desouky, N., Shoueir, K., El-Mehasseb, I., & El-Kemary, M. (2022). Synthesis of silver nanoparticles using bio valorization coffee waste extract: Photocatalytic flow-rate performance, antibacterial activity, and electrochemical investigation. *Biomass Conversion and Biorefinery*, 13(17), 15871–15885. https://doi.org/10.1007/s13399-021-02256-5

[20] Shim, S., & Yoo, H. S. (2020). The application of mucoadhesive chitosan nanoparticles in nasal drug delivery. *Marine Drugs*, 18(12), 605. https://doi.org/10.3390/md18120605

[21] Ajmal, M., Muslim, F., Ali, S. A., & Shah, M. R. (2024). Synthesis of bio-based nanolubricant using bimetallic nanoparticles to mitigate the surface deterioration of low carbon steel for metalforming industry. *Materials Chemistry and Physics*, 327, 129913. https://doi.org/10.1016/j.matchemphys.2024.129913

[22] Mondal, A., & Mukhopadhyay, C. (2024). Solvent-free microwave reactions towards significant organic transformations: A green approach. *Tetrahedron Green Chem*, 4, 100054. https://doi.org/10.1016/j.tgchem.2024.100054

[23] El-Araby, R. (2024). Biofuel production: Exploring renewable energy solutions for a greener future. *Biotechnology for Biofuels and Bioproducts*, 17(1). https://doi.org/10.1186/s13068-024-02571-9

[24] Khosravi, A., Zarepour, A., Iravani, S., Varma, R. S., & Zarrabi, A. (2024). Sustainable synthesis: Natural processes shaping the nanocircular economy. *Environmental Science Nano*, 11(3), 688–707. https://doi.org/10.1039/d3en00973d

[25] Mittal, P., Sharma, H., Kapoor, R., Gautam, R. K., Garg, N., & Dhankhar, S. (2024). Biobased nanomaterials in biomedical applications. *Biobased Nanomaterials*, 141–171. https://doi.org/10.1007/978-981-97-0542-9_6

5 Green Synthesis of Gold Nanorods
Implications for Sustainable Clean Environment

Zendesha Stephen Mbalaha, Intisar Albander, and Peverga Rex Jubu

5.1 BACKGROUND

Gold nanorods (AuNRs) are anisotropic nanoparticles with unique optical characteristics. Gold nanorods absorb incident light due to the surface plasmon resonance of oscillating surface electrons. Surface plasmon resonance is the phenomenon by which excited free conducting electrons oscillate in tandem with the incident light, leading to optical absorption. The optical absorption of AuNRs uniquely occurs at the transverse and longitudinal axes, respectively. The optical absorption at the transverse axis of AuNRs is fixed, while that at the longitudinal axis can be tuned to near-infrared wavelengths by varying synthetic parameters. The longitudinal absorption of AuNRs is very sensitive to their aspect ratio and the surrounding medium. Following optical absorption, AuNRs convert photon energy to photothermal energy via a non-radiative process. Thus, AuNRs are essential nanoparticles of interest in various biomedical applications such as drug delivery, detection of diseases, and photothermal therapy because of their deep tissue penetration and tunable surface plasmon resonance in the biological transmission window (Lage et al., 2021; Mbalaha et al., 2019). The deployment of AuNRs in biomedical applications requires the synthesis of stable AuNRs. Wet chemistry methods such as seedless and silver-assisted seed-mediated growth methods are often used for the synthesis of gold nanorods. For example, the silver-assisted seed-mediated growth method is used very often due to the high yield of AuNRs and flexibility of tuning the size of AuNRs to the required dimension with ease. However, the toxicity of the synthetic chemical reagents associated with wet chemistry methods to human health and the safety of the ecosystem is very concerning. Furthermore, the cost of equipment and chemical reagents places a huge financial burden on researchers. Thus, alternative methods are desirable.

Green synthesis of nanoparticles is inspiring because it is environmentally friendly and very cheap due to the abundance of the raw materials that can be sourced locally. Green synthesis is a method that utilizes plant extracts and microorganisms for the synthesis of nanoparticles. Extracts of plants are composed of natural phytochemical compounds such as flavonoids, polyphenols, proteins, fatty acids, vitamins, and polysaccharides which facilitate the reduction of the precursor material and stabilize the nanostructure (Zambonino et al., 2021). Figure 5.1 describes the green synthesis of AuNRs, where extracts of plants are subjected to heating to a certain temperature to facilitate the reduction of gold (III) ions to gold, or enzymes of microorganisms induce intracellular reduction of gold (III) ions to gold atoms to grow nanorods. The UV-vis extinction spectrum shows two unique peaks, illustrating absorption at the transverse and longitudinal axes, respectively. Tremendous successes have been achieved for the green synthesis of stable gold nanospheres using plant extracts and microorganisms (Donga, Ram, & Chanda, 2020; Jha et al., 2017). However, green synthesis of stable anisotropic gold nanoparticles can be very challenging because plant-based extracts or microorganisms may not be very efficient in contrast to cetyl trimethylammonium bromide (CTAB) in providing a soft template for the evolution of anisotropic shapes.

5.2 HISTORICAL PERSPECTIVE OF THE GREEN SYNTHESIS OF AUNRS

To the best of our knowledge, the green synthesis of AuNRs was first reported in 2003. The leaves of the geranium plant (*Pelargonium graveolens*) were used as a reducing agent for the green synthesis of AuNRs (Shankar et al., 2003). Boiled extract of geranium leaves was added to aqueous solution of chloroauric acid. After 1 hour of reaction time, it was observed that gold nanoparticles of different shapes, including rods, flat sheets, and triangles, were formed (Shankar et al., 2003). Even though the reaction time of this approach was short, the gold nanorods obtained were large and the yield was low relative to other shapes. The longitudinal absorption of the as-prepared AuNRs was 610 nm below the biological transmission window, thus rendering the AuNRs unattractive for biomedical applications in the biological transmission window.

Intracellular fabrication of AuNRs in the cyanobacterium *Nostoc ellipsosporum* was demonstrated in 2012 whereby a solution of gold (III) ions was up taken by the cyanobacteria (Parial et al., 2012). AuNRs were produced after exposing the cyanobacterium to a solution of gold for

DOI: 10.1201/9781003644439-5

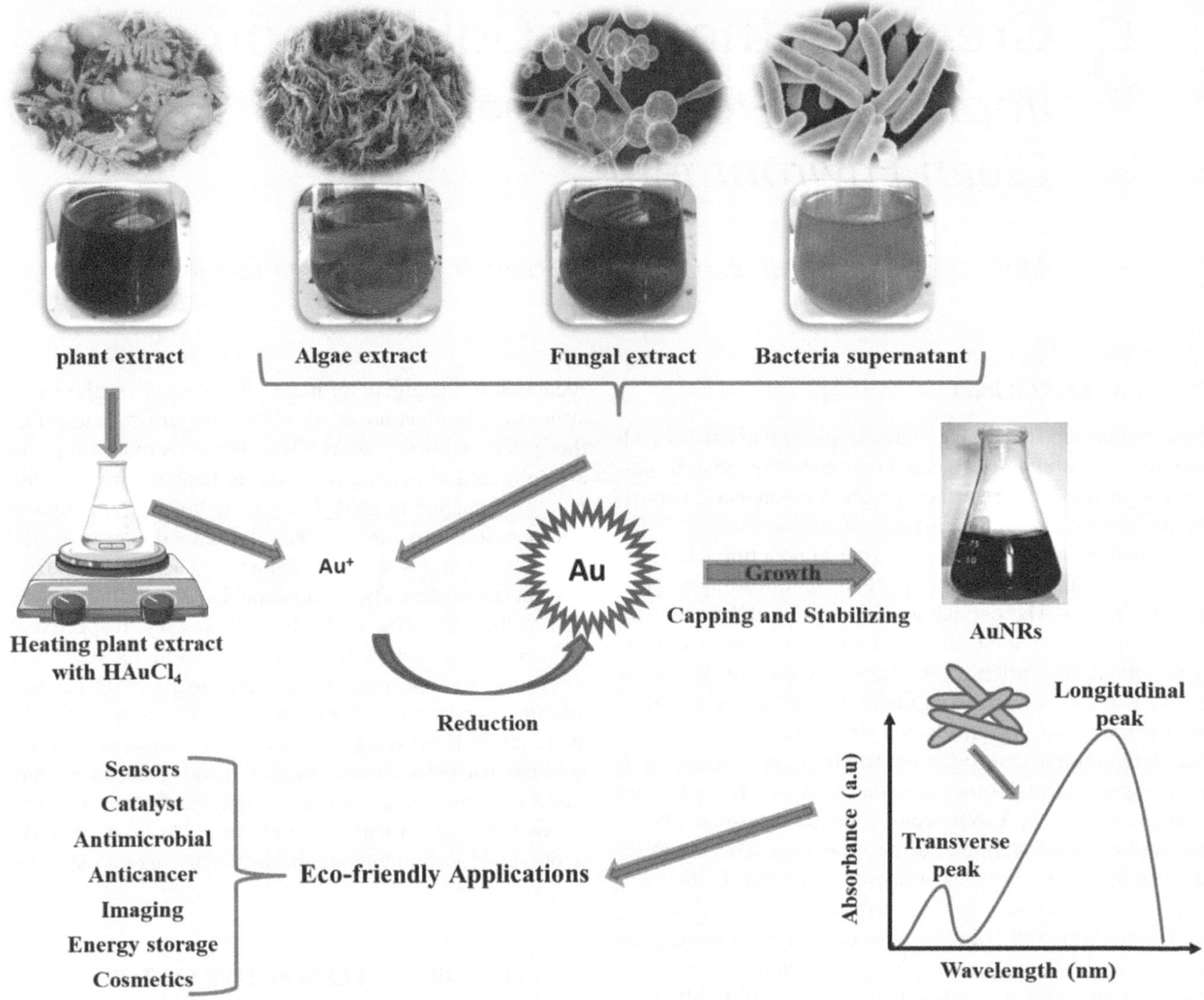

FIGURE 5.1 Green synthesis of gold nanorods from plant extracts and microorganisms; UV-vis spectrum and eco-friendly applications of green-synthesized gold nanorods.

48 hours at 20°C. The cyanobacterium enzymes reduced gold (III) ions to gold atoms, while intracellular proteins facilitated the stabilization of the nanorod structure. Gold nanorods of various sizes ranging from 137 to 209 nm in length and 33 to 68 nm in diameter were obtained. The AuNRs were large with low yield. It is worth noting that the secretion of intracellular enzymes and proteins could be susceptible to the physiological conditions of cells, raising doubts about the reproducibility of AuNRs by a microorganism-based biosynthesis approach. Moreover, the biosynthesis of metal nanoparticles using microorganisms as nanofactories is very slow relative to that with plant extracts (Nishanthi et al., 2019). Furthermore, control over the size of AuNRs could be very difficult to achieve in contrast to wet chemistry methods.

By 2014, citric acid and aniline were found to enhance the synthesis of AuNRs. It was shown that AuNRs of high aspect ratio with spherical-like and nanofibrous structures could be produced by varying the molar ratio of citric acid to aniline in the aqueous solution of chloroauric acid (Mondal et al., 2014). Additionally, extracts of cocoa powder were utilized as a reducing and stabilizing agent for the green synthesis of anisotropic gold nanoparticles (Fazal et al., 2014). Optimization of green-synthesized AuNRs with longitudinal surface plasmon resonance coinciding with the biological transmission window was achieved by varying the amount of the extracts of cocoa powder, aqueous solution of chloroauric acid, and temperature (Fazal et al., 2014). However, the yield of AuNRs relative to other shapes with extracts of cocoa powder was poor, casting doubts on the contribution of the as-prepared AuNRs by this approach in biomedical applications.

Optimization of the longitudinal absorption of green-synthesized AuNRs to the near infrared was demonstrated in 2016. A seedless approach was exploited for the green synthesis of AuNRs using a natural polyphenolic compound, resveratrol as a reducing agent, aqueous solution of gold, and CTAB surfactant (Wang et al., 2016). The AuNRs were monodispersed, and their longitudinal absorption was tuned to 950 nm.

In 2017, it was shown that AuNRs can be fabricated in the presence of CTAB and without CTAB using the extracts of *Olea europaea* fruit and *Acacia nilotica* husk (Awad et al., 2017). The plant extracts were added to a solution containing chloroauric acid and CTAB, and the final mixture was subjected to stirring at 45°C for 13 minutes until the final solution became colorless, marking the formation of AuNRs. Different sizes (short and long) of AuNRs were observed, and the yield was low. The longitudinal absorption was approximately 577 nm. In the absence of CTAB, monodispersed rod shapes were formed. This approach successfully produced AuNRs with longitudinal absorption in the visible region with a short reaction time. However, the optical absorption of the AuNRs in the visible region could potentially limit their biomedical application *in vivo*.

Recently, resveratrol (RSV), gallic acid (GA), and extract of *Stryphnodendron obovatum* were used as reducing agents for the synthesis of AuNRs by a seedless approach (Lage et al., 2021). Three separate growth solutions containing chloroauric solution, silver nitrate, and CTAB were prepared. To each mixture, RSV, GA, and extract of *Stryphnodendron obovatum* were added, followed by the addition of sodium borohydride. The samples containing RSV and GA were heated to 70°C, while the sample with extract of *Stryphnodendron obovatum* was heated to 50°C. The three samples produced stable AuNRs with phenolic compounds on their surfaces. The longitudinal absorption of AuNRs synthesized using RSV, GA, and extract of *Stryphnodendron obovatum* were 657, 687, and 655 nm, respectively. Though AuNRs were biosynthesized, this approach is compromised by several concerns such inhomogeneous morphology of the resulting AuNRs, the formation of a large quantity of undesirable nanospheres as a byproduct and most critically, the use of highly toxic chemicals such as CTAB and sodium borohydride (De Lima Oliveira et al., 2022).

Green synthesis of AuNRs with RSV phenolic compound using a similar seedless approach was demonstrated (Carmo et al., 2024). The study found the presence of the CTAB and RSV mixture on the surface of the AuNRs, and the RSV-coated AuNRs exhibited enhanced antifungal activity against *Candida albicans* in contrast to the gold nanospheres. The presence of CTAB surfactant on the surface of AuNRs raises cytotoxic concerns for their utilization in biomedical applications.

Extract of the peels of sour guava was used as a reducing agent for the green synthesis of AuNRs via a seedless approach for which the growth solution containing an aqueous solution of gold and CTAB mixture was subjected to heating at 50°C. Then silver nitrate, hydrochloric acid (HCl), and aqueous extract of sour guava peels were added to the mixture and stirred for 1 hour before cold sodium borohydride was added and stored in the oven at 50°C (Patiño-González et al., 2025). It was observed that the nanoparticles were heterogeneously composed of nanorods, nanospheres, and nanocubes, as depicted from the absorption spectra and TEM image shown in Figure 5.2. The UV-vis spectrum shows that the absorbance intensity of the transverse band is almost at the same height with the longitudinal band, suggestive of a low yield of AuNRs.

The green synthesis of AuNRs by a seedless approach, as demonstrated in the previous studies, can be considered partial green synthesis because of its reliance on chemicals such as CTAB and silver nitrate for the evolution of the anisotropic shape (Carmo et al., 2024; Lage et al., 2021;

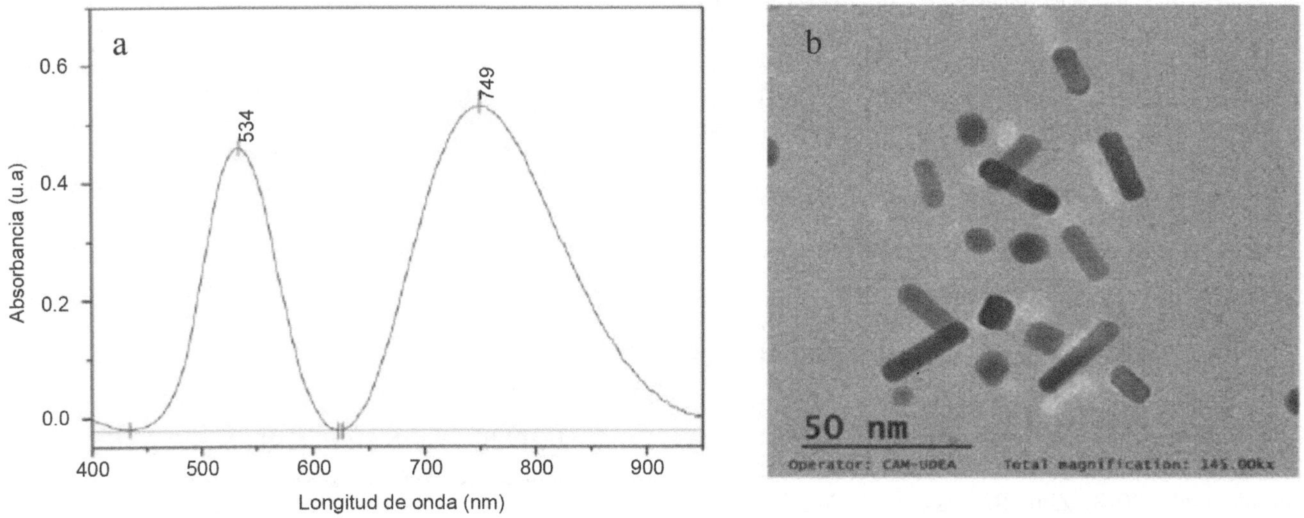

FIGURE 5.2 UV-vis spectra and TEM image of AuNRs synthesized by green chemistry.

(Patiño-González et al., 2025)

Patiño-González et al., 2025). These chemicals can be very harmful to humans and the environment. Achieving monodisperse and homogenous AuNRs without the use of CTAB and silver nitrate is a challenging task that may require further research. Furthermore, the presence of phytochemical compounds of plant origin on the surface of nanoparticles hinders the assembly of functional groups (Medeot et al., 2024). Therefore, functionalization of green-synthesized AuNRs with targeting ligands could be difficult. This could trigger nanoparticle instability *in vivo* capable of limiting broader biomedical applications of green-synthesized AuNRs. Another area of concern is the non-standardization of protocols for green synthesis of AuNRs.

5.3 BIOMEDICAL APPLICATIONS OF GREEN-SYNTHESIZED AUNRS

Gold nanorods are essential nanoparticles for biosensing, drug delivery, photothermal therapy of cancer and bio-imaging because of their biocompatibility and well-established surface chemistry for assembling functional ligands on their surface for drug delivery and targeting analytes (Mbalaha et al., 2024; Mbalaha et al., 2019; Wei et al., 2016). It has been demonstrated that upon laser irradiation, a mixture of green-synthesized anisotropic gold nanoparticles induced a photothermal effect on epidermoid carcinoma cell line A431 that led to the disruption of cell membranes and cell lysis (Fazal et al., 2014). Furthermore, green-synthesized anisotropic gold nanoparticles were found to be a better X-ray contrast agent than omnipaque in computed tomography imaging (Fazal et al., 2014).

The antibacterial and antifungal activities of green-synthesized AuNRs have been demonstrated (Awad et al., 2017; Carmo et al., 2024; Lage et al., 2021). Extracts of plants containing, polyphenols such as gallic acid and resveratrol intrinsically have anticancer and antifungal properties that could enhance the anticancer effect of green-synthesized AuNRs (Berman et al., 2017; Hadidi et al., 2024; Salehi et al., 2018). Although very effective for stabilization, the adhering surface layer of plants based compounds on the AuNRs hinder their subsequent chemical functionalization with targeting ligands, thus, restricting the use of green synthesized AuNRs in biomedical applications (Medeot et al., 2024).

5.4 MECHANISM OF THE ANTIBACTERIAL EFFECT OF GREEN-SYNTHESIZED AUNRS

Gold nanorods have been found to be very effective against several bacterial strains, such as *E. coli, S. aureus, P. aeruginosa*, and *B. subtilis* (Awad et al., 2017; Mahmoud et al., 2018; Zhu, Ramasamy, & Yi, 2014). The antibacterial effect of green-synthesized AuNRs can be influenced by various factors such as the particle size, concentration, surface charge, and intrinsic properties of plant phytochemical agents. For example, the small AuNRs could exhibit better antibacterial effects than the large AuNRs, possibly due to the larger surface area-to-volume ratio (Paesa et al., 2023).

The surface charge of small AuNRs of gold nanoparticles is a critical parameter for determining their interaction with the bacterial cell wall (De Lima Oliveira et al., 2022). The surface charge of green-synthesized gold nanoparticles is dependent on the intrinsic properties of the phenolic compounds embedded in the plant. For example, gallic acid, resveratrol, and tannins are naturally cationic; thus extracts from plants such as *Stryphnodendron obovatum* and *Acacia nilotica* rich in gallic acid, resveratrol, and tannins carry positive charges (Lage et al., 2021). For this reason, AuNRs synthesized using plant extracts rich in gallic acid, resveratrol, and tannins have a positive surface charge (Awad et al., 2017; Carmo et al., 2024; Lage et al., 2021). The bacterial cell wall, on the other hand, is negatively charged. The surface charge of AuNRs could facilitate their binding to the bacterial cell wall via electrostatic attraction. The interaction of the bacterial cell wall with AuNRs can cause damage and cell lysis, leading to cell death. In an aqueous solution, AuNRs could generate gold ions that may permeate the bacterial cell wall, causing the denaturation of bacterial proteins and DNA damage, leading to cell death. In addition, gold ions generate reactive oxygen species (ROS) that can induce oxidative stress, leading to cell membrane and DNA damage (Nishanthi et al., 2019; Paesa et al., 2023). Moreover, the antibacterial effect is an intrinsic property of some polyphenols such as resveratrol, gallic acid, and tannin (Borges et al., 2015; De Lima Oliveira et al., 2022; Fiori et al., 2013; Salehi et al., 2018). Polyphenols generate ROS that can damage vital cellular components of the bacteria (Hadidi et al., 2024). In addition, polyphenols can diffuse through the bacterial cell wall into the cytoplasm and disrupt the metabolic activities of the bacterial cell (Hadidi et al., 2024). For example, tanfloc, a derivative of tannin, has been found to inhibit the uptake of sugar and amino acid of the invading *Staphylococcus aureus* cell (De Lima Oliveira et al., 2022), while gallic acid disrupts intracellular functions such as adenosine triphosphate (ATP) synthesis (Hadidi et al., 2024). The study demonstrated that gold nanoparticles synthesized with tanfloc as a reducing and stabilizing agent exerted a faster, superior antibacterial effect against *Staphylococcus aureus* and biofilm in contrast to the gold nanoparticles and tanfloc alone (De Lima Oliveira et al., 2022). This suggests that green-synthesized AuNRs could work better in synergy with natural phytochemical compounds of plant origin against bacterial infectious diseases.

5.5 IMPLICATIONS OF GREEN SYNTHESIS OF AUNRS FOR A SUSTAINABLE CLEAN ENVIRONMENT

The increase in bio-wastes released to the environment has been a source of concern in recent years, and this can be attributed to the growth of the human population and increase of human activities. It has been reported recently that about 1.3 million tons of food is wasted annually (Flores-Contreras et al., 2024). Bio-waste can be either plant based or animal based. These bio-wastes emanate from several sources such as industries, schools, offices, markets, abattoirs, and household activities like cooking. The waste can pollute the environment and cause severe harm to human and animal health if not properly disposed of. Disposal of waste by burning in the open spaces, dumping on the streets, or discharging into the streams and rivers are commonly seen. Inappropriate methods of waste disposal (especially by burning) can trigger the emission of greenhouse gases capable of facilitating climate changes, with attendant consequences such as natural disaster, infectious diseases, and environmental contamination (Flores-Contreras et al., 2024). Wastes from peels of fruits, vegetables, and nuts can be reused or recycled for green synthesis of AuNRs and by extension sustain clean environment. For example, recently, peels of fruits and vegetables utilized for the green synthesis of AuNRs could reduce the amount of bio-waste that could have polluted the environment, thus fostering the sustenance of a clean and healthy environment (Patiño-González et al., 2025).

The interface between green-synthesized AuNRs and the environment can facilitate the removal of environmental pollutants and contaminants in water due to the large surface area-to-volume ratio and surface functionality of AuNRs for binding analytes (Abdelbary & Abdelfattah, 2020). Furthermore, polyphenols such as tannin found in plant extracts have the potentials of neutralizing contaminants in wastewater for a sustainable clean environment (De Lima Oliveira et al., 2022). The antimicrobial effect of green-synthesized AuNRs can be leverage for the preservation of food and food packaging materials against microbial growth. This may not only prolong the shelf life of packaged food, minimize food wastage, but reduce the amount of food packaging materials that litter the streets, dumpsites, or waterways, thereby mitigating environmental pollution (Flores-Contreras et al., 2024; Hadidi et al., 2024). Moreover, AuNRs could enhance the degradation of soil contaminants for a sustainable environment.

5.6 BENEFITS AND LIMITATIONS OF THE GREEN SYNTHESIS OF AUNRS

Green-synthesized AuNRs are very promising in biomedical applications and remediation of environmental pollution. The green synthesis of AuNRs offers numerous advantages such as eco-friendliness and low cost-. Green synthesized AuNRs are very promising in synergizing their antibacterial effect with that of plant extracts for overcoming antibiotic resistance. Additionally, this method is very promising in remediating environmental pollution. Despite the benefits, the green synthesis of AuNRs faces several significant challenges such as poor control over nanorod size, high yield of byproducts, difficulty of reproducibility and functionalization of the resulting nanorods.

5.7 CONCLUSION

The conventional methods of synthesizing AuNRs are based on wet chemistry. However, the chemical agents employed in conventional methods have been found to be unfriendly to the ecosystem. To protect the ecosystem from the hazards associated with wet chemistry–based methods, green synthesis approaches have evolved for the synthesis of AuNRs. Green synthesis approaches explore plant extracts for reducing the precursor gold metal for the growth and stabilization of AuNRs, while the microorganisms act as nanofactories where intracellular enzymes and proteins facilitate the formation and stabilization of AuNRs. Plant extracts enhance a faster formation of AuNRs in contrast to that by microorganism. Though heating is not required for the microorganism-based green synthesis of AuNRs, reproducibility of AuNRs based on microorganisms can be very challenging. The functionalization of green-synthesized AuNRs with targeting ligands very challengingand may require further investigation. By utilizing plant extracts and microorganisms for the green synthesis of AuNRs, the hazardous effects of chemical agents associated with wet chemistry on the ecosystem will be minimized. Green-synthesized AuNRs have the potentials of overcoming antibiotic resistance of some strains of bacteria agents for improving human health and sustenance of a clean environment. This chapter provides the groundwork for the exploration of green-synthesized AuNRs for improving human health and a sustainable clean environment.

REFERENCES

Abdelbary, S., & Abdelfattah, H. (2020). Modern trends in uses of different wastes to produce nanoparticles and their environmental applications. In M. Sen (Ed.), *Nanotechnology and the Environment* (p. 9). London: Intechopen Limited. https://doi.org/10.5772/intechopen.93315.

Awad, M. A., Hendi, A. A., Virk, P., Ortashi, K. M. O., Elobeid, M. A., Merghani, N. M., . . . Ibrahim, E. M. (2017). Preparation, characterization and antibacterial effects of eco-friendly gold nanorods. *Tropical Journal of Pharmaceutical Research*, *16*(2), 313–318. https://doi.org/10.4314/tjpr.v16i2.8.

Berman, A. Y., Motechin, R. A., Wiesenfeld, M. Y., & Holz, M. K. (2017). The therapeutic potential of resveratrol: A review of clinical trials. *NPJ Precision Oncology*, *1*(1), 35. https://doi.org/10.1038/s41698-017-0038-6.

Borges, F. A., Siguematsu, P. R., Herculano, R. D., & dos Santos, C. (2015). Novel sustained-release of Stryphnodendron obovatum leaves extract using natural rubber latex as carrier. *Journal of Basic and Applied Pharmaceutical Sciences*, *36*(3), 379–384.

Carmo, P. H. F. D., Lage, A. C. P., Garcia, M. T., Silva, N. S., Santos, D. A., Mylonakis, E., & Junqueira, J. C. (2024).

Resveratrol-coated gold nanorods produced by green synthesis with activity against Candida albicans. *Virulence*, *15*(1), 2416550. https://doi.org/10.1080/21505594.2024.2416550.

De Lima Oliveira, E. G., Vieira, S. A., Da Silva, F. A. G., Da Costa, M. M., Gomes, A. S. L., & De Oliveira, H. P. (2022). Synergistic antibacterial activity of green gold nanoparticles and tannin-based derivatives. *BioChem*, *2*(4), 269–279. https://doi.org/10.3390/biochem2040019.

Donga, S., Ram, B. G., & Chanda, S. (2020). Antimicrobial, antioxidant and anticancer activities of gold nanoparticles green synthesized using Mangifera indica seed aqueous extract. *Artificial Cells, Nanomedicine, and Biotechnology*, *48*(1), 1315–1325. https://doi.org/10.1080/21691401.2020.1843470.

Fazal, S., Jayasree, A., Sasidharan, S., Koyakutty, M., Nair, S. V., & Menon, D. (2014). Green synthesis of anisotropic gold nanoparticles for photothermal therapy of cancer. *ACS Applied Materials & Interfaces*, *6*, 8080–8089. https://doi.org/10.1021/am500302t.

Fiori, G. M. L., Fachin, A. L., Correa, V. S. C., Bertoni, B. W., Giuliatti, S., Amui, S. F., . . . Pereira, A. M. S. (2013). Antimicrobial activity and rates of tannins in stryphnodendron adstringens. *American Journal of Plant Sciences*, *4*, 2193–2198. https://doi.org/10.4236/ajps.2013.411272.

Flores-Contreras, E. A., González-González, R. B., Pizaña-Aranda, J. J. P., Parra-Arroyo, L., Rodríguez-Aguayo, A. A., Iñiguez-Moreno, M., . . . Melchor-Martínez, E. M. (2024). Agricultural waste as a sustainable source for nanoparticle synthesis and their antimicrobial properties for food preservation. *Frontiers in Nanotechnology*, *6*, 1346069. https://doi.org/10.3389/fnano.2024.1346069.

Hadidi, M., Liñán-Atero, R., Tarahi, M., Christodoulou, M. C., & Aghababaei, F. (2024). The potential health benefits of gallic acid: Therapeutic and food applications. *Antioxidants (Basel)*, *13*(8). https://doi.org/10.3390/antiox1308100.

Jha, P., Saraf, A., Rath, D., & Sharma, D. (2017). Green synthesis and antimicrobial property of gold nanoparticles: A review. *World Journal of Pharmaceutical and Medical Research*, *3*(8), 431–435.

Lage, A. C. P., Ladeira, L. O., Mosqueira, L., Paniago, R. M., Castilho, R. O., Amorim, J. M., . . . Faraco, A. A. G. (2021). Synthesis and characterization of gold nanorods using the natural products resveratrol, gallic acid, and a purified fraction of Stryphnodendron obovatum by seedless method. *Environmental Nanotechnology, Monitoring & Management*, *16*, 100473–100482. https://doi.org/10.1016/j.enmm.2021.100473.

Mahmoud, N. N., Alkilany, A. M., Khalil, E. A., & Al-Bakri, A. G. (2018). Nano-photothermal ablation effect of hydrophilic and hydrophobic functionalized gold nanorods on Staphylococcus aureus and Propionibacterium acnes. *Scientific Reports*, *8*(1), 6881. https://doi.org/10.1038/s41598-018-24837-7.

Mbalaha, Z. S., Craig, G., Birch, D. J. S., & Chen, Y. (2024). Aptamer functionalised small gold nanorod-based nanoprobes: Targeting EpCAM over-expressed cancer cells for photothermal therapy. *Austin Journal of Analytical and Pharmaceutical Chemistry*, *11*(1), 1170.

Mbalaha, Z. S., Edwards, P. R., Birch, D. J. S., & Chen, Y. (2019). Synthesis of small gold nanorods and their subsequent functionalization with hairpin single stranded DNA. *ACS Omega*, *4*(9), 13740–13746. https://doi.org/10.1021/acsomega.9b01200.

Medeot, C., Alsadig, A., D'Amico, F., Costantini, R., Campallegio, C. L., A, C., . . . Bonin, S. (2024). Green synthesis of gold nanoparticles: Pros and cons of natural compounds. *Nano Express*, *5*, 045020. https://doi.org/10.1088/2632-959X/ad9f85.

Mondal, S., Rana, U., Bhattacharjeeb, R. R., & Malik, S. (2014). One pot green synthesis of polyaniline coated gold nanorods and its applications. *RSC Advances*, *4*, 57282–57289. https://doi.org/10.1039/c4ra12080a.

Nishanthi, R., Malathi, S., John Paul, S., & Palani, P. (2019). Green synthesis and characterization of bioinspired silver, gold and platinum nanoparticles and evaluation of their synergistic antibacterial activity after combining with different classes of antibiotics. *Materials Science & Engineering C-Materials for Biological Applications*, *96*, 693–707. https://doi.org/10.1016/j.msec.2018.11.050.

Paesa, M., Remirez de Ganuza, C., Alejo, T., Yus, C., Irusta, S., Arruebo, M., . . . Mendoza, G. (2023). Elucidating the mechanisms of action of antibiotic-like ionic gold and biogenic gold nanoparticles against bacteria. *Journal of Colloid and Interface Science*, *633*, 786–799. https://doi.org/10.1016/j.jcis.2022.11.138.

Parial, D., Patra, H. K., Roychoudhury, P., Dasgupta, A. K., & Pal, R. (2012). Gold nanorod production by cyanobacteria—a green chemistry approach. *Journal of Applied Phycology*, *24*(1), 55–60. https://doi.org/10.1007/s10811-010-9645-0.

Patiño-González, M. C., Echeverri-Cuartas, C. E., Torijano-Gutiérrez, S. T., Naranjo-Rios, S. M., & AgudeloI, N. A. (2025). Optimization of the green synthesis of gold nanorods using aqueous extract of peeled sour guava as a source of antioxidants. *PLoS ONE*, *20*(1), e0313485. https://doi.org/10.1371/journal.pone.0313485.

Salehi, B., Mishra, A. P., Nigam, M., Sener, B., Kilic, M., Sharifi-Rad, M., . . . Sharifi-Rad, J. (2018). Resveratrol: A double-edged sword in health benefits. *Biomedicines*, *6*, 91–111. https://doi.org/10.3390/biomedicines6030091.

Shankar, S. S., Ahmad, A., Pasricha, R., & Sastry, M. (2003). Bioreduction of chloroaurate ions by geranium leaves and its endophytic fungus yields gold nanoparticles of different shapes. *Journal of Materials Chemistry*, *13*(7), 1822–1826. https://doi.org/10.1039/B303808B.

Wang, W., Li, J., Lan, S., Rong, L., Liu, Y., Sheng, Y., . . . Yang, B. (2016). Seedless synthesis of gold nanorods using resveratrol as a reductant. *Nanotechnology*, *27*(16), 165601. https://doi.org/10.1088/0957-4484/27/16/165601.

Wei, G., Yu, J., Wang, J., Gu, P., Birch, D. J. S., & Chen, Y. (2016). Hairpin DNA-functionalized gold nanorods for mRNA detection in homogenous solution. *Journal of Biomedical Optics*, *21*(9), 097001–097009. https://doi.org/10.1117/1.JBO.21.9.097001.

Zambonino, M. C., Quizhpe, E. M., Jaramillo, F. E., Rahman, A., Vispo, N. S., Jeffryes, C., & Dahoumane, S. A. (2021). Green synthesis of selenium and tellurium nanoparticles: Current trends, biological properties and biomedical applications. *International Journal of Molecular Sciences*, *22*, 989–1023. https://doi.org/10.3390/ijms22030989.

Zhu, Y., Ramasamy, M., & Yi, D. K. (2014). Antibacterial activity of ordered gold nanorod arrays. *ACS Applied Materials & Interfaces*, *6*, 15078–15085. https://doi.org/10.1021/am503153v.

6 Green Synthesis of Calcium Nanoparticles from *Momordica cymbalaria* and Their Applications in *Drosophila melanogaster* for Type 2 Diabetes Mellitus

Aishwarya H S, Mukta Dixit, Anjan P Kumar, Gurudatta B V, and Savithri Bhat

6.1 INTRODUCTION

Type 2 Diabetes Mellitus is the third major cause of mortality after cancer and cardiovascular diseases. Considering the high incidence rates, compounded with severe complications, including cardiovascular, renal, and neural impairments, there is a great impetus for developing effective therapeutic strategies for this metabolic disorder. The contemporary drugs for treating diabetes are basically designed to affect β-cells to initiate insulin secretion, of which the most conventional are the sulfonylurea drugs. However, the therapies often come with limited effectiveness in treating diabetes, which presents in a multilateral way. Moreover, most of these have side effects. Hence, a drug that exhibits multiple mechanisms of action and increased potency is necessary [1].

Since the Vedic period, medicinal plants have been an important source of bioactive substances in helping cure illnesses. *Momordica cymbalaria*, indigenous to the southern states of India, has gained the spotlight in research for its health benefits. Traditionally used because of its abortifacient and antimicrobial properties, MC has also gained significant attention for its antidiabetic potential. Various studies indicate that the fruit and extracts of MC exhibit significant hypoglycaemic activities in experimental models of diabetes. The various mechanisms of action attributed to its β-cell stimulation, enhancement of insulin secretion, and improvement of glycogen storage account for its activity. Saponins of this plant may play a pivotal role in reversing pancreatic β-cell atrophy, thereby increasing insulin secretion and alleviating hyperglycaemia. Also, by blocking adrenergic signalling, they may improve insulin sensitivity. A novel method to combat the increasing burden of diabetes has been proposed, which involves studying a triterpenoid saponin extracted from the roots of *Momordica cymbalaria* and its antidiabetic properties [2].

A research study was conducted on the synthesis of $CaCO_3$ nanoparticles ($CaCO_3$-NPs) using green extraction from *Ailanthus altissima* leaf extract, and their potential therapeutic applications were examined. Techniques such as Fourier Transform Infrared Spectroscopy (FTIR), Ultraviolet–visible spectroscopy (UV spectroscopy) and scanning electron microscopy (SEM) were carried out for the characterization of the obtained nanoparticles. Nanoparticles with antioxidant and anti-diabetic activity were examined through various types of assays. This establishes an advantage of *A. altissima* extracts in the synthesis of nanoparticles, where plant extracts can act as both reducing and stabilizing agents. Moreover, the study highlights the increasing focus on using green methods for nanoparticle synthesis, demonstrating the biomedical aspects of $CaCO_3$-NPs in drug delivery and the treatment of metabolic diseases [3].

A study highlights the growing interest in calcium carbonate nanoparticles ($CaCO_3$-NPs) as versatile nanocarriers in biomedical applications. Owing to their favourable properties, such as biocompatibility, low cytotoxicity, pH responsiveness, and cost-effectiveness, $CaCO_3$-NPs have emerged as promising candidates for therapeutic delivery. The review outlines their potential utility in three key areas: antimicrobial activity, drug delivery, and gene transfer. The emphasis on their pH-sensitive degradation and environmentally benign nature further supports their relevance in targeted and controlled therapeutic strategies. This work contributes to the broader understanding of inorganic nanoparticle systems in modern nanomedicine [4].

A work outlines the growing significance of calcium carbonate nanoparticles in cancer therapy, particularly as drug delivery systems. Due to their favourable properties, including biocompatibility, low toxicity, cost-effectiveness, and sensitivity to pH, $CaCO_3$-based nanocarriers have attracted attention for targeted tumour treatment. The paper discusses recent progress in the design and synthesis of these nanoparticles, highlighting their ability to function not only as drug carriers but also as sources of calcium and carbon dioxide, which may influence the tumour

DOI: 10.1201/9781003644439-6

microenvironment. It further examines the mechanisms through which these systems exert therapeutic effects and evaluates their potential in advancing tumour diagnosis and treatment strategies. The findings support the promising future of $CaCO_3$ nanoplatforms in biomedical applications, especially in oncology [5].

To carry out these studies, the diabetic model was chosen to be *Drosophila melanogaster.* Multiple studies have demonstrated that the effect of a high-sugar diet in the flies induces classic T2DM features, such as persistent hyperglycaemia, developmental and motor delays, insulin resistance and lipid accumulation [6]. This is accompanied by an increase in trehalose, glucose, glycogen, and Drosophila Insulin-Like Peptide 3/Adipokinetic Hormone (DILP3/AKH) expression [7, 8] detailed the AKH, target of Rapamycin (TOR), and insulin pathways involved in phenotype and treatment evaluation and how they are used to screen and test traditional Chinese medicine (TCM) compounds via dietary manipulation. *Drosophila* has been constantly used at early-stage drug evaluation, and [7] validated this by screening Dipeptidyl Peptidase-4 (DPP-4) to demonstrate evolutionary conservation.

Innovative and integrative approaches to understanding the growing global prevalence of diabetes are necessary to manage the disease. Recent studies underscore the significance of bridging metabolic phenotypes to genetic and environmental influences, specifically with inflammatory and antioxidant pathways [9]. The strain Canton S (CS) is a widely used wild-type strain, with a normal glucose lipid metabolism, often to establish a baseline response. CS flies, on long-term high sugar exposure, show transcriptional disturbances analogous to human T2DM [10]. On the other hand, vg strains are typically studied for development, gene regulation, and signal integration (Notch, Wingless, and Dpp pathways) and can be used to establish a control.

In this context, the present chapter aims to contribute to this expanding body of work by investigating the effects of green-synthesized $CaCO_3$ nanoparticles from *M. cymbalaria* on *Drosophila* to understand the hyperglycaemic and antioxidant performance of this novel formulation.

6.2 MATERIALS AND METHODOLOGY

6.2.1 Plant Collection

Specimens of *Momordica cymbalaria* plant were collected at the early flowering stage from agricultural regions in Andhra Pradesh, India. The seed and the root were made into powders to be used for extraction by air-drying them at 20 to 25°C for 7 days, after which they were powdered [11].

6.2.2 Extract Preparation

To extract and concentrate the bioactive compounds, the extraction method outlined by [12] was followed, with slight changes. A total of 10 grams of *M. cymbalaria* root and fruit (including seeds) powder was subjected to extraction using 100 mL of methanol. The mixture was continuously agitated for a duration of 5 hours using a magnetic stirrer to ensure uniform mixing and to enhance extraction efficiency. Following extraction, the mixture was filtered through a Whatman No. 1 filter paper to obtain a clear filtrate. The resulting purified methanolic extract was subsequently employed for the synthesis of nanoparticles [12, 13].

6.2.3 Green Synthesis of Calcium Nanoparticles

$CaCO_3$ nanoparticles were formulated using the green synthesis method, as [3]. detailed. A 25-mL aliquot of the methanolic extract was combined with 50 mL of $CaCl_2$ (5×10^{-2} M). The pH was adjusted to 8.5 with NaOH as necessary, and the reaction mixture was stirred magnetically for 2 h. Subsequently, Na_2CO_3 was added, prompting an immediate colour change indicative of nanoparticle nucleation. The sealed flasks were incubated for 2–3 days at 27°C to allow precipitation. Upon precipitation and a colour change from green to brown, indicating the formation of nanoparticles, the NPs were recovered by centrifugation at 8000 rpm. Upon subsequent washes with distilled water and ethanol, a white precipitate was extracted, confirming the removal of residual organics. Finally, a fine powder was obtained after leaving the precipitate to air-dry for 3 days at room temperature [3].

6.2.4 Culture of Fly Stock

The two types of *Drosophila melanogaster* strains used for this study were the Canton-S type, typically known as the wild type, and the Vestigial Wing Mutant, obtained from Dr. Gurudatta Baraka's laboratory, National Institute of Mental Health and Neurosciences (Bangalore, India).

6.2.4.1 Fly Maintenance and Diet Preparation

A standard diet of Bangalore FLY, Drosophila media preparation from NCBS, with optimized measurements for the flies, included 37.5 g corn flour, 50 g jaggery, 12 g yeast, 15 g malt extract, and 5 g agar in 350 mL of distilled water. The diet was autoclaved to eliminate any microbial contaminants, upon which 0.5 mL of orthophosphoric acid, 2 mL of propionic acid, and 0.35 g of methyl paraben were added as preservatives. The diet could be preserved and stable for 2 months under refrigeration at 4°C.

6.2.4.2 Model System Development

To induce diabetes-like conditions in *Drosophila*, we used the [6] study. The flies were first subjected to a 24-hour starvation period, during which they were transferred from the standard diet to a starvation medium composed of 1% agar in distilled water to initiate mobilization of internal energy and create a clean metabolic baseline [6].

Following starvation, flies were transferred to the diabetic diet, prepared by increasing the jaggery concentration to threefold (150 g) while maintaining the base composition:

37.5 g corn flour, 12 g yeast, 15 g malt, and 5 g agar in 300–350 mL of distilled water. The mixture was autoclaved and subsequently supplemented with 0.5 mL orthophosphoric acid, 2 mL propionic acid, and 0.35 g methyl paraben. Continuous exposure to this high-sugar diet for 10 days resulted in the induction of diabetic-like phenotypes in flies.

6.2.4.3 Exposure of *Drosophila* to Nanoparticles

To evaluate the therapeutic effect, a treatment diet was prepared with the same composition as the diabetic diet, with the addition of nanoparticles at a concentration of 2 mL per gram of diet. After autoclaving, preservatives were added as described previously. Following 24 hours of starvation, diabetic-induced flies were transferred to the treatment diet and maintained for 1 day to assess the effect of nanoparticle supplementation. Biochemical assays were carried out with the haemolymph and tissue homogenate of the treated *Drosophila*.

6.2.5 Climbing Assay of *Drosophila*

To evaluate the effects of different sucrose diets on the flies, they were exposed to the respective diets for 15 days. Following this, their climbing ability was assessed using a climbing assay. A standard climbing assay protocol (negative geotaxis) per [14] was followed. Approximately ten flies from each diet group were transferred into a sealed 50 ml measuring cylinder. The flies were gently tapped to the bottom of the cylinder to ensure they started from a consistent position. Under ambient light conditions, their climbing activity was recorded for 2 minutes. The number of flies crossing the 17.5 cm mark was documented. This assay was repeated across all groups to quantify locomotor differences. Additionally, observations on reproductive timelines, generation times, and lifespans were recorded to correlate behavioural changes with physiological effects induced by the diets.

6.2.6 Biochemical Assays in Control and Treated *Drosophila*

The selection and adaptation of biochemical protocols were guided by established methods in *Biochemical Assays to Detect the Antioxidant Level in Drosophila melanogaster* by [15]. The chapter outlines the biochemical assays used to evaluate oxidative stress and therapeutic effects in *Drosophila melanogaster*, specifically catalase activity, 2,2-diphenyl-1-picrylhydrazyl (DPPH) radical scavenging, α-glucosidase inhibition, peroxidase activity, and nitric oxide levels [15].

6.2.6.1 DPPH Radical Scavenging Assay

The compound DPPH exists in a stable free radical form at room temperature (25°C) and exhibits a purple colour when dissolved in methanol. The fading of this colour indicates that reactive oxygen species (ROS) have interacted with the nitrogen-based free radicals of DPPH. By measuring the DPPH levels, the antioxidant capacity and radical scavenging activity can be quantified [16].

The antioxidant activity of the samples was evaluated using the DPPH (2,2-diphenyl-1-picrylhydrazyl) assay. A 0.2-mM DPPH solution was prepared by dissolving 7.9 mg of DPPH in 100 mL of methanol and vertexing thoroughly. The prepared solution was stored at 4°C, protected from light using aluminium foil.

Aliquots of 10 µL of the sample homogenate were transferred into a test tube. Each well was supplemented with 125 µL of DPPH solution. A volume of 50 µL of absolute methanol was added to each well to adjust the reaction volume (Table 6.1). The test tubes were incubated in the dark at room temperature for 20 minutes. The absorbance was measured at 517 nm using a UV spectrophotometer. The percentage of DPPH scavenging activity was calculated using the following formula (Equation 6.1).

$$\textbf{Percentage of scavenging} = \left(ABS_{control} - ABS_{sample} / ABS_{control}\right) \times 100 \quad (6.1)$$

Equation 6.1: DPPH scavenging percentage formula.

6.2.6.2 Catalase Assay

For the catalase assay, the preparation involved creating a 50-mM phosphate buffer (PB). First, 10.105 g of Na_2HPO_4 (50 mM) and 6.80 g of KH_2PO_4 (50 mM) were dissolved in 800 mL of MilliQ water, and the pH was adjusted to 7.4 using HCl or NaOH. The solution was diluted to a final volume of 1 L with MilliQ water. A hydrogen peroxide (H_2O_2) solution was prepared by mixing 0.75 mL of 30% H_2O_2 with 100 mL of phosphate buffer. This solution was freshly prepared and stored at 4°C prior to use. For the testing, 2.0 mL of the sample was pipetted into a cuvette with a 1-cm path length and allowed to equilibrate at 25°C for 5 minutes. Then 1.0 mL of the H_2O_2 solution was added to the sample (Table 6.2). Two blanks were prepared, one containing only phosphate buffer and the other containing phosphate buffer with H_2O_2. The absorbance of the sample was measured at 240 nm using a UV-visible spectroscope. The enzyme activity was calculated as follows: the volume activity (U/mL) was determined using the formula ΔA × 34.4 × DF, where ΔA represented the change in absorbance, 34.4 was the constant derived from the extinction coefficient of H_2O_2 at 240 nm (0.0436), and

TABLE 6.1
Procedure of Assaying DPPH

Components	Sample (in µL)	Conc. (in µg/mL)	DW (in µL)	DPPH (in µL)	Methanol (in µL)
Solvent	0	0	0	0	250
Sample	10	2.5	90	125	50
DPPH control	125	0.1 mM		125	125

TABLE 6.2
Procedure of Assaying Catalase

	Phosphate Buffer	Sample		H_2O_2 Solution
Test	–	2ml	Equilibrate the sample in the cuvette for 5 minutes at 25°C	1mL
Blank 1	2mL	–		–
Blank 2	2mL	–		1mL

DF was the dilution factor. The weight activity (U/mg) was calculated by dividing the volume activity (U/mL) by the catalase content (C, in mg/mL) of the sample.

6.2.6.3 Nitric Oxide Assay

Nitric oxide levels, in the form of nitrate and nitrite, were quantified using the Griess reaction method. For this, 1 ml of the sample was incubated with 1 mL of Griess reagent (a 1:1 mixture of 0.1% N-(1-naphthyl) ethylenediamine dihydrochloride and 1% sulphanilamide in 5% phosphoric acid) at room temperature for 20 minutes. The absorbance was measured at 540 nm using a spectrophotometer.

6.2.6.4 Glutathione Peroxidase Assay

To prepare the reagents for the glutathione peroxidase assay, a 50 mM sodium phosphate buffer was made by dissolving 2.76 g of NaH_2PO_4 and 2.04 g of Na_2HPO_4 in 80 mL of MilliQ water, adjusting the pH to 7.0 with HCl or NaOH, and diluting to 100 mL. Additional reagents included 10 mM sodium azide (6.5 mg in 10 mL MilliQ water), 4 mM reduced glutathione (0.012 g in 10 mL MilliQ water), and 0.25 mM H_2O_2 (5 μL stock H_2O_2 in 9995 μL of sodium phosphate buffer). Trichloroacetic acid (TCA) was prepared at 100 mM by dissolving 1.639 g in 100 mL MilliQ water, while 0.3 M sodium hydrogen phosphate required 4.258 g in 100 mL MilliQ water. For the assay, 0.05% 5,5′-Dithiobis(2-Nitrobenzoic acid) (DTNB) was prepared by dissolving 0.05 g in 100 mL of 1% sodium phosphate buffer with 0.1 M Ethylenediaminetetraacetic acid (EDTA), and 0.08 mM EDTA was made by dissolving 23.3 mg in 100 mL MilliQ water. The procedure involved homogenizing tissue in a buffer containing phosphate buffer, sodium azide, EDTA, H_2O_2, reduced glutathione, protein sample, and MilliQ water. The mixture was split into two tubes, incubated at 37°C for 90 and 180 minutes, and treated with TCA to precipitate proteins. After centrifugation at 10,000 rpm for 10 minutes at 4°C, the supernatant was collected and stored. For analysis, 200 μL of supernatant was mixed with 120 μL of sodium hydrogen phosphate and 100 μL of DTNB, with a blank prepared using sodium hydrogen phosphate and DTNB. Absorbance readings were taken at 412 nm using a microplate reader.

6.2.6.5 Glucosidase-Peroxidase Assay

To perform the Glucose Oxidase – Peroxidase (GOD-POD) assay using a standard kit, first, prepare the working reagent per the kit instructions. Label test tubes or microplate wells for the blank, glucose standard, and test samples. Add 1.0 mL of GOD-POD reagent to each test tube (or 200 μL per well in a microplate), followed by 10 μL of the sample, glucose standard, or distilled water (for the blank). Mix well and incubate at 37°C for 10–15 minutes or at room temperature for 20–30 minutes. After incubation, measure the absorbance at 505 nm or 540 nm using a spectrophotometer, ensuring the blank is used for baseline correction. The glucose concentration is calculated using the formula:

$$\text{Glucose} = (\text{Absorbance of Sample}/\text{Absorbance of Standard}) \times \text{Concentration of Standard}$$

where the standard concentration is provided in the kit. The intensity of the developed colour is directly proportional to the glucose concentration in the sample, allowing for accurate quantification.

6.3 RESULTS

6.3.1 Green Synthesis of Nanoparticles

During the green synthesis of calcium nanoparticles, a colour change in the reaction mixture is usually the initial sign of nanoparticle formation (Figure 6.1a and 6.1b). The transition results from the reduction and stabilization of calcium ions through phytochemicals in plant extracts, which initiate nucleation and surface interactions responsible for the formation of nanoparticles [17, 18]. While the precise colour will vary with plant species and synthesis conditions, such a colour change indicates the beginning of nanoscale development.

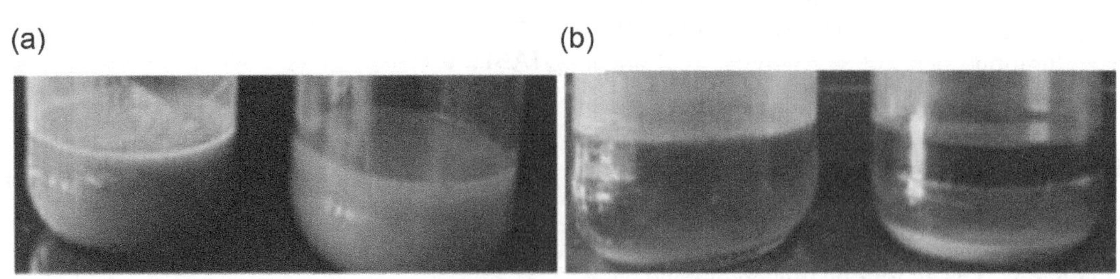

FIGURE 6.1 Synthesised calcium nanoparticles at: a) Day 1 and b) Day 5.

6.3.2 Climbing Assays in *Drosophila*

To evaluate the effect of the diet on the *Drosophila* locomotive ability, and in turn neurodegeneration, the climbing assay was performed. A significant decline in the motor performance of the flies can be seen with respect to the diabetic conditions in both wild and mutant types. The medians of both wild type and mutant flies on the normal diet are around 5.5, which drops drastically to 1 for flies on the diabetic diet (Figure 6.2).

6.3.3 DPPH Radical Scavenging Assay

The DPPH assay was employed on the haemolymph of the flies on varying diets to compare the ability of the haemolymph to neutralize free radicals through the measurement of scavenging activity percentage. Under normal diet conditions, the scavenging activities were the lowest in both genotypes, at around 45%. However, treatment with the $CaCO_3$ nanoparticles (CaNPs) from seed extract significantly increased scavenging activity to approximately 110% in WT and 115% in VS. Similarly, with CaNP root, both strains showed activities at around 120% (Figure 6.3). All treatments demonstrated a statistically significant increase in antioxidant activity compared to the normal control, with Root CaNP effectively reducing oxidative stress by neutralizing free radicals through hydrogen atom or electron donation, as reflected in the high scavenging percentage

6.3.4 Catalase Assay

The ability of catalase to break down hydrogen peroxide was tested by comparing the mutant and wild type strains on various diets to the catalase activity (U/mL) spectroscopically. In the mutant VS group, catalase activity was significantly reduced in diabetic flies (~0.012) compared to the control (~0.017; $p \leq 0.001$), seen as a slight decline in the wild type. However, treatment with green-synthesized calcium carbonate nanoparticles derived from seed extract notably restored catalase activity to the highest level observed (~0.023), significantly higher than both the diabetic and crude extract groups in the mutant, whereas the crude seed diet shows the highest activity for the wild-type strain (Figure 6.4).

6.3.5 Nitric Oxide Assay

To establish the effect of diabetic stress and plant-derived therapy on nitric oxide (NO) concentrations, NO activity was assessed in both the mutant and wild-type *Drosophila*. Diabetic flies from both genotypes showed an unprecedented rise in NO activity to around 3.5 U/mL in VS and 3.0 U/mL in WT compared to their respective control groups that had basal NO levels of ~0.8 U/mL. $CaCO_3$

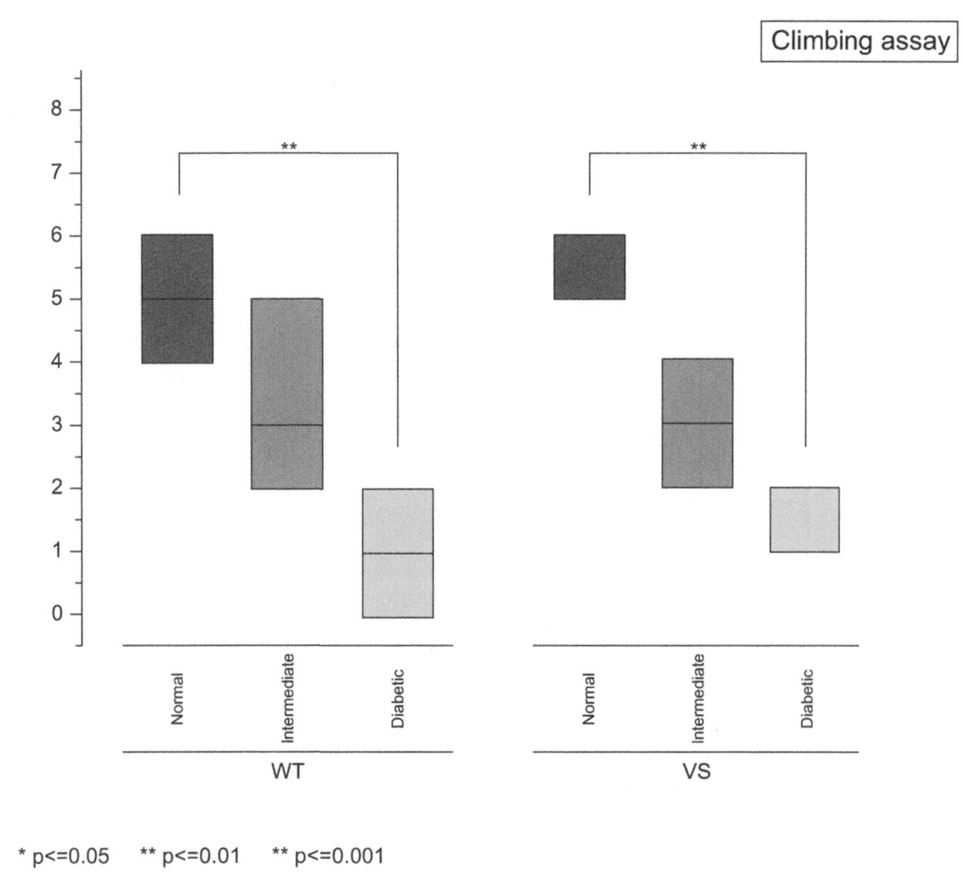

FIGURE 6.2 Effect of the diet on the two strains of *Drosophila* on their locomotive ability.

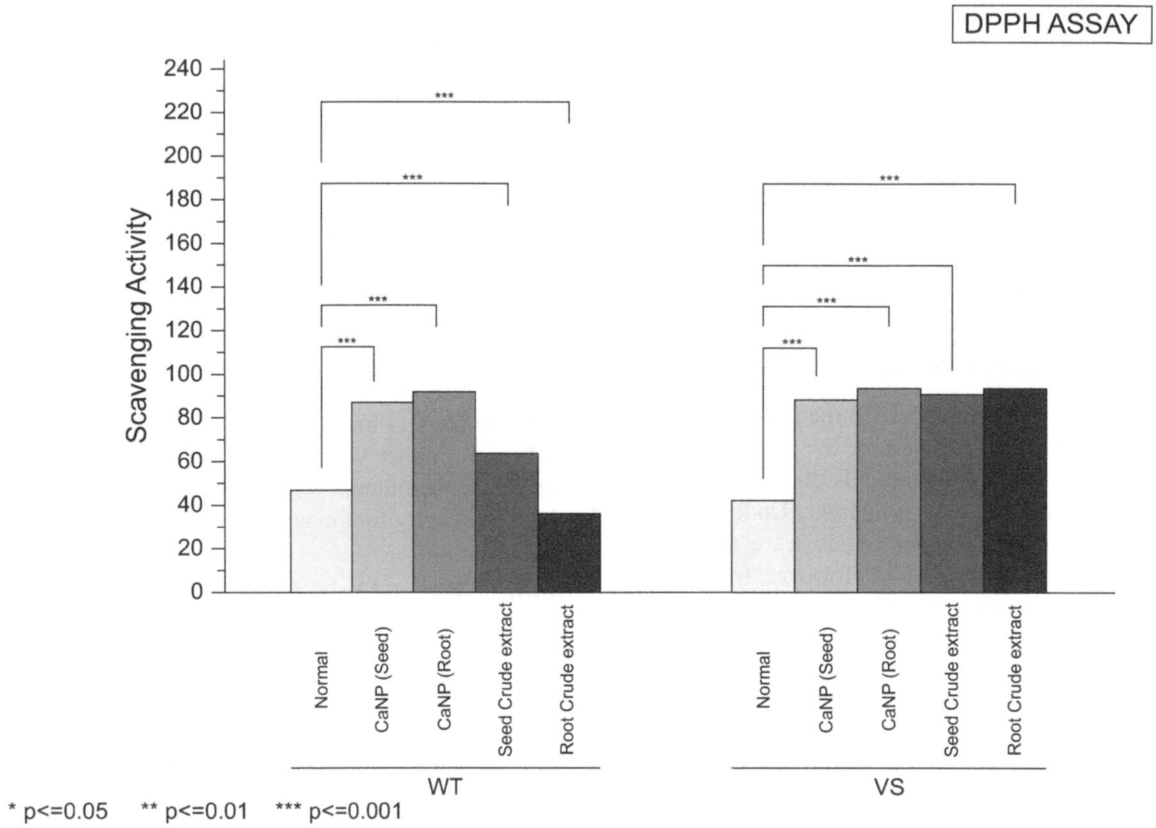

FIGURE 6.3 Comparative ability of the haemolymph to neutralize free radicals in the two strains of *Drosophila*.

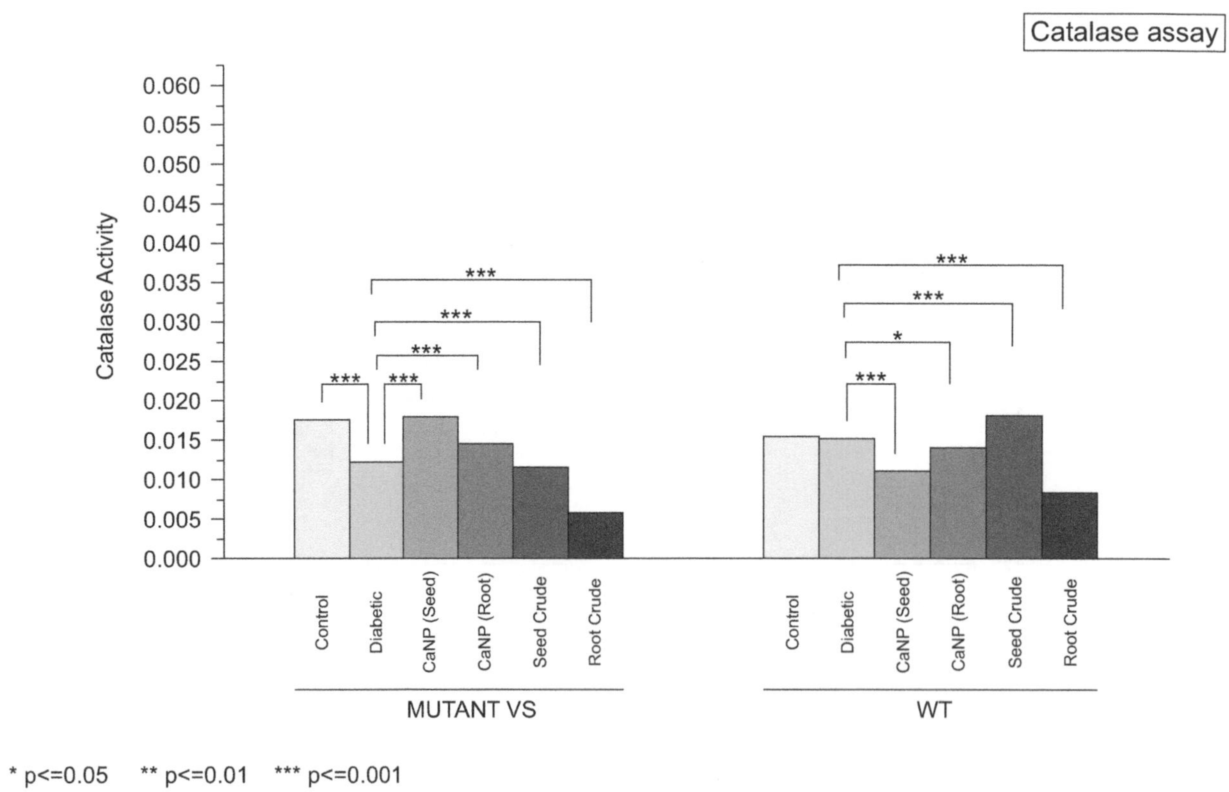

FIGURE 6.4 Ability of catalase to break down hydrogen peroxide in the two strains of *Drosophila*.

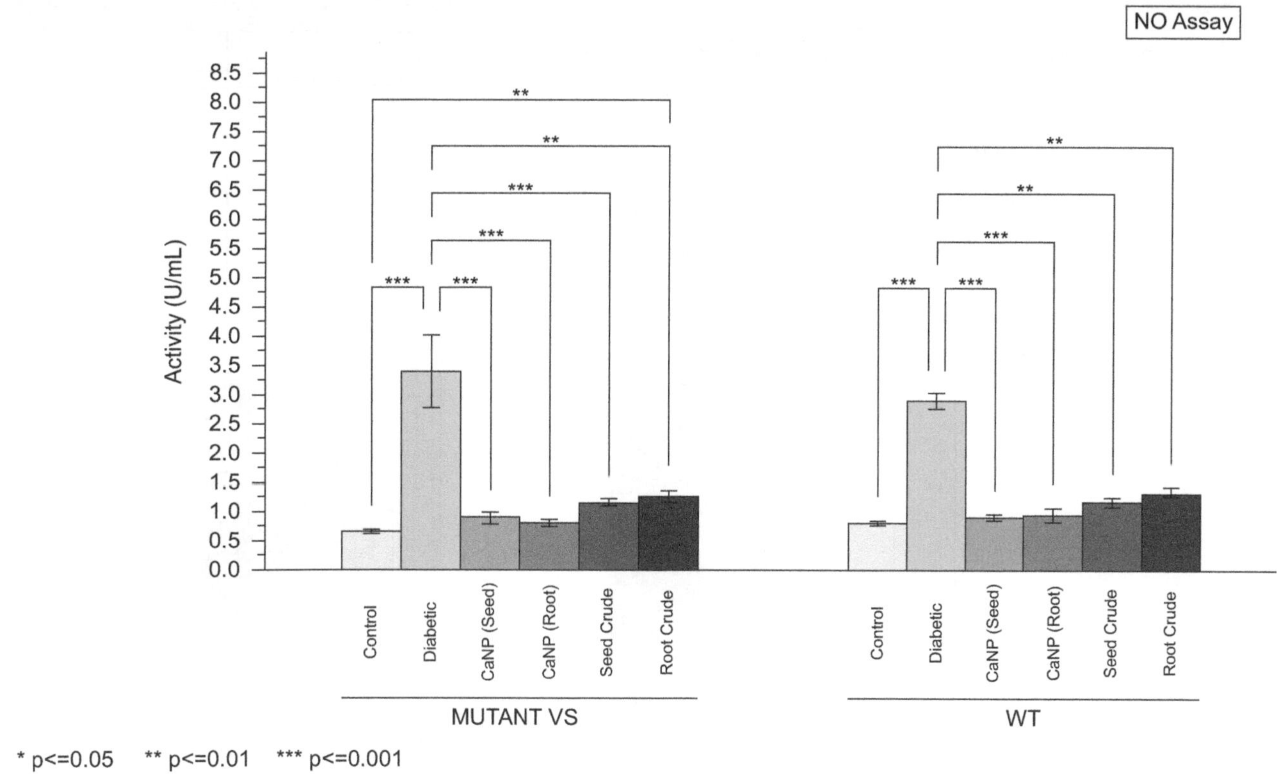

FIGURE 6.5 NO activity as assessed in the two strains of *Drosophila*.

nanoparticles isolated from seed extract decreased NO activity by ~0.9 U/mL in VS and ~1.0 U/mL in WT, reaching the normal value ($p \leq 0.001$) (Figure 6.5). CaNP (root) and crude extracts also decreased NO activity (~1.2–1.4 U/mL) but to a lesser extent than the seed-derived nanoparticles. All the treatments exhibited statistically significant reductions from diabetic conditions, with the most dramatic reduction found in CaNP (seed) groups.

6.3.6 Glutathione Assay

Glutathione (GSH) activity was measured to appraise the antioxidant reaction in diabetic and treated *Drosophila* in both wild-type and mutant strains. Diabetic flies in both genotypes significantly presented a reduction in the levels of GSH (~0.42 U/mL) relative to the control group (~0.70 U/mL in VS and WT), which points towards oxidative stress entailed with the diabetic state ($p \leq 0.05$). Treatment with $CaCO_3$ nanoparticles extracted from seed extract significantly improved GSH activity to ~0.60 U/mL in VS and WT flies ($p \leq 0.01$), close to control levels (Figure 6.6). CaNP (root) and crude extracts (seed and root) also enhanced GSH levels to ~0.55 U/mL, although the effect was slightly less significant. Statistical analysis verified that all treatments increased GSH activity significantly from the diabetic group, and the most reliable antioxidant effect was observed from CaNP (seed) across both genotypes. The findings indicate that CaNPs synthesized by green methods, especially from seed extracts can replenish glutathione levels in diabetic flies, implying lowered oxidative stress.

6.3.7 Glucosidase-Peroxidase Assay

To evaluate the antidiabetic efficacy of plant remedies, glucose content was determined through the GOD-POD assay in the mutant and wild-type *Drosophila* models. Diabetic flies showed a remarkable increase in glucose content to about 220 mg/dL in VS and 190 mg/dL in WT compared with their respective controls, which had normal levels of ~70 mg/dL ($p \leq 0.001$). $CaCO_3$ nanoparticles, specifically derived from seed extracts, significantly lowered glucose levels to ~75 mg/dL for both genotypes, nearly at control levels ($p \leq 0.001$). CaNP (root) was similarly effective. Although the crude seed and root extracts also resulted in significant glucose decreases (~85–90 mg/dL), their efficacy was somewhat less (Figure 6.7).

6.4 DISCUSSION

The present chapter demonstrates the therapeutic potential of green-synthesized calcium carbonate nanoparticles derived from *Momordica cymbalaria* root and seed extracts in alleviating diabetic and oxidative stress-induced impairments in *Drosophila melanogaster*. $CaCO_3$-NPs were selected for their inherent biocompatibility, reduced toxicity, and pH-responsive degradation profile, which make them suitable for biomedical applications, especially in targeted drug delivery [19]. The current research represents a novel approach, as no prior research has explored the synthesis of $CaCO_3$-NPs from *M. cymbalaria*, despite

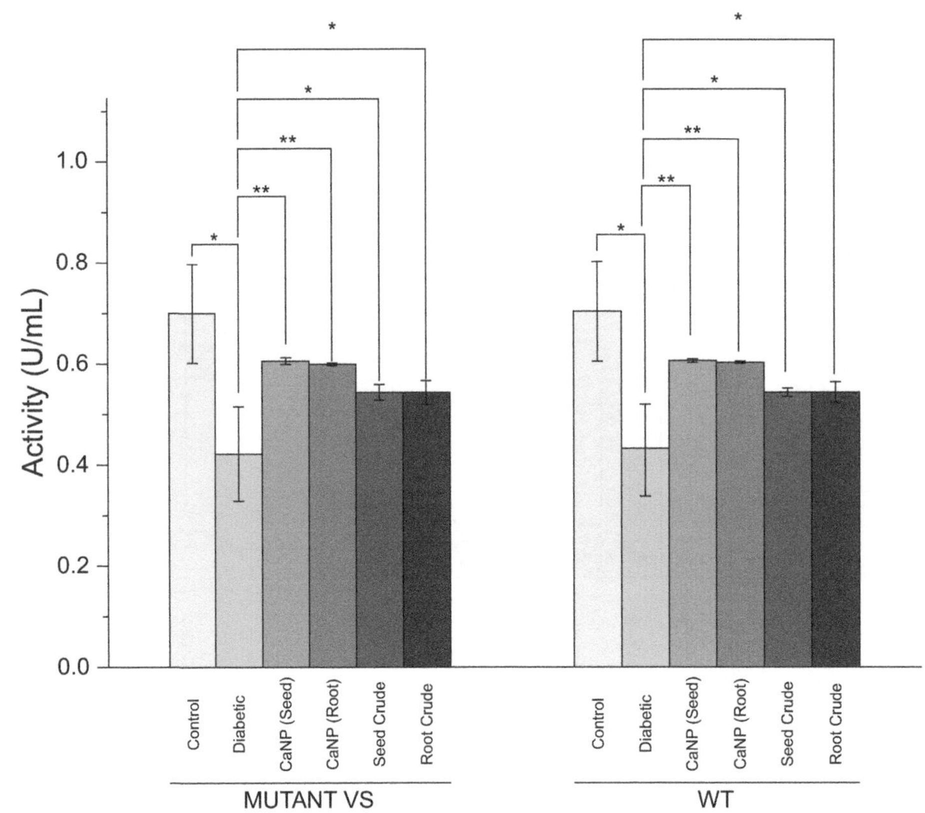

* p<=0.05 ** p<=0.01 *** p<=0.001

FIGURE 6.6 Glutathione activity as measured in the two strains of *Drosophila*.

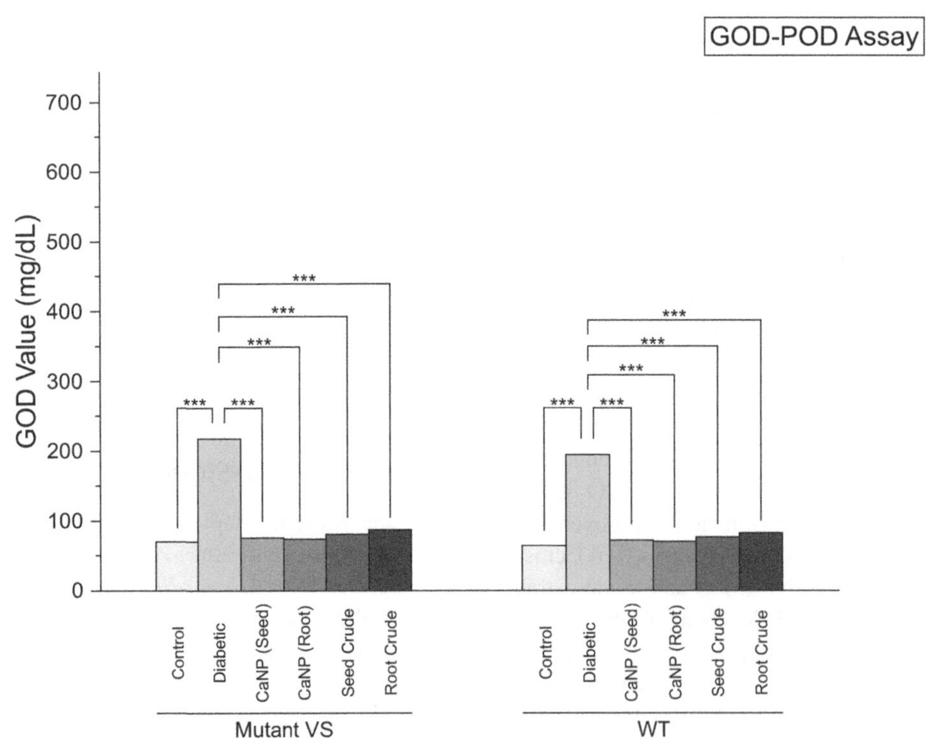

* p<=0.05 ** p<=0.01 *** p<=0.001

FIGURE 6.7 Glucose content as determined through the GOD-POD assay in the two strains of *Drosophila*.

its traditional use and demonstrated antidiabetic efficacy in other forms, such as silver nanoparticles [20].

Diabetic stress significantly reduced motor performance, as reflected in the climbing assay, which is indicative of neurodegeneration—an increasingly recognized complication of chronic hyperglycaemia. Treatment with CaNPs restored locomotor functions in both mutant and wild-type flies, suggesting neuroprotective effects possibly mediated by redox stabilization and reduced oxidative stress. These findings are consistent with curcumin-loaded $CaCO_3$-NPs studied in lead-induced neurotoxicity models, where improvements in catalase and SOD activities correlated with enhanced behaviour and oxidative recovery [21].

The antioxidant potential of CaNPs was clearly reflected in the DPPH, catalase, and glutathione assays. Diabetic flies showed decreased antioxidant activity and reduced levels of catalase and GSH, while CaNP treatments, particularly seed-derived ones, restored these to near-control levels. These results support the hypothesis that nanoformulation enhances the bioavailability of phytochemicals such as saponins and charantin, which are known for their ROS-scavenging and insulin-sensitizing properties [6, 17]. Many plants form calcium-oxalate crystals as part of their normal physiology, and this natural accumulation of calcium could potentially make calcium-rich species useful as a source material for synthesizing calcium-based nanoparticles [22].

The nitric oxide assay results reinforce the anti-inflammatory role of CaNPs. Elevated NO levels in diabetic flies, indicative of vascular inflammation and oxidative damage, were significantly reduced post-treatment, with seed CaNPs performing best. These findings mirror those of [23], who reported effective NO reduction using phytochemical-based nanocarriers in metabolic models.

A significant outcome of this study is the near-complete normalization of glucose levels in diabetic *Drosophila* treated with CaNPs, especially seed-derived nanoparticles. This aligns with earlier work by Kasetti et al. (2010) [24], where methanolic fruit extracts of *M. cymbalaria* reduced fasting glucose in diabetic rats by 66%. The superiority of the nanoparticle form over crude extracts suggests that nanoencapsulation not only enhances bioactive delivery but also supports synergistic pharmacological effects—combining the carrier efficiency of $CaCO_3$ with the intrinsic antidiabetic action of triterpenoid saponins and charantin [13, 18].

Furthermore, previous studies with $CaCO_3$-NPs synthesized using *Ailanthus altissima* demonstrated dual antioxidant and antidiabetic potential through α-glucosidase inhibition and hydrogen peroxide scavenging [3], further supporting our findings. Our study thus extends the scope of calcium carbonate nanocarriers, showing their adaptability to different therapeutic plant systems with favourable outcomes in metabolic disease models.

In summary, this study provides novel and robust evidence for the use of green-synthesized $CaCO_3$ nanoparticles from *M. cymbalaria*, especially seed-derived, in managing hyperglycaemia and oxidative stress. The observed benefits are likely a result of both improved bioactive compound delivery and the synergistic therapeutic effects of the phytochemicals and calcium carbonate carrier. These results support the potential of green nanomedicine in developing safe, targeted, and effective treatments for diabetes and related metabolic disorders.

6.5 CONCLUSION

This study demonstrates the significant antidiabetic and antioxidant efficacy of green-synthesized calcium carbonate nanoparticles derived from *Momordica cymbalaria* seed and root extracts using *Drosophila melanogaster* as a model organism. Diabetic flies showed marked impairments in motor ability, elevated oxidative stress markers (such as nitric oxide), reduced antioxidant enzyme activities (catalase and glutathione), and severe hyperglycaemia. Treatment with CaNPs, especially those synthesized from seed extracts, led to substantial recovery in locomotor performance, normalization of glucose levels, and restoration of antioxidant defence systems. The enhanced efficacy of CaNPs over crude extracts emphasizes the advantages of nanoencapsulation in improving the delivery and bioactivity of plant-derived therapeutics like saponins and charantin. CaNPs demonstrated excellent biocompatibility and redox-regulating capacity, suggesting a synergistic mechanism involving both the therapeutic properties of the plant and the carrier potential of $CaCO_3$.

These findings provide a strong foundation for the use of *M. cymbalaria*-based green nanomedicine in combating diabetes and oxidative stress-related complications. Further, mammalian models need to be explored to validate these results and the underlying molecular pathways. The assessment of long-term biosafety and clinical applicability of these nanoparticles will pave the way for advanced research in translational medicine and clinical pharmacology. In conclusion, CaNPs present a sustainable, plant-based, and multifunctional therapeutic solution for the effective management of metabolic disorders.

REFERENCES

[1] R. B. Koneri, S. Samaddar, S. M. Simi, and S. T. Rao, "Neuroprotective effect of a triterpenoid saponin isolated from Momordica cymbalaria Fenzl in diabetic peripheral neuropathy," *Indian J Pharmacol*, vol. 46, no. 1, pp. 76–81, 2014, doi: 10.4103/0253-7613.125179.

[2] R. Shanmugapriya and S. Poornima, "Detection of Charantin in the leaves and fruits of Momoridca tuberosa (Cogn) Roxb and Momordica dioica (Roxb Ex Wild) by Analytical HPTLC," vol. 4, no. 6, 2014.

[3] N. Samad *et al*., "A novel approach to assessing the antioxidant and anti-diabetic potential of synthesized calcium carbonate nanoparticles using various extracts of Ailanthus altissima," *Front Chem*, vol. 12, p. 1345950, 2024, doi: 10.3389/fchem.2024.1345950.

[4] R. B. S. M. N. Mydin, I. N. M. Zahidi, N. N. Ishak, N. S. S. N. Ghazali, S. Moshawih, and S. Siddiquee, "Potential of

Calcium Carbonate Nanoparticles for Therapeutic Applications," *Malaysian Journal of Medicine and Health Sciences*, vol. 14, pp. 201–206, Dec. 2018.

[5] C. Tan *et al.*, "Advanced CaCO3-derived delivery systems for bioactive compounds," *Advances in Colloid and Interface Science*, vol. 309, p. 102791, Nov. 2022, doi: 10.1016/j.cis.2022.102791.

[6] D. Chen, J. Yang, Z. Xiao, S. Zhou, and L. Wang, "A diet-induced type 2 diabetes model in Drosophila," *Sci China Life Sci*, vol. 64, no. 2, pp. 326–329, Feb. 2021, doi: 10.1007/s11427-020-1774-y.

[7] F. A. Lagunas-Rangel, S. Liao, M. J. Williams, V. Trukhan, R. Fredriksson, and H. B. Schiöth, "Drosophila as a Rapid Screening Model to Evaluate the Hypoglycemic Effects of Dipeptidyl Peptidase 4 (DPP4) Inhibitors: High Evolutionary Conservation of DPP4," *Biomedicines*, vol. 11, no. 11, p. 3032, Nov. 2023, doi: 10.3390/biomedicines11113032.

[8] W. S. Meshrif, I. M. El Husseiny, and H. Elbrense, "Drosophila melanogaster as a low-cost and valuable model for studying type 2 diabetes," *J Exp Zool A Ecol Integr Physiol*, vol. 337, no. 5, pp. 457–466, June 2022, doi: 10.1002/jez.2580.

[9] M. Cnop *et al.*, "RNA sequencing identifies dysregulation of the human pancreatic islet transcriptome by the saturated fatty acid palmitate," *Diabetes*, vol. 63, no. 6, pp. 1978–1993, June 2014, doi: 10.2337/db13-1383.

[10] O. T. Kayode, D. Rotimi, E. Okoh, M. Iyobhebhe, A. A. A. Kayode, and O. A. Ojo, "Novel ketogenic diet formulation improves sucrose-induced insulin resistance in canton strain Drosophila melanogaster," *J Food Biochem*, vol. 45, no. 9, p. e13907, Sept. 2021, doi: 10.1111/jfbc.13907.

[11] A. Krakowska-Sieprawska, A. Kiełbasa, K. Rafińska, M. Ligor, and B. Buszewski, "Modern Methods of Pre-Treatment of Plant Material for the Extraction of Bioactive Compounds," *Molecules*, vol. 27, no. 3, p. 730, Jan. 2022, doi: 10.3390/molecules27030730.

[12] M. K. Swamy, M. S. Akhtar, S. K. Mohanty, and U. R. Sinniah, "Synthesis and characterization of silver nanoparticles using fruit extract of Momordica cymbalaria and assessment of their in vitro antimicrobial, antioxidant and cytotoxicity activities," Dec. 2015, doi: 10.1016/j.saa.2015.07.009.

[13] I. Nasim, Z. Jabin, S. R. Kumar, and V. Vishnupriya, "Green synthesis of calcium hydroxide-coated silver nanoparticles using Andrographis paniculata and Ocimum sanctum Linn. leaf extracts: An antimicrobial and cytotoxic activity," *J Conserv Dent*, vol. 25, no. 4, pp. 369–374, 2022, doi: 10.4103/jcd.jcd_411_21.

[14] J. W. Gargano, I. Martin, P. Bhandari, and M. S. Grotewiel, "Rapid iterative negative geotaxis (RING): a new method for assessing age-related locomotor decline in *Drosophila*," *Experimental Gerontology*, vol. 40, no. 5, pp. 386–395, May 2005, doi: 10.1016/j.exger.2005.02.005.

[15] J. Bag and M. Mishra, "Biochemical Assays to Detect the Antioxidant Level in Drosophila melanogaster," in *Fundamental Approaches to Screen Abnormalities in Drosophila*, M. Mishra, Ed., New York, NY: Springer US, 2020, pp. 151–168. doi: 10.1007/978-1-4939-9756-5_13.

[16] "Gulcin, . and Alwasel, S.H. (2023) DPPH Radical Scavenging Assay. Processes, 11, Article No. 2248. - References - Scientific Research Publishing." Accessed: July 15, 2025. [Online]. Available: https://www.scirp.org/reference/referencespapers?referenceid=3864954

[17] D. A. Omoboyowa, M. D. Agoi, S. A. Shodehinde, O. A. Saibu, and J. A. Saliu, "Antidiabetes study of Spondias mombin (Linn) stem bark fractions in high-sucrose diet-induced diabetes in Drosophilamelanogaster," *J Taibah Univ Med Sci*, vol. 18, no. 4, pp. 663–675, Jan. 2023, doi: 10.1016/j.jtumed.2023.01.011.

[18] B. Ramola, N. C. Joshi, M. Ramola, J. Chhabra, and A. Singh, "Green Synthesis, Characterisations and Antimicrobial Activities of CaO Nanoparticles," *Oriental Journal of Chemistry*, vol. 35, no. 3, pp. 1154–1157, June 2019.

[19] P. Zhao, Y. Tian, J. You, X. Hu, and Y. Liu, "Recent Advances of Calcium Carbonate Nanoparticles for Biomedical Applications," *Bioengineering (Basel)*, vol. 9, no. 11, p. 691, Nov. 2022, doi: 10.3390/bioengineering9110691.

[20] M. Sundar, R. Gopalan, A. Nivetha, P. K. Seetharaman, and M. Selvaraj, "Phyto-Mediated Green Synthesis of Silver Nanoparticles Using an Aqueous Leaf Extract of Momordica cymbalaria: Antioxidant, Cytotoxic, Antibacterial, and Photocatalytic Properties," Feb. 2024.

[21] M. M. Mailafiya *et al.*, "Curcumin-loaded cockle shell-derived calcium carbonate nanoparticles: A novel strategy for the treatment of lead-induced hepato-renal toxicity in rats," *Saudi J Biol Sci*, vol. 27, no. 6, pp. 1538–1552, June 2020, doi: 10.1016/j.sjbs.2020.03.009.

[22] V. R. Franceschi and P. A. Nakata, "Calcium oxalate in plants: formation and function," *Annu Rev Plant Biol*, vol. 56, pp. 41–71, 2005, doi: 10.1146/annurev.arplant.56.032604.144106.

[23] K. A. Kelly, C. L. Heaps, G. Wu, V. Labhasetwar, and C. J. Meininger, "Nanoparticle-mediated delivery of tetrahydrobiopterin restores endothelial function in diabetic rats," *Nitric Oxide*, vol. 148, pp. 13–22, July 2024, doi: 10.1016/j.niox.2024.04.009.

[24] R. B. Kasetti *et al.*, "Antihyperglycemic and antihyperlipidemic activities of methanol:water (4:1) fraction isolated from aqueous extract of Syzygium alternifolium seeds in streptozotocin induced diabetic rats," *Food Chem Toxicol*, vol. 48, no. 4, pp. 1078–1084, Apr. 2010, doi: 10.1016/j.fct.2010.01.029.

7 Cow's Milk Silver Nanoparticles Synthesized at Various pH Combat Multidrug-Resistant *Pseudomonas aeruginosa*

Akshata G Athreya, Ismail Shareef M, and S M Gopinath

7.1 INTRODUCTION

Multi-drug resistance (MDR) is a growing global problem that threatens the effective treatment of bacterial infections. *P. aeruginosa* is an opportunistic pathogen that can cause a wide range of infections, including pneumonia, urinary tract infections, and sepsis. This bacterium has become a major concern due to its ability to develop resistance to multiple antibiotics. MDR *P. aeruginosa* infections are associated with high mortality rates, prolonged hospital stays, and increased healthcare costs. Several studies have shown that the prevalence of MDR *P. aeruginosa* infections is increasing worldwide, with some regions reporting resistance rates as high as 50–60% (Mongkolrattanothai et al., 2003; Gillet et al., 2002). MDR *P. aeruginosa* is a significant concern in healthcare settings, particularly in intensive care units (ICUs) and long-term care facilities. According to a study published in the *Journal of Antimicrobial Chemotherapy* in 2018, the prevalence of MDR *P. aeruginosa* infections in ICUs ranges from 14% to 47% (Miller et al., 2005). Several risk factors increase the likelihood of developing an MDR *P. aeruginosa* infection. These include prior antibiotic use, hospitalization, surgery, immunosuppression, and mechanical ventilation (Cosgrove et al., 2003; Kourtis et al., 2019). The development of resistance in *P. aeruginosa* is driven by a combination of intrinsic and acquired mechanisms, including the production of beta-lactamases, efflux pumps, and mutations in target sites (Glen & Lamont, 2021). Therefore, treatment of MDR *P. aeruginosa* infections is challenging, and there is a need for new antibiotics and alternative therapies/strategies.

The discipline of nanotechnology has a new line of defense in a time when antibiotics appear to be compromised and starting to give way to the mechanisms created by microorganisms. By altering the size of known antibacterial chemicals, this method aids in the reexamination of their properties, which in turn alters their effects (Thill et al., 2006). Due to AgNPs' robust antibacterial properties, especially when compared to other nanoparticles, there is a lot of opportunity for future research in this field. According to reports, AgNPs have been employed as antibacterial agents and have been derived from a variety of sources, including plants (Barros et al., 2018), microbes (Barros et al., 2018; Willner et al., 2006; Ashraf et al., 2014), enzymes (Willner et al., 2006; Ashraf et al., 2014), fruit juices (Hungund et al., 2015), and animal products (Lee et al., 2013; Hegazi et al., 2014; Rajapandiyan et al., 2011; Athreya et al., 2018; Karthika & Sevarkordiyone, 2015). Cow's milk was one of the possible sources that was underutilized, despite the fact that the AgNPs produced from it had effective antifungal and antibacterial characteristics when produced at neutral pH (Lee et al., 2013; Hegazi et al., 2014; Athreya et al., 2018). According to a number of studies, variables like temperature, pH, ion concentration, and reaction duration affect the synthesis of AgNPs. These factors have a significant impact on the antibacterial activity of AgNPs, as demonstrated by Marambio-Jones and Hoek in their study (Marambio-Jones & Hoek, 2010). The creation and characterization of nanoparticles are impacted by a number of factors, the most important being pH. They alter the charge of the metabolites, which affects redox reactions and the interaction between phytochemical capping agents and the metal ions. In addition, the acidity or basicity of the reaction medium influences nanoparticle formation and size, thereby impacting the overall stability of the nanoparticles (Halima & Archna, 2016). Determining the antibacterial potential of pasteurized cow's milk is the purpose of the current investigation. In contrast to other studies where the effect of pH on the antibacterial activity of AgNPs synthesized from plants and microbes as sources against non-MDR bacteria (Muthu & Priya, 2017; Khalil et al., 2013; Vanaja et al., 2013; Anigol et al., 2017; Kredy, 2018), this study examined the effects of pasteurized cow milk AgNPs synthesized across 3, 5, 7, 9, and 11 pH against MDR *P. aeruginosa* for the first time.

7.2 METHODS

7.2.1 Isolation of MDR from Clinical Sample

Infected patients' pus samples were taken from Kaade Multi Specialty Hospital, Bangalore. They were streaked at

DOI: 10.1201/9781003644439-7

a right angle on blood agar and incubated at 37°C for 24 to 48 hours. Colonies with the greatest colony-forming units (10^6 CFU/ml) were picked. In accordance with Bergey's manual, they were identified by looking at the colony morphology, Gram's staining, and biochemical tests (catalase, oxidase, indole, citrate, methyl red, and Voges Proskauer) (Holt et al., 1994). They were also subcultured on nutrient agar plates for further research. Required chemicals were purchased from Sigma Aldrich.

The 16S rRNA technique was used for molecular identification. The conventional technique was used to isolate genomic DNA (Athreya et al., 2020). The forward primer 5'-AGAGTTTGATCCTGGCTCAG-3 and the reverse primer 5'-ACGGCTACCTTGTTACGACTT-3 were used to amplify the 16S rRNA gene. A DNA ladder was used in 35 reaction cycles of PCR in a 50 µl total volume (0.1, 0.2, 0.3, 0.6, 1, 1.5, 2.0, 2.5, 3.0, 3.5 Kb). Sanger sequencing was used to sequence the DNA. The obtained sequence was tested using the NCBI GenBank database's basic local alignment for nucleotide sequence (BlastN) algorithm (www.ncbi.nlm.nih.gov). The identity of the bacteria was ascertained using sequences with the highest query coverage, identity, and score. ClustalW analysis was used to match the sequences of the isolated bacterial sequences with reference sequences from GenBank. The MEGA6 program was used to create the phylogenetic tree using the unweighted pair group method with arithmetic mean (UPGMA) approach. The Sequin tool was used to generate the accession number before the sequence was submitted to GenBank.

7.2.2 Synthesis and Characterization of AgNPs

In our earlier work, we reported the silver nitrate ($AgNO_3$)–based synthesis of AgNPs using pasteurized cow's milk at pH 7, along with a detailed description of the characterization using UV/VIS spectroscopy, scanning electron microscopy (SEM), energy dispersive spectroscopy (EDS), Fourier transform infrared spectroscopy (FTIR), and transmission electron microscopy (TEM) (Athreya et al., 2018). In the current investigation, AgNPs were made adopting the same procedure as in our earlier research but at several pH levels, including 3, 5, 9, and 11, by adding 1N NaOH and HCl, respectively. Research was carried out to determine the most probable pH that would promote antibacterial activity against the isolated MDR bacteria. To determine the spectrum shifts, UV/VIS spectroscopy was used. Transmission electron microscopy was used to determine how the size of AgNPs varied across the pH range.

7.2.3 Antibiotic Sensitivity

The Kirby-Bauer disc diffusion method, a test for antibiotic sensitivity, was carried out using 26 common antibiotics purchased from Hi Media (Table 7.1) (Bauer et al., 1966). Two hundred µl of an isolated bacterial culture were used to inoculate media plates with 20 ml of Mueller–Hinton agar (MHA). The media was covered with standard antibiotic discs, which were then incubated for 24 hours at 37°C (Anigol et al., 2017). The CLSI reference standard (Wayne, 2017) was used to interpret the inhibition zones for susceptibility and resistance patterns.

7.2.4 Antibacterial Activity

The antibacterial activity methodology was similar to an antibiotic sensitivity process (Section 7.2.3). Interpretation of the inhibition zones for susceptibility and resistance patterns was done according to the CLSI reference standard, as shown in Table 7.1. The experiment was done in triplicate. The data were pooled and statistically analyzed. Treatments for antibacterial activity are indicated in Table 7.2.

TABLE 7.1
Antibiotics Used in the Study and Their Standard Inhibition Zones (Habib et al., 2015)

				Standard Measures of Zones		
Antibiotic Class	Antibiotic	a-z Representation on Plates	Dosage (mcg)	Sensitive (mm)	Intermediate (mm)	Resistant (mm)
Chloramphenicol	Chloramphenicol C	a	30	18	13–17	12
Sulfonamides	Co-trimoxazole COT	b	25	16	11–15	10
Aminoglycosides	Amikacin-AK	c	30	17	15–16	14
	Gentamycin-GE	d	10	15	13–14	12
	Tobramycin-TO	e	10	15	13–14	12
Monobactams	Aztreonam-AT	f	30	21	18–20	17
Penicillin	Piperacillin-PI	g	100	21	15–20	14
Amino penicillins	Ampicillin-AMP	h	10	17	14–16	13
Antipseudomonal penicillins	Piperacillin/ tazobactam-PIT	i	100/10	21	18–20	17

TABLE 7.1 (*Continued*)
Antibiotics Used in the Study and Their Standard Inhibition Zones (Habib et al., 2015)

Antibiotic Class	Antibiotic	a-z Representation on Plates	Dosage (mcg)	Standard Measures of Zones: Sensitive (mm)	Intermediate (mm)	Resistant (mm)
Penicillin combination (polypeptides)	Colistin(methyl sulphonate)-CL	j	10	-	-	-
Combinatorial antibiotic	Ceftazidime/clavulanic acid CAC	k	30/10	-	-	-
Cephalosporin 1st gen	Cefazolin-CZ	l	30	23	20–22	19
Cephalosporin 2nd gen	Cefuroxime-CX	m	30	18	15–17	14
Cephalosporin 3rd gen	Ceftriaxone-CTR	n	30	23	20–22	19
	Cephoperazone Sulbactum-CFS	o	75	21	16–20	15
	Ceftazidime-CA	p	30	21	18–20	17
Cephalosporin 4th gen	Cefepime-CPM	q	30	25	19–24	18
	Cefixime-CFM	r	5	19	16–18	15
Carbapenem	Meropenem-MR	s	10	23	20–22	19
Fluoroquinolones 1st gen	Nalidixic acid-N	t	30	19	14–18	13
Fluoroquinolones 2nd gen	Ofloxacin-OF	u	5	16	13–15	12
	Norfloxacin-NX	v	10	17	13–16	12
	Ciprofloxacine-C	w	5	21	16–20	15
Tetracycline	Tetracycline-TE	x	30	15	12–14	11
	Amoxyclav-AM	y	30	18	14–17	13
Nitrofurans	Nitrofurantoin-NI	z	300	17	15–16	14

- Indicates standard zones not mentioned.

TABLE 7.2
Treatments Used for Antibacterial Studies of AgNPs Synthesized across 3.0, 5.0, 7.0, 9.0, and 11.0 pH against the Isolated MDR

Treatment No.	Treatment	Description	Labels Indicated in Fig
1	AgNPs	AgNPs synthesized at pH 3, 5, 7, 9, and 11–50 and 75 µl concentration	AgNPs synthesized across 3, 5, 7, 9, and 11 pH indicated by i, ii, iii, iv, v, respectively, across Set A to E 2, a-50 µl AgNPs 4,c-75 µl AgNPs
2	Positive control-antibiotics	Five antibiotics were chosen specifically for *P. aeruginosa*. Two criteria were considered for choosing antibiotics: (1) Antibiotics had to be in regular clinical use (2) All isolates of the group had to exhibit resistance (gentamycin, nalidixic acid, chloramphenicol, cefepime, piperacillin/tazobactam)	Ab1, Ab2, Ab3, Ab4, and Ab5
3	Synergistic activity: antibiotic + AgNPs	Respective antibiotics for *P. aeruginosa* + AgNPs pH 3, 5, 7, 9, and 11(50 and 75 µl)	3,b-50 µl AgNPs + respective antibiotics 5,d-75 µl AgNPs + respective antibiotics
4	Negative control	Best source + D.H2O	1

For treatment 3 (synergistic activity calculation), the *C = (B – A)/A formulae is applied, where A and B are the inhibitory zones (mm) determined for the antibiotic alone and AgNPs + antibiotics, respectively (Jyoti et al., 2016). The diameter of the disc (6 mm) was used for calculating purposes in the absence of a zone of inhibition (Habib et al., 2015).

7.2.5 In Vitro Cytotoxicity Study of AgNPs on Human Embryonic Kidney Cell Line

The in vitro cytotoxicity of silver nanoparticles, synthesized at different pH levels, was tested at various concentrations (500, 200, 100, 75, 50, and 25 μL). Silver nitrate ($AgNO_3$) was used as a positive control at concentrations of 10, 5, 2.5, and 1 mM. The tests were carried out on human embryonic kidney (HEK) cells, which were obtained from the National Centre for Cell Sciences (NCCS), Pune, India, to determine the toxic concentration of each substance. Each test sample was first suspended in Ham's F12 medium containing 2% heat-inactivated fetal bovine serum (FBS) to make a 10% v/v stock solution. The solutions were sterilized using a 0.22-micron syringe filter, and serial twofold dilutions were prepared for cytotoxicity testing. HEK cells were cultured in Ham's F12 medium with 10% heat-inactivated FBS, along with penicillin (100 IU/mL), streptomycin (100 μg/mL), and amphotericin B (5 μg/mL). They were incubated in a humidified atmosphere with 5% CO_2 at 37°C until they reached confluence. To passage the cells, they were detached using TPVG solution, which contains 0.2% trypsin, 0.02% EDTA, and 0.05% glucose in PBS. Stock cultures were maintained in 25 cm^2 flasks, and all cytotoxicity experiments were performed using 96-well microtiter plates (Tarsons India Pvt. Ltd., Kolkata, India).

All the reference chemicals and materials such as MTT (3-(4,5-dimethylthiazol-2-yl)-2,5-diphenyltetrazolium bromide), phosphate buffer saline (PBS), Ham's F12, trypsin, antibiotics and dimethyl sulfoxide (DMSO) were procured from HI Media. Fetal bovine serum was procured from Gibco. The biosafety cabinet for the study was procured from Ascension, Instrument, id: RRS/INS/CB/01.

The monolayer cell culture was trypsinized and adjusted to 100,000 cells/ml using Ham's F12 containing 10% FBS. After 24 h, the supernatant was flicked off, the monolayer was washed once with medium, and 100 μl of different test concentrations of test drugs were added. The plates were incubated at 37°C for 72 h in a 5% CO_2 atmosphere (CO_2 incubator, make: NUAIRE, instrument id: RRS/INS/CB/02), and microscopic examination was carried out (make: Motic China, instrument id: RRS/INS/CB/03). After 72 h, the drug solutions were discarded and 50 μl of MTT in PBS was added to each well. The absorbance was measured using a microplate reader, and the percentage growth inhibition was calculated using the standard formula and concentration of test drug needed to inhibit cell growth by 50% (CTC50).

7.3 RESULTS AND DISCUSSION

7.3.1 Isolation of MDR from Clinical Samples

In blood agar, colonies that had more than 10^6 CFU/ml were regarded as the main infection-causing organism. One such significant colony was chosen. It tested Gram negative upon Gram's staining. Colonies were irregular, greyish, smooth (freshly isolated samples), mucoid (if there was formation of slime), convex-flat, translucent-opaque, and β-hemolysis (in a few) according to colony morphological investigations. Biochemical testing indicated positive to catalase, oxidase, and citrate and negative to indole, methyl red, and Voges Proskauer. *Pseudomonas* sp. was determined to be the organism based on these findings. Further 16S rRNA gene amplification results revealed a single 1.5-Kb band (100 ng intensity). The organism's 16S rRNA sequencing revealed 100% to *P. aeruginosa* (Figure 7.1). The obtained sequence was submitted to GenBank, and the accession number generated was MH078568.

7.3.2 Antibiotic Sensitivity

Among 26 antibiotics tested for sensitivity, no antibiotics exhibited sensitivity, and they failed to show any zone of inhibition and thus were categorized under the resistant category according to the CLSI guidelines (Table 7.1, Figure 7.2).

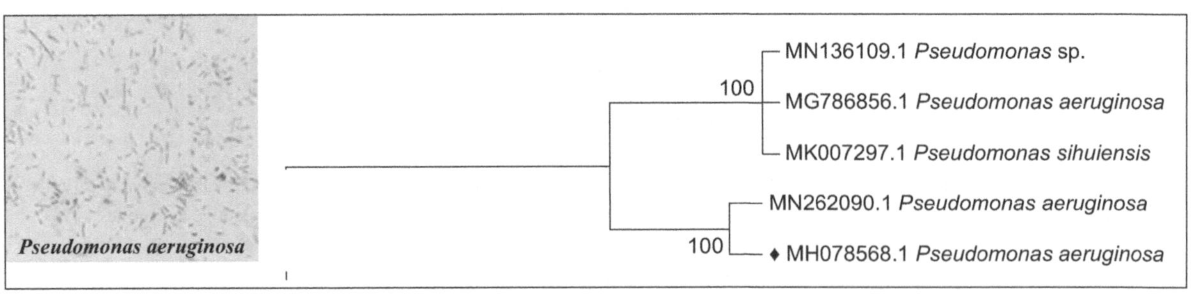

FIGURE 7.1 a. Gram staining of the bacterial isolates at 100× magnification. b. Phylogenetic tree depicting homology of isolated species in the study. Species in current study are marked with black diamonds.

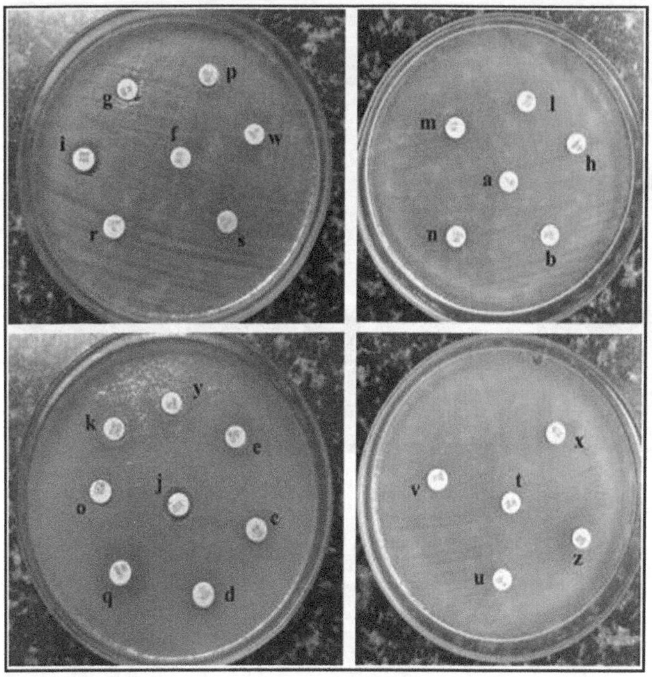

FIGURE 7.2 Antibiotic sensitivity pattern for *P. aeruginosa* (refer to Table 7.1 for a–z representation).

FIGURE 7.3 Flasks a, b, c, d, e represent AgNP synthesis at pH 3, 5, 7, 9, and 11, respectively (experimental flasks—formation of brown color observed); controls [Flask i: silver nitrate + D. water (Transparent solution); Flask ii: milk + D. water (white color)]

7.3.3 Synthesis and Characterization of AgNPs

In our earlier study (Athreya et al., 2018), we presented the results of the synthesis and characterization of AgNPs from pasteurized cow's milk at pH 7. In the current investigation, a shift in color from white to brown in experimental flasks a, b, c, d, and e (Figure 7.3) was observed, denoting the synthesis of AgNPs at pH values of 3, 5, 7, 9, and 11, respectively. The intensity of the brown color was seen to rise as the pH became more basic. In the instance of AgNPs generated from *Lawsonia inermis* extract across pH 4, 7, and 9, Kredy made a comparable observation (Kredy, 2018). According to Muthu and Priya's report on the synthesis of AgNPs from *Cassia auriculata* flowers at pH levels of 3, 4, 5, 6, 7, 8, and 9, AgNP synthesis occurred more quickly at neutral pH than acidic pH, which is consistent with the current findings (Muthu & Priya, 2017).

As pH increased from 3 to 9, UV-vis spectroscopy showed that absorbance increased, with sharp and narrow peaks appearing at pH 7 and 9. This might be because milk's functional groups are more bioavailable at this pH and were involved in the synthesis of AgNPs. Low pH led to the creation of a small, spreading SPR band, which suggested the formation of larger-sized nanoparticles. The production of unstable and agglomerated particles at pH 11 may be the cause of the decrease in absorbance [Figure 7.4]. The decrease in particle size seen as the pH becomes more basic could be the cause of the rise in absorbance. Whereas larger particles display higher scattering and have broader peaks towards longer wavelengths, smaller particles primarily absorb light and have sharp peaks close to shorter wavelengths. Similar findings were made in a study where AgNPs were generated from *Coleus aromaticus* leaf extract and olive leaf extract, respectively, across a range of pH (Khalil et al., 2013; Vanaja et al., 2013).

According to transmission electron microscopy, a pH adjustment had an impact on the size of the produced nanoparticles since it might alter the charge on organic molecules and biomolecules. According to Figure 7.5,

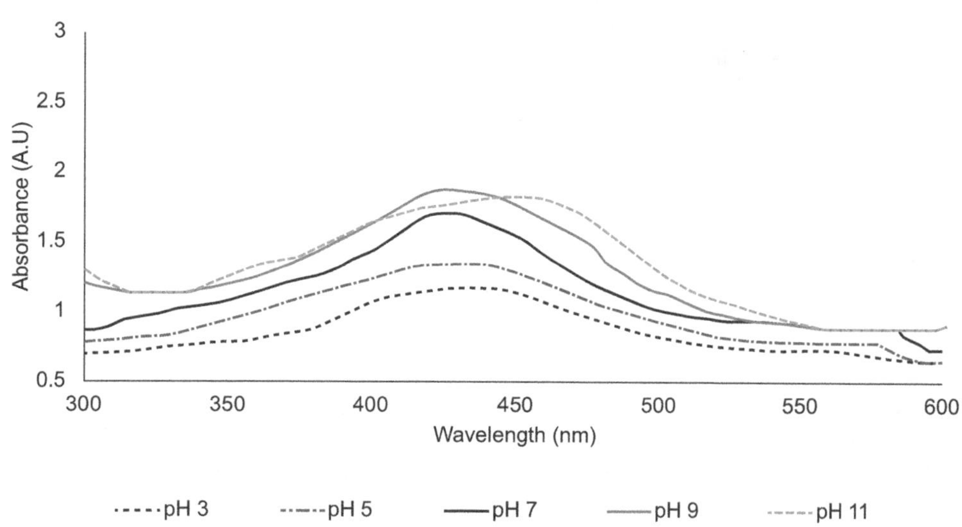

FIGURE 7.4 UV-vis spectroscopy of AgNPs synthesized at various pH values.

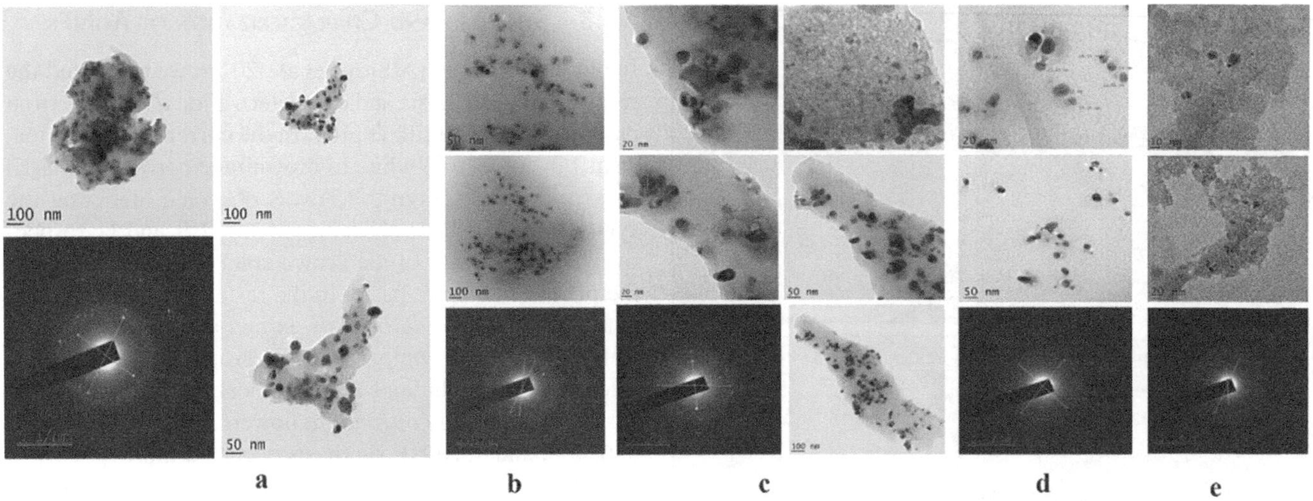

FIGURE 7.5 TEM images of AgNPs synthesized at pH 3, 5, 7, 9, and 11, represented by a, b, c, d, and e, respectively.

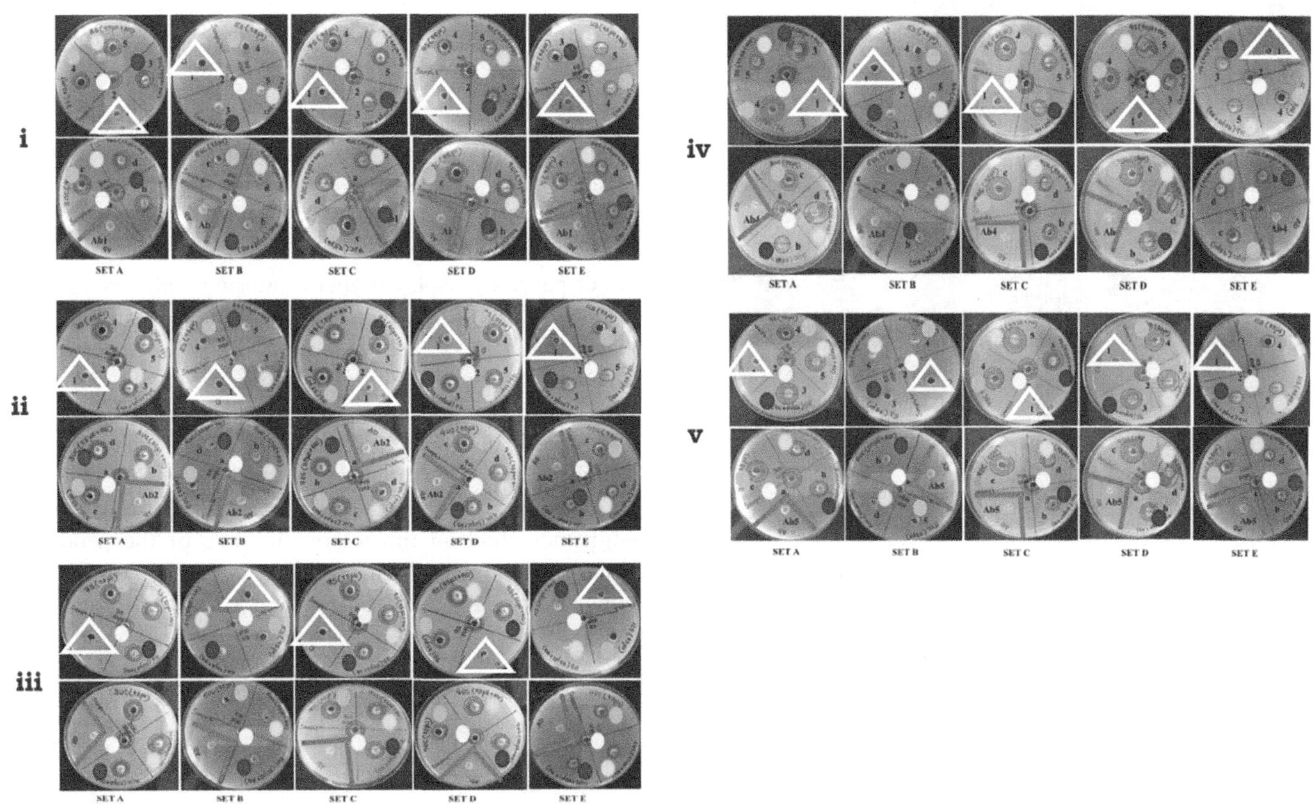

FIGURE 7.6 Antibacterial activity of AgNPs across 3, 5, 7, 9, 11 pH against MDR *P. aeruginosa* in Set A, B, C, D, E, respectively. Positive control: antibiotics [Discs represented as pink lines for Ab1(GEN), Ab2(NA), Ab3(C), Ab4(CPM), Ab5(PIT) in Figure 7.6] in i, ii, iii, iv, v, respectively, across Sets A to E; 1: negative control (pasteurized cow milk + D.H_2O, represented as white triangles in Figure 7.6).

which is consistent with the results reported in the case of the spectral shifts, the particle size reduced from 100 to 10–20 nm when pH changed from 3 to 11. Similar findings were made by Muthu et al., who demonstrated that the size of AgNPs (produced from *Cassia auriculata* flower extract) decreased at high pH. Khalil et al. made similar discoveries in another investigation, finding that particle size decreased from pH 3 to pH 8 in all cases while maintaining a consistent spherical shape. According to their investigation, the alkaline pH environment improved AgNPs' ability to reduce and stabilize (Khalil et al., 2013).

7.3.4 Antibacterial Activity

- 2, a-50 µl AgNPs
- 4,c-75 µl AgNPs
- 3,b-50 µl AgNPs + respective antibiotics
- 5,d-75 µl AgNPs + respective antibiotics

The antibacterial study was performed to find the capability of AgNPs synthesized across pH to inhibit the MDR *P. aeruginosa*. Table 7.3 and Figure 7.6 provide the details of inhibition zones and fold increases recorded for the same.

AgNPs: AgNPs at 50 µl exhibited inhibition zones (mm ± S.D) of 12.62 ± 0.379, 12.49 ± 1.10, and 12.73 ± 0.43 at pH 3,7, and 9, respectively. AgNPs at 75 µl displayed inhibition zones of 13.44 ± 0.60, 12.82 ± 1.27, and 13.44 ± 1.24 at pH 3, 7, and 9, respectively. At pH 11, AgNPs at both concentrations exhibited an inhibition zone of 7.11 ± 0.72, which was less than the zones recorded at pH 3, 7, and 9. AgNPs at both concentrations failed to show any zone of inhibition at pH 5.0. It was observed that inhibition zones obtained with 75-µl AgNPs were greater than 50-µl AgNPs by 0.8-fold. For MDR *P. aeruginosa*, AgNPs synthesized at pH 9.0 exhibited the maximum antibacterial activity, followed by pH 7.0 and pH 3.0.

Antibiotics: All five antibiotics tested failed to exhibit any zones.

It was observed that the MDR *P. aeruginosa* exhibited resistance to all antibiotics tested, whereas it showed sensitivity to the synthesized AgNPs at all pH except pH 5.

TABLE 7.3
Inhibition Zones Recorded by AgNPs Synthesized at pH 3, 5, 7, 9, 11 against MDR *P. aeruginosa* (mm ± S.D)

Treatments	pH 3.0	pH 5.0	pH 7.0	pH 9.0	pH 11.0
Pasteurized cow milk + D.H2O	0	0	0	0	0
50 µl AgNPs	12.49 ± 0.379	0	12.62 ± 1	12.73 ± 0.43	7.11 ± 0.
Ab (GEN)	0	0	0	0	0
50 µl AgNPs + GEN	12.65 ± 0.588	0	12.49 ± 0.	12.81 ± 0.77	7.10 ± 1.
Fold increase	3.44	0	3.33	3.55	0.40
Ab (NA)	0	0	0	0	0
50 µl AgNPs + NA	12.98 ± 0.260	0	13.00 ± 0.10	12.75 ± 0.24	7.20 ± 0.55
Fold increase	3.68	0	3.69	3.51	0.40
Ab (C)	0	0	0	0	0
50 µl AgNPs + C	13.01 ± 0.273	0	13.50 ± 0.50	12.73 ± 0.16	7.20 ± 0.98
Fold increase	3.70	0	4.06	3.50	0.40
Ab (CPM)	0	0	0	0	0
50 µl AgNPs +CPM	12.64 ± 2.10	0	13.50 ± 0.50	13.50 ± 1.35	7.40 ± 1.1
Fold increase	3.43	0	4.06	4.06	0.52
Ab (PIT)	0	0	0	0	0
50 µl AgNPs + PIT	12.65 ± 0.816	0	14.55 ± 0.45	14.58 ± 0.44	7.43 ± 1.0
Fold increase	3.44	0	4.88	4.90	0.53
75 µl AgNPs	12.82 ± 0.609	0	13.25 ± 1.27	13.44 ± 1.24	7.11 ± 0.73
Ab (GEN)	0	0	0	0	0
75 µl AgNPs + GEN	13.00 ± 1.01	0	13.0 ± 0.56	13.45 ± 0.86	7.43 ± 1.48
Fold increase	3.69	0	3.69	4.02	0.53
Ab (NA)	0	0	0	0	0
75 µl AgNPs + NA	13.25 ± 0.281	0	13.85 ± 0.68	13.52 ± 0.42	7.00 ± 0.95
Fold increase	3.87	0	4.32	4.07	0.36
Ab (C)	0	0	0	0	0
75 µl AgNPs + C	15.08 ± 0.21	0	13.68 ± 0.69	13.5 ± 0.42	7.00 ± 0.85
Fold increase	5.31	0	4.19	4.06	0.36
Ab (CPM)	0	0	0	0	0
75 µl AgNPs + CPM	15.22 ± 1.33	0	14.05 ± 0.08	15.66 ± 1.72	7.50 ± 1.1
Fold increase	5.43	0	4.48	5.81	0.56
Ab (PIT)	0	0	0	0	0
75 µl AgNPs + PIT	15.08 ± 0.885	0	15.08 ± 0.09	15.83 ± 0.74	7.50 ± 1.23
Fold increase	5.31	0	5.31	5.96	0.56

Note: All assays were performed in triplicate, and standard deviations were noted.

Negative control (pasteurized cow milk + D.H2O): No inhibition zones were recorded.

Synergistic activity of AgNPs: AgNPs synthesized at pH 3.0 (50 µl) showed the highest synergy with chloramphenicol, with a 3.70-fold increase, followed by nalidixic acid with 3.68 and cefepime, piperaciliin/tazobactam, and gentamycin with a 3.44-fold increase. A similar pattern in synergy was obtained on treatment with 75-µl AgNPs, where the highest increase was with chloramphenicol at 5.43-fold, followed by cefepime and piperaciliin/tazobactam with 5.31. AgNPs with nalidixic acid and gentamycin produced increases of 3.87- and 3.69-fold, respectively.

AgNPs synthesized at pH 5.0 (50 and 75 µl) exhibited no synergy with all five antibiotics. The reason for this could be attributed to the fact that no inhibition zones were observed with AgNPs alone at this pH.

AgNPs synthesized at pH 7.0 increased the efficacy of piperacilin/tazobactam by displaying a maximum increase of 4.88-fold and 5.31-fold, followed by cefepime, showing an increase of 4.06-fold and 4.48-fold at 50- and 75-µl concentrations, respectively. Further, chloramphenicol and nalidixic acid exhibited a 4.06-fold and 3.69-fold increase with 50-µl AgNPs and a 4.19-fold and 4.32-fold increase with 75-µl AgNPs. Both concentrations of AgNPs with gentamycin displayed the lowest increases, 3.33-fold and 3.69-fold, respectively.

pH 9.0 synthesized AgNPs at 50 µl with antibiotics showed the highest increase of 4.90-fold with piperacilin/tazobactam, followed by cefepime, chloramphenicol, nalidixic acid, and gentamycin, with increases of 4.06-, 4.06-, 3.69-, and 3.33-fold, respectively. A similar trend was observed on the interaction of AgNPs with antibiotics at a higher concentration of 75 µl, where a 20% increase in efficacy was observed.

AgNPs synthesized at pH 11.0 at both concentrations exhibited a very low synergy of less than 0.55 to all antibiotics.

Overall it can be observed that wherever antibiotics have failed to display any inhibition zones, AgNPs synthesized at pH 3, 7, and 9 produced clear inhibition zones against MDR *P. aeruginosa*. AgNPs when combined with antibiotics produced inhibition zones greater than zones observed due to individual effects. This indicated clearly that antibiotic efficacy was increased in the presence of AgNPs (Table 7.3).

7.3.5 In Vitro Cytotoxicity Studies of AgNPs on Human Embryonic Kidney Cells

AgNPs from pasteurized cow milk at pH 7.0 and 9.0 showed strong antibacterial activity, followed by pH 3.0, against isolated MDR *P. aeruginosa* at 50- and 75-µl concentrations. To assess biocompatibility, an in vitro cytotoxicity evaluation of the biosynthesized AgNPs was conducted using an MTT assay, in which HEK cells were exposed to varying concentrations of the test substances. The results showed a dose-response activity (Table 7.4, Figure 7.7). CTC50 (µl/ml) of AgNPs synthesized at pH 3.0 was found to be 85.37 ± 0.6. AgNPs synthesized at pH 3.0 (50 and 75 µl) displayed cytotoxicity of 27.09 ± 0.5 and 41.92 ± 0.5, which indicated they were non-toxic.

TABLE 7.4
Cytotoxic Properties of AgNPs Synthesized at pH 3, 7, and 9 against HEK Cell Line

Sl. No.	Name of Test Compound	Conc. (µl/ml)	% Cytotoxicity	CTC50 (µl/ml)
1.	AgNPs (pH 3.0)	500	90.79 ± 0.7	85.37 ± 0.6
		200	89.20 ± 0.5	
		100	61.41 ± 0.5	
		75	41.92 ± 0.5	
		50	27.09 ± 0.5	
		25	17.37 ± 0.5	
	AgNPs (pH 7.0)	500	91.66 ± 0.2	61.91 ± 0.1
		200	91.44 ± 0.3	
		100	90.15 ± 0.3	
		75	74.55 ± 0.3	
		50	27.65 ± 0.6	
		25	19.88 ± 1.4	
	AgNPs (pH 9.0)	500	91.75 ± 0.1	56.16 ± 0.3
		200	91.31 ± 0.1	
		100	90.32 ± 0.5	
		75	77.35 ± 0.5	
		50	41.05 ± 0.7	
		25	28.00 ± 1.2	

AgNPs synthesized at pH 7.0 and 9.0 exhibited CTC50 of 61.91 ± 0.1 and 56.16 ± 0.3 µl/ml, respectively. The cytotoxicity of 50-µl AgNPs at pH 7.0 and pH 9.0 was 27.65 ± 0.6 and 41.05 ± 0.7 µl/ml, well within the CTC50 range and considered non-toxic. On the other hand, 75-µl AgNPs displayed 74.55 ± 0.3 and 77.35 ± 0.5 µl/ml at pH 7.0 and 9.0, respectively, which was above the CTC50 range and considered toxic. Overall, the outcome indicated 50-µl AgNPs were safe at 3, 7, and 9 synthesized pH and could be considered effective antibacterial agents against MDR and non-MDR bacteria.

7.4 CONCLUSION

The antibacterial activity of AgNPs against MDR *P. aeruginosa* was investigated in this study. AgNPs were produced and described at different pH levels and were found to exhibit antibacterial properties and work synergistically with selected antibiotics at pH 3, 7, and 9 to combat MDR. The study highlights the importance of pH in demonstrating antibacterial activity, as alkaline pH facilitated the production of small-sized nanoparticles. In vitro cytotoxicity assays revealed that a 50-µl concentration of nanoparticles was non-toxic and safe. However, further research involving purification of AgNPs and in vivo trials is needed for

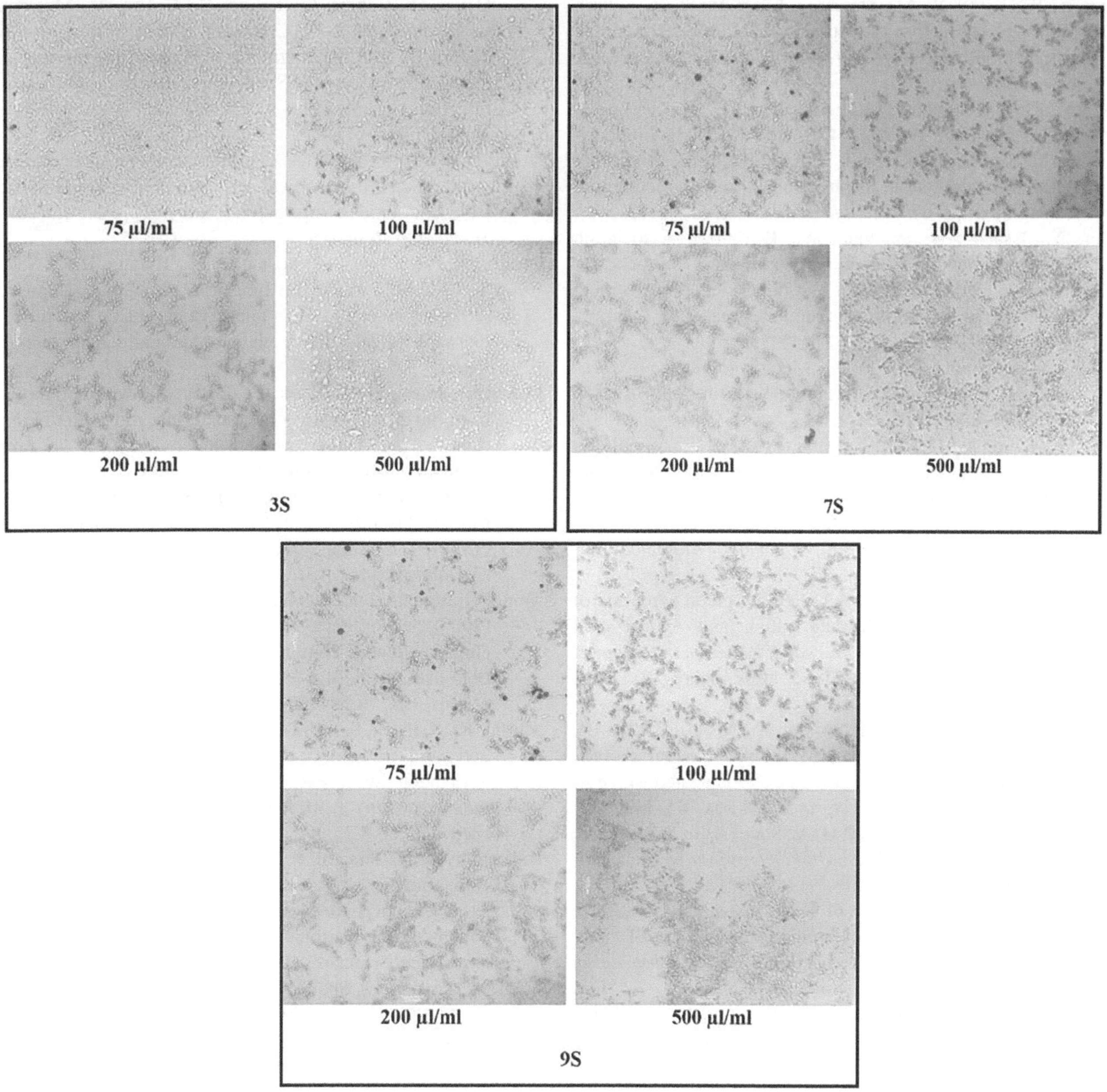

FIGURE 7.7 Cytotoxic effect of AgNPs synthesized at pH 3, 7, and 9, labeled in figures as 3S, 7S, and 9S, respectively, on HEK cell line.

it to be considered a viable therapeutic antibacterial agent against MDR bacteria.

REFERENCES

Anigol, L. B., Charantimath, J. S., & Gurubasavaraj, P. M. (2017). Effect of concentration and pH on the size of silver nanoparticles synthesized by green chemistry. *Organic & Medicinal Chemistry International Journal*, 3(5). https://doi.org/10.23880/OMCIJ-MS-ID-555622

Ashraf, S., Chatha, M. A., Ejaz, W., Janjua, H. A., & Hussain, I. (2014). Lysozyme-coated silver nanoparticles for differentiating bacterial strains on the basis of antibacterial activity. *Research Letters*, 9, 565.

Athreya, A. G., Shareef, M. I., & Gopinath, S. M. (2018). Antibacterial activity of silver nanoparticles isolated from cow's milk, hen's egg white, and lysozyme: A comparative study. *Arabian Journal for Science and Engineering*. https://doi.org/10.1007/s13369-018-3685-1

Athreya, A. G., Shareef, M. I., & Gopinath, S. M. (2020). Silver nanoparticles from cow's milk to combat multidrug-resistant gram-negative bacteria from clinical isolates. *Proceedings of the Indian National Academy-Biological Sciences*. https://doi.org/10.1007/s40011-019-01160-3

Barros, C. H. N., Fulaz, S., Stanisic, D., & Tasic, L. (2018). Biogenic nanosilver against multidrug-resistant bacteria (MDRB). *Antibiotics*, 7, 69. https://doi.org/10.3390/antibiotics7030069

Bauer, A. W., Kirby, W. M., Sherris, J. C., & Turck, M. (1966). Antibiotic susceptibility testing by a standardized single disk method. *American Journal of Clinical Pathology*, 45, 493–496.

Cosgrove, S. E., Sakoulas, G., Perencevich, E. N., Schwaber, M. J., Karchmer, A. W., & Carmeli, Y. (2003). Comparison of mortality associated with methicillin-resistant and methicillin-susceptible Staphylococcus aureus bacteremia: A meta-analysis. *Clinical Infectious Diseases*, 36, 53–59.

Gillet, Y., Issartel, B., & Vanhems, P. (2002). Association between Staphylococcus aureus strains carrying gene for Panton-Valentine leukocidin and highly lethal necrotizing pneumonia in young immunocompetent patients. *The Lancet*, 359, 753–759.

Glen, K. A., & Lamont, I. L. (2021). β-lactam resistance in *Pseudomonas aeruginosa*: Current status, future prospects. *Pathogens*, 10(12), 1638.

Habib, F., Rind, R., Durani, N., Bhutto, A. L., Buriro, R. S., Tunio, A., Aijaz, N., Lakho, S. A., Bugti, A. G., & Shoaib, M. (2015). Morphological and cultural characterization of Staphylococcus aureus isolated from different animal species. *Journal of Applied & Environmental Biological Sciences*, 5(2), 15–26.

Halima, R., & Archna. (2016). A review on green synthesis of silver nanoparticles, characterization and optimization parameters. *International Journal of Research in Engineering & Technology*, 5(15), 49–52. https://doi.org/10.15623/ijret.2016.0527010

Hegazi, A., Hamdy Elshazly, E., Abdou, A. M., Abd Allah, F., Eman, H., & Abdel-Rahman, M. (2014). Potential antibacterial properties of silver nanoparticles conjugated with cow and camel milks. *Global Veterinaria*, 12(6), 745–749.

Holt, J. G., Krieg, N. R., Sneath, P. H. A., Staley, J. T., & Williams, S. T. (1994). *Bergey's Manual of Determinative Bacteriology*. Williamsons & Wilkins.

Hungund, B. S., Dhulappanavar, G. R., & Ayachit, N. H. (2015). Comparative evaluation of antibacterial activity of silver nanoparticles biosynthesized using fruit juices. *Journal of Nanomedicine & Nanotechnology*, 6, 271.

Jyoti, K., Baunthiyal, M., & Singh, A. (2016). Characterization of silver nanoparticles synthesized using Urtica dioica Linn. leaves and their synergistic effects with antibiotics. *Journal of Radiation Research & Applied Sciences*, 9, 217–227.

Karthika, R., & Sevarkordiyone, S. P. (2015). Synthesis and characterization of silver nanoparticles using aqueous extract of goat faecal pellets. *International Journal of Current Science and Research*, 1(1), 1–7.

Khalil, M., Ismail, E. H., El-Baghdady, K. Z., & Mohamed, D. (2013). Green synthesis of silver nanoparticles using olive leaf extract and its antibacterial activity. *Arabian Journal of Chemistry*. https://doi.org/10.1016/j.arabjc.2013.04.007

Kourtis, A. P., Hatfield, K., Baggs, J., Mu, Y., See, I., Epson, E., Nadle, J., Kainer, M. A., Dumyati, G., Petit, S., Ray, S. M., David, H., Capers, C., Ewing, H., Coffin, N., McDonald, L. C., Jernigan, J., & Cardo, D. (2019). Vital signs: Epidemiology and recent trends in methicillin-resistant and methicillin-susceptible Staphylococcus aureus bloodstream infections—United States. *Morbidity and Mortality Weekly Report*, 68(9), 214–219.

Kredy, H. M. (2018). The effect of pH and temperature on the green synthesis and biochemical activities of silver nanoparticles from Lawsonia inermis extract. *Journal of Pharmaceutical Sciences & Research*, 10(8), 2022–2026.

Lee, K. J., Park, S. H., Govarthanan, M., Hwang, P. H., Seo, Y. S., Cho, M., Lee, W. H., Lee, J. Y., & Kamala-Kannan, S. (2013). Synthesis of silver nanoparticles using cow milk and their antifungal activity against phytopathogens. *Materials Letters*, 105, 128–131.

Marambio-Jones, C., & Hoek, E. (2010). A review of the antibacterial effects of silver nanomaterials and potential implications for human health and the environment. *Journal of Nanoparticle Research*, 12(5), 1531–1551.

Miller, L. G., Perdreau-Remington, F., & Rieg, G. (2005). Necrotizing fasciitis caused by community-associated methicillin-resistant Staphylococcus aureus in Los Angeles. *New England Journal of Medicine*, 352, 1445–1453.

Mongkolrattanothai, K., Boyle, S., Kahana, M. D., & Daum, R. S. (2003). Severe Staphylococcus aureus infections caused by clonally related community-acquired methicillin-susceptible and methicillin-resistant isolates. *Clinical Infectious Diseases*, 37, 1050–1058.

Muthu, K., & Priya, S. (2017). Green synthesis, characterization, and catalytic activity of silver nanoparticles using Cassia auriculata flower extract separated fraction. *Spectrochimica Acta Part A: Molecular and Biomolecular Spectroscopy*, 179, 66–72.

Rajapandiyan, K., Shanthi, S., Murugan, A. M., Muthu, G. A., & Ranjit Singh, A. J. A. (2011). Azadirachta indica—cow urine extract, a novel controlling agent towards clinically significant multidrug-resistant pathogens. *Journal of Applied Pharmaceutical Science*, 1(10), 107–113.

Thill, A., Zeyons, O., Spalla, O., Chauvat, F., Rose, J., & Auffan, M. (2006). Cytotoxicity of CeO_2 nanoparticles for Escherichia coli: Physicochemical insight of the cytotoxicity mechanism. *Environmental Science & Technology*, 40(6), 151–156.

Vanaja, M., Rajeshkumar, S., Paulkumar, K., Gnanajobitha, G., Malarkodi, C., & Annadurai, G. (2013). Kinetic study on green synthesis of silver nanoparticles using Coleus aromaticus leaf extract. *Advances in Applied Science Research*, 4(3), 50–55.

Wayne, P. A. (2017). CLSI. Performance standards for antibiotic disc susceptibility testing (27th ed.). *CLSI Supplement M100*, 37(1).

Willner, I., Baron, R., & Willner, B. (2006). Growing metal nanoparticles by enzymes. *Advanced Materials*, 18, 1109–1120.

8 Eco-Friendly Synthesis of Nanoparticles

Gurushantha K, Prashanth Kumar H P, Roopashree B N, and Shashidhar S

8.1 INTRODUCTION

Nanoparticles constitute extremely small particles between 1 and 100 nanometers (nm) in dimensions. A nanometer stands as the smallest unit possible when measuring a meter because it represents a single billionth of a meter. These small particles demonstrate properties that are vastly different from those of bulk particles sharing the same substance.

Nanoscale materials often reveal unique combinations of physical properties alongside chemical and biological properties. Nanoparticles generally outperform bulk materials in terms of strength as well as chemical reactivity and optical properties. Nanoparticulates exhibit dual properties which make them valuable for various applications from electronics to medicine and energy storage and environmental clean-up work [1].

Nanoparticles interact with biological systems through mechanisms that bigger particles cannot because of their small sizes and large surface areas, thus creating opportunities for research in fields such as food packaging, cancer treatment, and medical imaging [2].

The many industrial sectors, including electronics, environmental clean-up, and medicine utilize nanoparticles because of their specific physical and chemical characteristics. The typical methods to synthesize nanoparticles often result in three main problems: harmful chemical use, high energy requirements, and significant negative environmental consequences. Eco-friendly synthesis or green synthesis provides an effective answer to these difficulties since it continues to gain wide acceptance [3].

Eco-friendly synthesis produces nanoparticles through natural and benign agents including bacterial, fungal, algal, and plant extracts which replace conventional traditional hazardous methods. The approach aims to resolve the growing need for powerful nanomaterials through the development of earth-friendly economical and sustainable production processes for nanoparticles [4].

1. Important strategies in the synthesis of eco-friendly nanoparticles: biological approaches (green synthesis):
   a. Nanoparticles can be synthesized using flavonoids, alkaloids, polyphenols, and other biomolecules present in plant extracts, which reduce metal ions into nanoparticles.
   b. Nanoparticles result from microbial methods which incorporate bacteria, fungi, and algae as the main production agents. The reduction of metal ions receives help from metabolites and enzymes which microorganisms naturally produce. Successful nanoparticle production involves fungi strains like *Fusarium* and *Aspergillus* along with bacteria which include *Bacillus subtilis* and *Escherichia coli.*
   c. The synthesis of nanoparticles benefits from algae as well as seaweeds because these organisms contain abundant protein and polysaccharide components that serve as sustainable renewable sources. Specific seaweeds have proved useful for producing metal nanoparticles together with metal oxide nanoparticles.
2. Enzyme-mediated synthesis: Nanoparticles of equivalent metals can be produced by enzymatic reduction of metal ions, which offers a beneficial alternative instead of using extreme chemicals or harsh processing methods.
3. Plant-mediated synthesis vs. chemical synthesis:
   a. Toxic chemicals, high temperatures, and pressure are frequently needed in traditional chemical methods of synthesizing nanoparticles, which can be hazardous for the environment and human health.
   b. On the other hand, plant-based techniques (or other biological techniques) do not require hazardous chemicals and function at room temperature and pressure. Plant extracts are also abundant, renewable, and frequently inexpensive.
4. Green solvents: Green solvents like water or natural organic solvents are frequently used in eco-friendly synthesis in place of toxic solvents, which lowers the amount of hazardous byproducts and environmental contamination.

Amal Mohamed Al-Mohaimeed reported the multifunctional eco-friendly synthesis of ZnO nanoparticles using aqueous oat extract in his article titled “Multifunctional Eco-Friendly Synthesis of ZnO Nanoparticles in Biomedical

DOI: 10.1201/9781003644439-8

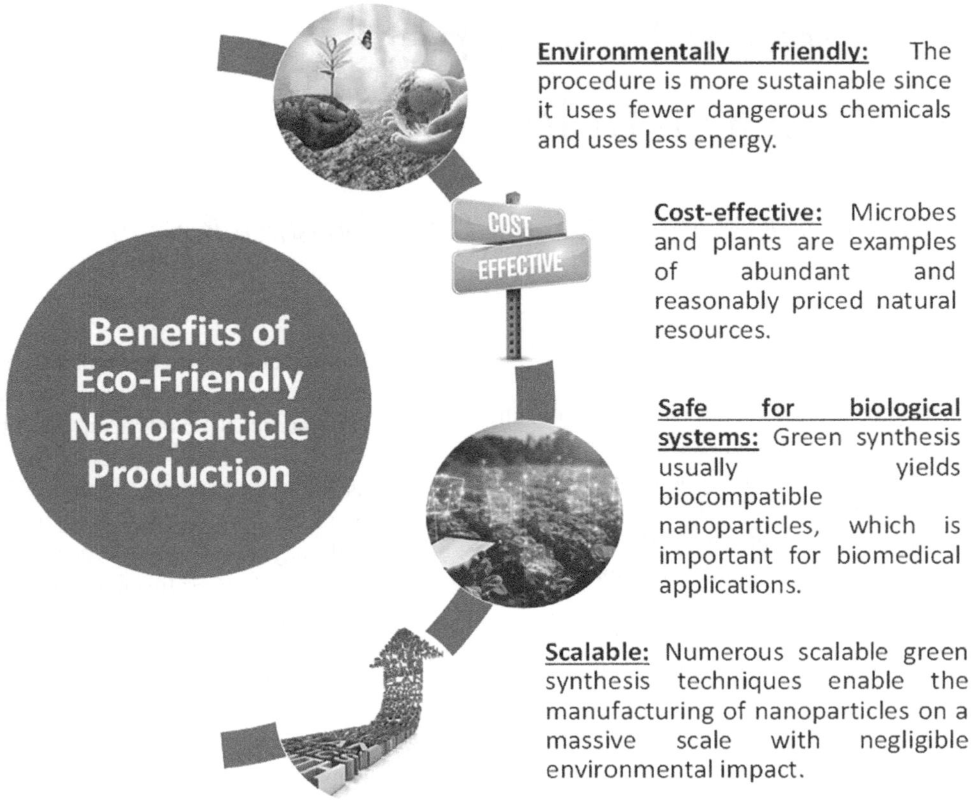

FIGURE 8.1 Benefits of eco-friendly nanoparticle production.

TABLE 8.1
Precursors Employed for the Synthesis of Metal Nanoparticles and Their Applications

Sl. No.	Nanoparticle	Precursor Used	Applications	Year	Reference
1	ZnO NPs	Aqueous oat extract	Biomedical applications	2022	[5]
2	Au NPs	*Salix alba L.* leaves extract	Biomedical and pharmaceutical applications	2022	[6]
3	NiO NPs	*Punica granatum* L. juice extract	Methyl orange degradation	2020	[7]
4	Co_3O_4 NPs	*Coriandrum sativum and Allium sativumare* seeds and cloves extracts	Rhodamine-B degradation	2021	[8]
5	M-Cu NPs	*Mentha pulegium* leaf extract	Antibacterial and cytotoxic activities	2021	[9]
6	Ag NPs	*Hypericum perforatum L.* extract	-	2022	[10]
7	FeO NPs	*Cassia occidentalis L.* flower extract	Nano-priming agent for boosting seed germination in rice	2020	[11]
8	CuO NPs	Biological extract of *Cynomorium coccineum*	Methylene blue degradation	2019	[12]
9	La_2O_3 NPs	*Eucalyptus globulus* leaf extracts	Effective biomedical applications	2020	[13]
10	$Co(OH)_2$ and Co_3O_4 NPs	Extract of *Litchi chinensis*	Various applications	2020	[14]

Applications." The synthesis of ZnO nanoparticles using aqueous oat extract is described in this work as being environmentally beneficial. Excellent antibacterial activity was demonstrated by the sophisticated electrochemical and optical properties of green-synthesized ZnO NPs, which also played a significant role in pharmacological measurements.

Similarly, in another review paper titled "Green Synthesis of Gold Nanoparticles: An Eco-Friendly Approach," Poornima Budime Santhosh reported the eco-friendly synthesis of gold nanoparticles from *Salix alba L.* leaf extract for biomedical and pharmaceutical applications. An overview of the developments in the synthesis of AuNPs using various biological resources, along with examples, is given in this article, together with information on their significant uses in biomedicine. Particular attention is paid to the production of AuNPs from various medicinal plants and their numerous uses in immunological responses and antibacterial and anti-inflammatory processes. Furthermore, a thorough discussion is given on the uses of AuNPs in cancer theragnostics, including contrast imaging, medication delivery, hyperthermia, and cancer therapies. Additionally, this study highlights the significance of the green synthesis technique and go over the benefits, difficulties, and future possibilities in this area.

For environmental bioremediation process, Azeez A. Barzinjy described the green and eco-friendly synthesis of nickel oxide nanoparticles in his research article titled "Green and Eco-Friendly Synthesis of Nickel Oxide Nanoparticles and Its Photocatalytic Activity for Methyl Orange Degradation." In this work, the degradation of methyl orange, a dangerous environmental pollutant in water, is examined using the catalytic potential of green-produced nickel oxide nanoparticles (NiO NPs). The antioxidant content of pomegranate juice extract from *Punica granatum L.* was used to prepare NiO NPs under ambient circumstances, and their capacity to reduce biological pollutants was thoroughly examined. This procedure is completely environmentally friendly and devoid of any dangerous or poisonous solvents [7].

In another article titled "Eco-Friendly Synthesis and Characterization of Cobalt Oxide Nanoparticles by *Sativum* Species and Its Photo-Catalytic Activity," G. Asha showed the eco-friendly synthesis of cobalt oxide nanoparticles using *Coriandrum sativum* and *Allium sativumare* seeds and clove extracts. Rhodamine-B dye was used to study photocatalytic breakdown in the presence of ultraviolet light. Co_3O_4-C exhibited higher dye degradation efficiency than the Co_3O_4-A sample when the two samples were compared for the removal of organic dyes, which are used to purify water and clean the environment [8].

Maryam Gholami reported the synthesis of copper nanoparticles for antibacterial activity in his article titled "Eco-Friendly Synthesis of Copper Nanoparticles Using *Mentha pulegium* Leaf Extract Characterisation, Antibacterial and Cytotoxic Activities." *Mentha pulegium*, a flowering plant in the *Lamiaceae* family, was effectively used as an extract to create green copper nanoparticles (M-Cu NPs). Potential antibacterial activity was demonstrated by the plant extract and synthesized Cu NPs against the bacterial pathogens *Klebsiella pneumonia, Bacillus cereus, Escherichia coli,* and *Staphylococcus aureus.* Against *B. cereus* bacteria, a maximum zone of inhibition of 12 mm was observed. M-Cu NPs were tested for cell cytotoxicity on healthy human lymphocytes using methyl thiazolyl diphenyltetrazolium bromide (MTT). Up to 25 μg/mL of M-Cu NPs, nearly 98% of healthy human peripheral blood cells were found to survive, demonstrating the NPs strong antibacterial activity against *S. aureus* and *B. cereus* [9].

In the review paper titled "Uncovering the Phytochemical Basis and the Mechanism of Plant Extract-Mediated Eco-Friendly Synthesis of Silver Nanoparticles Using Ultra-Performance Liquid Chromatography Coupled with a Photo Diode Array and High-Resolution Mass Spectrometry," Matam Pradeep reported silver nanoparticle synthesis. Ag+ ions were not reduced by the pure compounds of the xanthone and phloroglucinol families (mangiferin and hyperforin), the low polar component of *H. perforatum* extract, or rich in these compounds. On the other hand, the extracts polar fraction, which was high in flavonoids and naphthodianthrones, as well as the pure compounds of these classes, namely hypericin, pseudohypericin, protohypericin, quercetin, and kaempferol-3-glucoside, was able to decrease Ag^+ ions. All things considered, the study makes a substantial contribution to our knowledge of the environmentally friendly synthesis of NPs using plant extracts [10].

Shadma Afzal described the ecofriendly synthesis of phytochemical-capped iron oxide nanoparticles for application in agriculture in the paper titled "Eco-Friendly Synthesis of Phytochemical-Capped Iron Oxide Nanoparticles as Nano-Priming Agent for Boosting Seed Germination in Rice (*Oryza sativa L*)." FeO NPs primed at 20 and 40 mg/L effectively increased germination and seedling vigor in rice seeds as compared to hydro-primed control and ferrous sulphate ($FeSO_4$) priming. Biophysical measures including root length and dry weight were stimulated by up to 50% in the seeds primed with 20 mg/L FeO NPs. At the same concentration, significant stimulation of the sugar and amylase content was also seen. Comparing the antioxidant enzyme activity to $FeSO_4$ priming and control, there was a considerable increase. Fe, Zn, K, Ca, and Mn concentrations in seeds were also analyzed using inductively coupled plasma mass spectrometry (ICP-MS). The seed priming method represents an understandable and novel strategy that may improve α-amylase activity, iron uptake, and ROS generation, resulting in higher levels of soluble sugar to promote seedling growth and increase the seed germination rate, respectively. This study presents phytochemical-capped FeO NPs as a powerful nano-priming agent that can promote the germination of naturally aged rice seedlings [11].

Biosorption activity is highlighted in the paper titled "Eco-Friendly Synthesis of *Cynomorium coccineum* Extract for Controlled Production of Copper Nanoparticles for Sorption of Methylene Blue Dye." Nouha Sebeia reported utilizing copper sulfate as a starting material. Copper nanoparticles were synthesized utilizing a biological extract of *Cynomorium coccineum.* Using methylene blue as an adsorbate, the adsorption properties of the micro-adsorbents were investigated, and at room temperature, the biosorption capacity was 64 mg/g [12].

G Maheshwaran reported the eco-friendly synthesis of lanthanum oxide nanoparticles for anti-inflammation activity in his article titled "Eco-Friendly Synthesis of Lanthanum Oxide Nanoparticles by *Eucalyptus globulus* Leaf Extracts for Effective Biomedical Applications." This study's primary goal was to use La_2O_3 as a possible inhibitory medication to regulate different types of inflammation and diabetes illnesses [13].

Principal scientist and investigator Damian C. Onwudiwe reported the synthesis of cobalt hydroxide and cobalt oxide NPs in the paper titled "Eco-Friendly Synthesis, Structural Properties and Morphology of Cobalt Hydroxide and Cobalt

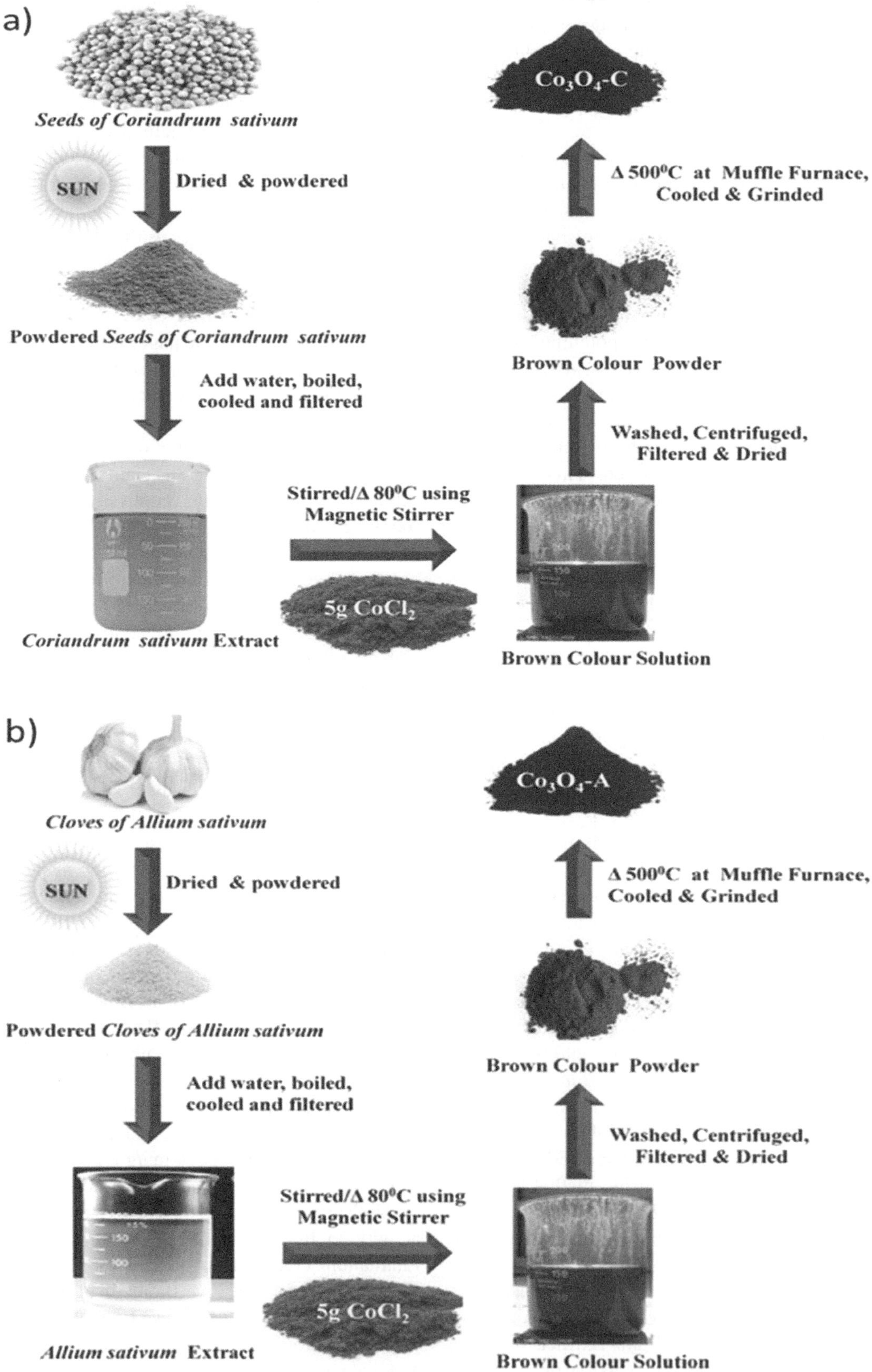

FIGURE 8.2 Material synthesis process that includes (a) the extraction of the necessary phytocomponents from the fruit peels of *Litch chinensis* and (b) the synthesis of $Co(OH)_2$ and Co_3O_4 nanoparticles.

Reprinted with permission [14].

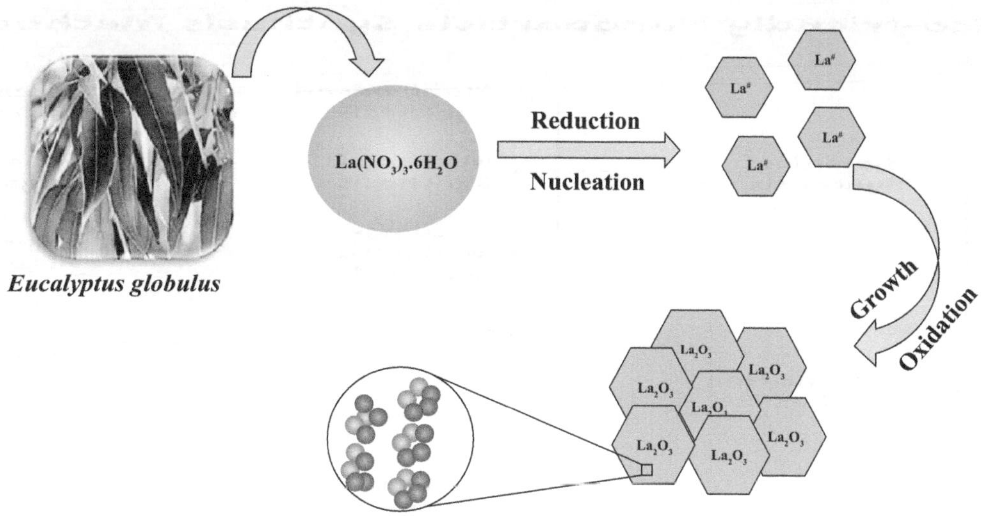

FIGURE 8.3 Synthesis scheme.
Reprinted with permission [13].

Oxide Nanoparticles Using Chinensis." The production of $Co(OH)_2$ and Co_3O_4NPs, which have a variety of potential uses, may be accomplished efficiently, simply, quickly, and affordably using this environmentally friendly method that doesn't require the use of any chemical reagents [14].

8.2 SYNTHESIS

8.2.1 General Method of Synthesis of Eco-Friendly Nanoparticles

Green chemistry concepts are used in the synthesis of eco-friendly nanoparticles in order to minimize the toxicity and environmental impact of traditional nanoparticle production techniques. The goal of eco-friendly techniques is to reduce or do away with the usage of dangerous chemicals, solvents, and energy-intensive procedures [15].

General techniques for the environmentally friendly production of nanoparticles are listed in the following sections.

8.2.1.1 Biological (Green) Synthesis

This process develops nanoparticles by using biological creatures such as fungi, microbes, plants, and enzymes. The biological agents are frequently regarded as economical and environmentally safe.

- *Plant-based synthesis:* Secondary metabolites from plants, including polyphenols, flavonoids, and terpenoids, serve as reducing agents for the production of nanoparticles.

  Common plant materials include extracts from leaves, flowers, fruits, and more.

  Example: Leaf extracts from *Azadirachta indica,* or neem, were used for producing silver nanoparticles.
- *Microbial synthesis:* Metal ions can be reduced into nanoparticles by specific bacteria, fungi, and algae. Scalability and the capacity to generate nanoparticles under mild conditions are two benefits of these microbes.

  Example: Gold nanoparticles are synthesized using either *Fusarium oxysporum* or *Bacillus subtilis.*
- *Enzyme-mediated synthesis*: Eco-friendly reduction of metal ions into nanoparticles can be catalyzed by enzymes like reductases. This approach can be more carefully managed and is quite particular.

  Example: The production of silver nanoparticles using *Aspergillus niger* enzymes.

8.2.1.2 Solvent-Free or Low-Solvent Synthesis

Nanoparticles are produced using this technique without the use of hazardous organic solvents. For instance, a dry or semi-dry synthesis method frequently involves a reducing agent and makes use of heat, pressure, or microwave energy.

- *Mechanochemical synthesis:* In a ball mill, solid precursors are broken down into nanoparticles using mechanical energy. This approach uses fundamental equipment and avoids solvents.
  Example: Zinc oxide powder is crushed with a reducing agent to produce zinc oxide nanoparticles.
- *Solid-state synthesis:* This method creates nanoparticles by heating and mixing solid reactants, frequently with green solvents like water or supercritical CO_2 as reaction medium. It uses less energy and doesn't include dangerous chemicals.

  Example: Heating a combination of iron salts with a green solvent to produce iron oxide nanoparticles.

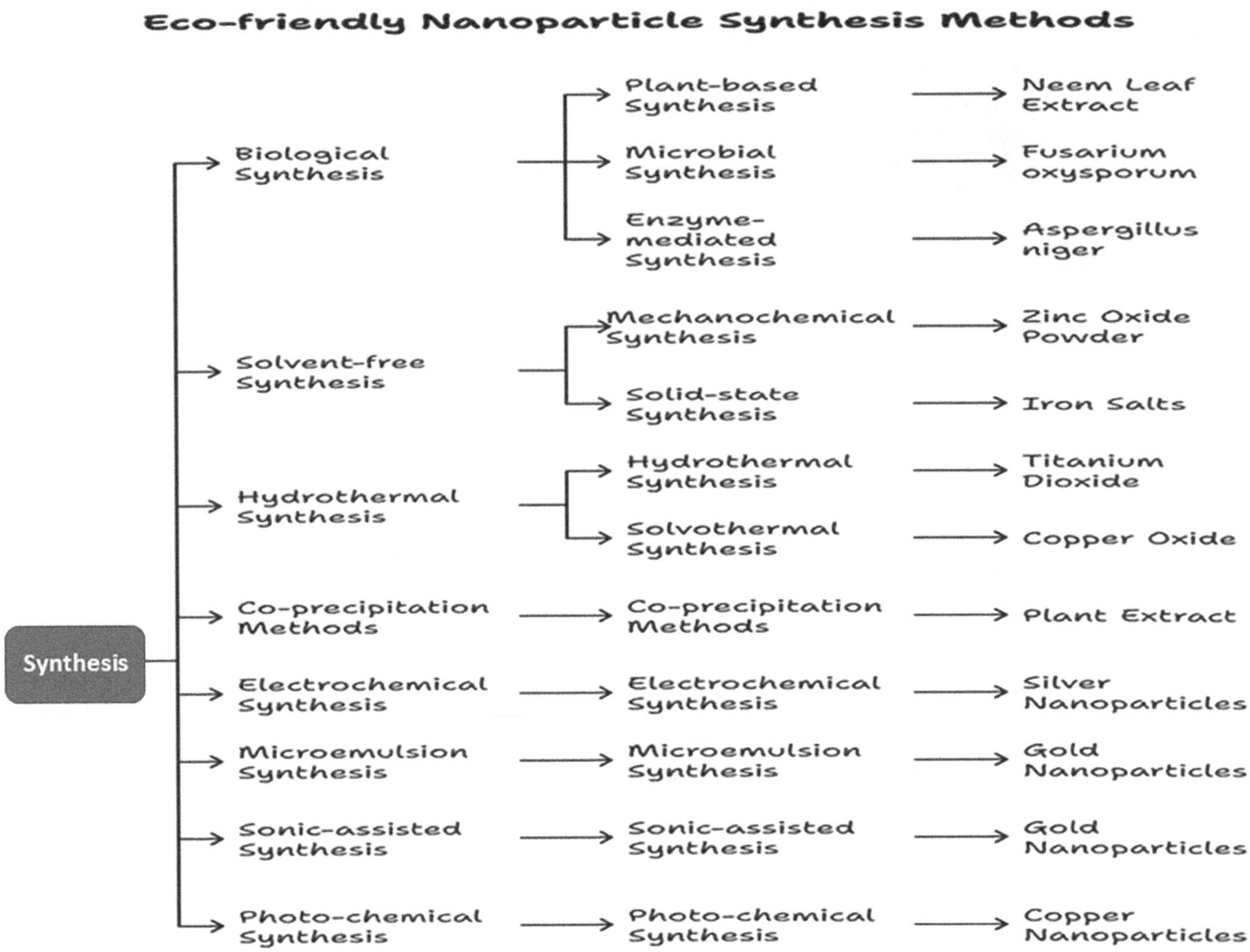

FIGURE 8.4 Synthesis of eco-friendly nanoparticles flow chart.

8.2.1.3 Hydrothermal and Solvothermal Synthesis

This technique uses regulated high-temperature and high-pressure conditions to create nanoparticles using either water (hydrothermal) or other eco-friendly solvents (solvothermal).

- *Hydrothermal synthesis:* At high temperatures and pressures, nanoparticles are created in a sealed vessel using water as a solvent. This method can produce distinct nanoparticles with special qualities.

  Example: Synthesis of titanium dioxide nanoparticles via hydrothermal synthesis.
- *Solvothermal synthesis:* Comparable to hydrothermal synthesis but using different solvents like ethanol or green solvents that might be better for the environment.

  Example: Copper oxide nanoparticle synthesis with an eco-friendly solvent.

8.2.1.4 Co-Precipitation Methods

This is the process of combining two or more precursor solutions, which causes a solid to precipitate and produce nanoparticles. Toxic chemicals can be eliminated by employing natural materials or aqueous solutions as reducing agents in green co-precipitation techniques.

Example: Iron oxide nanoparticle co-precipitation with a plant extract acting as a stabilizing factor.

8.2.1.5 Electrochemical Synthesis

This process produces nanoparticles by reducing metal ions in an aqueous solution using an electric current. The size and shape of nanoparticles can be regulated by adjusting variables including temperature, time, and current density.

Example: Environmentally friendly electrochemical technologies that use few or no hazardous chemicals to synthesize silver nanoparticles.

8.2.1.6 Microemulsion and Reverse Micelles

This technique minimizes the use of hazardous organic solvents by forming microemulsions or reverse micelles out of ecologically benign surfactants or stabilizing agents that serve as nanoreactors for the controlled synthesis of nanoparticles.

Example: Synthesis of gold nanoparticles in an aqueous microemulsion with a non-toxic surfactant.

8.2.1.7 Sonic-Assisted or Ultrasound-Assisted Synthesis

In order to promote the creation of nanoparticles in an environmentally benign way, ultrasound waves can cause cavitation in the solution, which produces high energy. In general, the technique uses less energy and doesn't use any dangerous chemicals.

Example Production of gold or silver nanoparticles by sonication with liquids derived from plants.

8.2.1.8 Photo-Chemical Synthesis

Light energy functions as the reduction power to turn metal ions into nanoparticles. The method employs either artificial light sources or natural light sources together with environmentally friendly chemicals during operation. The production of copper nanoparticles through green synthesis requires reducing agents obtained from plant extracts which function under sunlight illumination conditions. Some advantages of eco-friendly nanoparticle synthesis are that the process of green synthesis produces minimal toxicity, which lowers the environmental and health threats associated with manufacturing and using nanoparticles. These methods require less energy for operation compared to traditional procedures because they operate at mild pressures and temperatures.

Synthesis through green methods regularly uses biodegradable natural solvents together with biodegradable reducing agents and stabilizers. The procedures demonstrate excellent scalability characteristics because they enable smooth industrial scale-up operations without harmful environmental consequences. Plants combined with microbes together with natural solvents serve as renewable natural resources for environmentally friendly nanoparticle synthesis methods to produce nanoparticles that generate minimal environmental strain.

Ecological procedures based on nanotechnology present possible solutions to replace traditional chemical practices that are risky to the environment.

Applications: Environmentally friendly synthesized nanoparticles have been used in numerous fields.

- Medicine: Antimicrobial agents, cancer treatment, and drug delivery
- Environmental remediation
- The removal of contaminants and heavy metals from the air and water
- Agriculture: Nano-fertilizers and pest control
- Energy: Solar cells, energy storage and sensors

8.3 CHALLENGES AND FUTURE DIRECTIONS

Standardization: Benefits of Eco-Friendly Nanoparticle Production Environmentally friendly. Understanding mechanisms: The procedure is more sustainable since it uses fewer dangerous chemicals and uses less energy. Cost-effective Microbes and plants are examples of abundant and reasonably priced natural resources. Safe for biological systems Green synthesis usually yields biocompatible nanoparticles, this is important for biomedical applications. Scalable Numerous scalable green synthesis techniques enable the manufacturing of nanoparticles on a massive scale with negligible environmental impact.

- Standardization: Maintaining uniformity in nanoparticle size, shape, and functionality is one of the difficulties of green synthesis.
- Understanding mechanisms: The biological and molecular factors underlying green synthesis techniques require further investigation.
- Scalability: Scaling these techniques for industrial usage presents hurdles, including cost, time, and consistency, even though small-scale production is possible.

Eco-friendly nanoparticle synthesis is a promising area that combines sustainability and nanotechnology. This field of study has a great deal of promise for developing novel answers to global problems as the world moves toward more sustainable and greener practices [16].

8.4 CONCLUSION

To sum up, environmentally friendly nanoparticle synthesis is a viable and sustainable substitute for conventional techniques, which frequently call for hazardous chemicals and energy-intensive procedures. Utilizing natural resources like microbes, plant extracts, and other biogenic resources, the green synthesis of nanoparticles not only lessens the impact on the environment but also creates new opportunities for applications in industries like electronics, environmental remediation, and medicine. In addition to reducing waste and energy use, this strategy supports the growing need for more sustainable, greener technologies. To completely understand the underlying mechanics of nanoparticle generation and optimize these techniques for large-scale production, more research is necessary. Ultimately, environmentally friendly synthesis has the potential to completely transform nanotechnology, making it safer and more widely available for a variety of uses while also promoting environmental preservation.

REFERENCES

1. Bihal, R., et al. (2023). Entomopathogenic fungi: An eco-friendly synthesis of sustainable nanoparticles and their nanopesticide properties. *Microorganisms*, *11*(6), 1617. https://doi.org/10.3390/microorganisms11061617
2. Bhardwaj, B., et al. (2020). Eco-friendly greener synthesis of nanoparticles. *Advanced Pharmaceutical Bulletin*, *10*(4), 566. https://doi.org/10.34172/apb.2020.066
3. Priya, N., et al. (2021). Green synthesis: An eco-friendly route for the synthesis of iron oxide nanoparticles. *Frontiers in Nanotechnology*, *3*, 655062. https://doi.org/10.3389/fnano.2021.655062
4. Alahdal, F. A., et al. (2022). Eco-friendly synthesis of zinc oxide nanoparticles as nanosensor, nanocatalyst and antioxidant agent using leaf extract of *P. austro-arabica*. *OpenNano*, *8*, 100067. https://doi.org/10.1016/j.onano.2022.100067
5. Al-Mohaimeed, A. M., Al-Onazi, W. A., & El-Tohamy, M. F. J. M. (2022). Multifunctional eco-friendly synthesis of ZnO nanoparticles in biomedical applications. *Molecules*, *27*(2), 579. https://doi.org/10.3390/molecules27020579
6. Santhosh, P. B., Genova, J., & Chamati, H. J. C. (2022). Green synthesis of gold nanoparticles: An eco-friendly approach. *Chemistry*, *4*(2), 345–369. https://doi.org/10.3390/chemistry4020023
7. Barzinjy, A. A., et al. (2020). Green and eco-friendly synthesis of nickel oxide nanoparticles and its photocatalytic activity for methyl orange degradation. *Journal of Materials Science: Materials in Electronics*, *31*(14), 11303–11316. https://doi.org/10.1007/s10854-020-03627-5
8. Asha, G., et al. (2022). Eco-friendly synthesis and characterization of cobalt oxide nanoparticles by *Sativum* species and its photo-catalytic activity. *Materials Today: Proceedings*, *48*, 486–493. https://doi.org/10.1016/j.matpr.2021.08.292
9. Gholami, M., et al. (2022). Eco-friendly synthesis of copper nanoparticles using *Mentha pulegium* leaf extract: Characterisation, antibacterial and cytotoxic activities. *Materials Technology*, *37*(10), 1523–1531. https://doi.org/10.1080/10667857.2021.2019705
10. Pradeep, M., et al. (2021). Uncovering the phytochemical basis and the mechanism of plant extract-mediated eco-friendly synthesis of silver nanoparticles using ultra-performance liquid chromatography coupled with a photodiode array and high-resolution mass spectrometry. *ACS Sustainable Chemistry & Engineering*, *10*(1), 562–571. https://doi.org/10.1021/acssuschemeng.1c06179
11. Afzal, S., et al. (2021). Eco-friendly synthesis of phytochemical-capped iron oxide nanoparticles as nano-priming agent for boosting seed germination in rice (*Oryza sativa* L.). *Environmental Science and Pollution Research*, *28*, 40275–40287. https://doi.org/10.1007/s11356-021-13718-1
12. Sebeia, N., et al. (2020). Eco-friendly synthesis of *Cynomorium coccineum* extract for controlled production of copper nanoparticles for sorption of methylene blue dye. *Arabian Journal of Chemistry*, *13*(2), 4263–4274. https://doi.org/10.1016/j.arabjc.2020.01.025
13. Maheshwaran, G., et al. (2021). Eco-friendly synthesis of lanthanum oxide nanoparticles by *Eucalyptus globulus* leaf extracts for effective biomedical applications. *Materials Letters*, *283*, 128799. https://doi.org/10.1016/j.matlet.2020.128799
14. Onwudiwe, D. C., et al. (2020). Eco-friendly synthesis, structural properties and morphology of cobalt hydroxide and cobalt oxide nanoparticles using extract of *Litchi chinensis*. *Nano-Structures & Nano-Objects*, *23*, 100470. https://doi.org/10.1016/j.nanoso.2020.100470
15. Nayantara, K. P., & Kaur, P. (2018). Biosynthesis of nanoparticles using eco-friendly factories and their role in plant pathogenicity: A review. *Biotechnology Research and Innovation*, *2*, 63–73. https://doi.org/10.1016/j.biori.2018.09.001
16. Ansari, S. M., et al. (2020). Eco-friendly synthesis, crystal chemistry, and magnetic properties of manganese-substituted $CoFe_2O_4$ nanoparticles. *ACS Omega*, *5*(31), 19315–19330. https://doi.org/10.1021/acsomega.0c01939

9 Nanocatalyst-Driven Sustainable Electrochemical Machining

Role of Gold Nanoparticle Size in Improving Dissolution Efficiency

Ritesh Kumar Upadhyay, Gopisetti Srinivasarao, Shyam S. Majhi, and Susanta S. Mahapatra

9.1 INTRODUCTION

ECM is increasingly recognized for its ability to provide precise, stress-free machining of conductive materials without tool wear. This nontraditional machining process operates through anodic dissolution, where the workpiece serves as the anode and the tool functions as the cathode, both submerged in an electrolyte that completes the circuit (Acharya et al., 1986). Upon application of a potential difference, metal ions from the anode dissolve into the electrolyte through electrochemical reactions. The efficiency of the process, especially the MRR, is regulated by factors such as current density, electrolyte concentration, mobility of ions, and gas evolution on the surfaces of the electrodes (Landolt et al., 2003). A fundamental difficulty in ECM is the deposition of passive oxide films on the anode surface, which impedes the process of dissolution (Pourgharibshahi and Lambert, 2016). These passivating layers decrease surface conductivity and retard anodic reactions by hindering OH^- ion access and charge transfer, particularly under conditions where oxygen evolution becomes important (Liu et al., 2017). Researchers have explored various electrolytes, including NaCl, KCl, $CaCl_2$, and $NaNO_3$, along with additives like H_2O_2 and reducing agents such as nano zero-valent iron (NZVI) and copper sulfate, to enhance dissolution and surface quality (Kumaar et al., 2022; Upadhyay et al., 2016, 2017, 2025b). Recent research has tended to aim towards the betterment of anodic dissolution of difficult-to-machine metals like 20MnCr5 hard-wear alloy steel used across the automotive and industrial sectors. Although earlier research has shown the positive contribution of AuNPs in ORR, which results in enhanced MRR and improved surface finishes, this research explores a new aspect: the effect of particle size on ECM performance (Upadhyay et al., 2025a).

AuNPs exhibit excellent catalytic activity towards the ORR in alkaline electrolytes like sodium nitrate ($NaNO_3$) and hence are promising agents to rupture passive oxide films during ECM (Corti et al., 2005). In the present work, two distinct size ranges of AuNPs, 28–40 and 150–180 nm, were incorporated into the $NaNO_3$ electrolyte to study their effect on the ECM of 20MnCr5 steel. Small-sized AuNPs are found to be more active since they possess a greater surface area-to-volume ratio and consequently more active sites to catalyze the ORR and therefore enhance the dissolution rate by passive oxide layer inhibition or disruption. The electrolyte, during ECM, carries out the transport of ions and is also involved in electrochemical reactions. Sodium nitrate dissociates into Na^+ and NO_3^- ions, while water dissociates into H^+ and OH^- ions (Mukherjee et al., 2007).

At the anode (workpiece), metal atoms oxidize and dissolve into the electrolyte, while at the cathode (tool), reduction reactions occur. The presence of AuNPs influences these reactions by catalyzing ORR, which can suppress the formation of passivating oxide layers on the workpiece surface. The electrochemical dissolution reactions are presented in Equations (9.1) and (9.2) (Liu et al., 2022).

$$NaCl \rightarrow Na^+ + Cl^- \tag{9.1}$$

and H_2O as

$$H_2O \rightarrow H^+ + OH^- \tag{9.2}$$

The alloy steel 20MnCr5 workpiece undergoes an oxidation reaction in which its atoms are ionized, as shown in Equations (9.3) and (9.4) (Peng et al., 2024).

$$Fe^0 \rightarrow Fe^{2+} + 2e^- \tag{9.3}$$

and

$$Fe^0 \rightarrow Fe^{3+} + 3e^- \tag{9.4}$$

DOI: 10.1201/9781003644439-9

At the anode, hydroxyl ions (OH^-) participate in the evolution of oxygen, as shown in Equation (9.5)

$$4OH^- \rightarrow O_2 \uparrow + 2H_2O + 4e^- \quad (9.5)$$

This oxygen evolution reaction becomes predominant when metal dissolution does not occur. In NaCl solution, oxygen evolution is relatively limited, whereas in $NaNO_3$ solution, it becomes the dominant process. At potentials > 0 V, active dissolution of Fe^{2+} occurs in NaCl, whereas at current density > 10 A/cm^2, transpassive dissolution of Fe^{3+} takes place. Potentials > 1.8 V and lower current density lead to the evolution of oxygen in $NaNO_3$ electrolyte (Lohrengel et al., 2016).

The following reactions occur at the cathode: sodium ion reduction (Equation 9.6), reaction of sodium with water (Equation 9.7), and hydrogen ion reduction to hydrogen gas (Equation 9.8).

Reactions at Cathode

$$Na^+ + e^- \rightarrow Na \quad (9.6)$$
$$Na + H_2O \rightarrow NaOH + H^+ \quad (9.7)$$
$$2H^+ + 2e^- \rightarrow H_2 \quad (9.8)$$

AuNPs enable this process by speeding up the ORR, thereby reducing the accumulation of oxygen on the anode and minimizing oxide film formation tendencies. This results in a more stable current discharge, increased MRR, and better surface topography. In addition, using AuNPs of smaller size is likely to generate an enhanced ECM performance than larger particles (Rutkowska et al., 2015). This is because smaller particles enhance the kinetics of ORR and improve electrolyte conductivity between the inter-electrode gap. Larger AuNPs, however, while still catalytically active, have lower kinetics and lower dispersion uniformity, leading to less efficient passivating layer disruption. With an analysis of these effects, research reveals not just the significance of AuNP-aided ECM but, more significantly, the critical role of nanoparticle size in determining machining performance (Shvpunov et al., 2015). The findings turn out to be pivotal in realizing particle–electrolyte interactions as well as extending research potential in particle-engineered electrolytes for improving the efficiency of ECM, particularly of difficult-to-cut materials like alloy steels. This study is an original addition to the literature in quantifying the impact of nanoparticle size on dissolution behavior in ECM, a concept previously not fully explored in the literature.

9.2 MATERIALS AND PROCEDURES

9.2.1 Materials

The AuNP synthesis in this work uses a chemical reduction process, where sodium borohydride ($NaBH_4$) is used as the main reducing agent and trisodium citrate ($Na_3C_6H_5O_7$) as the capping (stabilizing) agent (Strobl and Scherson, 2021). This pair is commonly used because it is easy to use and can generate small, evenly distributed AuNPs. $HAuCl_4$ (chloroauric acid) is used as the gold precursor. In this process, $NaBH_4$ donates electrons necessary to reduce Au^{3+} ions from the $HAuCl_4$ solution to neutral gold atoms (Au^0), which nucleate to create nanoparticles. The citrate ions, although not participating directly in the reduction, are important in preventing the particles from aggregating by stabilizing their surfaces via electrostatic repulsion (Tang et al., 2008).

9.2.2 Gold Nanoparticle Synthesis Procedures

The synthesis procedure is carried out under controlled low-temperature conditions to ensure a slow and uniform reduction process, which favors the formation of monodisperse nanoparticles. To prepare, an aqueous solution of 0.05 mol/L $NaBH_4$ is prepared in ice-cold water. Low temperatures are particularly important in this case, since $NaBH_4$ is thermally unstable and degrades quickly at room temperature (Zhang and Bo-Qing, 2013). Cooling retards the reaction, enabling more controlled nucleation and growth of nanoparticles. A 5.0-mL stock solution of $HAuCl_4$ (with about 0.2 g/100 mL gold) is prepared. This solution is then diluted 10 times to lower the concentration of gold ions to control the size of the particles during reduction. To the diluted gold precursor, 2 mL of 1% trisodium citrate ($Na_3C_6H_5O_7$) solution is added. Citrate adsorbs onto the nanoparticle surface, inhibiting particle coalescence and allowing a stable colloidal dispersion. The solution is kept in an ice bath to keep the reaction temperature low throughout. Another 1 mL of 1% citrate solution is added to the $HAuCl_4$ solution before reduction. The cold $HAuCl_4$ citrate mixture is then stirred constantly with a magnetic stirrer to keep the mixture uniform. After the system is well mixed, $NaBH_4$ solution is added slowly. In less than 5 minutes of the addition of $NaBH_4$, the solution exhibits a visible change in color from pale yellow to pink, marking the nucleation of tiny AuNPs. With continued stirring, the color steadily changes to wine red, which is characteristic of well-dispersed AuNPs of average size 28–180 nm. This color shift is due to the phenomenon of surface plasmon resonance (SPR), where the conduction electrons on the surface of a nanoparticle, in unison, oscillate upon exposure to light. An excess of $NaBH_4$ is employed to facilitate full reduction of all gold ions and to facilitate homogeneous nucleation and particle size regulation (Leon-Fernandez et al., 2023). The fast provision of electrons by excess borohydride enhances the reduction kinetics and reduces secondary nucleation events that may cause polydispersity. After stabilizing the characteristic red color, which signals the formation of AuNPs, a sample is taken for subsequent characterization and analysis. The detailed reaction scheme for AuNP formation is shown in Figure 9.1.

9.2.3 Characterization of Synthesized Nano Zero-Valent Iron Particles

AuNP synthesis was characterized by transmission electron microscopy (TEM) to determine the size and shape. The measurement was taken by a JEM-1400 model from JEOL/USA, Inc., which provides high-resolution images to visually analyze the morphology of the nanoparticles in detail. Figures 9.2 (a) and (b) are the TEM images of AuNPs with two different ranges of sizes: 28–40 and 150–180 nm, respectively. These images confirm the effective synthesis of size-controlled nanoparticles and help determine their structural characteristics, which are critical to studying their performance in electrochemical machining applications.

9.2.4 Oxygen Reduction Reaction Scheme

Oxygen reduction reaction is a multi-step, intricate electrochemical process whose efficacy relies on catalyst type, electrode material, and electrolyte composition. In the current study, the MRR during ECM of 20MnCr5 steel was investigated using two size ranges of AuNPs: 28–40 and 150–180 nm. The findings indicate that large AuNPs in a simple $NaNO_3$ electrolyte significantly improve the ORR pathway, leading to the enhanced MRR. This is attributed to the higher catalytic activity of large nanoparticles, which facilitates more efficient electron transfer during the

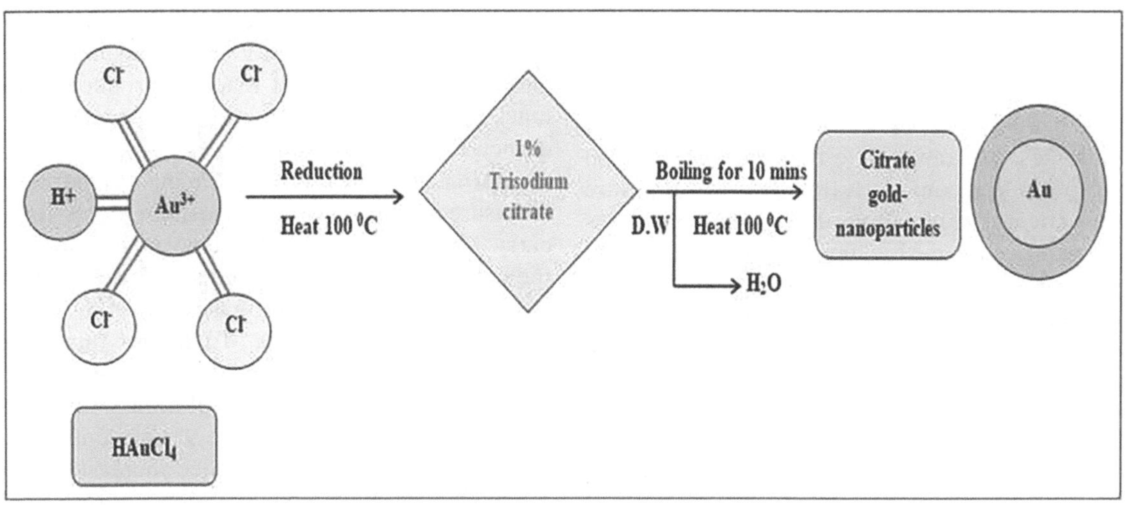

FIGURE 9.1 AuNP formation reaction scheme.

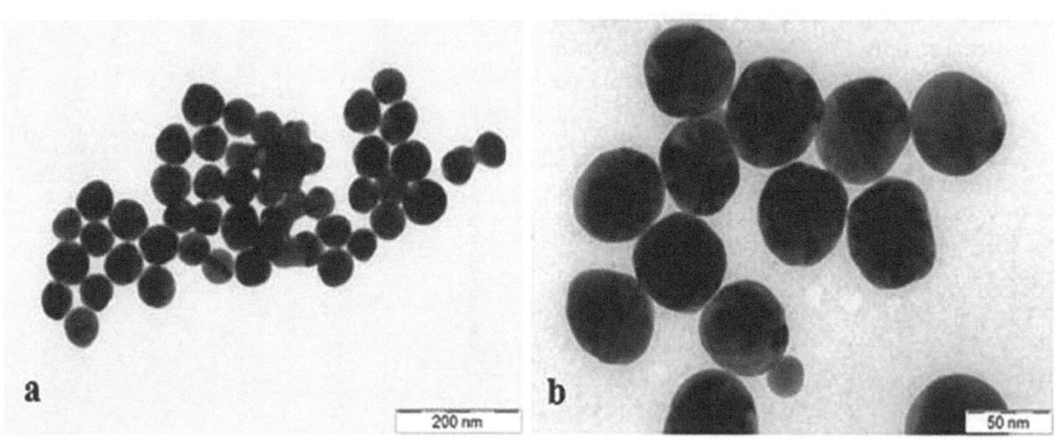

FIGURE 9.2 TEM images of gold nanoparticles with size ranges (a) 28–40 nm and (b) 150–180 nm.

process. The ORR can proceed via two principal pathways, as shown in Equations (9.9) to (9.10) (Guo et al., 2013):

$$4H^+ + O_2 + 4e^- \rightarrow 2H_2O \quad (9.9)$$
$$2H^+ + O_2 + 2e^- \rightarrow H_2O_2 \quad (9.10)$$

9.3 EXPERIMENTATION

9.3.1 Material Selection for ECM Experiments

The experimental setup for investigating the ECM characteristics was carefully designed to mimic controlled machining conditions. The ECM setup included a rectangular flow cell, tool feed system, and electrolyte circulation system to ensure a uniform and continuous supply of electrolyte ions through the inter-electrode gap. For the analysis, a rectangular sample of 20MnCr5 alloy steel with high wear resistance commonly used in automobile components was selected. The tool electrode (cathode) used here was brass with a cylindrical cross-section diameter of 12 mm, owing to its excellent electrical conductivity and corrosion resistance properties, making it a highly sought-after material to utilize as ECM cathodes. The electrolyte flow system was also designed to provide a consistent flow rate and pressure and thereby efficient elimination of reaction byproducts and also to maintain the desired temperature in the inter-electrode gap.

The design allowed for a stable and controlled condition for the ECM process where the material could be removed with high precision and surface finish quality. Two types of electrolytes were employed: a standard aqueous $NaNO_3$ solution and a $NaNO_3$ solution mixed with AuNPs to enhance the electrochemical dissolution process. The concentration of AuNPs was carefully controlled at 0.25 g per 100 cc of electrolyte for each trial. AuNPs are said to enhance the kinetics of the oxygen reduction reaction and reduce the formation of passive oxide films, thereby enhancing MRR and surface finish. The tool was constructed with a 4-mm bore inner hole to enable direct supply of electrolyte into the thin inter-electrode gap (0.012 mm). This provides uniform distribution of the electrolyte; avoids stagnation areas; and enhances flushing of reaction products, heat, and gases evolved during anodic dissolution. This kind of thin inter-electrode gap is essential for supporting the high current density required for effective machining. The complete setup and spatial arrangements of the components are schematically shown in Figure 9.3, which facilitates visual understanding of the ECM mechanism employed in this study.

9.3.2 Electrical Conductance and pH of Electrolyte

In ECM, the electrolyte serves as a critical medium that facilitates ion transfer between the tool and the workpiece, enabling controlled material removal through anodic dissolution. To achieve optimal performance, the electrolyte should possess high electrical conductivity to facilitate efficient current transmission across the narrow gap between the electrodes (Kumar and Pabla, 2021). Additionally, low viscosity supports smooth circulation and effective flushing of reaction products, while resistance to passive film formation is essential to avoid hindrance to the dissolution process. A key parameter used to evaluate electrolyte performance is its conductance (measured in mS), which directly reflects its ability to carry electric current. This is primarily influenced by the concentration of dissolved salts, such as $NaNO_3$, and the presence of dissolved gases that may participate in electrochemical reactions. To assess this relationship, an experimental study was conducted where the electrical conductivity of $NaNO_3$ solutions was measured at various concentrations. Figure 9.4 presents the observed trend, showing that conductance increases sharply with increasing $NaNO_3$ concentration, particularly up to 20 g/L. This steep rise indicates a significant improvement in ion availability and mobility within this range. However, beyond 20 g/L, the increase in conductivity becomes more gradual, suggesting saturation effects

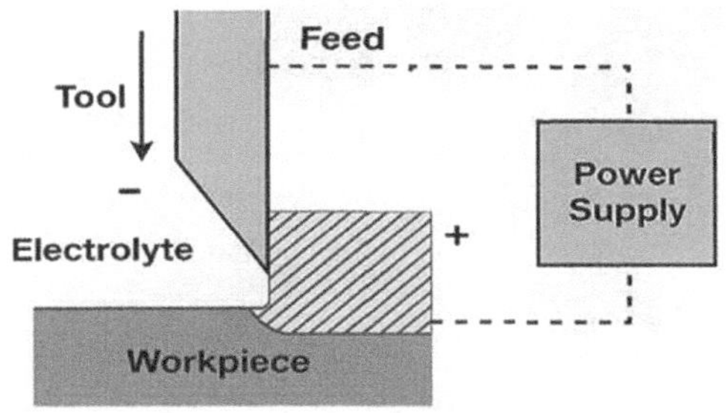

FIGURE 9.3 Experimental setup.

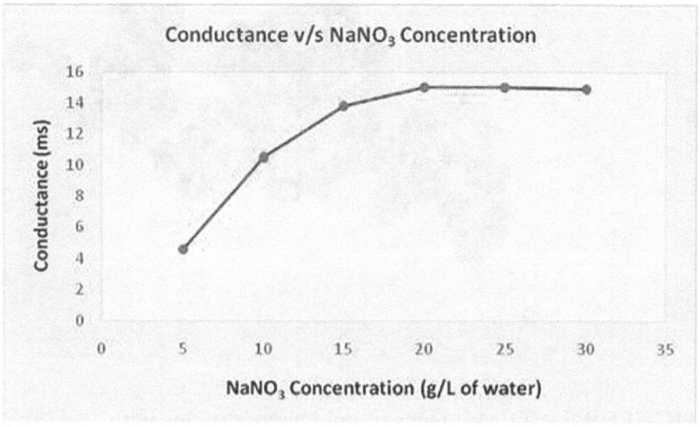

FIGURE 9.4 Conductance vs $NaNO_3$ concentration.

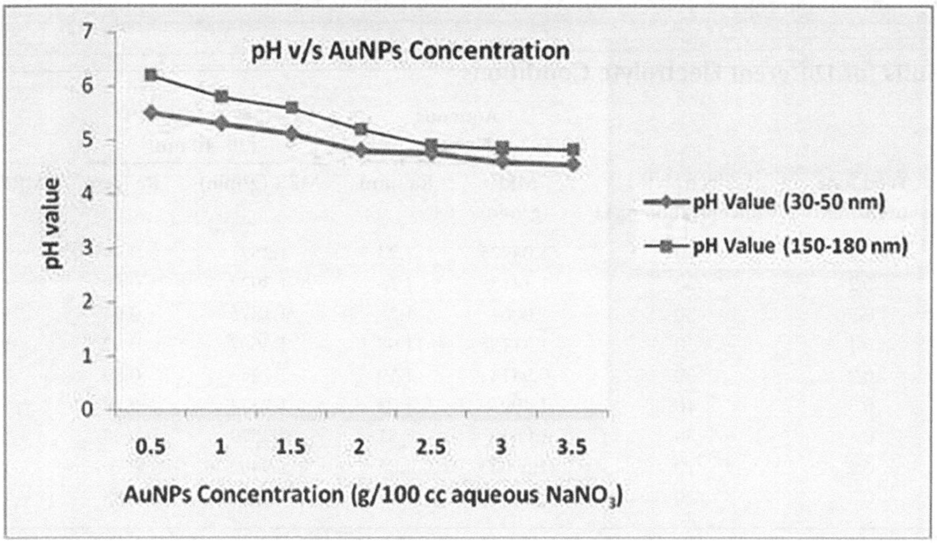

FIGURE 9.5 pH vs AuNP concentration (g/100 mL aqueous $NaNO_3$).

or reduced ion mobility due to increased ionic interactions. Therefore, 20 g/L was identified as the optimal concentration for further experimentation, offering a balance between conductivity and solution manageability. Furthermore, the study explored the influence of AuNPs added to the $NaNO_3$ solution at a concentration of 0.25 g per 100 mL. As illustrated in Figure 9.5, this composition resulted in a favorable pH value, which supports efficient anodic dissolution. However, it was observed that the reduction potential introduced by the AuNPs had a more pronounced effect on enhancing the dissolution rate than pH adjustment alone. This highlights the role of the electrolyte's reducing nature, provided by the electrocatalytic behavior of AuNPs, in accelerating the electrochemical reactions and improving overall machining efficiency.

9.3.3 Process Parameters

In ECM, the efficiency and accuracy of machining are greatly affected by key process parameters, including voltage, tool feed rate, and electrolyte concentration. These factors directly affect the anodic dissolution rate and, consequently, the MRR (Prakash et al., 2022; Weber et al., 2015). In the present study, various combinations of these parameters were investigated to evaluate their effect on the MRR of 20MnCr5 alloy using an AuNP-enhanced $NaNO_3$ electrolyte. The addition of AuNPs improves electrolyte conductivity and promotes the oxygen reduction reaction, reducing passive film formation (Shao et al., 2016). This systematic approach aids in optimizing ECM conditions for enhanced MRR and improved machining performance of alloy steel. Table 9.1 presents the machining conditions for the analysis.

TABLE 9.1
Process Parameters for Analysis

Parameters	Range/Specifications
Applied voltage	12V–20 V
Tool feed rate	0.1–0.3 mm/min
Electrolyte concentration	10–30 g/L

9.3.4 Quantification of MRR and Surface Topography

In this study, electrochemical machining was conducted using nine distinct parameter combinations based on the Taguchi L9 orthogonal array, a statistical method for experimental design aimed at minimizing variability. The selection of the L9 orthogonal array was based on our study's design, which involved three control factors, each at three levels. Hence, nine distinct parameter combinations were tested to assess their effects on the MRR and surface roughness of 20MnCr5 alloy steel. The weights of the 20MnCr5 specimens before and after machining were measured on a precision weighing machine. For each run, the MRR was calculated by finding the weight difference over machining time per Equation (9.11). This method allowed for systematic analysis of the effect of each parameter on the performance characteristics of ECM (Sekar and Marappan, 2008). While the primary focus was on understanding the effects of individual parameters on the ECM process, a comprehensive optimization of parameters was not within the scope of this study. Future research will aim to build upon these findings to achieve full parameter optimization.

$$MRR = \frac{\text{Initial weight} - \text{Final weight}}{\text{Total time}} \tag{9.11}$$

TABLE 9.2
MRR and Ra Results for Different Electrolyte Conditions

				Aqueous $NaNO_3$		$NaNO_3$ + AuNPs (28–40 nm)		$NaNO_3$ + AuNPs (150–180 nm)	
SN	Voltage (V)	Feed Rate (mm/min)	Electrolyte Concentration (g/L)	MRR (g/min)	Ra (μm)	MRR (g/min)	Ra (μm)	MRR (g/min)	Ra (μm)
1	12	0.1	10	1.04975	1.23	1.2597	0.94	1.1547	1.05
2	12	0.2	20	1.14155	1.16	1.3927	0.89	1.2499	0.99
3	12	0.3	30	1.0064	1.28	1.1875	0.94	1.0969	1.02
4	16	0.1	20	1.31155	1.08	1.5607	0.82	1.4427	0.92
5	16	0.2	30	0.9418	1.19	1.2243	0.89	1.0501	1.01
6	16	0.3	10	1.4093	1.03	1.7334	0.78	1.5149	0.88
7	20	0.1	30	1.0353	1.31	1.2579	0.98	1.1285	1.11
8	20	0.2	10	1.59035	1.05	1.9402	0.79	1.7509	0.89
9	20	0.3	20	1.6235	1.11	2.005	0.83	1.7534	0.94

Table 9.2 presents the MRR results obtained under the planned machining conditions using the selected electrolytes.

9.4 RESULTS AND DISCUSSION

The purpose of this study was to investigate the influence of incorporating AuNPs in an aqueous $NaNO_3$ electrolyte while conducting ECM tests and determine their impact on MRR and surface roughness. Particular focus was on how particle size affects the machining performance. AuNPs of two sizes, 28–40 and 150–180 nm, were compared to a baseline aqueous $NaNO_3$ electrolyte. The results from experiments at different operating conditions, such as voltage, feed rate, and concentration of electrolyte, are discussed, and the subsequent discussion clarifies the observed trends and implications.

9.4.1 Material Removal Rate

Under all the test conditions, it was noted that the introduction of AuNPs made a notable improvement in the MRR over the reference aqueous $NaNO_3$ electrolyte. For example, under the test conditions of 20 V, 0.3 mm/min feed rate, and 20 g/L electrolyte concentration, the MRR went up from 1.6235 g/min using pure $NaNO_3$ to 2.005 g/min when 28–40 nm AuNPs were added to the electrolyte. This is an impressive enhancement of about 23.4%. When even bigger AuNPs of 150–180 nm were employed using the same conditions, the MRR were 1.7534 g/min, which, while enhanced, exhibited a slightly lower increase of about 14.3% from the baseline electrolyte. This improvement trend was evident in all the experimental conditions. The introduction of the smaller-sized nanoparticles (28–40 nm) always led to the greatest increase in MRR, while the larger nanoparticles (150–180 nm) also enhanced the MRR but less. The main cause of the improvement found in MRR is the better electrical conductivity and enhanced current distribution caused by the introduction of AuNPs. These particles with their superior conductivity increase ionic mobility in the electrolyte and result in more rapid anodic dissolution of the workpiece material and thus greater material removal rates. However, when the particle size was increased to the range of 150–180 nm, the extent of MRR improvement was slightly reduced. This decline can be explained by several factors. First, larger nanoparticles have a reduced surface area-to-volume ratio, offering less surface area to support the conductivity improvements. Second, the tendency of bigger nanoparticles to agglomerate leads to uneven distribution within the electrolyte, causing localized variations in current density. Last, larger particles may partially obstruct the smooth flow of the electrolyte between the tool and the workpiece, adversely impacting the efficiency of the ECM process. This trend is presented in Figure 9.6, which indicates that the nanoparticle size plays a crucial role in determining the level of improvement in MRR during ECM using nanoparticle-enriched electrolytes.

9.4.2 Surface Roughness

Surface roughness (Ra) is a crucial parameter that directly indicates the quality of machining in electrochemical processes. In this study, it was consistently observed across all experimental conditions that the addition of AuNPs to the aqueous $NaNO_3$ electrolyte significantly reduced the Ra values, leading to improved surface finishes. Under the condition of 20 V, 0.3 mm/min feed rate, and 20 g/L electrolyte concentration, the Ra value for pure $NaNO_3$ electrolyte was measured at 1.11 μm. When 28–40 nm AuNPs were added, the roughness decreased markedly to 0.83 μm, demonstrating a notable 24% improvement. With the inclusion of larger AuNPs (150–180 nm), the Ra value was 0.94 μm, reflecting an improvement of approximately 15% compared to the baseline. As depicted in Figure 9.7, this consistent

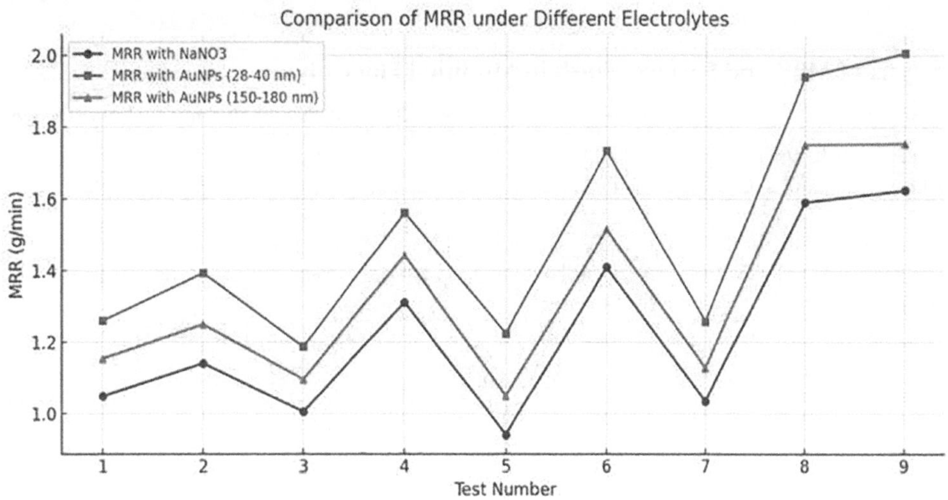

FIGURE 9.6 Material removal rate variations under different test conditions.

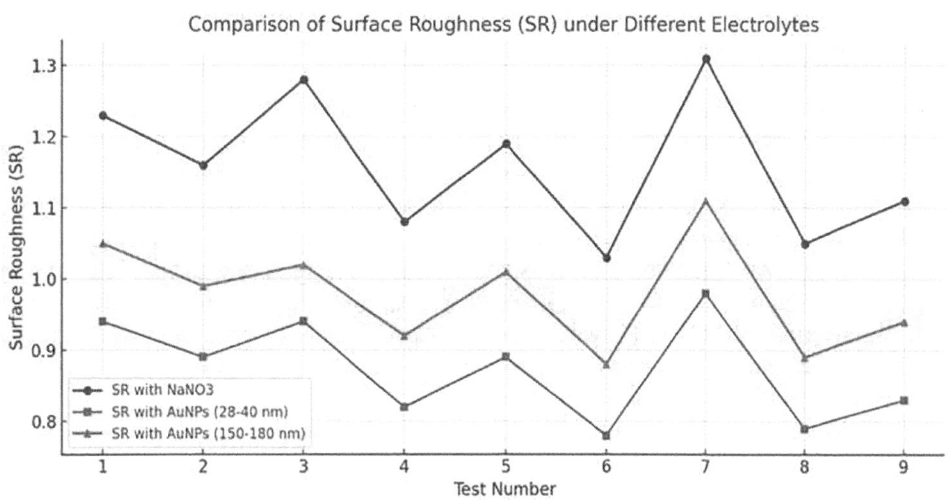

FIGURE 9.7 Surface roughness variations under different test conditions.

trend was observed across different voltages and feed rates, with smaller nanoparticles (28–40 nm) delivering superior surface finishes in all cases. The improved surface quality can be attributed to several factors. First, the presence of smaller nanoparticles contributes to a stable anodic dissolution process by creating a more uniform and stable electric field, which helps prevent localized micro-arcing events responsible for surface defects. Second, nanoparticles facilitate enhanced flushing of machining debris from the inter-electrode gap, reducing the chances of particle deposition and secondary reactions that can deteriorate the surface quality. Last, the improved conductivity of the electrolyte, aided by the nanoparticles, ensures smoother ion movement, promoting uniform material removal and contributing to finer surface finishes. When larger nanoparticles (150–180 nm) were used, although the surface quality still improved, the level of improvement was slightly reduced compared to smaller particles. This is primarily due to slightly disturbed flow patterns within the electrolyte caused by the larger particles and less effective stabilization of the electric field, owing to their lower surface-to-volume ratios.

Overall, the study clearly demonstrates that smaller nanoparticles are more effective in achieving better surface quality during electrochemical machining operations. Table 9.3 presents a comparative analysis of different electrolyte conditions.

9.4.3 Influence of Process Parameters

The impact of critical process parameters, namely voltage, feed rate, and electrolyte concentration on ECM performance, was systematically analyzed in conjunction with the addition of AuNPs to the electrolyte. Voltage plays a critical role by serving as the primary driving force for anodic

TABLE 9.3
Improvement Analysis of MRR and Surface Finish in Modified Electrolytes

SN	Description	Aqueous $NaNO_3$	$NaNO_3$ + AuNPs (28–40 nm)	$NaNO_3$ + AuNPs (150–180 nm)
1	Maximum achieved MRR (g/min)	1.6235	2.005	1.7534
2	Minimum achieved Ra (μm)	1.03	0.78	0.88
3	Test conditions	20 V	20 V	20 V
	For maximum MRR	0.3 mm/min	0.3 mm/min	0.3 mm/min
		20 g/L	20 g/L	20 g/L
4	For minimum Ra	16 V	16 V	16 V
		0.3 mm/min	0.3 mm/min	0.3 mm/min
		10 g/L	10 g/L	10 g/L
5	% improvement over reference electrolyte	Reference	23.4% in MRR, 24% in Ra	14.3% in MRR, 15% in Ra

dissolution. As the applied voltage increases, a corresponding rise in the MRR is observed across all test cases. This is because higher voltages enhance the electrochemical reactions at the workpiece surface. However, with pure $NaNO_3$ electrolyte, an undesirable side effect was noted: surface roughness increased at higher voltages as a result of increased micro-arcing and localized over-dissolution. Interestingly, the addition of AuNPs reduced this effect by stabilizing the electric field and facilitating uniform dissolution, thereby maintaining lower Ra values even at higher voltages. Feed rate also had a significant impact. Increased feed rates caused a reduction in MRR due to the decreased residence time between the workpiece and the tool, restricting the range of electrochemical reactions. However, the improved conductivity and enhanced ion transport provided by AuNPs reversed this effect, keeping the electrochemical dissolution efficient even at increased feed rates. Concentration of electrolyte was another critical aspect. An increase in the concentration of $NaNO_3$ enhanced the solution conductivity, thus enhancing MRR and surface finish. Incorporation of AuNPs further magnified these advantages, particularly at 20 and 30 g/L concentrations, where more significant enhancements were noted. Thus, clearly, a fine balance of the process parameters, greater but controlled voltage, middle-range feed rates, suitable concentration of the electrolyte, and application of thoroughly dispersed smaller-sized AuNPs all join together to create the best performance of ECM for both the material removing rate and the quality of surface.

9.4.4 Microstructure of Machined Surfaces

The SEM micrographs in Figure 9.8 reveal the significant role of AuNPs in improving the surface finish during ECM of 20MnCr5 steel. When machined with pure aqueous $NaNO_3$ (Figure 9.8a), the surface is covered with a dense, solid oxide layer formed due to the strong oxidizing nature of $NaNO_3$. This oxidation leads to erosion and surface roughening. However, when small-sized AuNPs (28–50 nm) were introduced into the electrolyte (Figure 9.8b), the surface morphology improved markedly. The microstructure appears less precipitated, and carbide particles (visible as blackish regions) are more distinctly distributed within the matrix, indicating minimal surface damage. The enhanced surface finish can be attributed to the electrocatalytic properties of smaller AuNPs, which actively inhibit the oxidation of Fe^{2+} to Fe^{3+}. This suppression prevents the formation of excessive oxide byproducts, thereby reducing the roughness and improving the quality of the machined surface. In comparison, larger AuNPs (150–180 nm) also contributed to surface improvement (Figure 9.8c), but to a lesser extent, due to reduced catalytic activity. Thus, smaller-sized AuNPs are more effective in achieving better surface finishes by promoting stable anodic dissolution and minimizing secondary oxidation reactions during ECM.

9.5 CONCLUSION

This research concludes that the introduction of AuNPs into an aqueous $NaNO_3$ electrolyte in ECM has a very positive effect. The inclusion of AuNPs increases the MRR and enhances the surface finish quality over the use of a normal electrolyte in isolation. The research indicates that the nanoparticle size has an important role to play in the level of improvement. Smaller nanoparticles, sized between 28–40 nm, provide greater performance compared to larger-sized nanoparticles of size 150–180 nm. Smaller-sized nanoparticles are more efficient because they provide a larger surface area-to-volume ratio, enhancing the electrolyte conductivity. Improved conductivity ensures that anodic dissolution occurs at a faster rate, resulting in increased MRR. Further, smaller nanoparticles disperse more homogeneously inside the electrolyte, having a constant and homogeneous electric field between the workpiece and the tool. This homogeneity contributes to getting a smoother material removal process, thereby producing surfaces with lower roughness values. On the contrary, larger particles, although also useful, will agglomerate more readily. Agglomeration leads to non-uniform conductivity and impairs the smooth flow of the electrolyte, reducing material removal efficiency and

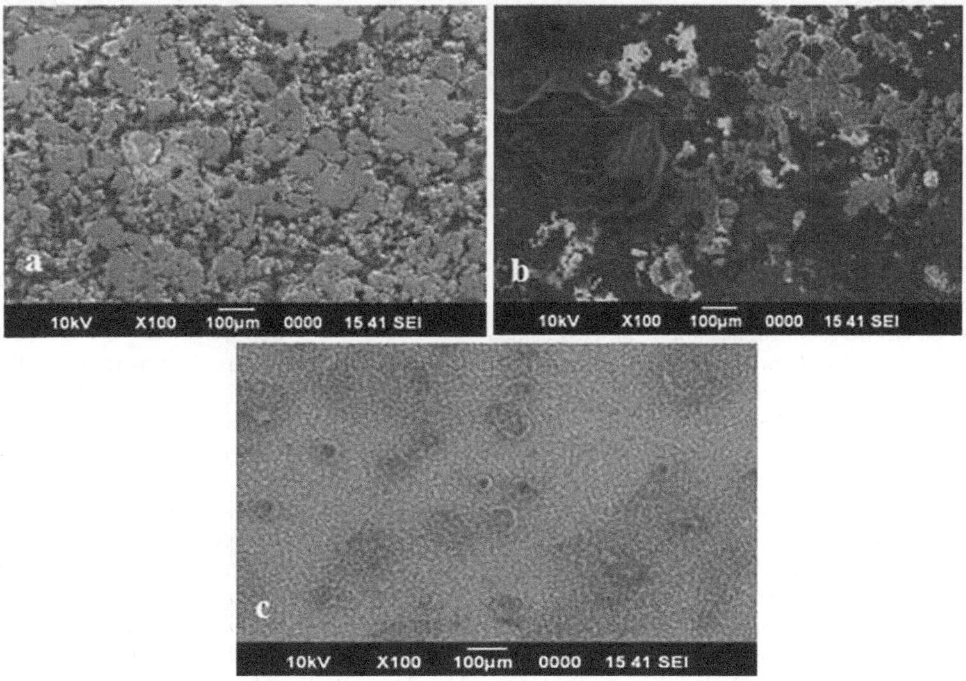

FIGURE 9.8 SEM micrographs of surfaces machined using: (a) aqueous $NaNO_3$, (b) aqueous $NaNO_3$ with AuNPs (28–50 nm), and (c) aqueous $NaNO_3$ with AuNPs (150–180 nm).

improvement in surface quality by a minute amount. They can also get in the way of the inter-electrode gap, giving rise to local irregularities while machining.

These findings imply that precise selection of nanoparticle size is essential in optimizing ECM processes. The research shows that nanoparticle-assisted ECM, especially with smaller AuNP sizes, has great potential for applications where high-precision machining with outstanding surface finishes are required, such as aerospace, biomedical, and micro-tool manufacturing industries. In general, this technique might result in more efficient and cost-effective machining methods in the future.

REFERENCES

Acharya, B. G., Jain, V. K., Batra, J. L. (1986). Multi-objective optimization of the ECM process. *Precision Engineering*, **8(2)**, 88–96. https://doi.org/10.1016/0141-6359(86)90091-7

Corti, C. W., Holliday, R. J., Thompson, D. T. (2005). Commercial aspects of gold catalysis. *Applied Catalysis A: General*, **291(1–2)**, 253–261. https://doi.org/10.1016/j.apcata.2005.01.051

Guo, S., Zhang, S., Sun, S. (2013). Tuning nanoparticle catalysis for the oxygen reduction reaction. *Angewandte Chemie International Edition*, **52(33)**, 8526–8544. https://doi.org/10.1002/anie.201207186

Kumar, A., Pabla, B. S. (2021). Review on optimized process parameters of electrochemical machining and its variants. *Materials Today: Proceedings*, **46**, 10854–10860. https://doi.org/10.1016/j.matpr.2021.01.807

Kumaar, J. R. K. V., Thanigaivelan, R., Soundarrajan, M. (2022). Effect of different electrolytes on electrochemical micro-machining of SS 316L. *Chemical Industry and Chemical Engineering Quarterly*, **28(4)**, 329–337. https://doi.org/10.2298/CICEQ211204007V

Landolt, D., Chauvy, P. F., Zinger, O. J. E. A. (2003). Electrochemical micromachining, polishing and surface structuring of metals: Fundamental aspects and new developments. *Electrochimica Acta*, **48(20–22)**, 3185–3201. http://doi.org/10.1016/S0013-4686(03)00368-2

Leon-Fernandez, L. F., Caballero-Ortiz, A., Martinez-Mora, O., Fransaer, J., Dominguez-Benetton, X. (2023). Mechanism and kinetics of gold recovery and Au nanoparticle synthesis by Gas-Diffusion Electrocrystallization (GDEx). *Electrochimica Acta*, **460**, 142592. https://doi.org/10.1016/j.electacta.2023.142592

Liu, W., Ao, S., Li, Y., Liu, Z., Zhang, H., Manladan, S. M., Wang, Z. (2017). Effect of anodic behavior on electrochemical machining of TB6 titanium alloy. *Electrochimica Acta*, **233**, 190–200. http://doi.org/10.1016/j.electacta.2017.03.025

Liu, Y., Huang, S., Peng, S., Zhang, H., Wang, L., Wang, X. (2022). Novel Au nanoparticles-inlaid titanium paper for PEM water electrolysis with enhanced interfacial electrical conductivity. *International Journal of Minerals, Metallurgy and Materials*, **29(5)**, 1090–1098. https://doi.org/10.1007/s12613-022-2452-1

Lohrengel, M. M., Rataj, K. P., Münninghoff, T. J. E. A. (2016). Electrochemical machining—mechanisms of anodic dissolution. *Electrochimica Acta*, **201**, 348–353. http://doi.org/10.1016/j.electacta.2015.12.219

Mukherjee, S. K., Kumar, S., Srivastava, P. K. (2007). Effect of electrolyte on the current-carrying process in electrochemical machining. *Proceedings of the Institution of Mechanical Engineers, Part C: Journal of Mechanical Engineering Science*, **221(11)**, 1415–1419. https://doi.org/10.1243/09544062JMES355

Peng, Y., Lin, Y., Xia, R., Dai, Z., Zhang, W., Liu, W. (2024). Electrochemical investigation of chloride ion-induced breakdown of passive film on P110 casing steel surface in simulated pore solution: Behavior and critical value determination. *Metals*, **14(1)**, 93. https://doi.org/10.3390/met14010093

Pourgharibshahi, M., Lambert, P. (2016). The role of indium in the activation of aluminum alloy galvanic anodes. *Materials and Corrosion*, **67(8)**, 857–866. https://doi.org/10.1002/maco.201508685

Prakash, S. O., Jeyakumar, M., Gandhi, B. S. (2022). Parametric optimization on electro chemical machining process using PSO algorithm. *Materials Today: Proceedings*, **62**, 2332–2338. http://doi.org/10.1016/j.matpr.2022.04.141

Rutkowska, I. A., Zoladek, S., Kulesza, P. J. (2015). Polyoxometallate-assisted integration of nanostructures of Au and ZrO2 to form supports for electrocatalytic PtRu nanoparticles: Enhancement of their activity toward oxidation of ethanol. *Electrochimica Acta*, **162**, 215–223. https://doi.org/10.1016/j.electacta.2014.12.098

Sekar, K. R., Marappan, R. (2008). Experimental investigations into the influencing parameters of electrochemical machining of AISI 202. *Journal of Advanced Manufacturing Systems*, **7(2)**, 337–343. http://doi.org/10.1142/S0219686708001486

Shao, M., Chang, Q., Dodelet, J. P., Chenitz, R. (2016). Recent advances in electrocatalysts for oxygen reduction reaction. *Chemical Reviews*, **116(6)**, 3594–3657. https://doi.org/10.1021/acs.chemrev.5b00462

Shypunov, I., Kongi, N., Kozlova, J., Matisen, L., Ritslaid, P., Sammelselg, V., Tammeveski, K. (2015). Enhanced oxygen reduction reaction activity with electrodeposited Ag on manganese oxide–graphene supported electrocatalyst. *Electrocatalysis*, **6**, 465–471. https://doi.org/10.1007/s12678-015-0266-x

Strobl, J. R., Scherson, D. (2021). On the mechanism of the oxygen reduction reaction on Au (poly) in aqueous alkaline electrolytes: A critical reassessment. *The Journal of Physical Chemistry C*, **125(25)**, 13862–13870. https://doi.org/10.1021/acs.jpcc.1c02683

Tang, W., Lin, H., Kleiman-Shwarsctein, A., Stucky, G. D., McFarland, E. W. (2008). Size-dependent activity of gold nanoparticles for oxygen electroreduction in alkaline electrolyte. *The Journal of Physical Chemistry C*, **112(28)**, 10515–10519. http://doi.org/10.1021/jp710929n

Upadhyay, R. K., Chakraborty, A. K., Majhi, S. S., Singh, A. C., Kumar, B., Yadav, N. (2025a). Influences of redox electrolyte containing AuNPs on microscopic surface structure, material removal and surface roughness of 20MnCr5 steel alloy during electrochemical machining. *Surface Science and Technology*, **3(1)**, 11. https://doi.org/10.1007/s44251-025-00074-9

Upadhyay, R. K., Kumar, A., Srivastava, P. K. (2016). High rate anodic dissolution of stainless steel 316 (SS316) using nano zero valent iron as reducing agent. *Journal of Applied Science and Engineering*, **19(1)**, 47–52. http://doi.org/10.6180/jase.2016.19.1.06

Upadhyay, R. K., Kumar, A., Srivastava, P. K. (2017). Experimental investigations of catalytic effect of Cu2+ during anodic dissolution of iron in NaCl electrolyte. *Proceedings of the Institution of Mechanical Engineers, Part B: Journal of Engineering Manufacture*, **231(13),** 2408–2415. https://doi.org/10.1177/0954405416629865

Upadhyay, R. K., Majhi, S. S., Mahapatra, S. S., Yadav, N., Chakraborty, A. K., Singh, A. C., Kumar, B. (2025b). A comparative study on material removal rate and surface roughness during electrochemical machining of 100Cr6 steel in oxidizing and reducing electrolytic environments. *Proceedings of the Institution of Mechanical Engineers, Part E: Journal of Process Mechanical Engineering*, 09544089251316751. https://doi.org/10.1177/09544089251316751

Weber, O., Weinmann, M., Natter, H., Bähre, D. (2015). Electrochemical dissolution of cast iron in NaNO3 electrolyte. *Journal of Applied Electrochemistry*, **45**, 591–609. http://doi.org/10.1007/s10800-015-0809-0

Zhang, G. R., Bo-Qing, X. U. (2013). Nano-size effect of Au catalyst for electrochemical reduction of oxygen in alkaline electrolyte. Chinese *Journal of Catalysis*, **34(5)**, 942–948. https://doi.org/10.1016/S1872-2067(12)60546-4

10 Characterization and Applications of Green Nanoparticles

Liny. P, Jagadish H Patil, Ismail Shareef M, and Ansh Varkar

10.1 INTRODUCTION

Green chemistry, when coupled with nanotechnology—an interdisciplinary field dealing with materials at the nanoscale (1–100 nm)—enables transformative innovations in material science, medicine, energy, and manufacturing (Ahmed *et al.*, 2022). Green nanomaterials, synthesized via eco-friendly and sustainable processes, are revolutionizing fields ranging from medicine to environmental remediation. These nanoparticles are produced using plant extracts, microorganisms, and other natural agents, eliminating the need for toxic chemicals and harsh conditions (Baharlouei *et al.*, 2022). Their biocompatibility, functional versatility, and reduced toxicity make them ideal candidates for a range of biomedical applications, from targeted drug delivery to cancer therapy and wound healing. To harness their full potential, precise characterization is essential to understand their structural, morphological, and functional properties (Farhan, 2023).

In this chapter, we will focus on the characterization techniques used to assess nanoparticles synthesized using green chemistry principles. The focus is also on the methods that preserve the principles of green chemistry while ensuring accuracy and reproducibility. This chapter explores characterization techniques of green nanoparticles and the diverse biomedical applications of green nanoparticles, focusing on their synthesis, mechanisms of action, and real-world uses. It also explores the environmental applications of green chemistry principles in nanotechnology, highlighting key roles in pollution remediation, water purification, air quality control, and sustainable energy.

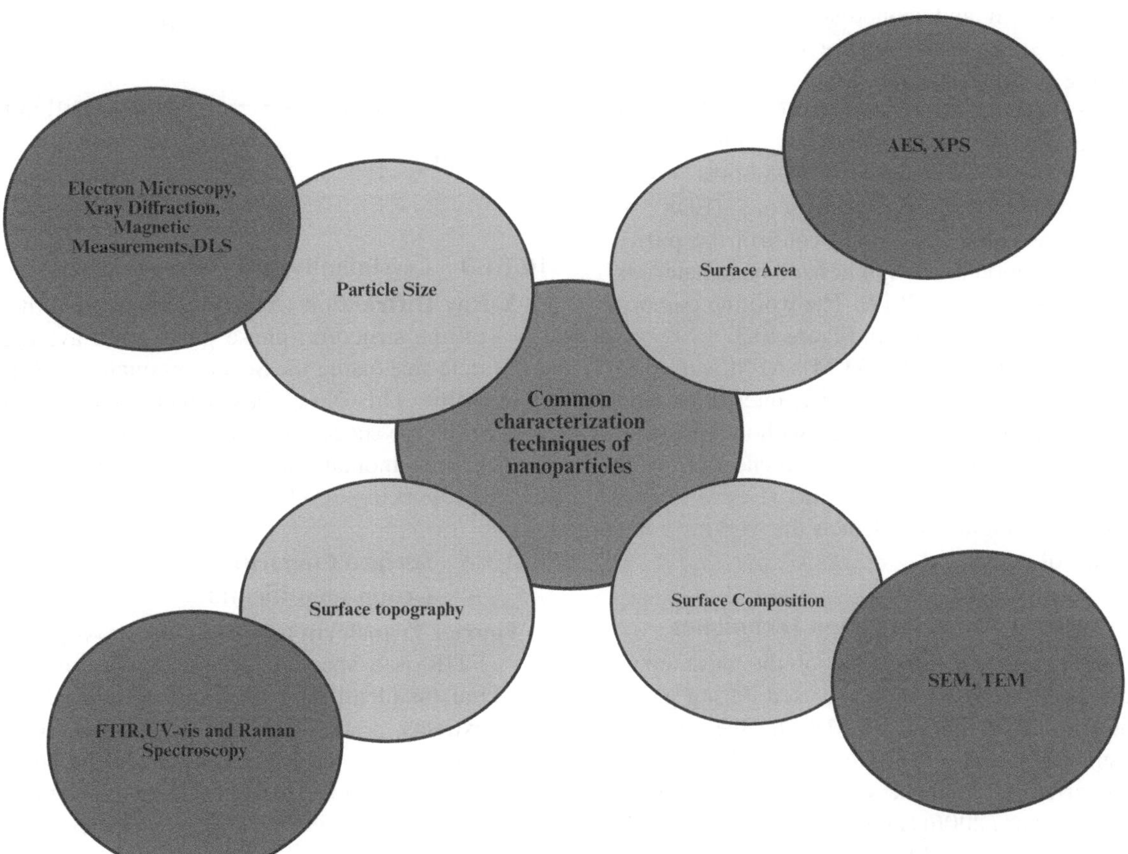

FIGURE 10.1 Common characterization techniques for nanoparticles.

DOI: 10.1201/9781003644439-10

10.1.1 Advanced Characterization Techniques for Green Nanomaterials

The rise of green chemistry has led to the development of green nanomaterials, which indicates nanostructures synthesized using natural extracts, microorganisms, or biodegradable materials in place of toxic chemicals. While their sustainable synthesis marks a critical shift, it is the characterization of these materials that defines their functionality, safety, and performance (Fatimah *et al.*, 2022). Characterization is key to determining properties like size, morphology, surface chemistry, crystallinity, and stability—factors that directly influence their behavior in biological, environmental, or industrial applications (Halagali *et al.*, 2025). The precise characterization of green nanomaterials is fundamental to understanding their structure–function relationships. Traditional characterization methods are adapted and complemented with novel approaches to ensure sustainability and minimize waste.

10.1.1.1 Morphological and Structural Characterization

Transmission Electron Microscopy (TEM): TEM provides high-resolution imaging to determine the shape, size, and crystallinity of green nanomaterials at the atomic or molecular scale. This is particularly useful in verifying whether green synthesis yields uniform and monodisperse nanoparticles (Halagali *et al.*, 2025). Figure 10.2 indicates the working components of TEM.

Scanning Electron Microscopy (SEM): SEM offers surface imaging of nanoparticles with detailed topographical information. Combined with energy-dispersive x-ray spectroscopy (EDS), it enables elemental analysis to confirm the purity and composition of the synthesized nanomaterials (Leyva-Porras *et al.*, 2019). The working components of SEM are shown in Figure 10.3.

Atomic Force Microscopy (AFM): AFM is used to assess surface roughness, particle interaction, and three-dimensional morphology without requiring conductive coatings, making it a greener option for soft or bio-based nanomaterials (Leyva-Porras *et al.*, 2019). Figure 10.4 depicts the working components of AFM.

10.1.1.2 Optical Characterization Techniques

UV-Visible Spectroscopy: One of the most common spectroscopic techniques used during green nanoparticle synthesis. It detects the surface plasmon resonance (SPR) band, which indicates nanoparticle formation, size, and concentration (Liqiang *et al.*, 2006) (Figure 10.5).

Photoluminescence (PL) Spectroscopy: PL is used to understand electron-hole recombination processes in semiconducting nanomaterials and detect defects, making it valuable for photocatalytic applications (Liqiang *et al.*, 2019).

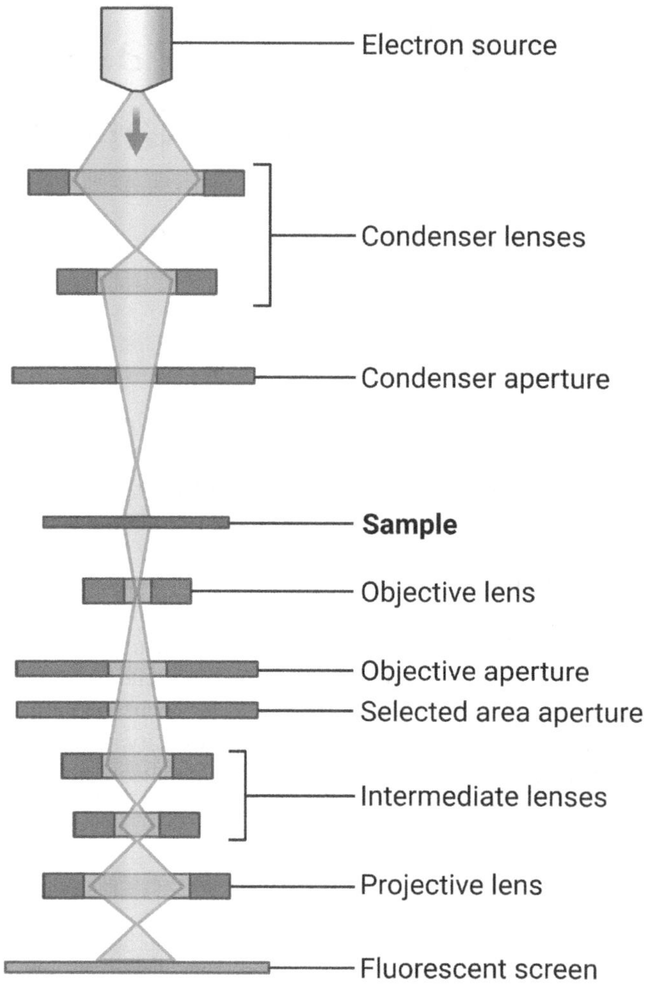

FIGURE 10.2 Transmission electron microscopy.

10.1.1.3 Crystallinity and Phase Analysis

X-Ray Diffraction (XRD): XRD reveals the crystalline structure, phase purity, and average particle size (using the Scherrer equation) (Raj *et al.*, 2016). This X-ray–based technique helps determine the success of synthesis and the stability of green nanomaterials over time. Figure 10.7 shows the working of XRD.

10.1.1.4 Surface Chemistry and Functional Group Identification

Fourier Transform Infrared Spectroscopy (FTIR): FTIR is a spectroscopic technique. It is essential for identifying functional groups from plant extracts, proteins, or polysaccharides involved in capping or reducing metal ions during green synthesis. It validates the eco-friendly synthesis and surface modification of nanoparticles (Rabiee *et al.*, 2022).

Raman Spectroscopy: Raman complements FTIR by providing vibrational information, especially for carbon-based nanomaterials or those with complex organic matrices (Rabiee *et al.*, 2022).

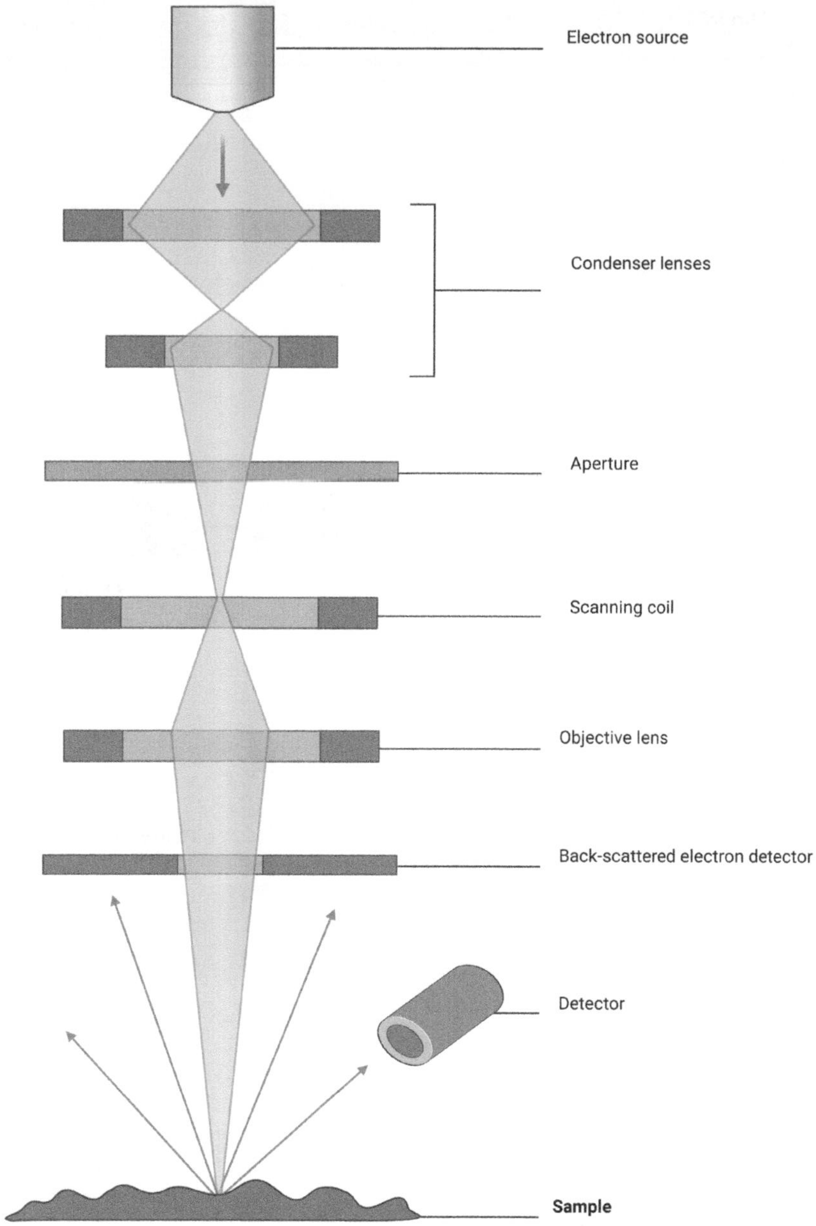

FIGURE 10.3 Working components of scanning electron microscopy.

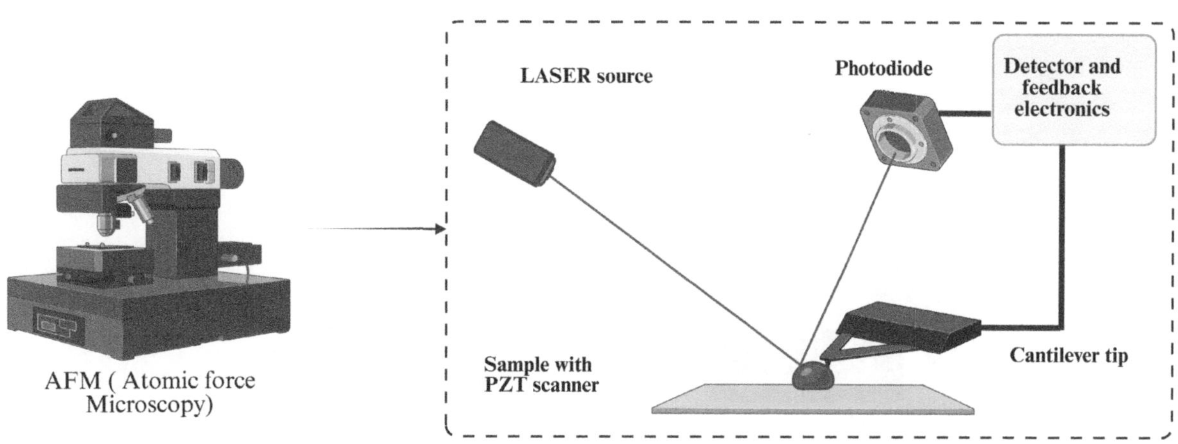

FIGURE 10.4 Working components of atomic force microscopy.

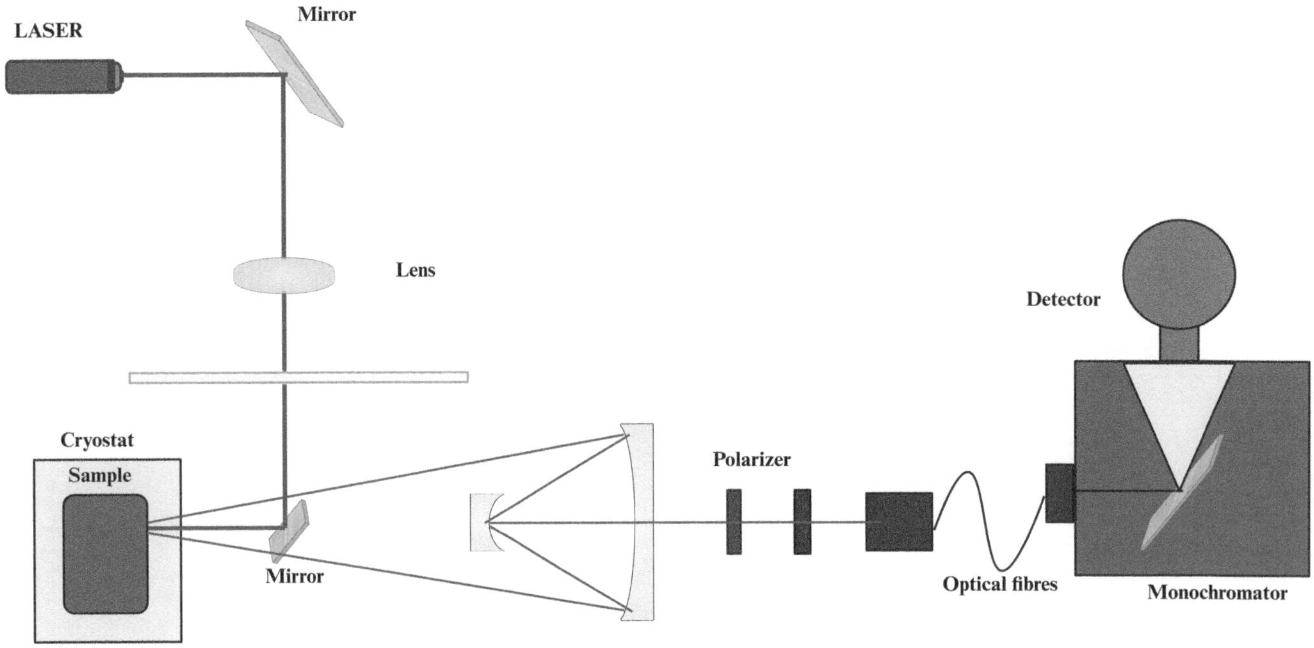

FIGURE 10.5 Components of UV-vis spectroscopy

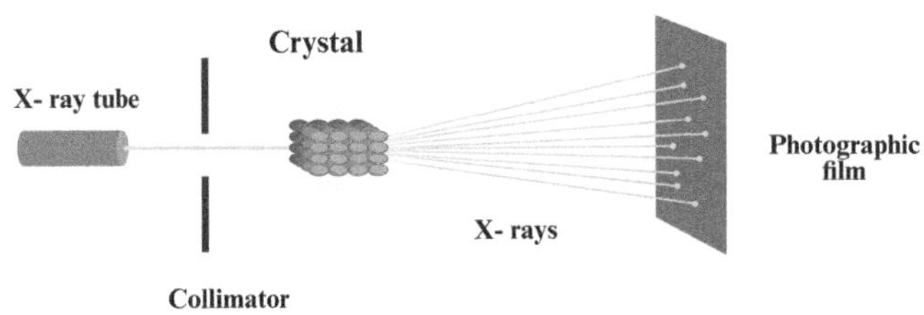

FIGURE 10.6 Photoluminescence spectroscopy.

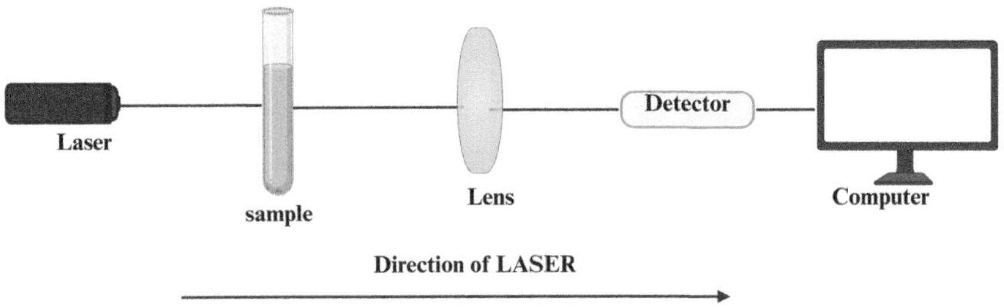

FIGURE 10.7 X-ray diffraction setup.

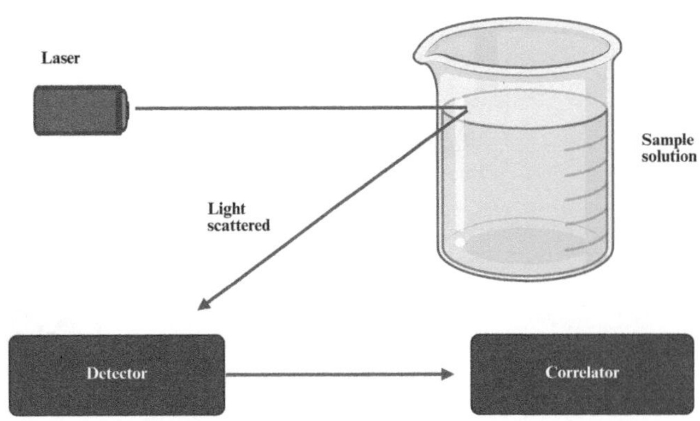

FIGURE 10.8 Working components of Fourier transform infrared spectroscopy.

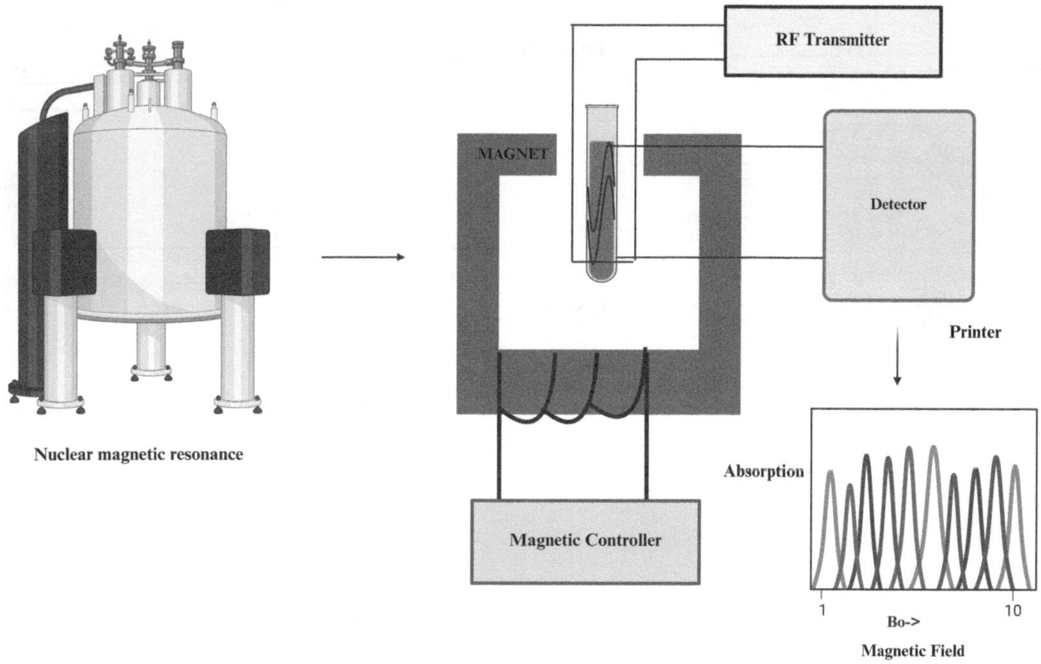

FIGURE 10.9 Working components of Raman spectroscopy.

10.1.1.5 Surface Charge and Stability

Zeta Potential Analysis: Zeta potential measures the surface charge of nanoparticles in suspension, providing insight into colloidal stability. A high absolute zeta potential (> ±30 mV) usually indicates stable dispersions, critical for biomedical or environmental use (Srivastava *et al.*, 2022).

10.1.1.6 Size Distribution and Hydrodynamic Behavior

Dynamic Light Scattering (DLS): DLS measures the hydrodynamic diameter and size distribution of nanoparticles in solution. This technique is noninvasive and suitable for characterizing biogenic nanoparticles under physiological conditions. It also gives information about the polydispersity index (PDI), critical for biomedical applications. The working components of DLS are shown in Figure 10.10.

Zeta potential and the DLS technique not only confirm the successful synthesis of green nanomaterials but also influence the design of application-specific nanoparticles (Halagali *et al.*, 2025).

10.1.1.7 Thermal and Chemical Stability

Thermogravimetric Analysis (TGA): TGA assesses thermal stability and the decomposition behavior of green nanomaterials, which are important for storage and application under various temperature conditions (Farhan, 2023).

Differential Scanning Calorimetry (DSC): DSC evaluates phase transitions, melting points, and heat capacities, helping in the formulation of nanomaterials for controlled drug release or industrial use (Farhan, 2023).

10.1.1.8 Spectroscopic Fingerprinting

X-Ray Photoelectron Spectroscopy (XPS): XPS is an X-ray–based technique used to determine the elemental composition and chemical states at the surface of nanomaterials. This is crucial for understanding surface passivation, oxidation states, and functionalization. The electron numbers reflect the element's proportion. Binding energy can be calculated, as the energy of the X-ray used to excite the electron from a core orbital is a known quantity, by using the equation (Shah *et al.*, 2019):

$$E_{binding} + E_{photon} - E_{kinetic} = \phi \tag{10.1}$$

where $E_{binding}$ is the energy of the emitted electron, E_{photon} is the X-ray photon energy used, $E_{kinetic}$ is the kinetic energy of the emitted electron, and Φ is the work function.

XPS can also be used for detecting contaminants in a sample.

Nuclear Magnetic Resonance (NMR): It is used primarily for characterizing organic ligands, capping agents, and the structure of soft nanomaterials. NMR supports the elucidation of molecular interactions on nanoparticle surfaces (Midekessa *et al.*, 2020).

10.1.1.9 Emerging Characterization Techniques

Synchrotron-Based Techniques: High-resolution X-ray absorption fine structure (XAFS) and X-ray

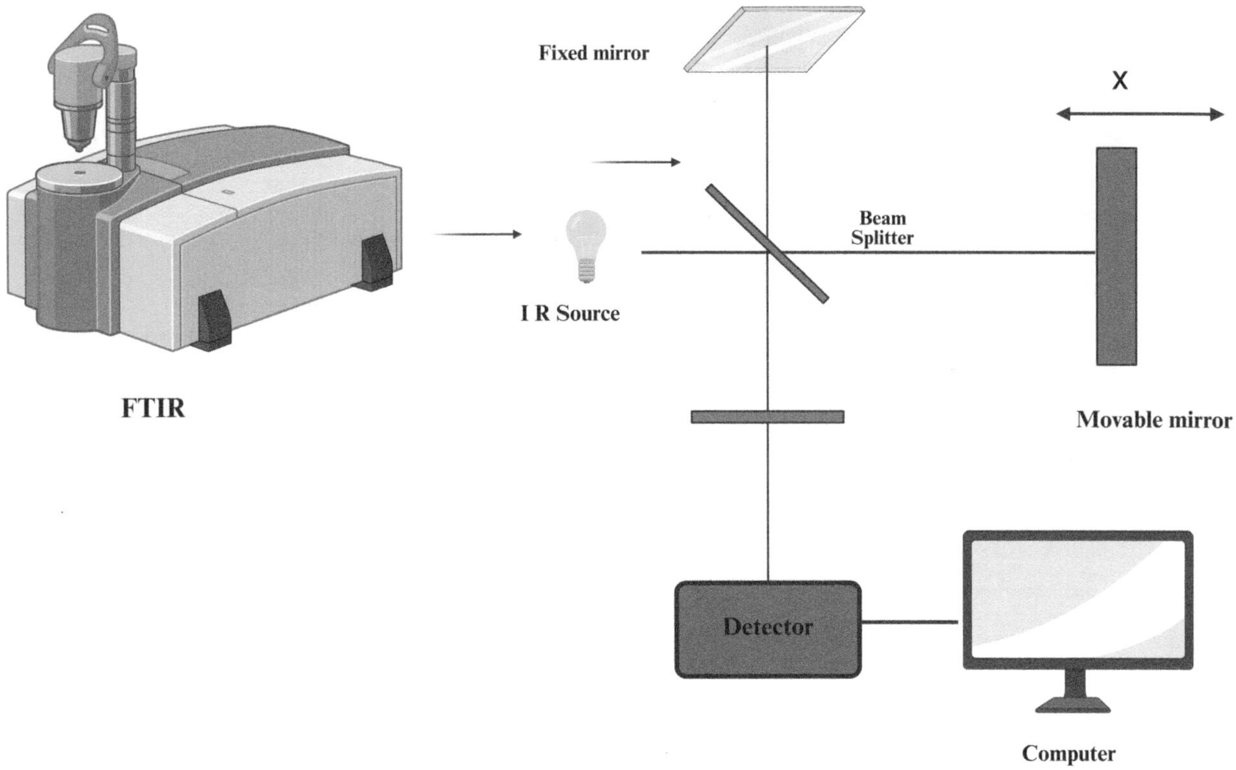

FIGURE 10.10 Working components of dynamic light scattering.

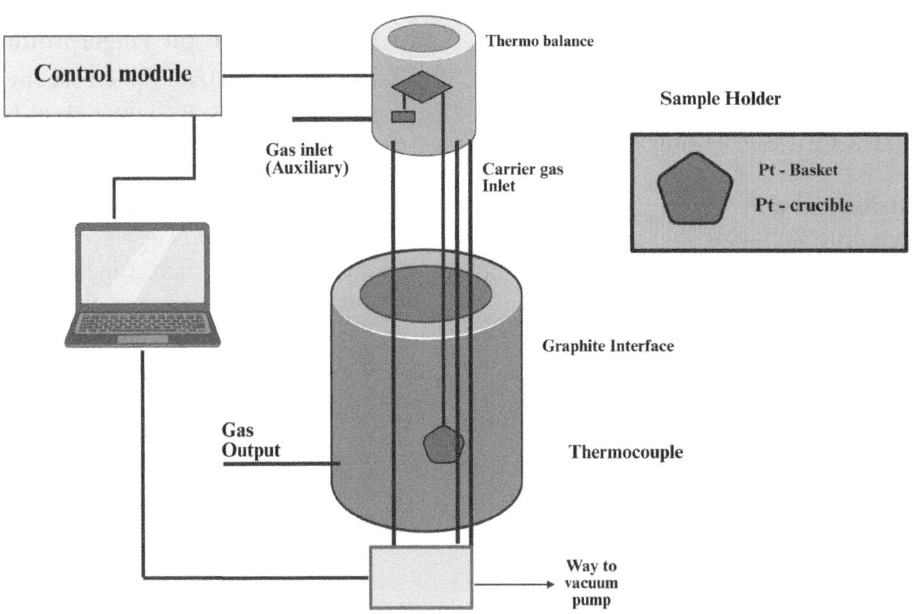

FIGURE 10.11 Thermogravimetric analysis.

fluorescence (XRF) at synchrotron facilities enable element-specific insights into oxidation states and coordination environments (Pinho *et al.*, 2022).

- **In Situ and Real-Time Monitoring**: New tools like in situ TEM or UV-vis allow monitoring of nanoparticle synthesis and growth mechanisms in real time, ensuring reproducibility and deeper understanding of green synthesis pathways.

10.1.2 Biomedical Applications of Green Nanoparticles

The biomedical field demands materials that are both effective and safe. Conventional nanoparticles, while useful, often involve hazardous reagents in their production, raising concerns over cytotoxicity and environmental impact (Ruan *et al.*, 2023). Green nanoparticles overcome these

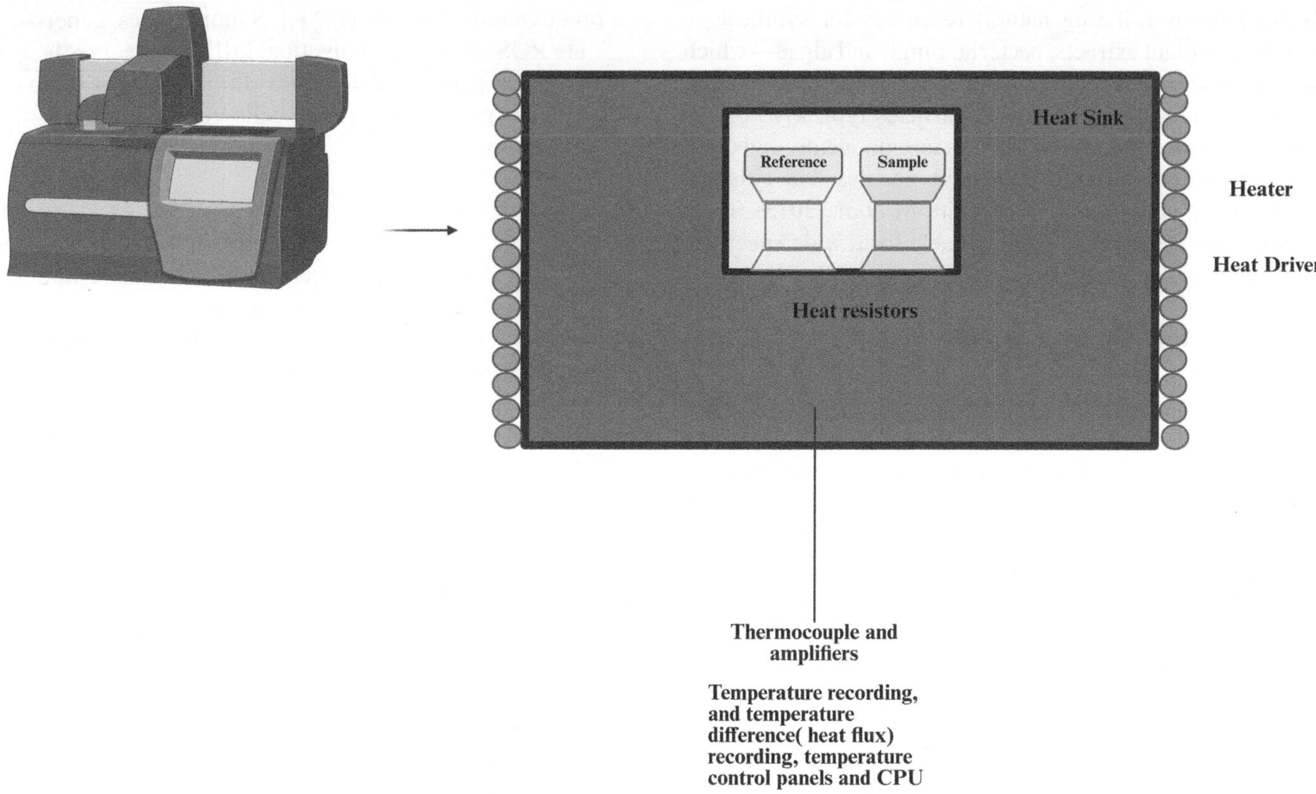

FIGURE 10.12 Working components of differential scanning calorimetry.

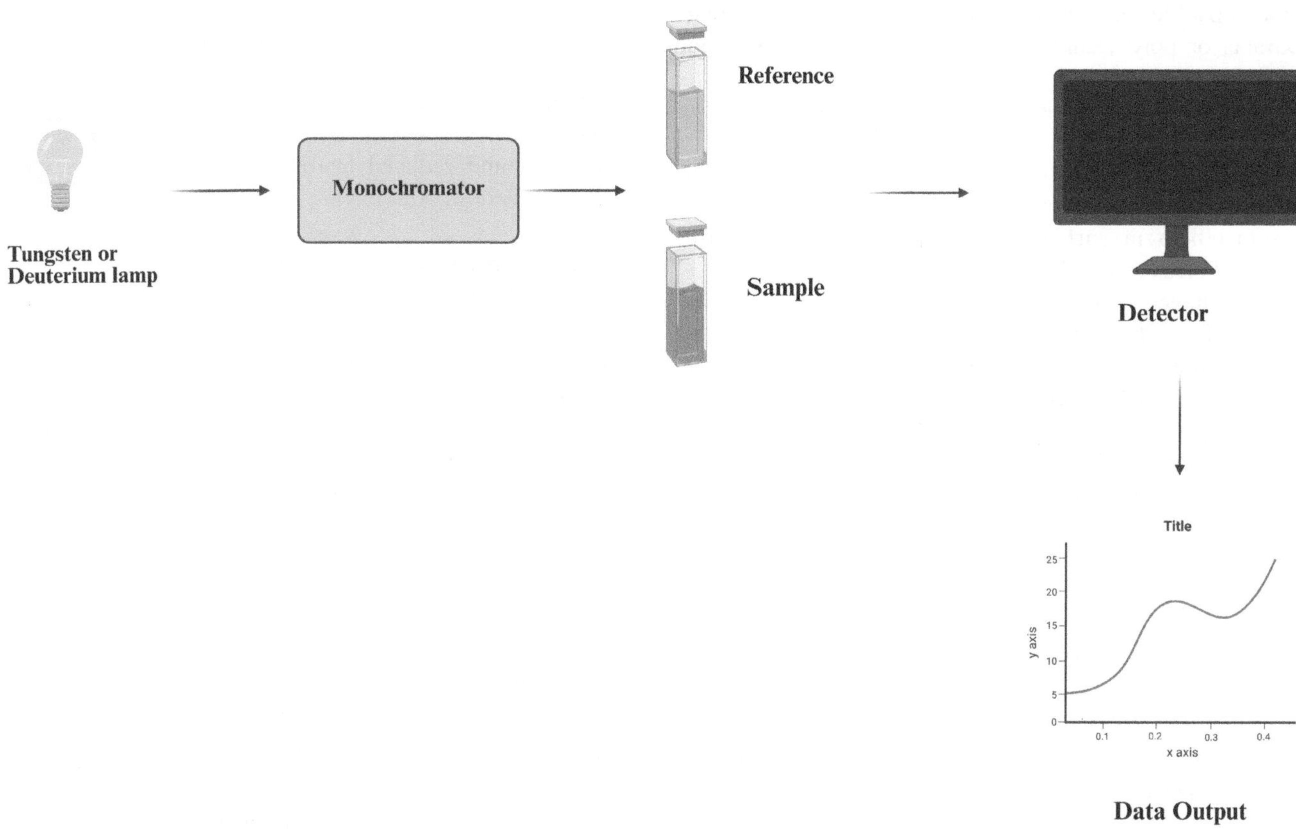

FIGURE 10.13 Working elements of nuclear magnetic resonance.

limitations by utilizing natural resources for synthesis—commonly plant extracts, bacteria, fungi, and algae—which act as reducing and stabilizing agents.

These naturally derived nanoparticles typically possess enhanced biocompatibility, are less immunogenic, and often come pre-functionalized with biomolecules that enhance their biological interactions (Tibbitt, M *et al.*, 2015). Their unique properties make them suitable for a wide spectrum of biomedical applications.

10.1.2.1 Antimicrobial Agents

Silver, gold, and copper nanoparticles synthesized via green routes demonstrate strong antimicrobial activity against drug-resistant strains. One of the most well-established uses of green nanoparticles is in antimicrobial therapy. Green-synthesized metallic nanoparticles—especially silver (AgNPs), gold (AuNPs), and zinc oxide (ZnO)—have shown strong activity against a wide range of pathogens, including antibiotic-resistant strains. The mode of action includes disruption of bacterial cell membranes, generation of reactive oxygen species (ROS), and inhibition of DNA replication and protein synthesis (Ruan *et al.*, 2023).

Silver nanoparticles synthesized using neem or tea extract exhibit broad-spectrum antimicrobial properties. ZnO nanoparticles have been incorporated into bandages and wound dressings to prevent infection.

10.1.2.2 Drug Delivery Systems

Biocompatible and biodegradable coatings from plant extracts or polysaccharides enhance circulation time and targeting. Gold nanoparticles, in particular, are used for photothermal therapy and drug conjugation. Green nanoparticles offer excellent platforms for targeted drug delivery, improving the therapeutic index of drugs while minimizing side effects. Their natural capping agents (e.g., flavonoids, tannins) help in surface functionalization and targeting (Leyva-Porras *et al.*, 2019).

Advantages include enhanced solubility and bioavailability of poorly soluble drugs; they also show controlled and sustained drug release and target-specific delivery (e.g., ligand–receptor interactions).

Examples such as curcumin-loaded green nanoparticles enhance bioavailability and are being studied for anti-inflammatory and anticancer effects. Gold nanoparticles synthesized with honey are being explored for targeted delivery of anticancer drugs due to their ease of functionalization (Ruan *et al.*, 2023).

10.1.2.3 Cancer Diagnosis and Therapy

Cancer treatment demands highly targeted approaches with minimal side effects. Green nanoparticles contribute to both diagnostic imaging and therapy (Ruan *et al.*, 2023).

Therapeutic roles include:

- Photothermal therapy (PTT): Green AuNPs absorb near-infrared light and convert it to heat, selectively destroying tumor cells.
- Photodynamic therapy (PDT): Nanoparticles generate ROS upon light activation, killing cancer cells.
- Chemotherapeutic delivery: Targeted delivery reduces toxicity to healthy cells via ligand-receptor interactions.

Diagnostic roles include:

- MRI and CT contrast agents: Green iron oxide nanoparticles enhance imaging clarity.
- Biosensors: Functionalized nanoparticles detect cancer biomarkers with high specificity.

Examples include green-synthesized gold nanoparticles using pomegranate peel extract, which have shown cytotoxic effects on breast cancer cell lines and are being explored in clinical studies.

10.1.2.4 Wound Healing and Tissue Engineering

Nanoparticles derived from natural polymers like chitosan or alginate promote healing due to their anti-inflammatory and antimicrobial properties. The regenerative potential of green nanoparticles is gaining interest in wound care and tissue engineering. The mechanism in wound healing leads to the promotion of fibroblast migration and collagen deposition, reduces microbial infection, and stimulates angiogenesis (Pasieczna-Patkowska *et al.*, 2025).

Applications include nanoformulated hydrogels with green silver or zinc oxide nanoparticles and biopolymer scaffolds for skin grafts embedded with green nanoparticles to enhance tissue regeneration.

Examples include chitosan-silver nanocomposites synthesized with aloe vera, which show enhanced wound contraction and reduced healing time in animal models (Baharlouei *et al.*, 2022).

10.1.2.5 Anti-Inflammatory and Antioxidant Applications

Chronic inflammation and oxidative stress are underlying factors in many diseases. Green nanoparticles often carry antioxidant phytochemicals from their source materials, giving them therapeutic potential. Some benefits include reduction in inflammatory cytokine levels, scavenging of free radicals, and protection of cells from oxidative damage (Ahmed *et al.*, 2022).

Examples include green selenium nanoparticles synthesized using garlic or turmeric extracts that exhibit strong anti-inflammatory and antioxidant effects in vitro and in vivo. Gold nanoparticles from ginger extract reduce oxidative stress in models of neurodegenerative diseases.

10.1.2.6 Antiviral Applications

The ongoing search for antiviral agents has highlighted green nanoparticles for their ability to inhibit viral entry and replication. The mechanism includes blocking virus–host cell receptor interactions, then disrupting viral envelopes, followed by modulating host immune responses (Halagali *et al.*, 2025).

Examples include silver nanoparticles from eucalyptus extract, which have shown inhibition of influenza and herpes simplex viruses. Green AuNPs functionalized with antiviral agents are being explored for use against SARS-CoV-2.

10.1.3 Environmental Applications of Green Chemistry in Nanotechnology

Traditional industrial practices have contributed significantly to environmental degradation through excessive use of hazardous chemicals, unsustainable resource extraction, and generation of toxic waste. When combined with nanotechnology, green chemistry allows for the development of novel materials that operate efficiently at the nanoscale, often with enhanced reactivity and selectivity (Soni *et al.*, 2022). Green nanomaterials synthesized using natural reducing agents such as plant extracts, bacteria, and fungi are biodegradable, non-toxic, and capable of addressing environmental challenges across multiple domains.

10.1.3.1 Water Purification and Wastewater Treatment

Removal of Heavy Metals and Toxic Ions: Green-synthesized nanoparticles like iron oxide, magnetite, and silver can adsorb or reduce heavy metals such as lead, arsenic, mercury, and chromium from contaminated water. These nanoadsorbents can also remove dye and pathogens from wastewater effectively. One example is green iron oxide nanoparticles synthesized using tea extract, which have been used to remove arsenic from groundwater.

Photocatalytic Degradation of Organic Pollutants: Metal oxide nanoparticles such as TiO_2 and ZnO, synthesized through green methods, serve as photocatalysts to break down organic pollutants (e.g., dyes, pharmaceuticals) under sunlight. The mechanism of action indicates that upon exposure to light, these nanoparticles generate reactive oxygen species, which degrade complex organic molecules into harmless byproducts (Nagajyothi *et al.*, 2020).

Antimicrobial Disinfection: Silver nanoparticles, synthesized using plant extracts, are effective in inactivating a wide range of microbial contaminants in water. AgNPs derived from neem leaf extract show strong antibacterial activity against *E. coli* and *Salmonella*, common pathogens in drinking water (Nagajyothi *et al.*, 2020).

10.1.3.2 Air Pollution Control

Removal of Volatile Organic Compounds (VOCs): Green nanomaterials are used in coatings for catalytic converters and air filters, helping reduce VOCs and NOx emissions, as they can adsorb or catalytically decompose VOCs from industrial emissions. Functionalized carbon nanotubes (CNTs) derived from agricultural waste show high surface area and reactivity for VOC capture.

Catalytic Converters and Air Filters: Green nanoparticles like ceria (CeO_2) and platinum-group metals have been used in automotive and industrial catalytic converters to convert CO, NOx, and hydrocarbons into less harmful gases. Green synthesis of these nanoparticles reduces reliance on hazardous precursors and lowers production emissions.

Particulate Matter (PM) Control: Nanofibrous filters incorporating green nanoparticles can trap fine particulate matter (PM2.5 and PM10), improving indoor and outdoor air quality.

10.1.3.3 Soil Remediation and Land Restoration

In Situ Stabilization of Contaminants: Green zero-valent iron (nZVI) nanoparticles, produced using eco-friendly reducing agents (e.g., grape seed extract), can immobilize or reduce or degrade contaminants/heavy metals and organic pollutants in soils. nZVI is used in mining sites to stabilize lead and cadmium, reducing leaching into groundwater.

Degradation of Pesticides and Herbicides: Photocatalytic nanomaterials like green-synthesized TiO_2 degrade persistent agrochemicals in agricultural soils under solar light, restoring soil health and preventing groundwater contamination (Nagajyothi *et al.*, 2020).

10.1.3.4 Sustainable Energy Solutions

Green Nanomaterials in Solar Cells: Biogenic nanoparticles are incorporated into solar cells to improve their efficiency by enhancing light absorption and charge transfer. Silver and gold nanoparticles synthesized from fruit extracts are incorporated into dye-sensitized solar cells (DSSCs) to boost performance. Biogenic nanoparticles are used in hydrogen production through water splitting as eco-friendly catalysts (Soni *et al.*, 2022).

Hydrogen Production and Fuel Cells: Green metal nanoparticles act as catalysts in hydrogen production via water splitting and are being studied for use in proton exchange membrane (PEM) fuel cells. This shows a reduced reliance on rare or toxic metals by using biosynthesized, cost-effective alternatives.

10.1.3.5 Green Nanotechnology in Waste Management

Degradation of Plastics and Polymers: Enzyme-functionalized nanoparticles that are produced through green synthesis can help break down synthetic polymers into biodegradable components.

E-Waste Recycling: Green chemistry methods using benign solvents and plant-mediated nanoparticles can recover valuable metals (e.g., gold, palladium) from electronic waste (Sharma *et al.*, 2022).

Overall, the advantages of green chemistry in environmental nanotechnology indicate eco-friendly synthesis that avoids toxic reagents, protecting ecosystems. The process is cost effective due to the use of renewable biological materials. It highlights enhanced biocompatibility and reduced long-term toxicity. It is in alignment with the circular economy through recyclability and biodegradability.

10.1.4 Industrial Applications and Case Studies of Green Chemistry in Nanotechnology

The 21st century has seen an increasing demand for sustainable technological advancement. Green chemistry, aimed at designing products and processes that minimize the use and generation of hazardous substances, offers a roadmap to reduce the environmental footprint of the chemical industry. Meanwhile, nanotechnology—manipulating matter at the scale of atoms and molecules—promises revolutionary breakthroughs in efficiency, functionality, and performance.

When combined, green chemistry and nanotechnology provide opportunities to revolutionize industrial practices. Green nanotechnology emphasizes not only the safe design and use of nanomaterials but also cleaner production processes, resource conservation, and lifecycle sustainability. This section includes industries integrating the principles of green chemistry into the design, synthesis, and application of nanomaterials (Pinho *et al.*, 2022). Through key case studies across diverse sectors—including energy, medicine, electronics, and materials—this section illustrates the transformative impact and challenges of implementing green nanotechnology in real-world industrial contexts

10.1.4.1 Green Chemistry Principles in Nanotechnology

Before delving into industrial examples, it's important to understand how green chemistry principles are applied in nanotechnology (Soni *et al.*, 2022):

Prevention of waste: Using atom-efficient synthetic pathways to minimize byproducts, thus avoiding waste generation in nanoparticle synthesis. Monitoring nanoparticle synthesis to avoid hazardous byproducts.

Safer solvents and auxiliaries: Using water, supercritical CO_2, or ionic liquids instead of volatile organic solvents.

Energy efficiency: Utilizing processes that occur at ambient temperatures and pressures.

Renewable feedstocks: Deriving starting materials from plant-based, bio-based, or recycled sources.

Degradable materials: Designing nanoparticles that are biodegradable or recyclable, like engineering nanomaterials that degrade into non-toxic components.

Less hazardous synthesis: Using non-toxic precursors in nanoparticle production.

Designing safer chemicals: Creating nanomaterials with low toxicity profiles.

Atom economy: Maximizing the incorporation of all materials into the final product.

Reduce derivatives: Minimizing the use of modifying agents that require additional reagents.

Inherently safer chemistry: Reducing accident risks in nanomanufacturing environments.

Catalysis: Using nanocatalysts to reduce the energy and materials needed for chemical reactions (Milovanovic *et al.*, 2025).

These principles have guided several successful industrial implementations across various sectors.

10.1.4.2 Industrial Applications

10.1.4.2.1 Green Nanomaterials in Electronics Manufacturing

Synthesis of conductive inks for printed electronics: HP developed silver nanoparticles for use in printed electronics using a water-based synthesis method that eliminated the need for toxic solvents like ethylene glycol. The process also leveraged microwave-assisted synthesis, significantly reducing energy consumption and reaction time. The effect includes a reduction in waste solvents. There are lower operational costs due to decreased energy use and increased worker safety (Farhan *et al.*, 2023).

10.1.4.2.2 Pharmaceutical Nanocarriers Using Biodegradable Polymers

Nanoparticle drug delivery systems for cancer therapy: Innovations include the adoption of PLGA (poly(lactic-co-glycolic acid))—a biodegradable and biocompatible polymer—as the carrier material for encapsulating drugs like paclitaxel. Nanoparticle synthesis employed supercritical CO_2 as a solvent, eliminating toxic organic solvents. Outcomes include enhanced drug targeting with minimal systemic toxicity, reduced solvent disposal costs, and increased regulatory acceptance due to biocompatibility (Baharlouei *et al.*, 2022).

10.1.4.2.3 Bio-Based Nanocomposites in the Automotive Industry

Lightweight, durable components for vehicle interiors: Green innovation includes Ford incorporating cellulose nanofibers derived from agricultural waste (e.g., wheat straw, kenaf) into polymer composites. These nanomaterials were produced via low-energy mechanical processes, and the final composites were recyclable. Benefits include weight reduction of up to 30%, improving fuel efficiency, use of renewable feedstocks, and enhanced life cycle sustainability of vehicle parts (Liu *et al.*, 2010).

10.1.4.2.4 Catalytic Nanoparticles in Petrochemical Refining

Hydrocracking catalysts for fuel production: BASF developed zeolite-based nanocatalysts with high surface areas that improve hydrocarbon cracking efficiency. The catalysts are manufactured using aqueous routes and recycled rare earth elements. Impacts include increased product yield, lower process temperatures, saving energy, and reducing reliance on virgin raw materials (Farhan, 2023).

10.1.4.2.5 Green Synthesis of Iron Oxide Nanoparticles for Water Purification

Providing affordable water purification solutions for rural areas: Iron oxide nanoparticles were synthesized using plant extracts (e.g., neem, tulsi) in a one-pot aqueous method. These particles were incorporated into household water filters to remove arsenic and microbes. Sustainability features include the use of plant-based reducing agents (green synthesis); low-cost, scalable production methods; and zero toxic waste generation. The outcomes of the method include safe drinking water, local job creation in manufacturing, and global recognition of green nanotech applications in public health (Ahmed *et al.*, 2023).

10.1.4.2.6 Agriculture: Nanofertilizers and Pesticides

Plant-based or microbial routes are considered to synthesize the nanoparticles for nanofertilizers and nanopesticides. Biodegradable nanoencapsulation ensures controlled release by minimizing the leaching of nutrients and pesticides into water bodies. It leads to improved crop yield with lower input, reduced environmental runoff, and enhanced sustainability of agricultural practices.

10.1.4.2.7 Medicine and Healthcare: Biodegradable Nanocarriers

Biodegradable nanocarriers are used in drug delivery, imaging, and diagnostics as they are biocompatible. Biodegradable polymers such as PLGA and chitosan are used as nanoparticle carriers. Supercritical CO_2 as a solvent is considered in the encapsulation process. Microwave-assisted green synthesis of metal oxide nanoparticles is used for bioimaging. These lead to reduced use of hazardous organic solvents, better patient safety through safer materials, and lower manufacturing waste and cost (Tibbitt *et al.*, 2015).

10.1.4.2.8 Green Synthesis in Cosmetics

Green-synthesized zinc oxide and titanium dioxide nanoparticles are increasingly used in sunscreens as eco-friendly alternatives to conventional UV filters, as it prevents prevent coral bleaching and reduce potential health risks. Sustainability features include the use of plant extract–mediated nanoparticle synthesis, better biodegradable formulations, and reduced environmental toxicity in aquatic ecosystems.

10.1.4.2.9 Cellulose Nanofibers in Manufacturing Industries

Bio-based nanocomposites: Ford has pioneered the use of cellulose nanofibers derived from agricultural waste to produce lightweight car components. These nanocomposites not only reduce vehicle weight (enhancing fuel economy) but also avoid petroleum-based polymers, aligning with green chemistry principles. This leads to 20–30% weight savings and the use of renewable feedstocks. It also indicates enhanced recyclability and a reduced carbon footprint (Liu *et al.*, 2010; Soni *et al.*, 2022; Milovanovic *et al.*, 2025).

10.2 CHALLENGES AND LIMITATIONS

Despite promising developments, the integration of green chemistry in nanotechnology faces several barriers; for example, scaling up synthesis while maintaining reproducibility is difficult without precise control and characterization. Many green synthesis routes remain inefficient at industrial scale (Sharma *et al.*, 2022). Heterogeneity of biological extracts can lead to variable surface chemistry. Biological variability in plant extracts can affect nanoparticle consistency. Batch-to-batch variability exists due to differences in natural sources. Hence, standardization of protocols is essential for comparing results across laboratories.

Cost: Bio-based feedstocks and green solvents can be more expensive or less available.

Regulation and Standardization: Lack of standardized testing for environmental fate and toxicity of green nanoparticles. Regulatory hurdles in clinical translation. Lack of clear guidelines for environmental use and disposal of green nanomaterials.

Technology Transfer: Difficulty translating lab-scale methods into commercial production due to equipment or expertise gaps.

Maximizing the potential of green nanotechnology in industry lies in the development of standardized synthesis protocols. This is done by expanding in situ techniques to minimize sample preparation and preserve the natural state of nanomaterials. Advanced modeling tools and AI can aid in designing more efficient green synthesis pathways, such as the use of AI and machine learning to predict nanoparticle behavior; developing portable, cost-effective characterization tools suitable for field or clinical settings; integration with wearable medical devices for real-time monitoring and delivery; and real-time sensors and environmental monitoring tools using green nanosensors and considering the policy frameworks supporting green innovations at the local and global levels (Halagali *et al.*, 2025).

10.3 CONCLUSION

Nanoparticles synthesized by green methods are more than an innovation—they are a necessity. By integrating sustainability with advanced nanoscience, green nanomaterials offer powerful tools for addressing environmental challenges like pollution, resource depletion, and energy scarcity. As research continues to evolve, these eco-friendly nanotechnologies are poised to play a pivotal role in shaping a cleaner, safer, and more resilient world. Their versatility, coupled with inherent biocompatibility, makes them ideal for a broad array of applications—from treating infections to fighting cancer and promoting tissue regeneration. With continued research and interdisciplinary collaboration, green nanotechnology is poised to become a cornerstone of next-generation biomedical innovations. Advanced characterization is the backbone of green nanotechnology, providing the data necessary to tailor nanomaterials for specific applications while maintaining sustainability. As instrumentation continues to evolve, our ability to design safer, smarter, and more efficient green nanomaterials will grow, propelling the field toward real-world impact.

The convergence of green chemistry and nanotechnology holds transformative potential for sustainable industrial development. Through innovative synthesis methods, renewable materials, and environmentally benign processing, companies across electronics, pharmaceuticals, automotive, and chemical sectors are already showcasing the tangible benefits of this synergy. While challenges remain, ongoing research, policy support, and industrial commitment will play a pivotal role in steering nanotechnology towards a greener future.

REFERENCES

Ahmed, S. F., Mofijur, M., Rafa, N., Chowdhury, A. T., Chowdhury, S., Nahrin, M., & Ong, H. C. (2022). Green approaches in synthesising nanomaterials for environmental nanobioremediation: Technological advancements, applications, benefits and challenges. *Environmental Research*, *204*, 111967. https://doi.org/10.1016/j.envres.2021.111967

Baharlouei, P., & Rahman, A. (2022). Chitin and chitosan: Prospective biomedical applications in drug delivery, cancer treatment, and wound healing. *Marine Drugs*, *20*(7), 460. https://doi.org/10.3390/md20070460

Farhan, M. (2023). Catalytic converters: Enhancing green chemistry and their technological implications. https://doi.org/10.35248/2161-0398.25.15.420

Fatimah, S., Ragadhita, R., Al Husaeni, D. F., & Nandiyanto, A. B. D. (2022). How to calculate crystallite size from x-ray diffraction (XRD) using Scherrer method. *ASEAN Journal of Science and Engineering*, *2*(1), 65–76. https://doi.org/10.17509/ajse.v2i1.37647

Halagali, P., Nayak, D., Rathnanand, M., Tippavajhala, V. K., Sharma, H., & Biswas, D. (2025). Synergizing sustainable green nanotechnology and AI/ML for advanced nanocarriers: A paradigm shift in the treatment of neurodegenerative diseases. In *The Neurodegeneration Revolution* (pp. 373–397). Academic Press. https://doi.org/10.1016/B978-0-443-28822-7.00017-9

Leyva-Porras, C., Cruz-Alcantar, P., Espinosa-Solís, V., Martínez-Guerra, E., Pinon-Balderrama, C. I., Compean Martínez, I., & Saavedra-Leos, M. Z. (2019). Application of differential scanning calorimetry (DSC) and modulated differential scanning calorimetry (MDSC) in food and drug industries. *Polymers*, *12*(1), 5. https://doi.org/10.3390/polym12010005

Liqiang, J., Yichun, Q., Baiqi, W., Shudan, L., Baojiang, J., Libin, Y., & Jiazhong, S. (2006). Review of photoluminescence performance of nano-sized semiconductor materials and its relationships with photocatalytic activity. *Solar Energy Materials and Solar Cells*, *90*(12), 1773–1787. https://doi.org/10.1016/j.solmat.2005.11.010

Liu, C. J., Burghaus, U., Besenbacher, F., & Wang, Z. L. (2010). Preparation and characterization of nanomaterials for sustainable energy production. https://doi.org/10.1021/nn102420c

Midekessa, G., Godakumara, K., Ord, J., Viil, J., Lättekivi, F., Dissanayake, K., . . . & Fazeli, A. (2020). Zeta potential of extracellular vesicles: Toward understanding the attributes that determine colloidal stability. *ACS Omega*, *5*(27), 16701–16710. https://doi.org/10.1021/acsomega.0c02051

Milovanovic, J., Banjanac, K., Nikolić, J., Nikodinović-Runić, J., & Prlainović, N. Ž. (2025). The organic-functionalized silica nanoparticles as lipase carriers for biocatalytic application: Future perspective in Biodegradation. *Catalysts*, *15*(1), 54. https://doi.org/10.3390/catal15010054

Nagajyothi, P. C., Prabhakar Vattikuti, S. V., Devarayapalli, K. C., Yoo, K., Shim, J., & Sreekanth, T. V. M. (2020). Green synthesis: Photocatalytic degradation of textile dyes using metal and metal oxide nanoparticles-latest trends and advancements. *Critical Reviews in Environmental Science and Technology*, *50*(24), 2617–2723. https://doi.org/10.1080/10643389.2020.1743675

Pasieczna-Patkowska, S., Cichy, M., & Flieger, J. (2025). Application of fourier transform infrared (FTIR) spectroscopy in characterization of green synthesized nanoparticles. *Molecules*, *30*(3), 684. https://doi.org/10.3390/molecules30030684

Patra, J. K., & Baek, K. H. (2014). Green nanobiotechnology: Factors affecting synthesis and characterization techniques. *Journal of Nanomaterials*, *2014*(1), 417305. https://doi.org/10.1155/2014/417305

Pinho, B., Zhang, K., Hoye, R. L., & Torrente-Murciano, L. (2022). Importance of monitoring the synthesis of light-interacting nanoparticles—a review on in situ, ex situ, and online time-resolved studies. *Advanced Optical Materials*, *10*(14), 2200524. https://doi.org/10.1002/adom.202200524

Rabiee, N., Ahmadi, S., Akhavan, O., & Luque, R. (2022). Silver and gold nanoparticles for antimicrobial purposes against multi-drug resistance bacteria. *Materials*, *15*(5), 1799. https://doi.org/10.3390/ma15051799

Raj, D. R., Prasanth, S., Vineeshkumar, T. V., & Sudarsanakumar, C. (2016). Surface plasmon resonance based fiber optic dopamine sensor using green synthesized silver nanoparticles. *Sensors and Actuators B: Chemical*, *224*, 600–606. https://doi.org/10.1016/j.snb.2015.10.106

Ruan, H., Aulova, A., Ghai, V., Pandit, S., Lovmar, M., Mijakovic, I., & Kádár, R. (2023). Polysaccharide-based antibacterial coating technologies. *Acta Biomaterialia*, *168*, 42–77. https://doi.org/10.1016/j.actbio.2022.12.041

Shah, K. W., & Li, W. (2019). A review on catalytic nanomaterials for volatile organic compounds VOC removal and their applications for healthy buildings. *Nanomaterials*, *9*(6), 910. https://doi.org/10.3390/nano9060910

Sharma, A., Manocha, D., Arora, A., Kumar, A., Srivastava, C. M., Rawat, V., . . . & Rao, G. K. (2022). Applications of green nanomaterials in electronic and electrical industries. In *Green Nanomaterials for Industrial Applications* (pp. 397–421). Elsevier. https://doi.org/10.1016/B978-0-12-823296-5.00016-2

Soni, R. A., Rizwan, M. A., & Singh, S. (2022). Opportunities and potential of green chemistry in nanotechnology. *Nanotechnology for Environmental Engineering*, *7*(3), 661–673. https://doi.org/10.1007/s41204-022-00233-5

Srivastava, S., & Bhargava, A. (2022). *Green Nanoparticles: The Future of Nanobiotechnology* (pp. 1–13). Singapore: Springer. https://doi.org/10.1007/978-981-16-7106-7

Tibbitt, M. W., Rodell, C. B., Burdick, J. A., & Anseth, K. S. (2015). Progress in material design for biomedical applications. *Proceedings of the National Academy of Sciences*, *112*(47), 14444–14451. https://doi.org/10.1073/pnas.1508520112

11 Advance Characterization and Applications for Green Nanomaterials

J. Preethi, P.K. Praveen Kumar, Srinidhi Varadharajan, and M. Kevin

11.1 CH01_INTRODUCTION

Nanotechnology, a science integrated with technology will be a prominent tool in upcoming years (Huston, DeBella, DiBella, and Gupta 2021). Nano systems embraced with active and passive Nano assemblies & small-scale molecular Nano systems are widely used across various scientific fields (Malik, Muhammad, and Waheed 2023). However, the conventional synthesis methods pose serious environmental and health risks therefore it has led to the blending of green chemistry principles, which prioritize the use of natural resources and energy-efficient processes (Yuan, Xu, Wang, and Yao 2015). This shift is essential as the chemicals utilized in these convectional processes causes environmental pollution due to the presence of hazardous chemicals (Asghar, Hussain, Nguyen, Ali, Hussain, Junejo, and Ali 2024). Plant extract-mediated nanoparticle synthesis, such as using *Withania somnifera* (Ashwagandha) extracts, Boerhavia offers a simpler, expandable, and sustainable solution, prohibiting the complexities of microbial synthesis, reduced cell viability, under particular growth requirements. These methods are sustainable and can be synthesized under mild conditions such as ambient temperature and neutral pH which can be applied to treat hepatocellular carcinoma and type 2 diabetes mellitus. Green synthesis is carried out utilizing biodegradable materials like plant extracts and plant-based primary and secondary metabolite products. This approach helps alleviate the environmental impact while at the same time enhancing biocompatibility and providing sustainable solutions in environmental remediation, medicine and industrial sectors (Devatha and Thalla 2018; De, Singh, Guin, and Barthwal 2023; Chaudhary, Mahapatra, Pandey, Patel, Dubey, Kumar, and Mukherjee 2024). For instance, the structure of nanoparticles must be meticulously constructed and geometry-optimized to accurately replicate their size and shape before molecular docking simulations can predict their binding modes (Kar, Oriola, and Oyedeji 2024). Thus, integrating green synthesis with computational modeling not only advances the precision and safety of nanomaterial applications but also paves the way for innovative, eco-friendly therapeutic strategies in modern medicine. In this book chapter, overall characterization and applications of Green Nanochemistry is mentioned in the Figure 11.1

11.2 CH_02 BACKGROUND_ADVANCED CHARACTERIZATION TECHNIQUES FOR GREEN SYNTHESIS

Some of the advanced characterization techniques used in green synthesis include spectroscopic methods such as UV-Vis spectroscopy, microscopic methods like Scanning Electron Microscopy and Transmission Electron Microscopy, X-ray Diffraction (XRD), Nuclear Magnetic Resonance (NMR) and particle size and stability analysis using Dynamic Light Scattering (DLS) and Zeta potential. The advanced characterization techniques for green synthesis are shown in Figure 11.2

11.3 CH02_MICROSCOPIC METHODS_ SCANNING ELECTRON MICROSCOPE (SEM)

Nanoparticles morphology, topography and chemical composition are analyzed using Scanning Electron Microscopy (SEM). A beam of electrons is passed through the sample and high-resolution images of the particles surface are produced. For example, the identification of silica nanoparticles can be confirmed through SEM imaging by their characteristic spherical morphology and distinct surface roughness. Its magnifications range from low levels to several hundred thousand times hence, SEM is used to accurately visualize micro- and Nano scale structures. Additionally, the detection of secondary electrons, backscattered electrons, and x-rays, helps in detailed analysis of topography and elemental composition of nanoparticles (Kumar, Pavithra, and Naushad 2019).

11.4 CH02_MICROSCOPIC METHODS_ TRANSMISSION ELECTRON MICROSCOPE (TEM)

TEM has the ability to measure and visualize the inward structure of the nanoparticles (Sri, Chauhan, Lakshmi, Thakar, and Solanki 2022) mentioned that Selected Area Electron Diffraction (SAED) applied in conjunction with Transmission Electron Microscopy (TEM), is an important technique for examining the crystalline characteristics of nanoparticles. SAED patterns analysis is important in identifying specific

DOI: 10.1201/9781003644439-11

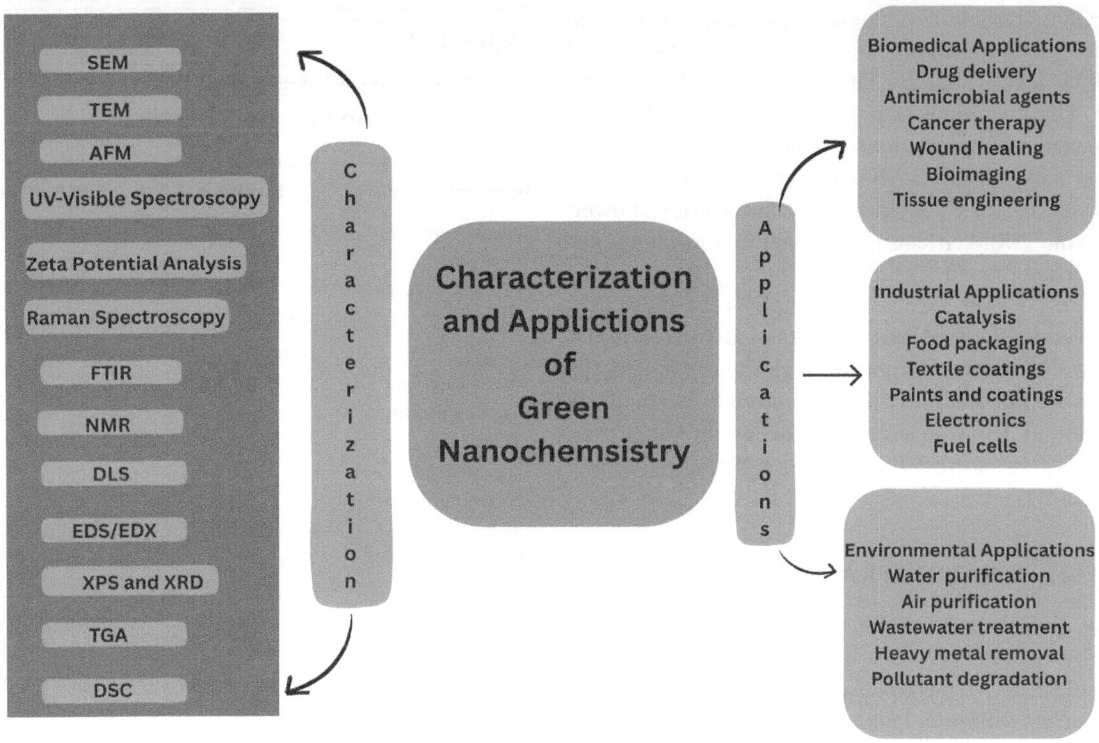

FIGURE 11.1 Characterization and Applications of Green Nanochemistry

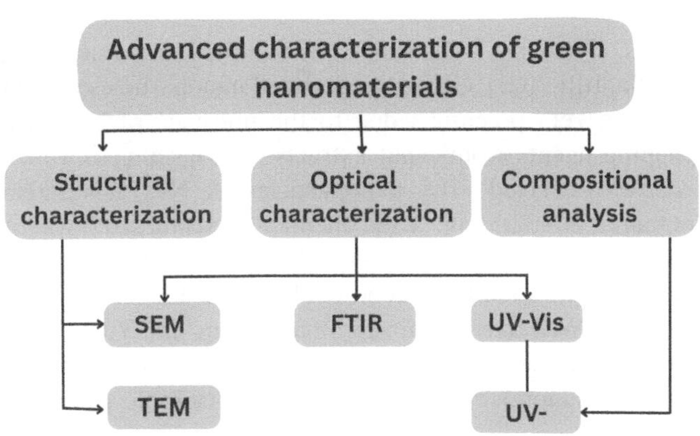

FIGURE 11.2 Advance Characterization of Green Nanomaterials

crystallographic phases, assess the degree of crystallinity within the sample and determination of lattice planes based on inter planar spacing (d-spacing), (Braidy, Béchu, de Souza Terra, and Patience 2020). It also helps validate the green synthesis of nanoparticles by confirming nanoparticle size and ensuring uniform morphology, which plays an important role in applications like drug delivery and antimicrobial activity (Rauwel, Küünal, Ferdov, and Rauwel 2015).

11.5 CH02_SPECTROSCOPIC METHODS_ UV-VIS SPECTROMETER

Nanoparticles size, shape and distribution are analyzed based on surface plasma resonance confirmation (Mahmudin, Suharyadi, Utomo, and Abraha 2015). Key parameters such as band gap energy are confirmed based on the absorption spectrum, therefore optical and electronic properties can be studied (Chen, Yang, Wu, and Lin 2018). It is also useful for monitoring the progress of nanoparticle synthesis and for optimizing reaction conditions.

11.6 CH02_SPECTROSCOPIC METHODS_ FOURIER TRANSFORM INFRARED SPECTROSCOPY (FTIR)

FTIR identifies the functional groups that are present in the green nanoparticle. (Pasieczna-Patkowska, Cichy, and Flieger 2025). It also provides insight into their stabilization, formation and structural properties. The chemical integrity of the green nanoparticles can also be assessed using FTIR (Medina-Ramirez, Bashir, Luo, and Liu 2009), while spectral features offer valuable information on nanoparticle composition, green synthesis mechanism, and surface chemistry which thus helps to ensure a sustainable synthesis process (Sharma, Tyagi, Vikal, Devi, Gautam, and Singh 2022).

11.7 CH02_SPECTROSCOPIC METHODS_NUCLEAR MAGNETIC RESONANCE (NMR)

Nuclear magnetic resonance (NMR) is a spectroscopic technique which absorbs the atomic nuclei and re-emit electromagnetic radiation in magnetic field and it is used to characterize the conformational dynamics of nanomaterial at the atomic level (Sharma, Pandey, Ghatak, Watal, and Rai 2018). This technique is becoming important because of its wide range of applications, including structural characterization and analysis of dynamic interactions between species under various conditions (Carter, Harley, Power, and Augustine 2005). It can also be applied to characterize phase behavior and surface adsorption phenomena of colloids in suspension. One such example is the pulse gradient spin-echo (PGSE) NMR technique. Diffusion NMR spectroscopy is another important technique used for its effective resolution of large colloidal materials, particularly vesicles formed from macromolecular amphiphiles and nanoparticles (Valentini, Vaccaro, Rehor, Napoli, Hubbell, and Tirelli 2004).

11.8 CH02_X-RAY DIFFRACTION (XRD)

The X-ray diffraction (XRD) technique that was first described by German physicist Max von Laue, is a powerful and nondestructive method used for characterizing crystalline materials (Sharma, Pandey, Ghatak, Watal, and Rai 2018). Number of phrases, size and characteristics of the material along with diffraction peaks can be studied using XRD analysis. (Liu, Wang, Chang, and Chen 1995) Nanocatalyst can also be characterized using XRD (Amin, Iram, Iqbal, Saeed, Raza, and Alam 2013).

11.9 CH02_PARTICLE SIZE ANALYSIS_ DYNAMIC LIGHT SCATTERING (DLS)

Dynamic Light Scattering (DLS) is spectroscopic technique used for characterizing nanoparticles by measuring the size distribution in liquid suspensions (Jia, Li, Gao, Yang, and Kanaev 2023). The size of the particles in the range of 1 to 1000 nanometers can be analyzed by DLS, which predicts polydispersity, nanoparticle aggregation, and stability in various environments. Hydrodynamic size of green-synthesized nanoparticles is analyzed based on time dependent fluctuations in scattered light intensity caused by the Brownian motion of particles in suspension (Hassan, Rana, and Verma 2015). Due to a quicker decay in the intensity autocorrelation function, smaller particles move faster, whereas larger particles move more slowly leading to a slower decay (Babick 2020). This principle allows for precise measurement of the nanoparticle size distribution within the colloidal system.

11.10 CH02_NANOPARTICLE STABILITY_ZETA POTENTIAL

Zeta potential is used to analyze the stability and surface charge of the nanoparticle. It indicates electrostatic repulsion between nanoparticles. An absolute value ≥ ±30 mV denotes strong colloidal stability due to effective repulsive forces. (Pochapski, Carvalho dos Santos, Leite, Pulcinelli, and Santilli 2021). Green-synthesized nanoparticles exhibit negative zeta potentials due to the presence of bioactive capping agents from plant extracts. For instance, *Cynara scolymus*-derived silver nanoparticles (AgNPs) has a zeta potential of –32.3 mV, while chemically synthesized water-based AgNPs has (–31.9 to −43.49 mV). Such negative surface charges enhance biocompatibility for biomedical use. (Ahmad, Wei, Syed, Tahir, Rehman, Khan, and Yuan 2017) The techniques used in different applications of green synthesis of nanomaterials are summarized in Table 11.1

TABLE 11.1
Techniques used for green synthesis and characterization of nanomaterials

Techniques Used	Source of Green Synthesis	References
Spectroscopic Methods	Parts of Plants	(Iravani 2011)
UV-Vis Spectroscopy, FTIR	Leaves, Roots, Flowers & Fruits	(Jadoun, Arif, Jangid, and Meena 2021)
Microscopic analysis	Micro Organisms	(Shafey 2020)
Scanning Electron Microscope, Transmission Electron Microscope	Bacteria, Fungi, Algae, Yeast.	(Das, and Brar 2013)
Nuclear Magnetic Resonance X-Ray Diffraction	Enzymes	(Bahrulolum, Nooraei, Javanshir, Tarrahimofrad, Mirbagheri, Easton, and Ahmadian 2021)
Dynamic Light Scattering Technique, Zeta Potential Analysis Technique	Laccases, Lipases and Amylases.	(Romano, Vitiello, Gallucci, Di Girolamo, Cattaneo, Poli, and Di Donato 2022) (Otari, Patil, Nadaf, Ghosh, and Pawar 2014)

11.11 CH03_BACKGROUND_ BIOMEDICAL APPLICATION OF GREEN NANOMATERIALS

Nano-medicine delivery system offers meticulous drug delivery methods. Techniques such as media milling, high-pressure homogenization, and micro emulsion templates have been triumphantly applied to produce Nano suspensions (Rahuman, Dhandapani, Narayanan, Palanivel, Paramasivam, Subbarayalu, Thangavelu, and Muthupandian 2022). Nanoparticles (NPs) are remarkably promising in biomedical applications due to their innate antioxidant, antibacterial, and antimicrobial properties, making them suitable for targeted drug delivery, wound healing, infection control, and tissue engineering (Shreyash, Bajpai, Khan, Vijay, Tiwary, and Sonker 2021). Nano suspensions synthesized using flash Nano precipitation enhances scalability,

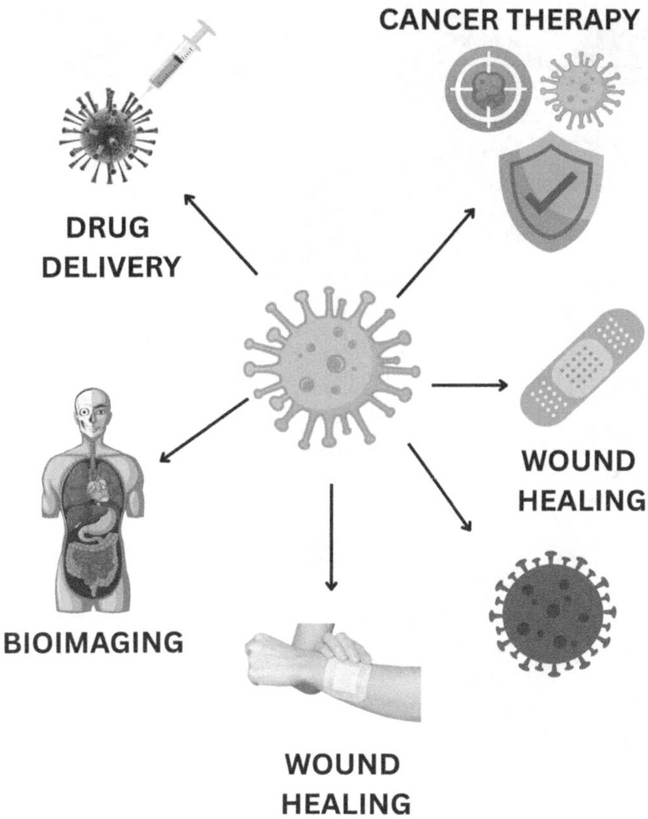

FIGURE 11.3 Biomedical applications of nanomaterials

solubility and encapsulation efficiency of the green Nano drugs (Zhu, Shao, Guo, Feng, Chen, and Yang 2021).

Biomedical applications of nanomaterials are shown in Figure 11.3

11.12 CH03_BIOMEDICAL APPLICATIONS_ DRUG DELIVERY

An eminent fusion of ancient medicines specifically Ayurveda are blended with modern science such as Nano technology are applied to synthesize Green Nano Bio based medicines. (Kaushik, Singh, and Tyagi 2023) It can be used as oral, parenteral as well as pulmonary drugs. (Ravichandran 2009). Nanoparticles are synthesized with specific surface properties to target the diseased cells. (Yusuf, Almotairy, Henidi, Alshehri, and Aldughaim 2023)

11.13 CH03_BIOMEDICAL APPLICATIONS_ ANTI-MICROBIAL ACTIVITY

The global demand of antibiotic resistance due to public threat has urged the research for alternative antimicrobial strategies (Ehtesabi, Fayaz, Hosseini-Doabi, and Rezaei 2023). Silver nanoparticles (AgNPs) exhibit potent antibacterial activity which is drug resistant to many drugs. It also helps to treat inflammatory conditions (Natarajan, Singh, Singh, Singh, Pandey, and Singh 2024; Dhaka, Mali, Sharma, and Trivedi 2023)

11.14 CH03_BIOMEDICAL APPLICATIONS_ WOUND HEALING

Green-synthesized nanoparticle bandages produced using phytochemicals are used as an Antibiotic Adjuvants to treat Bio-film related infections and diabetic wound (Khatami, Varma, Zafarnia, Yaghoobi, Sarani, and Kumar 2018; Kungwani, Panda, Mishra, Chavda, Shukla, Vikhe, Sharma, Mohanta, and Sharifi-Rad 2024)

11.15 CH03_BIOMEDICAL APPLICATIONS_ TISSUE ENGINEERING

Polymer based biomaterials specifically Chitosan has many biological properties such as Anti-Bacterial and Anti-fungal activity is used in osteochondrial scaffolds. (Malafaya, Pedro, Peterbauer, Gabriel, Redl, and Reis 2005; Beulah, Jinu, Ghorbanpour, and Venkatachalam 2019)

11.16 CH03_BIOMEDICAL APPLICATIONS CANCER THERAPY

Citrus sinensis silver Nanoparticles are used to treat Lung cancer due to its anti-microbial and anti-cancer properties (Gupta, Choudhary, and Sundaram 2025). Microorganisms based therapy stimulates the immune response and exhibits tumor specific effects (Huang, Guo, Wang, and Shao 2022) Medicinal plants such as *Boerhavia* and Ashwagandha are mostly used in cancer treatment (Krishnamoorthy, Muthukumaran, Maheswaran, and Sukumaran 2017)

11.17 CH03_BIOMEDICAL APPLICATIONS_BIOSENSORS

Biosensors like electrochemical, optical and plant variable biosensors are utilized in smart-farming (Bharti, Jain, and Chauhan 2024). Nano bio-sensors are utilized to identify the seed viability, shell life of the agricultural products and also the nutrient requirements (Dar, Qazi, and Pirzadah 2020).

11.18 CH04_ENVIRONMENTAL APPLICATIONS OF GREEN CHEMISTRY IN NANOTECHNOLOGY

Green chemistry utilizing nanotechnology is applied in waste water treatment and pollution control (Khan 2019)

The 12 principles of green chemistry are shown in Figure 11.4.

Environmental applications of green nanomaterials are shown in Figure 11.5

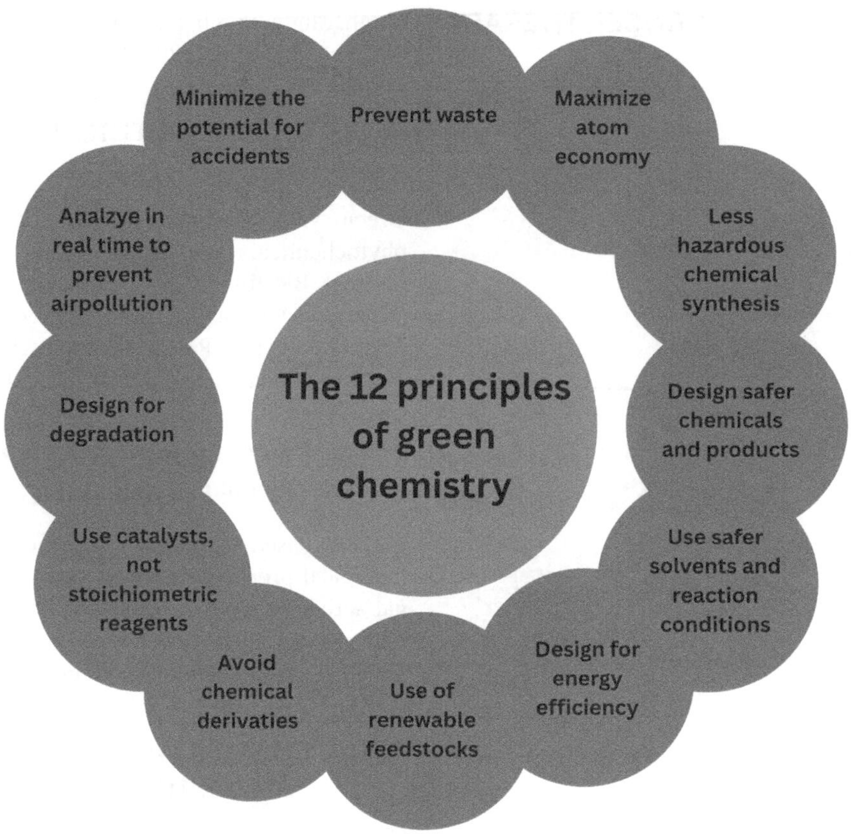

FIGURE 11.4 Twelve principles of green chemistry

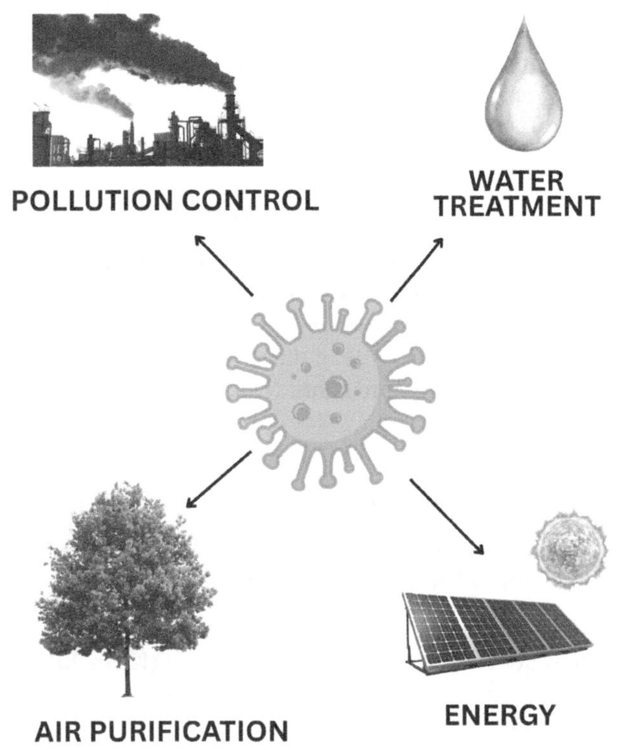

FIGURE 11.5 Environmental applications of green nanomaterials

11.19 CH04_ENVIRONMENTAL APPLICATIONS_POLLUTION CONTROL AND REMEDIATION

Nano-photo catalyst & Nano-adsorbents are applied in waste water treatment to eliminate the pollutants (Mehta 2021). Air pollutants are identified and deteriorated applying green nanoparticle due to their large surface area and increased energy (Sharma, Tyagi, Vikal, Devi, Gautam, and Singh 2022). Microbial Nano remediation technique eradicates the soil pollutants (Khan 2019). Green nanoparticles with various shapes and morphologies are utilized as adsorbents and photocatalysts to detect and eradicate various pollutants in water. (Singh, De, Guin, and Tomar 2023)

11.20 CH05_BACKGROUND_INDUSTRIAL APPLICATIONS AND CASE STUDIES

Synergistic effects of plant and microorganisms with metals are remarkable due to their physiochemical properties. Henceforth these Nanoparticles are applied across various sectors including Health care, agricultural and Bioremediation. (Bhardwaj, Singh, Kumar, Kumar, and Budhwar 2020; Joudeh and Linke 2022).

11.21 CH05_INDUSTRIAL APPLICATIONS IN THE AGRIFOOD SECTOR

Global threat in agrifood sector is environmental degradation. (Hendrawan, Rinanti, Fachrul, Tazkiaturrizki, Marendra, and Zahra 2024). These issues are expected to Amplify with rising population demands and the looming threat of global food insecurity. Consequently, there is an immediate need to transition from traditional farming practices to innovative, sustainable technologies. In this context, nanotechnology—particularly green nanotechnology—emerges as a promising solution (Cristofoli, Lima, Tchonkouang, Quintino, and Vieira 2023). It has the opportunity to transform various aspects of agriculture by dealing with eco-friendly, efficient products. Phosphorus-containing hydroxyapatite nanoparticles (nHAPs) were synthesized as a unique Nano-fertilizer using an environmentally friendly green synthesis method. Pomegranate peel extract (PPE) and coffee ground extract (CE) served as viable reducing and stabilizing agents in the synthesis process (Abdelmigid, Morsi, Hussien, Alyamani, Alhuthal, and Albukhaty 2022). Nanoparticles can selectively stimulate specific micronutrients in different aspects of crop performance and nutritional quality, providing a targeted approach to optimize agricultural productivity.

11.22 CH05_INDUSTRIAL APPLICATIONS IN FOOD SECTOR

Phytochemicals encapsulated metal Nanoparticles specifically silver nanoparticles serve as natural antimicrobial agents for food preservation (Singh, Tiwari, and Dubey 2021; Jadhav, Pawar, Choudhari, Topare, Raut-Jadhav, Bokil, and Khan 2023). Microbial tagged metal Nanoparticles, offers an eco-friendly strategy to control foodborne pathogens (Himaja, Archana, Kodaparthi, and Merugu 2025). For instance, green synthesis of SeNPs using plant extracts are used as reducing and stabilizing agents (Ndwandwe, Malinga, Kayitesi, and Dlamini 2021).

11.23 CH05_CASE STUDIES_ GREEN NANOPARTICLES IN WASTEWATER TREATMENT

Plant and microbial loaded nanoparticles are applied in eliminating heavy metals due to its biocompatibility. Carbon based nanoparticles and metal-based nanoparticles are effective in removal of heavy metals and toxic microorganisms (Sikiru, Abiodun, Sanusi, Sikiru, Soleimani, Yekeen, and Haslija, 2022) (Devatha, Thalla, and Katte 2016). Heavy metals such as cadmium (Cd^{2+}) and lead (Pb^{2+}) ions are removed using metal-based nanoparticles specifically freshwater algal extract loaded silver nanoparticles which is less than 15nm (Negi, Singh, and Rawat 2021).

11.24 CH06_CONCLUSION

Green nanomaterial is environmentally friendly and beneficial compared to chemical synthesis method. Their characterization uses advanced techniques to analyze their structural, morphological, and functional features for optimized performance. Green synthesis methods utilize plant extracts, microorganisms, and agricultural waste to produce nanomaterials with specific size control, enhanced surface reactivity, and a significantly lower environmental impact than traditional chemical synthesis techniques. A green nanoparticle in drug delivery systems provides targeted cancer therapy with minimal off-target effects, while their antimicrobial properties enhance wound healing and infection control. Nano-fertilizers and biosensors boost crop yields and reduce pesticide use, catalytic nanoparticles aid in environmental remediation, and Nano composites and Nano-catalysts magnify fuel efficiency and renewable energy storage.

ACKNOWLEDGEMENT

I acknowledge Mr. S. Bharathi Kumar, a native English Language expert from the Department of Humanities and Social Sciences, Sri Venkateswara College of Engineering, Sriperumbudur, for thoroughly and extensively editing the book chapter for grammar, punctuation, spelling, and overall style.

REFERENCES

Abdelmigid, H. M., Morsi, M. M., Hussien, N. A., Alyamani, A. A., Alhuthal, N. A., and Albukhaty, S. (2022). Green synthesis of phosphorous-containing hydroxyapatite nanoparticles (nHAP) as a novel nano-fertilizer: Preliminary assessment on pomegranate (Punica granatum L.). *Nanomaterials*, *12*(9), 1527.

Ahmad, A., Wei, Y., Syed, F., Tahir, K., Rehman, A. U., Khan, A., . . . and Yuan, Q. (2017). The effects of bacteria-nanoparticles interface on the antibacterial activity of green synthesized silver nanoparticles. *Microbial Pathogenesis*, *102*, 133–142.

Amin, M., Iram, F., Iqbal, M. S., Saeed, M. Z., Raza, M., and Alam, S. (2013). Arabinoxylan-mediated synthesis of gold and silver nanoparticles having exceptional high stability. *Carbohydrate Polymers*, *92*(2), 1896–1900.

Asghar, N., Hussain, A., Nguyen, D. A., Ali, S., Hussain, I., Junejo, A., and Ali, A. (2024). Advancement in nanomaterials for environmental pollutants remediation: A systematic review on bibliometrics analysis, material types, synthesis pathways, and related mechanisms. *Journal of Nanobiotechnology*, *22*(1), 26.

Babick, F. (2020). Dynamic light scattering (DLS). In *Characterization of Nanoparticles*. Elsevier, pp. 137–172.

Bahrulolum, H., Nooraei, S., Javanshir, N., Tarrahimofrad, H., Mirbagheri, V. S., Easton, A. J., and Ahmadian, G. (2021). Green synthesis of metal nanoparticles using microorganisms and their application in the agrifood sector. *Journal of Nanobiotechnology*, *19*, 1–26.

Beulah, P., Jinu, U., Ghorbanpour, M., and Venkatachalam, P. (2019). Green engineered chitosan nanoparticles and its biomedical applications—An overview. *Advances in Phytonanotechnology*, 329–341.

Bhardwaj, B., Singh, P., Kumar, A., Kumar, S., and Budhwar, V. (2020). Eco-friendly greener synthesis of nanoparticles. *Advanced Pharmaceutical Bulletin*, *10*(4), 566.

Bharti, A., Jain, U., and Chauhan, N. (2024). From lab to field: Nano-biosensors for real-time plant nutrient tracking. *Plant Nano Biology*, 100079.

Braidy, N., Béchu, A., de Souza Terra, J. C., and Patience, G. S. (2020). Experimental methods in chemical engineering: Transmission electron microscopy—TEM. *The Canadian Journal of Chemical Engineering*, *98*(3), 628–641.

Carter, R. S., Harley, S. J., Power, P. P., and Augustine, M. P. (2005). Use of NMR spectroscopy in the synthesis and characterization of air-and water-stable silicon nanoparticles from porous silicon. *Chemistry of Materials*, *17*(11), 2932–2939.

Chaudhary, S., Mahapatra, B., Pandey, V., Patel, C., Dubey, M., Kumar, M., and Mukherjee, S. (2024). Ultrasensitive detection of Hg 2+ ions with CVD grown MoS 2 functionalized MgZnO/CdZnO HEMT. *IEEE Sensors Journal*, vol. 25, no. 2, pp. 2048–2055, 15 Jan.15, 2025, doi: 10.1109/JSEN.2024.3504842.

Chen, J. L., Yang, P. C., Wu, T., and Lin, Y. W. (2018). Determination of mercury (II) ions based on silver-nanoparticles-assisted growth of gold nanostructures: UV–Vis and surface enhanced Raman scattering approaches. *Spectrochimica Acta Part A: Molecular and Biomolecular Spectroscopy*, *199*, 301–307.

Cristofoli, N. L., Lima, A. R., Tchonkouang, R. D. N., Quintino, A. C., and Vieira, M. C. (2023). Advances in the food packaging production from agri-food waste and by-products: Market trends for a sustainable development. *Sustainability*, *15*(7), 6153.

Dar, F. A., Qazi, G., and Pirzadah, T. B. (2020). Nano-biosensors: NextGen diagnostic tools in agriculture. In *Nanobiotechnology in Agriculture: An Approach Towards Sustainability*. Cham: Springer International Publishing, pp. 129–144.

Das, R. K., and Brar, S. K. (2013). Plant mediated green synthesis: Modified approaches. *Nanoscale*, *5*(21), 10155–10162.

De, A., Singh, N. B., Guin, M., and Barthwal, S. (2023). Water purification by green synthesized nanomaterials. *Current Pharmaceutical Biotechnology*, *24*(1), 101–117.

Devatha, C. P., and Thalla, A. K. (2018). Green synthesis of nanomaterials. In *Synthesis of Inorganic Nanomaterials*. Woodhead Publishing, pp. 169–184.

Devatha, C. P., Thalla, A. K., and Katte, S. Y. (2016). Green synthesis of iron nanoparticles using different leaf extracts for treatment of domestic waste water. *Journal of Cleaner Production*, *139*, 1425–1435.

Dhaka, A., Mali, S. C., Sharma, S., and Trivedi, R. (2023). A review on biological synthesis of silver nanoparticles and their potential applications. *Results in Chemistry*, *6*, 101108.

Ehtesabi, H., Fayaz, M., Hosseini-Doabi, F., and Rezaei, P. (2023). The application of green synthesis nanoparticles in wound healing: A review. *Materials Today Sustainability*, *21*, 100272.

Gupta, S., Choudhary, D. K., and Sundaram, S. (2025). Green Synthesis and characterization of silver nanoparticles using Citrus sinensis (Orange peel) extract and their antidiabetic, antioxidant, Antimicrobial and Anticancer Activity. *Waste and Biomass Valorization*, *16*(3), 1101–1114.

Hassan, P. A., Rana, S., and Verma, G. (2015). Making sense of Brownian motion: Colloid characterization by dynamic light scattering. *Langmuir*, *31*(1), 3–12.

Hendrawan, D. I., Rinanti, A., Fachrul, M. F., Tazkiaturrizki, Minarti, A., Marendra, S. M. P., and Zahra, L. A. (2024). Addressing algal bloom and other ecological issues caused by microalgae biomass conversion technology. In *Algae as a Natural Solution for Challenges in Water-Food-Energy Nexus: Toward Carbon Neutrality*. Singapore: Springer Nature Singapore, pp. 373–431.

Himaja, N., Archana, Y., Kodaparthi, A., and Merugu, R. (2025). Green nanotechnology: Harnessing plant-based nanoparticles for sustainable food preservation. In *Plant-Based Nanoparticle Synthesis for Sustainable Agriculture*. CRC Press, pp. 277–288.

Huang, X., Guo, H., Wang, L., and Shao, Z. (2022). Engineered microorganism-based delivery systems for targeted cancer therapy: A narrative review. *Biomaterials Translational*, *3*(3), 201.

Huston, M., DeBella, M., DiBella, M., and Gupta, A. (2021). Green synthesis of nanomaterials. *Nanomaterials*, *11*(8), 2130.

Iravani, S. (2011). Green synthesis of metal nanoparticles using plants. *Green Chemistry*, *13*(10), 2638–2650.

Jadhav, R., Pawar, P., Choudhari, V., Topare, N., Raut-Jadhav, S., Bokil, S., and Khan, A. (2023). An overview of antimicrobial nanoparticles for food preservation. *Materials Today: Proceedings*, *72*, 204–216.

Jadoun, S., Arif, R., Jangid, N. K., and Meena, R. K. (2021). Green synthesis of nanoparticles using plant extracts: A review. *Environmental Chemistry Letters*, *19*(1), 355–374.

Jia, Z., Li, J., Gao, L., Yang, D., and Kanaev, A. (2023). Dynamic light scattering: A powerful tool for *in situ* nanoparticle sizing. *Colloids and Interfaces*, *7*(1), 15.

Joudeh, N., and Linke, D. (2022). Nanoparticle classification, physicochemical properties, characterization, and applications: A comprehensive review for biologists. *Journal of Nanobiotechnology*, *20*(1), 262.

Kar, P., Oriola, A. O., and Oyedeji, A. O. (2024). Molecular docking approach for biological interaction of green synthesized nanoparticles. *Molecules*, *29*(11), 2428.

Kaushik, A., Singh, R. K., and Tyagi, P. K. (2023). Green synthesized nanoparticle-based drug delivery: Recent trends and future prospects. *Precision Nanomedicine*, *6*(4), 1109–1131.

Khan, S. H. (2019). Green nanotechnology for the environment and sustainable development. In *Green Materials for Wastewater Treatment*. Cham: Springer International Publishing, pp. 13–46.

Khatami, M., Varma, R. S., Zafarnia, N., Yaghoobi, H., Sarani, M., and Kumar, V. G. (2018). Applications of green synthesized Ag, ZnO and Ag/ZnO nanoparticles for making clinical antimicrobial wound-healing bandages. *Sustainable Chemistry and Pharmacy*, *10*, 9–15.

Krishnamoorthy, P. K., Muthukumaran, S., Maheswaran, A., and Sukumaran, P. (2017). Isolation, purification and characterization of Boeravinone B from Boerhaavia diffusa Linn. *International Research Journal of Pharmacy*, *8*, 140–144.

Kumar, P. S., Pavithra, K. G., and Naushad, M. (2019). Characterization techniques for nanomaterials. In *Nanomaterials for Solar Cell Applications*. Elsevier, pp. 97–124.

Kungwani, N. A., Panda, J., Mishra, A. K., Chavda, N., Shukla, S., Vikhe, K., Sharma, G., Mohanta, Y. K., and Sharifi-Rad, M. (2024). Combating bacterial biofilms and related drug resistance: Role of phyto-derived adjuvant and nanomaterials. *Microbial Pathogenesis*, 106874.

Liu, D., Wang, Q., Chang, H. L. M., and Chen, H. (1995). Variant structure in metal-organic-chemical-vapor-deposition-derived SnO2 thin films on sapphire (0001). *Journal of Materials Research*, *10*(6), 1516–1522.

Mahmudin, L., Suharyadi, E., Utomo, A. B. S., and Abraha, K. (2015). Optical properties of silver nanoparticles for surface plasmon resonance (SPR)-based biosensor applications. *Journal of Modern Physics*, *6*(8), 1071–1076.

Malafaya, P. B., Pedro, A. J., Peterbauer, A., Gabriel, C., Redl, H., and Reis, R. L. (2005). Chitosan particles agglomerated scaffolds for cartilage and osteochondral tissue engineering approaches with adipose tissue derived stem cells. *Journal of Materials Science: Materials in Medicine*, *16*, 1077–1085.

Malik, S., Muhammad, K., and Waheed, Y. (2023). Nanotechnology: A revolution in modern industry. *Molecules*, *28*(2), 661. https://doi.org/10.3390/molecules28020661

Medina-Ramirez, I., Bashir, S., Luo, Z., and Liu, J. L. (2009). Green synthesis and characterization of polymer-stabilized silver nanoparticles. *Colloids and Surfaces B: Biointerfaces*, *73*(2), 185–191.

Mehta, K. P. (2021). Application of nano technology in waste water treatment. In *Climate Change and Water Security: Select Proceedings of VCDRR 2021*. Singapore: Springer, pp. 423–432.

Natarajan, A., Singh, K. R. B., Singh, P., Singh, J., Pandey, S. S., and Singh. R. P. (2024). Silver nanoparticles based functional materials for anti-bacterial and antiviral applications. In *Antibacterial and Antiviral Functional Materials, Volume 2*. American Chemical Society, pp. 185–219.

Ndwandwe, B. K., Malinga, S. P., Kayitesi, E., and Dlamini, B. C. (2021). Advances in green synthesis of selenium nanoparticles and their application in food packaging. *International Journal of Food Science and Technology*, *56*(6), 2640–2650.

Negi, S., Singh, V., and Rawat, J. (2021). Green synthesis of silver nanoparticles using microalgal extract and its application in metal ion removal from aqueous solution, *Journal of Experimental Biology and Agricultural Sciences*, *9*(2), 214–230.

Otari, S. V., Patil, R. M., Nadaf, N. H., Ghosh, S. J., and Pawar, S. H. (2014). Green synthesis of silver nanoparticles by microorganism using organic pollutant: Its antimicrobial and catalytic application. *Environmental Science and Pollution Research*, *21*, 1503–1513.

Pasieczna-Patkowska, S., Cichy, M., and Flieger, J. (2025). Application of Fourier Transform Infrared (FTIR) spectroscopy in characterization of green synthesized nanoparticles. *Molecules*, *30*(3), 684.

Pochapski, D. J., Carvalho dos Santos, C., Leite, G. W., Pulcinelli, S. H., and Santilli, C. V. (2021). Zeta potential and colloidal stability predictions for inorganic nanoparticle dispersions: Effects of experimental conditions and electrokinetic models on the interpretation of results. *Langmuir*, *37*(45), 13379–13389.

Rahuman, H., Beevi, H., Dhandapani, R., Narayanan, S., Palanivel, V., Paramasivam, R., Subbarayalu, R., Thangavelu, S., and Muthupandian, S. (2022). Medicinal plants mediated the green synthesis of silver nanoparticles and their biomedical applications. *IET Nanobiotechnology*, *16*(4), 115–144.

Rauwel, P., Küünal, S., Ferdov, S., and Rauwel, E. (2015). A review on the green synthesis of silver nanoparticles and their morphologies studied via TEM. *Advances in Materials Science and Engineering*, *1*, 682749.

Ravichandran, R. (2009). Nanoparticles in drug delivery: Potential green nanobiomedicine applications. *International Journal of Green Nanotechnology: Biomedicine*, *1*(2), B108–B130.

Romano, I., Vitiello, G., Gallucci, N., Di Girolamo, R., Cattaneo, A., Poli, A., and Di Donato, P. (2022). Extremophilic microorganisms for the green synthesis of antibacterial nanoparticles. *Microorganisms*, *10*(10), 1885.

Shafey, A. M. E. (2020). Green synthesis of metal and metal oxide nanoparticles from plant leaf extracts and their applications: A review. *Green Processing and Synthesis*, *9*(1), 304–339.

Sharma, G., Pandey, S., Ghatak, S., Watal, G., and Rai, P. K. (2018). Potential of spectroscopic techniques in the characterization of green nanomaterials. In *Nanomaterials in Plants, Algae, and Microorganisms*. Academic Press, pp. 59–77.

Sharma, K., Tyagi, S., Vikal, S., Devi, A., Gautam, Y. K., and Singh, B. P. (2022). Green nanomaterials for remediation of environmental air pollution. In *Handbook of Green and Sustainable Nanotechnology: Fundamentals, Developments and Applications*. Cham: Springer International Publishing, pp. 1–26.

Shreyash, N., Bajpai, S., Khan, M. A., Vijay, Y., Tiwary, S. K., and Sonker, M. (2021). Green synthesis of nanoparticles and their biomedical applications: A review. *ACS Applied Nano Materials*, *4*(11), 11428–11457.

Sikiru, S., Abiodun, O. J. A., Sanusi, Y. K., Sikiru, Y. A., Soleimani, H., Yekeen, N., and Haslija, A. B. A. (2022). A comprehensive review on nanotechnology application in wastewater treatment a case study of metal-based using green synthesis. *Journal of Environmental Chemical Engineering*, *10*(4), 108065.

Singh, B. K., Tiwari, S., and Dubey, N. K. (2021). Essential oils and their nanoformulations as green preservatives to boost food safety against mycotoxin contamination of food commodities: A review. *Journal of the Science of Food and Agriculture*, *101*(12), 4879–4890.

Singh, N. B., De, A., Guin, M., and Tomar, R. (2023). Applications of green nanomaterials in environmental remediation. In Shanker, U., Hussain, C. M., and Rani, M. (eds) *Handbook of Green and Sustainable Nanotechnology*. Cham: Springer.

Sri, S., Chauhan, D., Lakshmi, G. B. V. S., Thakar, A., and Solanki, P. R. (2022). MoS2 nanoflower based electrochemical biosensor for TNF alpha detection in cancer patients. *Electrochimica Acta*, *405*, 139736.

Valentini, M., Vaccaro, A., Rehor, A., Napoli, A., Hubbell, J. A., and Tirelli, N. (2004). Diffusion NMR spectroscopy for the characterization of the size and interactions of colloidal matter: The case of vesicles and nanoparticles. *Journal of the American Chemical Society*, *126*(7), 2142–2147.

Yuan, H. X., Wang, B., and Yao, X. (2015). A new dominance relation-based evolutionary algorithm for many-objective optimization. *IEEE Transactions on Evolutionary Computation*, *20*(1), 16–37.

Yusuf, A., Almotairy, A. R. Z., Henidi, H., Alshehri, O. Y., and Aldughaim, M. S. (2023). Nanoparticles as drug delivery systems: A review of the implication of nanoparticles' physicochemical properties on responses in biological systems. *Polymers* (Basel), *15*(7), 1596.

Zhu, Z., Shao, C., Guo, Y., Feng, J., Chen, C., and Yang, H. (2021). Facile pathway to generate agrochemical nanosuspensions integrating super-high load, eco-friendly excipients, intensified preparation process, and enhanced potency. *Nano Research*, *14*, 2179–2187.

12 Advanced Characterization Techniques for Green Nanomaterials

G.K. Prashanth, Srilatha Rao, H.S. Lalithamba, N.P. Bhagya, K.V. Rashmi, and Mithun Kumar Ghosh

12.1 INTRODUCTION

Nanotechnology has revolutionized the scientific landscape, offering materials with extraordinary physical, chemical, and biological properties owing to their nanoscale dimensions. Among the various approaches to synthesizing nanomaterials, green synthesis has gained substantial attention as a sustainable, non-toxic, and environmentally benign method. Unlike conventional physical and chemical routes that often involve high energy consumption, hazardous reagents, or harsh conditions, green synthesis leverages biological systems—including plant extracts, microorganisms, enzymes, and polysaccharides—to reduce and stabilize nanomaterials [1–5].

Green nanomaterials have found extensive applications in environmental remediation, biomedical sciences, catalysis, energy harvesting, and food safety due to their eco-friendly synthesis and functional surface chemistry imparted by bioactive compounds. However, the performance and functionality of these nanomaterials are intimately linked to their physicochemical characteristics—such as size, shape, crystallinity, surface area, chemical composition, band gap energy, and surface charge. Hence, advanced characterization techniques are indispensable for unveiling these critical properties and for establishing the relationship between synthesis parameters and material performance [6–10].

This chapter focuses on the systematic characterization of green nanomaterials using a wide array of sophisticated analytical tools. The objective is to provide readers with a detailed understanding of each technique's working principles, the type of information it delivers, and its relevance in the context of green synthesis. Key characterization techniques covered include structural techniques like X-ray diffraction (XRD) and Raman spectroscopy; morphological techniques such as scanning and transmission electron microscopy (SEM and TEM); surface and chemical analysis methods, including FT-IR, XPS, and EDX; optical tools such as UV-vis spectroscopy and photoluminescence (PL); thermal methods like TGA and DSC; and electrochemical techniques, including cyclic voltammetry (CV) and electrochemical impedance spectroscopy (EIS).

The chapter also features case studies of plant-based green nanomaterial syntheses demonstrating how these methods are used in practice to verify material characteristics and evaluate their suitability for planned uses. Including new characterization techniques and real-time analytical tools provides a forward-looking perspective on the changing area of green nanomaterial research.

Readers will have a complete knowledge by the end of this chapter of how enhanced characterization supports the logical design, manufacture, and use of green nanomaterials in many scientific and industrial domains.

12.2 MORPHOLOGICAL CHARACTERIZATION

It is crucial to grasp the dimensional qualities, surface features, and morphological traits of green nanomaterials. Their size, shape, surface texture, degree of aggregation, and physicochemical and biological performance in medication delivery, environmental cleanup, biosensing, and catalysis all interact to produce nanomaterials.

Often, the shape of green nanomaterials is determined by the kind of bio-reducing and stabilizing chemicals present in plant extracts or microbial cultures. These biomolecules can regulate particle development, which produces a range of morphologies and surface roughnesses, including spherical, rod-like, flower-shaped, triangular, and irregular. Thus, morphological study provides significant knowledge on the stability of nanoparticles; their surface-to-volume ratio, which directly influences reactivity; and the effectiveness of the synthesis process.

12.2.1 SEM

An often-used technique, SEM looks at the shape and surface structure of nanomaterials. A concentrated electron beam projected across the specimen's surface produces high-resolution photographs. Their interactions with the material's atoms produce signals like secondary electrons, backscattered electrons, and distinctive X-rays. Detecting these transmitted signals then generates thorough topographical pictures. Surface roughness, material composition, and electron density in the sample [11, 12] are among the elements affecting image contrast and sharpness.

In the characterization of green nanomaterials, SEM is essential for evaluating the particle size, shape, porosity, and surface texture of materials produced utilizing plant extracts, microbial cultures, or biodegradable polymers.

DOI: 10.1201/9781003644439-12

Depending on the reducing and stabilizing chemicals used during synthesis, these materials often show unusual morphologies, including spherical, rod-like, flower-like, or porous architectures. SEM examination of biosynthesized zinc oxide nanoparticles produced by *Moringa oleifera* leaf extract revealed a regular hexagonal form with a low degree of aggregation [13]. Such morphological characteristics directly affect the photocatalytic, antibacterial, and adsorption capabilities of the materials. A schematic diagram of SEM is given in Figure 12.1.

12.2.2 TEM

An advanced characterization method, TEM offers in-depth knowledge of the internal structure, shape, particle size, and crystallinity of nanomaterials at atomic to nanoscale accuracy. Unlike SEM, which captures surface topography, TEM allows electrons to transmit through an ultra-thin specimen, generating high-resolution images that reveal the internal microstructure. The technique is indispensable for green nanomaterials, where detailed structural analysis is essential to confirm nanoscale features influenced by eco-friendly synthesis routes [14–17].

Green synthesis often yields nanoparticles capped with organic moieties derived from plant extracts or microbial metabolites, which influence shape and size at the nucleation and growth stages. TEM can directly visualize these nanoscale features and detect the lattice fringes of crystalline materials, aiding in the determination of interplanar spacing and crystallite orientation. For instance, gold nanoparticles synthesized using *Ocimum sanctum* (holy basil) leaf extract were shown to be spherical with sizes ranging from 10–25 nm, as observed under TEM, and exhibited clear lattice fringes confirming their crystalline nature [18]. Selected area electron diffraction (SAED) patterns obtained during TEM analysis offer complementary data on the crystalline phases present. In the case of green-synthesized silver nanoparticles using *Azadirachta indica* extract, SAED rings corresponded to the face-centered cubic structure of metallic silver, validating the phase purity [19]. High-resolution TEM (HRTEM) combined with EDX mapping within the same system allows for simultaneous imaging and elemental analysis at the nanoscale. This dual capability is particularly useful in studying composite or doped nanomaterials. For example, in the biosynthesis of ZnO-Fe_2O_3 nanocomposites using *Aloe vera* extract, HRTEM confirmed the heterojunction interface, while EDX confirmed elemental distribution [20]. The working principles of TEM are given in Figure 12.2.

12.2.3 Atomic Force Microscopy

Atomic force microscopy (AFM) is a high resolution surface characterization technique that provides three-dimensional (3D) topographical information about nanomaterials at the atomic and nanoscale levels. AFM works by mechanically scanning a sharp probe (or tip) across the sample's surface, in contrast to electron-based imaging methods like SEM or TEM. The forces between the probe tip and the sample surface—such as van der Waals, electrostatic, and capillary interactions—cause deflections in the cantilever,

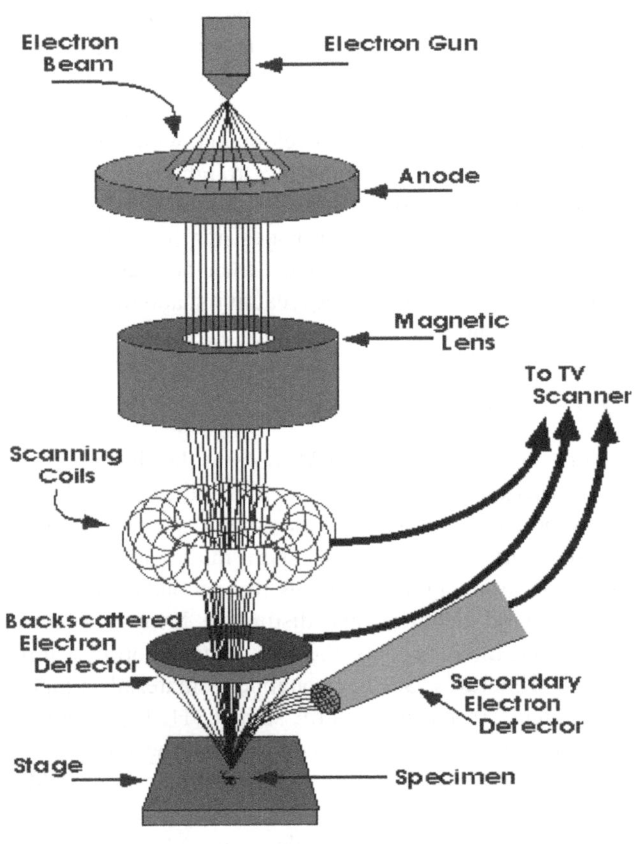

FIGURE 12.1 Schematic diagram of scanning electron microscopy.

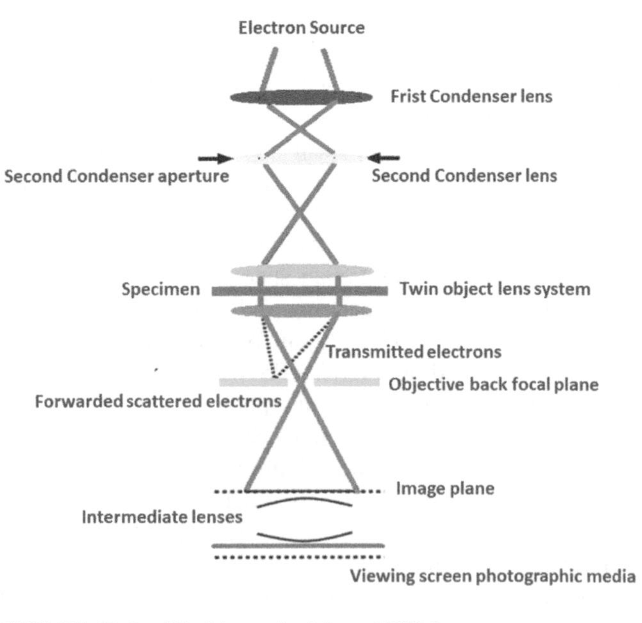

FIGURE 12.2 Working principles of TEM.

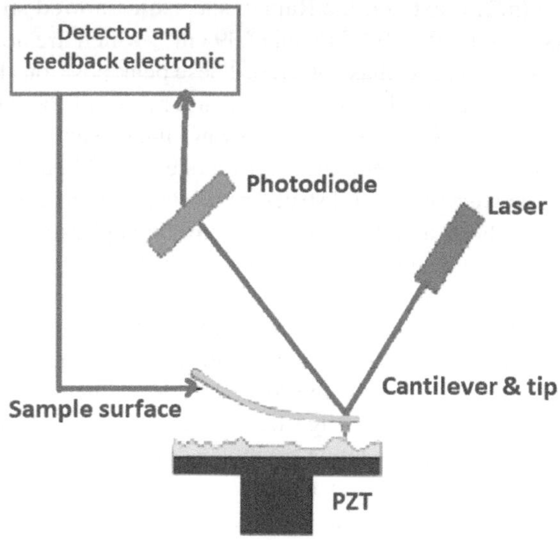

FIGURE 12.3 Working principles of AFM.

which are detected by a laser beam reflected into a photodetector. These deflections are converted into topographic maps, offering detailed information on surface roughness, texture, and particle height [21–25].

Since it is non-destructive and may operate in liquid, vacuum, or ambient settings, AFM is especially good at describing green nanomaterials. Its assistance allows for the analysis of soft, bio-functionalized, and hydrated surfaces—all of which are common in green-synthesized nanoparticles. AFM showed spherical particles averaging 18 nm in height and confirmed consistent surface distribution when *Carica papaya* leaf extract was used to generate silver nanoparticles [26]. The surface roughness values obtained by AFM can influence how nanomaterials interact with light, biomolecules, and catalytic reactants, helping link morphology with functional performance. The working principles of AFM are depicted in Figure 12.3.

12.3 STRUCTURAL CHARACTERIZATION

12.3.1 XRD

In the study of green nanomaterials, structural characterization of crystalline materials depends highly on XRD. XRD relies on the constructive interference of monochromatic X-rays with periodic atomic planes in a crystalline sample. X-rays hitting a crystalline material scatter its atoms and interplanar distances in particular directions. Each crystalline phase's unique diffraction patterns are recorded as peaks at various angles (2θ). The resulting XRD pattern [27–32] exposes important information about the material's crystal structure, phase composition, crystallite size, lattice strain, and degree of crystallinity. The calculation of crystallite size—which may be estimated using the Scherrer equation—is among the most useful uses of XRD in nanomaterials:

$$D = (K\lambda)/(\beta\cos\theta)$$

where λ is the X-ray wavelength, β is the full width at half maximum (FWHM) of the diffraction peak, θ is the Bragg angle, D is the average crystallite size, and K is the form factor (usually 0.9). This formula provides a reliable estimate of nanoscale dimensions and helps distinguish between nanoparticles with similar morphologies but different crystalline sizes. However, it is important to note that XRD measures crystallite size—not particle size—so complementary techniques like TEM are often used in conjunction with it.

In the context of green nanomaterials, XRD plays a vital role in confirming the formation and crystalline nature of nanoparticles synthesized using eco-friendly methods, such as plant extracts, bacteria, fungi, or algae. For instance, the XRD pattern revealed strong peaks at 2θ values of 31.7°, 34.4°, 36.2°, 47.5°, 56.6°, 62.8°, and 67.9° in the green synthesis of zinc oxide nanoparticles (ZnO NPs) using *Aloe vera* extract. These values correspond to the (100), (002), (101), (102), (110), (103), and (112) planes of the hexagonal wurtzite ZnO crystal structure (JCPDS Card No 36–1451). The great crystallinity and phase purity of the ZnO nanoparticles produced using an environmentally friendly method were validated by these diffraction peaks [33]. The working principles of XRD are given in Figure 12.4.

12.3.2 FT-IR

FT-IR is a powerful analytical technique widely employed to investigate chemical bonding, surface functional groups, and molecular interactions in green nanomaterials. It works on the principle of absorption of infrared (IR) radiation by chemical bonds in a molecule, causing them to vibrate at characteristic frequencies. These vibrations correspond to specific functional groups, and the resulting IR spectrum serves as a molecular fingerprint of the material. FT-IR is especially valuable in green nanotechnology for identifying the organic components involved in the synthesis and stabilization of nanoparticles, particularly those derived from plant extracts, microbial cultures, or natural polymers [34–39].

In green nanomaterials, FT-IR is widely used to confirm the presence of natural capping and reducing agents

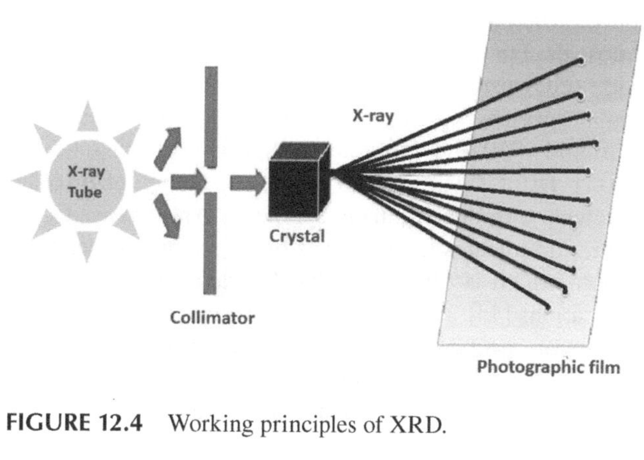

FIGURE 12.4 Working principles of XRD.

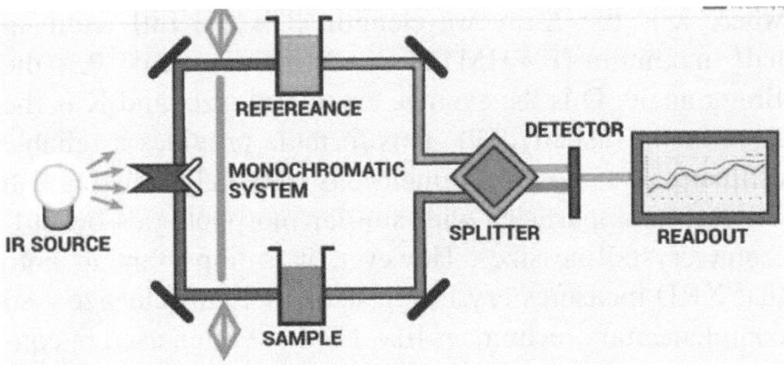

FIGURE 12.5 Working principles of FT-IR.

derived from biological sources such as plant extracts, microbes, or algae. These biomolecules—such as polyphenols, flavonoids, terpenoids, and proteins—act as stabilizers and reducing agents during nanoparticle formation. For instance, in the green synthesis of silver nanoparticles (Ag NPs) using *Ocimum sanctum* (holy basil) leaf extract, FT-IR spectra revealed significant peaks around 3420 and 1635 cm^{-1}, corresponding to O–H and N–H stretching and C=O stretching vibrations, respectively. These signals confirmed the involvement of hydroxyl and amide groups in the nanoparticle synthesis process [40]. Figure 12.5 depicts the working principles of FTIR.

12.3.3 Raman-Spectroscopy

A non-destructive vibrational spectroscopic method called Raman spectroscopy can reveal molecular-level details on the chemical and structural makeup of materials. When monochromatic light, usually from a laser source, interacts with phonons, molecular vibrations, or other excitations in the system, it scatters inelastically. Raman spectroscopy is a supplementary approach to FT-IR because it is sensitive to changes in the polarizability of molecular bonds, unlike infrared spectroscopy, which analyses changes in dipole moment. Because it can detect minute variations in molecular structure, crystallinity, chemical bonding, and the presence of functional groups, this technique is especially useful for characterizing green nanomaterials [41–43].

In the characterization of green nanomaterials, Raman spectroscopy is particularly valuable because it provides non-destructive, label-free chemical analysis even at the nanoscale. For example, in the green synthesis of graphene oxide (GO) using *Mangifera indica* (mango) leaf extract, Raman spectroscopy revealed characteristic D and G bands at around 1350 and 1580 cm^{-1}, respectively. The intensity ratio (I_D/I_G) was used to assess the degree of disorder or defects introduced during synthesis. An increased I_D/I_G ratio indicates higher levels of defects and functionalization, which is often desirable for catalysis and sensing applications [44].

Raman spectroscopy is also widely used in studying metal oxide nanoparticles such as TiO_2 and ZnO. In the case of green-synthesized TiO_2 nanoparticles using *Azadirachta indica* (neem) extract, the Raman spectrum showed strong peaks at 144, 197, 399, 515, and 639 cm^{-1}, which are attributed to the anatase phase of TiO_2. These peak positions help confirm the crystallographic phase and purity of the nanomaterial [45]. Similarly, in green ZnO nanostructures, the presence of a prominent E_2 (high) mode at ~437 cm^{-1} confirms the wurtzite crystal structure of ZnO, which is critical for their photocatalytic and optoelectronic applications.

For noble metal nanoparticles, such as gold (AuNPs) or silver, Raman spectroscopy can be further enhanced using a technique called surface-enhanced Raman scattering (SERS). When molecules are adsorbed on the surface of such nanostructures, the local electromagnetic field is amplified, leading to a significant increase in the Raman signal—enabling detection of molecules at even picomolar concentrations. For example, green-synthesized AgNPs using *Ocimum basilicum* extract showed excellent SERS enhancement for detecting dye pollutants like rhodamine B, making them ideal for environmental sensing applications [46]. The working principles of Raman spectroscopy are presented in Figure 12.6.

12.4 OPTICAL CHARACTERIZATION

12.4.1 UV-Vis Spectroscopy

UV-vis spectroscopy is one of the most widely used techniques in the characterization of green nanomaterials, particularly due to its simplicity, rapid analysis, and effectiveness in confirming nanoparticle formation. The technique measures the absorption of ultraviolet or visible light by a material, typically in the wavelength range of 200–800 nm. The absorption occurs due to electronic transitions within the molecules or nanomaterials, especially the excitation of valence electrons from the ground state to higher energy levels. Such absorption frequently produces surface plasmon resonance (SPR), a distinctive property of metal and metal oxide nanoparticles that offers information on the size, shape, concentration, and dispersion of the particles [47–49].

The main technique for tracking the effective reduction of metal ions into nanoparticles by biological agents like plant extracts or microbial secretions in the context of green synthesis is UV-vis spectroscopy. For instance, the synthesis and colloidal stability of silver nanoparticles made from *Azadirachta indica* leaf extract are confirmed by the presence of a characteristic SPR absorption band at about 420–450 nm [50]. Depending on their size and shape, gold nanoparticles show SPR bands between 520 and 540 nm. Changes in nanoparticle properties can be directly linked to shifts in peak location, bandwidth, and intensity in UV-vis spectra, providing a quick evaluation during production and storage.

Using Tauc's approach, UV-vis spectroscopy is also used to measure the optical band gap of semiconductor nanomaterials. For instance, ZnO or TiO_2 nanoparticles that are green synthesized frequently exhibit distinct absorption

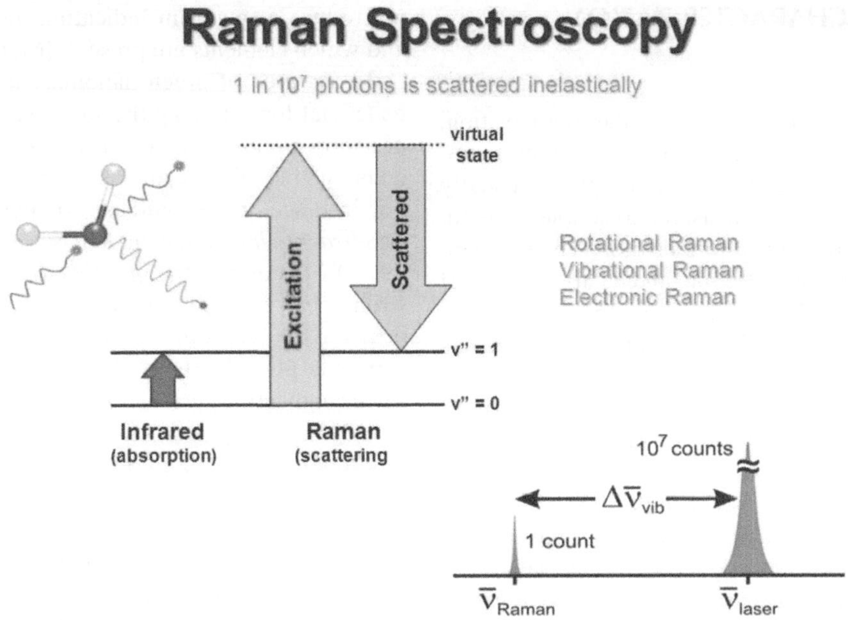

FIGURE 12.6 Working principles of Raman spectroscopy.

edges in the UV range (~350–390 nm), and their estimated band gap can reveal whether they are suitable for photocatalytic or optoelectronic uses. Following green alteration, variations in the band gap energy may indicate better light absorption capabilities or enhanced electron transition behavior [51].

12.4.2 PL Spectroscopy

A potent optical method for examining the electrical structure, defect states, and charge carrier dynamics of green nanomaterials—particularly semiconductors and quantum dots—is PL spectroscopy. A material's electrons are stimulated to higher energy states when it absorbs photons with a particular energy. The PL spectrum records the light emitted by these electrons as they relax back to their ground states. The nature, strength, and location of these emission peaks reveal important information on band gap energies, defect levels, and the recombination processes of photogenerated electron–hole pairs [52, 53].

Commonly produced using environmentally safe methods employing plant extracts or microbiological agents, metal oxide nanoparticles—such as ZnO, TiO_2, and CeO_2—are characterized by PL spectroscopy, which is extensively utilized in green nanotechnology. Green-synthesized ZnO nanoparticles, for instance, often exhibit visible emissions in the blue or green area and strong UV emission close to 380 nm (near-band-edge emission), which are ascribed to surface defects such as zinc interstitials or oxygen vacancies (Singh et al., 2020) [54].

PL study is also necessary for evaluating the photocatalytic efficacy of green nanomaterials. Generally speaking, lower PL intensity, which is usually a sign of less electron–hole recombination, is linked with higher photocatalytic activity. For example, a quenched or extended PL signal in green-synthesized TiO_2 nanoparticles employed to decompose organic pollutants would suggest improved charge separation and higher surface activity resulting from biofunctionalization using plant-based compounds. [55]

12.4.3 Band Gap Analysis (Tauc Plot)

Determining the optical band gap is absolutely essential for evaluating semiconductor nanomaterials, especially those meant for energy, optoelectronic, or photocatalytic applications. Among the most common methods for estimating the band gap is the Tauc plot study, which relies on UV-visible absorption spectroscopic data. The Tauc connection [56–58] links the absorption coefficient (α), incident photon energy (hv), and optical band gap (E_g):

$$(\alpha h\nu)^n = A(h\nu - E_g)$$

where A is a constant, hv is the photon energy, and n is defined by the type of electronic transition (for example, n = ½ for direct allowed transitions, n = 2 for indirect allowed transitions). The band gap energy is calculated by extrapolating the linear part of the curve to the photon energy axis (where $(\alpha h\nu)^n = 0$) from a plot of $(\alpha h\nu)^n$ versus hv. This strategy works especially well for nanomaterials made using environmentally friendly processes, such those mediated by microbes or plants, because the electrical characteristics are frequently influenced by surface and quantum effects. Green-synthesized ZnO nanoparticles, for instance, usually have band gaps between 3.1 and 3.3 eV, which are somewhat different from the bulk value because of quantum confinement or surface modification by biomolecules [59].

12.5 THERMAL CHARACTERIZATION

12.5.1 TGA

A common technique for evaluating the composition, breakdown behavior, and thermal stability of green nanomaterials is TGA. In a controlled atmosphere, usually air, nitrogen, or argon, TGA measures a sample's weight change as a function of temperature or time. Their uses in biomedicine, environmental cleanup, and catalysis [60–62] are particularly enhanced by knowledge of how nanomaterials made using ecologically benign techniques—such as plant extracts or microbial agents—react to heat.

In the production of green nanomaterials, TGA is used to verify the presence and measure the quantity of organic moieties—like capping agents or residual plant compounds—on the surface of nanoparticles. Though often utilized as stabilizing agents, these organic residues' thermal decomposition can offer significant insight on the material's composition. For instance, the evaporation of adsorbed moisture causes biosynthesized silver or zinc oxide nanoparticles to initially lose weight below 150°C. This is followed by a second stage of decomposition in the 200–500°C range, which is caused by the breakdown of phytochemicals or other biomolecules used in the synthesis [63].

12.6 SURFACE AND ELEMENTAL ANALYSIS

12.6.1 XPS

XPS is a surface-sensitive analytical technique that can be used to determine the elemental composition, oxidation states, and chemical environment of atoms within green nanomaterials. In order to perform XPS, a sample is exposed to monochromatic X-ray radiation, which causes photoelectrons to be ejected off the surface of the material. The binding energy of the elements present can be ascertained by measuring the kinetic energy of these released electrons, providing comprehensive information on the chemical state at a depth of around 5–10 nm [64, 65].

In green nanotechnology, XPS is particularly useful for verifying the presence of phytochemicals, functional groups, and dopants introduced during biosynthesis. Feature XPS examination of green-synthesized ZnO nanoparticles using *Moringa oleifera* extract revealed Zn $2p_{3/2}$ and Zn $2p_{1/2}$ peaks about 1022 and 1045 eV, respectively. Also discovered were O 1s peaks close to 530 eV, which imply surface hydroxyl groups and lattice oxygen [66].

12.6.2 EDX

Often used with electron microscopy methods like TEM or SEM, EDX—also known as EDS or EDAX—is an elemental analysis tool. It helps to evaluate and determine the elemental makeup of nanomaterials. Atoms in the sample are excited by a concentrated electron beam, which then releases distinct X-rays. They are then located and analyzed to produce a spectrum indicating their relative abundance and which elements are present. [67, 68].

In the area of green nanomaterials, EDX is especially beneficial for verifying the successful synthesis and purity of nanoparticles generated utilizing plant extracts, bacteria, or biodegradable polymers under eco-friendly conditions. For instance, green-synthesized silver nanoparticles using *Azadirachta indica* (neem) leaf extract can be characterized by EDX to confirm the presence of elemental silver, with a strong peak typically observed around 3 keV. Other minor peaks in the spectrum may correspond to residual elements from the plant biomolecules, such as carbon, oxygen, or sulfur, indicating surface functionalization or capping [69].

EDX also plays a crucial role in confirming the incorporation of dopants or additional components in composite or alloy nanostructures. In ZnO–Ag nanocomposites synthesized through green methods, EDX can clearly differentiate between Zn, O, and Ag signals, validating the formation of the desired composite and providing semi-quantitative analysis of the elemental distribution.

12.6.3 Brunauer–Emmett–Teller Surface Area Analysis

One popular technique for figuring out the precise surface area of porous and nanostructured materials is Brunauer–Emmett–Teller (BET) surface area analysis. The physical adsorption of gas molecules—usually nitrogen—onto a material's surface at cryogenic temperatures is the basis of this method. The BET theory makes it possible to calculate the surface area from adsorption isotherms and extends the Langmuir adsorption model to multilayer adsorption. The specific surface area is calculated from the linear part of the BET plot, which shows the quantity of gas adsorbed as a function of relative pressure [70].

In the context of green nanomaterials, BET analysis is essential for evaluating materials synthesized through environmentally benign methods, especially those intended for applications in catalysis, adsorption, and energy storage. Materials synthesized using plant extracts, agricultural waste, or biopolymers often exhibit high surface areas due to the templating or stabilizing effect of biomolecules during synthesis. For instance, green-synthesized ZnO nanoparticles using *Calotropis gigantea* leaf extract demonstrated a BET surface area of 45.6 m^2/g, which is significantly higher than that of bulk ZnO, enhancing their photocatalytic activity under sunlight [71].

Usually, a bigger surface area suggests more active zones for interactions for uses such dye degradation, pollutant adsorption, or antibiotic action. Combining BET analysis with the Barrett–Joyner–Halenda (BJH) method produces data on pore volume and pore size distribution. This is crucial for tailoring mesoporous or microporous structures in green nanomaterials either for drug delivery or selective adsorption. BET research thus helps to connect surface area with functional performance and offers significant information on the textural qualities of green nanoparticles. When paired with techniques like SEM, TEM, and XRD, it provides a comprehensive understanding of the shape, structure, and reactivity of the material.

12.7 ELECTROCHEMICAL CHARACTERIZATION

12.7.1 CV

Evaluating the redox behavior, electrochemical stability, and charge transfer characteristics of a nanomaterial using CV is a fundamental electrochemical method. After being linearly swept between two boundaries, CV reverses the potential of a working electrode, hence generating a cycle. The voltammogram, or voltage–current curve, reveals the capacitive or faradaic nature of the material, the oxidation and reduction processes at the electrode surface, and the reversibility of electron transfer reactions [72, 73].

CV is very useful for evaluating electrochemical activity in green nanomaterials, especially for sensors, supercapacitors, and catalytic processes. Green-synthesized metal oxide nanoparticles, such as NiO or FeO_3, made from plant extracts usually exhibit improved redox behavior owing surface functionalization by phytochemicals. A CV study of green-synthesized CuO nanoparticles using extract from *Ocimum sanctum*, for instance, showed obvious redox peaks related to the Cu^{20}/Cu^{i} pair, therefore suggesting the electrochemical activity and their application in energy storage devices. [74].

The peak current response and the form of the CV curve provide clues about diffusion-controlled processes and the kinetics of electron transfer. While a rectangular CV profile indicates electric double-layer capacitance, redox peaks in supercapacitor materials suggest pseudocapacitive behavior. The effect of scan rate in CV studies also helps to determine the charge storage mechanism, which is crucial for designing efficient energy devices using sustainable nanomaterials.

12.7.2 EIS

EIS is a strong technique in electrochemical systems for investigating the interfacial properties and charge transfer resistance of materials. It tracks the current response produced by delivering a small sinusoidal AC voltage to a material or device over a broad frequency range. Examining the impedance—a complex interplay of resistance and capacitance—helps one to understand the many physicochemical processes taking place at the electrode–electrolyte interface, including charge transfer, double-layer formation, ion transport, and reaction kinetics [75, 76].

EIS is needed to assess the performance of green nanomaterials in applications such electrocatalysts, batteries, and supercapacitors. Often, materials made using ecologically friendly techniques, including metal oxides or carbon-based nanomaterials generated from plant waste or biomass, benefit from enhanced surface functions and porosity. These characteristics greatly influence the EIS response, as they can greatly enhance ion diffusion and lower charge transfer resistance. For example, an EIS investigation of green-synthesized activated carbon made from coconut shells revealed a linear tail at low frequencies implying efficient ion diffusion and capacitive behavior, as well as a short semicircular arc in the Nyquist plot, indicating low charge transfer resistance [77].

12.8 ADDITIONAL TECHNIQUES

12.8.1 Zeta-Potential Analysis

Zeta-potential analysis is a significant tool for evaluating the surface charge and colloidal stability of nanomaterials spread in a liquid environment. It computes the electrostatic potential at the sliding plane of particles using electrophoretic light scattering, which shows how strongly particles resist or attract one another in suspension. Usually, a high absolute value of zeta-potential—either positive or negative—indicates strong particle repulsion, which suggests good dispersion and stability. On the other hand, low numbers indicate a tendency to agglomerate [78, 79].

In the area of green nanomaterials, zeta-potential study is crucial for assessing the stability of nanoparticles produced from plant extracts, biodegradable polymers, or microorganisms. Often, capping chemicals or biomolecules used in green synthesis methods alter the surface charge by attaching to the surfaces of nanoparticles and stabilizing them sterically or electrostatically. Silver nanoparticles produced from *Azadirachta indica* (neem) leaf extract, for example, have shown a zeta-potential of about –32 mV, implying better stability and even dispersion in water [80].

12.8.2 Dynamic Light Scattering

A strong method for ascertaining the hydrodynamic size and size distribution of suspended nanoparticles is dynamic light scattering (DLS), sometimes called photon correlation spectroscopy or quasi-elastic light scattering. It tracks variations in the intensity of scattered light brought on by particle Brownian motion. The Stokes-Einstein equation is used to transform the acquired diffusion coefficients into particle size. Unlike microscopic techniques that study particles in a dry state, DLS provides significant insights into the behavior of nanoparticles in their original colloidal form [81, 82].

When it comes to green nanomaterials, DLS is especially crucial for comprehending how particle size and colloidal stability in aqueous environments are impacted by the use of natural reducing and stabilizing agents, such as plant extracts, polysaccharides, or microbial secretions. Green-produced nanoparticles frequently have a hydrodynamic diameter that is marginally greater than what electron microscopy can see because of the solvation layer and surface-bound proteins. For instance, the DLS measurement of 65 nm for gold nanoparticles made with *Hedychium coronarium* leaf extract was larger than the TEM measurement of 40 nm, suggesting that biomolecules were capping the particles [83].

Therefore, a variety of analytical methods, including TEM, AFM, XRD, FT-IR, UV-vis, PL, EDX, BET, EIS, zeta potential, DLS, and many more, can be used to better understand green nanomaterials [84–101].

12.9 CONCLUSION AND FUTURE PROSPECTS

Green nanomaterials, which provide sustainable and environmentally benign substitutes for a range of technical, environmental, and biological applications, have paved the way for a new era in nanoscience. In order to comprehend the basic characteristics of these nanomaterials, such as their shape, structural integrity, surface chemistry, thermal stability, optical behavior, electrochemical performance, and colloidal characteristics, advanced characterization techniques are essential. Techniques such as SEM, TEM, AFM, XRD, FT-IR, Raman spectroscopy, UV-vis, PL, TGA, XPS, EDX, BET, CV, EIS, zeta potential, DLS, and others collectively provide a holistic understanding of the physicochemical nature and functional capabilities of green nanomaterials.

These characterization tools are essential for both confirming the effectiveness of green synthesis routes and customizing material attributes for particular uses, such as medicine delivery, energy storage, environmental cleanup, and photocatalysis. Moreover, the ability to correlate structure–property relationships through such techniques is vital for rational design and scalable production of green nanomaterials with reproducible and enhanced performance.

Looking ahead, the field of green nanotechnology demands the integration of in-situ and real-time characterization methods to monitor nanoparticle formation and transformation in dynamic environments. The advancement of hybrid techniques (such as correlative microscopy) and machine learning-assisted data interpretation can significantly accelerate the material discovery process. Additionally, there is a growing need for standardization of characterization protocols specifically tailored for bio-derived and plant-based nanomaterials to ensure consistency, reproducibility, and regulatory compliance, especially for biomedical and environmental applications.

In conclusion, while green nanomaterials offer a promising solution to the sustainability challenge, their true potential can only be realized through the continued innovation and application of advanced, precise, and complementary characterization methods. These techniques not only deepen our understanding of material behavior but also pave the way for next-generation applications where functionality and environmental compatibility go hand in hand.

REFERENCES

1. Malik, S., Muhammad, K. and Waheed, Y., 2023. Nanotechnology: A revolution in modern industry. *Molecules*, *28*(2), p.661. https://doi.org/10.3390/molecules28020661
2. Singh, N.B., Kumar, B., Usman, U.L. and Susan, M.A.B.H., 2024. Nano revolution: Exploring the frontiers of nanomaterials in science, technology, and society. *Nano-Structures & Nano-Objects*, *39*, p.101299. https://doi.org/10.1016/j.nanoso.2024.101299
3. Kumar, R.D.S., Kumar, L.H., Jeeva Roshini, S.K., Varghese, J. and Singh, L., 2024, December. Nanotechnology in mechanical engineering-a review. In *Materials Science Forum* (Vol. 1143, pp.47–70). Trans Tech Publications Ltd. https://doi.org/10.4028/p-abdB7G
4. Muneer, N., 2023. Emerging materials and technologies for the next generation of engineering solutions. *Liberal Journal of Language and Literature Review*, *1*(1), pp.39–46. https://doi.org/10.1007/978-3-031-72955-3_4
5. Hornyak, G.L., Moore, J.J., Tibbals, H.F. and Dutta, J., 2018. *Fundamentals of Nanotechnology*. CRC Press. https://doi.org/10.1201/9781315222561
6. Sharma, A., Manocha, D., Arora, A., Kumar, A., Srivastava, C.M., Rawat, V., Kim, H., Verma, M. and Rao, G.K., 2022. Applications of green nanomaterials in electronic and electrical industries. In *Green Nanomaterials for Industrial Applications* (pp. 397–421). Elsevier. https://doi.org/10.1016/B978-0-12-823296-5.00016-2
7. Gómez-López, P., Puente-Santiago, A., Castro-Beltrán, A., do Nascimento, L.A.S., Balu, A.M., Luque, R. and Alvarado-Beltrán, C.G., 2020. Nanomaterials and catalysis for green chemistry. *Current Opinion in Green and Sustainable Chemistry*, *24*, pp.48–55. https://doi.org/10.1016/j.cogsc.2020.03.00
8. Dutta, V., Devasia, J., Chauhan, A., VL, V., Jha, A., Nizam, A., Lin, K.Y.A. and Ghotekar, S., 2022. Photocatalytic nanomaterials: Applications for remediation of toxic polycyclic aromatic hydrocarbons and green management. *Chemical Engineering Journal Advances*, *11*, p.100353. https://doi.org/10.1016/j.ceja.2022.100353
9. Noah, N.M. and Ndangili, P.M., 2022. Green synthesis of nanomaterials from sustainable materials for biosensors and drug delivery. *Sensors International*, *3*, p.100166. https://doi.org/10.1016/j.sintl.2022.100166
10. Verma, C., Verma, D.K., Berdimurodov, E., Barsoum, I., Alfantazi, A. and Hussain, C.M., 2024. Green magnetic nanoparticles: A comprehensive review of recent progress in biomedical and environmental applications. *Journal of Materials Science*, *59*(2), pp.325–358. https://doi.org/10.1007/s10853-023-08914-5
11. Rasool, S., Faheem, M., Hanif, U., Bahadur, S., Taj, S., Liaqat, F., Pereira, L., Liaqat, I., Shaheen, S., Shuaib, M. and Gulzar, S., 2022. Toxicological effects of the chemical and green ZnO NPs on Cyprinus carpio L. observed under light and scanning electron microscopy. *Microscopy Research and Technique*, *85*(3), pp.848–860. https://doi.org/10.1002/jemt.23954
12. Rozina, Ahmad, M., Zafar, M., Yousaf, Z., Ullah, S.A., Sultana, S. and Bibi, F., 2022. Identification of novel, nonedible oil seeds via scanning electron microscopy as potential feedstock for green synthesis of biodiesel. *Microscopy Research and Technique*, *85*(2), pp.708–720. https://doi.org/10.1002/jemt.23942
13. Singh, J., Dutta, T., Kim, K.H., Rawat, M., Samddar, P. and Kumar, P., 2021. Green synthesis of metals and their oxide nanoparticles: Applications for environmental remediation.

Journal of Nanobiotechnology, *19*(1), pp.1–24. https://doi.org/10.1186/s12951-018-0408-4
14. Su, D., 2017. Advanced electron microscopy characterization of nanomaterials for catalysis. *Green Energy & Environment*, *2*(2), pp.70–83. https://doi.org/10.1016/j.gee.2017.02.001
15. Rauwel, P., Küünal, S., Ferdov, S. and Rauwel, E., 2015. A review on the green synthesis of silver nanoparticles and their morphologies studied via TEM. *Advances in Materials Science and Engineering*, *2015*(1), p.682749. https://doi.org/10.1155/2015/682749
16. Anjum, D.H., 2016, August. Characterization of nanomaterials with transmission electron microscopy. In *IOP Conference Series: Materials Science and Engineering* (Vol. 146, No. 1, p.012001). IOP Publishing. https://doi.org/10.1088/1757-899X/146/1/012001
17. Banala, R.R., Nagati, V.B. and Karnati, P.R., 2015. Green synthesis and characterization of Carica papaya leaf extract coated silver nanoparticles through X-ray diffraction, electron microscopy and evaluation of bactericidal properties. *Saudi Journal of Biological Sciences*, *22*(5), pp.637–644. https://doi.org/10.1016/j.sjbs.2015.01.007
18. Bar, H., Bhui, D.K., Sahoo, G.P., Sarkar, P. and Misra, A., 2009. Green synthesis of silver nanoparticles using Ocimum sanctum (Tulsi) leaf extract and screening its antimicrobial activity. *Journal of Materials Science*, *44*(19), pp.5063–5070. https://doi.org/10.1016/j.bbrep.2020.100848
19. Ahmed, S., Ahmad, M., Swami, B.L. and Ikram, S., 2016. A review on plants extract mediated synthesis of silver nanoparticles for antimicrobial applications: A green expertise. *Journal of Advanced Research*, *7*(1), pp.17–28. https://doi.org/10.1016/j.jare.2015.02.007
20. Kumar, P., Senthamil Selvi, S., Govindaraju, M. and Sivakumar, R., 2020. Green synthesis and characterization of $ZnO–Fe_2O_3$ nanocomposite using Aloe vera extract and its photocatalytic activity. *Environmental Nanotechnology, Monitoring & Management*, *14*, p.100326. https://doi.org/10.1016/j.enmm.2020.100326
21. Kim, J., Yhim, W.B., Park, J.W., Lee, S.H., Kim, T.Y., Cha, S.H., Kim, H.S., Jang, H.L., Cho, M., Park, Y. and Cho, S., 2016. Gallotannin-capped gold nanoparticles: Green synthesis and enhanced morphology of AFM images. *Journal of Nanoscience and Nanotechnology*, *16*(6), pp.5991–5998. https://doi.org/10.1166/jnn.2016.10965
22. Moosa, A., Ridha, A.M. and Allawi, M.H., 2015. Green synthesis of silver nanoparticles using spent tea leaves extract with atomic force microscopy. *International Journal of Current Engineering and Technology*, *5*(5), pp.3233–3241.
23. Demir-Yilmaz, I., Guiraud, P. and Formosa-Dague, C., 2021. The contribution of Atomic Force Microscopy (AFM) in microalgae studies: A review. *Algal Research*, *60*, p.102506. https://doi.org/10.1016/j.algal.2021.102506
24. Mišić Radić, T., Vukosav, P., Čačković, A. and Dulebo, A., 2023. Insights into the morphology and surface properties of microalgae at the nanoscale by atomic force microscopy (AFM): A review. *Water*, *15*(11), p.1983.https://doi.org/10.3390/w15111983
25. Fuoco, D., 2012. A new method for characterization of natural zeolites and organic nanostructure using atomic force microscopy. *Nanomaterials*, *2*(1), pp.79–91. https://doi.org/10.3390/nano2010079
26. Hoseinzadeh, E., Makhdoumi, P., Taha, P., Hossini, H. and Stelling, J., 2019. Green synthesis of silver nanoparticles using *Carica papaya* and evaluation of their antifungal activity. *Applied Nanoscience*, *9*(3), pp.403–411. https://doi.org/10.1007/s13204-018-0783-4.
27. Kumar, M., Ranjan, R., Dandapat, S., Srivastava, R. and Sinha, M.P., 2022. XRD analysis for characterization of green nanoparticles: A mini review. *Global Journal of Pharmacy and Pharmaceutical Sciences*, *10*(1), p.555779.
28. Banala, R.R., Nagati, V.B. and Karnati, P.R., 2015. Green synthesis and characterization of Carica papaya leaf extract coated silver nanoparticles through X-ray diffraction, electron microscopy and evaluation of bactericidal properties. *Saudi Journal of Biological Sciences*, *22*(5), pp.637–644. https://doi.org/10.1016/j.sjbs.2015.01.007
29. Ahmad, A., Pervaiz, M., Ramzan, S., Kiyani, M.Z., Khan, A., Ahmad, I. and Asiri, A.M., 2021. Role of XRD for nanomaterial analysis. In *Nanomedicine Manufacturing and Applications* (pp. 149–161). Elsevier. https://doi.org/10.1016/B978-0-12-820773-4.00008-1
30. Khan, H., Yerramilli, A.S., D'Oliveira, A., Alford, T.L., Boffito, D.C. and Patience, G.S., 2020. Experimental methods in chemical engineering: X-ray diffraction spectroscopy—XRD. *The Canadian Journal of Chemical Engineering*, *98*(6), pp.1255–1266. https://doi.org/10.1002/cjce.23747
31. Thakar, M.A., Jha, S.S., Phasinam, K., Manne, R., Qureshi, Y. and Babu, V.H., 2022. X ray diffraction (XRD) analysis and evaluation of antioxidant activity of copper oxide nanoparticles synthesized from leaf extract of Cissus vitiginea. *Materials Today: Proceedings*, *51*, pp.319–324. https://doi.org/10.1016/j.matpr.2021.05.410
32. Singh, V., Shrivastava, A. and Wahi, N., 2015. Biosynthesis of silver nanoparticles by plants crude extracts and their characterization using UV, XRD, TEM and EDX. *African Journal of Biotechnology*, *14*(33), pp.2554–2567. https://doi.org/10.5897/AJB2015.14692
33. Rajendran, R.L., Balasubramanian, V. and Palanisamy, S., 2020. Green synthesis of ZnO nanoparticles using Aloe vera leaf extract and evaluation of their antimicrobial and antioxidant activities. *Materials Science in Semiconductor Processing*, *118*, p.105282. https://doi.org/10.1016/j.mssp.2020.105282
34. Pasieczna-Patkowska, S., Cichy, M. and Flieger, J., 2025. Application of Fourier Transform Infrared (FTIR) spectroscopy in characterization of green synthesized nanoparticles. *Molecules*, *30*(3), p.684. https://doi.org/10.3390/molecules30030684
35. Nikolić, G.S., Cakić, M.D., Glišić, S., Cvetković, D.J., Mitić, Ž.J. and Marković, D.Z., 2017. Study of green nanoparticles and biocomplexes based on exopolysaccharide by modern Fourier transform spectroscopy. In *Fourier Transforms-High-tech Application and Current Trends*. IntechOpen. https://doi.org/10.5772/65776
36. Vasanth, V., Senguttuvan, K., Murugesh, K.A., Swathiga, G., Arasakumar, E., Kumar, S.R., Navaneetha Krishnan, S., SJ, M.P. and Nitish, G., 2023. Phytochemical analysis of green synthesised S. platensis mediated TiO2 NPs through Fourier transform infrared spectroscopy (FT-IR). *International Journal of Statistics and Applied Mathematics*, *8*(6S), pp.642–647.

37. Khan, S.A., Khan, S.B., Khan, L.U., Farooq, A., Akhtar, K. and Asiri, A.M., 2018. Fourier transform infrared spectroscopy: Fundamentals and application in functional groups and nanomaterials characterization. In *Handbook of Materials Characterization* (pp.317–344). https://doi.org/10.1007/978-3-319-92955-2_9
38. Faghihzadeh, F., Anaya, N.M., Schifman, L.A. and Oyanedel-Craver, V., 2016. Fourier transform infrared spectroscopy to assess molecular-level changes in microorganisms exposed to nanoparticles. *Nanotechnology for Environmental Engineering*, *1*, pp.1–16. https://doi.org/10.1007/s41204-016-0001-8
39. Díaz-Visurraga, J., Daza, C., Pozo, C., Becerra, A., von Plessing, C. and García, A., 2012. Study on antibacterial alginate-stabilized copper nanoparticles by FT-IR and 2D-IR correlation spectroscopy. *International Journal of Nanomedicine*, pp.3597–3612. https://doi.org/10.2147/IJN.S32648
40. Ahmed, S., Saifullah, Ahmad, M., Swami, B.L. and Ikram, S., 2016. Green synthesis of silver nanoparticles using Azadirachta indica aqueous leaf extract. *Journal of Radiation Research and Applied Sciences*, *9*(1), pp.1–7. https://doi.org/10.1016/j.jrras.2015.06.006
41. Ge, J., Li, Y., Wang, J., Pu, Y., Xue, W. and Liu, X., 2016. Green synthesis of graphene quantum dots and silver nanoparticles compounds with excellent surface enhanced Raman scattering performance. *Journal of Alloys and Compounds*, *663*, pp.166–171. https://doi.org/10.1016/j.jallcom.2015.12.055
42. Li, Y.S. and Church, J.S., 2014. Raman spectroscopy in the analysis of food and pharmaceutical nanomaterials. *Journal of Food and Drug Analysis*, *22*(1), pp.29–48. https://doi.org/10.1016/j.jfda.2014.01.003
43. Pasieczna-Patkowska, S., Cichy, M. and Flieger, J., 2025. Application of Fourier Transform Infrared (FTIR) spectroscopy in characterization of green synthesized nanoparticles. *Molecules*, *30*(3), p.684. https://doi.org/10.3390/molecules30030684
44. Khan, Z., Hussain, J.I. and Hashmi, A.A., 2019. Green synthesis of graphene oxide using Mangifera indica leaf extract and its application in the removal of methylene blue dye. *Journal of Environmental Chemical Engineering*, *7*(3), p.103100. https://doi.org/10.1016/j.jece.2019.103100
45. Srinivasan, R., Santhosh, R. and Jeyaraj, B., 2020. Green synthesis of TiO_2 nanoparticles using Azadirachta indica leaf extract and its photocatalytic activity. *Materials Today: Proceedings*, *33*, pp.3536–3540. https://doi.org/10.1016/j.matpr.2020.03.048
46. Kaviya, S., Santhanalakshmi, J., Viswanathan, B., Muthumary, J. and Srinivasan, K., 2011. Biosynthesis of silver nanoparticles using Citrus sinensis peel extract and its antibacterial activity. *Spectrochimica Acta Part A: Molecular and Biomolecular Spectroscopy*, *79*(3), pp.594–598. https://doi.org/10.1016/j.saa.2011.03.037
47. Rizwan, M. and Gwenin, C., 2021. Nanomaterials in renewable energy: UV-Visible spectroscopy characterization and applications. In *Nano Tools and Devices for Enhanced Renewable Energy* (pp. 103–120). Elsevier. 10.1016/B978-0-12-821709-2.00017-7
48. Kumar, A., Meenakshi, Saxena, N., Seema, K., Srivastava, A., Rajak, J., Singh, D., Singh, S.V. and Singh, L., 2023. Physico-chemical characterization of green synthesized nanomaterials by UV-visible spectroscopy. *Current Nanomaterials*, *8*(2), pp.102–109. https://doi.org/10.2174/2405461507666220721115604
49. Shukla, A.K. and Iravani, S., 2016. Green synthesis and spectroscopic characterization of nanoparticles. *Nanoscience in Food and Agriculture*, *1*, pp.65–99. https://doi.org/10.1007/978-3-319-39303-2_3
50. Ahmed, S., Ahmad, M., Swami, B.L. and Ikram, S., 2016. A review on plants extract mediated synthesis of silver nanoparticles for antimicrobial applications: A green expertise. *Journal of Advanced Research*, *7*(1), pp.17–28. https://doi.org/10.1016/j.jare.2015.02.007
51. Kaviya, S., Santhanalakshmi, J., Viswanathan, B., Muthumary, J. and Srinivasan, K., 2011. Biosynthesis of silver nanoparticles using Citrus sinensis peel extract and its antibacterial activity. *Spectrochimica Acta Part A: Molecular and Biomolecular Spectroscopy*, *79*(3), pp.594–598. https://doi.org/10.1016/j.saa.2011.03.040
52. Goswami, S., Nandy, S., Deuermeier, J., Marques, A.C., Nunes, D., Patole, S.P., Costa, P.M., Martins, R. and Fortunato, E., 2018. Green nanotechnology from waste carbon–polyaniline composite: Generation of wavelength-independent multiband photoluminescence for sensitive ion detection. *Advanced Sustainable Systems*, *2*(1), p.1700137. https://doi.org/10.1002/adsu.201700137
53. Peng, W.Q., Cong, G.W., Qu, S.C. and Wang, Z.G., 2006. Synthesis and photoluminescence of ZnS: Cu nanoparticles. *Optical Materials*, *29*(2–3), pp.313–317. https://doi.org/10.1016/j.optmat.2005.10.00
54. Singh, P., Kim, Y.J., Zhang, D. and Yang, D.C., 2020. Biological synthesis of nanoparticles from plants and microorganisms. *Trends in Biotechnology*, *34*(7), pp.588–599. https://doi.org/10.1016/j.tibtech.2016.02.006
55. Pushpamalini, T., Keerthana, M., Sangavi, R., Nagaraj, A. and Kamaraj, P., 2021. Comparative analysis of green synthesis of TiO2 nanoparticles using four different leaf extract. *Materials Today: Proceedings*, *40*, pp.S180–S184. https://doi.org/10.1016/j.matpr.2020.08.438
56. Johannes, A.Z., Pingak, R.K. and Bukit, M., 2020, April. Tauc Plot Software: Calculating energy gap values of organic materials based on Ultraviolet-Visible absorbance spectrum. In *IOP Conference Series: Materials Science and Engineering* (Vol. 823, No. 1, p.012030). IOP Publishing. https://doi.org/10.1088/1757-899X/823/1/012030
57. Buniyamin, I., Akhir, R.M., Asli, N.A., Khusaimi, Z. and Rusop, M., 2022. Green synthesis of tin oxide nanoparticles by using leaves extract of Chromolaena Odorata: The effect of different thermal calcination temperature to the energy band gap. *Materials Today: Proceedings*, *48*, pp.1805–1809. https://doi.org/10.1016/j.matpr.2021.09.117
58. Jubu, P.R., Obaseki, O.S., Nathan-Abutu, A., Yam, F.K., Yusof, Y. and Ochang, M.B., 2022. Dispensability of the conventional Tauc's plot for accurate bandgap determination from UV–vis optical diffuse reflectance data. *Results in Optics*, *9*, p.100273. https://doi.org/10.1016/j.rio.2022.100273
59. Abegunde, S.M., Olasehinde, E.F. and Adebayo, M.A., 2024. Green synthesis of ZnO nanoparticles using Nauclea latifolia fruit extract for adsorption of Congo red. *Hybrid Advances*, *5*, p.100164. https://doi.org/10.1016/j.hybadv.2024.100164
60. Rami, J.M., Patel, C.D., Patel, C.M. and Patel, M.V., 2021. Thermogravimetric analysis (TGA) of some synthesized

metal oxide nanoparticles. *Materials Today: Proceedings*, *43*, pp.655–659. https://doi.org/10.1016/j.matpr.2020.12.554

61. Loganathan, S., Valapa, R.B., Mishra, R.K., Pugazhenthi, G. and Thomas, S., 2017. Thermogravimetric analysis for characterization of nanomaterials. In *Thermal and Rheological Measurement Techniques for Nanomaterials Characterization* (pp. 67–108). Elsevier. https://doi.org/10.1016/B978-0-323-46139-9.00004-9
62. Otabor, G.O., Ikhuoria, E.U., Onaifo, J.O., Ifijen, I.H., Ehigie, A.C. and Aigbodion, A.I., 2025. Comparative thermogravimetric and differential thermal analysis of biosynthesized MgO-nanoparticles using Chromolaena odorata, Hevea brasiliensis, and Elaeis guineensis leaf extracts. *Tropical Journal of Chemistry*, *1*(1), pp.1–7. https://doi.org/10.71148/tjoc/v1i1.1
63. Kumar, V., Yadav, S.K. and Yadav, A., 2020. Thermogravimetric analysis and characterization of green synthesized zinc oxide nanoparticles using Ocimum sanctum leaf extract. *Materials Today: Proceedings*, *33*, pp.1235–1240.
64. Khan, Z.U.H., Khan, A., Chen, Y.M., Shah, N.S., Khan, A.U., Muhammad, N., Tahir, K., Shah, H.U., Khan, Z.U., Shakeel, M. and Nadeem, M., 2018. Enhanced antimicrobial, anti-oxidant applications of green synthesized AgNPs-an acute chronic toxicity study of phenolic azo dyes & study of materials surface using X-ray photoelectron spectroscopy. *Journal of Photochemistry and Photobiology B: Biology*, *180*, pp.208–217. https://doi.org/10.1016/j.jphotobiol.2018.02.01
65. Susi, T., Pichler, T. and Ayala, P., 2015. X-ray photoelectron spectroscopy of graphitic carbon nanomaterials doped with heteroatoms. *Beilstein Journal of Nanotechnology*, *6*(1), pp.177–192. https://doi.org/10.3762/bjnano.6.17
66. Matinise, N., Fuku, X.G., Kaviyarasu, K., Mayedwa, N. and Maaza, M.J.A.S.S., 2017. ZnO nanoparticles via Moringa oleifera green synthesis: Physical properties & mechanism of formation. *Applied Surface Science*, *406*, pp.339–347. https://doi.org/10.1016/j.apsusc.2017.01.219
67. Mishra, R.K., Zachariah, A.K. and Thomas, S., 2017. Energy-dispersive X-ray spectroscopy techniques for nanomaterial. In *Microscopy Methods in Nanomaterials Characterization* (pp. 383–405). Elsevier. https://doi.org/10.1016/B978-0-323-46141-2.00012-2
68. Scimeca, M., Bischetti, S., Lamsira, H.K., Bonfiglio, R. and Bonanno, E., 2018. Energy Dispersive X-ray (EDX) microanalysis: A powerful tool in biomedical research and diagnosis. *European Journal of Histochemistry: EJH*, *62*(1), p.2841. https://doi.org/10.4081/ejh.2018.2841
69. Ahmed, S., Saifullah, Ahmad, M., Swami, B.L. and Ikram, S., 2016. Green synthesis of silver nanoparticles using Azadirachta indica aqueous leaf extract. *Journal of Radiation Research and Applied Sciences*, *9*(1), pp.1–7. https://doi.org/10.1016/j.jrras.2015.06.006
70. Mohan, V.B., Jayaraman, K. and Bhattacharyya, D., 2020. Brunauer–Emmett–Teller (BET) specific surface area analysis of different graphene materials: A comparison to their structural regularity and electrical properties. *Solid State Communications*, *320*, p.114004. https://doi.org/10.1016/j.ssc.2020.114004
71. Raliya, R., Tarafdar, J.C. and Biswas, P., 2013. Enhancing the mobilization of native phosphorus in the hydroponic system using ZnO nanoparticles synthesized by green method. *Journal of Nanoparticle Research*, *15*, p.1559. https://doi.org/10.1021/acs.jafc.5b05224
72. Surendra, B.S., 2018. Green engineered synthesis of Ag-doped CuFe2O4: Characterization, cyclic voltammetry and photocatalytic studies. *Journal of Science: Advanced Materials and Devices*, *3*(1), pp.44–50. https://doi.org/10.1016/j.jsamd.2018.01.005
73. Shashanka, R., Jayaprakash, G.K., BG, P., Kumar, M. and Kumara Swamy, B.E., 2022. Electrocatalytic determination of ascorbic acid using a green synthesised magnetite nanoflake modified carbon paste electrode by cyclic voltammetric method. *Materials Research Innovations*, *26*(4), pp.229–239. https://doi.org/10.1080/14328917.2021.1945795
74. Asha, S.C., Mahesh, B., Ravikumar, C.R., Chamaraja, N.A. and Murthy, H.A., 2024. Green synthesis of calcium oxide nanoparticles using Ocimum sanctum leaf extracts: Photocatalytic and electrochemical sensor applications. *Journal of Materials Science: Materials in Electronics*, *35*(26), p.1738. https://doi.org/10.1007/s10854-024-13374-x
75. Patel, R., Vinchurkar, M., Patkar, R., Naik, T., Adami, A., Giacomozzi, F., Ramesh, R., Pramanick, B., Lorenzelli, L. and Baghini, M.S., 2023. Green synthesized silver nanoparticles functionalized interdigitated electrodes for bacterial sensing using non-faradaic electrochemical impedance spectroscopy. *Micro and Nano Engineering*, *21*, p.100231. https://doi.org/10.1016/j.mne.2023.100231
76. Hu, Y., Hua, S., Li, F., Jiang, Y., Bai, X., Li, D. and Niu, L., 2011. Green-synthesized gold nanoparticles decorated graphene sheets for label-free electrochemical impedance DNA hybridization biosensing. *Biosensors and Bioelectronics*, *26*(11), pp.4355–4361. https://doi.org/10.1016/j.bios.2011.04.037
77. Glogic, E., Kamali, A.K., Keppetipola, N.M., Alonge, B., Kumara, G.A., Sonnemann, G., Toupance, T. and Cojocaru, L., 2022. Life cycle assessment of supercapacitor electrodes based on activated carbon from coconut shells. *ACS Sustainable Chemistry & Engineering*, *10*(46), pp.15025–15034. https://doi.org/10.1021/acssuschemeng.2c03239
78. Lunardi, C.N., Gomes, A.J., Rocha, F.S., De Tommaso, J. and Patience, G.S., 2021. Experimental methods in chemical engineering: Zeta potential. *The Canadian Journal of Chemical Engineering*, *99*(3), pp.627–639. https://doi.org/10.1002/cjce.23914
79. Varenne, F., Botton, J., Merlet, C., Vachon, J.J., Geiger, S., Infante, I.C., Chehimi, M.M. and Vauthier, C., 2015. Standardization and validation of a protocol of zeta potential measurements by electrophoretic light scattering for nanomaterial characterization. *Colloids and Surfaces A: Physicochemical and Engineering Aspects*, *486*, pp.218–231. https://doi.org/10.1016/j.colsurfa.2015.08.044
80. Ahmed, S., Ahmad, M., Swami, B.L. and Ikram, S., 2016. A review on plants extract mediated synthesis of silver nanoparticles for antimicrobial applications: A green expertise. *Journal of Advanced Research*, *7*(1), pp.17–28. https://doi.org/10.1016/j.jare.2015.02.007
81. Rahdar, A., Amini, N., Askari, F. and Susan, M.A.B.H., 2019. Dynamic light scattering: A useful technique to characterize nanoparticles. *Journal of Nanoanalysis*, *6*(2), pp.80–89.
82. Marucco, A., Aldieri, E., Leinardi, R., Bergamaschi, E., Riganti, C. and Fenoglio, I., 2019. Applicability and

limitations in the characterization of poly-dispersed engineered nanomaterials in cell media by dynamic light scattering (DLS). *Materials*, *12*(23), p.3833. https://doi.org/10.3390/ma12233833

83. Ghosh, M.K., et al., 2024. Sustainable synthesis of gold nanoparticles using Hedychium coronarium leaf extract: CT-DNA and protein binding, cytotoxicity and docking studies. *Journal of Molecular Structure*, *1301*, p.135218. https://doi.org/10.1016/j.inoche.2024.113401
84. Prashanth, G.K., Prashanth, P.A., Bora, U., Gadewar, M., Nagabhushana, B.M., Ananda, S., Krishnaiah, G.M. and Sathyananda, H.M., 2015. In vitro antibacterial and cytotoxicity studies of ZnO nanopowders prepared by combustion assisted facile green synthesis. *Karbala International Journal of Modern Science*, *1*(2), pp.67–77. https://doi.org/10.1016/j.kijoms.2015.10.007
85. Prashanth, G.K., Prashanth, P.A., Nagabhushana, B.M., Ananda, S., Krishnaiah, G.M., Nagendra, H.G., Sathyananda, H.M., Rajendra Singh, C., Yogisha, S., Anand, S. and Tejabhiram, Y., 2018. Comparison of anticancer activity of biocompatible ZnO nanoparticles prepared by solution combustion synthesis using aqueous leaf extracts of Abutilon indicum, Melia azedarach and Indigofera tinctoria as biofuels. *Artificial Cells, Nanomedicine, and Biotechnology*, *46*(5), pp.968–979. https://doi.org/10.1080/21691401.2017.1351982
86. Prashanth, G.K., Prashanth, P.A., Singh, P., Nagabhushana, B.M., Shivakumara, C., Krishnaiah, G.M., Nagendra, H.G., Sathyananda, H.M. and Chaturvedi, V., 2020. Effect of doping (with cobalt or nickel) and UV exposure on the antibacterial, anticancer, and ROS generation activities of zinc oxide nanoparticles. *Journal of Asian Ceramic Societies*, *8*(4), pp.1175–1187. https://doi.org/10.1080/21870764.2020.1824328
87. Prashanth, G.K., Rao, S., Gadewar, M., Lalithamba, H.S., Swamy, M.M. and Sowmyashree, A.S., 2024, April. Cultivating eco-friendly nanomaterials: Mimosa pudica-enhanced zinc oxide nanoparticles for enhanced health and environmental well-being. *Journal of Physics: Conference Series*, *2748*(1), p.012001. https://doi.org/10.1088/1742-6596/2748/1/012001
88. Prashanth, G.K., Giresha, A.S., Lalithamba, H.S., Aman, M., Rao, S., Ravindra, K.N., Gadewar, M., Bhagya, N.P., Swamy, M.M. and Chaturvedi, V., 2025. Sustainable biofabrication of Ni/Mn co-doped ZnO nanoparticles using Simarouba glauca leaf extract: Evaluation of non-cytotoxic, anti-carcinogenic, anti-tubercular, anti-bacterial properties, anti-oxidant and hyaluronidase inhibition activities. *Inorganic Chemistry Communications*, *171*, p.113592. https://doi.org/10.1016/j.inoche.2024.113592
89. Prashanth, G.K., Sathyananda, H.M., Prashanth, P.A., Gadewar, M., Mutthuraju, M., Prabhu, S.B., Nagabhushana, B.M., Shivakumara, C., Rao, S. and Mohanty, D., 2022. Controlled synthesis of Ag/CuO nanocomposites: Evaluation of their antimycobacterial, antioxidant, and anticancer activities. *Applied Physics A*, *128*(7), p.614. https://doi.org/10.1007/s00339-022-05748-x
90. Malini, B.R., Prashanth, G.K., Jayasimha, H.N., Prashanth, P.A., Rao, S. and Lalithamba, H.S., 2024. Green synthesis of undoped and Ag-doped NiO nanoparticles: Evaluation and their synergetic effect on antimicrobial, anticancer, and antioxidant activities. *Inorganic Chemistry Communications*, *170*, p.113155. https://doi.org/10.1016/j.inoche.2024.113155
91. Banger, A., Kumari, A., Jangid, N.K., Jadoun, S., Srivastava, A. and Srivastava, M., 2025. A review on green synthesis and characterisation of copper nanoparticles using plant extracts for biological applications. *Environmental Technology Reviews*, *14*(1), pp.94–126.
92. Lalithamba, H.S., Prashanth, G.K., Latha, H.K.E., Siddekha, A. and Nagendra, G., 2025. Cerium oxide nanoparticles: Sustainable synthesis and diverse applications in electrical properties, catalysis and biomedicine. *Chemical Papers*, *79*(1), pp.193–209. https://doi.org/10.1007/s11696-024-03770-2
93. Prashanth, G.K., Gadewar, M., Rao, S., Lalithamba, H.S., Prashant, S.H., Mahadevaswamy, M., Bhagya, N.P. and Nagabhushana, B.M., 2024. Sustainable *Mirabilis jalapa*-derived zinc oxide nanoparticles: Synergistic bioactivity in antimicrobial, anticancer, and antioxidant. In *Advanced Materials in Engineering Applications* (pp. 329–335). CRC Press. https://doi.org/10.1201/9781003545941
94. Swamy, M.M., Giresha, A.S., Rao, S., Prashanth, G.K., Aman, M., Ravindra, K.N., Shree, S.S., Chougala, M.B., Lalithamba, H.S. and Akolkar, H.K.N., 2025. Eco-friendly development of Leucas aspera-derived MoO3 nanoparticles: Corrosion studies and multifunctional applications in medicine, agriculture, and industry. *Applied Physics A*, *131*(1), pp.1–21. https://doi.org/10.1007/s00339-024-08199-8
95. Malini, B.R., Prashanth, G.K., Prashanth, P.A., Lalithamba, H.S., Rao, S. and Bhagya, N.P., 2024. Multifunctional Ag/CuO nanocomposites: Synthesis, antimycobacterial efficacy, and cytotoxicity against colon cancer cells. *Transition Metal Chemistry*, pp.1–17. https://doi.org/10.1007/s11243-024-00623-5
96. Siddekha, A., HS, L. and GK, P., 2024. Plant-mediated synthesis of biocompatible ZnO NPs: Comprehensive characterisation and investigation of their catalytic, corrosion inhibition, antibacterial, scavenging and anti-oncogenic activities. *Materials Research Innovations*, pp.1–11. https://doi.org/10.1080/14328917.2024.2408128
97. Bhagya, N.P., Rao, S., Prashanth, G.K., Sowmyashree, A.S., Lalithamba, H.S., Yashodha, S.R. and Gadewar, M., 2025. Exploring the multifaceted potential of ZrO2 nanoparticles: Sustainable synthesis, anticancer properties, and electrochemical insight. *Journal of the Australian Ceramic Society*, pp.1–13. https://doi.org/10.1007/s41779-025-01177-0
98. Prashanth, G.K., Prashanth, P.A., Ramani, M., Ananda, S., Nagabhushana, B.M., Krishnaiah, G.M., Nagendra, H.G., Sathyananda, H.M. and Singh, C.R., 2019. Comparison of antimicrobial, antioxidant and anticancer activities of ZnO nanoparticles prepared by lemon juice and citric acid fueled solution combustion synthesis. *BioNanoScience*, *9*, pp.799–812. https://doi.org/10.1007/s12668-019-00670-8
99. Sathyananda, H.M., Prashanth, P.A., Prashanth, G.K., Dileep, M.S., Boselin Prabhu, S.R., Nagabhushana, B.M., Shivakumara, C. and Nagendra, H.G., 2022. Evaluation of antimycobacterial, antioxidant, and anticancer activities of CuO nanoparticles through cobalt doping. *Applied Nanoscience*, *12*, pp.79-86. https://doi.org/10.1007/s13204-021-02156-0

100. Gadewar, M., Prashanth, G.K., Babu, M.R., Dileep, M.S., Prashanth, P.A., Rao, S., Mahadevaswamy, M., Ghosh, M.K., Singh, N., Mandotra, S.K. and Chauhan, A., 2024. Unlocking nature's potential: Green synthesis of ZnO nanoparticles and their multifaceted applications—A concise overview. *Journal of Saudi Chemical Society, 28*(1), p.101774. https://doi.org/10.1016/j.jscs.2023.101774

101. Krishna, P.G., Ananthaswamy, P.P., Trivedi, P., Chaturvedi, V., Mutta, N.B., Sannaiah, A., Erra, A. and Yadavalli, T., 2017. Antitubercular activity of ZnO nanoparticles prepared by solution combustion synthesis using lemon juice as bio-fuel. *Materials Science and Engineering: C, 75*, pp.1026–1033. https://doi.org/10.1016/j.msec.2017.02.093

13 Applications of Nanoparticles
A Revolutionary Approach in the Food Industry

Shashank Vasanthkumar, Chandan Kumar Sahu, and Ravi-Kumar Kadeppagari

13.1 INTRODUCTION

Over the past few decades, nanotechnology has focused on designing, producing, and applying materials at the nanoscale (1–100 nm). These materials have unique characteristics like high surface energy, enhanced adsorption capacity, and stronger biological activity (Khan et al., 2017). Nanomaterials can be broadly categorized based on either their composition or structure. Composition-wise, they include metallic, organic, carbon-based, mineral-based, boron nitride, and silicon-based materials. Structurally, they are classified into various forms such as nanorods, nanowires, nanofibers, nanotubes, nanosheets, quantum dots, and aerogels (Gómez-Gómez et al., 2023). When these materials are reduced to the nanoscale, their physical and chemical properties, such as strength, diffusivity, and solubility, improve significantly. As a result, they become lighter, more stable, and more durable in chemical, mechanical, and kinetic terms (Ashfaq et al., 2022). Because of these properties, nanotechnology is widely applied in various fields, including agriculture, environmental science, medicine, and food technology (Chudasama & Goyary, 2024). In the food industry, nanotechnology has led to the development of innovative products with enhanced texture, taste, stability, and overall quality. Therefore, nanotechnology changes the traditional methods of food product production, packaging, and transportation and has a beneficial effect on consumers' attitudes (Biswas et al., 2022).

The nanotechnology market in the food industry has grown to a billion dollars, with the majority invested in food packaging, particularly in health-enhancing products, NP coatings, and beverages (Vasile, 2018). Increases in consumer concerns about food quality and health benefits are inspiring researchers to find alternative ways to enhance food quality without disturbing the nutritional content of the product (Singh et al., 2017). The demand for non-toxic nanoparticle-based materials has increased in the food industry (Onyeaka et al., 2022). An active role is played by NPs in detecting foodborne illnesses, formulating personalized nutrition plans, and ensuring sustainable food production through nanoencapsulation techniques (De Oliveira Mallia et al., 2021). Nanocapsules can be used as gelating and viscosity improving agents. They can also be used as nutritional supplements or food additives to cover or modify undesirable flavor, taste, or smell, and they can be used to support the better distribution of insoluble additives without adding surfactants or emulsifiers (Ayyaril et al., 2023). Smart nutrient delivery systems driven by nanobased technologies will enable the precise and controlled release of essential nutrients, which will enhance the bioavailability and effectiveness of nutrients (Singh et al., 2023).

Encapsulated food additives, preservatives, flavoring agents, food ingredients, and antimicrobial compounds are employed to improve the nutritional quality and enhance the shelf life, aroma, and texture of food products (Biswas et al., 2022). Nanoparticles, particularly Cu/CuO, Ag, MgO, TiO_2, ZnO, carbon dots, mesoporous particles, and graphene, were used in food packaging (Biswas et al., 2022; Suvarna et al., 2022). Among them, antimicrobial polymeric films or encapsulated antimicrobial compounds extend the shelf life of the products by controlling microbial growth. They can also improve barrier properties by reducing the oxygen level in packed food items (Chawla et al., 2021).

After further research, nanosensors were developed that can convert physical parameters into signals for detection and analysis. This approach is used in the food industry to identify microbial infection or contaminants in food products throughout processing, packaging, storage, and transport (Awlqadr et al., 2024). However, the application of NPs in food products certainly raises concerns for humans. According to some studies, oral consumption of nanoparticle-enhanced food or ingestion of NPs that have transported from packaging has been linked to DNA damage, protein denaturation, stimulation of oxidative stress, and other biological consequences of the host leading to gastrointestinal tract (GIT) and secondary organ damage (Wang et al., 2021). Hence, there is a need for close inspection before applying NPs in food products, and investigating the toxicity and food safety must be a major practice before applying NPs in food items (Gupta et al., 2024a). Nevertheless, they are successfully used in food quality monitoring, food packaging, and shelf-life improvement. This chapter will focus on these applications and a few new trends.

 DOI: 10.1201/9781003644439-13

13.2 NANOPARTICLES IN FOOD PACKAGING

Food deteriorates when it is exposed to oxygen, moisture, and gases. These factors will increase spoilage and lead to decreases in nutritive value, taste, texture, and shelf life. Therefore, the development of good packaging material becomes essential to combat these problems (Ahari & Soufiani, 2021). These materials not only provide physical barriers like temperature resistance, mechanical damage, and microbial contamination but also help in regulating internal conditions by limiting exposure to oxygen and other spoilage-induced gases (Chawla et al., 2021). Recent innovations are focusing on the utilization of nanoparticles in the development of polymer-based nanocomposites with good barrier and antimicrobial properties (Rashid et al., 2024). Natural biopolymers like chitosan, starch, cellulose, and alginate are used for coating fresh produce in thin layers, and they showed excellent results in extending the shelf life by reducing moisture, delaying oxidation, and preventing microbial contamination (Priya et al., 2023; Karnwal et al., 2025). Combining biopolymers with nanomaterials during food packaging will not only solve food preservation challenges but also support a sustainable approach and increase opportunities in food industries (Ahmad et al., 2024). It is also important for researchers to concentrate on barrier properties and antimicrobial compounds and also detect contaminants in packaging material (Gupta et al., 2024b). Food preservation depends on a well-organized food handling chain, which includes harvesting, processing, storage, and transportation, where packaging plays an important role for food safety and preservation (Amit et al., 2017).

13.3 TYPES OF FOOD PACKAGING

Food packaging is divided into two groups, active packaging and smart packaging. Active packaging materials focus on preserving food items by directly interacting with food and its surroundings. This kind of approach focuses on controlling the rate of microbial contaminants in the food (Malhotra et al., 2015). The shelf life of food products is increased by controlling the growth of microbes by using packaging materials that are incorporated with NPs with antimicrobial properties. In addition, intrinsic factors are maintained by functional agents like oxygen scavengers, carbon dioxide regulators, and moisture controllers that also help in preserving organoleptic properties by maintaining nutritional qualities (Fadiji et al., 2023). Smart packaging materials are used to maintain the internal conditions of packed food items. For example, sensors incorporated with NPs are used to detect physiological changes like temperature, pH, gas composition, spoilage, contaminants, and food pathogens in the food product (Jafarzadeh et al., 2024).

13.3.1 Active Packaging Materials

Active packaging material plays an important role in controlling spoilage, decreasing microbial contamination, and preserving the nutritional qualities of the food product (Ahmed et al., 2022; Majid et al., 2016). It is achieved by maintaining physiological changes and releasing functional compounds to control microbial growth (Du et al., 2023). According to the literature, functional agents like NPs, water vapor regulators, antimicrobial compounds, and oxygen absorbers are added to enhance the mechanical properties of polymer-based packaging material that helps in extending the shelf life of food products (Upadhyay et al., 2024). For example, functional agents such as NPs like gold, silver, zinc oxide (ZnO), and titanium dioxide (TiO_2) were added by direct or indirect contact to control microbial growth in food items. The NPs present on food surfaces bind to the bacterial cell wall and cause apoptosis (Wang et al., 2022a).

Among the different types of NPs, silver NPs (AgNPs) are gaining more importance for their broad-spectrum antimicrobial properties. AgNPs are not only effective against pathogenic bacteria and fungi but also can inhibit viruses like monkey pox (Duncan, 2011). Their mode of action is to act on the microbial cell wall, leading to an imbalance in cellular respiration by increasing the ROS, thus leading to apoptosis. According to the literature, researchers have explained the importance of AgNPs as a good functional agent to control foodborne pathogens. For example, Rhim et al. (2014) developed AgNP-based agar films with antimicrobial activity on *E. coli* and *L. monocytogene.* Similarly, Mohammed et al. (2009) developed AgNP-based sodium alginate active packaging film that has significant antimicrobial properties on *E. coli* and *S. aureus.* Alternatively, Damm et al. (2008) synthesized polyamide AgNP-nano- and microcomposites with an 80% inhibition rate on *E. coli.*

In further research, fresh-cut fruits and vegetables were preserved by using different types of active packaging films containing AgNPs. For example, An et al. (2008) developed a nanocomposite film using AgNPs and polyvinyl pyrrolidone (PVP) to extend the shelf life of asparagus spears for 25 days under refrigerated conditions. In addition, Mohammed et al. (2009) used AgNP-based sodium alginate active packaging film to increase the shelf life of carrots and pears. Kraśniewska et al. (2020) explained the significant benefits of biopolymer-based materials that were incorporated with AgNPs for food packaging application. Fasihnia et al. (2017) also supported the use of organoclay NPs for nanocomposite films as a promising approach for antimicrobial packaging. Dias et al. (2013) developed antimicrobial films with carbon nanotubes and allyl isothiocyanate for packing cooked and shredded chicken meat. The films in storage extended the shelf life of meat up to 40 days by controlling the microbial contamination, oxidation, and color change of the product.

13.3.2 Smart Packaging Materials

Smart packaging materials are nanosensors used for detecting food contaminants. These sensors are made up of NPs that are highly sensitive and capable of detecting a wide range of physical, chemical, and biological contaminants during food processing, preservation, transportation, and storage (Chelliah et al., 2021). In addition, they can also monitor physical parameters like temperature, humidity, and gas levels (Javaid et al., 2021). A smart packaging material also contains visual indicators or labels like temperature-time indicators, freshness detectors, and integral monitors that can detect whether the packaged food item has been exposed to any contaminants or physically damaged (Chiu et al., 2024). This method helps in controlling the food waste and product quality throughout the supply chain (Shruti et al., 2024). On further advancements, nanobarcodes NP-based tags were developed to store product-specific data, providing a secure and traceable amount of information for identifying the authenticity of the food products (Bajpai et al., 2018).

Nanosensors are bio-analytical devices made up of nanostructured materials and biological receptors (Huang et al., 2021). In food industries, these sensors are advantageous for being user friendly, cost effective, high accuracy, and less time consuming for detecting contaminants (Tarannum et al., 2024). The main advantage of these devices is that they can detect trace amounts of foreign particles called analytes (Malik et al., 2023). These sensors are made up of different types of metal and non-metal NPs, metal oxides, nanorods, nanowires, nanofibers, and carbon tubes. A few gas sensors are developed from metallic NPs like gold, palladium, and platinum (Gil et al., 2025). These materials bring an advantage by increasing their surface area and enhancing their optical and electrical properties. The sensors have the ability to exhibit a color change to detect food spoilage and gases released during the process (Wang et al., 2022b).

13.4 NANOCOMPOSITES AS FOOD PACKAGING MATERIALS

Nanocomposites are developed by combining NPs with polymers to increase the mechanical properties and strength and gain resistance to external stress. These materials have antimicrobial properties that help in protecting food items from spoilage (Basavegowda & Baek, 2021). In the food industry, they are used in protecting food items from mechanical stress and temperature changes and also extending the shelf life of products. Kim et al. (2020) developed ZnO nanocomposites with significant antioxidant and antimicrobial properties to extend the shelf life of food products. Nanocomposites known for significant gas barrier properties are used in packaging carbonated beverages to avoid carbon dioxide leakage. They are made from nanoclays like aegis, durethan, nanocar, and imperium. These materials are transparent, have low density and good surface properties, and are biodegradable (Wu et al., 2021).

Durethan contains polyamide, which helps in packaging fruit juices (Biswas et al., 2022). Aegis and imperium nanoclay acts as an oxygen scavenger and is used to pack carbonated drinks (Flanagan & Singh, 2006). In addition, nanocar, with good gas barrier properties, is used in the manufacturing of plastic beer bottles (Pradhan et al., 2015). The market has different types of nanocomposite-based commercialized products that can benefit food preservation, packaging, and extending the shelf life of food products. For example, guard IN is used as an ethylene gas scavenger for controlling ripening in fruits and vegetables (Wei et al., 2020). Top screen DS13, with excellent barrier properties and biodegradability, is used in the packaging of various fresh fruits, dairy products, and dried food products (Mason et al., 2006). Nano-Cream PAC is a nanocomposite-based coating material used for volatile organic compounds and other impurities in food industry. This property helps in maintaining the freshness and flavor of food products during storage (Bordes et al., 2009).

13.5 NANOTECHNOLOGY IN FOOD QUALITY MONITORING

Nanosensors are sensitive to gases like ammonia, hydrogen, sulfur dioxide, hydrogen sulfide, and nitrogen oxide (Kumar et al., 2020). They can also detect foreign substances like pathogens, adulterants, toxins, and pesticidal residues (Abedi-Firoozjah et al., 2023). Compared to conventional sensors, nanosensors are more sensitive and selective. They are made up of sensing and electronic component that process data (Dhahi et al., 2024). The sensing component detects changes in physiological parameters like temperature, pH, light, gas, and chemical constituents of the food item; then it converts them into electrical signals that can be detected by electronic components for interpretation (Thakur & Ragavan, 2012).

Carbon nanotubes; super paramagnetic nanoparticles; and metallic nanoparticles like gold, copper, silver, platinum, and zinc have been used to develop electrochemical sensors and biosensors that can identify toxins in food products (Gong et al., 2023). Among those toxins, aflatoxin is known to be a potent carcinogen found as contaminant in a variety of food items (Dadmehr & Korouzhdehi, 2023). Radoi et al. (2008) applied a combination of supermagnetic NPs and gold nanoprobes to detect aflatoxin M1 in milk. Additionally, Sharma et al. (2010) developed gold NPs that can detect aflatoxin B1 in food samples. Similarly, Zhou et al. (2009) developed a gold NP-based immunochromatographic assay to detect brevetoxin in seafoods. Zamolo et al. (2012) also developed gold-based immunoassays for detecting botulinum neurotoxin type B in processed food items.

Microfluid sensors, being highly sensitive, are used to detect trace amounts of toxins in water-based samples (Zhang et al., 2024). In industries, food packaging facilities are installed with nanosensors to detect contaminants, reducing the time for testing in the lab. These sensors provide knowledge about freshness in food to consumers by indicating a visible color change. Commonly, gas, time, and temperature detectors are used in packaging industries (Heo & Lim, 2024). On further advancement, array biosensors, nanoparticle-based sensors, nanocantilevers, nanotest strips, and electronic noses are used (Pradhan et al., 2015). An electronic nose sensor mimics the human sense of smell, indicating spoilage gases released from contaminated food items. Similarly, there are also electronic tongue sensors that indicate food spoilage by changing color when they come into contact with spoiled food items (Roy & Yadav, 2022).

In addition, nanobarcode-based technology is used in the food industry for food quality assessment. Biometric sensors are also applied in the food industry (Neethirajan & Jayas, 2011). They are constructed from proteins and synthetic membranes to detect mycotoxins and other toxic compounds (Mohammad et al., 2022). Reflective interferometry sensors are another type of sensor capable of detecting *E. coli* contamination in packaged foods. In this method, when an *E. coli* protein present on a silicon chip comes into contact with pathogens from food items, interaction and binding take place, resulting in the release of scattering signals. Then all the light scattering signals caused by interactions with mitochondrial proteins are analyzed using digital imaging to identify the spoilage (Wanekaya et al., 2006).

13.6 NANOPARTICLES AS ANTIMICROBIAL AGENTS AND SHELF-LIFE IMPROVEMENT

Food is an essential part of human life and is prone to biological contamination during the manufacturing process. Hence, there is a great demand in the market for improving food quality by developing novel antimicrobial compounds with a beneficial effect on food products (Mafe et al., 2024). Food products, whether refined or raw and derived from plant or animal sources, require preservation to maintain quality throughout their storage period. NPs play an emerging role in food preservation (Lisboa et al., 2024). They can either be incorporated into packaging material or directly mixed with food products. A literature review highlights their application in preserving a wide range of food items like fresh-cut fruits, vegetables, dairy products, meat, and baked items (Chadha et al., 2022).

Food is generally contaminated by foodborne pathogens, and they will affect human health, shelf life, and economy. Hence, food manufacturers use synthetic preservatives and antibiotics to control microbial growth in food (Farid et al., 2023). However, excessive and inappropriate application of antibiotics is causing the development of antimicrobial resistance in bacteria. Therefore, researchers are focusing on using NPs as antimicrobial nanostructures and considering them as an alternative approach to synthetic compounds (Moradi et al., 2023). Also, NPs act as good carriers of bioactive compounds and protect these compounds against adverse conditions during the manufacturing and storage process so that bioactive compounds will act against pathogenic bacteria (Marín et al., 2024). The efficiency of these materials is based on the release rate of antimicrobial compounds on the food matrix or surfaces during microbial contact (Chawla et al., 2021).

There are different types of NPs that are used in the food industry based on their specific functions. Metal-based NPs, including silver, zinc oxide, and copper oxide, are extensively used in food packaging, edible coatings, and preservation to minimize spoilage and contamination (Adeyemi & Fawole, 2023). Among these, AgNPs are most commonly utilized in food industries because they can easily act on the cell membrane by adhesion and degrade lipopolysaccharides, resulting in the loss of the membrane structure in food pathogens (More et al., 2023). In addition, these AgNPs bind to electron donor groups (N_2, O_2, and S_2) and prevent DNA replication and ATP synthesis. Also, when the concentration of Ag^+ increases, cytoplasm will shrink, leading to cell wall rupture, followed by protein and nucleic acid degradation and eventually cell death (Dakal et al., 2016) (Figure 13.1). Other metal-based NPs kill pathogens by producing reactive

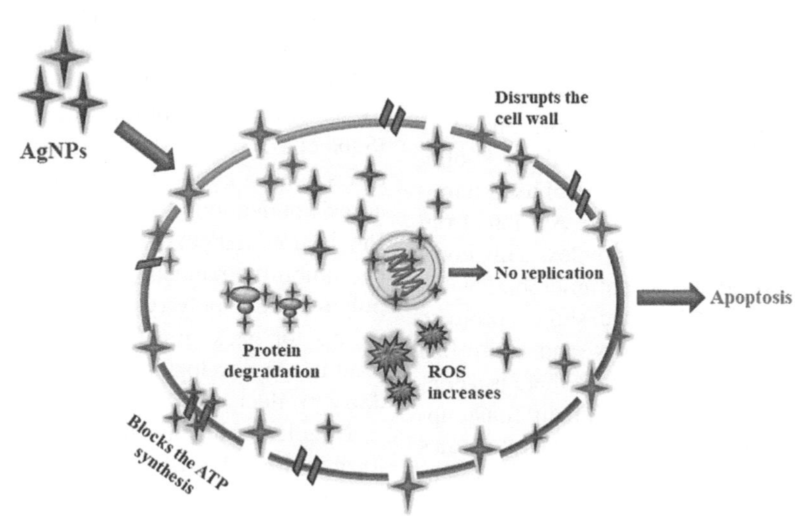

FIGURE 13.1 The mechanism of antibacterial activity of AgNPs.

oxygen species, which cause oxidative cellular damage, resulting in the disruption of the bacterial membrane and a decrease in the concentration of intracellular ATP (Slavin et al., 2017). Organic NPs like chitosan exhibit antimicrobial properties by binding to bacterial surfaces and disrupting their structures, while lipid-based nanoemulsions enable controlled release of antimicrobial compounds like essential oils (Kravanja et al., 2019).

Carbon-based NPs, including graphene oxide and carbon nanotubes, physically damage bacterial membranes and have strong adsorption properties (Sani et al., 2025). Earlier reports have proved that NP application by layering, immobilizing, inclusing, and conformational changes on top of packaging material would have beneficial effects on increasing the shelf life of food products (Dash et al., 2022). For example, Arfat et al. (2016) developed an antimicrobial nanocomposite film made from silver-copper bimetallic nanoparticles and fish skin gelatin that was effective against *L. monocytogenes* and *S. enterica* pathogens. Similarly, harmful bacteria present on food surfaces were controlled by using TiO_2 as a photocatalytic disinfectant (Fujishima et al., 2000). Tripathi et al. (2009) developed chitosan-based NPs with antimicrobial activity against *E.coli*, *S. aureus*, and *B. subtilis* that also increased the shelf life of tomatoes. This is due to the interaction between the negatively charged cell membrane and positively charged chitosan surface, leading to a breakdown of membrane permeability and release of intracellular components.

Sarwar et al. (2018) developed antimicrobial films made from silver nanocomposites, nanocellulose, and polyvinyl alcohol with antimicrobial properties against both methicillin-resistant *Staphylococcus aureus* (MRSA) and the DH5-alpha strain of *Escherichia coli.* Valerini et al. (2018) developed a Al-doped ZnO-based nanostructure, and the coating demonstrated enhanced antimicrobial activity against *E. coli.* In addition, Lu et al. (2017) synthesized an antimicrobial nanoemulsion with particles measuring 100 nm by treating citral with ultrasonic power. It acted as an effective antimicrobial agent in the food industry.

Yang et al. (2010) developed a nanomaterial for preserving fruits like strawberries (*Fragaria ananassa* Duch.) by using polyethylene and nanopowders like silver, kaolin, rutile TiO_2, and anatase TiO_2. Johnston (2010) used nanostructured calcium silicate (NCS) to capture Ag^+ ions from a solution, and it formed an NCS-Ag complex. This complex was successfully applied as an antimicrobial agent in food preservation. Though NPs have shown promise in enhancing the stability, bioavailability, and controlled release of active compounds in food systems, there is a concern regarding health risks associated with their consumption (Atanda et al., 2025). Therefore, comprehensive studies are necessary to evaluate the prolonged toxicity, bioaccumulation, and potential environmental impact of nanoencapsulated foods to ensure their safe application in the food industry (De Oliveira Mallia et al., 2021).

13.7 NEW TRENDS

Nanomaterials play an important role in the food sector for various applications, and they can exist in powder or liquid form, with different physical and chemical properties (Joudeh & Linke, 2022). Different types of nanomaterials are NPs, nanocapsules, nanoclays, nanoemulsions, nanotubes, nanofibers, and nanolaminates (Khan et al., 2022; Baraketi & Khwaldia, 2024). Food-grade biopolymeric NPs are derived from proteins and polysaccharides, and they are produced using different techniques such as emulsification, desolvation, coacervation, and electrospray drying. These methods were used for encapsulating micronutrients to biopolymers and delivering to the host (Pateiro et al., 2021). Synthetic polymers such as polylactic acid (PLA) or polylactic-co-glycolic acid (PLGA) or natural polymers such as chitosan, agarose, fibrin, and collagen were used as carriers for the encapsulation of biomolecules like vitamins and proteins (Song et al., 2018). Since NPs have a larger surface area, they have gained application in bio-nanosensing where immobilized enzymes are attached on NP surfaces in order to increase the efficiency of enzymes and control the diffusion rate of the substrates (Fu et al., 2024).

Nanoemulsions are mixtures of two or more liquids that are difficult to combine and structured as dispersed droplets with a diameter of 500 nm or less (Mushtaq et al., 2023). For example, when water droplets are dispersed in oil, it is called a water-in-oil emulsion (W-O), and when oil droplets are dispersed in water, it is called an oil-in-water emulsion (O-W). Since they can be either hydrophobic or hydrophilic, depending on their composition and formulation, they are used to deliver active substances like nutraceuticals, flavoring, coloring agents, and antimicrobial compounds into food products (Ho et al., 2021). Additionally, biodegradable coatings and packaging smart films are developed from these emulsions to increase the shelf life of food products (Thakur et al., 2024).

Nanoclays are complex clay crystallites structures made up of layered mineral silicates that are cost effective and environmentally friendly (Perera et al., 2023). Based on shape and chemical composition, they are classified into different types, including bentonite, montmorillonite, hectorite, smectite, kaolinite, halloysite, illite, and chlorite (Saba et al., 2018). Nanoclay has shown good barrier properties against gases and moisture and is used for food packaging applications (Nath et al., 2021; Perera et al., 2023). In addition, nanoclay–polymer composites were developed by combining nanoclays with polymers such as olyamides, polyurethane, polystyrene, polyolefns, epoxy resins, and nylons for packaging meats and cereals. They were also used in the development of coating solutions for fruits and dairy products (Mohammad et al., 2022; Vasile, 2018).

Nanolaminates are made from food-grade materials to form multiple layers, providing thickness of less than 1 nm with good barrier properties (Thakur et al., 2024). Generally, they are made from biomolecules like proteins, lipids, and polysaccharides (Aswathanarayan & Vittal, 2019).

They are used to increase the shelf life of vegetables, fruits, candies, meats, chocolates, and bakery products. In addition, they provide significant functional and sensory properties such as flavor, colors, antioxidant, and antimicrobial properties to food materials (Mohammad et al., 2022).

Nanofibers and nanotubes are a type of polymer with good barrier properties and size measuring from 10 to 1000 nm (Goyal et al., 2015). Nanofibers are synthesized by the electrospinning technique, where a solution is passed through a spinneret with a small capillary orifice. Then a strong electric field is applied to generate thin, solid polymer strands called nanofibers (Guarino et al., 2017). On other hand, globular proteins from milk called A-lactalbumin are used for the development of nanotubes. These were used in an enzyme immobilization process. The proteins in nanotubes have the ability to withstand pasteurization temperatures (72°C for 40 sec) and can also be used as carriers for encapsulation process (Mohammad et al., 2022). Nanoencapsulation is a technique used for encapsulating bioactive compounds to increase the shelf life and functional and nutritional quality of food products without affecting the organoleptic properties (Awuchi et al., 2022).

Functional foods play an important role for their therapeutic applications in the human body. They are developed from bioactive compounds that are difficult to absorb in the host (Konstantinidi & Koutelidakis, 2019). Functional foods were developed from the nanodelivery system (NDS) of active substances (Su et al., 2022). An NDS is a carrier that can transport bioactives that are sensitive to physiological changes in the gut. Therefore, they form micelles, liposomes, emulsions, and nanocapsules and are transported inside the gut for better absorption and increase the health benefits of the host (Lv et al., 2024). Additionally, NDSs can control and target the release of bioactives inside the host (Patra et al., 2018). NDSs are advantageous for having characteristics like small size, pH sensitivity, and ionic behavior. They also help in the absorption and adhesion of bioactives in the area of the gastrointestinal tract (Rezagholizade-Shirvan et al., 2024).

Nanomicelles consist of a colloidal dispersion system. The particle size measures around 5 to 100 nm (Bose et al., 2021). They are made up of a polymeric micellar hydrophobic core and hydrophilic shell. They play a major role in delivering insoluble drugs and bioactives to a target site (Li et al., 2022). Nanomicelles also improve bioactive characteristics like solubility, stability, and absorption, thereby increasing bioavailability and contributing to the overall benefits of the host (Altemimi et al., 2024). Du et al. (2019) synthesized a nanomicelle from egg yolk peptides to use as an emulsifier for pickering milk. Chen et al. (2019) developed curcumin-based nanomicelles by linking them with hydroxyl starch to increase the rate of solubility. This nanoformulation also provided protection against light and high temperatures, which are known to degrade curcumin. Similarly, Mohan et al. (2012) reported that recombinant casein micelles in milk can act as carriers for hydrophobic molecules such as curcumin, vitamin A, vitamin D, and docosahexaenoic acid. Therefore, the selection of nanomicelle systems in foods is based on particle nature, loading capacity, physiochemical behavior, sensory properties, and economical benefits (Janik et al., 2023). Once nanomicelles are loaded with functional bioactive nutrients such as saponins, polyphenols, and lipids, they can be fortified into food products or beverages in either powder or liquid form (Su et al., 2022). This approach not only enhances the nutritional quality of the product but also brings economic and health benefits.

Nanoliposomes are nanostructured lipid carriers (NLCs) measuring 10–100 nm in size. Their main aim is to increase the bioavailability of active compounds and decrease the toxic levels in the GIT (Motsoene et al., 2023). They help in delivering antioxidants, lipids, proteins, minerals, and vitamins. For example, Tamjidi et al. (2017) synthesized NLC with astaxanthin for developing functional beverages with hydrophobic nutraceuticals with commercial value. Alternatively, Han et al. (2021) synthesized DHA-NLC with soy lecithin and β-sitosterol to use as a food stabilizer. The presence of polar heads in liposomes has a slight disadvantage in the host. Since it can be absorbed and digested by epithelial cells through enzyme hydrolysis, it is quite difficult to understand the hydrolysis and toxicity levels in the intestine and in other organs. In addition, the presence of the oil phase in NLC cannot be digested by enzymes, which leads to deposition of NLC in organs and causes toxicity.

Nanocapsules are nanosized polymers that are dispersed easily and can be suspended in water to form a homogeneous solution. In functional foods, this nanocapsule contains functional bioactives for effective delivery in the GIT of the human body (Deng et al., 2020). Generally these materials are non-toxic, edible, degradable, and biocompatible with the host for food applications. The walls of these capsules are made from soluble gums, carbohydrates, and proteins. Prakash et al. (2018) explained the functional application of plant essential oil nanocapsules as a new trend for food preservatives. Reddy et al. (2019) used β-cyclodextrin or whey protein to encapsulate coffee bean oil to form nanocapsules that prevent rancidity and can be used as a food flavoring agent. Liu et al. (2021) synthesized nanocapsule containing lactic acid bacteria with sodium alginate that can escape GIT barriers and reach the intestine as a food additive for beneficial effects. Therefore, various studies have demonstrated the utilization of nanocapsules in preserving food, enhancing flavor, and delivering bioactives. They hold the potential to transform the food industry by enhancing the shelf life and nutritional value of food products.

Nanoemulsions, as discussed earlier, serve as effective carriers for delivering large quantities of bioactive compounds by enhancing their dispersion in both oil and water phases (Preeti et al., 2023). This approach significantly increases the bioavailability of the compounds and allows more efficient delivery and greater nutritional effects in food and beverage applications (Choi & McClements, 2020). Rao and McClements (2011) synthesized a food-grade nanoemulsion with lemon oil and sucrose monopalmitate for functional food applications. Dasgupta et al. (2015) developed a vitamin E nanoemulsion with mustard oil with high antioxidant and antifungal properties that can be used in the production of fruit juices, which extends their shelf life, nutritional content, and stability.

Joung et al. (2016) prepared curcumin nanoemulsions for controlling fat oxidation in milk. Spernath et al. (2002) synthesized lycopene microemulsions by using ethoxylated sorbitan anhydride esters as a new carrier material for delivering natural food supplements. Therefore, nanoemulsions are an additional option to deliver active substances by increasing their bioavailability (Wilson et al., 2021). Since these emulsions can easily pass through tissues and the GIT barrier, they accumulate in tissues after delivering bioactives. There are not many studies provided on the reaction of accumulated nanoemulsions in tissues, predicting toxicity levels (Jacob et al., 2024).

13.8 CONCLUSIONS

Nanotechnology has brought many benefits to the food industry. It provides an opportunity for controlling, detecting, and monitoring contaminants like microbes, toxins, pesticides, and chemicals (Rahmati et al., 2020). Overall, only a few NPs are in early stages of development and have been applied to food items. In further development, creating biodegradable nanocomposites has found wide application in food packaging (Ghosh et al., 2025). Using nanosensors to detect food pathogens has become a simpler and less time-consuming process. Though NPs have wide range of beneficial effects, they also have potential risks related to their application in the food industry (Valenzuela-Amaro et al., 2023). Hence, a wide range of research must be conducted and NPs evaluated for toxicity levels before incorporating them into foods. Since NPs are in contact with food material, there is a chance of transmitting NPs into food items, which can cause health risks. Even though NPs have good antimicrobial and antioxidant properties, they can still cause DNA damage, apoptosis, uncontrolled cell responses and cell division, changes in cell mobility, and cytotoxicity (Yu et al., 2020). Therefore, the toxicity of NPs depends on the type of packaging material and absorption rate of NPs in to food items. According to the literature, the health risks associated with absorption of NPs are a growing concern, and nanobased food items may have a negative impact on health due to intake, bioaccumulation, and toxic effects (Xuan et al., 2023).

The World Health Organization (WHO), United Nations, Food and Agricultural Organization (FAO), and European Union (EU) have highlighted the need for rules and safety measures for using NPs in food (Allan et al., 2021). In response, the European Food Safety Authority (EFSA) has updated its guidelines for monitoring the risks of using NPs in the feed and food chain by considering both animal and human health (Xavier et al., 2021). In India, food safety regulations are not sufficient to monitor the safety of NPs. At present, rules associated with nanobased foods and packaging are not sufficient because the potential risk and toxicity levels are not properly understood (Kumari et al., 2023). Currently, there are no standard operating protocols for safety assessment of NPs, and available safety data comes with confidential industrial studies, limiting transparency. Therefore, further research is required to determine the potential toxicity, and it provides a great opportunity for food scientists and researchers to resolve all safety aspects of NPs in food applications.

Nanotechnology plays a significant role in all food sectors for extending the shelf life of food products. Different types of nanomaterials have specific applications in the food industry either as food additives or functional components. Presently, NPs have been applied as antioxidant and antimicrobial agents for food packaging, coating solutions, and development of functional foods. They also help the food industry by using nanosensors to detect pathogens and their toxins, nutritional content, and organoleptic properties of food. With such wide applications, there is a risk for human health if NP concentration exceeds regulatory levels. Therefore, as nanotechnology continues to evolve, its applications in food packaging, quality monitoring, and shelf-life extension will grow, and its future depends on safe and sustainable food systems. Continued research and innovations are essential to fully harness the benefits of NPs while ensuring their safety for consumers.

REFERENCES

Abedi-Firoozjah, R., Ebdali, H., Soltani, M., Abdolahi-Fard, P., Heydari, M., Assadpour, E., Azizi-Lalabadi, M., Zhang, F., & Jafari, S. M. (2023). Nanomaterial-based sensors for the detection of pathogens and microbial toxins in the food industry; a review on recent progress. *Coordination Chemistry Reviews*, 500, 215545. https://doi.org/10.1016/j.ccr.2023.215545

Adeyemi, J. O., & Fawole, O. A. (2023). Metal-based nanoparticles in food packaging and coating technologies: A review. *Biomolecules*, 13(7), 1092. https://doi.org/10.3390/biom13071092

Ahari, H., & Soufiani, S. P. (2021). Smart and active food packaging: Insights in novel food packaging. *Frontiers in Microbiology*, 12. https://doi.org/10.3389/fmicb.2021.657233

Ahmad, K., Li, Y., Tu, C., Guo, Y., Yang, X., Xia, C., & Hou, H. (2024). Nanotechnology in food packaging with implications for sustainable outlook and safety concerns. *Food Bioscience*, 58, 103625. https://doi.org/10.1016/j.fbio.2024.103625

Ahmed, M. W., Haque, M. A., Mohibbullah, M., Khan, M. S. I., Islam, M. A., Mondal, M. H. T., & Ahmmed, R. (2022). A review on active packaging for quality and safety of foods: Current trends, applications, prospects and challenges. *Food Packaging and Shelf Life*, 33, 100913. https://doi.org/10.1016/j.fpsl.2022.100913

Allan, J., Belz, S., Hoeveler, A., Hugas, M., Okuda, H., Patri, A., Rauscher, H., Silva, P., Slikker, W., Sokull-Kluettgen, B., Tong, W., & Anklam, E. (2021). Regulatory landscape of nanotechnology and nanoplastics from a global perspective. *Regulatory Toxicology and Pharmacology*, 122, 104885. https://doi.org/10.1016/j.yrtph.2021.104885

Altemimi, A. B., Farag, H. A. M., Salih, T. H., Awlqadr, F. H., Al-Manhel, A. J. A., Vieira, I. R. S., & Conte-Junior, C. A. (2024). Application of nanoparticles in human nutrition: A review. *Nutrients*, 16(5), 636. https://doi.org/10.3390/nu16050636

Amit, S. K., Uddin, M. M., Rahman, R., Islam, S. M. R., & Khan, M. S. (2017). A review on mechanisms and commercial aspects of food preservation and processing. *Agriculture & Food Security*, 6(1). https://doi.org/10.1186/s40066-017-0130-8

An, J., Zhang, M., Wang, S., & Tang, J. (2008). Physical, chemical and microbiological changes in stored green asparagus spears as affected by coating of silver nanoparticles-PVP. *LWT*, 41(6), 1100–1107. https://doi.org/10.1016/j.lwt.2007.06.019

Arfat, Y. A., Ahmed, J., Hiremath, N., Auras, R., & Joseph, A. (2016). Thermo-mechanical, rheological, structural and antimicrobial properties of bionanocomposite films based on fish skin gelatin and silver-copper nanoparticles. *Food Hydrocolloids*, 62, 191–202. https://doi.org/10.1016/j.foodhyd.2016.08.009

Ashfaq, A., Khursheed, N., Fatima, S., Anjum, Z., & Younis, K. (2022). Application of nanotechnology in food packaging: Pros and Cons. *Journal of Agriculture and Food Research*, 7, 100270. https://doi.org/10.1016/j.jafr.2022.100270

Aswathanarayan, J. B., & Vittal, R. R. (2019). Nanoemulsions and their potential applications in food industry. *Frontiers in Sustainable Food Systems*, 3. https://doi.org/10.3389/fsufs.2019.00095

Atanda, S. A., Shaibu, R. O., & Agunbiade, F. O. (2025). Nanoparticles in agriculture: Balancing food security and environmental sustainability. *Discover Agriculture*, 3(1). https://doi.org/10.1007/s44279-025-00159-x

Awlqadr, F. H., Altemimi, A. B., Qadir, S. A., Salih, T. a. H., Alkanan, Z. T., AlKaisy, Q. H., Mohammed, O. A., & Hesarinejad, M. A. (2024). Emerging trends in nano-sensors: A new frontier in food safety and quality assurance. *Heliyon*, 11(1), e41181. https://doi.org/10.1016/j.heliyon.2024.e41181

Awuchi, C. G., Morya, S., Dendegh, T. A., Okpala, C. O. R., & Korzeniowska, M. (2022). Nanoencapsulation of food bioactive constituents and its associated processes: A revisit. *Bioresource Technology Reports*, 19, 101088. https://doi.org/10.1016/j.biteb.2022.101088

Ayyaril, S. S., Shanableh, A., Bhattacharjee, S., Rawas-Qalaji, M., Cagliani, R., Shabib, A. G., & Khan, M. I. (2023). Recent progress in micro and nano-encapsulation techniques for environmental applications: A review. *Results in Engineering*, 18, 101094. https://doi.org/10.1016/j.rineng.2023.101094

Bajpai, V. K., Kamle, M., Shukla, S., Mahato, D. K., Chandra, P., Hwang, S. K., Kumar, P., Huh, Y. S., & Han, Y. (2018). Prospects of using nanotechnology for food preservation, safety, and security. *Journal of Food and Drug Analysis*, 26(4), 1201–1214. https://doi.org/10.1016/j.jfda.2018.06.011

Baraketi, S., & Khwaldia, K. (2024). Nanoparticles from agri-food by-products: Green technology synthesis and application in food packaging. *Current Opinion in Green and Sustainable Chemistry*, 49, 100953. https://doi.org/10.1016/j.cogsc.2024.100953

Basavegowda, N., & Baek, K. (2021). Advances in functional Biopolymer-Based nanocomposites for active food packaging applications. *Polymers*, 13(23), 4198. https://doi.org/10.3390/polym13234198

Biswas, R., Alam, M., Sarkar, A., Haque, M. I., Hasan, M. M., & Hoque, M. (2022). Application of nanotechnology in food: Processing, preservation, packaging and safety assessment. *Heliyon*, 8(11), e11795. https://doi.org/10.1016/j.heliyon.2022.e11795

Bordes, P., Pollet, E., & Averous, L. (2009). Nano-biocomposites: Biodegradable polyester/nanoclay systems. *Progress in Polymer Science*, 34(2), 125–155. https://doi.org/10.1016/j.progpolymsci.2008.10.002

Bose, A., Burman, D. R., Sikdar, B., & Patra, P. (2021). Nanomicelles: Types, properties and applications in drug delivery. *IET Nanobiotechnology*, 15(1), 19–27. https://doi.org/10.1049/nbt2.12018

Chadha, U., Bhardwaj, P., Selvaraj, S. K., Arasu, K., Praveena, S., Pavan, A., Khanna, M., Singh, P., Singh, S., Chakravorty, A., Badoni, B., Banavoth, M., Sonar, P., & Paramasivam, V. (2022). Current trends and future perspectives of nanomaterials in food packaging application. *Journal of Nanomaterials*, 2022(1). https://doi.org/10.1155/2022/2745416

Chawla, R., Sivakumar, S., & Kaur, H. (2021). Antimicrobial edible films in food packaging: Current scenario and recent nanotechnological advancements—a review. *Carbohydrate Polymer Technologies and Applications*, 2, 100024. https://doi.org/10.1016/j.carpta.2020.100024

Chelliah, R., Wei, S., Daliri, E. B., Rubab, M., Elahi, F., Yeon, S., Jo, K. H., Yan, P., Liu, S., & Oh, D. H. (2021). Development of nanosensors based intelligent packaging systems: Food quality and medicine. *Nanomaterials*, 11(6), 1515. https://doi.org/10.3390/nano11061515

Chen, S., Wu, J., Tang, Q., Xu, C., Huang, Y., Huang, D., Luo, F., Wu, Y., Yan, F., Weng, Z., & Wang, S. (2019). Nanomicelles based on hydroxyethyl starch-curcumin conjugates for improved stability, antioxidant and anticancer activity of curcumin. *Carbohydrate Polymers*, 228, 115398. https://doi.org/10.1016/j.carbpol.2019.115398

Chiu, I., Ye, H., Aayush, K., & Yang, T. (2024). Intelligent food packaging for smart sensing of food safety. *Advances in Food and Nutrition Research*, 215–259. https://doi.org/10.1016/bs.afnr.2024.06.006

Choi, S. J., & McClements, D. J. (2020). Nanoemulsions as delivery systems for lipophilic nutraceuticals: Strategies for improving their formulation, stability, functionality and bioavailability. *Food Science and Biotechnology*, 29(2), 149–168. https://doi.org/10.1007/s10068-019-00731-4

Chudasama, M., & Goyary, J. (2024). Nanostructured materials in food science: Current progress and future prospects. *Next Materials*, 5, 100206. https://doi.org/10.1016/j.nxmate.2024.100206

Dadmehr, M., & Korouzhdehi, B. (2023). Nanobiosensors for aflatoxin B1 detection, current research trends and future outlooks. *Microchemical Journal*, 194, 109344. https://doi.org/10.1016/j.microc.2023.109344

Dakal, T. C., Kumar, A., Majumdar, R. S., & Yadav, V. (2016). Mechanistic basis of antimicrobial actions of silver nanoparticles. *Frontiers in Microbiology*, 7. https://doi.org/10.3389/fmicb.2016.01831

Damm, C., Münstedt, H., & Rösch, A. (2008). The antimicrobial efficacy of polyamide 6/silver-nano- and microcomposites. *Materials Chemistry and Physics*, 108(1), 61–66. https://doi.org/10.1016/j.matchemphys.2007.09.002

Dasgupta, N., Ranjan, S., Mundra, S., Ramalingam, C., & Kumar, A. (2015). Fabrication of food grade vitamin E nanoemulsion by low Energy Approach, Characterization and its application. *International Journal of Food Properties*, 19(3), 700–708. https://doi.org/10.1080/10942912.2015.1042587

Dash, K., Deka, P., Bangar, S., Chaudhary, V., Trif, M., & Rusu, A. (2022). Applications of inorganic nanoparticles in food Packaging: A comprehensive review. *Polymers*, 14(3), 521. https://doi.org/10.3390/polym14030521

Deng, S., Gigliobianco, M. R., Censi, R., & Di Martino, P. (2020). Polymeric nanocapsules as nanotechnological alternative for drug delivery system: Current status, challenges and opportunities. *Nanomaterials*, 10(5), 847. https://doi.org/10.3390/nano10050847

De Oliveira Mallia, J., Galea, R., Nag, R., Cummins, E., Gatt, R., & Valdramidis, V. (2021). Nanoparticle food applications and their toxicity: Current trends and needs in risk assessment strategies. *Journal of Food Protection*, 85(2), 355–372. https://doi.org/10.4315/jfp-21-184

Dhahi, T. S., Dafhalla, A. K. Y., Tayfour, O. E., Mubarakali, A., Alqahtani, A. S., Ahmed, A. E. T., Elobaid, M. E., Adam, T., & Gopinath, S. C. B. (2024). Advances in nano sensors for monitoring and optimal performance enhancement in photovoltaic cells. *iScience*, 27(4), 109347. https://doi.org/10.1016/j.isci.2024.109347

Dias, M. V., De Fátima F Soares, N., Borges, S. V., De Sousa, M. M., Nunes, C. A., De Oliveira, I. R. N., & Medeiros, E. A. A. (2013). Use of allyl isothiocyanate and carbon nanotubes in an antimicrobial film to package shredded, cooked chicken meat. *Food Chemistry*, 141(3), 3160–3166. https://doi.org/10.1016/j.foodchem.2013.05.148

Du, H., Sun, X., Chong, X., Yang, M., Zhu, Z., & Wen, Y. (2023). A review on smart active packaging systems for food preservation: Applications and future trends. *Trends in Food Science & Technology*, 141, 104200. https://doi.org/10.1016/j.tifs.2023.104200

Du, Z., Li, Q., Li, J., Su, E., Liu, X., Wan, Z., & Yang, X. (2019). Self-Assembled egg yolk peptide micellar nanoparticles as a versatile emulsifier for Food-Grade Oil-in-Water pickering nanoemulsions. *Journal of Agricultural and Food Chemistry*, 67(42), 11728–11740. https://doi.org/10.1021/acs.jafc.9b04595

Duncan, T. V. (2011). Applications of nanotechnology in food packaging and food safety: Barrier materials, antimicrobials and sensors. *Journal of Colloid and Interface Science*, 363(1), 1–24. https://doi.org/10.1016/j.jcis.2011.07.017

Fadiji, T., Rashvand, M., Daramola, M. O., & Iwarere, S. A. (2023). A review on antimicrobial packaging for extending the shelf life of food. *Processes*, 11(2), 590. https://doi.org/10.3390/pr11020590

Farid, N., Waheed, A., & Motwani, S. (2023). Synthetic and natural antimicrobials as a control against food borne pathogens: A review. *Heliyon*, 9(6), e17021. https://doi.org/10.1016/j.heliyon.2023.e17021

Fasihnia, S. H., Peighambardoust, S. H., & Peighambardoust, S. J. (2017). Nanocomposite films containing organoclay nanoparticles as an antimicrobial (active) packaging for potential food application. *Journal of Food Processing and Preservation*, 42(2), e13488. https://doi.org/10.1111/jfpp.13488

Flanagan, J., & Singh, H. (2006). Microemulsions: A potential delivery system for bioactives in food. *Critical Reviews in Food Science and Nutrition*, 46(3), 221–237. https://doi.org/10.1080/10408690590956710

Fu, Y., Liu, T., Wang, H., Wang, Z., Hou, L., Jiang, J., & Xu, T. (2024). Applications of nanomaterial technology in biosensing. *Journal of Science Advanced Materials and Devices*, 9(2), 100694. https://doi.org/10.1016/j.jsamd.2024.100694

Fujishima, A., Rao, T. N., & Tryk, D. A. (2000). Titanium dioxide photocatalysis. *Journal of Photochemistry and Photobiology C Photochemistry Reviews*, 1(1), 1–21. https://doi.org/10.1016/s1389-5567(00)00002-2

Ghosh, S., Mandal, R. K., Mukherjee, A., & Roy, S. (2025). Nanotechnology in the manufacturing of sustainable food packaging: A review. *Discover Nano*, 20(1). https://doi.org/10.1186/s11671-025-04213-x

Gil, M., Rudy, M., Duma-Kocan, P., & Stanisławczyk, R. (2025). Electronic sensing technologies in food quality assessment: A comprehensive literature review. *Applied Sciences*, 15(3), 1530. https://doi.org/10.3390/app15031530

Gómez-Gómez, A. L., Martínez-Ayala, A. L., Del Rosario Moguel-Concha, D., Borges-Martínez, J. E., De Jesús Perea-Flores, M., & Dávila-Ortiz, G. (2023). Relationship of nanomaterials' structure based on their application in the food industry: Physicochemical and Techno-Functional characteristics. *Applied Sciences*, 13(12), 7167. https://doi.org/10.3390/app13127167

Gong, Z., Huang, Y., Hu, X., Zhang, J., Chen, Q., & Chen, H. (2023). Recent progress in electrochemical Nano-Biosensors for detection of pesticides and mycotoxins in foods. *Biosensors*, 13(1), 140. https://doi.org/10.3390/bios13010140

Goyal, R., Macri, L. K., Kaplan, H. M., & Kohn, J. (2015). Nanoparticles and nanofibers for topical drug delivery. *Journal of Controlled Release*, 240, 77–92. https://doi.org/10.1016/j.jconrel.2015.10.049

Guarino, V., Bonadies, I., & Ambrosio, L. (2017). *Fabrication of Nanofibers and Nanotubes for Tissue Regeneration and Repair.* Elsevier eBooks (pp. 205–228). https://doi.org/10.1016/b978-0-08-100803-4.00008-5

Gupta, R. K., Gawad, F. a. E., Ali, E. A., Karunanithi, S., Yugiani, P., & Srivastav, P. P. (2024a). Nanotechnology: Current applications and future scope in food packaging systems. *Measurement Food*, 13, 100131. https://doi.org/10.1016/j.meafoo.2023.100131

Gupta, R. K., Guha, P., & Srivastav, P. P. (2024b). Investigating the toxicological effects of nanomaterials in food packaging associated with human health and the environment. *Journal of Hazardous Materials Letters*, 100125. https://doi.org/10.1016/j.hazl.2024.100125

Han, C., Yang, C., Li, X., Liu, E., Meng, X., & Liu, B. (2021). DHA loaded nanoliposomes stabilized by β-sitosterol: Preparation, characterization and release in vitro and vivo. *Food Chemistry*, 368, 130859. https://doi.org/10.1016/j.foodchem.2021.130859

Heo, W., & Lim, S. (2024). A review on gas indicators and sensors for smart food packaging. *Foods*, 13(19), 3047. https://doi.org/10.3390/foods13193047

Ho, T. M., Abik, F., & Mikkonen, K. S. (2021). An overview of nanoemulsion characterization via atomic force microscopy. *Critical Reviews in Food Science and Nutrition*, 62(18), 4908–4928. https://doi.org/10.1080/10408398.2021.1879727

Huang, X., Zhu, Y., & Kianfar, E. (2021). Nano Biosensors: Properties, applications and electrochemical techniques.

Journal of Materials Research and Technology, 12, 1649–1672. https://doi.org/10.1016/j.jmrt.2021.03.048

Jacob, S., Kather, F. S., Boddu, S. H. S., Shah, J., & Nair, A. B. (2024). Innovations in nanoemulsion technology: Enhancing drug delivery for oral, parenteral, and ophthalmic applications. *Pharmaceutics*, 16(10), 1333. https://doi.org/10.3390/pharmaceutics16101333

Jafarzadeh, S., Yildiz, Z., Yildiz, P., Strachowski, P., Forough, M., Esmaeili, Y., Naebe, M., & Abdollahi, M. (2024). Advanced technologies in biodegradable packaging using intelligent sensing to fight food waste. *International Journal of Biological Macromolecules*, 261, 129647. https://doi.org/10.1016/j.ijbiomac.2024.129647

Janik, M., Hanula, M., Khachatryan, K., & Khachatryan, G. (2023). Nano-/Microcapsules, liposomes, and micelles in polysaccharide carriers: Applications in food technology. *Applied Sciences*, 13(21), 11610. https://doi.org/10.3390/app132111610

Javaid, M., Haleem, A., Singh, R. P., Rab, S., & Suman, R. (2021). Exploring the potential of nanosensors: A brief overview. *Sensors International*, 2, 100130. https://doi.org/10.1016/j.sintl.2021.100130

Johnston, C. T. (2010). Probing the nanoscale architecture of clay minerals. *Clay Minerals*, 45(3), 245–279. https://doi.org/10.1180/claymin.2010.045.3.245

Joudeh, N., & Linke, D. (2022). Nanoparticle classification, physicochemical properties, characterization, and applications: A comprehensive review for biologists. *Journal of Nanobiotechnology*, 20(1). https://doi.org/10.1186/s12951-022-01477-8

Joung, H. J., Choi, M., Kim, J. T., Park, S. H., Park, H. J., & Shin, G. H. (2016). Development of food-grade curcumin nanoemulsion and its potential application to food beverage system: Antioxidant property and in vitro digestion. *Journal of Food Science*, 81(3). https://doi.org/10.1111/1750-3841.13224

Karnwal, A., Kumar, G., Singh, R., Selvaraj, M., Malik, T., & Tawaha, A. R. M. A. (2025). Natural biopolymers in edible coatings: Applications in food preservation. *Food Chemistry X*, 25, 102171. https://doi.org/10.1016/j.fochx.2025.102171

Khan, I., Saeed, K., & Khan, I. (2017). Nanoparticles: Properties, applications and toxicities. *Arabian Journal of Chemistry*, 12(7), 908–931. https://doi.org/10.1016/j.arabjc.2017.05.011

Khan, Y., Sadia, H., Shah, S. Z. A., Khan, M. N., Shah, A. A., Ullah, N., Ullah, M. F., Bibi, H., Bafakeeh, O. T., Khedher, N. B., Eldin, S. M., Fadhl, B. M., & Khan, M. I. (2022). Classification, synthetic, and characterization approaches to nanoparticles, and their applications in various fields of nanotechnology: A review. *Catalysts*, 12(11), 1386. https://doi.org/10.3390/catal12111386

Kim, I., Viswanathan, K., Kasi, G., Thanakkasaranee, S., Sadeghi, K., & Seo, J. (2020). ZNO nanostructures in active Antibacterial food packaging: Preparation methods, antimicrobial mechanisms, safety issues, future prospects, and challenges. *Food Reviews International*, 38(4), 537–565. https://doi.org/10.1080/87559129.2020.1737709

Konstantinidi, M., & Koutelidakis, A. E. (2019). Functional foods and bioactive compounds: A review of its possible role on weight management and obesity's metabolic consequences. *Medicines*, 6(3), 94. https://doi.org/10.3390/medicines6030094

Kraśniewska, K., Galus, S., & Gniewosz, M. (2020). Biopolymers-based materials containing silver nanoparticles as active packaging for food applications—a review. *International Journal of Molecular Sciences*, 21(3), 698. https://doi.org/10.3390/ijms21030698

Kravanja, G., Primožič, M., Knez, Ž., & Leitgeb, M. (2019). Chitosan-based (nano)materials for novel biomedical applications. *Molecules*, 24(10), 1960. https://doi.org/10.3390/molecules24101960

Kumar, M., Agrawal, A. V., Moradi, M., & Yousefi, R. (2020). *Nanosensors for Gas Sensing Applications*. Elsevier eBooks (pp. 107–130). https://doi.org/10.1016/b978-0-12-818821-7.00006-3

Kumari, R., Suman, K., Karmakar, S., Mishra, V., Lakra, S. G., Saurav, G. K., & Mahto, B. K. (2023). Regulation and safety measures for nanotechnology-based agri-products. *Frontiers in Genome Editing*, 5. https://doi.org/10.3389/fgeed.2023.1200987

Li, L., Zeng, Y., Chen, M., & Liu, G. (2022). Application of nanomicelles in enhancing bioavailability and biological efficacy of bioactive nutrients. *Polymers*, 14(16), 3278. https://doi.org/10.3390/polym14163278

Lisboa, H. M., Pasquali, M. B., Anjos, A. I. D., Sarinho, A. M., De Melo, E. D., Andrade, R., Batista, L., Lima, J., Diniz, Y., & Barros, A. (2024). Innovative and sustainable food preservation techniques: Enhancing food quality, safety, and environmental sustainability. *Sustainability*, 16(18), 8223. https://doi.org/10.3390/su16188223

Liu, Z., Chen, X., & Guo, H. (2021). Preparation and properties of a novel sodium alginate microcapsule. *Journal of Physics Conference Series*, 1893(1), 012005. https://doi.org/10.1088/1742-6596/1893/1/012005

Lu, W., Huang, D., Wang, C., Yeh, C., Tsai, J., Huang, Y., & Li, P. (2017). Preparation, characterization, and antimicrobial activity of nanoemulsions incorporating citral essential oil. *Journal of Food and Drug Analysis*, 26(1), 82–89. https://doi.org/10.1016/j.jfda.2016.12.018

Lv, Y., Li, W., Liao, W., Jiang, H., Liu, Y., Cao, J., Lu, W., & Feng, Y. (2024). Nano-Drug delivery systems based on natural products. *International Journal of Nanomedicine*, 19, 541–569. https://doi.org/10.2147/ijn.s443692

Mafe, A. N., Edo, G. I., Makia, R. S., Joshua, O. A., Akpoghelie, P. O., Gaaz, T. S., Jikah, A. N., Yousif, E., Isoje, E. F., Igbuku, U. A., Ahmed, D. S., Essaghah, A. E. A., & Umar, H. (2024). A review on food spoilage mechanisms, food borne diseases and commercial aspects of food preservation and processing. *Food Chemistry Advances*, 5, 100852. https://doi.org/10.1016/j.focha.2024.100852

Majid, I., Nayik, G. A., Dar, S. M., & Nanda, V. (2016). Novel food packaging technologies: Innovations and future prospective. *Journal of the Saudi Society of Agricultural Sciences*, 17(4), 454–462. https://doi.org/10.1016/j.jssas.2016.11.003

Malhotra, B., Keshwani, A., & Kharkwal, H. (2015). Antimicrobial food packaging: Potential and pitfalls. *Frontiers in Microbiology*, 6. https://doi.org/10.3389/fmicb.2015.00611

Malik, S., Singh, J., Goyat, R., Saharan, Y., Chaudhry, V., Umar, A., Ibrahim, A. A., Akbar, S., Ameen, S., & Baskoutas, S. (2023). Nanomaterials-based biosensor and their applications: A review. *Heliyon*, 9(9), e19929. https://doi.org/10.1016/j.heliyon.2023.e19929

Marín, V. N., Buccini, D. F., Da Silva, V. G., Soliz, I. A. F., & Franco, O. L. (2024). Nanoformulations of bioactive compounds derived from essential oils with antimicrobial activity. *Nano Trends*, 100070. https://doi.org/10.1016/j.nwnano.2024.100070

Mason, T. G., Wilking, J. N., Meleson, K., Chang, C. B., & Graves, S. M. (2006). Nanoemulsions: Formation, structure, and physical properties. *Journal of Physics Condensed Matter*, 18(41), R635–R666. https://doi.org/10.1088/0953-8984/18/41/r01

Mohammad, Z. H., Ahmad, F., Ibrahim, S. A., & Zaidi, S. (2022). Application of nanotechnology in different aspects of the food industry. *Discover Food*, 2(1). https://doi.org/10.1007/s44187-022-00013-9

Mohammed Fayaz, A., Balaji, K., Girilal, M., Kalaichelvan, P. T., & Venkatesan, R. (2009). Mycobased synthesis of silver nanoparticles and their incorporation into sodium alginate films for vegetable and fruit preservation. *Journal of Agricultural and Food Chemistry*, 57(14), 6246–6252. https://doi.org/10.1021/jf900337h

Mohan, A., Jurat-Fuentes, J., & Harte, F. (2012). Binding of vitamin A by casein micelles in commercial skim milk. *Journal of Dairy Science*, 96(2), 790–798. https://doi.org/10.3168/jds.2012-5777

Moradi, F., Ghaedi, A., Fooladfar, Z., & Bazrgar, A. (2023). Recent advance on nanoparticles or nanomaterials with anti-multidrug resistant bacteria and anti-bacterial biofilm properties: A systematic review. *Heliyon*, 9(11), e22105. https://doi.org/10.1016/j.heliyon.2023.e22105

More, P. R., Pandit, S., De Filippis, A., Franci, G., Mijakovic, I., & Galdiero, M. (2023). Silver nanoparticles: Bactericidal and mechanistic approach against drug resistant pathogens. *Microorganisms*, 11(2), 369. https://doi.org/10.3390/microorganisms11020369

Motsoene, F., Abrahamse, H., & Kumar, S. S. D. (2023). Multifunctional lipid-based nanoparticles for wound healing and antibacterial applications: A review. *Advances in Colloid and Interface Science*, 321, 103002. https://doi.org/10.1016/j.cis.2023.103002

Mushtaq, A., Wani, S. M., Malik, A., Gull, A., Ramniwas, S., Nayik, G. A., Ercisli, S., Marc, R. A., Ullah, R., & Bari, A. (2023). Recent insights into nanoemulsions: Their preparation, properties and applications. *Food Chemistry X*, 18, 100684. https://doi.org/10.1016/j.fochx.2023.100684

Nath, D., R, S., Pal, K., & Sarkar, P. (2021). Nanoclay-based active food packaging systems: A review. *Food Packaging and Shelf Life*, 31, 100803. https://doi.org/10.1016/j.fpsl.2021.100803

Neethirajan, S., & Jayas, D. S. (2011). Nanotechnology for the food and bioprocessing industries. *Food and Bioprocess Technology*, 4(1), 39–47. https://doi.org/10.1007/s11947-010-0328-2

Onyeaka, H., Passaretti, P., Miri, T., & Al-Sharify, Z. T. (2022). The safety of nanomaterials in food production and packaging. *Current Research in Food Science*, 5, 763–774. https://doi.org/10.1016/j.crfs.2022.04.005

Pateiro, M., Gómez, B., Munekata, P. E. S., Barba, F. J., Putnik, P., Kovačević, D. B., & Lorenzo, J. M. (2021). Nanoencapsulation of promising bioactive compounds to improve their absorption, stability, functionality and the appearance of the final food products. *Molecules*, 26(6), 1547. https://doi.org/10.3390/molecules26061547

Patra, J. K., Das, G., Fraceto, L. F., Campos, E. V. R., Del Pilar Rodriguez-Torres, M., Acosta-Torres, L. S., Diaz-Torres, L. A., Grillo, R., Swamy, M. K., Sharma, S., Habtemariam, S., & Shin, H. (2018). Nano based drug delivery systems: Recent developments and future prospects. *Journal of Nanobiotechnology*, 16(1). https://doi.org/10.1186/s12951-018-0392-8

Perera, K. Y., Hopkins, M., Jaiswal, A. K., & Jaiswal, S. (2023). Nanoclays-containing bio-based packaging materials: Properties, applications, safety, and regulatory issues. *Journal of Nanostructure in Chemistry*, 14(1), 71–93. https://doi.org/10.1007/s40097-023-00525-5

Pradhan, N., Singh, S., Ojha, N., Shrivastava, A., Barla, A., Rai, V., & Bose, S. (2015). Facets of nanotechnology as seen in food processing, packaging, and preservation industry. *BioMed Research International*, 2015, 1–17. https://doi.org/10.1155/2015/365672

Prakash, B., Kujur, A., Yadav, A., Kumar, A., Singh, P. P., & Dubey, N. (2018). Nanoencapsulation: An efficient technology to boost the antimicrobial potential of plant essential oils in food system. *Food Control*, 89, 1–11. https://doi.org/10.1016/j.foodcont.2018.01.018

Preeti, N., Sambhakar, S., Malik, R., Bhatia, S., Harrasi, A. A., Rani, C., Saharan, R., Kumar, S., Geeta, N., & Sehrawat, R. (2023). Nanoemulsion: An emerging novel technology for improving the bioavailability of drugs. *Scientifica*, 2023, 1–25. https://doi.org/10.1155/2023/6640103

Priya, K., Thirunavookarasu, N., & Chidanand, D. (2023). Recent advances in edible coating of food products and its legislations: A review. *Journal of Agriculture and Food Research*, 12, 100623. https://doi.org/10.1016/j.jafr.2023.100623

Radoi, A., Targa, M., Prieto-Simon, B., & Marty, J. (2008). Enzyme-Linked Immunosorbent Assay (ELISA) based on superparamagnetic nanoparticles for aflatoxin M1 detection. *Talanta*, 77(1), 138–143. https://doi.org/10.1016/j.talanta.2008.05.048

Rahmati, F., Hosseini, S. S., Safai, S. M., Lajayer, B. A., & Hatami, M. (2020). New insights into the role of nanotechnology in microbial food safety. *3 Biotech*, 10(10). https://doi.org/10.1007/s13205-020-02409-9

Rao, J., & McClements, D. J. (2011). Food-grade microemulsions, nanoemulsions and emulsions: Fabrication from sucrose monopalmitate & lemon oil. *Food Hydrocolloids*, 25(6), 1413–1423. https://doi.org/10.1016/j.foodhyd.2011.02.004

Rashid, A. B., Haque, M., Islam, S. M. M., & Labib, K. R. U. (2024). Nanotechnology-enhanced fiber-reinforced polymer composites: Recent advancements on processing techniques and applications. *Heliyon*, 10(2), e24692. https://doi.org/10.1016/j.heliyon.2024.e24692

Reddy, M. N. P., Ishwarya, S. P., & Anandharamakrishnan, C. (2019). Nanoencapsulation of roasted coffee bean oil in whey protein wall system through nanospray drying. *Journal of Food Processing and Preservation*, 43(3), e13893. https://doi.org/10.1111/jfpp.13893

Rezagholizade-Shirvan, A., Soltani, M., Shokri, S., Radfar, R., Arab, M., & Shamloo, E. (2024). Bioactive compound encapsulation: Characteristics, applications in food systems, and implications for human health. *Food Chemistry X*, 24, 101953. https://doi.org/10.1016/j.fochx.2024.101953

Rhim, J., Wang, L., Lee, Y., & Hong, S. (2014). Preparation and characterization of bio-nanocomposite films of agar and silver nanoparticles: Laser ablation method. *Carbohydrate Polymers*, 103, 456–465. https://doi.org/10.1016/j.carbpol.2013.12.075

Roy, M., & Yadav, B. K. (2022). Electronic nose for detection of food adulteration: A review. *Journal of Food Science and Technology*, 59(3), 846–858. https://doi.org/10.1007/s13197-021-05057-w

Saba, N., Jawaid, M., & Asim, M. (2018). *Nanocomposites with Nanofibers and Fillers from Renewable Resources*. Elsevier eBooks (pp. 145–170). https://doi.org/10.1016/b978-0-08-102177-4.00007-0

Sani, M. A., Khezerlou, A., Rezvani-Ghalhari, M., McClements, D. J., & Varma, R. S. (2025). Advanced carbon-based nanomaterials: Application in the development of multifunctional next-generation food packaging materials. *Advances in Colloid and Interface Science*, 339, 103422. https://doi.org/10.1016/j.cis.2025.103422

Sarwar, M. S., Niazi, M. B. K., Jahan, Z., Ahmad, T., & Hussain, A. (2018). Preparation and characterization of PVA/nanocellulose/Ag nanocomposite films for antimicrobial food packaging. *Carbohydrate Polymers*, 184, 453–464. https://doi.org/10.1016/j.carbpol.2017.12.068

Sharma, A., Matharu, Z., Sumana, G., Solanki, P. R., Kim, C., & Malhotra, B. (2010). Antibody immobilized cysteamine functionalized-gold nanoparticles for aflatoxin detection. *Thin Solid Films*, 519(3), 1213–1218. https://doi.org/10.1016/j.tsf.2010.08.071

Shruti, A., Bage, N., & Kar, P. (2024). Nanomaterials based sensors for analysis of food safety. *Food Chemistry*, 433, 137284. https://doi.org/10.1016/j.foodchem.2023.137284

Singh, A. K., Pal, P., Pandey, B., Goksen, G., Sahoo, U. K., Lorenzo, J. M., & Sarangi, P. K. (2023). Development of "smart foods" for health by nanoencapsulation: Novel technologies and challenges. *Food Chemistry X*, 20, 100910. https://doi.org/10.1016/j.fochx.2023.100910

Singh, T., Shukla, S., Kumar, P., Wahla, V., Bajpai, V. K., & Rather, I. A. (2017). Application of nanotechnology in food science: Perception and overview. *Frontiers in Microbiology*, 8. https://doi.org/10.3389/fmicb.2017.01501

Slavin, Y. N., Asnis, J., Häfeli, U. O., & Bach, H. (2017). Metal nanoparticles: Understanding the mechanisms behind antibacterial activity. *Journal of Nanobiotechnology*, 15(1). https://doi.org/10.1186/s12951-017-0308-z

Song, R., Murphy, M., Li, C., Ting, K., Soo, C., & Zheng, Z. (2018). Current development of biodegradable polymeric materials for biomedical applications. *Drug Design Development and Therapy*, 12, 3117–3145. https://doi.org/10.2147/dddt.s165440

Spernath, A., Yaghmur, A., Aserin, A., Hoffman, R. E., & Garti, N. (2002). Food-Grade microemulsions based on nonionic emulsifiers: Media to enhance lycopene solubilization. *Journal of Agricultural and Food Chemistry*, 50(23), 6917–6922. https://doi.org/10.1021/jf025762n

Su, Q., Zhao, X., Zhang, X., Wang, Y., Zeng, Z., Cui, H., & Wang, C. (2022). Nano functional food: Opportunities, development, and future perspectives. *International Journal of Molecular Sciences*, 24(1), 234. https://doi.org/10.3390/ijms24010234

Suvarna, V., Nair, A., Mallya, R., Khan, T., & Omri, A. (2022). Antimicrobial nanomaterials for food packaging. *Antibiotics*, 11(6), 729. https://doi.org/10.3390/antibiotics11060729

Tamjidi, F., Shahedi, M., Varshosaz, J., & Nasirpour, A. (2017). Stability of astaxanthin-loaded nanostructured lipid carriers in beverage systems. *Journal of the Science of Food and Agriculture*, 98(2), 511–518. https://doi.org/10.1002/jsfa.8488

Tarannum, N., Gautam, A., Chauhan, T., & Kumar, D. (2024). Nanomaterial based sensors for detection of food contaminants: A prospect. *Sensing Technology*, 2(1). https://doi.org/10.1080/28361466.2024.2373196

Thakur, D., Rana, P., Singh, S. K., Bakshi, M., Kumar, S., & Singh, S. (2024). Nanoemulsion edible coating for shelf-life improvement and quality control in perishable products. *Plant Nano Biology*, 100114. https://doi.org/10.1016/j.plana.2024.100114

Thakur, M. S., & Ragavan, K. V. (2012). Biosensors in food processing. *Journal of Food Science and Technology*, 50(4), 625–641. https://doi.org/10.1007/s13197-012-0783-z

Tripathi, S., Mehrotra, G., & Dutta, P. (2009). Physicochemical and bioactivity of cross-linked chitosan–PVA film for food packaging applications. *International Journal of Biological Macromolecules*, 45(4), 372–376. https://doi.org/10.1016/j.ijbiomac.2009.07.006

Upadhyay, P., Zubair, M., Roopesh, M. S., & Ullah, A. (2024). An overview of advanced antimicrobial food packaging: Emphasizing antimicrobial agents and Polymer-Based films. *Polymers*, 16(14), 2007. https://doi.org/10.3390/polym16142007

Valenzuela-Amaro, H. M., Aguayo-Acosta, A., Meléndez-Sánchez, E. R., De La Rosa, O., Vázquez-Ortega, P. G., Oyervides-Muñoz, M. A., Sosa-Hernández, J. E., & Parra-Saldívar, R. (2023). Emerging applications of nanobiosensors in pathogen detection in water and food. *Biosensors*, 13(10), 922. https://doi.org/10.3390/bios13100922

Valerini, D., Tammaro, L., Di Benedetto, F., Vigliotta, G., Capodieci, L., Terzi, R., & Rizzo, A. (2018). Aluminum-doped zinc oxide coatings on polylactic acid films for antimicrobial food packaging. *Thin Solid Films*, 645, 187–192. https://doi.org/10.1016/j.tsf.2017.10.038

Vasile, C. (2018). Polymeric nanocomposites and nanocoatings for food Packaging: A review. *Materials*, 11(10), 1834. https://doi.org/10.3390/ma11101834

Wanekaya, A. K., Chen, W., Myung, N. V., & Mulchandani, A. (2006). Nanowire-based electrochemical biosensors. *Electroanalysis*, 18(6), 533–550. https://doi.org/10.1002/elan.200503449

Wang, K., Lin, X., Zhang, M., Li, Y., Luo, C., & Wu, J. (2022a). Review of electrochemical biosensors for food safety detection. *Biosensors*, 12(11), 959. https://doi.org/10.3390/bios12110959

Wang, M., Li, S., Chen, Z., Zhu, J., Hao, W., Jia, G., Chen, W., Zheng, Y., Qu, W., & Liu, Y. (2021). Safety assessment of nanoparticles in food: Current status and prospective. *Nano Today*, 39, 101169. https://doi.org/10.1016/j.nantod.2021.101169

Wang, S., Alenius, H., El-Nezami, H., & Karisola, P. (2022b). A new look at the effects of engineered ZnO and TiO2 nanoparticles: Evidence from transcriptomics studies. *Nanomaterials*, 12(8), 1247. https://doi.org/10.3390/nano12081247

Wei, H., Seidi, F., Zhang, T., Jin, Y., & Xiao, H. (2020). Ethylene scavengers for the preservation of fruits and vegetables: A review. *Food Chemistry*, 337, 127750. https://doi.org/10.1016/j.foodchem.2020.127750

Wilson, R. J., Li, Y., Yang, G., & Zhao, C. (2021). Nanoemulsions for drug delivery. *Particuology*, 64, 85–97. https://doi.org/10.1016/j.partic.2021.05.009

Wu, Y., Liang, Y., Mei, C., Cai, L., Nadda, A., Van Le, Q., Peng, Y., Lam, S. S., Sonne, C., & Xia, C. (2021). Advanced nanocellulose-based gas barrier materials: Present status and prospects. *Chemosphere*, 286, 131891. https://doi.org/10.1016/j.chemosphere.2021.131891

Xavier, M., Parente, I. A., Rodrigues, P. M., Cerqueira, M. A., Pastrana, L., & Gonçalves, C. (2021). Safety and fate of nanomaterials in food: The role of in vitro tests. *Trends in Food Science & Technology*, 109, 593–607. https://doi.org/10.1016/j.tifs.2021.01.050

Xuan, L., Ju, Z., Skonieczna, M., Zhou, P., & Huang, R. (2023). Nanoparticles-induced potential toxicity on human health: Applications, toxicity mechanisms, and evaluation models. *MedComm*, 4(4). https://doi.org/10.1002/mco2.327

Yang, F., Li, H., Li, F., Xin, Z., Zhao, L., Zheng, Y., & Hu, Q. (2010). Effect of nano-packing on preservation quality of fresh strawberry (*Fragaria ananassa* Duch. cv Fengxiang) during storage at 4°C. *Journal of Food Science*, 75(3). https://doi.org/10.1111/j.1750-3841.2010.01520.x

Yu, Z., Li, Q., Wang, J., Yu, Y., Wang, Y., Zhou, Q., & Li, P. (2020). Reactive oxygen species-related nanoparticle toxicity in the biomedical field. *Nanoscale Research Letters*, 15(1). https://doi.org/10.1186/s11671-020-03344-7

Zamolo, V. A., Valenti, G., Venturelli, E., Chaloin, O., Marcaccio, M., Boscolo, S., Castagnola, V., Sosa, S., Berti, F., Fontanive, G., Poli, M., Tubaro, A., Bianco, A., Paolucci, F., & Prato, M. (2012). Highly sensitive electrochemiluminescent nanobiosensor for the detection of palytoxin. *ACS Nano*, 6(9), 7989–7997. https://doi.org/10.1021/nn302573c

Zhang, Y., Li, J., Jiao, S., Li, Y., Zhou, Y., Zhang, X., Maryam, B., & Liu, X. (2024). Microfluidic sensors for the detection of emerging contaminants in water: A review. *The Science of the Total Environment*, 929, 172734. https://doi.org/10.1016/j.scitotenv.2024.172734

Zhou, Y., Pan, F., Li, Y., Zhang, Y., Zhang, J., Lu, S., Ren, H., & Liu, Z. (2009). Colloidal gold probe-based immunochromatographic assay for the rapid detection of brevetoxins in fishery product samples. *Biosensors and Bioelectronics*, 24(8), 2744–2747. https://doi.org/10.1016/j.bios.2009.01.034

14 Industrial Applications of Green-Synthesized Nanoparticles

Venkatesh Sivalingam, Gaurav Rajasekhar, Bandaru Manideep, Asha P Johnson, Jeevan Kumar M, Poornima DS, and Lokesh Ravi

14.1 INTRODUCTION

14.1.1 Green Synthesis of Nanoparticles and Their Importance in Pharmaceutical Industry

Nanoparticles (NPs) are minute particles with dimensions in the nanometer range (1 nm = 10^{-9} m). Due to their nanoscale size, they exhibit distinctive physicochemical properties, including thermal, optical, electronic, and magnetic characteristics. Typically, metal-based NPs are synthesized through chemical methods using either a top-down or bottom-up approach, each with its own advantages and limitations. The most widely utilized metal-based NPs include gold (Au-NPs), silver (Ag-NPs), zinc oxide (ZnO-NPs), platinum (Pt-NPs), and other metallic nanostructures (1).

Chemical synthesis methods have been reported to exhibit toxicity and pose environmental hazards. Additionally, these methods often yield low nanoparticle production, making them an inefficient approach for large-scale synthesis. Various industrial techniques for NP fabrication include ball milling, lithography, laser ablation, sol-gel processing, pyrolysis, and chemical vapor deposition (CVD), among others To achieve non-toxic, cost-effective, and environmentally sustainable NP synthesis, researchers have increasingly focused on green synthesis. This approach utilizes biological systems such as plants, microorganisms, and cells to facilitate NP formation (2), as shown in Figure 14.1.

Compared to chemical synthesis, green synthesis offers multiple advantages, including non-toxicity, high production efficiency, well-defined morphology, environmental sustainability, and scalability, as shown in Figure 14.2. Several factors influence NP biosynthesis, such as pH, reactant concentration, reaction duration, and temperature, as these parameters aid in optimizing NP properties. Temperature plays a crucial role in controlling NP size and surface morphology, while pH affects structural characteristics and regulates nucleation center formation. Karade et al. highlighted that reaction time significantly impacts the synthesis of magnetic nanoparticles (MNPs) (3) Characterization techniques are essential for determining the physicochemical properties of synthesized NPs. Advanced analytical instruments are categorized into physical and chemical characterization tools, including UV-vis spectroscopy, X-ray diffraction (XRD), atomic force microscopy (AFM), and Fourier transform infrared spectroscopy (FTIR), among others (4). Strong scientific evidence suggests that green-synthesized NPs exhibit superior pharmacological activity compared to those produced via physicochemical methods. Phytosynthesized NPs demonstrate remarkable stability,

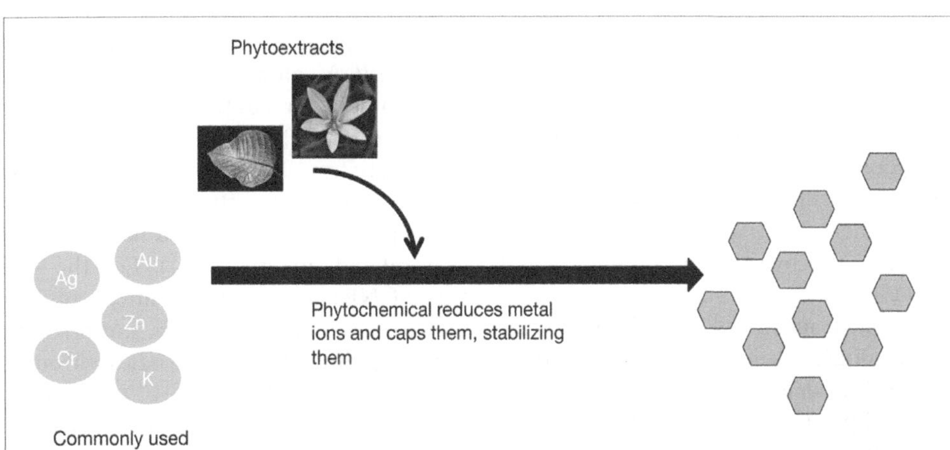

FIGURE 14.1 Typical procedure of green synthesis of metal nanoparticles using biological extracts.

DOI: 10.1201/9781003644439-14

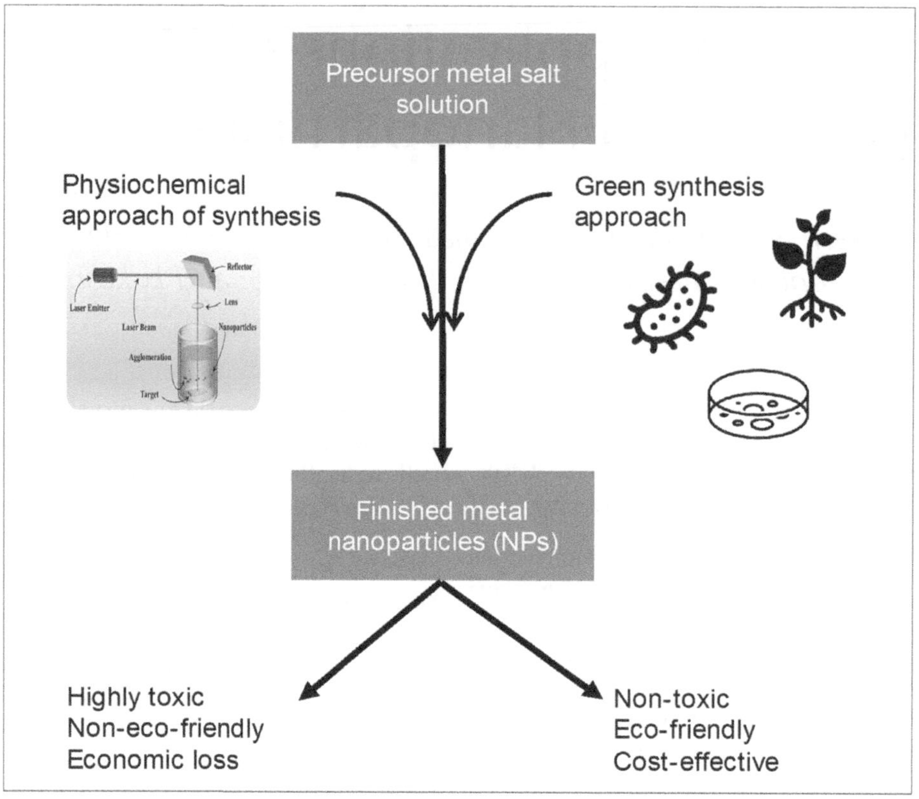

FIGURE 14.2 Schematic representation of difference between chemical synthesis and green synthesis of nanoparticles.

enabling them to exhibit crucial therapeutic properties, such as antimicrobial, antioxidant, antidiabetic, and insecticidal activities, making them highly relevant in biomedical applications (1).

14.1.2 Phytochemical-Derived Metal Nanoparticles and Their Applications

Several plants, including mango (*Mangifera indica*) leaves and tulsi (*Ocimum tenuiflorum*), have been employed for green NP synthesis (5, 6). Botanically, plants produce a diverse array of secondary metabolites, known as phytochemicals, which possess medicinal properties. Notable examples include flavonoids, terpenoids, alkaloids, glycosides, anthocyanins, and phenolics. These naturally occurring compounds facilitate the cost-effective, environmentally friendly, and non-toxic synthesis of metal-based NPs by serving as reducing and stabilizing agents (7).

Phytochemical-mediated metal NPs hold significant pharmaceutical applications, particularly in green drug synthesis and targeted drug delivery systems. Among the most widely synthesized plant-derived metal NPs are gold and silver. Recent studies indicate that wound dressings incorporating Ag-NPs, chitosan, and gelatin have effectively reduced infection rates and accelerated vascularization and scar tissue formation, demonstrating their potential in advanced wound healing therapies (8, 9). It is evident to say that nanoparticles synthesized using green synthesis has immense potential in various sectors, hence displaying unique applications as shown in Figure 14.3.

14.1.2.1 Green-Synthesized Ag-NPs and Their Applications

In vitro studies have demonstrated that polyphenol-enriched nanoparticles synthesized from *Cornus mas* effectively inhibited NF-κB activation in macrophage (10). Polyphenols derived from the white willow tree (*Salix alba*), along with salicin, exhibit potent anti-inflammatory and antibacterial properties and have been successfully utilized in the synthesis of gold and silver nanoparticles. The intrinsic antimicrobial activity of silver enhances its efficiency in pharmaceutical applications, particularly in wound healing and the treatment of skin-related infections (11). For the successful integration of NPs into pharmaceutical applications, it is crucial to assess their toxicity, antimicrobial efficacy, and physicochemical characteristics through comprehensive evaluation. Ag-NPs release ionic silver (Ag^+), which inactivates bacterial enzymes, disrupts bacterial DNA replication, depletes intracellular ATP levels, compromises cytoplasmic membrane integrity, and ultimately induces apoptosis. In addition to its antimicrobial activity, silver possesses diverse medicinal properties and has

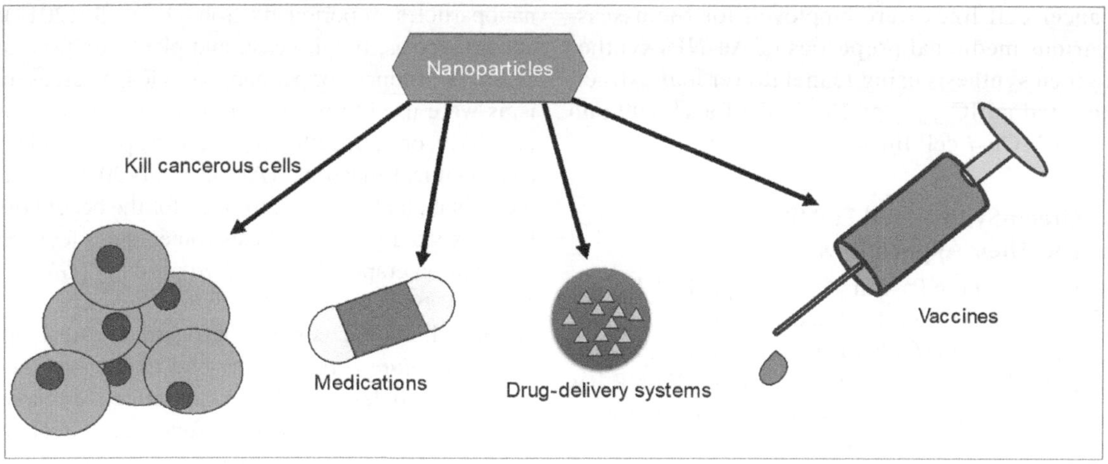

FIGURE 14.3 Applications of green-synthesized nanoparticles in the pharmaceutical industry.

TABLE 14.1
Plants along with Their Phytoextracts and Metal Nanoparticles Synthesized with Their Pharmaceutical Application(s) (15–23)

Name of Organism	Extract	Metal Nanoparticle Synthesized	Pharmaceutical Application
Punica granatum	Fruit, peels	$AgNO_3$, $HAuCl_4$, Gold	Antibacterial activity
Phyllanthus amarus	Leaves	Gold	Anticancer activity
Allium cepa	Bulbs	$AgNO_3$	Anticancer activity
Mentha piperita	Leaves	Gold	Antibacterial activity
Ocimum sanctum	Leaves	$AgNO_3$	Antimicrobial and anticancer activity
Costus speciosus	Callus culture	$AgNO_3$	Antibacterial activity
Calotropis procera	Milky latex	Copper	Antitumor activity
Arachis hypogaea	Leaves	Cr_2O_3	Antibacterial activity
Memecylon umbellatum	Leaves	$AgNO_3$, $HAuCl_4$	Antimicrobial activity
Solanum xanthocarpum	Berries	ZnO	Antioxidant activity
Achillea wilhelmsii	Flowers	Silver	Antimicrobial activity
Trianthema decandra	Roots	Silver and gold	Antimicrobial activity
Caesalpinia pulcherrima	Flowers	Silver	Antiviral, antimicrobial, antioxidant, analgesic
Tagetes erecta	Flowers	Gold	Antioxidant activity
Putranjiva roxburghii	Leaves	Silver	Anticancer activity

historically been employed in the treatment of epilepsy, venereal diseases, acne, and leg ulcers (12, 13). Dhrutika et al. (2013) reported that Ag-NPs release Ag^{3+} ions, which further contribute to their antimicrobial efficacy (14). Ag-NPs synthesized using ethanolic and hydroalcoholic plant extracts demonstrated antitubercular activity at two different minimum inhibitory concentrations (MICs), with the highest effective concentrations being 50 and 75 μg/mL (4). Some literature reports on phyto-synthesized Ag-NPs are summarized in Table 14.1.

14.1.2.2 Green-Synthesized Au-NPs and Their Applications

In addition to silver nanoparticles, gold nanoparticles also hold significant pharmaceutical potential. These materials exhibit diverse properties, including eco-friendliness, biocompatibility, and other functional characteristics, thereby ensuring improved safety for drug-delivery applications. Different plants such as *Citrus maxima, Piper longum,* and *Chenopodium album* were utilized to synthesize gold nanoparticles efficiently and in different sizes. These nanoparticles enhance antibacterial efficacy and drug delivery efficiency, particularly in cancer therapy (24–27). Lysenko et al. investigated the synthesis of gold-silicon dioxide carrier NPs, which significantly reduced lonicera gold ion toxicity and demonstrated strong antiviral activity against adenoviruses. Additionally, Au-NPs have been explored for targeting the dengue virus by inhibiting viral RNA replication. Furthermore, hybrid-coated Ag-NPs conjugated with domain III of the viral envelope protein (EDIII) exhibited enhanced inhibitory activity against the dengue virus. H1299 and

MCF-7 cancer cell lines were employed for the assessment of various medicinal properties of Au-NPs synthesized via green synthesis using fennel flower leaf extract, which exhibited an $IC_{50\ value}$ of 10 $\mu g ml^{-1}$ for H1299 and 25 $\mu g ml^{-1}$ for MCF-7 cell lines.

14.1.2.3 Green-Synthesized Fe-NPs and Their Applications

Synthesis of green Fe-NPs can be accomplished by utilizing different plant species, including thyme (*Thymus vulgaris*), water hyacinth (*Eichhornia crassipes*), and sensitive plants like *Mimosa pudica*. Phyto-derived Fe-NP has antioxidant activity and ferric-reducing antioxidant power. MCF-7 and AMJ-13 cancer cell lines were treated using Fe-NPs synthesized from flat-crown (*Albizia adianthifolia*) leaf extract, which caused apoptosis in the cancer cell lines. Nanoparticles synthesized using sugar-apple leaf extract demonstrated significant cytotoxicity against HepG2 cancer cells, while nanoparticles synthesized from the leaf extract of babchi (*Psoralea corylifolia*) exhibited strong antitumor activity against Caki-2 cells (28).

14.1.2.4 Green-Synthesized Se-NPs and Their Applications

Fruits, seeds, fruit peels, and leaf extracts are used for preparing selenium nanoparticles. The pharmaceutical applications of phyto-derived selenium nanoparticles are vast. It is a biotherapeutic agent without side effects and has physiological responses such as antioxidant, antitumor, and immunoregulatory functions. Assays such as Ferric Reducing Antioxidant Power (FRAP) assay, 2-2′-Azino-bis(3-ethylbenzothiazoline-6-sulfonic acid) assay (ABTS assay), 2-2-Diphenyl-1-picrylhydrazyl (DPPH) assay were conducted on cultivated garlic bud–synthesized selenium nanoparticles and displayed efficient antioxidant response, and the same was true of tea extract. Horseshoe geranium (*Pelargonium zonale*) leaf extract exhibited significant antifungal and antibacterial activity. Bombay ebony (*Diospyros montana*) leaf extract displayed antibacterial activity as well as anticancer and antioxidant activity. Oseltamivir combined with selenium nanoparticles was highly active and exhibited antiviral activity against enteroviruses, and in astrocytoma cells, it reduced ROS production (4). *Ulva fasciata* extract has strong antibacterial activity and anticancer activity. *Solanum nigrum* selenium nanoparticles exhibited efficient inhibitory activity against gram-positive and gram-negative bacteria and as an anticancer and antioxidant. Small selenium nanoparticles attach to cancer cells, induce cell death, and show cytotoxicity activity against cancer cells (4).

14.1.2.5 Green-Synthesized Cu-NPs and Their Applications

Copper nanoparticles (Cu-NPs) synthesized via green synthesis are also eco-friendly and cost-effective. *Penicillium* species were heavily employed to synthesize and stabilize copper nanoparticles, reported by Soheyla et al. (2012). Bacteria, actinomycetes, fungi, yeast, and plants are used as reducing agents to produce copper nanoparticles, whereas microorganisms were used to produce inorganic materials either extracellularly or intracellularly with properties like biological and chemical materials (Amrut et al. 2010). A spectrum of medicinal plants can be exploited for the benefit of producing Cu-NPs via a green synthesis route, and a few species, such as common grapes (*Vitis vinifera*), fire lily (*Gloriosa superba L.*), and even jackfruit (*Arabotrys odoratissimus*), can be employed. In the context of pharmaceutical applications, clove (*Syzgium aromaticum*) bud extract–synthesized copper nanoparticles have better and effective response due to antimicrobial and antioxidant functions. Green-synthesized copper nanoparticles can be used as a therapeutic option, as in therapy for microorganisms with antibiotic resistance, as the copper nanoparticles provide various medical and antimicrobial characteristics. Several studies have demonstrated that copper nanoparticles exhibit strong antiviral properties; the NPs have been shown to inhibit viral protein activity effectively. Additionally, Cu-NPs have exhibited significant cytotoxicity against cancer cell lines, such as human skin carcinoma cells (B16F10), within 24 hours.

14.1.2.6 Green-Synthesized TiO_2-NPs and Their Applications

Titanium dioxide nanoparticles (TiO_2-NPs) are widely utilized in water and air purification (Parthasarthi and Thilagavathi, 2009). However, in pharmaceutical applications, TiO_2-NPs possess a huge number of medicinal as well as antimicrobial properties, particularly when synthesized using *Nyctanthes* leaf extracts. Despite these advantages, the synthesis of TiO_2-NPs using *Nyctanthes* remains underexplored (Sundrarajan and Gowri, 2011) (4, 29).

14.1.3 Therapeutic Approach and Drug Delivery Systems

The primary goal of therapeutic drugs is to reach the target site of infection or disease and exert their intended curative effect. However, achieving this is often challenging due to several factors, including low bioavailability, drug lipophilicity, physiological and chemical barriers, limited drug distribution volume, and off-target effects. Most conventional drugs are administered either orally or via injectable routes. Regardless of the mode of administration, many drugs fail to exhibit optimal therapeutic efficacy due to poor absorption and distribution. Nanoparticles address these limitations due to their nanoscale size, which facilitates enhanced cellular uptake and bioavailability, making them highly efficient drug delivery agents.

14.1.3.1 Biological-Based NPs in the Role of Drug Delivery

Beyond metal-based nanoparticles, other green-synthesized nanocarrier systems for drug delivery include carbon nanotubes, micelles, dendrimers, polymeric nanoparticles, polymerosomes, liposomes, and lipid-coated polymeric nanoparticles. Biopolymeric nanoparticles play a crucial role in drug delivery systems and come in various types. For instance, chitosan nanoparticles have been used for drug delivery in the buccal cavity. Alginate-based chitosan-coated nanoparticles, developed by Costa et al., were designed to enhance ocular epithelium's absorption of daptomycin for antibacterial effects [25]. Xanthan gum was investigated by Menzel et al. as a nasal release excipient, where Cys-MNA was conjugated to xanthan gum to improve muco-adhesion and degradation resistance [26]. Cellulose-based nanoparticles have been employed to modulate drug release by altering drug solubility and gelation properties. Cellulose nanocrystals, through hydrogen bonding interactions, prolong drug release, leading to a controlled and sustained drug release profile (28).

14.1.3.1.1 Liposomes

Liposomes are extensively utilized in pharmaceutical industries as drug delivery systems. These spherical vesicles, composed of phospholipids and steroids, mimic biological membranes, facilitating efficient drug incorporation and targeted delivery (29). Their key advantages include molecular stabilization, compatibility with both hydrophobic and hydrophilic drugs, enhanced biodistribution, and biodegradability. Various types of liposomal nanoparticles include:

1. **Conventional liposomes**—composed of a lipid bilayer encapsulating an aqueous core, capable of forming cationic, anionic, or neutral phospholipids with cholesterol.
2. **PEGylated liposomes**—modified with polyethylene glycol (PEG) to enhance circulation time and create steric stabilization.
3. **Ligand-targeted liposomes**—functionalized with ligands such as antibodies and peptides for targeted drug delivery.
4. **Theranostic liposomes**—multifunctional nanoparticles used for simultaneous targeting, imaging, and therapeutic applications [30].

14.1.3.1.2 Polymeric Micelles

In an aqueous environment, polymeric micelles self assemble into core-shell structures. The hydrophobic core efficiently encapsulates poorly soluble drugs such as camptothecin, docetaxel, and paclitaxel, while the hydrophilic shell stabilizes the nanostructure. These micelles, with the help of the EPR effect, preferentially accumulate in tumor locations, reducing off-target interactions while improving bioavailability and therapeutic efficiency. Polymeric micelles are widely used for delivering both cancer and ophthalmic drugs.

14.1.3.1.3 Dendrimers

Dendrimers, highly branched three-dimensional macromolecules, are another promising class of drug delivery vehicles. While effective, their cationic surface groups can induce cytotoxicity. However, green-synthesized dendrimers exhibit enhanced biocompatibility. Once administered, these nanoparticles execute their intended therapeutic function and are subsequently degraded. Being biodegradable, they pose minimal risk to biological systems and the environment. Some biopolymeric nanoparticles can also be metabolized by enzymatic pathways, as they mimic cellular properties such as membrane composition, lipophilicity, and receptor–ligand interactions.

14.1.3.1.4 Biogenic Silica

Biogenic silica, derived from plants, algae, and certain animal tissues, serves as a natural source of silica nanoparticles. Silica is primarily obtained in the form of ortho-silicic acid from water or soil, making plant biomass a viable precursor for synthesizing silica nanoparticles [39]. Horsetail plants (*Equisetum* spp.) are particularly effective biogenic silica producers, as they generate hydroalcoholic extracts with notable analgesic, anti-inflammatory, hepatoprotective, antifungal, and antimutagenic properties.

14.2 GREEN SYNTHESIS OF NANOPARTICLES IN THE TEXTILE INDUSTRY

14.2.1 Introduction

The green synthesis of nanoparticles is a biological approach that utilizes fungi, bacteria, and plant extracts or enzymes as reducing and stabilizing agents instead of hazardous chemicals. In the textile industry, these nanoparticles are employed for antimicrobial fabric protection, wastewater treatment, self-cleaning, self-healing properties, and photocatalytic activity. As the textile industry continues to expand, increasing volumes of industrial wastewater effluents are discharged into natural water bodies, necessitating effective remediation strategies. ZnO_2 (zinc oxide) and TiO_2 (titanium oxide) nanoparticles were synthesized using a direct precipitation and hydrolysis protocol to investigate their photocatalytic activity in the degradation of Congo red dye. A schematic representation of nanoparticle synthesis for the textile industry is provided in Figure 14.4. Some of the applications of green-synthesized nanoparticles in the textile industry are summarized in Figure 14.5.

14.2.2 Synthesis and Characterization of ZnO and TiO_2 Nanoparticles

The crystallite size of the nanoparticles was determined using X-ray diffraction data and analyzed through various models, including the size-strain plot (SSP), least squares linear multiple scattering effects (LSLMSE), and Halder–Wagner models. Among these methods, the SSP model yielded the most reliable crystallite size for ZnO nanoparticles, estimated at 45.31 nm. In contrast, the Halder–Wagner model indicated a crystallite size of 2.44 nm for TiO_2 nanoparticles. Scanning electron microscopy (SEM)

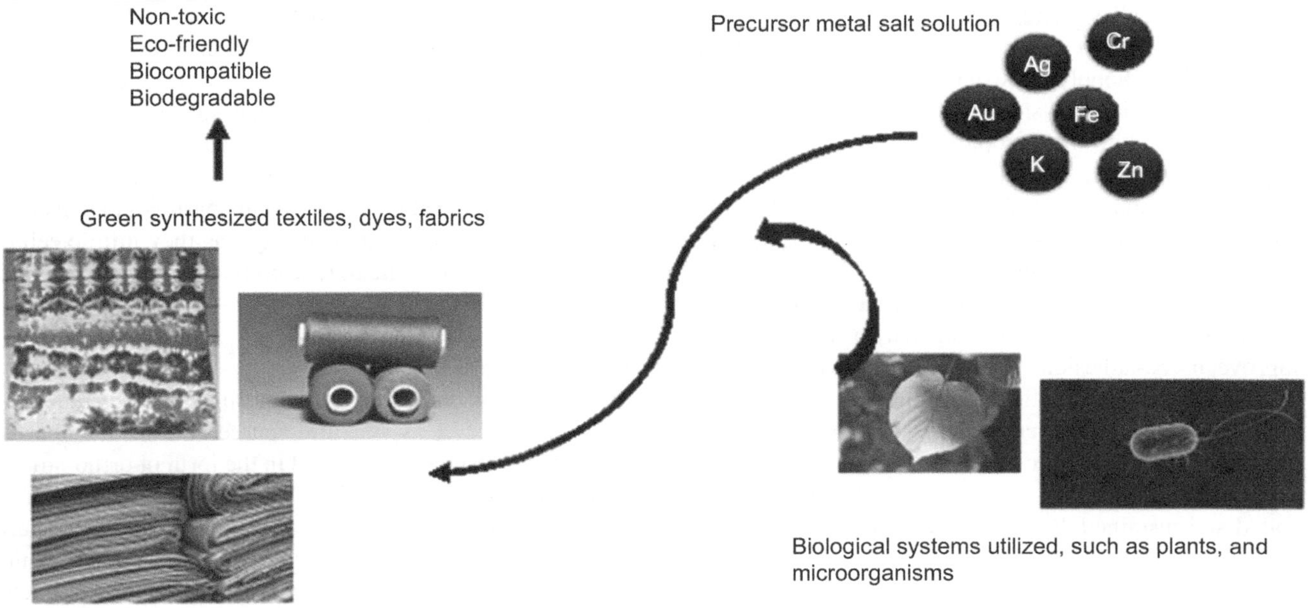

FIGURE 14.4 Typical procedure of producing green-synthesized textiles.

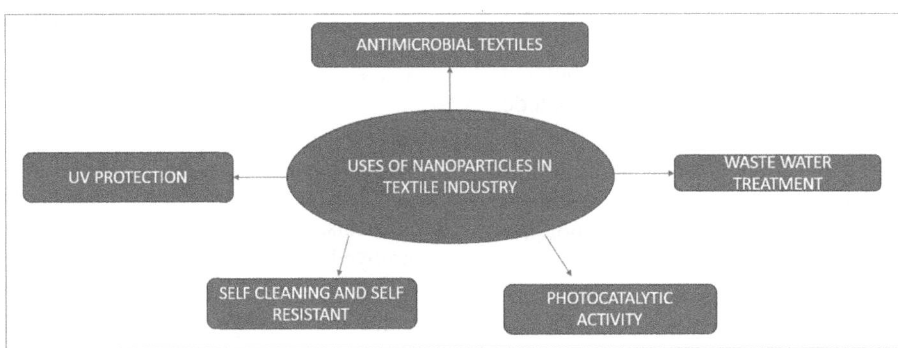

FIGURE 14.5 Uses of nanoparticles in textile industries.

imaging confirmed that both ZnO_2 and TiO_2 nanoparticles exhibited a predominantly spherical morphology, with particle dimensions measuring below 151 nm. Furthermore, Fourier transform infrared spectroscopy analysis validated the successful synthesis of ZnO_2 and TiO_2 nanoparticles.

14.2.3 Photocatalytic Activity

The photocatalytic activity of ZnO_2 and TiO_2 nanoparticles was evaluated using Congo red dye under various process parameters, including catalyst dosage, reaction time, initial dye concentration, radical scavenging ability, and reusability. The highest degradation efficiency (90%) was achieved after 180 minutes using 0.2 g of catalyst with a 20-ppm dye concentration at pH 9 (31, 32). Lokesh Ravi et al. proved green synthesis of AgCuO bimetallic nanomaterial using *Cordia sebestena* leaf extract and its application in methylene blue (MB) degradation. Phytochemical analysis confirmed the presence of flavonoids, tannins, saponins, phenols, and alkaloids, which play a crucial role in nanoparticle synthesis. SEM, EDX, and DLS analysis revealed that the AgCuO nanomaterial exhibits sharp needle-like clusters (~671 nm) with a high surface-to-volume ratio, enhancing catalytic efficiency.

AgCuO demonstrated remarkable MB degradation, converting the dye into a pale green solution. GC-MS analysis elucidated the degradation mechanism, wherein the sulfur atom in MB undergoes nucleophilic attack, leading to its fragmentation into biodegradable derivatives (N, N-dimethylcyclohexa-2,5-dienamine and N,N-dimethylcyclohexa-2,5-diene-1,4-diamine). Furthermore, the nanomaterial exhibited high reusability, retaining efficiency over five cycles, and displayed significant antibacterial activity against industrial pathogens. These findings position AgCuO as a potential dual-functional material for wastewater remediation and microbial decontamination in industrial effluent treatment (33). Industrial organic dyes, particularly those from the textile industry, are major environmental pollutants, posing severe ecological and health hazards due to their high toxicity.

14.2.3.1 Challenges in TiO_2 Photocatalysis

TiO_2 exhibits low sensitivity to visible light because of its large band gap (~3.2 eV for the anatase phase), and its photo-induced electron–hole pairs recombine rapidly, limiting its photocatalytic efficiency. Among various degradation techniques, heterogeneous photocatalysis using titanium dioxide is highly effective. To enhance performance, silver nanoparticles are incorporated into TiO_2, forming a Schottky barrier, which improves light absorption, minimizes electron-hole recombination, and enhances photocatalytic efficiency. Recent research highlights silver-doped TiO_2 nanocomposites as a promising solution for organic dye degradation, contributing to sustainable wastewater treatment (34, 35).

14.2.4 Silk-Based Biopolymers in Textiles

Silk is a fibrous biopolymer with a rich historical significance in the textile industry. It consists of two proteins, sericin and fibroin, which account for 20–30% of the silk mass. Sericin has long been considered a waste byproduct when extracting silk. However, silk sericin (Sser), a protein derived from silkworm cocoons, has gained attention due to its antioxidant properties, cellular development properties, and low immunogenicity. Its characteristics include morphology, specific structural features, molecular weight variations, layered composition, and gene sequences.

The functional groups present in the amino acid composition of sericin facilitate its applications in wound healing, regenerative medicine, drug delivery, textiles, environmental protection, and energy-related fields. It can be utilized in various forms, such as hydrogels, scaffolds, films, and conduits, making it a versatile biomaterial with significant multi-functional applications (36, 37).

14.2.4.1 Green Nanomaterials in Smart Textiles

Recently, smart textiles have garnered substantial attention for their potential in biomedical and environmental monitoring. The integration of green nanomaterials into smart textiles enhances their functionality, sustainability, and performance. Current research focuses on the synthesis, characterization, and application of green nanomaterials in the advancement of smart textile technology.

The use of green nanomaterials in smart textiles presents several challenges and limitations, including scalability, durability, stability, cost-effectiveness, and regulatory concerns. Ensuring biocompatibility and environmental sustainability while enhancing antimicrobial efficacy remains a key area of research. Future advancements should focus on eco-friendly synthesis methods, enhanced functionalization, and improved integration techniques to develop sustainable, antimicrobial smart textiles.

Nanomaterials play a critical role in the textile industry due to their unique physicochemical properties. Recently, plant-based nanoparticles have been explored for their antibacterial properties (39, 39). Aloe vera nanoparticles, synthesized from dried *Aloe vera* leaves using a ball milling method without chemical additives, exhibited an average size of 40 nm and UV absorption at 269 nm. To enhance fabric functionality, these nanoparticles were combined with chitosan and coated onto cotton fabrics using the pad-dry-cure method. The treated fabric demonstrated strong antibacterial activity against *E. coli* (22.05 mm) and *S. aureus* (27.17 mm), along with excellent UV protection (UPF 57.2) and high water resistance (contact angle of 155°). These findings suggest that *Aloe vera*–based nanomaterials could enhance fabric performance, making them suitable for protective clothing in healthcare and defense applications (40).

14.2.4.2 Antimicrobial Textiles

Maintaining a healthy body is essential, yet the rapid proliferation of harmful pathogens has intensified research into antimicrobial textiles. Microbial adherence to textiles can lead to serious clinical complications, necessitating the development of antimicrobial cellulosic fabrics that balance comfort, durability, and ease of care. Recent studies on organic antimicrobial agents for textile finishing classify them into natural agents (chitosan, cyclodextrins, natural dyes) and synthetic agents (quaternary ammonium salts, triclosan, halogenated phenols, metal-organic frameworks). These studies also examine their interaction with cellulose, antimicrobial mechanisms, and strategies to enhance durability and mechanical strength (41). Additionally, freshwater resources are increasingly strained because of climate change, population growth, water pollution, and inefficient resource management.

14.2.5 Water Purification via Textile-Based ISSG Systems

Interfacial solar steam generation (ISSG) systems, which utilize photothermal conversion for wastewater treatment and desalination, have emerged as a promising sustainable solution for addressing global water scarcity. Textile-based interfacial solar steam generation devices, including knitted fabrics, nonwoven, woven, and electro spun membranes, offer several advantages, such as high porosity, large surface area, flexibility, and durability. These properties make them cost-effective and widely applicable for water purification. Research is ongoing to enhance textile-based ISSG systems, focusing on performance improvements while addressing system limitations (42, 43).

14.2.5.1 Environmental Impact of Textile Dyes

Dyes and pigments play a crucial role in textile finishing, with common families including reactive, direct, dispersed, indigo, sulfur, and vat dyes. However, their presence in untreated wastewater poses severe environmental threats due to poor biodegradability, leading to aquatic ecosystem disruption. These pollutants contribute to eutrophication, reduced oxygen levels, turbidity, and odor issues while also causing long-term hazards such as bioaccumulation, mutagenicity, and carcinogenicity. Among the primary dyes used in the textile industry, azo and anthraquinone dyes are widely employed, but their release into wastewater introduces heavy metals, resulting in increased biochemical oxygen demand (BOD), chemical oxygen demand (COD), and overall water toxicity. New strategies are urgently needed to promote microbial bioremediation for the detoxification and biodegradation of these hazardous dyes.

Certain bacteria, fungi, and algae have demonstrated potential for dye decolorization and are already utilized in other textile industry processes. These microbes use enzymes such as peroxidase, laccase, and azoreductase to break down toxic dyes into less harmful compounds. Ongoing research aims to enhance microbial efficiency and optimize their bioremediation capabilities, paving the way for more sustainable wastewater treatment solutions in the textile industry (44, 45).

Streptomyces species exhibit significant potential for biotechnological applications, particularly in environmental bioremediation and the bioconversion of complex polymers. Their broad substrate adaptability and stability across diverse pH and temperature conditions make them valuable biocatalysts. Research focuses on *Streptomyces*-based technologies in the textile and pulp milling industries, emphasizing their role in dye decolorization, lignocellulosic biodegradation, and industrial waste treatment. Additionally, recent advancements and challenges in enhancing biodegradation efficiency for textile and pulp industry effluents are highlighted.

14.2.6 Sustainable Textile Recycling

The fashion industry's growth has led to significant environmental concerns, positioning textile recycling as a crucial solution for sustainability. Various chemical and mechanical recycling methods, such as pyrolysis, enzymatic hydrolysis, hydrothermal treatment, and ammonolysis, are employed, each with specific challenges and limitations. Mechanical recycling enables textile waste to be re-spun into yarn, while enzymatic hydrolysis is a promising approach under mild reaction conditions. Additionally, the Internet of Things (IoT) plays an emerging role in enhancing waste sorting and identification to improve recycling efficiency (46, 47).

14.2.7 COVID-19 and Advancements in Smart Textiles

The COVID-19 pandemic had a profound impact on the textile and apparel industry, leading to supply chain disruptions, decreased consumer demand, and financial constraints. However, it also accelerated digital transformation and innovation in functional textiles. Research on advanced and smart textiles in response to SARS-CoV-2 has focused on electrospun nanofibers and nanogenerators for real-time monitoring and biosensing applications.

Advancements in antiviral medical textiles have contributed to improved protection and pandemic control. However, challenges persist, particularly in the disposal of personal protective equipment (PPE). Ongoing research explores smart textile-based solutions that not only enhance safety and hygiene but also help reduce virus transmission, fostering a sustainable and resilient approach to future pandemic preparedness (48).

14.3 GREEN SYNTHESIS OF NANOPARTICLES IN THE FOOD INDUSTRY

Food is a vital source for sustaining life, and food safety is the primary objective when it comes to the food industry. Nanoparticles have been utilized in improving the quality of food but have led to harmful side effects. With the advent of green synthesis, the food industry has shifted its approach to a safer method of producing food with zero toxicity.

The demand has grown; keeping food safety as the highest priority, food packaging comes into the picture. Food packaging is an essential step prior to the transport of food to markets and stores. Environmental factors such as dust, pollen, and contaminants can degrade the preserved food if packaging is not done effectively. To avoid this issue, bioactive packaging, a technique which is based on the principles of active packaging, has been shown to impact consumer health. Packaged food is now being improved by inoculating natural active substances such as bioactive compounds. Shelf life is a growing concern when it comes to the storage of food. Additives are artificial substances, but green-synthesized nanoparticles as additives are less toxic and have displayed increased shelf life (49). The safety of food products and safer consumption have always been the ultimate goals of the food industry, and to achieve these, eco-friendly alternatives are a new focus in promoting better food quality.

14.3.1 Role of Green Nanoparticles in the Food Industry

Green synthesis uses biological extracts to synthesize nanoparticles. These biological extracts could be either plant extracts or microbes. Owing to the use of these organisms, green synthesis has less toxicity and is environmentally sustainable and safe for humans. Thus, green synthesis has gained a lot of attention in recent years. Feeding the ever-growing population is a challenge, but producing safe, consumable food is a bigger challenge (50). Addressing this issue is a key step in achieving it, and the nanotechnology industry has enhanced the food sector positively.

With the intervention of nanotechnology in the food sector, nanosensing and nanostructured ingredients are upcoming domains of interest. Nanotechnology in the food sector has been revolutionized with various technologies, which aim to provide precise data regarding the food quality, physiochemical properties, nutritional status, and shelf life. Nutritional enhancement through the adoption of green-synthesized nano-bio nutrient delivery systems is a promising approach, including the engineering of nanocarriers for targeted nutrient delivery and other emerging research innovations in the industry (51, 52). Biotechnology combines biological principles with physical and chemical approaches to obtain nanosized functional particles. Nanoparticles and their products have been implemented in various sectors such as medical, agriculture, food, electronics, and many other industries. Synthesis of these particles is primarily done by either chemical or physical methods.

The integration of nanotechnology with green chemistry and engineering principles has led to the emergence of 'green nanotechnology,' a field focused on developing environmentally friendly nanomaterials and processes, often using plant-based or renewable resources. Green nanoparticles-such as iron and iron oxide nanoparticles, are increasingly explored in the food industry; however, food technologists continue to evaluate their biocompatibility, safety, and consumption-related concerns.The integration of nanotechnology with green chemistry and engineering principles has led to the emergence of 'green nanotechnology,' a field focused on developing environmentally friendly nanomaterials and processes, often using plant-based or renewable resources. Green nanoparticles-such as iron and iron oxide nanoparticles, are increasingly explored in the food industry; however, food technologists continue to evaluate their biocompatibility, safety, and consumption-related concerns.

14.3.2 Application of Green-Synthesized Nanoparticles in the Food Industry

Iron nanoparticles exhibit wide range of applications like enzyme immobilization, food analysis, protein purification also used in active packaging, including the ability to eliminate oxygen and antimicrobial activity food processing also improve the in large aspects like smart delivery of nutrients, bio separation of proteins etc. also used in checking of biological and chemical contamination (53). Nanoscale devices are used to view biomolecules found in nature like proteins, DNA, and RNA.

14.3.3 Future Trends of Green-Synthesized Nanoparticles in the Food Industry

Nanotechnology has been positively improving the food sector, with more novel approaches yet to be discovered. Post-harvesting and processing are important steps undertaken to ensure food safety. At the molecular level, food undergoes a spectrum of biochemical changes, which, if altered, can cause great loss. Molecular and atomic-level alterations are new technology being implemented in the food sector for efficient food safety and consumption. With this technology, more precision and increased sustainability are observed, along with reduced cost. Site-specific targets of nutritionally bioactive compounds are being administered as nanocarriers encapsulating the target substance, benefiting human health and the administered site (54).

14.4 Green Synthesis of Nanoparticles in the Cosmetic Industry

Cosmetics are widely used by people of all ages, and these products are increasingly being formulated with the help of green nanoparticles, often referred to as "nano-cosmeceuticals." These minuscule particles enhance the delivery of active ingredients, allowing cosmetics to be more effective and longer lasting. Additionally, they extend the shelf life of skincare products. However, many cosmetics are not manufactured using environmentally sustainable methods, and they can contain harmful chemicals. These nanoparticles have the potential to penetrate the skin and interact with crucial proteins like elastin and collagen, which are vital for skin health. Chemically synthesized metallic nanoparticles may carry toxic substances, leading to gene mutations upon entering the skin (55). These harmful particles can spread beyond the skin, causing toxic effects on other parts of the body. Furthermore, their release into water systems can lead to environmental contamination. To mitigate these issues, the cosmetics industry is turning to green nanoparticles.

14.4.1 Cosmetics Synthesized Using Green Synthesis

Sunscreen products, for example, incorporate zinc oxide (ZnO) and titanium dioxide (TiO_2) nanoparticles as UV filters. Green synthesis of nanoparticles involves using plant extracts to produce these particles. Inorganic nanoparticles, preferred in cosmetics due to their hydrophilicity, biocompatibility, stability, and non-toxicity, are widely used. Both inorganic and organic nanoparticles are gaining popularity in cosmetic formulations, as they offer properties such as protection, cleansing, hydration, moisturizing, soothing, and firming (56). Nanoparticle-based carriers, which facilitate penetration of active ingredients through the skin, have become a significant focus in dermo-pharmaceutical products.

14.4.2 Importance of Green-Synthesized Nanoparticles in the Cosmetic Industry

Green nanoparticles offer several benefits, including anticancer, antioxidant, anti-inflammatory, antimicrobial, wound-healing, and anti-aging effects. Due to these properties, nanoparticles are increasingly employed in medical applications such as drug delivery, controlled release, diagnostics, and cosmetics (57). The small size, spherical shape, and high surface area of nanoparticles make them valuable across a variety of scientific disciplines, including biology, medicine, and engineering. Inorganic nanoparticles can be naturally occurring, such as iron oxide or silver, or synthesized in laboratories for specific applications in electronics, medicine, and materials science. Common metal-based nanoparticles include gold, silver, and iron oxide, while metal oxide nanoparticles include zinc oxide, iron oxide (Fe_2O_3), and aluminum oxide (Al_2O_3) (56, 57).

The skin, as the body's primary barrier against environmental factors, presents a significant challenge for delivering active ingredients effectively. Nanoparticles, especially those around 200–400 nm (for ZnO) and 150–300 nm (for TiO_2), can cause a white, powdery appearance on the skin, which may be more noticeable on darker skin tones. Despite their small size, these particles are undetectable by conventional microscopes and can provide a smoother, more transparent application (58). Green nanoparticles, considered environmentally friendly, offer a safe alternative to chemically manufactured nanoparticles used in cosmetics. The cosmetic industry is adopting these materials for more consistent, long-lasting results. Green nanotechnology is actively advancing, offering insights into the types, methods of generation, characterization, and properties of green nanomaterials. Green nanomaterials are highly regarded for their low cost, high efficiency, and minimal environmental impact (55).

Green nanoparticles are also used in the production of sustainable beauty products, such as anti-aging beauty masks made from chitin nanofibrils (CNs) and nano lignin (LG), derived from crustaceans and plant biomass. These materials are natural and environmentally sound. In vitro studies have shown that these beauty masks do not have toxic effects on keratinocytes or fibroblasts and promote collagen synthesis (59), which helps counteract skin aging. Silver nanoparticles synthesized with natural extracts have demonstrated antibacterial effects and reduced cytotoxicity in cell lines. The size and shape of Ag-NPs vary depending on the reducing agent and concentration of silver nitrate used during synthesis. Biosynthesized nanoparticles have proven effective in controlling oxidative stress, genotoxicity, and apoptosis-related changes. These nanoparticles, produced using environmentally friendly methods, offer a viable alternative to traditional chemical processes, which often involve toxic and hazardous substances (60).

Shape, size, and area influence the physiochemical properties of nanoparticles. Characterization of these particles is crucial for determining the morphology, and to achieve this, techniques such as FTIR, TEM, SEM, and AFM are employed. The cosmetic industry, which is continuously growing and increasingly focused on developing safer products, has begun adopting green synthesis approaches using plant- and animal-derived byproducts. Green-synthesized nanoparticles have also demonstrated significant potential in environmental applications, particularly in the mitigation of water pollution. Plant-based nanoparticles, for instance, have demonstrated the highest removal rates of pollutants, ranging from 80% to 90%, while microbe- and algae-based nanoparticles have also shown significant effectiveness, between 70% and 90%. Chitin/chitosan-based bio(nano) polymers have achieved a 70% removal rate (28).

14.4.3 Future Scope of Green Synthesis of Nanoparticles in the Cosmetic Industry

As awareness of personal appearance increases, cosmetics have evolved into a billion-dollar industry. Consumers are demanding products that are both effective and environmentally friendly. The use of human-made nanoparticles, which often involve harmful chemicals, has led to the discovery of green nanoparticles as a safer, more sustainable alternative. Derived from natural sources like algae, fungi, bacteria, and plants, green nanoparticles are synthesized under mild conditions and offer numerous benefits in cosmetics and other industries (59, 61).

14.5 GREEN SYNTHESIS OF NANOPARTICLES IN ELECTRONICS AND ENERGY

Nanoparticles are nanoscale structures with dimensions ranging from 1 to 100 nm. These minute particles exhibit diverse chemical, thermal, electronic, optical, plasmonic, and physical properties. Various types of nanoparticles have been synthesized from metals such as zinc oxide, iron, copper, selenium, silver, gold, titanium dioxide, and others. The synthesis of these metal-based nanoparticles primarily follows a chemical route. However, from an industrial standpoint, the synthesis process often lacks precision and efficiency, resulting in the uncontrolled formation of undesirable nanoparticles with inconsistent sizes, suboptimal physicochemical properties, and significant chemical wastage, leading to economic losses. Furthermore, chemical synthesis introduces several hazardous effects, including toxicity to the environment, humans, and animals. The accumulation of toxic substances in the soil can contaminate groundwater, potentially causing severe health issues in both plants and humans. If not managed effectively, metal nanoparticles may eventually pollute natural resources, posing a substantial threat to economic development and public health (62, 63).

14.5.1 Green Synthesis and Its Importance in the Electronics Industry

Recent research has shifted focus toward an innovative and efficient approach for synthesizing metal nanoparticles. One such method is green synthesis, which utilizes biological systems such as plants, algae, microorganisms, and cells to extract biomolecules that, in turn, facilitate the formation of metal nanoparticles without the involvement of synthetic chemicals. Green synthesis can occur either intracellularly, within living cells, or extracellularly, outside the cells.

Regardless of the pathway, the final product is entirely natural and non-toxic. From an economic perspective, green synthesis is highly cost effective, eliminating the need for substantial investments in chemical reagents. Moreover, living organisms are readily available, providing the advantage of accessibility without additional resource requirements. Green synthesis has extensive applications across various sectors, including biomedical sciences, electronics, pharmaceuticals, biotechnology, textiles, food, cosmetics, and other technological fields. The electronics industry encompasses a wide range of consumer products, such as televisions, computers, microchips, semiconductors, electrical switches, and more. The primary precursor materials for these products are metals and metalloids (64).

Nanotechnology plays a crucial role in the development of these products, as exemplified by nanochips, which are gaining significant attention in the electronics industry. The applications of nanotechnology in electronics include nanorods, nanofillers, nanochips, nanowires, nanocrystallites, and quantum dots, among others. However, as previously mentioned, chemical synthesis often results in products with lower efficiency and suboptimal properties. In contrast, nanoparticles synthesized through green synthesis have demonstrated superior electronic and optoelectronic properties, highlighting their effectiveness as a more efficient and sustainable alternative to conventional chemical methods. The dimensions of nanoparticles synthesized using plant extracts vary depending on factors such as the concentration of the extract and the synthesis temperature. The phytochemicals present in the extract primarily function as reducing and capping agents during nanoparticle formation. Various techniques are employed for green nanoparticle synthesis, including (1) the sol-gel technique, which utilizes a precipitation method to produce nanoparticles with the desired properties; (2) the hydrothermal method; and (3) the coprecipitation technique. Since the focus is on green synthesis, plant-derived extracts from stems, roots, and other phytochemicals rich plant parts, phytonutrients, and other bioactive compounds rich natural sources, are extensively used for nanoparticle preparation.

14.5.2 Applications of Green-Synthesized Nanoparticles in the Electronics Industry

Numerous scientific studies have highlighted the significance of green synthesis in the electronics industry. The applications of green-synthesized nanoparticles in electronics are vast. Zinc oxide exhibits high electron mobility and exceptional optoelectronic properties, making it an efficient donor material that can be easily synthesized. Researchers have reported the synthesis and characterization of ZnO nanoparticles using plant extracts from *Trifolium pratense* flowers, *Euphorbia jatropa* latex, and powdered *Mimosa pudica* and coffee. These ZnO nanoparticles have been investigated for their clinical applications, structural control, crystallite size regulation, and band gap energy evaluation. Additionally, ZnO nanoparticles were synthesized using a leaf extract of *Mimosa pudica*, in combination with aqueous zinc acetate dihydrate and NaOH, and subsequently characterized using various analytical techniques. The incorporation of phytoextracts in ZnO nanoparticle synthesis enhanced their suitability for use as photoanodes in dye-sensitized solar cells (65). *Haematoxylum brasiletto* powder was utilized for the synthesis of lead oxide (PbO) nanoparticles. These green-synthesized PbO nanoparticles exhibit significant potential for use in the fabrication of porous silicon photodetectors (66). Additionally, silica extracted from sugarcane biomass using organic acids, alkalis, and solvents has demonstrated promise in electronic applications such as photovoltaics, semiconductors, and nanoelectronics. Gas sensors designed for detecting hazardous gases—including NO, NO_2, N_2O, H_2S, CO, NH_3, CH_4, SO_2, and CO_2—are synthesized using phytoextracts and microorganisms (64).

14.5.3 Green Synthesis and Its Applications in Renewable Energy

Energy is another crucial aspect to consider, as every process involves energy consumption or dissipation. Industrial activities release substantial amounts of energy in the form of thermal, electrical, solar, and other energy types, as shown in Figure 14.8. The uncontrolled release of energy into the environment contributes to pollution and toxicity, ultimately leading to ecological damage. To mitigate these negative impacts, green synthesis presents a sustainable solution. In the energy sector, green synthesis follows key principles such as waste prevention, the production of environmentally safe chemicals, the utilization of natural resources, and the minimization of energy loss while improving overall energy efficiency (3). Polysaccharides, enzymes, proteins, phytochemicals, and phytonutrients function as reducing and capping agents to stabilize nanoparticles. Dextran, a biocompatible polysaccharide, has been employed as a stabilizing agent for nanoparticles. Several researchers have successfully synthesized dextran-derived nanoparticles from various microbial sources, such as *Bacillus anthracis* and *Weissella cibaria*.

Furthermore, mercaptoacetic acid-coated gold nanoparticles were synthesized by Esfahani et al. for the detection of phosphate ions in drinking water. Kokilavani et al. developed a colorimetric sensor utilizing tyrosine and a Cu/Ag-Dex nanocomposite for detecting Hg^{2+} ions in aqueous solutions (67). Renewable energy sources, which are naturally replenishable, offer a sustainable, non-toxic, and eco-friendly alternative to traditional energy systems.

14.5.4 Future Scope of Green-Synthesized Nanoparticles in Sustainable Energy

Nanoparticles play a crucial role in enhancing solar energy, geothermal energy, wind energy generation, and fuel cell efficiency. Long-term sustainability can be effectively achieved through green-synthesized nanoparticles. Numerous researchers have published scientific studies highlighting the role of green synthesis in producing metal nanoparticles for applications such as eco-friendly solar cells, photovoltaic cells, photosensitizers, and more. These nanoparticles are synthesized using phytochemicals, including chlorophyll, pigments, and protein scaffolds. Additionally, microorganisms have also been utilized for this purpose.

The pursuit of sustainable energy solutions creates novel opportunities for scientists to develop greener, more environmentally friendly strategies for energy consumption, storage, and recycling. Replacing conventional chemical synthesis methods with green synthesis is a fundamental step toward a more sustainable future. If immediate action is not taken, the continued reliance on environmentally harmful methods will contribute to the degradation of our planet Earth (63).

14.6 CONCLUSION

Although metals are essential in trace amounts for supporting human physiological functions, nanoparticle-based formulations are typically administered at very low concentrations to minimize cytotoxicity while enhancing therapeutic efficacy. In recent years, the pharmaceutical industry has increasingly embraced green synthesis methods that reduce hazardous byproducts and improve the pharmacological effectiveness of nanomedicines. These eco-friendly approaches not only mitigate the health and environmental risks associated with traditionally synthesized metal nanoparticles, but they also enhance the stability, consistency, and biocompatibility of pharmaceutical formulations, signaling a paradigm shift toward more sustainable drug development. Concurrently, innovations in nanotechnology are transforming the textile industry with sustainable solutions focused on environmental remediation, antimicrobial activity, and efficient waste management. The integration of bioremediation techniques, smart textile systems, and cutting-edge recycling technologies plays a critical role in reducing ecological harm and enhancing textile performance, propelling the sector toward circular economy models and low-impact manufacturing processes. In the food industry, the use of green-synthesized nanoparticles—particularly those based on iron oxide—represents a key advancement in sustainable, health-conscious food production and packaging. These biocompatible nanomaterials enhance food safety, extend shelf life, and improve nutrient bioavailability, which aligns with the rising consumer demand for clean-label, minimally processed, and ready-to-consume food products. Additionally, the integration of intelligent delivery systems enhances food integrity and promotes environmentally responsible packaging solutions. Similarly, in the cosmetic industry, the adoption of green nanoparticle synthesis has facilitated the development of safer, more sustainable, and highly effective personal care products. Derived from natural biological sources, these eco-friendly nanoparticles enhance the bioavailability and stability of active ingredients, offering multifunctional benefits such as anti-aging, antioxidant, and antimicrobial effects. Their low toxicity and compatibility with skin cater to the increasing consumer preference for sustainable, dermatologically safe beauty products. In conclusion, green synthesis of nanoparticles provides an environmentally benign, economically feasible alternative to conventional chemical synthesis methods. By harnessing biological resources such as plants, microorganisms, and natural extracts, this approach minimizes environmental harm and enhances the functional and physicochemical properties of nanoparticles. As it expands across diverse fields, including electronics, renewable energy, biomedicine, textiles, food, and cosmetics, green synthesis has emerged as a transformative strategy that aligns innovation with environmental responsibility, promoting a sustainable future for nanotechnology.

REFERENCES

1. Rafique M, Sadaf I, Rafique MS, Tahir MB. A review on green synthesis of silver nanoparticles and their applications. *Artif Cells Nanomed Biotechnol.* 2017;45(7):1272–1291.
2. Malik B, Pirzadah TB, Kumar M, Rehman RU. Biosynthesis of nanoparticles and their application in pharmaceutical industry. *In: Nanotechnology.* Singapore: Springer; 2017, pp. 235–252.
3. Karade VC, Dongale TD, Sahoo SC, Kollu P, Chougale AD, Patil PS, et al. Effect of reaction time on structural and magnetic properties of green-synthesized magnetic nanoparticles. *Journal of Physics and Chemistry of Solids.* 2018;120:161–166.
4. Vijayaram S, Razafindralambo H, Sun YZ, Vasantharaj S, Ghafarifarsani H, Hoseinifar SH, et al. Applications of green synthesized metal nanoparticles—a review. *Biol Trace Elem Res.* 2024;202(1):360–386.
5. Mayuri R Sawant,Yash S Chaudhari, Prashik P Jamdhade, Pramod S Shinde, Sanika S Pimpale, Sanchita R Kokate. Green Synthesis of Silver Nanoparticles Using Tulsi (Ocimum Sanctum) and Alo vera Leaf Extract and its Characterization. *Int J Adv Res Sci Commun Technol.* 2024;4(4):197–204.
6. Dhivyabharathi V, Bhavadharani S, Shameena S, Gomathi M. Green synthesis of silver nanoparticles using Mangifera indica leaves and Rose petals extract and their characterization; 2023;2(1):16–22.
7. Liu L, Yu C, Ahmad S, Ri C, Tang J. Preferential role of distinct phytochemicals in biosynthesis and antibacterial activity of silver nanoparticles. *J Environ Manage.* 2023;344:118546.
8. Khafaga DSR, El-Khawaga AM, Elfattah Mohammed RA, Abdelhakim HK. Green synthesis of nano-based drug delivery systems developed for hepatocellular carcinoma treatment: A review. *Mol Biol Rep.* 2023;50(12): 10351–10364.
9. Vijapur LS, Shalavadi M, Desai AR, Hiremath JN, Gudigennavar AS, Shidramshettar SL, et al. Wound healing potential of green synthesized silver nanoparticles of

Glycyrrhiza glabra linn root extract: A preclinical study. *J Trace Elem Miner.* 2025;11:100214.

10. Crisan D, Scharffetter-Kochanek K, Crisan M, Schatz S, Hainzl A, Olenic L, et al. Topical silver and gold nanoparticles complexed with *Cornus mas* suppress inflammation in human psoriasis plaques by inhibiting NF-κB activity. *Exp Dermatol.* 2018;27(10):1166–1169.
11. Majeed S, Khanday M. Green synthesis of silver nanoparticles using bark extract of Salix alba and its antimicrobial effect against bacteria isolated from dental plaque. *Orient J Chem.* 2016;32(3):1611–1618.
12. Dakal TC, Kumar A, Majumdar RS, Yadav V. Mechanistic basis of antimicrobial actions of silver nanoparticles. *Front Microbiol.* 2016;7.
13. Xu L, Wang YY, Huang J, Chen CY, Wang ZX, Xie H. Silver nanoparticles: Synthesis, medical applications and biosafety. *Theranostics.* 2020;10(20):8996–9031.
14. Malik B, Pirzadah TB, Kumar M, Rehman RU. Biosynthesis of nanoparticles and their application in pharmaceutical industry. *In: Metabolic Engineering for Bioactive Compounds.* Singapore: Springer; 2017, pp. 331–349.
15. Jayabalan J, Paramasivam S, Jayaseelan C, Paramanathan B, Mani G, Jang HT. Phytogenic zinc oxide nanoparticles (ZnO NPs) using Solanum xanthocarpum fruit extract: Its evaluation of photocatalytic activity, molecular docking and biomedical applications. *Inorg Chem Commun.* 2025;173:113792.
16. Ramesh C, Kumar KM, Senthil M, Ragunathan V. Antibacterial activity of Cr2O3 nanoparticles against E.coli; Reduction of chromate ions by Arachis hypogaea leaves. Arch Appl Sci Res. 2012;4(4):1894–1900.
17. Arunachalam K, Annamalai S, Hari S. One-step green synthesis and characterization of leaf extract-mediated biocompatible silver and gold nanoparticles from Memecylon umbellatum. *Int J Nanomedicine.* 2013;1307.
18. Chahrdoli A, Qalekhani F, Ghowsi M, Nemati H, Shokoohinia Y, Fattahi A. Achillea wilhelmsii C. Koch mediated blood compatible silver nanoparticles. *Mater Today Commun.* 2020;25:101577.
19. Moteriya P, Chanda S. Biosynthesis of silver nanoparticles formation from Caesalpinia pulcherrima stem metabolites and their broad spectrum biological activities. *J Genet Eng Biotechnol.* 2018;16(1):105–113.
20. Young Park S, Park K, Oh JW, Park G. Gold nanoparticle encoded with marigold (Tagetes erecta L.) suppressed hyperglycemia-induced senescence in retinal pigment epithelium via suppression of lipid peroxidation. *Arab J Chem.* 2023;16(10):105120.
21. Young Park S, Park K, Oh JW, Park G. Gold nanoparticle encoded with marigold (Tagetes erecta L.) suppressed hyperglycemia-induced senescence in retinal pigment epithelium via suppression of lipid peroxidation. *Arab J Chem.* 2023;16(10):105120.
22. Fakhra S, Shilpa S. Synthesis of silver nanoparticles using Costusspeciosus and study of its anti-microbial properties against urinary tract pathogens. 2014;3(1):248–252.
23. Ahmad N, Ansari MA, Al-Mahmeed A, Joji RM, Saeed NK, Shahid M. Biogenic silver nanomaterials synthesized from Ocimum sanctum leaf extract exhibiting robust antimicrobial and anticancer activities: Exploring the therapeutic potential. *Heliyon.* 2024;10(15): e35486.
24. Patil MP, Bayaraa E, Subedi P, Piad LLA, Tarte NH, Kim GD. Biogenic synthesis, characterization of gold nanoparticles using Lonicera japonica and their anticancer activity on HeLa cells. *J Drug Deliv Sci Technol.* 2019;51:83–90.
25. Dwivedi AD, Gopal K. Biosynthesis of silver and gold nanoparticles using Chenopodium album leaf extract. *Colloids Surf A Physicochem Eng Asp.* 2010;369(1–3):27–33.
26. Faiz N, Sivaswamy V, Rohinikumar S. Green synthesis of gold nanoparticles using eucalyptus and piper longum and antimicrobial activity evaluation. *Indian J Dent Res.* 2025;35(4):429–432.
27. Yuan C, Huo C, Gui B, Cao W. Green synthesis of gold nanoparticles using *Citrus maxima* peel extract and their catalytic/antibacterial activities. *IET Nanobiotechnol.* 2017;11(5):523–530.
28. Salem SS, Fouda A. Green synthesis of metallic nanoparticles and their prospective biotechnological applications: An overview. Biol Trace Elem Res. 2021;199(1):344–370.
29. Malik B, Pirzadah TB, Kumar M, Rehman RU. Biosynthesis of nanoparticles and their application in pharmaceutical industry. *In: Metabolic Engineering for Bioactive Compounds.* Singapore: Springer; 2017, pp. 331–349.
30. Gavanaroudi SB. Investigation of green chemistry approaches in the synthesis of pharmaceuticals. *EJCMPR.* 2024;3(4):1221–1237.
31. Tamanna NJ, Sahadat Hossain Md, Tabassum S, Bahadur NM, Ahmed S. Easy and green synthesis of nano-ZnO and nano-TiO2 for efficient photocatalytic degradation of organic pollutants. *Heliyon.* 2024;10(17):e37469.
32. Magdy G, Aboelkassim E, Abd Elhaleem SM, Belal F. A comprehensive review on silver nanoparticles: Synthesis approaches, characterization techniques, and recent pharmaceutical, environmental, and antimicrobial applications. *Microchem J.* 2024;196:109615.
33. Lokesh Ravi, Venkatesh Selvaraj, Saranya Shankar, Ranjitha Dhevi V. Sundar, Gayathri Segaran, Suganya Kumaresan, Venkatesh Sadhana. Methylene blue degradation by AgCuO bimetallic nanomaterial, green synthesized using cordia sebestena leaves. 2020;13(7): 3122–3128.
34. Ramesh N, Lai CW, Johan MR Bin, Mousavi SM, Badruddin IA, Kumar A, et al. Progress in photocatalytic degradation of industrial organic dye by utilising the silver doped titanium dioxide nanocomposite. *Heliyon.* 2024;10(24):e40998.
35. Sandin G, Peters GM. Environmental impact of textile reuse and recycling—A review. *J Clean Prod.* 2018;184: 353–365.
36. Yuan Y, Nasri M, Manayi A, Zhang J, Wu C, Jeon TJ, et al. Sericin coats of silk fibres, a degumming waste or future material? *Mater Today Bio.* 2024;29:101306.
37. Saad M, El-Samad LM, Gomaa RA, Augustyniak M, Hassan MA. A comprehensive review of recent advances in silk sericin: Extraction approaches, structure, biochemical characterization, and biomedical applications. *Int J Biol Macromol.* 2023;250:126067.
38. Popescu M, Ungureanu C. Green nanomaterials for smart textiles dedicated to environmental and biomedical applications. *Materials.* 2023;16(11):4075.
39. Stone C, Windsor FM, Munday M, Durance I. Natural or synthetic—how global trends in textile usage threaten freshwater environments. *Sci Total Environ.* 2020;718:134689.

40. Subramani K, Kolathupalayam Shanmugam B, Rangaraj S, Palanisamy M, Periasamy P, Venkatachalam R. Screening the UV-blocking and antimicrobial properties of herbal nanoparticles prepared from *Aloe vera* leaves for textile applications. *IET Nanobiotechnol.* 2018;12(4):459–465.
41. Emam HE. Antimicrobial cellulosic textiles based on organic compounds. *3 Biotech.* 2019;9(1):29.
42. Saeed HAM, Kazimoto VV, Xu W, Yang H. The application of textile materials in interfacial solar steam generation for water purification and desalination. *Polymers (Basel).* 2024;16(6):793.
43. Yaseen DA, Scholz M. Textile dye wastewater characteristics and constituents of synthetic effluents: A critical review. Int. J. Environ. Sci. Technol. (2019)16:1193–1226.
44. Berradi M, Hsissou R, Khudhair M, Assouag M, Cherkaoui O, El Bachiri A, et al. Textile finishing dyes and their impact on aquatic environs. *Heliyon.* 2019;5(11):e02711.
45. Das S, Cherwoo L, Singh R. Decoding dye degradation: Microbial remediation of textile industry effluents. *Biotechnol Notes.* 2023;4:64–76.
46. Cuebas-Irizarry MF, Grunden AM. *Streptomyces* spp. as biocatalyst sources in pulp and paper and textile industries: Biodegradation, bioconversion and valorization of waste. *Microb Biotechnol.* 2024;17(1).
47. Adenan NH, Lim YY, Ting ASY. Removal of triphenylmethane dyes by Streptomyces bacillaris: A study on decolorization, enzymatic reactions and toxicity of treated dye solutions. *J Environ Manage.* 2022;318:115520.
48. Ivanoska-Dacikj A, Oguz-Gouillart Y, Hossain G, Kaplan M, Sivri Ç, Ros-Lis JV, et al. Advanced and smart textiles during and after the COVID-19 pandemic: Issues, challenges, and innovations. *Healthcare.* 2023;11(8):1115.
49. Safa Baraketi, Khaoula Khwaldia. Nanoparticles from agrifood by-products: Green technology synthesis and application in food packaging. *Current Opinion in Green and Sustainable Chemistry.* 2024;49:100953.
50. Fakhar Islam FS, MA, AA, MH, MAK, SAS, AOK. Applications of green technologies-based approaches for food safety enhancement: A comprehensive review. [cited 2025 Mar 29]; Available from: https://www.tandfonline.com/doi/full/10.1080/19430871003684440#abstract
51. Ravichandran R. Nanotechnology applications in food and food processing: Innovative green approaches, opportunities and uncertainties for global market. [cited 2025 Mar 29]; Available from: https://www.tandfonline.com/doi/full/10.1080/19430871003684440#abstract
52. Tsekhmistrenko S, Bityutskyy V, Tsekhmistrenko O, Merzlo S, Tymoshok N, Melnichenko A, Polishchuk S, Demchenko A, Yakymenko I. Bionanotechnologies: Synthesis of metals' nanoparticles with using plants and their applications in the food industry: A review. *Journal of Microbiology, Biotechnology and Food Sciences.* 2021;10(6):e1513.
53. Góral D, Marczuk A, Góral-Kowalczyk M, Koval I, Andrejko D. Application of iron nanoparticle-based materials in the food industry. *Materials.* 2023;16(2):780.
54. Shende S, Rajput VD, Gade A, Minkina T, Fedorov Y, Sushkova S. Metal-based green synthesized nanoparticles: Boon for sustainable agriculture and food security. *IEEE Transactions on NanoBioscience.* 2022;21(1):44–54.
55. Rehman N, Pandey A. Green nanomaterials revolution in cosmetic products and skin treatment. *Green Nanoarchitectonics.* 2022: 10.1201/9781003318606-12.
56. Mascarenhas-Melo F, et al. Inorganic nanoparticles in dermopharmaceutical and cosmetic products: Properties, formulation development, toxicity, and regulatory issues. *Eur J Pharm Biopharm.* 2023;192:25–40.
57. Paiva-Santos AC, et al. Plant-mediated green synthesis of metal-based nanoparticles for dermopharmaceutical and cosmetic applications. *Int J Pharm.* 2021;597:120311.
58. Mascarenhas-Melo F, et al. Inorganic nanoparticles in dermopharmaceutical and cosmetic products: Properties, formulation development, toxicity, and regulatory issues. *Eur J Pharm Biopharm.* 2023;192:25–40.
59. Morganti P, et al. Green nanotechnology serving the bioeconomy: Natural beauty masks to save the environment. *Cosmetics.* 2016;3(4):41.
60. Arroyo GV, et al. Green synthesis of silver nanoparticles for application in cosmetics. *J Environ Sci Health A Tox Hazard Subst Environ Eng.* 2020;55(11):1304–1320.
61. Nhani GBB, Di Filippo LD, de Paula GA, Mantovanelli VR, da Fonseca PP, Tashiro FM, Monteiro DC, Fonseca-Santos B, Duarte JL, Chorilli M. High-tech sustainable beauty: Exploring nanotechnology for the development of cosmetics using plant and animal by-products. *Cosmetics.* 2024;11(4):112.
62. Pechyen C, Tangnorawich B, Toommee S, Marks R, Parcharoen Y. Green synthesis of metal nanoparticles, characterization, and biosensing applications. *Sens Int.* 2024;5:100287.
63. Parveen K, Banse V, Ledwani L. Green synthesis of nanoparticles: Their advantages and disadvantages. AIP Conf Proc. 2016;1724:020048.
64. Dadkhah M, Tulliani JM. Green synthesis of metal oxides semiconductors for gas sensing applications. *Sensors.* 2022; 22(13):4669.
65. Balogun SW, James OO, Sanusi YK, Olayinka OH. Green synthesis and characterization of zinc oxide nanoparticles using bashful (Mimosa pudica), leaf extract: A precursor for organic electronics applications. *SN Appl Sci.* 2020;2(3):504.
66. Salih AA, Abad WK, Fadaam SA, Hussein BH. Fabrication of lead oxide nanoparticles by green synthesis method for photovoltaic applications. *Dig J Nanomater Biostruct.* 2023;18(4):1225–1233.
67. Madani M, Hosny S, Alshangiti DM, Nady N, Alkhursani SA, Alkhaldi H, et al. Green synthesis of nanoparticles for varied applications: Green renewable resources and energy-efficient synthetic routes. *Nanotechnol Rev.* 2022;11(1):731–759.

15 Toxicity Assessment of Nanomaterials

Molecular Mechanisms and Assessment of Nanotoxicity

Sharmistha Dutta and Hadagali Ashoka

15.1 INTRODUCTION

In recent decades, advancement in nanotechnology has received much attention in the field of industrial and medical sciences, which has brought about a revolution in the design and synthesis of nanostructures for both industrial and medical applications (Sabella et al., 2014). These particles are essentially classified as "ultrafine" particles (UFPs), typically defined by a diameter of less than 100 nanometers. Nanoparticles (NPs) possess the unique properties of small size (1–100 nm); shape; greater surface area-to-volume ratio; and different electronic, magnetic, optical, and mechanical properties that make them desirable candidates of interest in various fields and applications, including biomedical applications, drug delivery systems, gene delivery, and therapeutic applications (Berube et al., 2007; Zhang et al., 2018b). The very same properties that make NPs such a desirable and appealing candidate raise concerns regarding their health and environmental impacts (Zhang et al., 2018a; Isibor et al., 2024; Xuan et al., 2023). Evidence of nanoparticle-induced *in vivo* and *in vitro* toxicity has surfaced from extensive study and research spanning the last two decades. Nanoparticle toxicity is multifaceted, and therefore gaining insight into the cellular, molecular mechanisms of toxicity induction helps us to assess the routes and therefore design methodologies of toxicity assessment for designing safer nanomaterials (Oberdörster, 2010; Odaudu & Akinsiku, 2022).

15.2 ROLE OF PHYSICOCHEMICAL PROPERTIES OF NPS IN NANOTOXICITY

The physicochemical characteristics of nanoparticles and their composites affect biological systems and their kinetics. The physico-chemical properties of metal and metal oxide NPs can be determined by various attributes like size, shape, architecture, crystallinity, agglomeration tendency, and composition (J. Li et al., 2015). Nanosize (<100 nm) and large surface areas of nanomaterials (NMs) are shown to correspond with the theoretical prediction for higher toxicities as compared to their corresponding bulk forms (Pareek et al., 2018).

The sizes of NPs determine their uptake mechanism by endocytosis, pinocytosis, phagocytosis, or direct diffusion and its bioaccumulation in cell organelles, which triggers aggregation and induces toxicity (Wu et al., 2025; Awashra & Młynarz, 2023). The shape of nanoparticles, namely whether they are rods, wires, or spheres, is a particularly critical parameter to affect cellular uptake and site-specific drug delivery and is therefore a significant factor in determining the potency and magnitude of the effect together with its toxicity (Demir, 2021; Sengul & Asmatulu, 2020).

Research indicates that asymmetrical nanoparticles such as rods, dumbbells, and bipyramids generally demonstrate greater cytotoxicity than their spherical counterparts. Studies have shown *in vivo* plasma biodistribution, histopathological alterations, and pharmacokinetics are notably influenced by the shape and size of AuNPs.(Wu et al., 2025; Djurišić et al., 2015; Schaeublin et al., 2011).

The key characteristic of nanoparticles is their high surface area-to-volume ratio, which enhances their chemical reactivity. This, however, also leads to increased generation of reactive oxygen species (ROS). This elevated reactivity is a primary contributor to their inherent toxicity, as the larger surface area offers more opportunities for interactions with biological systems. Moreover, studies have demonstrated that the physico-chemical properties of NPs modulate their dynamic interaction with *in vivo* biomolecules and cellular organelles, thereby leading to toxicity (Sabella et al., 2014; Buchman et al., 2019). When exposed to plasma or culture media (biological fluid), NPs tend to associate with a range of proteins, which organize to form the "protein corona" (protein coating) (Lundqvist et al., 2008; Lesniak et al., 2012). Understanding the dynamic formation of a protein corona around nanoparticles on exposure to biological fluid, which is a key molecular event, strongly impacts the biological response, including toxicity behavior, and is therefore crucial to standardizing toxicological tests (Maiorano et al., 2010). Systematic study of this protein corona reveals unique functional properties deeply associated with nanoparticle types. This protein corona formed around the NPs may modulate cellular uptake in regard to quality and quantity of substances in the cytoplasm, induce protein misfolding pathways which may thereby trigger inflammation, DNA damage or even induce apoptosis (Wang et al., 2013; Fatima et al., 2024).

DOI: 10.1201/9781003644439-15

15.3 MECHANISMS OF TOXICITY

The key mechanisms by which nanoparticles induce toxicity, like lysosomal toxicity, mitochondrial dysfunction, ER stress, oxidative stress, inflammation, and autophagy, are illustrated in Figure 15.1.

15.3.1 Lysosomal Machinery-Induced Intracellular Toxicity

This kind of cytotoxicity, also known as the lysosome-enhanced Trojan Horse (LETH) effect, is mediated and enhanced by lysosomes (E.-J. Park et al., 2010). When NPs are internalized by cells (endocytosis) and consequently the lysosome, the same protective machinery of the lysosomes that was meant for foreign bodies elicits the intracellular release of toxic ions (e.g., Ag^{+}, Cd^{2+}, $Fe^{2+/3+}$, $Au^{1+/3+}$ ions) in the cytoplasm driven by the acidic environment of the lysosome, leading to toxicity in cells (Service, 2010; Yao et al., 2019). Ultimately, this leads to elevated ROS levels, triggering apoptosis as well as damage to DNA and cellular membranes. If the lysosomal pathway slows down NP degradation in cases of less acidic pH, the toxic effects of the NPs get reduced to a large extent or even mitigated. This mechanism is partially influenced by experimental conditions, the composition of biological fluids, and the specific characteristics of the cell lines used. Alterations of lysosomal functionality caused by NPs have been described, particularly concerning ion channel proteins such as V-ATPase, TRPM2, CLC-7, and LAMPs (Tubatsi et al., n.d.; Feng et al., 2024).

Another approach of reducing the LETH effect is by coating the surface of the NPs with biocompatible materials like a silica shell surrounding Fe_3O_4 NPs. This approach considerably reduces ion leakage from the NPs in acidic conditions while maintaining the original physico-chemical characteristics of the NPs (same size, surface charge, and zeta potential). Silica or SiO_2 is resistant to acidic conditions and naturally found in numerous tissues and is therefore a nontoxic compound, which is efficiently excreted through the urine (J.-H. Park et al., 2009). Since the main cause of toxicity induced by metal-containing NPs is via their corrosion and ion leakage in the acidic lysosomal environment, it is now possible to devise several strategies to engineer safer NPs while maintaining their characteristic and powerful functionalities.

15.3.2 Mitochondrial Dysfunction and NP-Induced OS and Apoptosis

The mitochondria is the center for the generation of energy as well as free radicals, and these free radicals can lead to various complications (Nalika & Parvez, 2015). The mitochondria are not only the center of energy transduction and ATP production of the cell but play essential roles in cell metabolism, ROS generation via free radicals, redox regulation, calcium homeostasis, cell cycle progression, and apoptosis (Zheng et al., 2023). ROS production in cells through oxidative phosphorylation is normal and is controlled by enzymatic antioxidants like SOD, CAT, and GPx

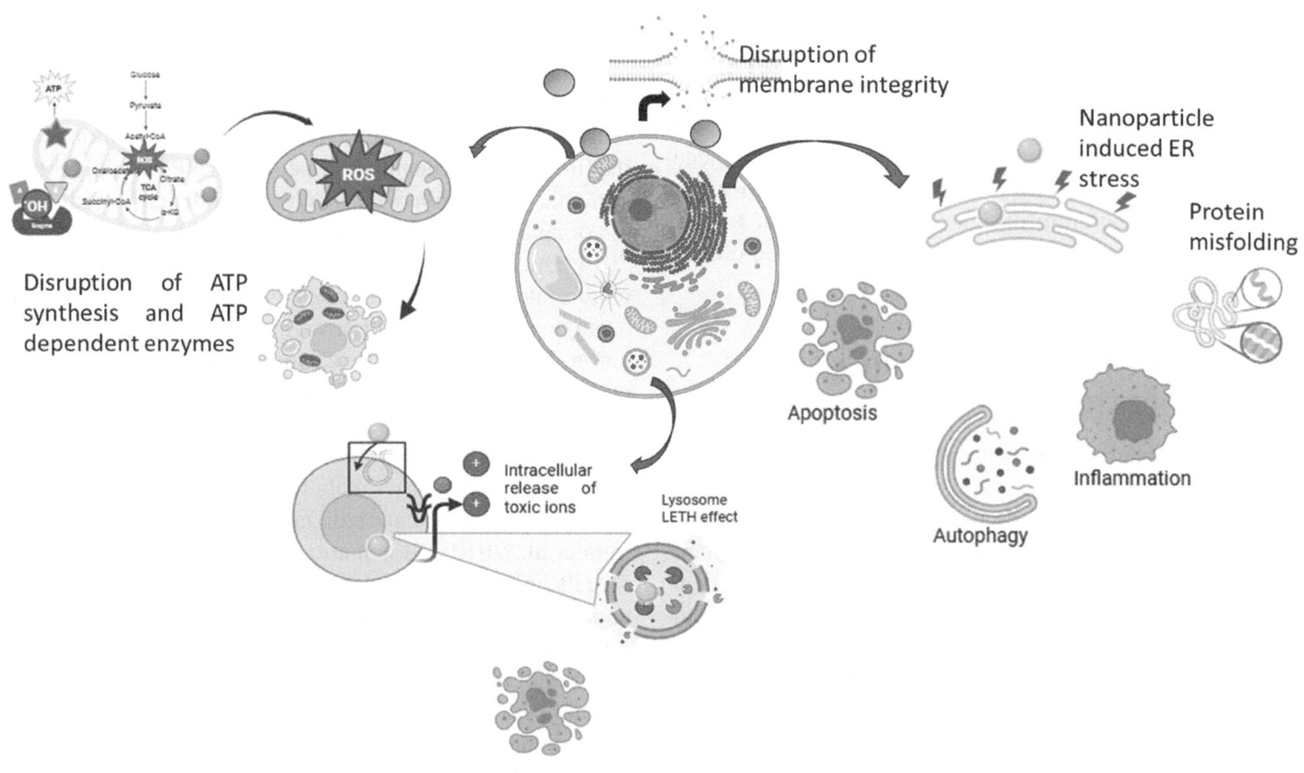

FIGURE 15.1 Mechanisms underlying the induction of nanotoxicity.

and non-enzymatic antioxidants like vitamin C, vitamin E, and carotenoids when well within optimal levels. However, strong oxidative potential induced by external agents like internalized NPs leads to overproduction of ROS beyond the cellular scavenging potential, thereby leading to altered cascades, signaling pathways, mutagenesis, and protein oxidative carboxylation, which eventually culminates in autophagy, necrosis, and pyroptosis (Buck et al., 2025; Patrón-Romero et al., 2022).

Thus, mitochondria are the major target of NP exposure, leading to major concerns about NP safety (Fujioka et al., 2014). Once internalized, NPs have been demonstrated to cause alterations of the mitochondrial endothelial layer, mitochondrial structure, and morphology, together with cascading reactions, leading to disruptions in mitochondrial dynamics and biogenesis. The effects of CdTe-QDs (metal-based nanoparticles < 10 nm) on mitochondrial oxidative phosphorylation were demonstrated by changes in the levels and activities of the enzymes involved in the electron transport chain (Nguyen et al., 2015; Malik et al., 2022; Manuja et al., 2021). It was even observed that NPs were capable of collapsing mitochondrial membrane potential, elevating Ca^{2+} levels, thus resulting in mitochondrial-mediated apoptosis (Y. Li et al., 2011). NPs not only lead to impairments in ATP synthesis and cellular respiration but also impair the activities of three ATP-dependent enzymes (Na^+/K^+-ATPase, Ca^{2+}-ATPase, and Ca^{2+}/Mg^{2+}-ATPase) (Guo et al., 2018).

NPs are self-oxidative in nature, but they are also capable of inducing pro-oxidant effects via ROS generation involving the electron transport chain and activation of NADPH-like enzyme systems (Driscoll et al., 2001). It has been demonstrated that ultrafine particles and cationic NPs are capable of inducing elevated cellular ROS production, GSH depletion, and toxic death without inflammation (Xia et al., 2006; Manke et al., 2013).

High levels of ROS in the mitochondria can result in damage to membrane phospholipids, inducing mitochondrial membrane depolarization (Lenaz, 2001). This damage to mitochondrial membrane permeability can be easily confirmed by lactate dehydrogenase (LDH) assay (Fujioka et al., 2014).

15.3.3 OS and Inflammation

Oxidative stress is primarily linked to mitochondrial dysfunction (Sharma et al., 2018), while autophagy is mediated through lysosomal pathways (He et al., 2013). Neurotoxicity frequently arises when both these processes are disrupted, as mitochondrial oxidative stress and lysosomal impairment act together to drive neuronal damage.

In response to oxidative stress, when there are excess ROS, proinflammatory mediators are released through nuclear factor-B, mitogen-activated protein kinase (MAPK), and phosphoinositide 3-kinase (PI3-K) (Poljak-Blaži et al., 2010). This indicates that oxidative stress and inflammation are mutually interconnected (Allen & Tresini, 2000).

Experimental data have also shown that metal and metal-derived NPs are also capable of interacting with nuclear proteins and DNA, causing oxidative deterioration of biological macromolecules. This results in elevated levels of oxidative stress that affect the levels and functions of numerous redox-sensitive receptors and genes (e.g., growth factor receptors and transcription factors) to induce apoptosis (programmed cell death). It has been demonstrated that metal-based NPs cause diminished cell viability, increased ROS, and apoptosis by first inducing these conditions through NPs. Then these stressed cells are rescued by reversing these changes by treating cells with N-acetylcysteine (NAC), which suggests that ROS generated by NPs are responsible for cell apoptosis (Suthar et al., 2023). As inflammation and oxidative stress are the most widely described mechanisms underlying nanomaterial-mediated toxicity, a growing number of studies suggest that autophagy could be a potential new mechanism explaining, at least in part, the toxicity of nanomaterials (Cohignac et al., 2014).

15.3.4 Autophagy

Autophagy is a cellular catabolic process in which organelles and macromolecules are delivered to lysosomes for degradation, and it occurs continuously at basal levels in most cells (Li & Ju, 2018). Autophagy involves the invagination of the lysosomal membrane, which leads to the sequestration and degradation of cytosolic components into the lysosome. The resulting degradation products are recycled to maintain nutrient and energy homeostasis. This basal level of autophagy increases under nanomaterial-induced OS, and studies have suggested the involvement of autophagy in turn in modulation of OS and inflammatory responses (Zhou et al., 2022).

The involvement of autophagy in these pathologies can be associated with its role in the modulation of oxidative stress and inflammatory responses (Khalili Fard et al., 2015). As inflammation and oxidative stress are the most widely described mechanisms underlying nanomaterial-mediated toxicity, a growing number of studies suggest that autophagy could be a potential new mechanism explaining, at least in part, the toxicity of nanomaterials. NPs have been demonstrated to increase cathepsin D protein levels, suggesting their involvement in autophagy (Cohignac et al., 2014). Recent studies have also suggested that many of the commonly used NPs like ZnO NPs initiate autophagy and pyroptosis simultaneously, a highly inflammatory programmed cell death (Pei et al., 2023). However, their interrelation was found to be intriguing. Autophagy, which is generally characterized by the production of autophagosome and increased expressions of autophagy-related proteins, seemed to inhibit the process of pyroptosis by degrading inflammasome (e.g., NLRP3) and preventing gasdermin D protein activation (Gao et al., 2022).

15.3.5 Alteration of Cell Signaling Pathways

In this context of cellular signaling disruption, three groups of MAP kinase cascades (extracellular signal-regulated kinases [ERKs], p38 mitogen-activated protein kinase, and c-Jun N-terminal kinases [JNKs]) are of prime importance. Redox-sensitive transcription factors such as NFκB and Nrf-2 have also shown special significance and have been studied in this context. Metal-based NPs and their oxides over long-term exposure have been shown to inhibit ERK signaling pathway producing excessive ROS which interrupt in cell cycle progression stages and chromosomal segregation (Huang et al., 2017). They also interact with membrane receptors, leading to ROS production and activating protein kinase involving signaling pathways (Marano et al., 2011). Again, ROS dependence is manifested by genetic factors that activate these pathways (Zia et al., 2023).

15.3.6 Endoplasmic Reticulum Stress and Dysfunction

The endoplasmic reticulum (ER) is an important site for protein synthesis, glycosylation, lipid and Ca^{2+} storage, protein folding, and secretion, as well nascent protein transport, thereby playing a critical role in cell and body's homeostasis. When the structure and function of the ER are destroyed due to NM-induced toxicity, it leads to a cascade of disruptive mechanisms, leading to toxicity, apoptosis, autophagy, and inflammation (Habas et al., 2021). Nanoparticles can interfere with proper protein folding in the ER, leading to an accumulation of misfolded proteins through protein misfolding and unfolded protein response (UPR). These responses involve activation of key pathways such as the PERK-mediated signaling cascade, activating transcription factor 6 (ATF6), and inositol-requiring enzyme 1 (IRE1) (B. Li et al., 2022; Song et al., 2023). UPR produces double-edged outputs. On the one hand, it restores homeostasis through transcription and translation in cells undergoing moderate ER stress, and on the other hand, it triggers apoptosis in cells undergoing excess ER stress (Ghosh et al., 2019). NPs of metals and metal oxides have been reported to activate ER stress response pathways, which results in elevated eukaryotic translation initiation factor 2 (eIF2α), caspase-12, and C/BEP homologous protein (CHOP) expression followed by apoptosis. NP-induced endoplasmic reticulum stress (ERS) not only plays a role in cell and tissue injury but is also linked to neuronal death, leading to neurodegenerative diseases (Yi et al., 2017; Chiang & Nicol, 2022). ERS intervenes, and its cross-talk with multiple inflammatory pathway gives rise to NLRP3 inflammasome (NOD-, LRR-, and pyrin domain-containing protein 3), a pattern recognition receptor, which is the key component of the innate immune system. This can initiate and aggravate chronic inflammatory diseases (W. Li et al., 2020; Tang et al., 2022).

15.3.7 Effects on Cell Morphology, Membrane Integrity, and Cytoskeletal Integrity

Studies have reported that metal-based NPs like Cu, Zn, CuO, and Cu-Zn composite caused hemolysis in RBCs. The membrane damage was caused by the metal release process at the cell membrane lipid bilayer surface. The asymmetry of the lipid component of the membrane paves way for the NP entry and interaction (Karlsson et al., 2013). The toxicity of the plasma membrane is also dependent on the particle size of the NPs: the smaller the NP, the greater the surface area, the greater the interaction, and the higher its potency of toxicity. The lipid composition of the membrane's outer leaflet(layer) plays a crucial in nanoparticle-induced toxicity (Bigdelou et al., 2020).

Studies have shown that NPs like AgNPs induce toxicity through degradation of cytoskeletal components such as β-tubulin and filamentous actin (F-actin) (Grady et al., 2017). NPs inhibited neurite outgrowth, reduced cell viability, disrupted both pre-synaptic (synaptophysin) and post-synaptic (PSD-95) protein expression, and induced mitochondrial dysfunction, ultimately leading to premature apoptosis (Xu et al., 2013). Also, the elasticity of the cell membrane is significantly modified by exposure to NPs (Co, Ti), thereby altering the cell conformation and functionality (Perni et al., 2018). Various studies with inorganic NPs like Ag, Au, Si, Ti, SiO_2, TiO, and Fe_2O_3 have shown their capacity to permeate the BBB, remodel cytoskeletal structures, and induce electrophysiological changes, thereby hindering neurodevelopment (Singh, 2015). One of the most significant mechanisms linking nanoparticles to neurotoxicity appears to involve alterations in NMDA receptor function, disruptions in ion channel activity, and activation of the Rho signaling pathway. Oxidation of the thiol group of tubulin monomers and perturbations of microtubule polymerization have been the result of NP-induced OS (Kang et al., 2016).

15.4 NANOTOXICITY EVALUATION TECHNIQUES

Toxicity refers to the adverse effects of a chemical on the body and can be assessed quantitatively (e.g., LD_{50}), qualitatively through classifications such as low, moderate, or highly toxic, or computationally using QSAR models, simulation approaches, and statistical analyses (Raies & Bajic, 2016; Sharma et al., 2024). Understanding the molecular mechanisms of NP-induced toxicity requires thorough evaluation of the toxicity profiles of both natural and engineered nanoparticles before conducting application-specific in vitro or in vivo studies. In-depth understanding of the NP-induced toxicity mechanism help us devise techniques for toxicity evaluation.

Among several evaluation techniques, we will be focusing here on *in silico, in vitro*, and *in vivo* methods that provide reliable results with zero side effects.

15.4.1 *In Silico*/Computational Approaches

In silico computational approaches facilitate assessment of toxicity of nanomaterials at low cost and without animal testing (von Ranke et al., 2022). Quantitative structure activity relationship (QSAR), molecular docking, and molecular dynamics (MD) simulations for absorption, distribution, metabolism, excretion, and toxicity (ADMET) property evaluation are the primary methods used for toxicological assessments. The development of QSAR models for NMs happens to be a little challenging due to a lack of adequate experimental data, but it still finds application through the toxicological profiling derived from physico-chemical and structural characteristics (Shityakov et al., 2017). Also, Cheminformatics modelling helps to identify important descriptors for cytotoxicity, HOMO-LUMO parameters, zeta potential, surface charge, size, and shape that play a crucial role in inducing cytotoxicity and genotoxicity (El Yamani et al., 2022). To develop more efficient models, integrating data from multiple databases with statistical analyses enables innovative computational approaches for predicting toxicological effects during the design of novel nanomaterials (Yang et al., 2022).

15.4.2 Omics Platforms

Whole-genome sequencing (WGS) or next-generation sequencing (NGS) has emerged as an efficient tool to decipher mutations in genomes and elucidate the toxic impact of nanomaterials downstream at the metabolic pathways. In recent years, it has helped researchers identify the role of microRNAs (miRNAs), as a small non-coding RNA, in inducing nanomaterial-induced stress (Yang et al., 2022). RNA is the prefinal step of protein formation and is well known to have lower stability than DNA. This makes it highly susceptible to the toxicological effects of nanomaterials, including alterations in RNA or its copy number. The same platform of NGS and a microarray is also used for whole transcriptomics analyses. Instead of animal models, different *in vitro* models like 3D cell cultures and organoids are being extensively used for transcriptomic profiling (Chen et al., 2021). Proteomics platforms follow to understand the complete protein profile of pre- and post-treated cells in terms of both quality and quantity. The techniques include mass spectrometry-based techniques, mass cytometry, fluorescence flow cytometry, enzyme-linked immunosorbent assay, and single-molecule array (SiMoA) assays (Ventura et al., 2022). Proteomics exploration helps give deeper insights into accumulation, membrane stability, and alterations in protein pathways related to NP-based effects on organisms (Tubatsi et al., n.d.). Metabolites are the end products of metabolic pathways whose alterations lead to disorders. Analysis of qualitative and quantitative changes in metabolites on exposure to NPs helps us to understand the toxic effects of these NPs. Thus, assessment of these metabolites is key to assessing the safety of NPs in human applications as well as the environment. This is achieved by many platforms, such as the classical autoanalyzer biochemical analytes or modern techniques of chromatography coupled with MS-based techniques and recently developed sensors. LC or GC-MS and NMR-based platforms are also employed extensively for metabolomics analyses. In addition, other platforms such as microarray, Fourier transform infrared (FTIR) spectroscopy, PCR-based WGA methods, and fluorescence sensor-based metabolite identification are also commonly used for nanotoxicity assessment. Single-cell omics techniques can assess omic profiles in different layers simultaneously in single cells by sequencing (CITE-seq) or antigen profiling (ASAP-seq) (Diamante et al., 2024).

15.4.3 *In Vitro* Nanocytotoxicity Assessment

In vitro methods are the quickest and first line of cytotoxicity assessment techniques, among which calorimetric assays like MTT, MTS, XTT, and LDH assessing cell viability are the most frequently used. These assays are used to measure the cellular metabolism by assessment of metabolic active cells (Kumar et al., 2017; 2024). 3-(4,5-dimethyl thiazol-2-yl)-2,5-diphenyltetrazolium bromide (MTT) is the most commonly used tetrazolium salt for *in vitro* toxicity assessment of nanoparticles. The MTT assay also produces insoluble formazan; therefore, assays such as such XTT or WST-1, which produce soluble dyes, are preferred, as the solubilization step can be avoided. The BrdU (bromodeoxyuridine) incorporation assay measures DNA synthesis during the S-phase of the cell cycle. It is sensitive in assessing cell proliferation rates (Ganguly et al., 2018). Upregulation or exhaustion of SOD and glutathione stores indicates imbalance due to oxidation by ROS or RNS and serves as evidence of oxidative stress (Xu et al., 2012). Superoxide dismutase activity is determined indirectly using nitro blue tetrazolium. It causes inhibition to oxidation of the colored substrate, generating superoxide through exogenous xanthine oxidase. The degree of inhibition of NBT reduction is inversely proportional to SOD activity. To evaluate cellular ROS, the production plasmid assay, DCFH-DA, and FOX-1 are mostly used (Griffo et al., 2023). DCFH-DA, a fluorogenic marker, is cleaved to a fluorescent DCFH molecule during elevated production of ROS. Therefore, the increase in fluorescence indicates the level of increased ROS production. Dihydroethidium (DHE) also works in a similar way, and these methods are extremely useful in providing information on NPs alone, irrespective of the environment in the cell (El Yamani et al., 2024).

Glutathione, an essential cellular antioxidant, becomes oxidized to form GSSG under conditions of oxidative stress. The GSH:GSSG ratio, a key indicator of cellular oxidative stress, is typically measured using enzymatic recycling assays like Ellman's reagent, HPLC, or LC-MS/MS.

Assays like the TUNEL assay, COMET assay, Caspase assay, Annexin V assay, CLSM, and FACS are used to detect apoptosis and necrosis (Huang et al., 2025). The Comet assay and Ames's assay are frequently used to assess

DNA damage and gene mutations, respectively, that occur due to exposure to nanomaterials or any chemical agent. Immunological tests help to assess the effect of NMs on immunological health; for example, the Episcreen kit is used for T-cell activation profiling, Epibase for MHC class II binding prediction, and REVEAL for memory T-cell response detection. These *in vitro* assessments provide a deeper insight into the extent of damage inflicted by NMs on exposure to cellular fluids (Lakhani et al., 2024).

Inflammation induced by nanoparticles can be detected using the enzyme-linked immunosorbent assay (ELISA), which can detect the presence of pro-inflammatory cytokines and chemokines released by those exposed cells. The most commonly assessed human inflammatory markers—such as interleukin-8 (IL-8), TNF-α, and IL-6—can be measured using either PCR or FACS analysis. PCR assays and microarrays are also used for profiling, detection, and expression of a targeted group of genes that are affected by nanoparticle exposure. This can happen directly or indirectly through perturbation of cell signaling cascades (Summer et al., 2024). In recent years, advanced techniques like the single-molecule array and microfluidic antibody capture chip (MACS chip)–based assay have emerged as quick assays for detecting highly sensitive biomarkers at single molecule resolution in toxicological studies. The droplet-based (drop-seq) system can detect subtle transcriptomic changes to give a mechanistic insight into how NPs can induce toxicity. The DBiT-seq technique helps assess localized toxicity, and single-cell genome long-read sequencing technology (SMOOTH-seq) is used for genomic-level damage detection due to prolonged exposure to NPs (Diamante et al., 2024). Such techniques assist in the detection and quantification of specific RNA or a mixture of total RNA detected by autoradiographic imaging, thereby helping us focus on the expression of a panel of genes.

15.4.4 *In Vivo* Nanocytotoxicity Assessment

Histopathological assessments happen to be one of the classical techniques which look at malformations in tissues and organs in diseased conditions, which can be induced by any chemical, drug, or NP exposure. Routine stains such as Gram stain, Lugol's iodine, hematoxylin and eosin (H&E), and hematoxylin phloxine saffron (HPS) are used to distinguish various tissue components, while specialized techniques like immunofluorescence are employed for more specific diagnostic purposes. Also, different organic molecules are detected using staining procedures (Abebe, 2023; Buragohain et al., 2024). To gain a more insightful understanding of NP-induced toxicity, *in vivo* studies need to be performed to give us a clearer understanding at the organism level before the manifestation of the disease or malfunction (Abebe, 2023).

Assessment methods of *in vivo* toxicity include biodistribution, clearance, hematology, serum chemistry, and histopathology, and the animal models are generally zebra fish, mice, rats, guinea pigs and rabbits (Kumar et al., 2017). Nanoparticles are detected in killed or live animals through radiolabels (Egbuna et al., 2021). Toxicity assessment involves quantitative parameters like LD_{50}, LC_{50}, EC_{50}, IC_{50}, lowest observed adverse effect level (LOAEL), and no observed adverse effect level (NOAEL). The most commonly used metric is 50% of lethal dose (LD_{50}), which represents the dose required to cause death in 50% of a test population, typically expressed in mg/kg of body weight. In the case of airborne or aquatic exposures, LC_{50} is used to denote the concentration that causes mortality to 50% of exposed organisms. IC_{50} and EC_{50}, however, are more widely used in *in vitro* studies, signifying the concentrations needed to inhibit any biological function or enzyme to produce half the maximal effect. In order to signify the dosage, tolerance metrics like NOAEL and LOAEL, which define the dose ranges at which no or the first adverse effects are observed, respectively, are used. These parameters are crucial for evaluating nanoparticle safety and guiding their safe application in biomedical and environmental contexts (Arivazhahan, 2022). Similar to this are several behavioral and cognitive tests that are routinely studied like the Morris water maze test, radial maze test, open field test, hole board test (HBT), elevated plus maze (EPM) test, object exploration for emotional status, and Porsolt swim test (PST) (Seo et al., 2023). The hot plate test and tail flick test for are used for detection of startle response and nociception (Cerqueira et al., 2023).

15.5 CONCLUSION

Nanomaterials have provided benefits across diverse fields, which has led to the formulation of a new branch of study, nanotechnology. Progressive enhancements in the field of nanotechnology and nanomedicine have led to the invention of novel nanomaterials and their intervention in the global biome, including humans and animals. Investigations focus on their beneficial effects, but nanomaterial exposure over a long period of time leads to bio-accumulation, thus leaving a toxic impact which cannot be neglected. Though some toxicity studies have been carried out, they only elucidate short-term effects. Toxicological studies still rely upon conventional methods that only provide snapshots of the greater toxicological effects, ranging from cell to organ to organism and community. A comprehensive toxicity analysis involving multiple models and integrated analytical methods of *in silico*, *in vitro*, and *in vivo* 3D models that mimic a near-physiological microenvironment can probably be beneficial to provide a near-perfect toxicological analysis of these nanomaterials before flooding the environment or releasing them as nanomedicines on the market.

REFERENCES

Abebe, M. S. (2023). Acute and subacute toxicity of rhamnus prinoides leaves on histopathology of liver, kidney, and brain tissues, and biochemical profile of rats. *Journal of Toxicology*, *2023*(1), 3105615. https://doi.org/10.1155/2023/3105615

Allen, R. G., & Tresini, M. (2000). Oxidative stress and gene regulation. *Free Radical Biology and Medicine*, *28*(3), 463–499. https://doi.org/10.1016/S0891-5849(99)00242-7

Arivazhahan, A. (2022). Principles of EC50, ED50, pD2 and pA2 values of drugs. In M. Lakshmanan, D. G. Shewade, & G. M. Raj (Eds.), *Introduction to Basics of Pharmacology and Toxicology: Volume 3: Experimental Pharmacology: Research Methodology and Biostatistics* (pp. 143–149). Springer Nature. https://doi.org/10.1007/978-981-19-5343-9_10

Awashra, M., & Młynarz, P. (2023). The toxicity of nanoparticles and their interaction with cells: An in vitro metabolomic perspective. *Nanoscale Advances*, *5*(10), 2674–2723. https://doi.org/10.1039/D2NA00534D

Berube, K. A., Balharry, D., Sexton, K. J., Koshy, L., & Jones, T. P. (2007). Combustion-derived nanoparticles: Mechanisms of pulmonary toxicity. *Clinical and Experimental Pharmacology and Physiology*, *34*(10), Article 10.

Bigdelou, P., Vahedi, A., Kiosidou, E., & Farnoud, A. M. (2020). Loss of membrane asymmetry alters the interactions of erythrocytes with engineered silica nanoparticles. *Biointerphases*, *15*(4), 041001. https://doi.org/10.1116/6.0000246

Buchman, J. T., Hudson-Smith, N. V., Landy, K. M., & Haynes, C. L. (2019). Understanding nanoparticle toxicity mechanisms to inform redesign strategies to reduce environmental impact. *Accounts of Chemical Research*, *52*(6), 1632–1642. https://doi.org/10.1021/acs.accounts.9b00053

Buck, A. C., Maarman, G. J., Dube, A., & Bardien, S. (2025). Mitochondria targeted nanoparticles for the treatment of mitochondrial dysfunction-associated brain disorders. *Frontiers in Bioengineering and Biotechnology*, *13*. https://doi.org/10.3389/fbioe.2025.1563701

Buragohain, L., Kumar, R., Kumar, A., & Ghosh, M. (2024). *Nanointerventions for Detections of Viral Livestock Diseases* (pp. 139–176). https://doi.org/10.1007/978-981-16-1610-5_5

Cerqueira, M. M. de F., Castro, M. M. L., Vieira, A. A., Kurosawa, J. A. A., Junior, F. L. do A., Mendes, F. de C. C. de S., & Sosthenes, M. C. K. (2023). Comparative analysis between Open Field and Elevated Plus Maze tests as a method for evaluating anxiety-like behavior in mice. *Heliyon*, *9*(4). https://doi.org/10.1016/j.heliyon.2023.e14522

Chen, Y., Dong, L., Deng, F., Cao, Y., Fu, Y., Zhu, M., Qin, G., Schultz, D., Shekh, K., & Tang, S. (2021). Integrated high-throughput small RNA and transcriptome sequencing unveil the shape-dependent toxicity of nano-alumina in rat astrocytes. *Environmental Sciences Europe*, *33*(1), 95. https://doi.org/10.1186/s12302-021-00540-9

Chiang, M.-C., & Nicol, C. J. B. (2022). GSH-AuNP anti-oxidative stress, ER stress and mitochondrial dysfunction in amyloid-beta peptide-treated human neural stem cells. *Free Radical Biology and Medicine*, *187*, 185–201. https://doi.org/10.1016/j.freeradbiomed.2022.05.025

Cohignac, V., Landry, M. J., Boczkowski, J., & Lanone, S. (2014). Autophagy as a possible underlying mechanism of nanomaterial toxicity. *Nanomaterials*, *4*(3), Article 3. https://doi.org/10.3390/nano4030548

Demir, E. (2021). A review on nanotoxicity and nanogenotoxicity of different shapes of nanomaterials. *Journal of Applied Toxicology*, *41*(1), 118–147. https://doi.org/10.1002/jat.4061

Diamante, G., Ha, S. M., Wijaya, D., & Yang, X. (2024). Single cell multiomics systems biology for molecular toxicity. *Current Opinion in Toxicology*, *39*, 100477. https://doi.org/10.1016/j.cotox.2024.100477

Djurišić, A. B., Leung, Y. H., Ng, A. M. C., Xu, X. Y., Lee, P. K. H., Degger, N., & Wu, R. S. S. (2015). Toxicity of metal oxide nanoparticles: Mechanisms, characterization, and avoiding experimental artefacts. *Small*, *11*(1), 26–44. https://doi.org/10.1002/smll.201303947

Driscoll, K. E., Howard, B. W., Carter, J. M., Janssen, Y. M. W., Mossman, B. T., & Isfort, R. J. (2001). Mitochondrial-derived oxidants and quartz activation of chemokine gene expression. In P. M. Dansette, R. Snyder, M. Delaforge, G. G. Gibson, H. Greim, D. J. Jollow, T. J. Monks, & I. G. Sipes (Eds.), *Biological Reactive Intermediates VI: Chemical and Biological Mechanisms in Susceptibility to and Prevention of Environmental Diseases* (pp. 489–496). Springer US. https://doi.org/10.1007/978-1-4615-0667-6_74

Egbuna, C., Parmar, V. K., Jeevanandam, J., Ezzat, S. M., Patrick-Iwuanyanwu, K. C., Adetunji, C. O., Khan, J., Onyeike, E. N., Uche, C. Z., Akram, M., Ibrahim, M. S., El Mahdy, N. M., Awuchi, C. G., Saravanan, K., Tijjani, H., Odoh, U. E., Messaoudi, M., Ifemeje, J. C., Olisah, M. C., . . . Ibeabuchi, C. G. (2021). Toxicity of nanoparticles in biomedical application: Nanotoxicology. *Journal of Toxicology*, *2021*(1), 9954443. https://doi.org/10.1155/2021/9954443

El Yamani, N., Mariussen, E., Gromelski, M., Wyrzykowska, E., Grabarek, D., Puzyn, T., Tanasescu, S., Dusinska, M., & Rundén-Pran, E. (2022). Hazard identification of nanomaterials: *In silico* unraveling of descriptors for cytotoxicity and genotoxicity. *Nano Today*, *46*, 101581. https://doi.org/10.1016/j.nantod.2022.101581

El Yamani, N., Rundén-Pran, E., Varet, J., Beus, M., Dusinska, M., Fessard, V., Moschini, E., Serchi, T., Cimpan, M. R., Lynch, I., & Vinković Vrček, I. (2024). Hazard assessment of nanomaterials using *in vitro* toxicity assays: Guidance on potential assay interferences and mitigating actions to avoid biased results. *Nano Today*, *55*, 102215. https://doi.org/10.1016/j.nantod.2024.102215

Fatima, A., Shehzad, A., Shahzad, R., Khan, S., & Al-Suhaimi, E. A. (2024). Chapter seven—Impact of nanoparticles on structural elements within the cells. In H. Tombuloglu, G. Tombuloglu, E. Al-Suhaimi, A. Baykal, & K. R. Hakeem (Eds.), *Molecular Impacts of Nanoparticles on Plants and Algae* (pp. 111–141). Academic Press. https://doi.org/10.1016/B978-0-323-95721-2.00001-4

Feng, Y., Fu, H., Zhang, X., Liu, S., & Wei, X. (2024). Lysosome toxicities induced by nanoparticle exposure and related mechanisms. *Ecotoxicology and Environmental Safety*, *286*, 117215. https://doi.org/10.1016/j.ecoenv.2024.117215

Fujioka, K., Hanada, S., Inoue, Y., Sato, K., Hirakuri, K., Shiraishi, K., Kanaya, F., Ikeda, K., Usui, R., Yamamoto, K., Kim, S. U., & Manome, Y. (2014). Effects of silica and titanium oxide particles on a human neural stem cell line: Morphology, mitochondrial activity, and gene expression of differentiation markers. *International Journal of Molecular Sciences*, *15*(7), Article 7. https://doi.org/10.3390/ijms150711742

Ganguly, P., Breen, A., & Pillai, S. C. (2018). Toxicity of nanomaterials: Exposure, pathways, assessment, and recent advances. *ACS Biomaterials Science & Engineering*, *4*(7), 2237–2275. https://doi.org/10.1021/acsbiomaterials.8b00068

Gao, W., Wang, X., Zhou, Y., Wang, X., & Yu, Y. (2022). Autophagy, ferroptosis, pyroptosis, and necroptosis in tumor immunotherapy. *Signal Transduction and Targeted Therapy*, *7*(1), 1–26. https://doi.org/10.1038/s41392-022-01046-3

Ghosh, R., Colon-Negron, K., & Papa, F. R. (2019). Endoplasmic reticulum stress, degeneration of pancreatic islet β-cells, and therapeutic modulation of the unfolded protein response in diabetes. *Molecular Metabolism*, *27*, S60–S68. https://doi.org/10.1016/j.molmet.2019.06.012

Grady, M. E., Parrish, E., Caporizzo, M. A., Seeger, S. C., Composto, R. J., & Eckmann, D. M. (2017). Intracellular nanoparticle dynamics affected by cytoskeletal integrity. *Soft Matter*, *13*(9), 1873–1880. https://doi.org/10.1039/C6SM02464E

Griffo, A., Bosco, N., Pagano, A., Balestrazzi, A., & Macovei, A. (2023). Noninvasive methods to detect reactive oxygen species as a proxy of seed quality. *Antioxidants*, *12*(3), Article 3. https://doi.org/10.3390/antiox12030626

Guo, C., Wang, J., Jing, L., Ma, R., Liu, X., Gao, L., Cao, L., Duan, J., Zhou, X., Li, Y., & Sun, Z. (2018). Mitochondrial dysfunction, perturbations of mitochondrial dynamics and biogenesis involved in endothelial injury induced by silica nanoparticles. *Environmental Pollution*, *236*, 926–936. https://doi.org/10.1016/j.envpol.2017.10.060

Habas, K., Demir, E., Guo, C., Brinkworth, M. H., & Anderson, D. (2021). Toxicity mechanisms of nanoparticles in the male reproductive system. *Drug Metabolism Reviews*, *53*(4), 604–617. https://doi.org/10.1080/03602532.2021.1917597

He, L. Q., Lu, J. H., & Yue, Z. Y. (2013). Autophagy in ageing and ageing-associated diseases. *Acta Pharmacologica Sinica*, *34*(5), 605–611.

Huang, Q., Xia, Q., Qi, Y., Shen, C., Li, J., Pan, Z., Liu, Y., Zhang, Y., Hu, Q., Huan, P., Wang, C., Wang, X., Qiao, C., Zheng, B., Zhao, X., Shi, Z., Yu, J., & Yang, X. (2025). Testicular toxicity assessment of chronic low-dose exposure to polystyrene nanoparticles in multiexposure models using drosophila. *ACS Applied Nano Materials*, *8*(5), 2269–2279. https://doi.org/10.1021/acsanm.4c06138

Huang, Y.-W., Cambre, M., & Lee, H.-J. (2017). The toxicity of nanoparticles depends on multiple molecular and physicochemical mechanisms. *International Journal of Molecular Sciences*, *18*(12), Article 12. https://doi.org/10.3390/ijms18122702

Isibor, P. O., Sunday, A. S., Buba, A. B., & Oyewole, O. A. (2024). Mechanism of nanoparticle toxicity. In P. O. Isibor, G. Devi, & A. A. Enuneku (Eds.), *Environmental Nanotoxicology: Combatting the Minute Contaminants* (pp. 103–120). Springer Nature Switzerland. https://doi.org/10.1007/978-3-031-54154-4_6

Kang, Y., Liu, J., Song, B., Feng, X., Ou, L., Wei, L., Lai, X., & Shao, L. (2016). Potential links between cytoskeletal disturbances and electroneurophysiological dysfunctions induced in the central nervous system by inorganic nanoparticles. *Cellular Physiology and Biochemistry*, *40*(6), 1487–1505. https://doi.org/10.1159/000453200

Karlsson, H. L., Cronholm, P., Hedberg, Y., Tornberg, M., De Battice, L., Svedhem, S., & Wallinder, I. O. (2013). Cell membrane damage and protein interaction induced by copper containing nanoparticles—Importance of the metal release process. *Toxicology*, *313*(1), 59–69. https://doi.org/10.1016/j.tox.2013.07.012

Khalili Fard, J., Jafari, S., & Eghbal, M. A. (2015). A review of molecular mechanisms involved in toxicity of nanoparticles. *Advanced Pharmaceutical Bulletin*, *5*(4), 447–454. https://doi.org/10.15171/apb.2015.061

Kumar, R., Lakhani, P., Kumar, A. Y. N., & Ghosh, M. (2024). Nanotoxicity and mechanisms. In M. Prasad, R. Kumar, M. Ghosh, S. M. Syed, & S. Chakravarti (Eds.), *Nanotechnology Theranostics in Livestock Diseases and Management* (pp. 699–741). Springer Nature. https://doi.org/10.1007/978-981-16-1610-5_29

Kumar, V., Sharma, N., & Maitra, S. S. (2017). In vitro and in vivo toxicity assessment of nanoparticles. *International Nano Letters*, *7*(4), 243–256. https://doi.org/10.1007/s40089-017-0221-3

Lakhani, N., Chowdhury, A., Akshita, Maurya, K., Kumari, K., Rose, M. K., Sindhu, S., & Ghosh, M. (2024). Nanotoxicity evaluation methods. In M. Prasad, R. Kumar, M. Ghosh, S. M. Syed, & S. Chakravarti (Eds.), *Nanotechnology Theranostics in Livestock Diseases and Management* (pp. 743–763). Springer Nature. https://doi.org/10.1007/978-981-16-1610-5_30

Lenaz, G. (2001). The mitochondrial production of reactive oxygen species: Mechanisms and implications in human pathology. *IUBMB Life*, *52*(3–5), 159–164. https://doi.org/10.1080/15216540152845957

Lesniak, A., Fenaroli, F., Monopoli, M. P., Åberg, C., Dawson, K. A., & Salvati, A. (2012). Effects of the presence or absence of a protein corona on silica nanoparticle uptake and impact on cells. *ACS Nano*, *6*(7), 5845–5857. https://doi.org/10.1021/nn300223w

Li, B., Zhang, T., & Tang, M. (2022). Toxicity mechanism of nanomaterials: Focus on endoplasmic reticulum stress. *Science of the Total Environment*, *834*, 155417. https://doi.org/10.1016/j.scitotenv.2022.155417

Li, J., Zhao, T., Chen, T., Liu, Y., Nam Ong, C., & Xie, J. (2015). Engineering noble metal nanomaterials for environmental applications. *Nanoscale*, *7*(17), 7502–7519. https://doi.org/10.1039/C5NR00857C

Li, W., Cao, T., Luo, C., Cai, J., Zhou, X., Xiao, X., & Liu, S. (2020). Crosstalk between ER stress, NLRP3 inflammasome, and inflammation. *Applied Microbiology and Biotechnology*, *104*(14), 6129–6140. https://doi.org/10.1007/s00253-020-10614-y

Li, Y., & Ju, D. (2018). The role of autophagy in nanoparticles-induced toxicity and its related cellular and molecular mechanisms. In Q. Saquib, M. Faisal, A. A. Al-Khedhairy, & A. A. Alatar (Eds.), *Cellular and Molecular Toxicology of Nanoparticles* (Vol. 1048, pp. 71–84). Springer International Publishing. https://doi.org/10.1007/978-3-319-72041-8_5

Li, Y., Sun, L., Jin, M., Du, Z., Liu, X., Guo, C., Li, Y., Huang, P., & Sun, Z. (2011). Size-dependent cytotoxicity of amorphous silica nanoparticles in human hepatoma HepG2 cells. *Toxicology in Vitro*, *25*(7), 1343–1352. https://doi.org/10.1016/j.tiv.2011.05.003

Lundqvist, M., Stigler, J., Elia, G., Lynch, I., Cedervall, T., & Dawson, K. A. (2008). Nanoparticle size and surface properties determine the protein corona with possible implications for biological impacts. *Proceedings of the National Academy of Sciences*, *105*(38), 14265–14270. https://doi.org/10.1073/pnas.0805135105

Maiorano, G., Sabella, S., Sorce, B., Brunetti, V., Malvindi, M. A., Cingolani, R., & Pompa, P. P. (2010). Effects of cell culture media on the dynamic formation of protein-nanoparticle complexes and influence on the cellular response. *ACS Nano*, *4*(12), 7481–7491. https://doi.org/10.1021/nn101557e

Malik, A., Khan, J. M., Alhomida, A. S., Ola, M. S., Alshehri, M. A., & Ahmad, A. (2022). Metal nanoparticles: Biomedical applications and their molecular mechanisms of toxicity. *Chemical Papers*, *76*(10), 6073–6095. https://doi.org/10.1007/s11696-022-02351-5

Manke, A., Wang, L., & Rojanasakul, Y. (2013). Mechanisms of nanoparticle-induced oxidative stress and toxicity. *BioMed Research International*, *2013*(1), 942916. https://doi.org/10.1155/2013/942916

Manuja, A., Kumar, B., Kumar, R., Chhabra, D., Ghosh, M., Manuja, M., Brar, B., Pal, Y., Tripathi, B. N., & Prasad, M. (2021). Metal/metal oxide nanoparticles: Toxicity concerns associated with their physical state and remediation for biomedical applications. *Toxicology Reports*, *8*, 1970–1978. https://doi.org/10.1016/j.toxrep.2021.11.020

Marano, F., Hussain, S., Rodrigues-Lima, F., Baeza-Squiban, A., & Boland, S. (2011). Nanoparticles: Molecular targets and cell signalling. *Archives of Toxicology*, *85*(7), 733–741. https://doi.org/10.1007/s00204-010-0546-4

Nalika, N., & and Parvez, S. (2015). Mitochondrial dysfunction in titanium dioxide nanoparticle-induced neurotoxicity. *Toxicology Mechanisms and Methods*, *25*(5), 355–363. https://doi.org/10.3109/15376516.2015.1020183

Nguyen, K. C., Rippstein, P., Tayabali, A. F., & Willmore, W. G. (2015). Mitochondrial toxicity of cadmium telluride quantum dot nanoparticles in mammalian hepatocytes. *Toxicological Sciences*, *146*(1), 31–42. https://doi.org/10.1093/toxsci/kfv068

Oberdörster, G. (2010). Safety assessment for nanotechnology and nanomedicine: Concepts of nanotoxicology. *Journal of Internal Medicine*, *267*(1), 89–105. https://doi.org/10.1111/j.1365-2796.2009.02187.x

Odaudu, O. R., & Akinsiku, A. A. (2022). Toxicity and cytotoxicity effects of selected nanoparticles: A review. *IOP Conference Series: Earth and Environmental Science*, *1054*(1), 012007. https://doi.org/10.1088/1755-1315/1054/1/012007

Pareek, V., Gupta, R., & Panwar, J. (2018). Do physico-chemical properties of silver nanoparticles decide their interaction with biological media and bactericidal action? A review. *Materials Science and Engineering: C*, *90*, 739–749. https://doi.org/10.1016/j.msec.2018.04.093

Park, E.-J., Yi, J., Kim, Y., Choi, K., & Park, K. (2010). Silver nanoparticles induce cytotoxicity by a Trojan-horse type mechanism. *Toxicology in Vitro*, *24*(3), 872–878. https://doi.org/10.1016/j.tiv.2009.12.001

Park, J.-H., Gu, L., von Maltzahn, G., Ruoslahti, E., Bhatia, S. N., & Sailor, M. J. (2009). Biodegradable luminescent porous silicon nanoparticles for in vivo applications. *Nature Materials*, *8*(4), 331–336. https://doi.org/10.1038/nmat2398

Patrón-Romero, L., Luque-Morales, P. A., Loera-Castañeda, V., Lares-Asseff, I., Leal-Ávila, M. Á., Alvelais-Palacios, J. A., Plasencia-López, I., & Almanza-Reyes, H. (2022). Mitochondrial dysfunction induced by zinc oxide nanoparticles. *Crystals*, *12*(8), Article 8. https://doi.org/10.3390/cryst12081089

Pei, X., Liu, D., Li, J., Li, L., Ding, X., Zhang, W., Li, Z., Xu, G., Li, C., & Li, D. (2023). TFEB coordinates autophagy and pyroptosis as hepatotoxicity responses to ZnO nanoparticles. *Science of the Total Environment*, *865*, 161242. https://doi.org/10.1016/j.scitotenv.2022.161242

Perni, S., Yang, L., Preedy, E. C., & Prokopovich, P. (2018). Cobalt and Titanium nanoparticles influence on human osteoblast mitochondrial activity and biophysical properties of their cytoskeleton. *Journal of Colloid and Interface Science*, *531*, 410–420. https://doi.org/10.1016/j.jcis.2018.07.028

Poljak-Blaži, M., Jaganjac, M., & Žarković, N. (2010). *Cell Oxidative Stress: Risk of Metal Nanoparticles* (K. D. Sattler, Ed.; pp. 16–17). London; New York (NY): CRC Press; Taylor & Francis. https://www.croris.hr/crosbi/publikacija/prilog-knjiga/42444

Raies, A. B., & Bajic, V. B. (2016). In silico toxicology: Computational methods for the prediction of chemical toxicity. *WIREs Computational Molecular Science*, *6*(2), 147–172. https://doi.org/10.1002/wcms.1240

Sabella, S., P. Carney, R., Brunetti, V., Ada Malvindi, M., Al-Juffali, N., Vecchio, G., M. Janes, S., M. Bakr, O., Cingolani, R., Stellacci, F., & Paolo Pompa, P. (2014). A general mechanism for intracellular toxicity of metal-containing nanoparticles. *Nanoscale*, *6*(12), 7052–7061. https://doi.org/10.1039/C4NR01234H

Schaeublin, N. M., Braydich-Stolle, L. K., Schrand, A. M., Miller, J. M., Hutchison, J., Schlager, J. J., & Hussain, S. M. (2011). Surface charge of gold nanoparticles mediates mechanism of toxicity. *Nanoscale*, *3*(2), 410–420. https://doi.org/10.1039/C0NR00478B

Sengul, A. B., & Asmatulu, E. (2020). Toxicity of metal and metal oxide nanoparticles: A review. *Environmental Chemistry Letters*, *18*(5), 1659–1683. https://doi.org/10.1007/s10311-020-01033-6

Seo, M. K., Jeong, S., Seog, D.-H., Lee, J. A., Lee, J.-H., Lee, Y., McIntyre, R. S., Park, S. W., & Lee, J. G. (2023). Effects of liraglutide on depressive behavior in a mouse depression model and cognition in the probe trial of Morris water maze test. *Journal of Affective Disorders*, *324*, 8–15. https://doi.org/10.1016/j.jad.2022.12.089

Service, R. F. (2010). Nanoparticle trojan horses gallop from the lab into the clinic. *Science*, *330*(6002), 314–315. https://doi.org/10.1126/science.330.6002.314

Sharma, N., Kurmi, B. D., Singh, D., Mehan, S., Khanna, K., Karwasra, R., Kumar, S., Chaudhary, A., Jakhmola, V., Sharma, A., Singh, S. K., Dua, K., & Kakkar, D. (2024). Nanoparticles toxicity: An overview of its mechanism and plausible mitigation strategies. *Journal of Drug Targeting*, *32*(5), 457–469. https://doi.org/10.1080/1061186X.2024.2316785

Shityakov, S., Roewer, N., Broscheit, J.-A., & Förster, C. (2017). *In silico* models for nanotoxicity evaluation and prediction at the blood-brain barrier level: A mini-review. *Computational Toxicology*, *2*, 20–27. https://doi.org/10.1016/j.comtox.2017.02.003

Singh, A. K. (2015). *Engineered Nanoparticles: Structure, Properties and Mechanisms of Toxicity*. Academic Press.

Song, X., Qiao, L., Dou, X., Chang, J., Zhang, Y., & Xu, C. (2023). Selenium nanoparticles alleviate deoxynivalenol-induced intestinal epithelial barrier dysfunction by regulating endoplasmic reticulum stress in IPEC-J2 cells. *Toxicology*, *494*, 153593. https://doi.org/10.1016/j.tox.2023.153593

Summer, M., Ashraf, R., Ali, S., Bach, H., Noor, S., Noor, Q., Riaz, S., & Khan, R. R. M. (2024). Inflammatory response of nanoparticles: Mechanisms, consequences, and strategies for mitigation. *Chemosphere*, *363*, 142826. https://doi.org/10.1016/j.chemosphere.2024.142826

Suthar, J. K., Rakesh, B., Vaidya, A., & Ravindran, S. (2023). Comprehensive analysis of titanium oxide nanoparticle size and surface properties on neuronal PC-12 cells: Unraveling cytotoxicity, dopaminergic gene expression, and acetylcholinesterase inhibition. *Journal of Xenobiotics*, *13*(4), Article 4. https://doi.org/10.3390/jox13040043

Tang, Y., Zhou, X., Cao, T., Chen, E., Li, Y., Lei, W., Hu, Y., He, B., & Liu, S. (2022). Endoplasmic reticulum stress and oxidative stress in inflammatory diseases. *DNA and Cell Biology*, *41*(11), 924–934. https://doi.org/10.1089/dna.2022.0353

Tubatsi, G., Kebaabetswe, L. P., & Musee, N. (n.d.). *Proteomic Evaluation of Nanotoxicity in Aquatic Organisms: A Review.* https://doi.org/10.1002/pmic.202200008

Ventura, C., Torres, V., Vieira, L., Gomes, B., Rodrigues, A. S., Rueff, J., Penque, D., & Silva, M. J. (2022). New "omics" approaches as tools to explore mechanistic nanotoxicology. In H. Louro & M. J. Silva (Eds.), *Nanotoxicology in Safety Assessment of Nanomaterials* (pp. 179–194). Springer International Publishing. https://doi.org/10.1007/978-3-030-88071-2_8

von Ranke, N. L., Geraldo, R. B., Lima dos Santos, A., Evangelho, V. G. O., Flammini, F., Cabral, L. M., Castro, H. C., & Rodrigues, C. R. (2022). Applying *in silico* approaches to nanotoxicology: Current status and future potential. *Computational Toxicology*, *22*, 100225. https://doi.org/10.1016/j.comtox.2022.100225

Wang, F., Yu, L., Monopoli, M. P., Sandin, P., Mahon, E., Salvati, A., & Dawson, K. A. (2013). The biomolecular corona is retained during nanoparticle uptake and protects the cells from the damage induced by cationic nanoparticles until degraded in the lysosomes. *Nanomedicine: Nanotechnology, Biology and Medicine*, *9*(8), 1159–1168. https://doi.org/10.1016/j.nano.2013.04.010

Wu, H., Tang, L., Dong, H., Zhi, M., Guo, L., Hong, X., Liu, M., Xiao, Y., & Zeng, X. (2025). Shape and size dependence of pharmacokinetics, biodistribution, and toxicity of gold nanoparticles. *Molecular Pharmaceutics*, *22*(1), 196–208. https://doi.org/10.1021/acs.molpharmaceut.4c00832

Xia, T., Kovochich, M., Brant, J., Hotze, M., Sempf, J., Oberley, T., Sioutas, C., Yeh, J. I., Wiesner, M. R., & Nel, A. E. (2006). Comparison of the abilities of ambient and manufactured nanoparticles to induce cellular toxicity according to an oxidative stress paradigm. *Nano Letters*, *6*(8), 1794–1807. https://doi.org/10.1021/nl061025k

Xu, F., Piett, C., Farkas, S., Qazzaz, M., & Syed, N. I. (2013). Silver nanoparticles (AgNPs) cause degeneration of cytoskeleton and disrupt synaptic machinery of cultured cortical neurons. *Molecular Brain*, *6*(1), 29. https://doi.org/10.1186/1756-6606-6-29

Xu, P., Xu, J., Liu, S., Ren, G., & Yang, Z. (2012). In vitro toxicity of nanosized copper particles in PC12 cells induced by oxidative stress. *Journal of Nanoparticle Research*, *14*(6), 906. https://doi.org/10.1007/s11051-012-0906-5

Xuan, L., Ju, Z., Skonieczna, M., Zhou, P.-K., & Huang, R. (2023). Nanoparticles-induced potential toxicity on human health: Applications, toxicity mechanisms, and evaluation models. *MedComm*, *4*(4), e327. https://doi.org/10.1002/mco2.327

Yang, Y., Huang, J., Sun, Q., Wang, J., Huang, L., Fu, S., Qin, S., Xie, X., Ge, S., Li, X., Cheng, Z., Wang, X., Chen, H., Zheng, B., & He, Y. (2022). microRNAs: Key players in plant response to metal toxicity. *International Journal of Molecular Sciences*, *23*(15), Article 15. https://doi.org/10.3390/ijms23158642

Yao, Y., Zang, Y., Qu, J., Tang, M., & Zhang, T. (2019). The toxicity of metallic nanoparticles on liver: The subcellular damages, mechanisms, and outcomes. *International Journal of Nanomedicine*, *14*, 8787–8804. https://doi.org/10.2147/IJN.S212907

Yi, S., Shi, W., Wang, H., Ma, C., Zhang, X., Wang, S., Cong, B., & Li, Y. (2017). Endoplasmic reticulum stress PERK-ATF4-CHOP pathway is associated with hypothalamic neuronal injury in different durations of stress in rats. *Frontiers in Neuroscience*, *11*. https://doi.org/10.3389/fnins.2017.00152

Zhang, J., Guo, W., Li, Q., Wang, Z., & Liu, S. (2018a). The effects and the potential mechanism of environmental transformation of metal nanoparticles on their toxicity in organisms. *Environmental Science: Nano*, *5*(11), 2482–2499. https://doi.org/10.1039/C8EN00688A

Zhang, W., Xiao, B., & Fang, T. (2018b). Chemical transformation of silver nanoparticles in aquatic environments: Mechanism, morphology and toxicity. *Chemosphere*, *191*, 324–334. https://doi.org/10.1016/j.chemosphere.2017.10.016

Zheng, Q., Liu, H., Zhang, H., Han, Y., Yuan, J., Wang, T., Gao, Y., & Li, Z. (2023). Ameliorating mitochondrial dysfunction of neurons by biomimetic targeting nanoparticles mediated mitochondrial biogenesis to boost the therapy of Parkinson's disease. *Advanced Science*, *10*(22), 2300758. https://doi.org/10.1002/advs.202300758

Zhou, X., Jin, W., Sun, H., Li, C., & Jia, J. (2022). Perturbation of autophagy: An intrinsic toxicity mechanism of nanoparticles. *Science of the Total Environment*, *823*, 153629. https://doi.org/10.1016/j.scitotenv.2022.153629

Zia, S., Islam Aqib, A., Muneer, A., Fatima, M., Atta, K., Kausar, T., Zaheer, C.-N. F., Ahmad, I., Saeed, M., & Shafique, A. (2023). Insights into nanoparticles-induced neurotoxicity and cope up strategies. *Frontiers in Neuroscience*, *17*. https://doi.org/10.3389/fnins.2023.1127460

16 Toxicity Assessment of Nanomaterials Using Organs-on-Chips and 3D Cell Culture Systems

Hemavathi Brijesh, Brijesh Ajjappala, Pavithra S, and Prithvika R

16.1 INTRODUCTION

Nanotechnology is a multidisciplinary approach that integrates tools and technologies to examine, alter, and control systems that fall under nanoscale range, thereby enabling their application in nanoscience field using nanomaterials. Nanomaterials (NMs) are substances possessing size ranging from 1 to 100 nm, synthesized from materials belonging to various classes that display unique physicochemical properties due to their nanoscale dimension. However, the widespread use of nanotechnology in agriculture, health, and consumer products has raised concerns about its potential harmful health effects. Nanotoxicology, at the intersection of toxicology and nanomaterial science, studies the health hazards of NMs, which can translocate through the circulatory and lymphatic systems into vital organs like the liver, spleen, heart, brain, and placenta. NMs can cross the placental barrier, accumulating in fetal organs, primarily affecting the lungs and heart, mimicking air pollutants, causing pulmonary and systemic effects (Xuan et al., 2023). Nanotoxicity mechanisms include cytotoxicity, genotoxicity, ROS production, oxidative stress, inflammation, apoptosis, and cancer. Even FDA-approved nanotherapeutics carry some toxicity risks. Traditional 2D cell culture system lacks complex 3D interactions and rely on immortalized cell lines, limiting the toxicity predictions. While animal models provide systemic toxicity insights, physiological differences hinder the precise molecular mechanism studies. Organ-on-a-chip systems bridge gaps between 2D cultures, animal models, and human studies, improving nanoparticle (NP) toxicity screening (Lu and Radisic, 2021). Human organoids, closely mimicking real organs, enhance in vitro biological studies. 3D cultures show more physiological relevance and differentiation potential than traditional models. Toxicology evaluation employing organoids and 3D cultures provides deeper insights into chemical toxicity and improves prediction accuracy for human responses (Park et al., 2020). In this chapter, we aim to present a summary of the analysis of toxicity caused by nanoparticles derived from nanomaterials employing organ-on-a-chip and human organoid model systems.

16.2 PHYSICOCHEMICAL PROPERTIES OF NMS INFLUENCING TOXICITY

Engineered NMs (ENMs) are widely used in various fields such as agriculture, medical, pharmaceutical, cosmetics, food, and electronics due to their potential benefits. Nanoparticles possess unique properties such as high surface area, high reactivity, and quantum effects, with distinct biological properties which make them technologically advantageous. ENM toxicity is not solely determined by their composition but by their physicochemical properties, including size, shape, solubility, surface charge, and chemical composition. Unraveling the role of these parameters during nanoparticle biological interactions is required for their design and implementation in nanotechnology. However, concerns related to impact of NM toxicity on living organisms, including humans and the environment, are alarming, as their production and application are at the peak. Table 16.1 summarizes the important physicochemical properties that influence NM toxicity (Fröhlich, 2012).

16.2.1 SIZE AND SHAPE

16.2.1.1 Nanoparticle Size

NP size plays a major role in determining their biological interactions in various reactions. NPs less than 100 nm in size can easily cross biological membranes and are then

TABLE 16.1
Physicochemical Properties of NMs That Influence Toxicity (Fröhlich, 2012)

Properties of Nanomaterials	Influence on Toxicity
Size and shape	Determines cellular uptake, clearance, and inflammatory responses
Surface charge	Affects interactions with cell membranes and protein corona formation
Solubility	Influences ion release, bioavailability, and persistence in tissues
Chemical composition	Governs inherent reactivity, degradability, and oxidative stress induction

DOI: 10.1201/9781003644439-16

taken up by cells through endocytosis or passive diffusion. Further, smaller NPs possess a high surface area-to-volume ratio, making them increasingly reactive. This results in oxidative stress, DNA damage, and inflammation due to excess production of reactive oxygen species. Additionally, the size of the NPs affects their biodistribution and clearance within the living system. For example, NPs with 10-nm size could be readily filtered by the kidneys, while larger NPs get trapped by the reticulo-endothelial system, particularly in the liver and spleen (Oberdörster et al., 2005). This also raises concern about small NPs crossing the blood-brain barrier, eventually leading to neurotoxicity. Silver (Ag) NPs (10 nM) are distributed efficiently, leading to elevated hepatobiliary toxicity due to their small size (Xuan et al., 2023).

16.2.1.2 Nanoparticle Shape and Aspect Ratio

NPs are fabricated into various shapes including nanotubes, nanospheres, and nanorods, which can influence toxicity in vivo (Liu et al., 2022). The shape of NPs, often expressed as an aspect ratio (length-to-width), will have a significant influence during their biological interactions. Spherical NPs are typically internalized more readily by cells, whereas high aspect-ratio shapes such as rods, wires, or fibers can induce mechanical disruption of membranes or interfere with cytoskeleton structures. For instance, carbon nanotubes have been observed to cause phagocytosis in macrophages due to their needle-like morphology, leading to chronic inflammation (Donaldson et al., 2006). Furthermore, NP shape also affects circulation time and immune system recognition. Rod-shaped or disk-like NPs may exhibit prolonged systemic circulation but also have a higher propensity for aggregation or accumulation in tissues, which could lead to long-term toxicity.

16.2.2 Surface Charge

16.2.2.1 Zeta Potential and Membrane Interaction

Surface charge, represented by zeta potential, governs electrostatic interactions between NPs and cellular components. The internalization of NPs occurs due to the strong attraction between the positively charged NPs and the negatively charged phospholipid bilayers of cells leading to the cellular toxicity through cell membrane damage (Fröhlich, 2012). NPs cationic property causes a change in membrane integrity, thus resulting in intracellular content leakage, which in turn triggers apoptosis or necrosis in the tissues. Conversely, negatively charged or neutral NPs show reduced cellular uptake and also enhance colloidal stability with reduced immunological reactions. Positively charged NMs exhibiting higher cellular internalization was attributed to their stronger electrostatic interaction with the cell membrane that was negatively charged (Liu et al., 2022). Such types of NPs often evade rapid clearance and prolong their circulation period, leading to increased systemic exposure.

16.2.2.2 Influence on Protein Corona Formation

NPs are coated with a protein layer, and this process is known as the "protein corona". The characteristics of corona formation mainly depend on the surface charge that also influences the recognition of NPs by cells and immune receptors (Monopoli et al., 2012). The surface of positively charged NPs is attracted to different types of plasma proteins when compared to neutral or negatively charged ones. This results in variation in the biodistribution, opsonization, cellular uptake, and subsequent toxicity of NPs (Bai et al., 2021).

16.2.3 Solubility and Dissolution

16.2.3.1 Ion Release and Acute Toxicity

The bioavailability and extent of NP toxicity mainly depends on their solubility. For instance, metal based NPs, such as silver (AgNPs), cadmium-based quantum dots, and zinc oxide (ZnO), release toxic metal ions upon dissolution and interact with the thiol groups in cells, causing the disruption of the mitochondrial activity, interfering with enzymatic function, eventually resulting in oxidative damage and cellular toxicity (Lok et al., 2006). For example, the release of Ag^{+}ions from silver NPs induces toxicity by binding to sulfur and phosphorus containing biomolecules, causing disruption in cellular respiration and DNA replication. Further, NPs such as nano-Ag and nano-CuO (copper oxide) are involved in the dissolution of toxic ions, causing severe damage to biological systems (Sukhanova et al., 2018).

16.2.3.2 Persistence and Chronic Effects

NPs with poor solubility may result in their accumulation in tissues over a prolonged period of time, leading to chronic effects. For instance, titanium dioxide (TiO_2) and cerium oxide (CeO_2) NPs, although considered chemically stable, have exhibited inflammation and tissue damage in organs due to retention over long period of time (Auffan et al., 2009). However, NP solubility is affected by various factors such as pH, salt concentration, and the presence of proteins, all of which influence dissolution kinetics and bioavailability.

16.2.4 Chemical Composition

16.2.4.1 Core Material and Intrinsic Toxicity

NP properties such as reactivity and biological interactions are mainly dependent on their core chemical makeup. Various levels of oxidative potential influencing ROS generation and cytotoxicity are exhibited with different types of metallic NPs such as CuO, iron oxide (Fe_2O_2), and TiO_2 (Lynch et al., 2007). Similarly, semiconductor-based quantum dots could release toxic heavy metals like cadmium or selenium under oxidative or photolytic conditions. Further, polymorphic forms of the same NP (e.g., anatase vs. rutile TiO_2) can exhibit different toxicity patterns due to their surface reactivity and photocatalytic activity differences. Nanotoxicity can also result due to organic contaminants like endotoxins or lipopolysaccharides (LPSs), which resist sterilization and bind to NPs. NPs conjugated with large amounts of endotoxins could activate Toll-like receptor 4 (TLR4) present on various immune cells, triggering proinflammatory cytokine release (Himly et al., 2020).

16.2.4.2 Surface Functionalization and Ligands

Surface coatings and functional groups significantly impact NP behavior. Functionalization with polyethylene glycol (PEG), for instance, can enhance hydrophilicity and reduce nonspecific protein binding, thereby minimizing toxicity and immune recognition. However, certain functional groups, particularly cationic amines, may increase cellular uptake at the expense of elevated cytotoxicity. Modification of NPs with various surface ligands can activate endocytosis and later navigate through cytoplasm and get terminated in endoplasmic reticulum and golgi complex (Blanco et al., 2015; Donahue et al., 2019). Impurities or byproducts from the synthesis process may also contribute to toxicity. Residual surfactants, metal catalysts, or solvents can provoke inflammatory or genotoxic responses if not adequately removed during manufacturing.

16.3 NM EXPOSURE ROUTES INTO HUMANS

16.3.1 Inhalation

Inhalation represents a critical route of exposure to airborne NMs, particularly in occupational environments such as manufacturing or research facilities. Upon inhalation, NMs can settle in along the linings of the respiratory tract, influenced by their size, shape, and surface chemistry. Ultrafine particles, measuring less than 100 nm, have the potential to penetrate deeply into the alveolar region and may even migrate into the bloodstream (Geiser and Kreyling, 2010). For example, in vivo studies conducted for respiratory toxicity studies using TiO_2 NPs showed they could induce fibrosis and lung inflammation in animal models (Ma et al., 2019).

16.3.2 Dermal Contact

Dermal exposure to NMs may occur during the application of cosmetics or sunscreens or while handling products containing NMs. The degree of skin penetration is affected by factors such as particle size, surface coating, and the condition of the skin. Research has indicated that certain NPs, including TiO_2, are capable of infiltrating the viable epidermis, particularly when the skin is compromised (Nohynek et al., 2008).

16.3.3 Ingestion

Ingestion of NMs can happen indirectly through contaminated food, water, or hand-to-mouth behaviors, notably in children. The ingested NMs may interact with the gastrointestinal tract, leading to intestinal barrier crossing, and enter systemic circulation (Bouwmeester et al., 2009). This type of exposure pathway is found in food packaging materials and food additives made up of NMs (EFSA Scientific Committee et al., 2018).

16.3.4 Injection and Medical Applications

Various medical applications such as drug delivery systems, imaging, and diagnostics cause the introduction of NMs intentionally into the human body directly. The introduction of NMs through intravenous, intramuscular, or subcutaneous injections causes the crossing of physiological barriers and therefore requires thorough biocompatibility evaluations (Rzigalinski and Strobl, 2009).

16.4 NM EXPOSURE ROUTES INTO THE ENVIRONMENT

NMs infiltrate the environment during synthesis, usage, and disposal during their life cycle. The released NMs are then transported via air, water, and soil, posing a threat to ecological systems (Gottschalk et al., 2009).

16.4.1 Airborne Release of NMs

NMs are released into the environment and reach the atmosphere as aerosols or particulate matter during their production, handling, or combustion stages (Keller et al., 2013). These airborne NMs then travel long distances and settle in soil and water bodies, causing widespread exposure in the environment.

16.4.2 Waterborne Release of NMs

The exposure route of NMs into the aquatic environments occurs through industrial and domestic wastewater discharge. For instance, AgNPs used in textile and personal care product production join the drain after the washing process, thus accumulating in sewage sludge (Mueller and Nowack, 2008). These NMs later impact aquatic organisms when the sewage sludge runs off to the aquatic environment, thus causing oxidative stress and inducing genotoxicity (Handy et al., 2012).

16.4.3 Soil Contamination by NMs

NM contamination in the soil is mostly due to the application of biosolids or wastewater used for irrigation purposes, resulting in the disruption of normal microflora and altered nutrient cycles. For instance, ZnO NPs have displayed effects on root growth and altered microbial enzyme activity within soil ecosystems contaminated with NPs (Raliya et al., 2015).

16.4.4 Bioaccumulation and Trophic Transfer of NMs

NMs could have a severe impact in the ecosystem when they are transferred across different trophic levels after bioaccumulation among various types of organisms. Both terrestrial and aquatic organisms such as plants, invertebrates, and fish readily accumulate carbon nanotubes and metal-based NMs, causing concern about their long-term ecological outcome.

16.5 TOXICITY MECHANISMS OF NMS

Current advancements in the field of NMs analytical techniques like transmission electron microscopy and atomic force microscopy have allowed researchers to examine NMs interaction with cells more precisely. The studies conducted have revealed that NMs can induce genotoxicity, oxidative stress, inflammation, and cytotoxicity in different cell types (Xuan et al., 2023).

16.5.1 NM-Induced ROS Accumulation

The balanced oxidative–antioxidant system is found to be disrupted by excess ROS accumulation, leading to a series of events such as oxidative stress, cellular damage, and diseases. Previous research found that blood–brain barrier dysfunction and oxidative stress occurred due to silica NP exposure (Liu et al., 2017). It was reported that polystyrene NPs (PS-NPs) induce ROS-mediated apoptosis in the fetal thalamus, indicating reproductive toxicity of NPs (Yang et al., 2022). These findings highlight the potential toxicity of various NPs through ROS generation, emphasizing the need for further evaluation of their biological impacts.

16.5.2 NM-Induced Mitochondrial Damage

NMs exposure could cause mitochondrial damage, leading to dysfunction, apoptosis, and altered cellular physiology. Studies show nanocomposites (NCs) disrupt mitochondrial Ca^{2+} buffering, leading to calcium overload and membrane damage. Huang et al. (2022a) found that ZnO-NPs reduced mitochondrial membrane potential and increased cytochrome C release, triggering apoptosis. Fu et al. (2022) reported that amino-functionalized polystyrene NPs induced mitochondrial damage in endothelial cells.

16.5.3 NM-Induced Inflammation

Inflammation is a response to harmful stimuli, but NPs like PS-NPs, silica NPs, and carbon NPs can induce inflammatory damage. Wu et al. (2022) found that PS-NPs activate the TLR4/NOX2 pathway, causing oxidative stress and Th1/Th2 imbalance in carp myocardial tissue. Tang et al. (2022) reported that PS nanoplastics exacerbated lipopolysaccharide-induced spleen inflammation via the ROS/MAPK pathway. Additionally, nano-TiO is linked to genotoxicity, membrane damage, and oxidative stress. An et al. (2018) reported that high-dose graphene oxide exposure caused intraocular inflammation, corneal apoptosis, and iris neovascularization in mice, highlighting the inflammatory risks posed by environmental NPs.

16.5.4 NM-Induced Apoptosis

Research shows that NP exposure induces apoptosis through various pathways. Xu et al. (2022) found that ZnO-NPs trigger endoplasmic reticulum (ER) stress via the Keap/Nrf2 pathway, causing apoptosis in ovarian cells. Similarly, Zhang et al. (2019) reported that silica NPs activate the LC3 and Bax pathways, while melatonin inhibits SNP-induced apoptosis. Further, Li et al. (2022) reported that nanoplastics could disrupt ion transport and induce apoptosis in Japanese giant short gill cells.

16.5.5 NM-Induced DNA Damage

DNA damage, a key factor in cancer and aging-related diseases, can be induced by NP exposure. TiO_2 NPs, with a mild cytotoxic nature, generate ROS and oxidative stress, leading to DNA damage. Ultra-small super-paramagnetic iron oxide NPs were found to disrupt the oxidative balance in the mouse heart, causing increased production of oxidative stress markers and DNA damage. Shi et al. (2015) found that TiO_2-NPs inhibit the Nrf2 pathway, inducing oxidative DNA damage in HepG2 cells and mice. Similarly, Mo et al. (2021) showed nickel NPs were found to cause DNA damage and impair repair mechanisms via the HIF-1α/miR-210/Rad52 pathway, inturn increasing the phosphorylation of key DNA damage response proteins like p53 and H2AX. These results were consistent with earlier research findings conducted on genetic instability caused by titania NPs.

16.6 MODE OF NM TOXICITY ASSESSMENT

16.6.1 Organ-on-a-Chip

16.6.1.1 Lung-on-a-Chip

Silica NP exposure induced higher cellular toxicity under cyclic motion, increasing ROS production and NP translocation compared to static conditions. This study highlighted NP toxicity in the lungs through breathing motion amplification, unlike traditional 2D culture system. An organ-on-a-chip model replicating the alveolus–capillary interface under cyclic mechanical strain was previously developed (Huh et al., 2012). This system consists of two polydimethylsiloxane (PDMS) microfluidic channels which are separated by a porous extracellular matrix (ECM) coated membrane. Human epithelial cells are cultured at the air–liquid interface, while microvascular endothelial cells are grown in the basal compartment. Mechanical strain is controlled via lateral side chambers with vacuum modulation to simulate breathing. To simulate airborne pollution model, 12 nm silica NPs are perfused into the alveolar compartment and left as a surface layer on the epithelial cells. This type of platform provides a true-to-type physiologically relevant model for studying nanoparticle-induced lung toxicity. In a recent study, a device was designed using an alveoli based epithelium-on-a-chip composed of lung cancer cells possessing an inbuilt sensor to monitor zinc-quantum dots conjugated with human serum albumin NP. The results revealed that a significant amount of NP internalization occurred during co-culturing of cancer cells (Meghani et al., 2020).

16.6.1.2 Heart-on-a-Chip

Previously, studies have evaluated NP toxicity using heart-on-a-chip systems, despite the link between ultraprocessed foods (UPFs) and cardiovascular disease. Ahn et al. (2018) used a cell-sheet culture system combined with bioelectronics to monitor cardiac contractions in real time. TiO_2 NP exposure reduced the contractile force and diminished the calcium propagation temporarily by disrupting the sarcomere structures, which are the fundamental contractile units responsible for heart function. Recent study demonstrated that the bioelectronic heart-on-a-chip model developed exhibited the ability to regulate the oxygen level by microfluidic channel. The researchers integrated the bioelectronic devices and successfully monitored acute hypoxia through the cardiac electrophysiology responses (Liu et al., 2020).

16.6.1.3 Liver-on-a-Chip

Li et al. (2019) cultured primary rat hepatocytes in a microfluidic system and observed higher albumin and urea production after 72 hours, indicating enhanced functionality. However, exposure to superparamagnetic iron oxide NPs significantly reduced these markers, suggesting potential liver injury. Esch et al. (2016) developed a gut-liver model using Caco-2/HT29-MTX cells for the gut and HepG2/C3A cells for the liver to simulate NP toxicity under ingestion conditions. A significant higher amount of aspartate aminotransferase (AST) release was found in the gut-liver chip when compared to the gut chip alone when treated with carboxylated PS NPs (50 nm), highlighting inter-organ crosstalk in NP-mediated toxicity. These results indicate that organ-on-a-chip models improve understanding of liver toxicity mechanisms.

16.6.2 NM Toxicity Assessment Using Organoid Model

Organoids are 3D cell aggregates derived from embryonic stem cells (ESCs), induced pluripotent stem cells (iPSCs), or adult stem cells, replicating physiological and genomic features of tissues or organs. They originate from diverse cell types, including cancer and primary tissue cells. Subject-specific or multi-organoid models offer fast, reliable nanotoxicity screening, surpassing conventional methods. Organoids enhance high-throughput in vitro toxicity assessments, mimicking in vivo tissue structures and microenvironments. They bridge gaps between animal and cell models, aiding embryonic organ development studies. Organoids have successfully modeled early development of the human brain, retina, heart, liver, kidney, lung, and intestines. Table 16.2 summary of organoid systems used for NM study.

A 3D proximal tubule kidney organoid was used to assess nephrotoxicity of hydroxylated generation-5 PAMAM dendrimer (G5-OH) and gold NPs (<6 nm). G5-OH dendrimer induced dose-dependent cytotoxicity, with 20% cell death at 0.8 mg/ml, but showed minimal lysosomal enzyme NAG upregulation. Kim-1 and TNF-α (tumor necrosis factor) were elevated at 0.675 mg/ml, indicating mild toxicity. Unlike nephrotoxic cisplatin, G5-OH caused limited cytokine elevation. Gold NPs adhered to the matrix, restricting toxicity. In vivo, G5-OH had minimal impact on kidney function, aligning with organoid results. The 3D model effectively mimics in vivo nephrotoxicity compared to 2D cultures (Astashkina et al., 2014).

NPs such as silicon dioxide (SiO_2) and TiO_2 were found to be cytotoxic when their toxic effect was analyzed using human colon organoids, which was similar to the in vivo result (Park et al., 2020). Further, one study demonstrated that accumulation of high levels of Ag-NPs (0.5 g/mL) in cerebral organoids exhibited reduced cell proliferation, shunting neuronal growth, and triggered the apoptosis suggesting that Ag NPs can be a latent congenital risk factor (Huang et al., 2022b).

Hepatocyte-based organoids and 3D monocultures are commonly used for nanomaterial-induced hepatotoxicity assessments, often overlooking nonparenchymal cells. Kermanizadeh et al. (2019) emphasized the critical role of Kupffer cells in hepatic immune responses and nanotoxicity regulation. Their 3D hepatocyte co-culture microtissue (MT) model exhibited higher nanotoxicity than monocultures when exposed to ZnO, Ag, multiwalled carbon nanotubes, and TiO_2, highlighting inter-individual variation in toxicity response. The model effectively assessed long-term nanotoxicity from repeated low-dose exposure, showing minimal cell death over 3 weeks, but showed aging effects beyond that period. Despite its advantages, the model had limitations, including failure of AST to indicate subtle toxicity and inability to mimic in vivo liver regeneration within a short recovery period. However, it remains a useful tool for studying nanomaterial-induced hepatotoxicity, particularly for prolonged exposure scenarios, but has constraints in completely replicating in vivo hepatic responses. Similarly, an organoid model derived from gastric cancer patients was used to assess the antitumor effect of nab-paclitaxel, a nano-drug used in various cancers. It showed the lowest IC50 compared to 5-FU and epirubicin in all three gastric cancer organoid models, with apoptosis increasing after 48 hours of exposure (Xiao et al., 2020).

The blood–brain barrier (BBB) blocks most conventional drugs, but nanodrugs can penetrate it to treat CNS disorders. However, NP-induced neurotoxicity requires proper evaluation. In vivo models differ from humans, and

TABLE 16.2
Nanomaterial Toxicity Assessment Using Various Organoid Models (Prasad et al., 2021)

Type of Organoid Used to Study Nanomaterial Toxicity	NPs Causing Toxicity	Models Used for Toxicity Study
Liver organoid	ZnO-NPs	Human hepatocyte (L02)
	Ti-NPs C Ag-NPs	Male Wistar rats
	Ag-NPs, ZnO-NPs	Primary liver cells of mice
	Cu-NPs	Male SD rats
	NiO-NPs	Male Wistar rats
	Silica	HepG2
	Iron oxide	Human hepatocellular carcinoma cells
Kidney organoid	ZnO-NPs	Human embryonic kidney (HEK293) cells
	Dendrimers	COS-7 (African green monkey kidney)
	CuO NPs	HEK293
	Fullerenes	HEK293
	QD	HEK293
Lung organoid	MWCNTs with carbonyl (CdO), carboxyl (COOH), hydroxyl (OH)	Human lung-tumor cell lines (H596, H446, and Calu-1)
	MWCNTs with pristine	Human embryonic lung fibroblasts (IMR-90)
	Smaller CNTs	Fetal lung tissue
	Cerium oxide(CeO_2) NPs	Human lung epithelial cells (BEAS-2B)
	Chitosan nanoparticle	fibroblasts)A549 lung epithelial cells
	PLGA-NPs	Lung bronchial epithelial cells (BEAS-2B cells)
	Liposomes	Male Han Wistar rats
Brain organoid	Aluminum oxide NPs	Human brain microvascular endothelial cells (HBMVEC)
	Cadmium telluride quantum dots	Human neuroblastoma cell
	Carbon (C60)-NPs	Neuronal human astrocyte
	Dendrimers	N2a (mouse neuroblastoma cells), mHippoE-18 (mouse embryonic hippocampal cells), primary neural cell cultures, intracerebroventricular injection in mice
	Copper(II)-conjugated phosphorus dendrimers	U87 (human glioblastoma–astrocytoma, epithelial-like)
	Fe_2O_3-NPs	Growing neuron cell line PC12
Cardiac organoid	TiO_2-NPs	Wistar albino rats, female ICR mice
	ZnO NPs	CD-ICR mice, Sprague–Dawley rats, Wistar albino rat
	MWCNTs	Male C57BL/6J mice
	Silica NPs	Sprague–Dawley rats, zebrafish, Wistar rats
	TiO_2 NPs, ultrafine titanium dioxide (UFTiO (2)	Rats
	Iron oxide NPs	Human cardiac microvascular endothelial cells (HCMECs)
	Ultra-small superparamagnetic iron oxide nanoparticles	Mice
Intestinal organoid	Zinc oxide	Human colon carcinoma cells
	Dendrimers (PAMAM)	Caco-2 (colon adenocarcinoma cells) SW480 (primary adenocarcinoma cells of colon)
	Copper(II)-conjugated phosphorus dendrimers	HCT116 (human colon cancer)
	ZnO-NPs	Human intestinal Caco-2 and SW480 cells
	CuO NPs	Rat small intestine epithelial cells (IEC-6)
	Liposomal 5-fluoro-20-deoxyuridine (FUdR)-dipalmitate	Mice

2D cultures fail to replicate neural complexity. Human brain organoids and spheroids offer effective alternatives for nanotoxicity assessment. iPSC-derived brain spheroids and 3D LUHMES dopaminergic neuron models were used to evaluate sodium citrate, polyethylene glycol-functionalized gold NPs, and polylactic acid (PLA) NPs. All NPs were internalized; however, AuNP uptake in LUHMES was delayed. Mitochondrial potential analysis showed acute NP response in brain spheroids, while LUHMES sensitivity increased with dose and time. Au-PEG strongly downregulated cytokines in LUHMES but had a moderate effect on brain spheroids. Both AuNPs and PLA NPs induced toxicity in a dose-dependent manner. The brain spheroid model, containing glia for neuronal support, was considered to be suitable for general neurotoxicity assessments (Leite et al., 2019).

16.7 CONCLUSION

Since the inception of toxicological research, well-plate assays and animal models have served as foundational methods for assessing NP toxicity. However, their limitations in accurately predicting human health outcomes have raised concerns. In response, the Food and Drug Administration (FDA) and other global regulatory agencies have turned their attention to organ-on-a-chip technology, which offers a more physiologically relevant in vitro environment for cells, tissues, and organs. This approach enhances cellular function and responsiveness to drug exposure, making it a promising tool for preclinical drug screening.

Organ-on-a-chip systems hold considerable promise for enhancing drug toxicity research, despite ongoing challenges such as material optimization, cell source selection, scalability, and validation. Providing consistent data that correlates to clinical outcomes is one of their main goals. This could lead to a major change in toxicity testing if it is successful. Despite their growing use in drug efficacy testing, their relevance in nanotoxicology is limited due to the complexities of NP behavior and lack of standardized methodologies for NP synthesis and administration. Simultaneously, organoid technology is rapidly evolving as a reliable platform for toxicity testing and drug screening, overcoming many of the shortcomings of previous models. The potential of organoid technology includes disease modeling and regenerative and personalized medicines. Overall, the combination of organoid models and nanotechnology holds enormous potential for accelerating the development of nano-based medications, vaccines, and diagnostics, hence expanding biomedicine's frontiers and driving precision healthcare innovation.

REFERENCES

Ahn, S., Ardoña, H. A. M., Lind, J. U., Eweje, F., Kim, S. L., Gonzalez, G. M., Liu, Q., Zimmerman, J. F., Pyrgiotakis, G., Zhang, Z., Beltran-Huarac, J., Carpinone, P., Moudgil, B. M., Demokritou, P., & Parker, K. K. (2018). Mussel-inspired 3D fiber scaffolds for heart-on-a-chip toxicity studies of engineered nanomaterials. *Analytical and Bioanalytical Chemistry*, *410*(24), 6141–6154. https://doi.org/10.1007/s00216-018-1106-7

An, W., Zhang, Y., Zhang, X., Li, K., Kang, Y., Akhtar, S., Sha, X., & Gao, L. (2018). Ocular toxicity of reduced graphene oxide or graphene oxide exposure in mouse eyes. *Experimental Eye Research*, *174*, 59–69. https://doi.org/10.1016/j.exer.2018.05.024

Astashkina, A. I., Jones, C. F., Thiagarajan, G., Kurtzeborn, K., Ghandehari, H., Brooks, B. D., & Grainger, D. W. (2014). Nanoparticle toxicity assessment using an in vitro 3-D kidney organoid culture model. *Biomaterials*, *35*(24), 6323–6331. https://doi.org/10.1016/j.biomaterials.2014.04.060

Auffan, M., Rose, J., Bottero, J. Y., Lowry, G. V., Jolivet, J. P., & Wiesner, M. R. (2009). Towards a definition of inorganic nanoparticles from an environmental, health and safety perspective. *Nature Nanotechnology*, *4*(10), 634–641. https://doi.org/10.1038/nnano.2009.242

Bai, X., Wang, J., Mu, Q., & Su, G. (2021). In vivo protein corona formation: Characterizations, effects on engineered nanoparticles' biobehaviors, and applications. *Frontiers in Bioengineering and Biotechnology*, *9*, 646708. https://doi.org/10.3389/fbioe.2021.646708

Blanco, E., Shen, H., & Ferrari, M. (2015). Principles of nanoparticle design for overcoming biological barriers to drug delivery. *Nature Biotechnology*, *33*(9), 941–951. https://doi.org/10.1038/nbt.3330

Bouwmeester, H., Dekkers, S., Noordam, M. Y., Hagens, W. I., Bulder, A. S., De Heer, C., Ten Voorde, S. E., Wijnhoven, S. W., Marvin, H. J., & Sips, A. J. (2009). Review of health safety aspects of nanotechnologies in food production. *Regulatory Toxicology and Pharmacology*, *53*(1), 52–62. https://doi.org/10.1016/j.yrtph.2008.10.008

Donahue, N. D., Acar, H., & Wilhelm, S. (2019). Concepts of nanoparticle cellular uptake, intracellular trafficking, and kinetics in nanomedicine. *Advanced Drug Delivery Reviews*, *143*, 68–96. https://doi.org/10.1016/j.addr.2019.04.008

Donaldson, K., Aitken, R., Tran, L., Stone, V., Duffin, R., Forrest, G., & Alexander, A. (2006). Carbon nanotubes: A review of their properties in relation to pulmonary toxicology and workplace safety. *Toxicological Sciences*, *92*(1), 5–22. https://doi.org/10.1093/toxsci/kfj130

EFSA Scientific Committee, Hardy, A., Benford, D., Halldorsson, T., Jeger, M. J., Knutsen, H. K., More, S., Naegeli, H., Noteborn, H., Ockleford, C., & Mortensen, A. (2018). Guidance on risk assessment of the application of nanoscience and nanotechnologies in the food and feed chain: Part 1, human and animal health. *EFSA Journal*, *16*(7), e05327. https://doi.org/10.2903/j.efsa.2018.5327

Esch, M. B., Ueno, H., Applegate, D. R., & Shuler, M. L. (2016). Modular, pumpless body-on-a-chip platform for the coculture of GI tract epithelium and 3D primary liver tissue. *Lab on a Chip*, *16*(14), 2719–2729. https://doi.org/10.1039/c6lc00461j

Fröhlich, E. (2012). The role of surface charge in cellular uptake and cytotoxicity of medical nanoparticles. *International Journal of Nanomedicine*, 5577–5591. https://doi.org/10.2147/IJN.S36111

Fu, Y., Fan, M., Xu, L., Wang, H., Hu, Q., & Jin, Y. (2022). Amino-functionalized polystyrene nano-plastics induce mitochondria damage in human umbilical vein endothelial cells. *Toxics*, *10*(5), 215. https://doi.org/10.3390/toxics10050215

Geiser, M., & Kreyling, W. G. (2010). Deposition and biokinetics of inhaled nanoparticles. *Particle and Fibre Toxicology*, *7*, 1–17. https://doi.org/10.1186/1743-8977-7-2

Gottschalk, F., Sonderer, T., Scholz, R. W., & Nowack, B. (2009). Modeled environmental concentrations of engineered nanomaterials (TiO2, ZnO, Ag, CNT, fullerenes) for different regions. *Environmental Science & Technology*, *43*(24), 9216–9222. https://doi.org/10.1021/es9015553

Handy, R. D., Cornelis, G., Fernandes, T., Tsyusko, O., Decho, A., Sabo-Attwood, T., Metcalfe, C., Steevens, J. A., Klaine, S. J., Koelmans, A. A., & Horne, N. (2012). Ecotoxicity test methods for engineered nanomaterials: Practical experiences and recommendations from the bench. *Environmental Toxicology and Chemistry*, *31*(1), 15–31. https://doi.org/10.1002/etc.706.

Himly, M., Geppert, M., Hofer, S., Hofstätter, N., Horejs-Höck, J., & Duschl, A. (2020). When would immunologists consider a nanomaterial to be safe? Recommendations for planning studies on nanosafety. *Small*, *16*(21), 1907483. https://doi.org/10.1002/smll.201907483

Huang, C., Lin, B., Chen, C., Wang, H., Lin, X., Liu, J., Ren, Q., Tao, J., Zhao, P., & Xu, Y. (2022a). Synergistic reinforcing of immunogenic cell death and transforming tumor-associated macrophages via a multifunctional cascade bioreactor for optimizing cancer immunotherapy. *Advanced Materials*, *34*(51), 2207593. https://doi.org/10.1002/adma.202207593.

Huang, Y., Guo, L., Cao, C., Ma, R., Huang, Y., Zhong, K., Gao, H., Huang, Y., & Bu, Q. (2022b). Silver nanoparticles exposure induces developmental neurotoxicity in hiPSC-derived cerebral organoids. *Science of the Total Environment*, *845*, 157047. https://doi.org/10.2147/IJN.S406184

Huh, D., Leslie, D. C., Matthews, B. D., Fraser, J. P., Jurek, S., Hamilton, G. A., Thorneloe, K. S., McAlexander, M. A., & Ingber, D. E. (2012). A human disease model of drug toxicity–induced pulmonary edema in a lung-on-a-chip microdevice. *Science Translational Medicine*, *4*(159), 159ra147–159ra147. https://doi.org/10.1126/scitranslmed.aau4555.

Keller, A. A., McFerran, S., Lazareva, A., & Suh, S. (2013). Global life cycle releases of engineered nanomaterials. *Journal of Nanoparticle Research*, *15*, 1–17. https://doi.org/10.1007/s11051-013-1692-4

Kermanizadeh, A., Berthing, T., Guzniczak, E., Wheeldon, M., Whyte, G., Vogel, U., Moritz, W., & Stone, V. (2019). Assessment of nanomaterial-induced hepatotoxicity using a 3D human primary multi-cellular microtissue exposed repeatedly over 21 days-the suitability of the in vitro system as an in vivo surrogate. *Particle and Fibre Toxicology*, *16*, 1–14. https://doi.org/10.1186/s12989-019-0326-0

Leite, P. E. C., Pereira, M. R., Harris, G., Pamies, D., Dos Santos, L. M. G., Granjeiro, J. M., Hogberg, H. T., Hartung, T., & Smirnova, L. (2019). Suitability of 3D human brain spheroid models to distinguish toxic effects of gold and poly-lactic acid nanoparticles to assess biocompatibility for brain drug delivery. *Particle and Fibre Toxicology*, *16*, 1–20. https://doi.org/10.1186/s12989-019-0307-3

Li, L., Gokduman, K., Gokaltun, A., Yarmush, M. L., & Usta, O. B. (2019). A microfluidic 3D hepatocyte chip for hepatotoxicity testing of nanoparticles. *Nanomedicine*, *14*(16), 2209–2226. https://doi.org/10.2217/nnm-2019-0086

Li, Y., Liu, Z., Jiang, Q., Ye, Y., & Zhao, Y. (2022). Effects of nanoplastic on cell apoptosis and ion regulation in the gills of Macrobrachium nipponense. *Environmental Pollution*, *300*, 118989. https://doi.org/10.1016/j.envpol.2022.118989

Liu, H., Bolonduro, O. A., Hu, N., Ju, J., Rao, A. A., Duffy, B. M., Huang, Z., Black, L. D., & Timko, B. P. (2020). Heart-on-a-chip model with integrated extra-and intracellular bioelectronics for monitoring cardiac electrophysiology under acute hypoxia. *Nano Letters*, *20*(4), 2585–2593. https://doi.org/10.1021/acs.nanolett.0c00076

Liu, X., Sui, B., & Sun, J. (2017). Blood-brain barrier dysfunction induced by silica NPs in vitro and in vivo: Involvement of oxidative stress and Rho-kinase/JNK signaling pathways. *Biomaterials*, *121*, 64–82. https://doi.org/10.1016/j.biomaterials.2017.01.006

Liu, Y., Zhu, S., Gu, Z., Chen, C., & Zhao, Y. (2022). Toxicity of manufactured nanomaterials. *Particuology*, *69*, 31–48. https://doi.org/10.1016/j.partic.2021.11.007

Lok, C. N., Ho, C. M., Chen, R., He, Q. Y., Yu, W. Y., Sun, H., Tam, P. K. H., Chiu, J. F., & Che, C. M. (2006). Proteomic analysis of the mode of antibacterial action of silver nanoparticles. *Journal of Proteome Research*, *5*(4), 916–924. https://doi.org/10.1021/pr0504079

Lu, R. X. Z., & Radisic, M. (2021). Organ-on-a-chip platforms for evaluation of environmental nanoparticle toxicity. *Bioactive Materials*, *6*(9), 2801.

Lynch, I., Cedervall, T., Lundqvist, M., Cabaleiro-Lago, C., Linse, S., & Dawson, K. A. (2007). The nanoparticle–protein complex as a biological entity; a complex fluids and surface science challenge for the 21st century. *Advances in Colloid and Interface Science*, *134*, 167–174. https://doi.org/10.1016/j.cis.2007.04.021

Ma, Y., Guo, Y., Ye, H., Huang, K., Lv, Z., & Ke, Y. (2019). Different effects of titanium dioxide nanoparticles instillation in young and adult mice on DNA methylation related with lung inflammation and fibrosis. *Ecotoxicology and Environmental Safety*, *176*, 1–10.

Meghani, N., Kim, K. H., Kim, S. H., Lee, S. H., & Choi, K. H. (2020). Evaluation and live monitoring of pH-responsive HSA-ZnO nanoparticles using a lung-on-a-chip model. *Archives of Pharmacal Research*, *43*, 503–513.

Mo, Y., Zhang, Y., Zhang, Y., Yuan, J., Mo, L., & Zhang, Q. (2021). Nickel nanoparticle-induced cell transformation: Involvement of DNA damage and DNA repair defect through HIF-1α/miR-210/Rad52 pathway. *Journal of Nanobiotechnology*, *19*, 1–20. https://doi.org/10.1186/s12951-021-01117-7

Monopoli, M. P., Aberg, C., Salvati, A., & Dawson, K. A. (2012). Biomolecular coronas provide the biological identity of nanosized materials. *Nature Nanotechnology*, *7*, 779–786. https://doi.org/10.1038/nnano.2012.207

Mueller, N. C., & Nowack, B. (2008). Exposure modeling of engineered nanoparticles in the environment. *Environmental Science & Technology*, *42*(12), 4447–4453. https://doi.org/10.1021/es7029637

Nohynek, G. J., Dufour, E. K., & Roberts, M. S. (2008). Nanotechnology, cosmetics and the skin: Is there a health risk?. *Skin Pharmacology and Physiology*, *21*(3), 136–149. https://doi.org/10.1159/000131078

Oberdörster, G., Maynard, A., Donaldson, K., Castranova, V., Fitzpatrick, J., Ausman, K., Carter, J., Karn, B., Kreyling, W., Lai, D., & Olin, S. (2005). Principles for characterizing the potential human health effects from exposure to nanomaterials: Elements of a screening strategy. *Particle and Fibre Toxicology*, *2*(1), 8. https://doi.org/10.1186/1743-8977-2-8

Park, S. B., Jung, W. H., Kim, K. Y., & Koh, B. (2020). Toxicity assessment of SiO2 and TiO2 in normal colon cells, in vivo and in human colon organoids. *Molecules*, *25*(16), 3594.

Prasad, M., Kumar, R., Buragohain, L., Kumari, A., & Ghosh, M. (2021). Organoid technology: A reliable developmental biology tool for organ-specific nanotoxicity evaluation. *Frontiers in Cell and Developmental Biology*, *9*, 696668. https://doi.org/10.3389/fcell.2021.696668

Raliya, R., Nair, R., Chavalmane, S., Wang, W. N., & Biswas, P. (2015). Mechanistic evaluation of translocation and physiological impact of titanium dioxide and zinc oxide nanoparticles on the tomato (*Solanum lycopersicum* L.) plant.

Metallomics, *7*(12), 1584–1594. https://doi.org/10.1039/c5mt00168d

Rzigalinski, B. A., & Strobl, J. S. (2009). Cadmium-containing nanoparticles: Perspectives on pharmacology and toxicology of quantum dots. *Toxicology and Applied Pharmacology*, *238*(3), 280–288. https://doi.org/10.1016/j.taap.2009.04.010

Shi, Z., Niu, Y., Wang, Q., Shi, L., Guo, H., Liu, Y., . . . & Zhang, R. (2015). Reduction of DNA damage induced by titanium dioxide nanoparticles through Nrf2 in vitro and in vivo. *Journal of Hazardous Materials*, *298*, 310–319. https://doi.org/10.1016/j.jhazmat.2015.05.043

Sukhanova, A., Bozrova, S., Sokolov, P., Berestovoy, M., Karaulov, A., & Nabiev, I. (2018). Dependence of nanoparticle toxicity on their physical and chemical properties. *Nanoscale Research Letters*, *13*, 1–21.

Tang, X., Fan, X., Xu, T., He, Y., Chi, Q., Li, Z., & Li, S. (2022). Polystyrene nanoplastics exacerbated lipopolysaccharide-induced necroptosis and inflammation via the ROS/MAPK pathway in mice spleen. *Environmental Toxicology*, *37*(10), 2552–2565. https://doi.org/10.1002/tox.23618

Wu, H., Guo, J., Yao, Y., & Xu, S. (2022). Polystyrene nanoplastics induced cardiomyocyte apoptosis and myocardial inflammation in carp by promoting ROS production. *Fish & Shellfish Immunology*, *125*, 1–8. https://doi.org/10.1016/j.fsi.2022.04.048

Xiao, X., Chen, W., Wei, Z. W., Chu, W. W., Lu, X. F., Li, B., Chen, H., Meng, S. J., Hao, T. F., Wei, J. T., & Zhang, C. H. (2020). The anti-tumor effect of nab-paclitaxel proven by patient-derived organoids. *OncoTargets and Therapy*, 6017–6025. https://doi.org/10.2147/OTT.S237431

Xu, Y., Zhao, Y., Liu, S., Lv, S., Chen, L., Wang, W., Feng, Y., Fu, F., & Xu, H. (2022). Zinc oxide particles can cause ovarian toxicity by oxidative stress in female mice model. *International Journal of Nanomedicine*, *17*, 4947. https://doi.org/10.2147/IJN.S373147

Xuan, L., Ju, Z., Skonieczna, M., Zhou, P. K., & Huang, R. (2023). Nanoparticles-induced potential toxicity on human health: Applications, toxicity mechanisms, and evaluation models. *MedComm*, *4*(4), e327. https://doi.org/10.1002/mco2.327

Yang, D., Zhu, J., Zhou, X., Pan, D., Nan, S., Yin, R., Lei, Q., Ma, N., Zhu, H., Chen, J., & Ding, Y. (2022). Polystyrene micro-and nano-particle coexposure injures fetal thalamus by inducing ROS-mediated cell apoptosis. *Environment International*, *166*, 107362. https://doi.org/10.1016/j.envint.2022.107362

Zhang, S. H., Wu, H. M., Li, S., Wang, M. Z., Fang, L., & Liu, R. Y. (2019). Melatonin enhances autophagy and decreases apoptosis induced by nanosilica in RAW264. 7 cells. *IUBMB Life*, *71*(7), 1021–1029. https://doi.org/10.1002/iub.2055

17 Zinc Oxide Nanoparticles
Uses in Food Additives and Packaging and Toxicological Profile

Snigdha Roy

17.1 INTRODUCTION

Zinc (Zn) is an essential element for human health because it plays a key role in different biological functions. For more than 300 enzymes, it acts as a vital cofactor, activates T-cells to fight against infections, protects cells from oxidative stress, repairs tissues, and helps in cell division and protein production. Considering all its beneficial activities, a prescribed amount of Zn supplements is necessary for human health. The body receives this crucial nutrient from different dietary sources like legumes, beans, lentils, whole grains, nuts, eggs, and meats.

Apart from these natural sources, the trace element is absorbed from various artificial resources like multivitamins, processed fortified foods, and Zn-only tablets. In these supplements, Zn metal is present in different salt forms like zinc gluconate, zinc sulfate, zinc citrate, and zinc oxide. Slowly, with progress in nanotechnological research findings, people have understood that the bioavailability of Zn is higher when it is present in nano-size than in the bulk form of conventional salts. Therefore, from 2000 onwards, the application of Zn and ZnO-based NPs has been spreading widely in various fields, including medicine, the food industry, cosmetics, electronics, agriculture, and environmental remediation. In the case of the food industry, ZnO NPs are mainly utilized as nutrient supplements, food additives, commercial packaging, stabilizers, food sensors, and antibacterial wrapping. It was initially assumed that the addition of ZnO NPs in different fortified food items would enhance their nutritional value and extend the shelf life of the products. The higher solubility and greater absorptivity of these NPs would also reduce the required amount of addition compared to the bulk material. Moreover, the antibacterial property of ZnO NPs was significantly exploited in packaging of food products to protect them from microbial erosions. But within a few years of application, various toxic effects of these NPs started to emerge from different experimental research works. This has raised the concern of multiple business houses and the health administrative. Safety and regulatory bodies are now concentrating more on assessing the harmful effects of Zn nanomaterials before providing any approval to these synthetic particles for becoming a food additive. This chapter is written with the aim to discuss the various toxic effects of ZnO NPs being added to different food products and used in packaging. It also attempts to explore the actual reasons behind their toxicity.

17.2 APPLICATION OF ZNO NANOPARTICLES IN FOOD INDUSTRIES

ZnO NPs are widely used as dietary supplements in various fortified food items like cereals, diary products, beverages, and energy bars. Approximately 10 mg of elemental Zn is present in 12.5 mg of ZnO NP. In the case of an adult, the recommended dietary allowance for Zn ranges from 8 to 11 mg/day. The National Institutes of Health (NIH) declares the upper limit for elemental Zn consumption is 40 mg/day for an adult. Following these safety regulations, food product manufacturing companies add this essential micronutrient into their products with caution (Table 17.1).

TABLE 17.1
Zn-Containing Food Products

Product Type	Product Name	Manufacturing Company	Content of Zn Micronutrients	Approved By
Dairy products	NIDO Fortified Full Cream Milk Powder	Nestle	6.5 mg/100 g	FSSAI
	Bear Brand Fortified Powdered Milk Drink	Nestle Philippines	2.4 mg/33 g serving size	IDCP
	NIDO Growing Up Formula	Nestle, UAE	5.4 mg/240 ml drink	FSSAI
	Kiddle Me Yogurt	Dairy Rich Food LLP, Kolhapur	2.99 mg/100 g	FSSAI
Cereals	Unifit Corn Flakes	Vark Snack Food Pvt. Ltd.	2.55 mg/100 g	FSSAI
Whole wheat	Better Nutrition Biofortified Atta	SNKR Millers and TG Pvt. Ltd.	3.5 mg/100 g	FSSAI

DOI: 10.1201/9781003644439-17

In breakfast cereals, wholegrain wheat, daliya, dairy products like infant formula, milk and yogurt, and beverages like health drinks and fortified juice, Zn micronutrients are added either in NP or other salt forms to enhance the nutritional benefits. As a nutritional supplement, ZnO NP-based tablets containing 10–25 mg of Zn/tablet are available in pharmacies. Not only that, confectionery items like cakes are fortified with uncalcined ZnO NPs to increase the bioavailability of the essential element in body fluids (Ince et al., 2023). The NP is mixed with the minced fish and meat at a concentration of 20 mg/kg to enhance their sensory and antimicrobial property. Apart from food fortifying agents, ZnO NPs are also used in packaging materials due to their antibacterial and UV-protective properties. They can resist the growth of different bacteria like *Salmonella* and *Listeria* in food items. Various light-sensitive foods like oils and dairy products can be preserved effectively by using ZnO NP–based active and smart packaging materials. NPs can act as sensor agents to detect food spoilage or the presence of other contamination.

17.3 SPECIAL PROPERTIES OF ZNO NPS AND THEIR BIOLOGICAL BENEFITS

Like all other NPs, ZnO NPs are very small in size (nanometer range) and have a large surface area. Due to a high surface area-to-volume ratio, these particles can exhibit extraordinary adsorption properties, better catalytic efficiency, and higher reactivity. This unique set of features has been widely utilized in various applications, including physical, chemical, pharmaceutical, and biological areas. Some of the main physicochemical properties due to which the particles are considered an attractive choice among other food additives are discussed in the following.

17.3.1 Bioactivity of ZnO NPs

ZnO NPs are sparingly soluble in aqueous environments and can release zinc ions (Zn^{2+}) (Scheme 17.1). These ions are biologically active because they can interact with enzymes and proteins and promote cell proliferation and tissue regeneration. Due to their small size at the nanoscale level, ZnO particles can easily cross biological barriers like the blood–brain barrier, making them an efficient nutrient and drug delivery agent. Moreover, NPs can influence cell signaling pathways and induce gene expression through binding to the cellular membrane followed by penetration into cells. Thus, ZnO NPs can serve as a better source of the essential micronutrient Zn inside the body compared to other conventional Zn salts.

17.3.2 Antibacterial and Antifungal Activity

ZnO NPs can also disrupt microbial membranes through the generation of various reactive oxygen species (ROS) like superoxide anions (O_2^-), superoxide radicals ($O_2\bullet$), hydroxyl radicals (•OH), and hydrogen peroxide (H_2O_2) under UV or visible light (Scheme 17.2) (Choudhury et al., 2016). These particles are equally effective to both gram-positive and gram-negative bacteria. Different fungi like *Candida albicans* and *Candida glabrata* can be effectively killed by the same NPs via direct interaction with their microbial membranes. Apart from these human pathogens, the growth of other infective species like *Aspergillus, Penicillium,* and *Rhizopus* that cause food spoilage and post-harvest decay in fruits and bread mold, respectively, can also be significantly inhibited by ZnO NPs (Sirelkhatim et al., 2017). Thus, the addition of these particles can extend the shelf life of various food products and prevent multiple foodborne diseases.

17.3.3 Antioxidant and Anticancer Potential

The radical scavenging capacity of ZnO NPs reduces the oxidative stress of cells considerably. Through the generation of ROS, these particles can act as an effective cytotoxic agent to destroy cancerous cells and tissues. Different studies have pointed out that they can noticeably disrupt the mitochondrial functions of the cell and trigger programmed cell death or apoptosis (Yan et al., 2024).

17.3.4 Stability

ZnO NPs mostly remain stable under various pH conditions and high temperatures. They don't easily break down while being stores for a longer time with food items and maintain their functional properties over time.

$$2\,ZnO \longrightarrow 2\,Zn + O_2$$
$$Zn \longrightarrow Zn^{2+} + 2e^-$$

SCHEME 17.1 Formation of Zn^{2+} ions from ZnO.

$$ZnO + h\nu\ (\text{UV light}) \longrightarrow e^-(CB)\ (\text{excited electron in the conduction band}) + h^+(VB)\ (\text{hole in the valence band})$$
$$O_2 + e^- \longrightarrow \bullet O_2^-\ (\text{Superoxide radical anion})$$
$$\bullet O_2^- + e^- + 2H^+ \longrightarrow H_2O_2\ (\text{Hydrogen peroxide})$$
$$H_2O_2 + e^- \longrightarrow {}^-OH + \bullet OH\ (\text{Hydroxyl radical})$$

SCHEME 17.2 Generation of ROS from ZnO.

17.3.5 Biodegradability

Inside biological systems, ZnO NPs are slowly dissolved in acidic or slightly neutral environments and dissociated into Zn^{2+} and oxygen. Zn^{2+} can be absorbed, utilized, or excreted from the human body, avoiding long-term accumulation and thereby reducing the chances of metal-mediated corrosive impacts. ZnO NPs also degrade in natural environments like soil and water through dissolution and reaction with organic and inorganic compounds, minimizing ecological impact.

17.3.6 UV Absorption and Photocatalytic Activity

ZnO NPs can absorb and scatter a wide range of ultraviolet light that makes them effective for food packaging. Food products prone to UV-induced degradation can be protected by mixing with these NPs. Moreover, microbes and different organic pollutants are destroyed by these tiny materials when present in smart and active packaging substances.

17.3.7 Cost-Effective, Easy Synthesis

ZnO NPs can be easily synthesized from various eco-friendly precursors following relatively inexpensive green synthetic methods (J et al., 2024).

17.3.8 Low Toxicity and GRAS Status

At small concentrations, ZnO NPs are considered biocompatible, with low toxicity (Dey et al., 2024). A controlled and limited Zn^{2+} ion release from the NPs is sufficient to ensure the conductance of several essential biological functions, like boosting the immune system and carrying out different enzymatic and developmental activities. By modulating the particle size, dose, and surface coating, the toxicity level of the NPs can be easily controlled and monitored. On this basis, ZnO NPs received GRAS status and are recognized as a safe nutritional supplement and food fortifier by the FDA only when used in small quantities.

17.4 ROLES IN FOOD PACKAGING

ZnO NPs also play a vital role in food packaging and preservation. The nanoscale size of the ZnO particles increases their surface reactivity manifold. The high surface area-to-volume ratio of the NPs ensures stronger interaction of ZnO with microbes and other pathogens of food items, resulting in effective inhibition of a wide range of bacteria, fungi, and viruses. Special properties like antimicrobial activity, UV protection, biodegradability, and the ability to enhance the material strength of polymeric substances are attributed to its unique functionality. ZnO NPs can inhibit the growth of different foodborne pathogens very effectively with the help of released ions and ROS. When present in packaging material, Zn^{2+} ions interfere with the microbial metabolism process, and the ROS disrupt the bacterial membranes, protecting food items from microbial decay. Owing to a wide band gap of 3.4 eV, ZnO NPs can efficiently absorb UV light and thus resist the UV light–induced food oxidation process. The permeability of oxygen, carbon dioxide, and moisture are significantly limited by these particles to reduce the growth of microbial species in edible products. The ability to absorb various volatile organic compounds (VOCs) like ethylene ensures the freshness of food items, especially for fruits and vegetables, by significantly decreasing their ripening rate. Moreover, the elasticity and thermal resistance of polymeric packaging materials like polyethylene and polylactic acid are much improved when integrated with ZnO NPs. On combining them with other biodegradable polymers like chitosan, starch, and polylactic acid, the particles can also provide an eco-friendly alternative to traditional plastic-based packaging substances. Thus, ZnO NPs, when present as an active ingredient of packaging material, can not only ensure a long shelf life of food products in fresh form but also offer a durable packaging system which is less prone to tearing and deformation (Kim et al., 2020).

17.5 RESEARCH FINDINGS ON ZNO NP TOXICITY AND MECHANISTIC EXPLANATIONS

In spite of many applications of ZnO NPs, toxicity concerns about this NPs became prominent when a few research observations showed that, compared to bulk ZnO, the nanosized materials can be harmful to human cells and other living organisms. Brunner et al. (Brunner et al., 2006) and Franklin et al. (Franklin et al., 2007) pointed out the cytotoxic effects of ZnO NPs to human fibroblast cells and aquatic algae in 2006 and 2007, respectively, when used in high concentrations. These initial revelations prompted future research works to focus on the identification of the toxicological effects of ZnO NPs and their mechanistic interpretations.

Although ZnO NP toxicity both in *in vitro* and *in vivo* conditions is still under inspection, a good amount of research works have already disclosed some of the detrimental impacts of the NPs, with their probable reasons (Table 17.2). One important factor was found to be the high dose of applied ZnO NP.

TABLE 17.2
Timeline of ZnO NP Toxicity Detection

Year	Observed Toxic Impacts	Possible Reasons
2008–2010	Cell membrane damage, uncontrolled absorption of NPs	Increased Zn^{2+} release
2011–2013	Inflammation and oxidative stress in gut cells, DNA damage	Increased Zn^{2+} release and overproduction of ROS
2012–2015	Accumulation in liver, kidneys, intestines of animal samples; alteration in gut microbiota; inflammation and organ damage	Increased Zn^{2+} release and overproduction of ROS

17.5.1 Overproduction of Zn^{2+}

Research has suggested that when NPs are mixed at higher concentrations in any fortified food products, they cause toxicity in *in vitro* conditions (Youn & Choi, 2022). Oxidative stress, DNA damage, inflammation, and apoptosis of cells have been observed as the immediate result of this toxic effect. Scientists have explained the possible cause behind the dose-dependent toxicity of ZnO NPs. At high concentrations, ZnO NP dissolve more in the environment, especially in lower-pH media, and promote the release of a lot of Zn^{2+}. The excessive amounts of this ion disturb cellular processes and damage enzymes and mitochondrial functionality. Some of these perturbations are discussed in the following.

17.5.1.1 Disruption of Enzyme Activity

Inside the body, different enzymes need specific metal ions to function properly. For example, cytochrome oxidase, one of the crucial enzymes of the electron transport system, requires iron (Fe) and copper (Cu); ribonucleotide reductase, which catalyzes the conversion of ribonucleotide to deoxyribonucleotide, needs cobalt (Co); and acetyl-CoA carboxylase contains manganese (Mn) as an essential component for fatty acid synthesis. Now, in the presence of excessive Zn^{2+}, it starts to compete with other metal ions and binds to various wrong positions of other enzymes (Scheme 17.3). As a result, these enzymes lose their required structural integrity and stop functioning properly. Zn^{2+} also binds to the thiol (–SH) group of the cysteine residue of many enzymes, causing conformational changes and inactivation. In this manner, overuse of ZnO NPs has been found to block several key pathways of energy production, including glycolysis and mitochondrial respiration.

17.5.1.2 Interaction with Membrane Proteins and Lipid Bilayers

Excess Zn^{2+} ions bind to the membrane ion channel proteins that block the normal ion transport phenomenon of Na^+/K^+-ATPase, causing ion imbalance inside the cell (Noh et al., 2015). Ultimately, this causes cell stress or swelling. The negatively charged phospholipids bilayer of the cell membrane can also combine with oppositely charged Zn^{2+} ions and become neutral. The neutralization destabilizes the membrane structure, making it leaky, rigid, and broken. The presence of excessive Zn^{2+} ions breaks the ion gradient across the cell membrane, without which the membrane potential collapses. As a result of this, the cell cannot transport nutrients or remove waste through its outer membrane, leading to an early death of the cell. In addition to this, the replacement of other necessary metal ions by Zn^{2+} ions makes the signal transduction system of the cell almost non-functional (Haase & Maret, 2010).

17.5.1.3 Organ Toxicity

ZnO NPs, if overused, can also damage the activity of specific organs, including the liver, kidneys, and brain. In the case of the liver, more alanine transaminase (ALT) and aspartate transaminase (AST) enzymes are produced and released into the bloodstream, which indicates an acute liver injury. Creatinine production gets higher in the kidney, and the structure of renal tubules is damaged by excessive ZnO NPs. Experimental study with animal models also revealed the possibility of serious neurotoxicity with this NP (Tian et al., 2015). Due to its very small size, it can easily pass through the blood–brain barrier, causing several behavioral changes in rodents (Rahdar et al., 2020). Not only that, the sperm morphology of the animal models is also altered, and a hormonal imbalance is observed. The size of the spleen becomes bigger, with an associated immune dysfunction. With a high dose of ZnO NPs, the abundance of *Lactobacillus bacterium* in the gut also decreases, and metabolic activity is critically disturbed.

17.5.1.4 Mitochondrial Damage, Cell Apoptosis, and Genotoxicity

With the increasing concentration of Zn^{2+} ions, it starts to accumulate inside the mitochondria and leads to disruption of the electron transport chain (ETC), causing a significant depletion in energy production. A high dose of this ion competes with iron or copper in Fe-S or Cu centers present in the ETC complexes. The activities of two key enzymes of the TCA cycle, α-ketoglutarate dehydrogenase and isocitrate dehydrogenase, are greatly hampered by Zn^{2+} ions. It binds with the thiol (-SH) group of these enzymes by replacing the other necessary ions Mg^{2+}. Protein p53 is also activated by these ions, which eventually leads to programmed cell death, or apoptosis. Mitochondrial permeability is considerably enhanced by the presence of Zn^{2+} ions, which activate the permeability transition pore, leading to the collapse of the membrane potential. Moreover, when the excessive ions start to interact with DNA or DNA repair enzymes, the genetic material damages leading to

$$\text{Enzyme-M (active form)} + Zn^{2+} \longrightarrow \text{Enzyme-Zn (inactive form)} + M^{2+}$$

$$\text{Enzyme-SH (active form)} + Zn^{2+} \longrightarrow \text{Enzyme-S-Zn (inactive form)} + H^{+}$$

SCHEME 17.3 Inhibition of enzymatic activity by excessive Zn^{2+}.

mutation, chromosomal abnormalities and even cancer. Cytokine release is also triggered by the high concentration of Zn^{2+} ions, causing additional oxidative stress and indirect damage of DNA.

17.5.2 Generation of Reactive Oxygen Species

Apart from all the Zn^{2+} ion–mediated disruptions of cellular, enzymatic, mitochondrial, and chromosomal activities, excessive ROS generated by this ion could also impact human health adversely. ROS can cause severe oxidative stress inside the body. The cell membrane; protein; DNA; and overall various organs of the body like the kidneys, lungs, and liver can be seriously damaged by increased levels of ROS.

17.5.2.1 Lipid Peroxidation

Lipid peroxidation is the oxidative degradation process of the lipid layer of the cell membranes. The lipid layer is mainly made up of polyunsaturated fatty acids. ROS generated by ZnO NPs abstract hydrogen atoms from these unsaturated fatty acids, leading to the formation of lipid radicals (Scheme 17.4) and other reactive aldehydes like malondialdehyde (MDA) and 4-hydroxynonenal (4-HNE). The layer thus loses its flexibility and structural integrity. Several holes are formed in the outer layer of the cell, making it permeable and dysfunctional and altering the cellular signaling process, and can even cause cell death.

17.5.2.2 Mutation

Radical species directly attack the single or double strand of the DNA helix (Scheme 17.5) and can break it or oxidize the DNA bases (Scheme 17.6), resulting in mutation and chromosomal aberrations that increase cancer risk.

17.5.2.3 Protein Damage

The protein chain also undergoes oxidation with ROS (Scheme 17.7) and becomes misfolded or fragmented. Such structurally distorted protein molecules lose their enzymatic activities, and thereby the cell signaling and metabolic pathways of the biological system completely collapse.

17.5.2.4 Energy Crisis

In the presence of excessive ROS, ATP production inside the mitochondria decreases significantly, leading to an acute energy crisis of the cell and its ultimate programmed apoptosis.

17.5.2.5 Chronic Disease

Increased ROS activity plays a crucial role in activating different immune cells like macrophages and neutrophils that can trigger chronic inflammations, autoimmune response, and fibrosis inside the body. Chronic diseases like asthma, atherosclerosis, rheumatoid arthritis, and inflammatory bowel diseases are promoted by increased concentrations of ROS.

17.6 CHANGED ROLES OF FOOD REGULATORY AUTHORITIES

With the growing applications of ZnO NPs in consumable food products, nutrient fortification, and antimicrobial packaging worldwide, regulatory authorities have increased their surveillance on the safety, hygiene, and nutritional value of these commercial products through formulating different rules and regulations. The food administrative bodies reassess their roles, policies, and risk assessment strategies to curb possible harmful effects of ZnO NPs when used as food additives or in packaging materials. Continuous scientific studies on the toxicity effects of these nanomaterials on biological systems and new findings and revelations provide great help for these authorities to reset their policy frameworks for the benefit of human health and well-being.

17.6.1 Revised Views of Food Regulatory Authorities

A better understanding of the toxicological impacts of NPs drives food authorities to focus more on evaluating the nano-specific risk assessment of various fortified food items and packaging. Authorities have already acknowledged the different behaviors of nano-sized ZnO particles compared to its bulk format. Considering the more toxic effects of ZnO NPs, they are now looking to develop more specific guidelines

$$LH + {}^{\bullet}OH \longrightarrow L^{\bullet} + H_2O$$

$$L^{\bullet} + O_2 \longrightarrow LOO^{\bullet}$$

$$LOO^{\bullet} + LH \longrightarrow L^{\bullet} + LOOH$$

LH = Lipid, $L^{\bullet}$ = Lipid Radical

$LOO^{\bullet}$ = Lipid peroxy radical

LOOH = Lipid hydroperoxide

SCHEME 17.4 Chain reactions of lipid peroxidation.

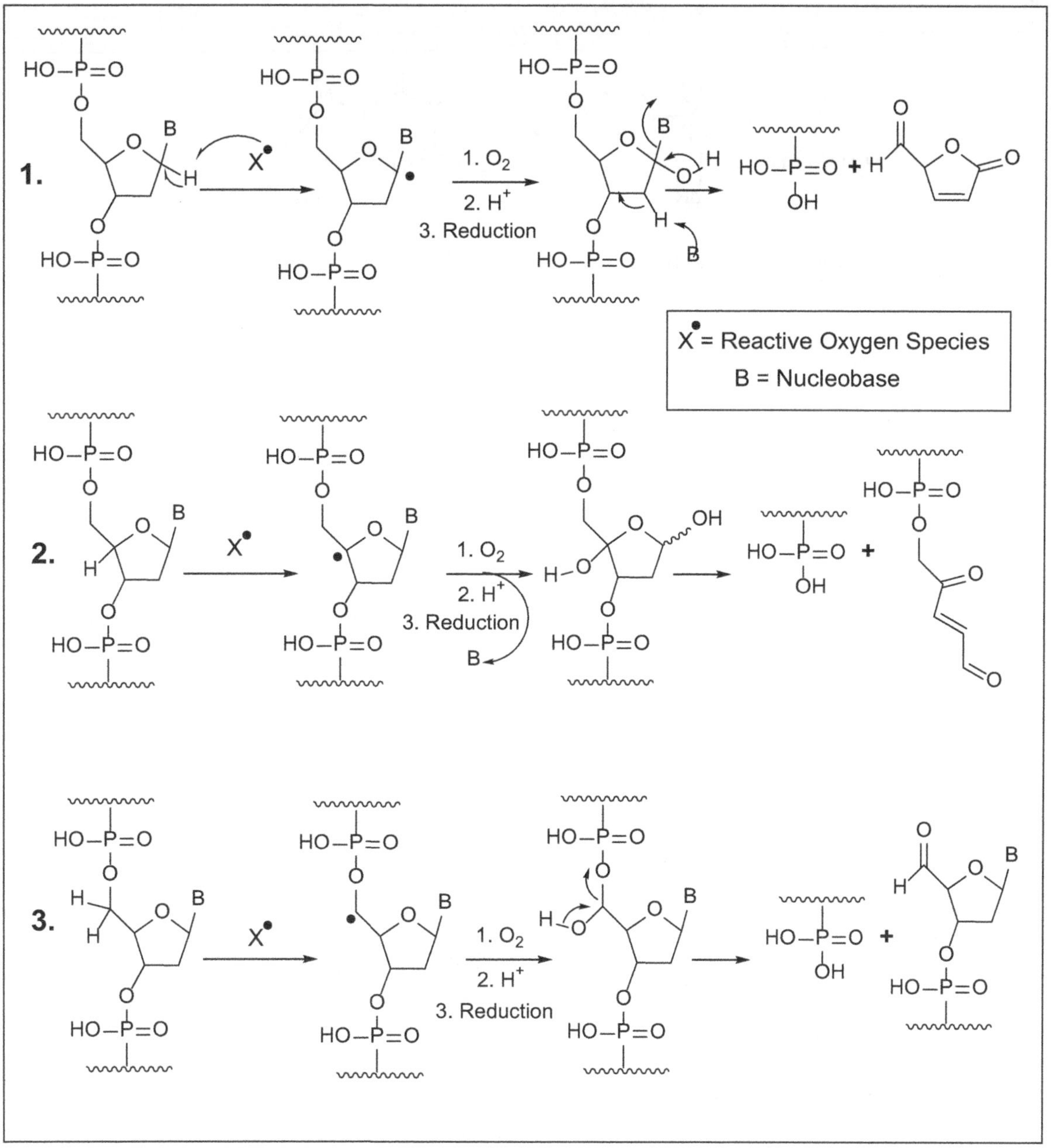

SCHEME 17.5 Breakage of DNA strand by reactive oxygen species.

which are different from the traditional generally recognized as safe (GRAS) regulations. As an effective step of this shift, the bioavailability and accumulation data of any NPs inside human tissues are now being questioned by authorities. The long-term exposure risks of the particles are being strictly evaluated. Regulatory bodies now require complete reactivity details of ZnO NPs on how they are interacting with the food matrix, digestive fluids of the body, and environmental pathways. The adoption of an eco-friendly synthetic methodology for NPs is continuously being promoted and encouraged by regulatory committees that can guide nanotech innovation towards a green and sustainable future. New surface technology such as nanoencapsulation and modification of particle coatings with biocompatible functionalities are continuously being encouraged by the authorities to reduce the toxic effects of NPs.

17.6.2 Requirements of Labeling

To maintain transparency and consumer safety, most of the regulatory bodies have instructed manufacturing and marketing companies to label their products clearly and mention the presence of nanoingredients such as ZnO ("zinc oxide [nano]") following a prescribed protocol. This information will help enhance consumer awareness about their product selection.

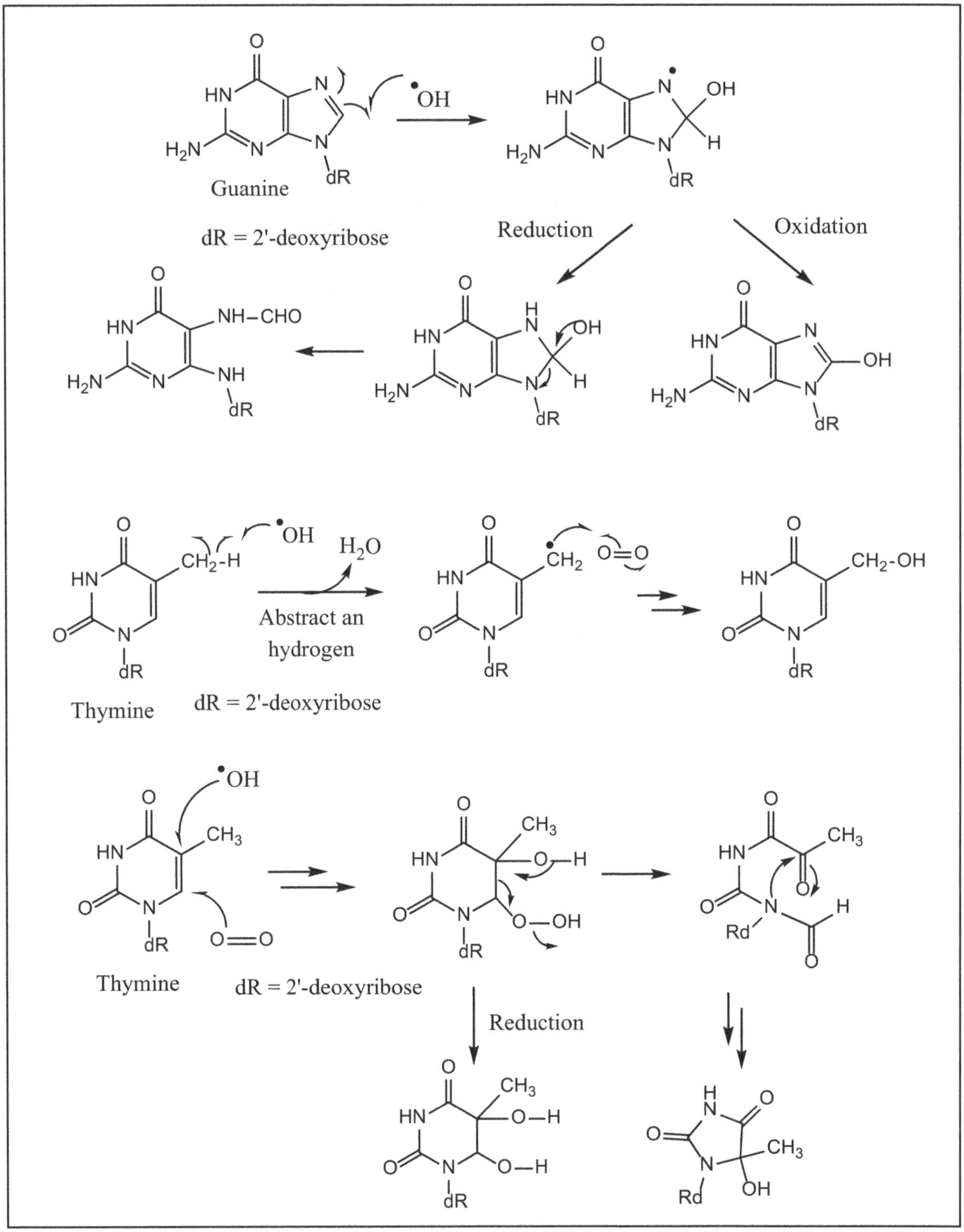

SCHEME 17.6 Oxidation and reduction of nucleobases by ROS.

17.6.3 Risk Communication and Surveillance

The regulatory institutions encourage food manufacturers to train their employees on the working principles of nanotechnology and how to maintain proper documents about nanoadditives and their toxicological data, including the biocompatibility, genotoxicity, and long-term health impact–related information. The authorities have further recommended setting a proper post-marketing monitoring surveillance procedure to track the health impacts of these fortified food products more precisely and recall items if any safety issue arises.

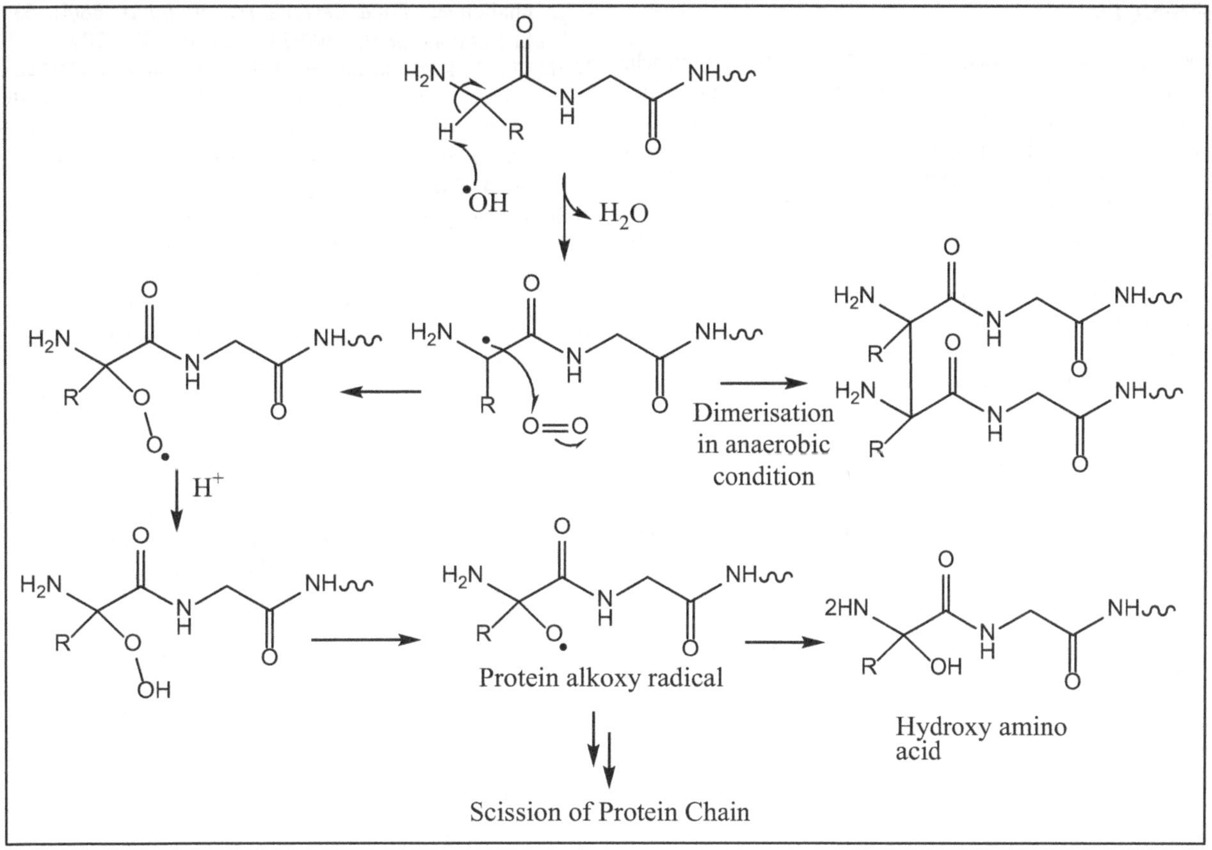

SCHEME 17.7 Cleavage of protein chain by ROS.

17.6.4 Guidelines of Different Food Authorities

The FDA, the food administrative body of the US, still recognizes bulk ZnO as GRAS, but for the nanoforms of the oxide, it has demanded a pre-market notification and safety data sheet. The European Food Safety Authority (EFSA) of the EU asks for the full physicochemical description of NPs, including the crystal structure, surface area, and nanoparticle size distribution from commercial companies to re-evaluate the particles as food additives. In the case of smart food packaging, it has demanded migration tests of NPs from the packaging of food items. The National Health Commission (NHC) of China restricted the use of nanoscale ZnO particles without proper safety assessment. In India, according to the guideline of Food Safety and Standards Authority of India (FSSAI), there is no official regulation of ZnO NPs to date. Although they have declared their complete alliance with the global nanotechnology frameworks and their safety data collection system. Thus, authorities of different countries are trying to impose unified rules and safety standards in connection with the use of ZnO NPs as food additives or packaging materials so that consistent regulations can be followed across borders.

17.7 CONCLUSION

ZnO NPs have shown promising applications in the food industry as nutritional additives to enhance zinc bioavailability and as an active ingredient in antimicrobial smart food packaging. Their unique physicochemical properties, including smaller size, high surface area, and enhanced reactivity, contribute to their functional effectiveness. However, these same properties can raise severe health concerns, especially when ingested or released from packaging into food in considerable amounts. Recent toxicological studies on ZnO NPs suggest that the particles may induce oxidative stress, cellular damage, mitochondrial dysfunction, genotoxicity, or bioaccumulation when used in high doses. Scientists have also highlighted the possible reasons for the harmful biological impacts of ZnO NPs, which are mainly because of the release of Zn^{2+} ions in high concentration and the generation of various reactive oxygen species. The degree of risk factors increases with the smaller particle size, acidic pH of the cellular environment, poor stability of uncoated nanoparticles, and longer exposure to UV light. Considering all these potential toxicities of ZnO NPs, global food regulatory institutions are moving towards developing a unified safety standard, comprehensive risk assessment strategies, and transparent labelling practices to ensure consumer safety and maintain quality standards. Responsible innovation of nanotechnology and its wise applications in food industry are also encouraged by the authorities to foster continuous growth in the field of food technology. Persistent research activities and cross-border regulatory cooperation will hopefully play a vital role in this aim in the future.

REFERENCES

Brunner, T. J., Wick, P., Manser, P., Spohn, P., Grass, R. N., Limbach, L. K., Bruinink, A., & Stark, W. J. (2006). In vitro cytotoxicity of oxide nanoparticles: Comparison to asbestos, silica, and the effect of particle solubility. *Environmental Science & Technology, 40*(14), 4374–4381. https://doi.org/10.1021/es052069i

Choudhury, S. R., Ordaz, J., Lo, C., Damayanti, N. P., Zhou, F., & Irudayaraj, J. (2016). ZnO nanoparticles induced reactive oxygen species promotes multimodal cyto- and epigenetic toxicity. *Toxicological Sciences*, kfw252. https://doi.org/10.1093/toxsci/kfw252

Dey, S., Mohanty, D. L., Divya, N., Bakshi, V., Mohanty, A., Rath, D., Das, S., Mondal, A., Roy, S., & Sabui, R. (2024). A critical review on Zinc oxide nanoparticles: Synthesis, properties and biomedical applications. *Intelligent Pharmacy.* https://doi.org/10.1016/j.ipha.2024.08.004

Franklin, N. M., Rogers, N. J., Apte, S. C., Batley, G. E., Gadd, G. E., & Casey, P. S. (2007). Comparative toxicity of nanoparticulate ZnO, bulk ZnO, and ZnCl2 to a freshwater microalga (pseudokirchneriella subcapitata): The importance of particle solubility. *Environmental Science & Technology, 41*(24), 8484–8490. https://doi.org/10.1021/es071445r

Haase, H., & Maret, W. (2010). *The Regulatory and Signaling Functions of Zinc Ions in Human Cellular Physiology.* CRC Press eBooks (pp. 181–212). https://doi.org/10.1201/9781420059984-c8

Ince, O. K., Aydogdu, B., & Ince, M. (2023). Development of new generation cakes fortified with zinc oxide nanoparticles. *Journal of Food Science and Technology, 61*(2), 290–299. https://doi.org/10.1007/s13197-023-05840-x

Kim, I., Viswanathan, K., Kasi, G., Thanakkasaranee, S., Sadeghi, K., & Seo, J. (2020). ZNO nanostructures in active Antibacterial food packaging: Preparation methods, antimicrobial mechanisms, safety issues, future prospects, and challenges. *Food Reviews International, 38*(4), 537–565. https://doi.org/10.1080/87559129.2020.1737709

Nandhini, J., Karthikeyan, E., & Rajeshkumar, S. (2024). Green synthesis of zinc oxide nanoparticles: Eco-friendly advancements for biomedical marvels. *Resources Chemicals and Materials, 3*(4), 294–316. https://doi.org/10.1016/j.recm.2024.05.001

Noh, S., Lee, S. R., Jeong, Y. J., Ko, K. S., Rhee, B. D., Kim, N., & Han, J. (2015). The direct modulatory activity of zinc toward ion channels. *Integrative Medicine Research, 4*(3), 142–146. https://doi.org/10.1016/j.imr.2015.07.004

Rahdar, A., Hajinezhad, M. R., Bilal, M., Askari, F., & Kyzas, G. Z. (2020). Behavioral effects of zinc oxide nanoparticles on the brain of rats. *Inorganic Chemistry Communications, 119*, 108131. https://doi.org/10.1016/j.inoche.2020.108131

Sirelkhatim, A., Mahmud, S., Seeni, A., Kaus, N. H. M., Ann, L. C., Bakhori, S. K. M., Hasan, H., & Mohamad, D. (2017). Review on Zinc oxide nanoparticles: Antibacterial activity and toxicity Mechanism. *Nano-Micro Letters, 7*(3), 219–242. https://doi.org/10.1007/s40820-017-0040-x

Tian, L., Lin, B., Wu, L., Li, K., Liu, H., Yan, J., Liu, X., & Xi, Z. (2015). Neurotoxicity induced by zinc oxide nanoparticles: age-related differences and interaction. *Scientific Reports, 5*(1), 16117. https://doi.org/10.1038/srep16117

Yan, Y., Huang, W., Lu, X., Chen, X., Shan, Y., Luo, X., Li, Y., Yang, X., & Li, C. (2024). Zinc oxide nanoparticles induces cell death and consequently leading to incomplete neural tube closure through oxidative stress during embryogenesis. *Cell Biology and Toxicology, 40*(1). https://doi.org/10.1007/s10565-024-09894-1

Youn, S., & Choi, S. (2022). Food additive zinc oxide nanoparticles: Dissolution, interaction, fate, cytotoxicity, and oral toxicity. *International Journal of Molecular Sciences, 23*(11), 6074. https://doi.org/10.3390/ijms23116074

18 Biomedical Applications of Green Nanoparticles

Emerging Trends and Therapeutic Potential

Suraj Prakash, Radha, and Sunil Puri

18.1 INTRODUCTION

The integration of nanotechnology with principles of sustainable chemistry has catalyzed a significant transformation in the field of biomedical sciences, primarily through green-synthesized nanoparticles (Khan 2020). Green-synthesized nanoparticles gained increasing attention with the shift from conventional nanoparticle synthesis techniques, which often involve the use of toxic reagents, stringent reaction conditions, high production costs, and high energy requirements (Fahim et al. 2024; Jain et al. 2024; Kirubakaran et al. 2025). Due to limitations in chemical synthesis, researchers are now shifting more towards sustainable approaches, and among them, green synthesis is getting wide attention. Out of various methods, the use of plant extracts for nanoparticle synthesis is more common because it is low cost and eco-friendly (Ahmad et al. 2022). In this method, the plant extract works as both a reducing and stabilizing agent, which helps in nanoparticle formation and gives better control over the size and shape. Although green synthesis of nanoparticles can also be carried out using microorganisms or other biological materials, the use of plant extracts is more frequently reported, primarily due to the presence of a wide range of phytochemicals. Medicinal plants are known to contain bioactive compounds such as flavonoids, phenolics, alkaloids, and terpenoids, which play a significant role not only in the reduction of metal ions but also in the stabilization of nanoparticles during synthesis (Abbas et al. 2022). These phytochemicals are also found to influence the physiochemical characteristics of synthesized nanoparticles, including shape, surface charge, and colloidal stability, which directly affect their behavior in biological systems (Abbasi et al. 2023). Studies have documented that green-synthesized nanoparticles exhibit better biological activity and reduced toxicity when compared with chemically synthesized nanoparticles (Ratan et al. 2020). Owing to these advantages, plant-based green-synthesized nanoparticles are being explored in various biomedical applications. For instance, gold and silver nanoparticles synthesized using plant extracts have been reported for their significant biological activity and surface plasmon resonance property (Muddapur et al. 2022). The smaller size, high surface area, and surface functionalization potential of green nanoparticles enhance their suitability in targeted drug delivery systems, thereby improving the therapeutic efficacy of drugs (Liu et al. 2023). Additionally, green nanoparticles exhibited broad-spectrum antimicrobial activity, including antibacterial, antifungal, and antiviral properties, which makes them a potential candidate in replacing conventional antimicrobial drugs, especially in conditions where drug resistance is emerging as a serious threat. In the field of cancer research, green-synthesized nanoparticles are gaining considerable attention due to their potential application in targeted therapy and photothermal treatment. Some studies have reported the design of such nanoparticles in a way that allows selective accumulation at the tumor site and facilitates intracellular localization within organelles such as mitochondria, thereby enhancing the anti-cancer efficacy of loaded drugs (Saminathan et al. 2022). Furthermore, green nanoparticles are also found to be promising candidates in biosensing applications owing to their unique optical and electrical properties which enable the sensitive and accurate detection of low concentrations of biomolecules (Harshita et al. 2023). In tissue engineering, they are added into scaffold materials, where they help in making the structure stronger and also affect how the cells behave. In case of imaging, they were found to act as contrast agents in different types of imaging techniques like fluorescence imaging, CT scan, and MRI (Shariati et al. 2023). The use of green nanotechnology in the medical field is increasing, as it combines natural sources with advanced nanoparticle-based tools to develop materials which are useful in diagnosis, treatment, and sensing (Shariati et al. 2023). Furthermore, studies on green-synthesized nanoparticles can play a more important role in future healthcare, disease control, and biomedical advancements. The main aim of this chapter is to examine the role of green-synthesized nanoparticles in drug delivery, antimicrobial activity, and cancer treatment.

18.2 ROLE IN TARGETED DRUG DELIVERY SYSTEMS

Various studies reported nanocarriers overcome several limitations linked with conventional drug delivery systems such as poor aqueous solubility, rapid degradation of drugs

DOI: 10.1201/9781003644439-18

in systemic circulation, and associated side effects (Prashar et al. 2024; Masane et al. 2025). Based on the nature of therapeutic requirements, different classes of nanocarriers have been developed and investigated for their potential applications. For instance, liposomes, polymeric nanoparticles, metallic nanoparticles such as gold, dendrimers, and carbon nanotubes have been extensively studied (Prashar et al. 2024; Masane et al. 2025). Nanosized vesicle liposomes are composed of lipid bilayers and are capable of encapsulating both hydrophilic and lipophilic drugs, thereby enhancing drug absorption and bioavailability (Prashar et al. 2024). Polymeric nanoparticles were reported to have biocompatibility, biodegradability, and tunable drug release properties which respond to stimuli such as temperature or pH changes (Kiran et al. 2023). Metallic nanocarriers such as gold nanoparticles are reported to show significant optical and physicochemical properties, making them suitable candidates for medical imaging and diagnostic applications (Georgeous et al. 2024). Dendrimers are hyperbranched, tree-like nanostructures with a large surface area, allowing attachment of multiple drug molecules and enhancement of drug solubility and therapeutic efficacy (Afzal et al. 2022). All these systems come with their own specific features, and the selection depends on what type of disease is being treated, what kind of drug is used, and where in the body the drug needs to reach. When these nanocarriers are combined with green-synthesized nanoparticles, the benefits may increase further due to better biocompatibility and lower toxicity. The design of nanocarriers generally depends on controlling factors like particle size, surface charge, and adding specific ligands which help in better targeting and uptake by cells. Specifically, the enhanced permeability and retention (EPR) effect promotes the passive accumulation of nanocarriers in tumor tissues, whereas surface ligands provide active targeting toward cancer cells (Cheng et al. 2023; Masane et al. 2025). These design features are important for improving drug pharmacokinetics and minimizing off-target toxicity (Figure 18.1).

Among their diverse applications, nanocarriers play a central role in enhancing the bioavailability of poorly soluble drugs. Natural compounds such as curcumin and emodin, which possess potent therapeutic effects but suffer from poor solubility and rapid metabolism, have benefited significantly from nanoformulations like liposomes and polymeric nanoparticles (Manzari-Tavakoli et al. 2024; Ai et al. 2025). Nanotechnology-based drug delivery systems have also improved the pharmacokinetics of anticancer agents; for instance, self-nanoemulsifying drug delivery systems have enhanced the oral bioavailability of hydrophobic drugs (Abdellatif et al. 2023; Shukla et al. 2023). These enhancements arise through multiple mechanisms, including solubilization of hydrophobic drugs within carrier matrices (Abdellatif et al. 2023), controlled and sustained drug release over prolonged periods (Mohapatra et al. 2024), and ligand-mediated targeting that ensures preferential uptake by diseased cells (Cheng et al. 2023). In therapeutic applications, nanocarriers have been employed to deliver anticancer drugs such as doxorubicin, paclitaxel, and cisplatin, reducing systemic toxicity and improving clinical outcomes (Tang et al. 2021; Masane et al. 2025). Surface functionalization with antibodies, peptides, or aptamers further allows for highly specific targeting to tumor cells (Cho et al. 2008; Sanità et al. 2020). Importantly, the integration of green nanotechnology utilizing plant-based extracts and eco-friendly synthesis methods has added a sustainable dimension to drug delivery. Green nanoparticles not only reduce environmental toxicity but are also reported to have enhanced biocompatibility and therapeutic efficacy. Their use in diagnostics, imaging, and therapy shows their potential for biomedical applications (Akram et al. 2023; Manzari-Tavakoli et al. 2024).

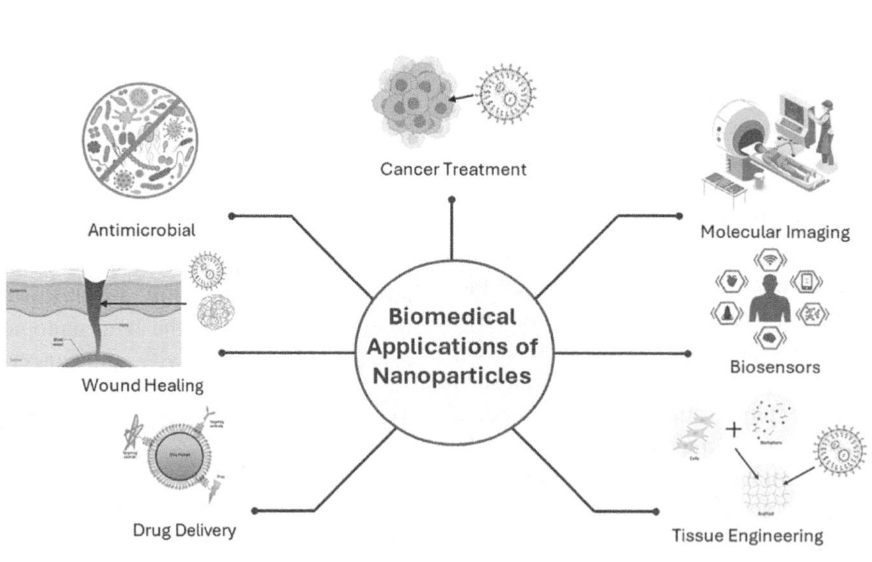

FIGURE 18.1 Biomedical applications of green-synthesized nanoparticles.

18.3 CANCER THERAPY

Cancer is one of the major global health issues, and conventional therapies are limited due to systemic toxicity, drug resistance, and poor targeting ability. Nanoparticles, especially those synthesized using a green chemistry approach, have shown promising potential in cancer therapy because of their selective targeting of tumors and inducing cytotoxicity in cancer cells without harming normal cells. The enhanced permeability and retention effect is a passive targeting mechanism which utilizes leaky blood vessels and poor lymphatic drainage in tumor regions. Due to EPR characteristics, nanoparticles accumulate more in tumor tissue and help increase the drug concentration at tumor sites and reduce systemic side effects. But EPR-based nanocarriers still face several clinical translation challenges due to heterogeneous tumor vasculature and variation in the EPR effect in different cancer types (Mah et al. 2014; Sharifi et al. 2022). Recent efforts were made to improve the physiochemical characteristics of nanoparticles like size, charge, and surface functionalization to increase tumor accumulation through the EPR effect. Nanoparticles with a size range between 10 and 200 nm are found to be more efficient for utilizing the EPR effect (Tian et al. 2022; Masane et al. 2025). Stimuli-responsive nanoparticles are also developed which can release drugs in specific environments such as acidic pH or heat to improve tumor targeting efficiency (Tian et al. 2022; Tian Yan et al. 2024). In active targeting, nanoparticles are usually modified on their surface with ligands which can bind to certain receptors that are more present on cancer cells. This kind of modification is known to improve the delivery of the drug at the site of cancer. Various researchers have used antibodies, short peptides, and ligands for this type of targeting, especially for biomarkers like Human Epidermal Growth Factor Receptor 2 (HER2), Epidermal Growth Factor Receptor (EGFR), and Cluster of Differentiation 44 (CD44) (Mah et al. 2014; Abdessalem and Adham 2025). Among green nanoparticles, silver and zinc oxide nanoparticles are often used after modifying them with biological molecules to achieve active targeting. For example, plant-extract–based zinc oxide nanoparticles were functionalized using hyaluronic acid, which can bind to CD44 markers present in many cancer cells (Todaria et al. 2024; Li et al. 2024). In another case, silver nanoparticles were loaded inside liposomes and designed in such a way that they released the drug only when the environment was acidic, like that inside tumors (Jayachandran et al. 2023; Al-Asiri et al. 2024).

Green-synthesized nanoparticles are known to have toxic effects on cancer cells in different ways like generation of reactive oxygen species (ROS) and damage to mitochondria and DNA. For example, silver nanoparticles prepared using plant extracts like *Euphorbia retusa* were found to increase ROS inside the cell, which caused the breakdown of the mitochondrial membrane and finally led to cell death (Al-Asiri et al. 2024; Todaria et al. 2024). Zinc oxide nanoparticles also showed similar toxic effects by changing the expression of genes related to apoptosis. Some studies reported that ZnO-NPs reduce the activity of anti-apoptotic gene Bcl-2 and increase pro-apoptotic genes like Bax and p53, which help in triggering programmed cell death in cancer (Aalami et al. 2020; Pouresmaeil et al. 2021; Efati et al. 2023). One of the important benefits of using green-synthesized nanoparticles is that they mostly act on cancer cells and cause less damage to normal cells. This selective nature is because of the unique conditions of cancer cells like high oxidative stress and altered metabolism. For instance, ZnO nanoparticles made using *Lepidium sativum* seed extract were toxic in a dose-dependent way to colorectal cancer cell lines like SW480, HT-29, and Caco-2 but did not show much toxicity to normal cells (Pouresmaeil et al. 2021; Efati et al. 2023). Similarly, silver nanoparticles that were packed in liposomes released the drug more at an acidic pH, which is commonly present inside tumors. At pH 5.5, these nanoparticles released their content more efficiently and helped in killing cancer cells selectively (Jayachandran et al. 2023; Al-Asiri et al. 2024). Apoptosis, or programmed cell death, is one of the major ways by which cancer cells are removed from the body. Green-synthesized nanoparticles are found to induce apoptosis through two main pathways, intrinsic and extrinsic. Studies have shown that pro-apoptotic proteins such as Bax and Bak become activated, leading to the disruption of the mitochondrial membrane in the intrinsic (mitochondrial) pathway. This allows cytochrome c to come out and activate caspase enzymes, which leads to cell death (Aalami et al. 2020; Efati et al. 2023; Al-Asiri et al. 2024). In the extrinsic pathway of apoptosis, specific surface receptors such as Fas and TNF-α are activated and were reported to play an important role in initiating programmed cell death. Upon activation, these receptors interact with adaptor proteins like FADD and subsequently activate caspase-8, which further triggers a downstream caspase cascade, leading to apoptosis (Wani et al. 2023; Todaria et al. 2024). Recent studies have showed involvement of microRNAs (miRNAs) in regulating this apoptotic pathway, particularly in cancer cells. For instance, zinc oxide nanoparticles synthesized using *Ceratonia siliqua* (carob) plant extract have been reported to downregulate the expression of cancer-associated miRNAs such as miR-155–5p, miR-203a-3p, and miR-223–3p in breast cancer cells. A decrease in expression of these miRNAs was found to be associated with suppression of anti-apoptotic gene expression and enhancement of pro-apoptotic genes, thereby promoting increased apoptosis in cancer cells (Aalami et al. 2020; Wani et al. 2023).

18.4 CELLULAR MODULATION EFFECTS

Green-synthesized nanoparticles gained attention for cellular modulation effects, especially in diseases such as cancer, wound healing, and immune responses. These nanoparticles interact with various cell types and induce specific biological responses, which makes them potential candidates for multiple biomedical applications. Studies reported that nanoparticles including silver, gold, and selenium have demonstrated cytotoxic effects against different cancer cell lines, primarily due to their ability to generate oxidative stress and induce apoptosis. One study showed silver nanoparticles synthesized with *Pimpinella anisum* extract showed dose-dependent cytotoxicity against Colo-205 colon cancer cells and were suggested as a promising agent for targeted cancer therapy (Fayed et al. 2024). Similarly, selenium nanoparticles formulated with carvacrol exhibited significant anticancer activity against MCF-7 breast cancer cells, which was attributed to the generation of reactive oxygen species and apoptosis induction (Othman et al. 2023). Silver nanoparticles from *Artemisia annua* helped to reduce liver and kidney damage in rats infected with resistant *Staphylococcus aureus*, suggesting that these nanoparticles may also improve immune function (Shaaban et al. 2024). They also increased the level of TNF-α and iNOS, which are markers of inflammation and immune response. Such findings show their possible use in immune-related therapies. In regenerative medicine, green-synthesized nanoparticles have also given good results. In wound healing, nanoparticles made from chitosan and medicinal plant extracts are often used in dressing materials. They support skin regeneration, control infections, and also help in forming new tissue and collagen (Punitha and Aruna 2022; Hajam et al. 2023). Because these nanoformulations are made using natural materials, they are usually considered safe and well tolerated, which is why many researchers have used them directly on wounds. From different studies, it appears that green nanoparticles affect a number of cellular activities, for example, destroying harmful cells, boosting immune function, or supporting tissue repair. Due to their natural origin and lower risk of toxicity, they are often seen as a better option for future use in medical treatments.

18.5 NANOPARTICLE-ENHANCED BIOMATERIALS

The use of green-synthesized nanoparticles in biomaterials is being explored more in recent times, as they seem useful for different medical purposes like drug delivery, wound healing, and even for making temporary implants. These nanoparticles are added to existing materials to improve their strength, compatibility with body cells, and overall function. In tissue engineering, for example, nanoparticles are mixed with biodegradable polymers such as PLLA and PDLLA. This combination helps cells attach better, grow faster, and form proper tissue structure (Punitha and Aruna 2022). Some studies also report that collagen-based nanomaterials combined with green nanoparticles are useful in healing bones and even in eye-related treatments, mainly because they don't cause strong immune reactions and work well with human tissues. Another major field is drug delivery. Plant-based nanoparticles and those from oils are used to develop drug carriers which can release medicine slowly and at the right site. These carriers increase drug effectiveness and reduce side effects. For example, anticancer drugs packed inside green nanoparticles were shown to target tumors better and avoid damage to healthy cells (Kumawat et al. 2024). Some plant-derived nanoparticles can also pass through biological barriers and fight drug resistance by reducing the expression of efflux transporters (Khatik 2022). In the case of wound healing and temporary medical implants, green nanoparticles again play a major role. Chitosan and chitin nanoparticles are widely used for wound dressing due to their antimicrobial action and support for skin recovery. Such materials are also used in short-term implants that release therapeutic agents and later break down safely inside the body (Punitha and Aruna 2022; Hajam et al. 2023). In all, nanoparticle-based biomaterials are a good example of combining eco-friendly synthesis with advanced medical applications. Their use in polymers, scaffolds, and drug systems is bringing new possibilities for safe, sustainable, and effective treatment options in modern medicine.

18.6 MOLECULAR RECOGNITION TECHNIQUES

In diagnostics, green-synthesized nanoparticles have been utilized for the detection of biomarkers with high sensitivity and specificity. A notable example includes gold nanoparticles synthesized from green tea leaves, which were integrated into optical fiber ball resonators to detect CD44 cancer antigens. These biosensors exhibited exceptional performance, with a limit of detection as low as 0.111 pM, showcasing the utility of green nanomaterials in non-invasive cancer diagnostics (Ashikbayeva et al. 2023). Silver nanoparticles have also been applied in molecular docking studies to predict drug-receptor interactions. Kaempferol-3-O-methyl ether, identified through such simulations, demonstrated high binding affinity to the epidermal growth factor receptor, suggesting its potential as a therapeutic candidate (Fayed et al. 2024). Targeted drug delivery systems benefit significantly from molecular recognition strategies enabled by green nanoparticles. By functionalizing the nanoparticle surfaces with ligands that recognize specific receptors overexpressed on diseased cells, selective drug delivery can be achieved. This kind of targeting helps in improving the effect of treatment while reducing the side effects on healthy tissues. Some studies reported that nanoparticles modified with ligands were used to target tumors through receptor-based uptake, helping the drug to enter cancer cells directly (Khatik 2022; Murugappan et al. 2024). Apart from cancer use, green-synthesized nanoparticles are also being explored for microbial infections using similar recognition mechanisms. Silver nanoparticles are among the most studied for their antibacterial effects. These particles can stick to bacterial surfaces, pass through the cell wall, and interfere with the internal structure and genetic material. In a few cases, they have also shown results against drug-resistant bacteria, which gives hope for using them in antibiotic-resistant infections (Mayegowda et al. 2023;

TABLE 18.1
Summary of Biomedical Applications of Green-Synthesized Nanoparticles

Type of Nanoparticle	Key Applications	References
Silver	Anticancer, antimicrobial, and immunomodulatory effects	Mayegowda et al. (2023), Fayed et al. (2024), Shaaban et al. (2024)
Gold	Biosensing, diagnostics, and targeted drug delivery	Khatik (2022), Ashikbayeva et al. (2023), Murugappan et al. (2024)
Selenium	Anticancer and apoptosis induction	Othman et al. (2023)
Chitosan	Wound dressing, tissue engineering, and drug delivery	Punitha and Aruna (2022), Asif et al. (2023), Hajam et al. (2023)
ZrO_2	Cancer inactivation	Tran et al. (2022)

Shaaban et al. 2024) (Table 18.1). So, the ability of green nanoparticles to bind specific biological molecules is useful not only in diagnosis and therapy but also for controlling harmful microbes. Because they are made from natural sources and are generally safe to use, such nanoparticles are now getting attention for future medical use where both sustainability and treatment efficiency are important.

18.7 CONCLUSION

Green-synthesized nanoparticles are now gaining attention as multi-use agents in areas like nanomedicine and biotechnology. Because of their origin from plants or other biological materials, they are now being studied for various medical uses. They are considered useful in areas like cancer therapy, wound healing, and immune system–related problems, mainly because they show lower toxicity and higher selectivity than many chemically made nanoparticles. When added to biomaterials, these nanoparticles can improve important features like strength, compatibility with body tissues, and support for healing. Because of this, they are being tested in drug delivery systems, tissue engineering scaffolds, and some types of implants. One useful property is that their surface can be changed easily, which makes them suitable for targeted treatment or diagnostic purposes. Some studies have also reported that green nanoparticles work against antibiotic-resistant bacteria, offering a possible solution in the fight against drug resistance. The use of green synthesis in nanomedicine not only supports safer approaches in treatment but also aligns with environmentally friendly methods. With more research, these biogenic nanoparticles may become an important part of future medical tools and therapies.

REFERENCES

A. H. Aalami, M. Mesgari, and A. Sahebkar, “Synthesis and Characterization of Green Zinc Oxide Nanoparticles with Antiproliferative Effects through Apoptosis Induction and MicroRNA Modulation in Breast Cancer Cells,” Bioinorganic Chemistry and Applications, vol. 2020, no. 1, pp. 1–17, Jan. 2020, doi: 10.1155/2020/8817110.

H. S. Abbas et al., “Prospects of Using Bioactive Compounds in Nanomaterials Surface Decoration and Their Biomedical Purposes,” International Nano Letters, vol. 12, no. 2, pp. 125–138, 2022, doi: 10.1007/s40089-021-00355-9.

R. Abbasi, G. Shineh, M. Mobaraki, S. Doughty, and L. Tayebi, “Structural Parameters of Nanoparticles Affecting Their Toxicity for Biomedical Applications: A Review,” Journal of Nanoparticle Research, vol. 25, no. 43, pp. 1–7, 2023, doi: 10.1007/s11051-023-05690-w.

M. M. Abdellatif, S. M. Ahmed, M. A. El-Nabarawi, and M. Teaima, “Nano-Delivery Systems for Enhancing Oral Bioavailability of Drugs,” International Journal of Applied Pharmaceutics, vol. 15, no. 1, pp. 13–19, 2023, doi: 10.22159/ijap.2023v15i1.46758.

M. A. Abdessalem and S. A. Adham, “Research Advancements in Nanoparticles and Cell-Based Drug Delivery Systems for the Targeted Killing of Cancer Cells,” Oncology Research, vol. 33, no. 1, pp. 27–44, 2025, doi: 10.32604/or.2024.056955.

O. Afzal et al., “Nanoparticles in Drug Delivery: From History to Therapeutic Applications,” Nanomaterials, vol. 12, no. 24, pp. 1–27, 2022, doi: 10.3390/nano12244494.

M. Ahmad et al., “Biosynthesized Silver Nanoparticles Using Polygonatum Geminiflorum Efficiently Control Fusarium Wilt Disease of Tomato,” Frontiers in Bioengineering and Biotechnology, vol. 10, 2022, doi: 10.3389/fbioe.2022.988607.

Z. Ai et al., “Advances in Nano Drug Delivery Systems for Enhanced Efficacy of Emodin in Cancer Therapy,” International Journal of Pharmaceutics: X, vol. 9, no. 2025, pp. 1–13, 2025, doi: 10.1016/j.ijpx.2024.100314.

M. Akram et al., “Exploring the Frontiers of Green Nanotechnology: Advancing Biomedicine, Herbonanoceuticals, Environment, and Sustainability,” Scicom Journal of Medical and Applied Medical Sciences, vol. 2, no. 1, pp. 18–30, Sept. 2023, doi: 10.54117/sjmams.v2i1.11.

W. Y. Al-Asiri et al., “Cytotoxic and Apoptotic Effects of Green Synthesized Silver Nanoparticles via Reactive Oxygen Species–Mediated Mitochondrial Pathway in Human Breast Cancer Cells,” Cell Biochemistry and Function, vol. 42, no. 7, p. e4113, 2024, doi: 10.1002/cbf.4113.

Z. Ashikbayeva, A. Bekmurzayeva, Z. Myrkhiyeva, N. Assylbekova, T. Sh. Atabaev, and D. Tosi, “Green-Synthesized Gold Nanoparticle-Based Optical Fiber Ball Resonator Biosensor for Cancer Biomarker Detection,” Optics & Laser Technology, vol. 161, pp. 1–15, 2023, doi: 10.1016/j.optlastec.2023.109136.

R. Asif et al., “Antimicrobial Potential, Drug Delivery and Therapeutic Applications of Bio-Nanoparticles in Medicine,” in Nanomaterials: The Building Blocks of Modern Technology, T. Khan, M. Jawaid, K. A. Ahmad, and B. Singh, Eds., Singapore: Springer Nature Singapore, 2023, pp. 115–130. doi: 10.1007/978-981-99-4149-0_6.

X. Cheng, Q. Xie, and Y. Sun, “Advances in Nanomaterial-Based Targeted Drug Delivery Systems,” Frontiers in

Bioengineering and Biotechnology, vol. 11, no. 2023, pp. 1–27, 2023, doi: 10.3389/fbioe.2023.1177151.

K. Cho, X. Wang, S. Nie, Z. (Georgia) Chen, and D. M. Shin, "Therapeutic Nanoparticles for Drug Delivery in Cancer," Clinical Cancer Research, vol. 14, no. 5, pp. 1310–1316, 2008, doi: 10.1158/1078-0432.CCR-07-1441.

Z. Efati, S. S. Shahangian, M. Darroudi, H. Amiri, S. I. Hashemy, and M. R. Aghamaali, "Green Chemistry Synthesized Zinc Oxide Nanoparticles in Lepidium Sativum L. Seed Extract and Evaluation of Their Anticancer Activity in Human Colorectal Cancer Cells," Ceramics International, vol. 49, no. 20, pp. 32568–32576, 2023, doi: 10.1016/j.ceramint.2023.07.221.

M. Fahim, A. Shahzaib, N. Nishat, A. Jahan, T. A. Bhat, and A. Inam, "Green Synthesis of Silver Nanoparticles: A Comprehensive Review of Methods, Influencing Factors, and Applications," JCIS Open, vol. 16, pp. 1–23, 2024, doi: 10.1016/j.jciso.2024.100125.

M. A. A. Fayed, I. A. Abdallah, I. Ahmad, H. Patel, and E. M. Abdou, "Green Nanotechnology for Targeted Drug Delivery: UPLC-ESI-MS/MS, In Vitro/In Silico Cytotoxic and Antibacterial Activity of Pimpinella Anisum L. and Its Silver Nanoparticles," Journal of Molecular Structure, vol. 1314, pp. 1–13, 2024, doi: 10.1016/j.molstruc.2024.138800.

J. Georgeous, N. AlSawaftah, W. H. Abuwatfa, and G. A. Husseini, "Review of Gold Nanoparticles: Synthesis, Properties, Shapes, Cellular Uptake, Targeting, Release Mechanisms and Applications in Drug Delivery and Therapy," Pharmaceutics, vol. 16, no. 10, pp. 1–32, 2024, doi: 10.3390/pharmaceutics16101332.

Y. A. Hajam, R. Rani, and P. Sharma, "Chapter 16: Green Synthesized Nanomaterials for Drug Delivery," in Synthesis of Bionanomaterials for Biomedical Applications, M. Ozturk, A. Roy, R. A. Bhat, F. Vardar-Sukan, and F. M. B. T.-S. of B. for B. A. Policarpo Tonelli, Eds., Elsevier, 2023, pp. 319–338. doi: 10.1016/B978-0-323-91195-5.00009-X.

Harshita, H.-F. Wu, and S. K. Kailasa, "Recent Advances in Nanomaterials-Based Optical Sensors for Detection of Various Biomarkers (Inorganic Species, Organic and Biomolecules)," Luminescence, vol. 38, no. 7, pp. 954–998, 2023, doi: 10.1002/bio.4353.

K. Jain, A. Takuli, T. K. Gupta, and D. Gupta, "Rethinking Nanoparticle Synthesis: A Sustainable Approach vs. Traditional Methods," Chemistry – An Asian Journal, vol. 19, no. 21, pp. 1–10, 2024, doi: 10.1002/asia.202400701.

P. Jayachandran et al., "Green Synthesized Silver Nanoparticle-Loaded Liposome-Based Nanoarchitectonics for Cancer Management: In Vitro Drug Release Analysis," Biomedicines, vol. 11, no. 217, pp. 1–18, 2023, doi: 10.3390/biomedicines 11010217.

S. H. Khan, "Green Nanotechnology for the Environment and Sustainable Development BT: Green Materials for Wastewater Treatment," Mu. Naushad and E. Lichtfouse, Eds., Cham: Springer International Publishing, 2020, pp. 13–46. doi: 10.1007/978-3-030-17724-9_2.

N. Khatik, "Green Synthesis of Nanomaterials and Their Utilization as Potential Vehicles for Targeted Cancer Drug Delivery," Advances in Pharmacology and Pharmacy, vol. 10, no. 2, pp. 114–121, 2022, doi: 10.13189/app.2022.100205.

P. Kiran, A. Kar, A. Banerjee, and A. Puri, "Advancement of Polymer-Based Nanocarrier System in Drug Delivery," in Nanomaterials in Healthcare, 1st ed., R. Srivastava, S. K. Debnath, and R. Prasad, Eds., CRC Press, 2023, ch. Advancemen, pp. 1–19. doi: 10.1201/9781003322368

D. Kirubakaran et al., "A Comprehensive Review on the Green Synthesis of Nanoparticles: Advancements in Biomedical and Environmental Applications," Biomedical Materials & Devices, 2025, doi: 10.1007/s44174-025-00295-4.

T. K. Kumawat, V. Kumawat, V. Sharma, M. Biyani, and N. Kumari, "Green Nanomaterials for Sustainable Targeted Drug Delivery Systems," in Sustainable Green Nanomaterials, 1st ed., S. Kulkarni and M. R. Goyal, Eds., Apple Academic Press, 2024, ch. Green Nano, pp. 1–33. doi: 10.1201/9781003501572-11

Y. Li et al., "Biodegradable Ferric Phosphate Nanocarriers with Tumor-Specific Activation and Glutathione Depletion for Tumor Self-Enhanced Ferroptosis and Chemotherapy," Advanced Functional Materials, vol. 34, no. 21, p. 2313540, May 2024, doi: 10.1002/adfm.202313540.

R. Liu et al., "Advances of Nanoparticles as Drug Delivery Systems for Disease Diagnosis and Treatment," Chinese Chemical Letters, vol. 34, no. 2, pp. 1–10, 2023, doi: 10.1016/j.cclet.2022.05.032.

L.-J. Mah, S. Tortorella, and T. C. Karagiannis, "Nano-Based Drug Delivery Modalities for the Treatment of Cancer: The Formulation of Tumour-Specific and Targeted Nanoparticles BT: Molecular Mechanisms and Physiology of Disease: Implications for Epigenetics and Health," in Molecular Mechanisms and Physiology of Disease, N. Maulik and T. Karagiannis, Eds., New York, NY: Springer New York, 2014, pp. 471–501. doi: 10.1007/978-1-4939-0706-9_18.

A. Manzari-Tavakoli, A. Babajani, M. M. Tavakoli, F. Safaeinejad, and A. Jafari, "Integrating Natural Compounds and Nanoparticle-Based Drug Delivery Systems: A Novel Strategy for Enhanced Efficacy and Selectivity in Cancer Therapy," Cancer Medicine, vol. 13, no. 5, pp. 1–19, 2024, doi: 10.1002/cam4.7010.

N. D. Masane, A. S. Rathod, V. G. Akhand, V. A. Katekar, and S. P. Deshmukh, "Nanoparticles Based Drug Delivery System for Cancer Therapy," GSC Advanced Research and Reviews, vol. 22, no. 1, pp. 223–237, 2025, doi: 10.30574/gscarr.2025.22.1.0014.

S. B. Mayegowda et al., "Green-Synthesized Nanoparticles and Their Therapeutic Applications: A Review," Green Processing and Synthesis, vol. 12, no. 1, pp. 1–17, 2023, doi: 10.1515/gps-2023-0001.

T. K. Mohapatra, P. Behera, P. Patel, and R. Pothal, "Nano Based Drug Delivery Systems: Emerging Trends and Future Prospects," International Journal of Pharmaceutical Research and Applications, vol. 9, no. 5, pp. 1198–1213, 2024, doi: 10.35629/4494-090511981213.

U. M. Muddapur et al., "Plant-Based Synthesis of Gold Nanoparticles and Theranostic Applications: A Review," Molecules, vol. 27, no. 4, pp. 1–22, 2022, doi: 10.3390/molecules27041391.

S. Murugappan, M. Pebam, S. A. Sankaranarayanan, and A. K. Rengan, "Drug-Delivery, Antimicrobial, Anticancerous Applications of Green Synthesized Nanomaterials," in Green Synthesis of Nanomaterials, 2024, pp. 131–168. doi: 10.1002/9781119900931.ch7.

M. S. Othman, S. M. Aboelnaga, O. A. Habotta, A. E. A. Moneim, and M. M. Hussein, "The Potential Therapeutic Role of Green-Synthesized Selenium Nanoparticles Using Carvacrol in Human Breast Cancer MCF-7 Cells," Applied

Sciences, vol. 13, no. 12, pp. 1–14, 2023, doi: 10.3390/app13127039.

V. Pouresmaeil, S. Haghighi, A. S. Raeisalsadati, A. Neamati, and M. Homayouni-Tabrizi, “The Anti-Breast Cancer Effects of Green-Synthesized Zinc Oxide Nanoparticles Using Carob Extracts,” 2021. doi: 10.2174/1871520620666200721132522.

D. Prashar, B. Sharma, S. Kumar, Namrata, L. Sharma, and N. Rani, “Exploring Advanced Drug Delivery Systems: Unvieling the Approaches, and Applications of Liposomes, Niosomes, Nanoparticles, and Monoclonal Antibodies,” Novel Drug Delivery System from Fundamental Principles to Advanced Applications, vol. 3., IIP Series, ch. Exploring, pp. 161–186, 2024, doi: 10.58532/nbennurndch9.

S. C. Punitha and G. Aruna, “Green Materials as Therapeutic Agents BT: Encyclopedia of Green Materials,” in Encyclopedia of Green Materials, C. Baskar, S. Ramakrishna, and A. Daniela La Rosa, Eds., Singapore: Springer Nature, 2022, pp. 1–13. doi: 10.1007/978-981-16-4921-9_61-1.

Z. A. Ratan et al., “Green Chemistry Synthesis of Silver Nanoparticles and Their Potential Anticancer Effects,” Cancers, vol. 12, no. 4, pp. 1–8, 2020, doi: 10.3390/cancers12040855.

A. Saminathan, M. Zajac, P. Anees, and Y. Krishnan, “Organelle-Level Precision with Next-Generation Targeting Technologies,” Nature Reviews Materials, vol. 7, no. 5, pp. 355–371, 2022, doi: 10.1038/s41578-021-00396-8.

G. Sanità, B. Carrese, and A. Lamberti, “Nanoparticle Surface Functionalization: How to Improve Biocompatibility and Cellular Internalization,” Frontiers in Molecular Biosciences, vol. 7, no. 2020, pp. 1–20, 2020, doi: 10.3389/fmolb.2020.587012.

M. T. Shaaban, S. H. Orabi, M. S. Abdel-Hamid, R. M. S. Korany, F. M. Alshehrei, and R. H. Elbawab, “Therapeutic Potential of Green-Synthesized Silver Nanoparticles: Combating Biofilms of Multidrug-Resistant Staphylococcus Aureus RM-Ph8 and Modulating the Immune Response in the Liver Tissue of Rats,” Veterinary World, vol. 17, no. 10, pp. 2211–2224, 2024, doi: 10.14202/vetworld.2024.2211-2224.

L. Shariati et al., “Advances in Nanobased Platforms for Cardiovascular Diseases: Early Diagnosis, Imaging, Treatment, and Tissue Engineering,” Environmental Research, vol. 238, no. 1, pp. 1–12, 2023, doi: 10.1016/j.envres.2023.116933.

M. Sharifi et al., “An Updated Review on EPR-Based Solid Tumor Targeting Nanocarriers for Cancer Treatment,” Cancers, vol. 14, no. 12, pp. 1–17, 2022, doi: 10.3390/cancers14122868.

E. Shukla, D. D. Kara, T. Katikala, and M. Rathnanand, “Self-Nanoemulsifying Drug Delivery Systems (SNEDDS) of Anti-Cancer Drugs: A Multifaceted Nanoplatform for the Enhancement of Oral Bioavailability,” Drug Development and Industrial Pharmacy, vol. 49, no. 1, pp. 1–16, 2023, doi: 10.1080/03639045.2023.2182124.

L. Tang, J. Li, Q. Zhao, T. Pan, H. Zhong, and W. Wang, “Advanced and Innovative Nano-Systems for Anticancer Targeted Drug Delivery,” Pharmaceutics, vol. 13, no. 8, pp. 1–30, 2021, doi: 10.3390/pharmaceutics13081151.

H. Tian et al., “Enhancing the Therapeutic Efficacy of Nanoparticles for Cancer Treatment Using Versatile Targeted Strategies,” Journal of Hematology & Oncology, vol. 15, no. 132, pp. 1–40, 2022, doi: 10.1186/s13045-022-01320-5.

H. Tian Yan et al., “Tumor Microenvironment Activated Mussel-Inspired Hollow Mesoporous Nanotheranostic for Enhanced Synergistic Photodynamic/Chemodynamic Therapy,” Journal of Colloid and Interface Science, vol. 665, no. 2024, pp. 188–203, 2024, doi: 10.1016/j.jcis.2024.03.133.

M. Todaria, D. Maity, and R. Awasthi, “Biogenic Metallic Nanoparticles as Game-Changers in Targeted Cancer Therapy: Recent Innovations and Prospects,” Future Journal of Pharmaceutical Sciences, vol. 10, no. 25, pp. 1–60, 2024, doi: 10.1186/s43094-024-00601-9.

T. Van Tran, D. T. C. Nguyen, P. S. Kumar, A. T. M. Din, A. A. Jalil, and D.-V. N. Vo, “Green Synthesis of ZrO2 Nanoparticles and Nanocomposites for Biomedical and Environmental Applications: A Review,” Environmental Chemistry Letters, vol. 20, no. 2, pp. 1309–1331, 2022, doi: 10.1007/s10311-021-01367-9.

A. K. Wani et al., “Targeting Apoptotic Pathway of Cancer Cells with Phytochemicals and Plant-Based Nanomaterials,” Biomolecules, vol. 13, no. 2, pp. 1–34, 2023, doi: 10.3390/biom13020194.

19 Metal Oxide Nanoparticles against Bacterial Biofilms
Emerging Trends and Technologies

Neeraj Kumari, Sunil Puri, and Radha

19.1 INTRODUCTION

Bacteria survive in nature by forming biofilms on various surfaces, both living and non-living, as long as these environments provide adequate nutrition and moisture (Høiby et al., 2011). A biofilm is a structured community of microorganisms embedded within a self-produced matrix composed of proteins, nucleic acids, and polysaccharides. This matrix provides structural stability and protects bacterial cells from external damage. Notably, bacterial biofilms exhibit strong resistance to phagocytosis, antibiotics, disinfectants, and both adaptive and innate immune responses (Khan et al., 2017). As a result, biofilm-associated bacteria contribute to chronic infections characterized by persistent inflammation and tissue damage (Costerton et al., 2003; Bjarnsholt et al., 2009). Biofilms can form on various surfaces within the human body, including teeth (Palmer et al., 2011), lungs (Bjarnsholt et al., 2009), prosthetic joints (Trampuz and Zimmerli, 2005; Del Pozo and Patel, 2009), intravenous catheters, stents (Reid, 1999; Stensballe et al., 2007), wounds (Bjarnsholt et al., 2008; Kirketerp-Møller et al., 2008), and the middle ear (Hall-Stoodley et al., 2006). The findings of various studies suggest that biofilms may develop multidrug resistance because antibiotics either fail to reach the biofilm in effective concentrations or their biological potential is reduced. Nutrient limitations and oxygen gradients within biofilm result in a reduction of bacterial metabolism, causing lower effectiveness of antibiotics (Taylor et al., 2012; Ramasamy et al., 2016). However, the growing resistance against antibiotics underscores the need to investigate new strategies to inhibit the formation of biofilm and eliminate bacterial infections. Therefore, limitations of traditional antibiotic therapies have driven the search for advanced solutions like material or drug development from nanotechnology. In recent years, nanoparticles (NPs) have emerged as a promising tool for biofilm control. Among them, metal oxide nanoparticles (MONPs) have gained attention, exhibiting significant inhibitory effects against gram-negative and gram-positive bacteria (Natan and Banin, 2017; Xiu et al., 2021). Furthermore, despite strong antimicrobial activity, challenges involving the use of nanoparticles for humans exist involving cytotoxicity, aggregation, and potential environmental toxicity that must be addressed to ensure safe biomedical applications. Additionally, the selection of appropriate NPs for clinical use remains difficult due to variation in synthesis and testing methodologies. The standardization of these processes is essential for ensuring safety and effectiveness in medical and industrial fields (Algadi et al., 2024). The present chapter discusses MONPs and provides a comparative analysis of different MONPs, highlighting their varying effects on biofilm formation and control. It also provides insights into optimized material synthesis and selection for applications like packaging material for food, coatings on medical devices, treatment of polluted water, and environment remediation. Thus, bridging fundamental research with applied sciences, the chapter highlights the transformative role of MONPs in biofilm prevention and treatment of biofilms along with outlining future pathways for safe and sustainable integration of nanotechnology for various applications.

19.2 METAL OXIDE NANOPARTICLES

Nanomaterials employed for bacterial biofilm management are categorized on the basis of their properties and chemical composition (Mahamuni-Badiger et al., 2020; Thambirajoo et al., 2021; Asma et al., 2022; Sheng et al., 2023). For example, several types of NPs such as metal oxide NPs (ZnO, CuO, TiO_2, Ag_2O), metal-based NPs (Cu-NP, Fe-NP, Ag-NP, Au-NP), carbon-based NPs (graphene nanoparticles and carbon nanotubes), and lipid-based NPs (ethosomes, solid lipid nanospheres, transfersomes, nanostructured lipid carriers, niosomes, and liposomes) have shown promising results in drug delivery because of their controlled-release properties (Algadi et al., 2024). Furthermore, NPs can also be categorized based on surface area, size, morphology, charge, and other physicochemical properties. For example, smaller NPs exhibit stronger antibacterial activity (Mohammadi et al., 2010; Seil and Webster, 2012a), while positively charged NPs bind effectively to the negatively charged bacterial surface, enhancing their bactericidal effect (Buzea et al., 2007; Mohammadi et al., 2011; Fellahi et al., 2013; Besinis et al., 2014). The flexibility of antibiofilm agents allows for surface modifications that target biofilms and facilitate their accumulation within biofilm structure (Seil and Webster, 2012a). Additionally, sharp edges, spikes, pillars, or protrusions on surface of NPs allow puncture and rupture of

DOI: 10.1201/9781003644439-19

bacterial cells by generating high local stress (Noreen et al., 2017). The exact mechanism of the antibacterial/antibiofilm effect of nanomaterials is still being investigated. NPs either inhibit or disrupt biofilm growth by generating ROS, damaging bacterial cell walls/membranes, and reducing bacterial adhesion to surfaces (Dizaj et al., 2014). Further, high surface area enhances the effectiveness of NPs in preventing biofilm formation (Xiu et al., 2021; Algadi et al., 2024). Thus, antimicrobial activity has been improved by combining different NPs with other carrier substances, leading to the creation of nanocomposites and nanocarriers and making effective drugs for treating biofilm infections (Muzammil et al., 2018; McNeilly et al., 2021).

Recently, green methods for MONP synthesis in particular have drawn significant attention due to their sustainable approach and eco-friendly nature (Singh et al., 2018). Generally, there are two synthesis approaches to obtain green NPs with the desired shape, size, and properties: top-down and bottom-up. The top-down approach involves a reduction in the size of the source material via physical or chemical methods and results in material with surface defects, influencing the properties of the NPs. In the bottom-up method, NPs are first formed and then assembled into the final material. This method is helpful in producing small metal NPs with uniform chemical composition (Thakkar et al., 2010). The green approach utilizes eco-friendly biological materials, including plant extracts, fungi, algae, and bacteria. Among these sources, plant extracts are widely employed for NP synthesis due to their low cost, simple process, and feasibility for large-scale production. Reaction factors, such as temperature, solvent type, pH, and pressure, play a key role in influencing the green synthesis technique (Soltys et al., 2021). In various studies, it has been observed that NPs synthesized via green methods show enhanced and effective antimicrobial activity. Doping and incorporation of additional material is further employed to tailor the properties of NPs for specific applications. In a study by Mahajan et al. (2019), Ag-doped cobalt ferrite NPs synthesized using plant extracts exhibited increased and improved antibacterial activity. Similarly, Ag-doped magnesium ferrite NPs using citric acid synthesized via a microwave-assisted combustion and sol-gel technique showed enhanced antimicrobial efficiency with increasing Ag concentration (Fantozzi et al., 2021). Further, the integration of MONPs with biopolymers and other functional materials to enhance the stability, reducing toxicity and improving biofilm disruption efficiency, has also been investigated in various studies.

19.3 ANTIBIOFILM MECHANISMS OF MONPS

As mentioned earlier, MONPs exhibit antimicrobial activity through multiple mechanisms, including direct interaction with bacterial cell walls, ROS generation, and the controlled release of metal ions, all of which contribute to biofilm inhibition and bacterial cell disruption. These mechanisms collectively lead to bacterial cell death by targeting essential cellular structures and functions (Raghunath and Perumal, 2017) (Figure 19.1). One of the primary ways MONPs exert their antibacterial effects is by directly damaging and compromising bacterial cell wall structural integrity. The surface charge and size of MONPs play a crucial role in their interaction with bacterial membranes. Since gram-positive and gram-negative bacteria have different membrane compositions, their binding efficiency and uptake vary. MONPs interact with bacterial

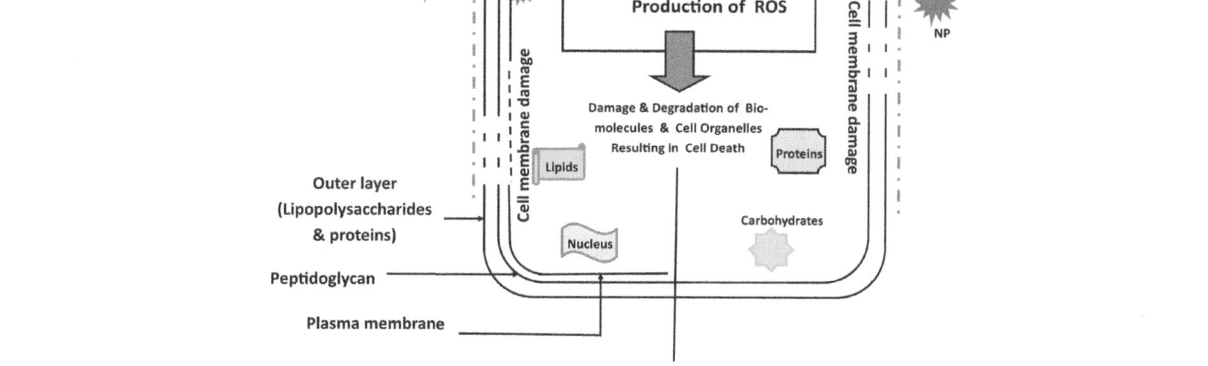

FIGURE 19.1 Mode of action of metal oxide nanoparticles on gram-positive and gram-negative bacteria (source and modified from Abo-zeid and Williams, 2020).

surfaces through electrostatic, hydrogen bonding, and van der Waals interactions (Parikh and Chorover, 2006). Lipopolysaccharides in gram-negative bacteria and lipoteichoic acids in gram-positive bacteria enable MONP binding, leading to endocytic uptake or nanoscale pore formation in the bacterial cell wall (Makin and Beveridge, 1996). These pores facilitate the entry of additional MONPs into the bacterial cytoplasm, where they interact with intracellular components such as lipids, enzymes, proteins, and DNA (Raghunath and Perumal, 2017). As a result, leakage of essential cellular contents occurs, ultimately leading to bacterial cell death.

Another major antibacterial mechanism of MONPs is ROS production, which includes superoxide anions, hydrogen peroxide, hydroxyl radicals, and organic hydroperoxides (Vatansever et al., 2013; Slavin et al., 2017). Although bacterial cells possess enzymatic (catalase, superoxide dismutase) and non-enzymatic (glutathione) antioxidant defenses, excessive ROS production overwhelms these systems, leading to oxidative stress and cellular dysfunction. MONPs release metal ions (e.g., Fe^{3+}, Cu^{2+}) both in the bacterial environment and inside the cytoplasm, triggering ROS formation (Pereira et al., 2012) (Table 19.1). These ROS damage essential biomolecules, including proteins, lipids, carbohydrates, and DNA, leading to bacterial dysfunction and ultimately cell death (Storz and Imlayt, 1999; Imlay, 2003). Some MONPs, such as TiO_2, can further enhance their antibacterial effects through photokilling, where UV light exposure increases ROS production, causing severe damage to bacterial membranes and genetic material (Chen and Wang, 2014). Cu_2O NPs exhibit particularly strong antibacterial effects due to redox cycling of Cu^+/Cu^{2+}, which sustains oxidative stress and disrupts bacterial survival (Meghana et al., 2015). Beyond ROS production, the release of metal ions (e.g., Cu^{2+}, Zn^{2+}, Fe^{3+}) disrupts bacterial metabolic pathways by interfering with enzymatic functions and redox homeostasis, ultimately leading to toxicity and bacterial death. Some metal ions can also directly interact with bacterial DNA, cross-linking and disrupting its helical structure, which inhibits replication and transcription (Gaballa and Helmann, 1998; Hood and Skaar, 2012; Skaar and Raffatellu, 2015; Raghunath and Perumal, 2017). Inside bacterial cells, MONPs undergo dissolution, creating a localized high concentration of toxic ions, further exacerbating bacterial damage. Additionally, MONPs induce genotoxic effects by damaging bacterial DNA, both chromosomal and plasmid, leading to DNA oxidation and fragmentation, impairing bacterial replication and survival (Giannousi et al., 2014). Studies have shown that exposure to MONPs, such as Cu_2O, can lead to significant DNA damage in bacteria like *E. coli* and *B. subtilis*. Overall, MONPs employ a combination of physical disruption, oxidative stress, metal ion toxicity, and genotoxicity to effectively kill bacterial cells (Table 19.1). These synergetic effects makes MONPs a promising tool for various applications, specifically as antimicrobial agents.

TABLE 19.1
Summary of Antibacterial Activity of Different MONPs Involving Target Bacteria; Mechanisms and Advantages of MONPs

Type of NP	Target Bacteria	Mechanisms of Antibacterial Action	Advantages of MONPs	References
CuO-NPs	MRSA, *E. coli*, oral *bacteria*, *B. subtilis*	ROS generation, Cu^{2+} release, physical interactions	Effective against antibiotic-resistant bacteria, antioxidant properties, eco-friendly synthesis	(Naskar and Kim, 2019; Shkodenko et al., 2020; Yang et al., 2021)
Fe_3O_4-NPs	*S. aureus, E. hirae, S. epidermidis, P. aeruginosa, E. faecalis, E. coli*	ROS generation, Trojan Horse effect, biofilm disruption	Catalytic action, antibacterial activity, antibiofilm effects	(Shkodenko et al., 2020; Franco et al., 2022)
TiO_2-NPs	*S. aureus, P. aeruginosa, E. coli*	ROS generation under UV, cell wall interaction, growth inhibition	Strong antibacterial activity, photocatalytic properties	(Hochvaldová et al., 2022; Franco et al., 2022)
ZnO-NPs	*E. coli, S. pneumoniae, S. aureus*	Cell membrane damage, ROS production, biofilm inhibition	Low toxicity, suppresses virulence, synergistic with antibacterials	(Shkodenko et al., 2020; Franco et al., 2022; Hamdan et al., 2023)
MgO-NPs	*E. coli, K. pneumoniae, R. solanacearum, S. aureus*	Spore interaction, ROS generation, defect-induced activity	Broad bacteriostatic action, effective at low doses, prolonged effect	(Naskar and Kim, 2019; Hochvaldová et al., 2022; Hamdan et al., 2023)
Al_2O_3-NPs	*E. coli, A. hydrophila, P. aeruginosa, P. putida*	Cell wall penetration, oxidative membrane breakdown	Broad-spectrum efficacy, MDR synergy	(Shkodenko et al., 2020; Balderrama-González et al., 2021; Hemeg, 2022)
Si/SiO_2-NPs	*S. aureus, E. coli, A. baumannii, C. freundii, Enterobacter* spp., *K. pneumoniae*	Cell interactions, ROS generation, biofilm prevention, antibiotic functionalization	Targeted delivery, reduced toxicity, MDR potential	Balderrama-González et al., 2021; Zhang et al., 2022; Mukherjee et al., 2023)
CaO-NPs	*E. coli, S. aureus, L. monocytogenes, S. typhimurium, R. solanacearum, B. cereus*	Superoxide generation, ROS-induced stress, membrane disruption	Enhanced with silver doping, eco-friendly	(Shkodenko et al., 2020; Hochvaldová et al., 2022; Hetta et al., 2023)

NPs: Nanoparticles; ROS: Reactive oxygen species; MDR: Multidrug resistance; UV: Ultraviolet

19.4 COMPARATIVE ANALYSIS OF DIFFERENT MONPS IN BIOFILM INHIBITION

Different investigations have been carried out to explore the effects of different MONPs against different bacterial strains. It was reported that the efficacy or inhibition rate of NPs varies with their characteristic properties, mainly depending on synthesis methods, size, and bacterial strains for which they are being tested.

19.4.1 Copper Oxide NPs

Copper oxide NPs have exhibited antagonistic effects against biofilm formation and promote eradication of preformed biofilms. The effect is mainly due to the toxicity of copper ions on cells associated with biofilm. For instance, in a study, it was reported that CuO NPs effectively inhibited biofilm formation against methicillin-resistant *S. aureus* (MRSA) and *E. coli*. The antibacterial activity of CuO NPs was observed to be the highest for MRSA, with a zone of inhibition (ZOI) of 22 mm, followed by 18 mm for *E. coli*. Additionally, CuO NPs exhibited higher toxicity than Fe_2O_3 NPs, displaying higher antibacterial activity and achieving complete eradication of bacterial biofilm within 4 days of exposure (Agarwala et al., 2014). At a 50-μg/mL dose, CuO NPs significantly inhibited the growth of total oral bacteria, suppressed the formation of biofilms of various model surfaces (like glass, acrylic dentures, and cultured human epithelial cells), and reduced the production of extracellular polysaccharides (EPSs) (Khan et al., 2013). In another study, CuO NPs displayed an inhibitory effect against *Ralstonia solanacearum* biofilms at concentrations of 125 and 250 μg/mL after periods of 24 and 72 hours of incubation, respectively (Chen et al., 2019).

Moreover, CuO NPs have also shown significant antibiofilm activity against other bacterial species, including *E. coli*, *B. megaterium*, *Acinetobacter* sp., *Klebsiella* sp., *Thiobacillus aquaesulis*, and *S. aureus*. The findings suggest that copper coating reacts readily with water, leading to release of copper ions, which in turn disrupt bacterial cell viability. However, the proposed mechanism involves interaction of CuO NPs with DNA or the cell wall of bacteria, leading to cell death, as confirmed by cell viability assays (Uthra et al., 2024). Overall, the findings suggest that CuO NPs can be considered a promising candidate for antibiofilm applications due to their effective antibacterial effect and broad spectrum activity against different bacterial strains.

19.4.2 Iron Oxide NPs

Iron oxide or Fe_3O_4 NPs (IONPs), referred to as magnetic NPs, have attracted considerable attention due to their unique magnetic properties, compatibility, and biomedical applications. Among these, their role as an antibiofilm agent is of particular interest. In a study, Fe_3O_4 NPs exhibited mild antibacterial activity with an effective dose of 10–20 mg/mL in a dose-dependent manner (Taylor and Webster, 2009; Tran et al., 2010; Margabandhu et al., 2015; Raghunath and Perumal, 2017). At higher concentrations, Fe_3O_4 NPs disrupted the architecture of biofilm and further destroyed cells embedded within the biofilm matrix. It has been observed that IONPs reduced biofilm formation by inhibiting the growth of *S. aureus*, *E. coli*, and *P. aeruginosa* (Thukkaram et al., 2014), as well as *S. epidermidis* (Taylor and Webster, 2009) and *Enterococcus hirae* (Gabrielyan et al., 2019). Besides the passive electrostatic effect of Fe_3O_4 NPs on biofilm, the penetration of these NPs is significantly enhanced in the presence of magnetic fields (Bandara et al., 2015; Geilich et al., 2017). This property allows NPs to exert mechanical disruption on biofilm structure, breaking and weakening the matrix integrity (Leuba et al., 2013; Raftery et al., 2014; Ranmadugala et al., 2017). However, due to these properties, IONPs are primarily used as carriers for biocides within biofilm. These enhance the efficacy of antibiotics, such as penicillin, kanamycin, streptomycin, cefotaxime, and erythromycin, against *S. aureus* biofilms (Niemirowicz et al., 2016). Fe_3O_4 NPs have also been conjugated with antifungal agents like nystatin and amphotericin B to improve their action against *Candida* sp. biofilms (Zhou et al., 2012). In recent studies, various biocide-conjugated Fe_3O_4 NPs have been developed to effectively target biofilms, enhancing drug efficacy through synergistic effects (Bandara et al., 2015; Geilich et al., 2017; Ranmadugala et al., 2017; Rumyantceva et al., 2019). Despites the limited studies on Fe_3O_4 NP antibacterial properties, their potential as an antibiofilm agent is significant, particularly in combination with antimicrobial drugs and irradiation with magnetic fields.

19.4.3 Aluminum Oxide NPs

Aluminum oxide or Al_2O_3 NPs are porous materials widely used for industrial and biomedical applications. These NPs possess excellent properties, including surface area, bio-inertness, mechanical strength, and resistance towards chemicals, and are cost effective and easily synthesized. It was observed that Al_2O_3 NPs exhibit weak antibacterial activity, inhibiting *E. coli* only at high concentrations (Sadiq et al., 2009). In another study, Al_2O_3 NPs were shown to be toxic to bacteria, with planktonic cells being more susceptible than biofilm-embedded cells when tested against *P. putida* and *Aeromonas hydrophila* in both planktonic and biofilm (Chrzanowska and Załęska-Radziwiłł, 2014). The MIC for *P. aeruginosa* biofilms was 1.6–3.2 mg/mL. At 2 mg/mL, it was observed that NPs resulted in significant inhibition of extended-spectrum-beta-lactamase (ESBL) and metallo-beta-lactamase (MBL), producing *P. aeruginosa* clinical isolates (Ansari et al., 2015). Overall, the antibiofilm and antimicrobial properties of Al_2O_3 NPs are weaker compared to other MONPs. However, due to limited efficacy, Al_2O_3 NPs are primarily useful in nanocomposites and biocide conjugates, where they can enhance antibacterial effects.

19.4.4 Zinc Oxide NPs

ZnO NPs are one of the most explored NPs due to their varied downstream applications. These are used due to their strong antimicrobial action, good biocompatibility, and ability to induce oxidative stress in bacterial cells. Studies have shown that the sub-MIC of ZnO NPs (3–12 µg/mL) exhibited significant antibiofilm activity against *S. pneumoniae*. The mechanism involved for this effect is reduction of cell adhesion to surfaces, thereby preventing the formation of biofilm (Bhattacharyya et al., 2018). The effective concentration of ZnO NPs typically ranges from 20–500 µg/mL (Hussain et al., 2019; Jamal et al., 2019). Happy et al. (2019) reported an inhibitory concentration of 50 µg/mL for *E. coli*. However, the efficacy of the NPs can be enhanced through physical treatments. For example, ultrasound-assisted treatment has resulted in enhanced cell penetration of NPs, leading to an increase (76%) in antibacterial activity against *S. aureus* due to generation of hydrogen peroxide (Seil and Webster, 2012b). Also, the treatment caused physical damage to the cell membrane, promoting the deeper penetration of ZnO NPs into bacterial cells (Wang et al., 2014). These NPs have also shown potential in controlling biofilm formation on dentures. They inhibit the growth of biofilm by *Rothia dentocariosa* and *R. mucilaginosa* through generation of Zn ions. The ions further interfere with the enzymatic activity of the DapE protein involved in peptidoglycan synthesis, leading to the failure of the formation of biofilm at initial stage (Khan et al., 2014). In conclusion, these findings highlight the strong antibiofilm potential of ZnO NPs, making them a promising candidate for application as coatings on devices and antibacterial materials.

19.4.5 Magnesium Oxide NPs

Magnesium oxide or MgO NPs are also a type of MONPs studied for their antibiofilm and antimicrobial properties. These NPs exhibited a broad spectrum of antimicrobial effect against spores, bacteria, and viruses at high concentrations, typically ranging from 100–1200 µg/mL (Blecher et al., 2011; Noori and Kareem, 2019; Maji et al., 2020). For *E.coli*, the MIC of MgO NPs was found to be 1 mg/mL (Nguyen et al., 2018). In another study, the antibiofilm activity of MgO NPs was evaluated against *E. coli* (250 µg/mL), *S. aureus* (500 µg/mL), and *K. pneumoniae* (125 µg/mL). The application of MgO NPs significantly reduced bacterial cell adhesion to plastic surfaces after incubation for 12 hours, thereby preventing biofilm development. In addition to inhibiting initial biofilm development, NPs have also been effective against mature biofilms. Treatment with 0.5 MIC led to a significant reduction in biofilm biomass (Hayat et al., 2018). Further evidence also supports that at 10 µg/mL, MgO NPs significantly inhibited *S. aureus* biofilm formation (Wong et al., 2020). Additionally, MgO NPs with 8-nm size demonstrated a strong inhibitory effect against *S. aureus* and *E. coli* biofilm formation (Makhluf et al., 2005) (Table 19.2). In conclusion, these findings suggest that MgO NPs possess substantial inhibitory activity in a dose-dependent manner.

19.4.6 Titanium Dioxide NPs

Titanium dioxide or TiO_2 has received considerable attention in various industrial applications. These NPs at concentrations ranging from 20–14000 µg/mL have been investigated for their antibiofilm activity against different bacterial strains (Daoud et al., 2005; Mahmoodi et al., 2006; Shah et al., 2008). These types of NPs have displayed an inhibition effect against both gram-positive and gram-negative bacteria, with the latter being more sensitive (Erkan et al., 2006; Landage et al., 2020). This difference in susceptibility might be due to the presence of a thick peptidoglycan layer in gram-positive bacteria, facilitating absorption of reactive radicals and preventing cell damage from oxidative stress (Pal et al., 2007). Generally, TiO_2 NPs exhibit antibiofilm activity by reducing cell adhesion and inducing the destruction of bacteria within biofilm. The action mechanism of NPs involves photocatalysis, which leads to the generation of ROS and lipid oxidation, leading to membrane damage (Shah et al., 2008; Dhandapani et al., 2012). It has been reported that these types of NPs effectively inhibited MRSA (Dhandapani et al., 2012) and *S. mitis* (Rajkumari et al., 2019). Specifically, these control the growth and biofilm formation of *S. mitis*, suggesting a potential application in oral health (Rajkumari et al., 2019). Additionally, at a concentration of 31.25 µg/mL, a reduction in *P. aeruginosa* biofilm development was observed, along with disruption of previously formed biofilms (Rajkumari et al., 2019). TiO_2 NPs result in a 40–50% reduction of *B. subtilis* and *E. coli* biofilm formation (Landage et al., 2020). Despite these effects, TiO_2 NPs do not exhibit significant bactericidal activity against certain antibiotic-resistant strains, such as *Cupriavidus metallidurans* CH34. This strain possesses high resistance to oxidative stress due to overexpression of membrane repair elements and protective components, which mitigate ROS-induced cell membrane damage (Simon-Deckers et al., 2009). Overall, TiO_2 NPs demonstrated strong antibiofilm and antibacterial properties, with applications in healthcare and environmental sanitation. However, their efficacy varies depending on bacterial species and resistance mechanisms.

19.4.7 Others

An Ag-SiO_2 NP-composite demonstrated prolonged antibacterial activity, with rapid initial Ag release at 24 h, followed by slower leaching at 48 and 72 h. The MONP was effective against MRSA and *E. coli*. UV irradiation was more effective as compared to filter sterilization in preserving its antibacterial properties due to a higher residual Ag concentration. Gauze impregnated with composite exhibited superior antibacterial activity compared to a commercial Ag-based dressing, highlighting its potential for wound care. Electron microscopy showed that released Ag ions interacted with bacterial cytoplasmic components, leading to membrane disruption. These findings suggest that Ag-SiO_2 is a promising material for wound dressing applications (Mosselhy et al., 2017). Altaf et al. (2021) reported that biosynthesized CeO_2-NPs from *Acorus calamus* extract inhibited bacterial biofilm formation by over 75% (Table 19.2). Confocal and electron microscopy confirmed that CeO_2-NPs effectively prevented bacterial colonization on solid surfaces. The NPs exhibited dose-dependent inhibition of preformed biofilms and significantly reduced EPS production.

TABLE 19.2
Comparison of Antibiofilm Activities of Different MONPs against Various Microbial Strains

Type of NP	Synthesis	Physical Properties	Target Bacteria	Key Finding	Reference
Ag-SiO_2-NPs	Stöber method	Spherical, size: 5 nm	MRSA, *E. coli*	Ag^+ interacts with bacterial cytoplasmic constituents, leading to membrane disruption at 250–500 µg/mL	(Mosselhy et al., 2017)
Ag-TiO_2-NPs	Chemical reduction	Spherical or quasi-spherical nanocrystals, size: 15–30 nm	*B. subtilis, C. albicans, S. aureus, E. coli, P. aeruginosa, K. pneumoniae*	ZOI: 0–11 mm, MIC: 150–1,200 µg/mL, MIC for biofilm inhibition: 9–5,000 µg/mL	(Lungu et al., 2014)
Nano-Al_2O_3 and Al_2O_3-NPs	Sigma-Aldrich	Size: 50 nm	*A. hydrophila*	With EPS: EC_{50} 709.68–1,000 mg/L; without EPS: EC_{50} 102.62–1,000 mg/L	(Chrzanowska and Załęska-Radziwiłł, 2014)
CeO_2-NPs	Purchased from Aladdin (Shanghai, China); sonochemical method	-	Periphytic biofilm	NPs cause ROS production, oxidative stress, and cell damage	(Tang et al., 2017)
CeO_2-NPs	Green synthesis	Spherical/ pseudo-spherical, size: 16.92 nm	*E. coli* ATCC 25922, *S. aureus* MTCC 3160, *P. aeruginosa* PAO1, *P. aeruginosa* PAO1	>75% biofilm inhibition	(Altaf et al., 2021)
CeO_2-CdO-NPs	Green synthesis, sol-gel	Platelet shaped, size: 10 nm	*P. aeruginosa* MTCC73, *S. pyogenes* MTCC26	ZOI: *P. aeruginosa* (33–38 mm), *S. pyogenes* (18–21 mm); ROS production	(Magdalane et al., 2017)
CuO-NPs	Chemical synthesized	Flower-like, size: ~25 nm	*Acinetobacter* sp., *E. coli, Klebsiella* sp., *S. aureus, B. megaterium, T. aquaesulis*	Interaction with bacterial DNA/ proteins leads to cell death; antibiofilm activity: 16–32 nm	(Uthra et al., 2024)
CuO-NPs	-	Size: 18.3 nm	*C. albicans, S. mutans, S. sobrinus, S. salivarius*	25–95% biofilm inhibition at 0–500 µg/mL, MIC: 64–128 µg/mL	(Ansarifard et al., 2021)
CuO-NPs	Nano Avijeh (Nanosav) Co.	Size: 40 nm	*C. albicans, C. krusei, C. glabrata*	ZOI: 0–21 mm	(Amiri et al., 2017)
CuO-NPs	Green synthesis	Semispherical, size: 10.7–30.9 nm	MRSA, *Acinetobacter* sp., *E. coli, K. pneumoniae*	MIC: 62.5–125 µg/mL, antibiofilm activity: 62.2–95% at 62.5–125 µg/mL	(Al-Jassani and Raheem, 2017)
CuO-NPs	Green synthesis	Spherical, size: 2–41 nm	*E. coli* ATCC 25922, *S. aureus* ATCC 43300	ZOI: 6–33 mm at 0.01–0.1 M	(Taran et al., 2017)
CuO-NPs	Gel-combustion method	Crystallite, size: 20–27 nm	*B. subtilis, E. coli, S. aureus, P. aeruginosa*	MIC: 20–75 µg/mL; ZOI: 10–24 mm	(Azam et al., 2012)
CuZnFe-NPs	From solutions of metals	Spherical, size: 42 nm	*E. coli, E. faecalis*	26–50% reduction in biofilm formation	(Alzahrani et al., 2018)
Fe_3O_4-NPs	Chemical synthesis	Spherical, 11–17 nm	*B. subtilis* (ATCC 6633)	Reduces biofilm population density while maintaining bacterial viability	(Ranmadugala et al., 2017)
Fe_3O_4-NPs	Wet chemical method	Size: <10nm	*S. aureus, P. aeruginosa, E. coli*	ZOI = 10–29 mm at 0.01–0.15 mg/mL	(Thukkaram et al., 2014)
Fe_3O_4-NPs	Co-precipitation method	Round, size: 10.64 nm	*E. hirae* ATCC 9790, *E. coli* BW 25113	Alters bacterial growth characteristics, ion transport, and enzymatic properties	(Gabrielyan et al., 2019)
MgO-NPs	Wet chemical method	Spherical, size: < 100 nm	*E. coli, K. pneumoniae, S. aureus*	ZOI: 17.5–35.5 mm, MBC: 500–1000 µg/mL, MIC: 125–500 µg/mL, biofilm inhibition: 25.3–49.8%	(Hayat et al., 2018)

(Continued)

TABLE 19.2 *(Continued)*
Comparison of Antibiofilm Activities of Different MONPs against Various Microbial Strains

Type of NP	Synthesis	Physical Properties	Target Bacteria	Key Finding	Reference
MgO-NPs	Sigma-Aldrich	Spherical, size: 50–100 nm	*R. solanacearum*	MIC = 200 μg/mL, MBC = 250 μg/mL, destruction of membrane integrity	(Cai et al., 2018)
MgO-NPs	Microwave radiation	Size: 8–23 nm	*E. coli, S. aureus*	Viability reduction: *S. aureus* (10–95%), *E. coli* (20–99.9%) at 1 mg/mL (1–4 hr)	(Makhluf et al., 2005)
MgO doped Li-NPs	Sol–gel method	Flake like, size: 20–60 nm	*E. coli* ATCC 25922	MIC: 500 μg/mL no bacterial colony for 24 hr	(Rao et al., 2013)
NiO-NPs	Green synthesis	Size: 10–20 nm	*S. aureus* (sensitive and resistant), *P. aeruginosa, E. coli*	MIC: 0.8–1.6 mg/mL, MBC: 1.6–3.2 mg/mL, biofilm inhibition: 1–83%	(Saleem et al., 2017)
TiO_2-NPs	Sol–gel method	-	*B. subtilis* 168	Exhibits anti-adhesive activity against *B. subtilis* biofilm	(Raie et al., 2017)
TiO_2-NPs	Sigma-Aldrich	Size: <50 nm	MRSA biofilm	ZOI = 11–14 mm (100–500 μg/mL)	(Jesline et al., 2015)
TiO_2-NPs	Green synthesis, by bacteria	Spherical, size: 10–30 nm	*B. subtilis* (FJ460362)	Formation of H_2O_2, resulting in destruction of microbes within biofilm	(Dhandapani et al., 2012)
Zn-doped CuO-NPs	Sonochemical method	Spherical/ elliptical, size: 20–95 nm	*E. coli* ATCC 25922, *P. mirabilis, S. aureus* ATCC 29213	High antibiofilm activity	(Shalom et al., 2017)
ZnMgO-NPs	Burning corresponding metal ribbons	Cubic, size: ~50 nm	*E. coli, B. subtilis*	Complete inactivation of *B. subtilis*, 80 % of inactivation *E. coli* growth	(Vidic et al., 2013)
ZnO-NPs	Wet chemical method	Spherical, size: 8–15 nm	*S. pneumoniae* MTCC 2672	MIC: 40 μg/mL, sub-MIC doses reduced biofilm formation at 12 μg/mL	(Bhattacharyya et al., 2018)
ZnO-NPs	Sigma-Aldrich	Size: <50 nm	*P. aeruginosa* PAO1	>95% biofilm inhibition	(Lee et al., 2014)
ZnO-NPs	Green synthesis	Triangles, hexagons, rods, rectangles, size: 10–50 nm	*E. coli, B. licheniformis, P. vulgaris B. pumilus*	ZOI: 20.3–26.3 mm, biofilm inhibition: 79–91%	(Ishwarya et al., 2018)
ZnO-NPs	Wet chemical method	Size: ~7–10 nm	Classical (O395) and ElTor (N16961) *V. cholerae*	Increased fluidity and depolarization of membrane, protein leakage, and DNA damage	(Sarwar et al., 2016)
ZnO-NPs	Green synthesis	Size: 65 nm	Strains from cystic fibrosis patients: 6; C-30 resistant strain (urine); PA14 (gallium resistant mutants): 2; PA14 C-30 (resistant mutant); environmental isolate: 4	Exhibits quorum quenching	(García-Lara et al., 2015)
ZnO-NPs	Green synthesis	Spherical/ hexagonal, size: 20–50 nm	MRSA	Poor biofilm formation on treatment with NPs (10 μg/mL)	(Vijayakumar et al., 2015)
Starch-capped ZnO-NPs	Wet chemical method	Spherical, size: 500 nm	*P. aeruginosa, S. aureus*	43–63% hydrophobicity inhibition, NPs disintegration and recalcitration of biofilm architecture	(Pati et al., 2014)
ZnO-NPs	Green synthesis	Size: 120–169 nm	*V. cholerae, E. coli* (ETEC)	Exhibits potential antibacterial activity	(Salem et al., 2015)
ZnO-NPs	Sigma-Aldrich	Size: <100 nm	MRSA	ZOI = 12–17 mm (100–500 μg/mL)	(Jesline et al., 2015)
ZnO-NPs	No information	Plates/spheres/ pyramids, size: 20 nm	*S. epidermidis* RP62A, *E. coli* UTI89 and MG1655, MRSA SH1000, *K. pneumoniae* LM21	Dose-dependent inhibition of growth of bacteria	(McGuffie et al., 2016)
ZnO-NPs	Green synthesis, microwave radiation	Hexagonal wurtzite, size: ~26 nm	*C. violaceum, P. aeruginosa, E. coli, L. monocytogenes*	27–93% reduction in biofilm biomass	(Al-Shabib et al., 2017)

TABLE 19.2 *(Continued)*
Comparison of Antibiofilm Activities of Different MONPs against Various Microbial Strains

Type of NP	Synthesis	Physical Properties	Target Bacteria	Key Finding	Reference
ZnO-NPs	Green synthesis, microwave radiation	Spherical, size: <50 nm	*C. violaceum, E. coli, P. aeruginosa, K. pneumoniae, S. marcescens, L. monocytogenes*	MIC: 50–400 μg/mL; biofilm inhibition: 16–84%	(Al-Shabib et al., 2018)
ZnO-NPs	Sol-gel method	Polygonal, size: 35 nm	*R. dentocariosa, R. mucilaginosa*	IC_{50} = 53–76 μg/mL, reduction in biofilm: 20–94% (50–200 μg/mL)	(Khan et al., 2014)
ZnO-NPs	Purchased at US Research Nanomaterials Co.	Size: 10–30 nm	*E. coli* PTCC 1399	33.4–50% reduction in biofilm formation	(Shakerimoghaddam et al., 2017)
ZnOAu-NPs	Green synthesis, photoreduction	Size: 0.77–2.3 nm	*S. aureus, E. coli*	Enhanced ROS generation, photocatalytic and antibacterial activity	(He et al., 2014)
α-Fe_2O_3-NPs	No information	Size: 2–540 nm	*P. aeruginosa* (PA01)	Small-sized NPs inhibited antimicrobial properties, while larger ones did not; 21.69–87.15 nm biofilm formation	Borcherding et al. (2014)

MRSA: Methicillin-resistant *Staphylococcus aureus*; EPS: Extracellular polysaccharides; ZOI: Zone of inhibition; ROS: Reactive oxygen species; NPs: Nanoparticles

19.5 APPLICATION OF MONPS IN VARIOUS SECTORS

19.5.1 Medical Sector

Bacterial infections and biofilm formation pose major challenges in implants, often leading to complications. MONPs have shown potential in preventing biofilm formation, particularly in dental and orthopedic applications. Beyene and Ghosh (2019) demonstrated that hydroxyapatite (HAP)-ZnO nanocomposites (90:10, 75:25, 60:40) effectively inhibited *S. aureus* and *E. coli*, with ZOI ranging from 3–16 mm and 1–11 mm, respectively. The MBC was 0.2 mg/mL for the 75:25 and 60:40 nanocomposites, with maximum biofilm inhibition (52% for *S. aureus*, 54% for *E. coli*) observed at a 60:40 ratio. These results suggest the potential of HAP-ZnO biomaterials for antimicrobial and antibiofilm applications in bone implants. Similarly, TiO_2 NPs incorporated into toothpaste enhanced antibacterial activity against dental plaque compared to conventional formulations. TiO_2 synthesized via wet chemical methods exhibited superior antimicrobial activity over commercial TiO_2 and bulk oxide NPs. A mouthwash containing TiO_2 NPs showed a ZOI of 2 cm against oral pathogens (Rimjhim and Thomas, 2016). Additionally, dental composite resins with nanoceria (CeO_2) demonstrated strong antibacterial activity against *S. mutans* (99.503%), *S. mitis* (99.441%), *S. aureus* (99.416%), and *Lactobacillus* spp. (99.233%) (Varghese et al., 2022). MONPs have also been explored in wound dressings and orthopedic implants. Krokowicz et al. (2015) reported that SiO_2 NPs exhibited antibacterial (ZOI: 0–8.5 mm) and antifungal (ZOI: 1.5–16 mm) activity when incorporated into wound dressings. Colon et al. (2006) observed that nanophase ZnO-NPs significantly reduced *S. epidermidis* colonies compared to microphase ZnO and TiO_2. When adjusted for surface area, bacterial colony counts decreased by 60% and 69%, while cell density dropped by 60% and 75%, respectively (Table 19.3). Nanophase ZnO-NPs exhibited lowest bacterial count, whereas microphase ZnO and TiO_2-NPs showed higher values. Beyond their antimicrobial effects, ZnO NPs also enhance osteoblast function and calcium mineral deposition, which are essential for the success of orthopedic implants. Their role in improving implant longevity and effectiveness underscores their potential in regenerative medicine and biomedical applications (Colon et al., 2006).

19.5.2 Food Packaging

In recent years, the utilization of metal oxide biocomposites has received attention in terms of sustainable material for food packaging. It offers a more advanced, eco-friendly alternative to traditional packaging materials. In an investigation, Mehmood et al. (2020) developed a nanocomposite film by incorporating IONPs with bovine gelation. The increasing concentration of IONPs up to 10% improved the film properties, while a 20% increase exhibited strong antimicrobial activity. This technique effectively extended the shelf life of fresh produce like grapes. In another study, Jayakumar et al. (2022) developed a biodegradable nanocomposite film using TiO_2 NPs, polyvinyl alcohol/starch, and elderberry extract. The addition of NPs improved tensile strength, moisture resistance, and antibacterial activity.

Additionally, the developed films functioned as pH indicators due to the presence of elderberry extract. Thus, when this film is used for cherry tomatoes, it reduced weight loss and microbial contamination and preserved the freshness of the produce for up to 22 days.

Liu et al. (2023) synthesized film by combining gelatin/chitosan with nano-ZnO and tea tree essential oil. The final developed film exhibited improved thermal stability with vapor resistance and antibacterial activity against *S. aureus* and *E. coli*, making it a potential material for active food packaging applications (Table 19.3). In another study, hydroxypropyl methylcellulose films containing Al_2O_3 and SiO_2 NPs showed efficacy against foodborne pathogens on chicken filets. At a particle size of approximately 80 nm and dose of 80 ppm, the films reduced the microbial load by 2–3 log^{10} CFU/cm^2 under refrigeration conditions for 15 days. The study highlights the potential of NP-based composite polymer film in enhancing safety and reducing contamination of perishable meat products. Overall, this will not only improve the structural and functional properties of packaging material but also help in increasing the shelf life of perishable products by inhibiting microbial growth. The incorporation of NPs, like TiO_2, ZnO, and SiO_2, has shown potential in maintaining and improving food quality, maximizing shelf life, and ensuring food safety and environmental sustainability, making them a viable solution for next-generation packaging material.

19.5.3 Water and Environmental Remediation

The increasing threat of multidrug-resistant bacteria in water systems has necessitated the development of innovative antimicrobial solutions. In this context, ZnO and CuO NPs have demonstrated significant antibiofilm efficacy against six highly biofilm-producing, multidrug-resistant bacterial strains: *Acinetobacter* sp., *Klebsiella* sp., *E. coli*, *S. aureus*, *B. megaterium*, and *T. aquaesulis*. Among them, ZnO exhibited superior antibacterial activity compared to CuO, with bacterial growth inhibition depending on both NP concentration and exposure time. Furthermore, antibiofilm assays on NP-coated glass specimens revealed significantly lower cell survivability compared to controls, aligning with ICP analysis results, which confirmed that Zn and Cu coatings released ions into the water. These findings suggest that the antibacterial action of NPs is driven by their interactions with bacterial DNA or cell wall proteins, ultimately causing cell death. This highlights their potential as a promising strategy for mitigating antibiotic-resistant pathogens, even in post-wastewater treatment applications (Uthra et al., 2024) (Table 19.3).

TABLE 19.3
Summary of Different MONP-Based Applications across Different Sectors

Type	Synthesis	Bacteria	Application	Reference
ZnO-NPs	Green synthesis	*S. aureus, E. coli*	Bone implant	(Beyene and Ghosh, 2019)
TiO_2-NPs	Wet chemical method	Dental samples (Savitha Dental Clinic, Jawaharlal Technological University)	Dental plaque	(Rimjhim and Thomas, 2016)
TiO_2-NPs	-	*S. mutans*	Dental disinfectant	(Besinis et al., 2014)
ZnO/TiO_2-NPs	Nanophase technologies	*S. epidermidis*	Bone implant	(Colon et al., 2006)
SiO_2-NPs	-	*S. aureus, P. mirabilis, E. faecalis, C. freundii, P. aeruginosa, K. oxytoca, E. coli*	Wound dressing	(Krokowicz et al., 2015)
ZnO-NPs	USA	*Streptococcus sobrinus*	Dental biofilm	(Aydin Sevinç and Hanley, 2010)
ZnO-NPs	-	-	Oral biofilm	(Garcia et al., 2021)
CeO2	Coprecipitation method	*S. mutans, S. mitis, S. aureus, Lactobacillus* spp.	Dental composite filler	(Varghese et al., 2022)
IO-NPs	Sigma Aldrich, (St. Louis, MO)	*E. coli, S. aureus*	Shelf-life enhancement	(Mehmood et al., 2020)
TiO_2-NPs	Beijing Beike 2D nanomaterials Co., Ltd.	-	Shelf-life enhancement	(Jayakumar et al., 2022)
ZnO-NPs	-	*S. aureus, E. coli*	Food packaging	(Liu et al., 2023)
CuO-NPs	-	Gram-negative and Gram-positive bacteria	Food packaging	(Hasanin and Youssef, 2022)
Al_2O_3/SiO_2-NPs	Beni-Suef University, Egypt	*B. cereus*, S. *aureus, S. typhimurium*	Packaging film	(Osman et al., 2019)
NiO-NPs	Co-precipitation	*S. aureus, E. coli*	Food packaging	(Momtaz et al., 2024)
NiO-NPs	Solution combustion method	*S. aureus, S. typhimurium*	Food packaging	(Marand et al., 2021)
CuO, ZnO-NPs	Chemical method	*Acinetobacter* sp., *E. coli, Klebsiella* sp., *S. aureus, B. megaterium, T. aquaesulis*	Environment remediation	(Uthra et al., 2024)
ZnO-NPs	-	*E. coli, S. typhimurium*	Water treatment	(Cruz et al., 2018)
ZnO-NPs	-	*E. coli, S. aureus, B. subtilis*	Water treatment	(Sapkota et al., 2011)
ZnO-NPs	-	*E. coli, S. aureus*	Water treatment	(Baruah et al., 2012)

NPs: Nanoparticles

Beyond direct antibacterial applications, ZnO nanorods have shown remarkable potential for water purification through visible light-mediated photocatalysis. Studies have demonstrated their effectiveness in inactivating waterborne bacteria, including *E. coli*, *S. aureus*, and *B. subtilis*. The reduced bacterial load or density under sunlight suggests that ZnO nanorods might be an alternative to an ex-situ water decontamination approach, particularly in areas with high sunlight exposure. Additionally, post-catalysis analysis confirmed that levels of dissolved ZnO nanorods remained non-lethal within safe limits, ensuring the safety of treated water. Furthermore, a novel technique based on endonuclease pretreatment provided evidence of potential DNA damage in bacterial cells, supporting the mechanism of action of ZnO NPs (Sapkota et al., 2011). Based on this concept, researchers have developed antibacterial water purifiers using ZnO nanorods grown on polyethylene fibers. This demonstrated the excellent efficacy of NPs in treating contaminated water with ~10^{10} CFU of *E. coli* and *S. aureus*. Under direct sunlight exposure, 99% of bacterial cells were immobilized, while under standard room conditions, about 80% of *E. coli* and 59% of *S. aureus* were effectively inactivated. This technology offers a cost-effective and sustainable solution for treating polluted water, specifically in disaster-affected regions where the availability of clean water is low but sunlight is abundant (Baruah et al., 2012). Generally, the antibacterial property or efficacy of ZnO nanorods is the result of two synergistic mechanisms: the slow release of ions through ZnO dissolution and the generation of free radicals via photocatalysis. These combined effects enhance bacterial inactivation, making ZnO-based materials a promising tool for eco-friendly water purification (Baruah et al., 2012). As an advancement in this field, the integration of nanotechnology with water treatment strategies could provide an effective and sustainable solution to address challenges of water contaminated with microorganisms.

19.6 TOXICOLOGICAL AND ENVIRONMENTAL CONSIDERATIONS

Despite the excellent role of MONPs in various applications, they have shown impacts on biological systems by causing oxidative stress, inflammatory response, and the risk of potential toxicity (Boyadzhiev et al., 2023; Summer et al., 2024). The small size and high surface reactivity contribute to the increased interactions of NPs with cells and tissues, resulting in various biological effects. The exposure of MONPs disrupts cellular functions, causing oxidative damage and alteration in metabolic processes (Tee et al., 2016; Horie and Tabei, 2021). They generate ROS, which induce protein damage, cellular stress, and lipid peroxidation. However, prolonged exposure may impair organ function and contribute to systemic toxicity. Further, NPs have been found to accumulate in different organs, potentially affecting normal physiological processes (Sarkar et al., 2014; Yarjanli et al., 2017; Chaudhary et al., 2019). Cellular organelles like mitochondria, endoplasmic reticulum (ER), and lysosomes have been structurally damaged. Mitochondrial stress can disrupt energy production, while ER stress may interfere with protein synthesis and lipid metabolism. The lysosomal accumulation of MONPs leads to membrane instability, potentially triggering cell death pathways (Stuparu-Cretu et al., 2023). Additionally, there are also environmental concerns if there is continuous use of MONPs in an ecosystem. Their release negatively impacts water and soil systems by altering nutrient cycles and microbial communities and also poses a risk to aquatic life forms. Besides this, these impacts also depend on various factors such as concentration of NPs and their size, surface properties, and environmental conditions (Wang et al., 2023). However, despite these known factors, there is still a gap in knowledge about the long-term consequences of MONPs. Further studies are still required to fully understand the interactions of MONPs with biological and environmental systems and to develop strategies for minimizing the extent of potential risks.

19.7 FUTURE PROSPECTS AND EMERGING TRENDS

The increasing challenges associated with infections have gained intensive attention and research into innovative antimicrobial strategies. MONPs have been observed as promising candidates due to their physicochemical properties and potent antibiofilm activities. The recent advancements and future trends in this field are multifaceted, encompassing the development of nanocomposites, synthesis via green methods, and further integration with photothermal therapies (Shkodenko et al., 2020; Algadi et al., 2024). One of the notable trends observed is engineering nanocomposites that integrate MONPs with other materials in order to enhance their antibiofilm efficacy. For instance, integrating MONPs with polymers or other NPs may result in synergistic effects, improving the biocompatibility and mechanical properties. Further, these composites have demonstrated significant potential in inhibiting biofilm formation and eradicating established biofilms, offering a potential tool for antimicrobial applications (Shkodenko et al., 2020; Franco et al., 2022; Sun et al., 2022).

Incorporating MONPs into photothermal therapy (PTT) represents another emerging approach. Various MONPs covert light energy into heat, effectively disrupting the structural integrity of the biofilm upon local irradiation. Thus, this method has shown promise in reducing biofilm viability, particularly when combined with antimicrobial agents, thereby enhancing overall treatment efficacy (Franco et al., 2022; Mammari and Duval, 2023). Green methods of NP synthesis using biological materials like plant extracts and others are gaining attention as an eco-friendly and sustainable alternative to traditional/conventional methods. This approach results in minimizing environmental impact while enhancing biocompatibility, with photosynthesized MONPs exhibiting significant inhibitory biofilm properties (Shkodenko et al., 2020; Franco et al., 2022; Sun et al., 2022). Furthermore, the functionalization of MONPs to control the generation of ROS has also been investigated to enhance their efficiency. By modifying the surface properties of NPs, one can modulate free radical generation, improving biofilm disruption while reducing cytotoxic effects on host tissues (Franco et al., 2022; Sun et al., 2022). However, despite these advancements in biomedical research regarding NPs, there are still challenges. The environmental impact and potential cytotoxic effects of utilizing NPs necessitate comprehensive biocompatibility assessments. Moreover, the scalability of synthesis and integration into medical devices/equipment requires further optimization. Future research should focus on studying antibiofilm mechanisms, optimizing evaluation methodologies, and validating the clinical efficacy of MONPs through in vivo models.

19.8 CONCLUSION

In nanotechnology, MONPs have gained attention as a promising tool in battling against bacterial biofilms, offering an alternative to conventional antibiotics that are increasingly failing due to microbial resistance. Depending on their physicochemical properties, such as shape, size, solubility, high surface area-to-volume ratio, and ability to generate ROS, MONPs can effectively disrupt or inhibit biofilm development. Among the different MONPs, ZnO, TiO_2, CuO, and AgO have exhibited strong antibiofilm and antibacterial activity through mechanisms such as ROS generation, membrane disruption, and interference with quorum sensing. The applications of these extend beyond medical therapeutics to industrial and environmental settings. The integration of MONPs in medical devices, packaging materials, wastewater treatment, and antimicrobial surface modifications has highlighted advancements in controlling biofilms. The green method of synthesis of MONPs involving plant extracts and biological materials further enhances their biocompatibility and sustainability, addressing concerns regarding their accumulation in the environment and biological systems. However, besides these practices, there is still a gap that needs to be addressed. The long-term impact of utilization of nanomaterials on the human body and ecosystem necessitates in-depth toxicological studies and regulatory frameworks. Additionally, standardizing synthesis methods to improve specificity, stability, and targeted delivery will be critical for effective implementation of MONPs. Future research should focus on developing functionalized MONPs with enhanced target specificity capabilities and synergistic effects with existing antimicrobial agents and investigating their role in modulating microbial resistance patterns. Overall, MONPs represent a transformative approach to biofilm management, bridging the gap between real-world applications and scientific studies. The continued development and responsible integration of nanotechnology and nanomaterials into the medical and industrial fields and ecosystems hold great potential in advancing sustainable antimicrobial strategies while mitigating the rising threat of antibiotic-resistant biofilms.

REFERENCES

Abo-zeid, Y., & Williams, G. R. (2020). The potential anti-infective applications of metal oxide nanoparticles: A systematic review. *Wiley Interdisciplinary Reviews: Nanomedicine and Nanobiotechnology*, *12*(2), e1592. https://doi.org/10.1002/wnan.1592

Agarwala, M., Choudhury, B., & Yadav, R. N. S. (2014). Comparative study of antibiofilm activity of copper oxide and iron oxide nanoparticles against multidrug resistant biofilm forming uropathogens. *Indian Journal of Microbiology*, *54*, 365–368. https://doi.org/10.1007/s12088-014-0462-z

Algadi, H., Alhoot, M. A., Al-Maleki, A. R., & Purwitasari, N. (2024). Effects of metal and metal oxide nanoparticles against biofilm-forming bacteria: A systematic review. *Journal of Microbiology and Biotechnology*, *34*(9), 1748–1756. https://doi.org/10.4014/jmb.2403.03029

Al-Jassani, M. J., & Raheem, H. Q. (2017). Anti-bacterial activity of CuO nanoparticles against some pathogenic bacteria. *International Journal of ChemTech Research*, *10*(2), 818–822.

Al-Shabib, N. A., Husain, F. M., Ahmed, F., Khan, R. A., Ahmad, I., Alsharaeh, E., . . . & Aliev, G. (2017). Erratum: Biogenic synthesis of Zinc oxide nanostructures from *Nigella sativa* seed: Prospective role as food packaging material inhibiting broad-spectrum quorum sensing and biofilm. *Scientific Reports*, *7*, 42266. https://doi.org/10.1038/srep42266

Al-Shabib, N. A., Husain, F. M., Hassan, I., Khan, M. S., Ahmed, F., Qais, F. A., . . . & Ahmad, I. (2018). Biofabrication of zinc oxide nanoparticle from *Ochradenus baccatus* leaves: Broad-spectrum antibiofilm activity, protein binding studies, and in vivo toxicity and stress studies. *Journal of Nanomaterials*, *2018*(1), 8612158. https://doi.org/10.1155/2018/8612158

Altaf, M., Manoharadas, S., & Zeyad, M. T. (2021). Green synthesis of cerium oxide nanoparticles using *Acorus calamus* extract and their antibiofilm activity against bacterial pathogens. *Microscopy Research and Technique*, *84*(8), 1638–1648. https://doi.org/10.1002/jemt.23724

Alzahrani, K. E., Niazy, A. A., Alswieleh, A. M., Wahab, R., El-Toni, A. M., & Alghamdi, H. S. (2018). Antibacterial activity of trimetal (CuZnFe) oxide nanoparticles. *International Journal of Nanomedicine*, 77–87. https://doi.org/10.2147/IJN.S154218

Amiri, M., Etemadifar, Z., Daneshkazemi, A., & Nateghi, M. (2017). Antimicrobial effect of copper oxide nanoparticles on some oral bacteria and candida species. *Journal of Dental Biomaterials*, *4*(1), 347.

Ansari, M. A., Khan, H. M., Alzohairy, M. A., Jalal, M., Ali, S. G., Pal, R., & Musarrat, J. (2015). Green synthesis of Al 2 O 3 nanoparticles and their bactericidal potential against clinical isolates of multi-drug resistant Pseudomonas aeruginosa. *World Journal of Microbiology and Biotechnology*, *31*, 153–164. https://doi.org/10.1007/s11274-014-1757-2

Ansarifard, E., Zareshahrabadi, Z., Sarafraz, N., & Zomorodian, K. (2021). Evaluation of antimicrobial and antibiofilm activities of copper oxide nanoparticles within soft denture liners against oral pathogens. *Bioinorganic Chemistry and Applications*, *2021*(1), 9939275. https://doi.org/10.1155/2021/9939275

Asma, S. T., Imre, K., Morar, A., Herman, V., Acaroz, U., Mukhtar, H., . . . & Gerlach, R. (2022). An overview of biofilm formation–combating strategies and mechanisms of action of antibiofilm agents. *Life*, *12*(8), 1110. https://doi.org/10.3390/life12081110

Aydin Sevinç, B., & Hanley, L. (2010). Antibacterial activity of dental composites containing zinc oxide nanoparticles. *Journal of Biomedical Materials Research Part B: Applied Biomaterials*, *94*(1), 22–31. https://doi.org/10.1002/jbm.b.31620

Azam, A., Ahmed, A. S., Oves, M., Khan, M. S., & Memic, A. (2012). Size-dependent antimicrobial properties of CuO nanoparticles against Gram-positive and-negative bacterial strains. *International Journal of Nanomedicine*, 3527–3535. https://doi.org/10.2147/IJN.S29020

Balderrama-González, A. S., Piñón-Castillo, H. A., Ramírez-Valdespino, C. A., Landeros-Martínez, L. L., Orrantia-Borunda, E., & Esparza-Ponce, H. E. (2021). Antimicrobial resistance and inorganic nanoparticles. *International Journal of Molecular Sciences*, *22*(23), 12890. https://doi.org/10.3390/ijms222312890

Bandara, H. M. H. N., Nguyen, D., Mogarala, S., Osiñski, M., & Smyth, H. D. C. (2015). Magnetic fields suppress *Pseudomonas aeruginosa* biofilms and enhance ciprofloxacin activity. *Biofouling*, *31*(5), 443–457. https://doi.org/10.1080/08927014.2015.1055326

Baruah, S., Jaisai, M., & Dutta, J. (2012). Development of a visible light active photocatalytic portable water purification unit using ZnO nanorods. *Catalysis Science & Technology*, *2*(5), 918–921. https://doi.org/10.1039/C2CY20033C

Besinis, A., De Peralta, T., & Handy, R. D. (2014). The antibacterial effects of silver, titanium dioxide and silica dioxide nanoparticles compared to the dental disinfectant chlorhexidine on *Streptococcus mutans* using a suite of bioassays. *Nanotoxicology*, *8*(1), 1–16. https://doi.org/10.3109/17435390.2012.742935

Beyene, Z., & Ghosh, R. (2019). Effect of zinc oxide addition on antimicrobial and antibiofilm activity of hydroxyapatite: A potential nanocomposite for biomedical applications. *Materials Today Communications*, *21*, 100612. https://doi.org/10.1016/j.mtcomm.2019.100612

Bhattacharyya, P., Agarwal, B., Goswami, M., Maiti, D., Baruah, S., & Tribedi, P. (2018). Zinc oxide nanoparticle inhibits the biofilm formation of *Streptococcus pneumoniae*. *Antonie Van Leeuwenhoek*, *111*, 89–99. https://doi.org/10.1007/s10482-017-0930-7

Bjarnsholt, T., Jensen, P. Ø., Fiandaca, M. J., Pedersen, J., Hansen, C. R., Andersen, C. B., . . . & Høiby, N. (2009). Pseudomonas aeruginosa biofilms in the respiratory tract of cystic fibrosis patients. *Pediatric Pulmonology*, *44*(6), 547–558. https://doi.org/10.1002/ppul.21011

Bjarnsholt, T., Kirketerp-Møller, K., Jensen, P. Ø., Madsen, K. G., Phipps, R., Krogfelt, K., . . . & Givskov, M. (2008). Why chronic wounds will not heal: A novel hypothesis. *Wound Repair and Regeneration*, *16*(1), 2–10. https://doi.org/10.1111/j.1524-475X.2007.00283.x

Blecher, K., Nasir, A., & Friedman, A. (2011). The growing role of nanotechnology in combating infectious disease. *Virulence*, *2*(5), 395–401. https://doi.org/10.4161/viru.2.5.17035

Borcherding, J., Baltrusaitis, J., Chen, H., Stebounova, L., Wu, C. M., Rubasinghege, G., . . . & Comellas, A. P. (2014). Iron oxide nanoparticles induce *Pseudomonas aeruginosa* growth, induce biofilm formation, and inhibit antimicrobial peptide function. *Environmental Science: Nano*, *1*(2), 123–132. https://doi.org/10.1039/C3EN00029J

Boyadzhiev, A., Wu, D., Avramescu, M. L., Williams, A., Rasmussen, P., & Halappanavar, S. (2023). Toxicity of metal oxide nanoparticles: Looking through the lens of toxicogenomics. *International Journal of Molecular Sciences*, *25*(1), 529. https://doi.org/10.3390/ijms25010529

Buzea, C., Pacheco, I. I., & Robbie, K. (2007). Nanomaterials and nanoparticles: Sources and toxicity. *Biointerphases*, *2*(4), MR17–MR71. https://doi.org/10.1116/1.2815690

Cai, L., Chen, J., Liu, Z., Wang, H., Yang, H., & Ding, W. (2018). Magnesium oxide nanoparticles: Effective agricultural antibacterial agent against *Ralstonia solanacearum*. *Frontiers in Microbiology*, *9*, 790. https://doi.org/10.3389/fmicb.2018.00790

Chaudhary, R. G., Bhusari, G. S., Tiple, A. D., Rai, A. R., Somkuvar, S. R., Potbhare, A. K., . . . & Abdala, A. A. (2019). Metal/metal oxide nanoparticles: Toxicity, applications, and future prospects. *Current Pharmaceutical Design*, *25*(37), 4013–4029. https://doi.org/10.2174/1381612825666191111091326

Chen, H., & Wang, L. (2014). Nanostructure sensitization of transition metal oxides for visible-light photocatalysis. *Beilstein Journal of Nanotechnology*, *5*(1), 696–710. https://doi.org/10.3762/bjnano.5.82

Chen, J., Mao, S., Xu, Z., & Ding, W. (2019). Various antibacterial mechanisms of biosynthesized copper oxide nanoparticles against soilborne *Ralstonia solanacearum*. *RSC Advances*, *9*(7), 3788–3799. https://doi.org/10.1039/c8ra09186b

Chrzanowska, N., & Załęska-Radziwiłł, M. (2014). The impacts of aluminum and zirconium nano-oxides on planktonic and biofilm bacteria. *Desalination and Water Treatment*, *52*(19–21), 3680–3689. https://doi.org/10.1080/19443994.2014.884528

Colon, G., Ward, B. C., & Webster, T. J. (2006). Increased osteoblast and decreased Staphylococcus epidermidis functions on nanophase ZnO and TiO2. *Journal of Biomedical Materials Research A*, *78*(3), 595–604. https://doi.org/10.1002/jbm.a.30789

Costerton, W., Veeh, R., Shirtliff, M., Pasmore, M., Post, C., & Ehrlich, G. (2003). The application of biofilm science to the study and control of chronic bacterial infections. *The Journal of Clinical Investigation*, *112*(10), 1466–1477. https://doi.org/10.1172/JCI20365

Cruz, G. J. F., Gómez, M. M., Solis, J. L., Rimaycuna, J., Solis, R. L., Cruz, J. F., . . . & Keiski, R. L. (2018). Composites of ZnO nanoparticles and biomass based activated carbon: Adsorption, photocatalytic and antibacterial capacities. *Water Science and Technology*, *2017*(2), 492–508. https://doi.org/10.2166/wst.2018.176

Daoud, W. A., Xin, J. H., & Zhang, Y. H. (2005). Surface functionalization of cellulose fibers with titanium dioxide nanoparticles and their combined bactericidal activities. *Surface Science*, *599*(1–3), 69–75. 10.1016/j.susc.2005.09.038

Del Pozo, J. L., & Patel, R. (2009). Infection associated with prosthetic joints. *New England Journal of Medicine*, *361*(8), 787–794. https://doi.org/10.1056/NEJMcp0905029

Dhandapani, P., Maruthamuthu, S., & Rajagopal, G. (2012). Biomediated synthesis of TiO2 nanoparticles and its photocatalytic effect on aquatic biofilm. *Journal of Photochemistry and Photobiology B: Biology*, *110*, 43–49. https://doi.org/10.1016/j.jphotobiol.2012.03.003

Dizaj, S. M., Lotfipour, F., Barzegar-Jalali, M., Zarrintan, M. H., & Adibkia, K. (2014). Antimicrobial activity of the metals and metal oxide nanoparticles. *Materials Science and Engineering: C*, *44*, 278–284. https://doi.org/10.1016/j.msec.2014.08.031

Erkan, A., Bakir, U., & Karakas, G. (2006). Photocatalytic microbial inactivation over Pd doped SnO2 and TiO2 thin films. *Journal of Photochemistry and Photobiology A: Chemistry*, *184*(3), 313–321. https://doi.org/10.1016/j.jphotochem.2006.05.001

Fantozzi, E., Rama, E., Calvio, C., Albini, B., Galinetto, P., & Bini, M. (2021). Silver doped magnesium ferrite nanoparticles: Physico-chemical characterization and antibacterial activity. *Materials*, *14*(11), 2859. https://doi.org/10.3390/ma14112859

Fellahi, O., Sarma, R. K., Das, M. R., Saikia, R., Marcon, L., Coffinier, Y., . . . & Boukherroub, R. (2013). The antimicrobial effect of silicon nanowires decorated with silver and copper nanoparticles. *Nanotechnology*, *24*(49), 495101. https://doi.org/10.1088/0957-4484/24/49/495101

Franco, D., Calabrese, G., Guglielmino, S. P. P., & Conoci, S. (2022). Metal-based nanoparticles: Antibacterial mechanisms and biomedical application. *Microorganisms*, *10*(9), 1778. https://doi.org/10.3390/microorganisms10091778

Gaballa, A., & Helmann, J. D. (1998). Identification of a zinc-specific metalloregulatory protein, Zur, controlling zinc transport operons in *Bacillus subtilis*. *Journal of Bacteriology*, *180*(22), 5815–5821. https://doi.org/10.1128/JB.180.22.5815-5821.1998

Gabrielyan, L., Hovhannisyan, A., Gevorgyan, V., Ananyan, M., & Trchounian, A. (2019). Antibacterial effects of iron oxide (Fe3O4) nanoparticles: Distinguishing concentration-dependent effects with different bacterial cells growth and membrane-associated mechanisms. *Applied Microbiology and Biotechnology*, *103*(6), 2773–2782. https://doi.org/10.1007/s00253-019-09653-x

Garcia, I. M., Balhaddad, A. A., Ibrahim, M. S., Weir, M. D., Xu, H. H., Collares, F. M., & Melo, M. A. S. (2021). Antibacterial response of oral microcosm biofilm to nano-zinc oxide in adhesive resin. *Dental Materials*, *37*(3), e182–e193. https://doi.org/10.1016/j.dental.2020.11.022

García-Lara, B., Saucedo-Mora, M. Á., Roldán-Sánchez, J. A., Pérez-Eretza, B., Ramasamy, M., Lee, J., . . . & García-Contreras, R. (2015). Inhibition of quorum-sensing-dependent virulence factors and biofilm formation of clinical and environmental *Pseudomonas aeruginosa* strains by ZnO nanoparticles. *Letters in Applied Microbiology*, *61*(3), 299–305. https://doi.org/10.1111/lam.12456

Geilich, B. M., Gelfat, I., Sridhar, S., van de Ven, A. L., & Webster, T. J. (2017). Superparamagnetic iron oxide-encapsulating polymersome nanocarriers for biofilm eradication. *Biomaterials*, *119*, 78–85. https://doi.org/10.1016/j.biomaterials.2016.12.011

Giannousi, K., Lafazanis, K., Arvanitidis, J., Pantazaki, A., & Dendrinou-Samara, C. (2014). Hydrothermal synthesis of copper based nanoparticles: Antimicrobial screening and interaction with DNA. *Journal of Inorganic Biochemistry*, *133*, 24–32. https://doi.org/10.1016/j.jinorgbio.2013.12.009

Hall-Stoodley, L., Hu, F. Z., Gieseke, A., Nistico, L., Nguyen, D., Hayes, J., . . . & Kerschner, J. E. (2006). Direct detection of bacterial biofilms on the middle-ear mucosa of children with chronic otitis media. *Jama*, *296*(2), 202–211. https://doi.org/10.1001/jama.296.2.202

Hamdan, H. F., Zulkiply, N., & Yahya, M. F. Z. R. (2023). Control strategies of *Staphylococcus aureus* and methicillin-resistant *Staphylococcus aureus* (MRSA) biofilms: A review. *Science Letters*, *17*(2), 33–49. https://doi.org/10.24191/sl.v17i2.22718

Happy, A., Soumya, M., Kumar, S. V., Rajeshkumar, S., Sheba, R. D., Lakshmi, T., & Nallaswamy, V. D. (2019). Phyto-assisted synthesis of zinc oxide nanoparticles using *Cassia alata* and its antibacterial activity against *Escherichia coli*. *Biochemistry and Biophysics Reports*, *17*, 208–211. https://doi.org/10.1016/j.bbrep.2019.01.002

Hasanin, M. S., & Youssef, A. M. (2022). Ecofriendly bioactive film doped CuO nanoparticles based biopolymers and reinforced by enzymatically modified nanocellulose fibers for active packaging applications. *Food Packaging and Shelf Life*, *34*, 100979. https://doi.org/10.1016/j.fpsl.2022.100979

Hayat, S., Muzammil, S., Rasool, M. H., Nisar, Z., Hussain, S. Z., Sabri, A. N., & Jamil, S. (2018). In vitro antibiofilm and anti-adhesion effects of magnesium oxide nanoparticles against antibiotic resistant bacteria. *Microbiology and Immunology*, *62*(4), 211–220. https://doi.org/10.1111/1348-0421.12580

He, W., Kim, H. K., Wamer, W. G., Melka, D., Callahan, J. H., & Yin, J. J. (2014). Photogenerated charge carriers and reactive oxygen species in ZnO/Au hybrid nanostructures with enhanced photocatalytic and antibacterial activity. *Journal of the American Chemical Society*, *136*(2), 750–757. https://doi.org/10.1021/ja410800y

Hemeg, H. A. (2022). Combatting persisted and biofilm antimicrobial resistant bacterial by using nanoparticles. *Zeitschrift für Naturforschung C*, *77*(9–10), 365–378. https://doi.org/10.1515/znc-2021-0296

Hetta, H. F., Ramadan, Y. N., Al-Harbi, A. I., A. Ahmed, E., Battah, B., Abd Ellah, N. H., . . . & Donadu, M. G. (2023). Nanotechnology as a promising approach to combat multidrug resistant bacteria: A comprehensive review and future perspectives. *Biomedicines*, *11*(2), 413. https://doi.org/10.3390/biomedicines11020413

Hochvaldová, L., Večeřová, R., Kolář, M., Prucek, R., Kvítek, L., Lapčík, L., & Panáček, A. (2022). Antibacterial nanomaterials: Upcoming hope to overcome antibiotic resistance crisis. *Nanotechnology Reviews*, *11*(1), 1115–1142. https://doi.org/10.1515/ntrev-2022-0059

Høiby, N., Ciofu, O., Johansen, H. K., Song, Z. J., Moser, C., Jensen, P. Ø., . . . & Bjarnsholt, T. (2011). The clinical impact of bacterial biofilms. *International Journal of Oral Science*, *3*(2), 55–65. https://doi.org/10.4248/IJOS11026

Hood, M. I., & Skaar, E. P. (2012). Nutritional immunity: Transition metals at the pathogen–host interface. *Nature Reviews Microbiology*, *10*(8), 525–537. https://doi.org/10.1038/nrmicro2836

Horie, M., & Tabei, Y. (2021). Role of oxidative stress in nanoparticles toxicity. *Free Radical Research*, *55*(4), 331–342. https://doi.org/10.1080/10715762.2020.1859108

Hussain, A., Oves, M., Alajmi, M. F., Hussain, I., Amir, S., Ahmed, J., . . . & Ali, I. (2019). Biogenesis of ZnO nanoparticles using *Pandanus odorifer* leaf extract: Anticancer and antimicrobial activities. *RSC Advances*, *9*(27), 15357–15369. https://doi.org/10.1039/c9ra01659g

Imlay, J. A. (2003). Pathways of oxidative damage. *Annual Reviews in Microbiology*, *57*(1), 395–418. https://doi.org/10.1146/annurev.micro.57.030502.090938

Ishwarya, R., Vaseeharan, B., Kalyani, S., Banumathi, B., Govindarajan, M., Alharbi, N. S., . . . & Benelli, G. (2018). Facile green synthesis of zinc oxide nanoparticles using *Ulva lactuca* seaweed extract and evaluation of their photocatalytic, antibiofilm and insecticidal activity. *Journal of Photochemistry and Photobiology B: Biology*, *178*, 249–258. https://doi.org/10.1016/j.jphotobiol.2017.11.006

Jamal, A., Awad, R., & Yusef, H. (2019). Evaluation of antimicrobial activity of ZnO nanoparticles against foodborne pathogens. *International Journal of Current Microbiology and Applied Sciences*, *8*, 2000–2025. https://doi.org/10.20546/ijcmas.2019.811.234

Jayakumar, A., Radoor, S., Kim, J. T., Rhim, J. W., Parameswaranpillai, J., Nandi, D., . . . & Siengchin, S. (2022). Titanium dioxide nanoparticles and elderberry extract incorporated starch based polyvinyl alcohol films as active and intelligent food packaging wraps. *Food Packaging and Shelf Life*, *34*, 100967. https://doi.org/10.1016/j.fpsl.2022.100967

Jesline, A., John, N. P., Narayanan, P. M., Vani, C., & Murugan, S. (2015). Antimicrobial activity of zinc and titanium dioxide nanoparticles against biofilm-producing methicillin-resistant *Staphylococcus aureus*. *Applied Nanoscience*, *5*, 157–162. https://doi.org/10.1007/s13204-014-0301-x

Khan, M. S. A., Altaf, M. M., & Ahmad, I. (2017). Chemical nature of biofilm matrix and its significance. In *Biofilms in Plant and Soil Health*, 151–177. https://doi.org/10.1002/9781119246329.ch9

Khan, S. T., Ahamed, M., Al-Khedhairy, A., & Musarrat, J. (2013). Biocidal effect of copper and zinc oxide nanoparticles on human oral microbiome and biofilm formation. *Materials Letters*, *97*, 67–70. https://doi.org/10.1016/j.matlet.2013.01.085

Khan, S. T., Ahamed, M., Musarrat, J., & Al-Khedhairy, A. A. (2014). Anti-biofilm and antibacterial activities of zinc oxide nanoparticles against the oral opportunistic pathogens *Rothia dentocariosa* and *Rothia mucilaginosa*. *European Journal of Oral Sciences*, *122*(6), 397–403. https://doi.org/10.1111/eos.12152

Kirketerp-Møller, K., Jensen, P. Ø., Fazli, M., Madsen, K. G., Pedersen, J., Moser, C., . . . & Bjarnsholt, T. (2008). Distribution, organization, and ecology of bacteria in chronic wounds. *Journal of Clinical Microbiology*, *46*(8), 2717–2722. https://doi.org/10.1128/JCM.00501-08

Krokowicz, L., Tomczak, H., Bobkiewicz, A., Mackiewicz, J., Marciniak, R., Drews, M., & Banasiewicz, T. (2015). In vitro studies of antibacterial and antifungal wound dressings comprising H2TiO3 and SiO2 nanoparticles. *Polish Journal of Microbiology*, *64*(2), 5. https://doi.org/10.33073/pjm-2015-020

Landage, K. S., Arbade, G. K., Khanna, P., & Bhongale, C. J. (2020). Biological approach to synthesize TiO2 nanoparticles using *Staphylococcus aureus* for antibacterial and anti-biofilm applications. *Journal of Microbiology & Experimentation*, *8*(1), 36–43. https://doi.org/10.15406/jmen.2020.08.00283

Lee, J. H., Kim, Y. G., Cho, M. H., & Lee, J. (2014). ZnO nanoparticles inhibit *Pseudomonas aeruginosa* biofilm formation and virulence factor production. *Microbiological Research*, *169*(12), 888–896. https://doi.org/10.1016/j.micres.2014.05.005

Leuba, K. D., Durmus, N. G., Taylor, E. N., & Webster, T. J. (2013). carboxylate functionalized superparamagnetic iron oxide nanoparticles (SPION) for the reduction of *S. aureus* growth post biofilm formation. *International Journal of Nanomedicine*, 731–736. https://doi.org/10.2147/IJN.S38256

Liu, J., Wang, Y., Liu, Y., Shao, S., Zheng, X., & Tang, K. (2023). Synergistic effect of nano zinc oxide and tea tree essential oil on the properties of soluble soybean polysaccharide films. *International Journal of Biological Macromolecules*, *239*, 124361. https://doi.org/10.1016/j.ijbiomac.2023.124361

Lungu, M., Gavriliu, Ş., Enescu, E., Ion, I., Brătulescu, A., Mihăescu, G., . . . & Chifiriuc, M. C. (2014). Silver–titanium dioxide nanocomposites as effective antimicrobial and antibiofilm agents. *Journal of Nanoparticle Research*, *16*, 1–15. https://doi.org/10.1007/s11051-013-2203-3

Magdalane, C. M., Kaviyarasu, K., Vijaya, J. J., Jayakumar, C., Maaza, M., & Jeyaraj, B. (2017). Photocatalytic degradation effect of malachite green and catalytic hydrogenation by UV–illuminated CeO2/CdO multilayered nanoplatelet arrays: Investigation of antifungal and antimicrobial activities. *Journal of Photochemistry and Photobiology B: Biology*, *169*, 110–123. https://doi.org/10.1016/j.jphotobiol.2017.03.008

Mahajan, P., Sharma, A., Kaur, B., Goyal, N., & Gautam, S. (2019). Green synthesized (*Ocimum sanctum* and *Allium sativum*) Ag-doped cobalt ferrite nanoparticles for antibacterial application. *Vacuum*, *161*, 389–397. https://doi.org/10.1016/j.vacuum.2018.12.021

Mahamuni-Badiger, P. P., Patil, P. M., Badiger, M. V., Patel, P. R., Thorat-Gadgil, B. S., Pandit, A., & Bohara, R. A. (2020). Biofilm formation to inhibition: Role of zinc oxide-based nanoparticles. *Materials Science and Engineering: C*, *108*, 110319. https://doi.org/10.1016/j.msec.2019.110319

Mahmoodi, N. M., Arami, M., Limaee, N. Y., & Tabrizi, N. S. (2006). Kinetics of heterogeneous photocatalytic degradation of reactive dyes in an immobilized TiO2 photocatalytic reactor. *Journal of Colloid and Interface SCIENCE*, *295*(1), 159–164. https://doi.org/10.1016/j.jcis.2005.08.007

Maji, J., Pandey, S., & Basu, S. (2020). Synthesis and evaluation of antibacterial properties of magnesium oxide nanoparticles. *Bulletin of Materials Science*, *43*, 1–10. https://doi.org/10.1007/s12034-019-1963-5

Makhluf, S., Dror, R., Nitzan, Y., Abramovich, Y., Jelinek, R., & Gedanken, A. (2005). Microwave-assisted synthesis of nanocrystalline MgO and its use as a bacteriocide. *Advanced Functional Materials*, *15*(10), 1708–1715. https://doi.org/10.1002/adfm.200500029

Makin, S. A., & Beveridge, T. J. (1996). The influence of A-band and B-band lipopolysaccharide on the surface characteristics and adhesion of *Pseudomonas aeruginosa* to surfaces. *Microbiology*, *142*(2), 299–307. https://doi.org/10.1099/13500872-142-2-299

Mammari, N., & Duval, R. E. (2023). Photothermal/photoacoustic therapy combined with metal-based nanomaterials for the treatment of microbial infections. *Microorganisms*, *11*(8), 2084. https://doi.org/10.3390/microorganisms11082084

Marand, S. A., Almasi, H., & Marand, N. A. (2021). Chitosan-based nanocomposite films incorporated with NiO nanoparticles: Physicochemical, photocatalytic and antimicrobial properties. *International Journal of Biological Macromolecules*, *190*, 667–678. https://doi.org/10.1016/j.ijbiomac.2021.09.024

Margabandhu, M., Sendhilnathan, S., Maragathavalli, S., Karthikeyan, V., & Annadurai, B. (2015). Synthesis characterization and antibacterial activity of iron oxide nanoparticles. *Global Journal of Bio-Science and BioTechnology*, *4*(4), 335–341. https://doi.org/10.3390/antibiotics10070884

McGuffie, M. J., Hong, J., Bahng, J. H., Glynos, E., Green, P. F., Kotov, N. A., . . . & VanEpps, J. S. (2016). Zinc oxide nanoparticle suspensions and layer-by-layer coatings inhibit staphylococcal growth. *Nanomedicine: Nanotechnology, Biology and Medicine*, *12*(1), 33–42. https://doi.org/10.1016/j.nano.2015.10.002

McNeilly, O., Mann, R., Hamidian, M., & Gunawan, C. (2021). Emerging concern for silver nanoparticle resistance in *Acinetobacter baumannii* and other bacteria. *Frontiers in Microbiology*, *12*, 652863. https://doi.org/10.3389/fmicb.2021.652863

Meghana, S., Kabra, P., Chakraborty, S., & Padmavathy, N. (2015). Understanding the pathway of antibacterial activity of copper oxide nanoparticles. *RSC Advances*, *5*(16), 12293–12299. https://doi.org/10.1039/C4RA12163E

Mehmood, Z., Sadiq, M. B., & Khan, M. R. (2020). Gelatin nanocomposite films incorporated with magnetic iron oxide nanoparticles for shelf life extension of grapes. *Journal of Food Safety*, *40*(4), e12814. https://doi.org/10.1111/jfs.12814

Mohammadi, G., Nokhodchi, A., Barzegar-Jalali, M., Lotfipour, F., Adibkia, K., Ehyaei, N., & Valizadeh, H. (2011). Physicochemical and anti-bacterial performance characterization of clarithromycin nanoparticles as colloidal drug delivery system. *Colloids and Surfaces B: Biointerfaces*, *88*(1), 39–44. https://doi.org/10.1016/j.colsurfb.2011.05.050

Mohammadi, G., Valizadeh, H., Barzegar-Jalali, M., Lotfipour, F., Adibkia, K., Milani, M., . . . & Nokhodchi, A. (2010). Development of azithromycin–PLGA nanoparticles: Physicochemical characterization and antibacterial effect against Salmonella typhi. *Colloids and Surfaces B: Biointerfaces*, *80*(1), 34–39. https://doi.org/10.1016/j.colsurfb.2010.05.027

Momtaz, M., Momtaz, E., Mehrgardi, M. A., Momtaz, F., Narimani, T., & Poursina, F. (2024). Preparation and characterization of gelatin/chitosan nanocomposite reinforced by NiO nanoparticles as an active food packaging. *Scientific Reports*, *14*(1), 519. https://doi.org/10.1038/s41598-023-50260-8

Mosselhy, D. A., Granbohm, H., Hynönen, U., Ge, Y., Palva, A., Nordström, K., & Hannula, S. P. (2017). Nanosilver–silica composite: Prolonged antibacterial effects and bacterial interaction mechanisms for wound dressings. *Nanomaterials*, *7*(9), 261. https://doi.org/10.3390/nano7090261

Mukherjee, A., Bose, S., Shaoo, A., & Das, S. K. (2023). Nanotechnology based therapeutic approaches: An advanced strategy to target the biofilm of ESKAPE pathogens. *Materials Advances*, *4*(12), 2544–2572. https://doi.org/10.1039/d2ma00846g

Muzammil, S., Hayat, S., Fakhar-E-Alam, M., Aslam, B., Siddique, M. H., Nisar, M. A., . . . & Wang, Z. (2018). Nanoantibiotics: Future nanotechnologies to combat antibiotic resistance. *Frontiers in Bioscience*, *10*(2), 352–374. https://doi.org/10.2741/e827

Naskar, A., & Kim, K. S. (2019). Nanomaterials as delivery vehicles and components of new strategies to combat bacterial infections: Advantages and limitations. *Microorganisms*, *7*(9), 356. https://doi.org/10.3390/microorganisms7090356

Natan, M., & Banin, E. (2017). From nano to micro: Using nanotechnology to combat microorganisms and their multidrug resistance. *FEMS Microbiology Reviews*, *41*(3), 302–322. https://doi.org/10.1093/femsre/fux003

Nguyen, N. Y. T., Grelling, N., Wetteland, C. L., Rosario, R., & Liu, H. (2018). Antimicrobial activities and mechanisms of magnesium oxide nanoparticles (nMgO) against pathogenic bacteria, yeasts, and biofilms. *Scientific Reports*, *8*(1), 16260. https://doi.org/10.1038/s41598-018-34567-5

Niemirowicz, K., Durnaś, B., Tokajuk, G., Głuszek, K., Wilczewska, A. Z., Misztalewska, I., . . . & Bucki, R. (2016). Magnetic nanoparticles as a drug delivery system that enhance fungicidal activity of polyene antibiotics. *Nanomedicine: Nanotechnology, Biology and Medicine*, *12*(8), 2395–2404. https://doi.org/10.1016/j.nano.2016.07.006

Noori, A. J., & Kareem, F. A. (2019). The effect of magnesium oxide nanoparticles on the antibacterial and antibiofilm properties of glass-ionomer cement. *Heliyon*, *5*(10). https://doi.org/10.1016/j.heliyon.2019.e02568

Noreen, Z., Ahmad, I., Siddiqui, F., Ziya, A. B., Abbas, T., & Bokhari, H. (2017). Size dependent structural, anti-bacterial and anti-biofilm properties of Er doped Li-Ni ferrites synthesized by the sol-gel auto-combustion route. *Ceramics International*, *43*(14), 10784–10790.

Osman, A. G., El-Desouky, A. I., Morsy, M. K., Aboud, A. A., & Mohamed, M. H. (2019). Impact of aluminum oxide and silica oxide nanocomposite on foodborne pathogens in chicken fillets. *European Journal of Nutrition and Food Safety*, *9*(2), 152–162. https://doi.org/10.9734/EJNFS/2019/v9i230054

Pal, A., Pehkonen, S. O., Liya, E. Y., & Ray, M. B. (2007). Photocatalytic inactivation of Gram-positive and Gram-negative bacteria using fluorescent light. *Journal of Photochemistry and Photobiology A: Chemistry*, *186*(2–3), 335–341. https://doi.org/10.1016/j.jphotochem.2006.09.002

Palmer, R. J., Darveau, R., Lamont, R. J., Nyvad, B., & Teles, R. P. (2011). Human oral bacterial biofilms: Composition, dynamics, and pathogenesis. *Biofilm infections*, 35–68. https://doi.org/10.1007/978-1-4419-6084-9_4

Parikh, S. J., & Chorover, J. (2006). ATR-FTIR spectroscopy reveals bond formation during bacterial adhesion to iron oxide. *Langmuir*, *22*(20), 8492–8500. https://doi.org/10.1021/la061359p

Pati, R., Mehta, R. K., Mohanty, S., Padhi, A., Sengupta, M., Vaseeharan, B., . . . & Sonawane, A. (2014). Topical application of zinc oxide nanoparticles reduces bacterial skin infection in mice and exhibits antibacterial activity by inducing oxidative stress response and cell membrane disintegration in macrophages. *Nanomedicine: Nanotechnology, Biology and Medicine*, *10*(6), 1195–1208. https://doi.org/10.1016/j.nano.2014.02.012

Pereira, M. C., Oliveira, L. C. A., & Murad, E. (2012). Iron oxide catalysts: Fenton and Fenton-like reactions—a review. *Clay Minerals*, *47*(3), 285–302. https://doi.org/10.1180/claymin.2012.047.3.01

Raftery, T. D., Kerscher, P., Hart, A. E., Saville, S. L., Qi, B., Kitchens, C. L., . . . & McNealy, T. L. (2014). Discrete nanoparticles induce loss of *Legionella pneumophila* biofilms from surfaces. *Nanotoxicology*, *8*(5), 477–484. https://doi.org/10.3109/17435390.2013.796537

Raghunath, A., & Perumal, E. (2017). Metal oxide nanoparticles as antimicrobial agents: A promise for the future. *International Journal of Antimicrobial Agents*, *49*(2), 137–152. https://doi.org/10.1016/j.ijantimicag.2016.11.011

Raie, D. S., Mhatre, E., Thiele, M., Labena, A., El-Ghannam, G., Farahat, L. A., . . . & Kovács, Á. T. (2017). Application of quercetin and its bio-inspired nanoparticles as anti-adhesive agents against *Bacillus subtilis* attachment to surface. *Materials Science and Engineering: C*, *70*, 753–762. https://doi.org/10.1016/j.msec.2016.09.038

Rajkumari, J., Magdalane, C. M., Siddhardha, B., Madhavan, J., Ramalingam, G., Al-Dhabi, N. A., . . . & Kaviyarasu, K. (2019). Synthesis of titanium oxide nanoparticles using *Aloe barbadensis* mill and evaluation of its antibiofilm potential against *Pseudomonas aeruginosa* PAO1. *Journal of Photochemistry and Photobiology B: Biology*, *201*, 111667. https://doi.org/10.1016/j.jphotobiol.2019.111667

Ramasamy, M., Lee, J. H., & Lee, J. (2016). Potent antimicrobial and antibiofilm activities of bacteriogenically synthesized gold–silver nanoparticles against pathogenic bacteria and their physiochemical characterizations. *Journal of Biomaterials Applications*, *31*(3), 366–378. https://doi.org/10.1177/0885328216646910

Ranmadugala, D., Ebrahiminezhad, A., Manley-Harris, M., Ghasemi, Y., & Berenjian, A. (2017). The effect of iron oxide nanoparticles on *Bacillus subtilis* biofilm, growth and viability. *Process Biochemistry*, *62*, 231–240. https://doi.org/10.1016/j.procbio.2017.07.003

Rao, Y., Wang, W., Tan, F., Cai, Y., Lu, J., & Qiao, X. (2013). Influence of different ions doping on the antibacterial properties of MgO nanopowders. *Applied Surface Science*, *284*, 726–731.

Reid, G. (1999). Biofilms in infectious disease and on medical devices. *International Journal of Antimicrobial Agents*, *11*(3–4), 223–226. https://doi.org/10.1016/s0924-8579(99)00020-5

Rimjhim, & Thomas, A. (2016). Analysis of antimicrobial activity of titanium dioxide nanoparticles on aerobic and anaerobic dental isolates. *International Journal of Scientific Research in Science, Engineering and Technology*, *2*(4), 815–822.

Rumyantceva, V., Rumyantceva, V., Koshel, E., & Vinogradov, V. (2019). Biocide-conjugated magnetite nanoparticles as an advanced platform for biofilm treatment. *Therapeutic Delivery*, *10*(4), 241–250. https://doi.org/10.4155/tde-2019-0011

Sadiq, I. M., Chowdhury, B., Chandrasekaran, N., & Mukherjee, A. (2009). Antimicrobial sensitivity of Escherichia coli to alumina nanoparticles. *Nanomedicine: Nanotechnology, Biology and Medicine*, *5*(3), 282–286. https://doi.org/10.1016/j.nano.2009.01.002

Saleem, S., Ahmed, B., Khan, M. S., Al-Shaeri, M., & Musarrat, J. (2017). Inhibition of growth and biofilm formation of clinical bacterial isolates by NiO nanoparticles synthesized from Eucalyptus globulus plants. *Microbial Pathogenesis*, *111*, 375–387. https://doi.org/10.1016/j.micpath.2017.09.019

Salem, W., Leitner, D. R., Zingl, F. G., Schratter, G., Prassl, R., Goessler, W., . . . & Schild, S. (2015). Antibacterial activity of silver and zinc nanoparticles against Vibrio cholerae and enterotoxic *Escherichia coli*. *International Journal of Medical Microbiology*, *305*(1), 85–95. https://doi.org/10.1016/j.ijmm.2014.11.005

Sapkota, A., Anceno, A. J., Baruah, S., Shipin, O. V., & Dutta, J. (2011). Zinc oxide nanorod mediated visible light photoinactivation of model microbes inwater. *Nanotechnology*, *22*(21), 215703. https://doi.org/10.1088/0957-4484/22/21/215703

Sarkar, A., Ghosh, M., & Sil, P. C. (2014). Nanotoxicity: Oxidative stress mediated toxicity of metal and metal oxide nanoparticles. *Journal of Nanoscience and Nanotechnology*, *14*(1), 730–743. https://doi.org/10.1166/jnn.2014.8752

Sarwar, S., Chakraborti, S., Bera, S., Sheikh, I. A., Hoque, K. M., & Chakrabarti, P. (2016). The antimicrobial activity of ZnO nanoparticles against Vibrio cholerae: Variation in response depends on biotype. *Nanomedicine: Nanotechnology, Biology and Medicine*, *12*(6), 1499–1509. https://doi.org/10.1016/j.nano.2016.02.006

Seil, J. T., & Webster, T. J. (2012a). Antimicrobial applications of nanotechnology: Methods and literature. *International Journal of Nanomedicine*, 2767–2781. https://doi.org/10.2147/IJN.S24805

Seil, J. T., & Webster, T. J. (2012b). Antibacterial effect of zinc oxide nanoparticles combined with ultrasound. *Nanotechnology*, *23*(49), 495101. https://doi.org/10.1088/0957-4484/23/49/495101

Shah, R. R., Kaewgun, S., Lee, B. I., & Tzeng, T. R. J. (2008). The antibacterial effects of biphasic brookite-anatase titanium dioxide nanoparticles on multiple-drug-resistant Staphylococcus aureus. *Journal of Biomedical Nanotechnology*, *4*(3), 339–348. https://doi.org/10.1166/jbn.2008.324

Shakerimoghaddam, A., Ghaemi, E. A., & Jamalli, A. (2017). Zinc oxide nanoparticle reduced biofilm formation and antigen 43 expressions in uropathogenic Escherichia coli. *Iranian Journal of Basic Medical Sciences*, *20*(4), 451. https://doi.org/10.22038/IJBMS.2017.8589.

Shalom, Y., Perelshtein, I., Perkas, N., Gedanken, A., & Banin, E. (2017). Catheters coated with Zn-doped CuO nanoparticles delay the onset of catheter-associated urinary tract infections. *Nano Research*, *10*, 520–533. https://doi.org/10.1007/s12274-016-1310-8

Sheng, Y., Chen, Z., Wu, W., & Lu, Y. (2023). Engineered organic nanoparticles to combat biofilms. *Drug Discovery Today*, *28*(2), 103455. https://doi.org/10.1016/j.drudis.2022.103455

Shkodenko, L., Kassirov, I., & Koshel, E. (2020). Metal oxide nanoparticles against bacterial biofilms: Perspectives and limitations. *Microorganisms*, *8*(10), 1545. https://doi.org/10.3390/microorganisms8101545

Simon-Deckers, A., Loo, S., Mayne-L'hermite, M., Herlin-Boime, N., Menguy, N., Reynaud, C., . . . & Carrière, M. (2009). Size-, composition-and shape-dependent toxicological impact of metal oxide nanoparticles and carbon nanotubes toward bacteria. *Environmental Science & Technology*, *43*(21), 8423–8429. https://doi.org/10.1021/es9016975

Singh, J., Dutta, T., Kim, K. H., Rawat, M., Samddar, P., & Kumar, P. (2018). 'Green' synthesis of metals and their oxide nanoparticles: Applications for environmental remediation. *Journal of nanobiotechnology*, *16*, 1–24. https://doi.org/10.1186/s12951-018-0408-4

Skaar, E. P., & Raffatellu, M. (2015). Metals in infectious diseases and nutritional immunity. *Metallomics*, *7*(6), 926–928. https://doi.org/10.1039/c5mt90021b

Slavin, Y. N., Asnis, J., Hńfeli, U. O., & Bach, H. (2017). Metal nanoparticles: Understanding the mechanisms behind antibacterial activity. *Journal of nanobiotechnology*, *15*, 1–20. https://doi.org/10.1186/s12951-017-0308-z

Soltys, L., Olkhovyy, O., Tatarchuk, T., & Naushad, M. (2021). Green synthesis of metal and metal oxide nanoparticles: Principles of green chemistry and raw materials. *Magnetochemistry*, *7*(11), 145. https://doi.org/10.3390/magnetochemistry7110145

Stensballe, J., Tvede, M., Looms, D., Lippert, F. K., Dahl, B., Tønnesen, E., & Rasmussen, L. S. (2007). Infection risk with nitrofurazone-impregnated urinary catheters in trauma patients: A randomized trial. *Annals of Internal Medicine*, *147*(5), 285–293. https://doi.org/10.7326/0003-4819-147-5-200709040-00002

Storz, G., & Imlayt, J. A. (1999). Oxidative stress. *Current Opinion in Microbiology*, *2*(2), 188–194. https://doi.org/10.1016/s1369-5274(99)80033-2

Stuparu-Cretu, M., Braniste, G., Necula, G. A., Stanciu, S., Stoica, D., & Stoica, M. (2023). Metal oxide nanoparticles in food packaging and their influence on human health. *Foods*, *12*(9), 1882. https://doi.org/10.3390/foods12091882

Summer, M., Ashraf, R., Ali, S., Bach, H., Noor, S., Noor, Q., . . . & Khan, R. R. M. (2024). Inflammatory response of nanoparticles: Mechanisms, consequences, and strategies for mitigation. *Chemosphere*, *363*, 142826. https://doi.org/10.1016/j.chemosphere.2024.142826

Sun, C., Wang, X., Dai, J., & Ju, Y. (2022). Metal and metal oxide nanomaterials for fighting planktonic bacteria and biofilms: A review emphasizing on mechanistic aspects. *International Journal of Molecular Sciences*, *23*(19), 11348. https://doi.org/10.3390/ijms231911348

Tang, J., Zhu, N., Zhu, Y., Zamir, S. M., & Wu, Y. (2017). Sustainable pollutant removal by periphytic biofilm via microbial composition shifts induced by uneven distribution of CeO2 nanoparticles. *Bioresource Technology*, *248*, 75–81. https://doi.org/10.1016/j.biortech.2017.07.064

Taran, M., Rad, M., & Alavi, M. (2017). Antibacterial activity of copper oxide (CuO) nanoparticles biosynthesized by Bacillus sp. FU4: Optimization of experiment design. *pharmaceutical sciences*, *23*(3), 198–206. https://doi.org/10.15171/PS.2017.30

Taylor, E. N., Kummer, K. M., Durmus, N. G., Leuba, K., Tarquinio, K. M., & Webster, T. J. (2012). Superparamagnetic iron oxide nanoparticles (SPION) for the treatment of antibiotic-resistant biofilms. *Small*, *8*(19), 3016–3027. https://doi.org/10.1002/smll.201200575

Taylor, E. N., & Webster, T. J. (2009). The use of superparamagnetic nanoparticles for prosthetic biofilm prevention. *International Journal of Nanomedicine*, 145–152. https://doi.org/10.2147/IJN.S5976

Tee, J. K., Ong, C. N., Bay, B. H., Ho, H. K., & Leong, D. T. (2016). Oxidative stress by inorganic nanoparticles. *Wiley Interdisciplinary Reviews: Nanomedicine and Nanobiotechnology*, *8*(3), 414–438. https://doi.org/10.1002/wnan.1374

Thakkar, K. N., Mhatre, S. S., & Parikh, R. Y. (2010). Biological synthesis of metallic nanoparticles. *Nanomedicine: Nanotechnology, Biology and Medicine*, *6*(2), 257–262. https://doi.org/10.1016/j.nano.2009.07.002

Thambirajoo, M., Maarof, M., Lokanathan, Y., Katas, H., Ghazalli, N. F., Tabata, Y., & Fauzi, M. B. (2021). Potential of nanoparticles integrated with antibacterial properties in preventing biofilm and antibiotic resistance. *Antibiotics*, *10*(11), 1338. https://doi.org/10.3390/antibiotics10111338

Thukkaram, M., Sitaram, S., Kannaiyan, S. K., & Subbiahdoss, G. (2014). Antibacterial efficacy of iron-oxide nanoparticles against biofilms on different biomaterial surfaces. *International Journal of biomaterials*, *2014*(1), 716080. https://doi.org/10.1155/2014/716080

Trampuz, A., & Zimmerli, W. (2005). Prosthetic joint infections: Update in diagnosis and treatment. *Swiss Medical Weekly*, *135*(1718), 243–243. https://doi.org/10.4414/smw.2005

Tran, N., Mir, A., Mallik, D., Sinha, A., Nayar, S., & Webster, T. J. (2010). Bactericidal effect of iron oxide nanoparticles on *Staphylococcus aureus*. *International Journal of Nanomedicine*, 277–283. https://doi.org/10.2147/ijn.s9220

Uthra, C., Nagaraj, K., Wadaan, M. A., Karuppiah, C., Maity, P., Baabbad, A., . . . & Kumar, P. (2024). Zinc and copper oxide nanoparticles: Pioneering antibacterial and antibiofilm strategies for environmental restoration against antibiotic-resistant bacteria. *Materials*, *17*(14), 3444. https://doi.org/10.3390/ma17143444

Varghese, E. J., Sihivahanan, D., & Venkatesh, K. V. (2022). Development of novel antimicrobial dental composite resin with nano cerium oxide fillers. *International Journal of Biomaterials*, *2022*(1), 3912290. https://doi.org/10.1155/2022/3912290

Vatansever, F., de Melo, W. C., Avci, P., Vecchio, D., Sadasivam, M., Gupta, A., . . . & Hamblin, M. R. (2013). Antimicrobial strategies centered around reactive oxygen species–bactericidal antibiotics, photodynamic therapy, and beyond. *FEMS Microbiology Reviews*, *37*(6), 955–989. https://doi.org/10.1111/1574-6976.12026

Vidic, J., Stankic, S., Haque, F., Ciric, D., Le Goffic, R., Vidy, A., . . . & Delmas, B. (2013). Selective antibacterial effects of mixed ZnMgO nanoparticles. *Journal of Nanoparticle Research*, *15*, 1–10. https://doi.org/10.1007/s11051-013-1595-4

Vijayakumar, S., Vinoj, G., Malaikozhundan, B., Shanthi, S., & Vaseeharan, B. J. S. A. P. A. M. (2015). *Plectranthus amboinicus* leaf extract mediated synthesis of zinc oxide nanoparticles and its control of methicillin resistant *Staphylococcus aureus* biofilm and blood sucking mosquito larvae. *Spectrochimica Acta Part A: Molecular and Biomolecular Spectroscopy*, *137*, 886–891. https://doi.org/10.1016/j.saa.2014.08.064

Wang, B., Zhang, Y., Mao, Z., Yu, D., & Gao, C. (2014). Toxicity of ZnO nanoparticles to macrophages due to cell uptake and intracellular release of zinc ions. *Journal of Nanoscience and Nanotechnology*, *14*(8), 5688–5696. https://doi.org/10.1166/jnn.2014.8876

Wang, Q., Zhang, P., Zhao, W., Li, Y., Jiang, Y., Rui, Y., . . . & Lynch, I. (2023). Interplay of metal-based nanoparticles with plant rhizosphere microenvironment: Implications for nanosafety and nano-enabled sustainable agriculture. *Environmental Science: Nano*, *10*(2), 372–392. https://doi.org/10.1039/D2EN00803C

Wong, C. W., Chan, Y. S., Jeevanandam, J., Pal, K., Bechelany, M., Abd Elkodous, M., & El-Sayyad, G. S. (2020). Response surface methodology optimization of mono-dispersed MgO nanoparticles fabricated by ultrasonic-assisted sol–gel method for outstanding antimicrobial and antibiofilm activities. *Journal of Cluster Science*, *31*(2), 367–389. https://doi.org/10.1007/s10876-019-01651-3

Xiu, W., Shan, J., Yang, K., Xiao, H., Yuwen, L., & Wang, L. (2021). Recent development of nanomedicine for the treatment of bacterial biofilm infections. *View*, *2*(1), 20200065. https://doi.org/10.1002/VIW.20200065

Yang, X., Chung, E., Johnston, I., Ren, G., & Cheong, Y. K. (2021). Exploitation of antimicrobial nanoparticles and their applications in biomedical engineering. *Applied Sciences*, *11*(10), 4520. https://doi.org/10.3390/app11104520

Yarjanli, Z., Ghaedi, K., Esmaeili, A., Rahgozar, S., & Zarrabi, A. (2017). Iron oxide nanoparticles may damage to the neural tissue through iron accumulation, oxidative stress, and protein aggregation. *BMC Neuroscience*, *18*, 1–12. https://doi.org/10.1186/s12868-017-0369-9

Zhang, Y., Lin, S., Fu, J., Zhang, W., Shu, G., Lin, J., . . . & Fu, H. (2022). Nanocarriers for combating biofilms: Advantages and challenges. *Journal of Applied Microbiology*, *133*(3), 1273–1287. https://doi.org/10.1111/jam.15640

Zhou, Y., Wang, G., Li, Y., Liu, Y., Song, Y., Zheng, W., . . . & Jia, J. (2012). In vitro interactions between aspirin and amphotericin B against planktonic cells and biofilm cells of *Candida albicans* and *C. parapsilosis*. *Antimicrobial Agents and Chemotherapy*, *56*(6), 3250–3260. https://doi.org/10.1128/AAC.06082-11

20 Ethical Issues in Nanotechnology and Sustainability

Narendra Kumar S, Shamantha Gowda T S, Manjunath Yadav S M, and Satheesh Babu G

20.1 ETHICAL CHALLENGES IN NANOTECHNOLOGY AND SUSTAINABILITY

Nanotechnology, which is the control of matter on an atomic and molecular scale, is a technology whose potential is only recently emerging as one of the world's most promising scientific disciplines in the 21st century. The characteristic size scale is between 1 and 100 nm, and in this regime, a wide range of new materials and systems with novel physicochemical properties that are considerably different from the properties of the corresponding bulk materials could be realized. These properties, for example, increased strength, conductivity, or chemical reactivity are rooted in quantum effects and high surface area at the nanoscale [1]. Because of its multidisciplinary character, the area of nanotechnology aggregates fundamental concepts from physics, chemistry, biology, and engineering, and as a consequence, this convergence promotes an enormous range of impactful device applications in sectors as such medicine, electronics, energy, and environmental science. In medicine, it enables the creation of nanoparticles which can seek out and obliterate diseased cells, minimizing side effects of treatments such as chemotherapy. In the electronics arena, it enables the creation of smaller and more powerful components that are used in faster and more efficient devices such as smartphones and computers [2].

The energy industry is literally powered by nanotechnology, with better solar film cells and battery technologies, as well as the environment via its use in enhanced filtration systems for drinking water and air [2]. This versatility places nanotechnology at the forefront of new technology and resources that can directly affect global challenges such as affordable healthcare, clean energy, and environmental protection. As new applications of AI are discovered, its potential is growing, playing a central role in the future of technology and society.

20.2 HISTORICAL BACKGROUND AND DEVELOPMENT

The term "nanotechnology" dates back to 1959 when physicist Richard Feynman gave his prophetic speech "There's Plenty of Room at the Bottom" during an American Physical Society meeting. Feynman thought of a world where machines could be manufactured atom by atom; he posited that scientists could manipulate individual atoms to construct creative devices, anticipating the development of ultra-small computers and even surgical instruments capable of operating within the human body. These assumptions set the intellectual framework for what is now referred to as nanotechnology [3]. It wasn't until the 1980s that the field had some practical impetus, triggered by the invention of two revolutionary instruments: the scanning tunnelling microscope (STM) in 1981 and the atomic force microscope (AFM) in 1986. The STM, developed by Binnig and Rohrer, uses quantum tunneling to image surfaces on the atomic scale and was awarded a Nobel Prize in physics in 1986. AFM is a device that uses a tiny cantilever to "feel" surfaces while manipulating atoms and molecules with nanometric accuracy; it was invented by Binnig, Quate, and Gerber [4]. These advances not only made it possible for scientists to visualize atomic structures but to rearrange them, marking a new stage in science from theoretical speculation to reality.

This feature has also propelled the synthesis and characterization of nanomaterials (e.g., carbon nanotubes and quantum dots) and has corresponded with an explosive increase in nanotechnology research and applications [5]. By the 1990s, it was seen as promising by governments and industries around the world, spurring programs such as the U.S. National Nanotechnology Initiative in 2000 that led to a speed up of the progress [6]. And today, Feynman's once-impossible fantasies are at the core of a global industry, with applications that range from consumer gadgets to state-of-the-art medical treatments, an evolution that took decades of progress that built on the theoretical groundwork he laid.

U.S. NNI Funding (Figure 20.1): Yearly funding distributed by the U.S. National Nanotechnology Initiative (NNI), with an initial level of \$0.27 billion in 2000, a maximum of \$2.16 billion in 2024, and then constant at \$1.40 billion since 2017 [7].

Global Nanotech Funding: Worldwide estimated nano-investment (including funding from the U.S., Europe, Asia, and other countries) increased from \$0.50 billion in 2000 to \$11.00 billion in 2024 [8]. The numbers reflect the growing global importance of nanotechnology continuing

DOI: 10.1201/9781003644439-20

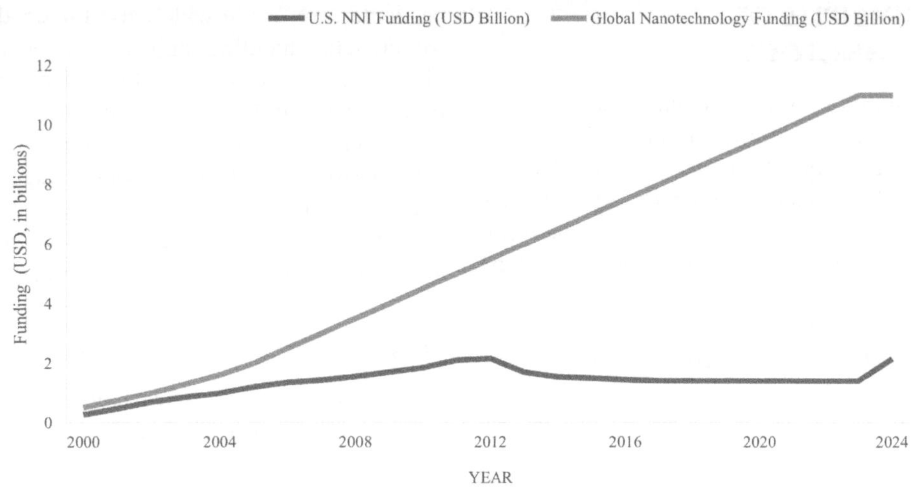

FIGURE 20.1 Funding for nanotechnology over the last two decades.

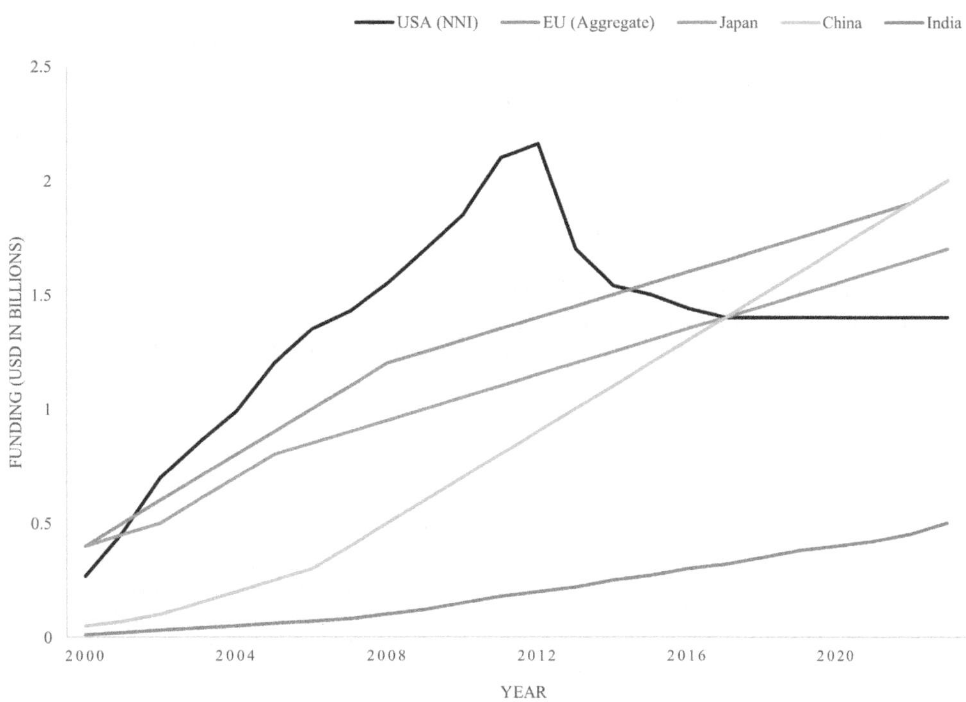

FIGURE 20.2 Fundings of countries for nanotechnology across the world over the last two decades.

to surge, although the U.S. began to move back down the steep slope of exponential growth after initial U.S. funding was introduced. Values are in USD and do not adjust for inflation. Government nanotechnology investments from 2000–2023 exhibit mixed trends [9].

The U.S. (NNI) (Figure 20.2) expanded from $0.27 billion to $2.16, billion peaking in 2012, to later level off, reaching $1.40 billion by 2017. The EU increased continuously from $0.40 billion to $2.00 billion as of 2023, overtaking the U.S. Japan also doubled, from $0.40 billion in 2000 to $0.80 billion in 2005, and then to $1.70 billion in 2023 [10, 11]. Chinese financing increased from $0.05 billion to $2.00 billion by 2023, motivated by post-2006 activities [12, 13]. India increased from $0.01 billion to $0.50 billion, facilitated by the Nano Mission in 2007. Numbers are not adjusted for inflation and are in US$, reflecting a variety of national interests. The funding trends from 2000 to 2023 reveal an early U.S. [14, 15] surge peaking in 2012, followed by stabilization, while the EU and China exhibit steady and rapid growth, respectively, overtaking the U.S. by 2023, with Japan showing consistent progress and India emerging gradually.

20.3 BASIC PRINCIPLES OF NANOTECHNOLOGY

Nanotechnology is based on very different principles than traditional material sciences, mainly because of the overwhelming importance of quantum mechanical effects at the nanometer scale. When materials are reduced to this size range (atomically or structurally) the new electronic, optical, and mechanical behavior of the material is affected by quantum effects related to electron confinement and wave particle duality. For example, in bulk, electrons are free to move, whereas at the nanoscale, constriction rules the way, creating a system with discrete energy levels that change electrical and light emission characteristics, as used in quantum dots for bright displays [6]. In addition, the very high surface-to-volume ratio of NPs will increase their reactivity and interaction with biological systems, which is essential for their biological application. For instance, a nanoparticle can possess as many atoms on its surface as half its total, whereas less than 1% of the atoms in bulk materials are on their surface, which enhances its catalytic activity and its response to environmental refinements [16]. This concept allows the construction of nanosystems with a specific focus, for example, liposomes for targeted drug therapy, which strive to encapsulate drugs and only release them at the target, thereby reducing systemic toxicity. Likewise, high-efficiency catalysts such as platinum nanoparticles employ added surface area to facilitate faster rates of chemical reactions in industrial applications, and new semiconductor materials exploit nanometer scale features to cram more transistors onto microchips, increasing computing speed [17]. This attribute, along with other distinctive features of nanotechnology, gives the field the ability to bridge diverse domains. For example, in energy applications, nanostructured solar cells can capture more sunlight than bulk materials, while in environmental science, nanofilters can remove pollutants with exceptionally high sensitivity [6]. Appreciating these principles quantum effects and surface dominance unlocks nanotechnology's potential to deliver smaller, smarter, and more efficient innovations than ever before and presents unique challenges in controlling and predicting nano-scale behavior.

20.4 CLASSIFICATION OF NANOMATERIALS

Nanomaterials are generally divided into four principal types.

20.4.1 Carbon-Based Nanomaterials

Fullerenes, carbon nanotubes (CNTs), and graphene display superior electrical conductivity and mechanical stability, rendering them critical in electronics and composite materials [18]. Their unique nanoscale structures enable high surface area-to-volume ratios, electron mobility, and thermal conductivity, which are essential for developing next-generation sensors, transistors, and energy storage devices. In particular, CNTs are widely used in conductive films, electromagnetic shielding, and flexible electronics, while graphene's two-dimensional lattice offers remarkable strength, transparency, and tunable electronic properties. Fullerenes, on the other hand, are explored in photovoltaic devices and drug delivery due to their hollow cage-like structure and chemical versatility. Together, these carbon-based nanomaterials are reshaping multiple high-tech industries, from aerospace to biomedical engineering [19].

20.4.2 Metallic Nanomaterials

Gold, silver, and platinum nanoparticles are extensively employed in a variety of applications due to their unique physicochemical properties such as tunable surface plasmon resonance, high surface reactivity, and biocompatibility [20]. Gold nanoparticles are particularly valued in biomedical imaging, drug delivery, and photothermal therapy owing to their ease of functionalization and stability in biological environments. Silver nanoparticles exhibit strong antimicrobial activity and are widely used in wound dressings, coatings for medical devices, and water purification systems. Platinum nanoparticles, known for their exceptional catalytic efficiency, play a crucial role in fuel cells, chemical sensing, and as anticancer agents. The versatility and multifunctionality of these metallic nanomaterials make them indispensable in fields ranging from nanomedicine to environmental remediation [21].

20.4.3 Nanoceramic Materials

Nanostructured titanium dioxide (TiO_2) and zinc oxide (ZnO) are widely utilized in environmental and biomedical applications due to their excellent photocatalytic and antimicrobial properties [22]. TiO_2 nanoparticles are prominent in air and water purification systems, where they degrade organic pollutants under UV or visible light through advanced oxidation processes. ZnO nanoparticles, in addition to their photocatalytic capabilities, exhibit strong antibacterial and antifungal activity, making them suitable for use in coatings, textiles, and packaging materials. Both materials are also incorporated into sunscreens and personal care products because of their ability to block harmful UV radiation. Furthermore, their chemical stability, low cost, and eco-friendliness contribute to their growing significance in sustainable nanotechnology solutions [22].

20.4.4 Polymeric Nanomaterials

Polymer structures at the nanoscale offer remarkable biocompatibility, tunable mechanical and chemical properties, and functional versatility, making them ideal for applications such as targeted drug delivery, gene therapy, and tissue engineering [23]. These nanomaterials can be engineered to respond to specific stimuli such as pH, temperature, or enzymes, enabling controlled and site-specific release of therapeutic agents. Biodegradable polymers like polylactic acid (PLA), poly(lactic-co-glycolic acid) (PLGA), and polyethylene glycol (PEG) are commonly used to design nanocarriers that minimize toxicity and enhance circulation time in the body. In tissue engineering, polymeric scaffolds support cell growth, differentiation, and tissue regeneration by mimicking the extracellular matrix. Overall, polymeric nanomaterials represent a cornerstone of advanced biomedical technologies due to their adaptability and safety profiles [24].

20.5 APPLICATIONS OF NANOTECHNOLOGY

Nanotechnology has been applied across diverse industries, yielding groundbreaking developments. Following is a detailed discussion of key applications.

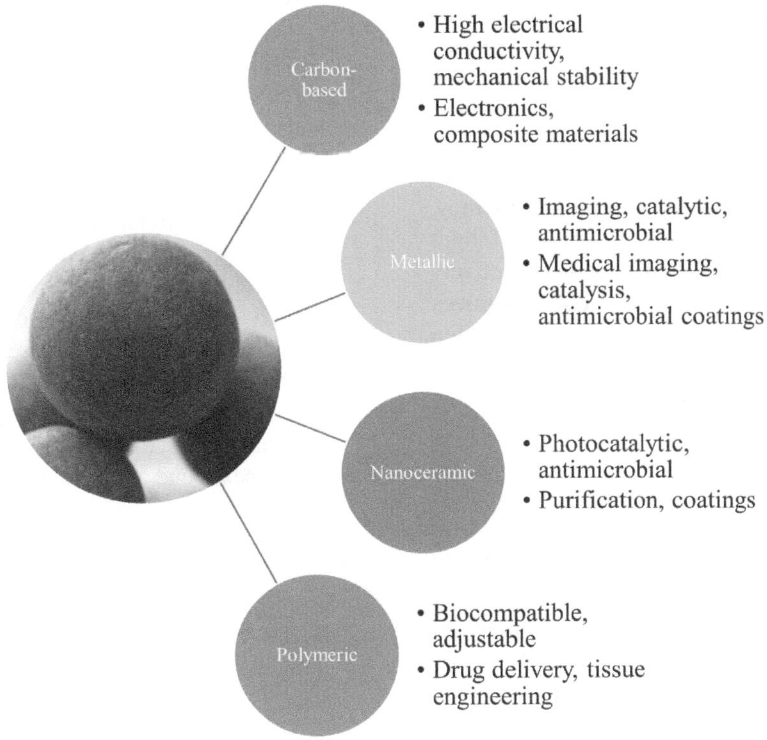

FIGURE 20.3 Classification of nanomaterials and their applications.

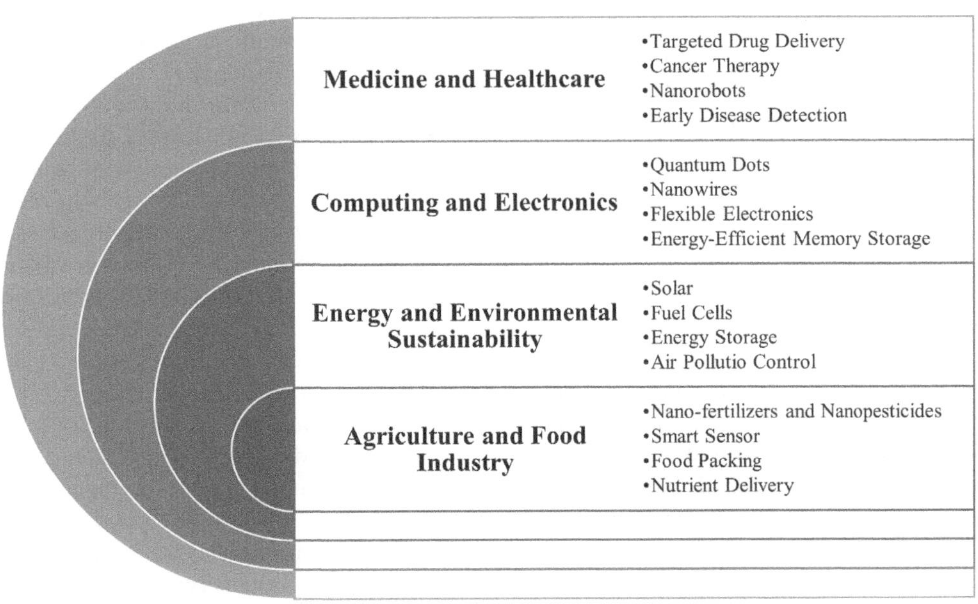

FIGURE 20.4 An overview of the application of nanotechnology.

20.5.1 Medicine and Healthcare

Nanotechnology has transformed the health sector by providing targeted drug delivery, cancer therapy, regenerative medicine, and high-tech diagnostics [21].

- *Targeted Drug Delivery:* Nanoparticles such as liposomes, dendrimers, and polymeric carriers are engineered to deliver therapeutic agents directly to diseased or affected cells, thereby improving drug efficacy and reducing systemic side effects. By exploiting surface ligands or responsive release mechanisms, these nanocarriers ensure precision delivery, particularly in diseases like cancer and inflammatory disorders [25].
- *Cancer Therapy:* Silver and gold nanoparticles are increasingly utilized in photothermal therapy, where they absorb specific wavelengths of light and convert it into localized heat to selectively destroy cancerous cells. This technique minimizes damage to adjacent healthy tissues and is often combined with imaging techniques for real-time monitoring and enhanced precision [3].
- *Nanorobots:* These microscopic, robot-like devices, often conceptual or in early-stage development, hold transformative potential in medicine. They are designed to perform minimally invasive tasks such as clearing arterial blockages, repairing damaged tissues, or delivering drugs at cellular resolution. Additionally, nanorobots may incorporate biosensors to continuously monitor physiological parameters and respond to changes in real time [4].
- *Early Disease Detection:* Nanosensors can detect biomarkers at extremely low concentrations, enabling early diagnosis of neurodegenerative diseases such as Alzheimer's and Parkinson's. These sensors leverage nanomaterials with high surface-to-volume ratios and sensitivity to specific biological molecules, improving diagnostic accuracy and allowing timely intervention [5].

20.5.2 Computing and Electronics

Nanotechnology has made miniaturization of electronics feasible, which would ultimately lead to faster, more efficient, and smaller transistors [6].

- *Quantum Dots:* Quantum dots are semiconductor nanocrystals that emit light at precise wavelengths when excited, making them highly valuable in next-generation LED and OLED display technologies. Their tunable emission spectra enhance color purity, brightness, and energy efficiency, significantly expanding the color gamut and improving visual performance in devices such as TVs, smartphones, and monitors [16].
- *Nanowires:* Due to their one-dimensional structure and quantum confinement effects, nanowires are at the forefront of advancements in nanoelectronics. They are critical in fabricating high-performance semiconductors, transistors, and logic gates, enabling faster data processing, reduced power consumption, and miniaturization of electronic components [17].
- *Flexible Electronics:* Flexible carbon nanotube (CNT) transistors, along with nanomaterials like gallium nitride (GaN) and graphene, are revolutionizing wearable and foldable technology. These materials offer exceptional mechanical flexibility, electrical conductivity, and thermal stability, allowing the development of bendable screens, stretchable sensors, and smart textiles that can conform to dynamic surfaces or human skin [18].
- *High-Density Nanoscale Memory:* Nanotechnology underpins the advancement of ultra-compact and high-speed memory devices. Nanoscale materials and architectures, such as resistive RAM (ReRAM), phase-change memory (PCM), and spintronic elements, enable dense data storage and faster computational speeds, forming the core of modern processors and next-generation computing systems [20].

20.5.3 Energy and Environmental Sustainability

To promote sustainable sources of energy and remediation of the environment to healthful living, nanotechnology is a fundamental science [22].

- *Solar:* Nanostructured photovoltaic materials, including quantum dots and perovskite nanoparticles, significantly enhance the efficiency of solar panels by improving light absorption and charge transport. These advanced materials enable broader spectral capture, including low-light conditions, and allow the development of thinner, lightweight, and flexible solar cells for a variety of applications [23].
- *Fuel Cells:* Nanocatalysts, particularly those composed of platinum or alloyed nanomaterials, dramatically increase the reaction rates and energy conversion efficiency in hydrogen fuel cells. These nanoscale catalysts lower activation energy and improve durability, contributing to the development of clean, sustainable, and scalable energy alternatives for transportation and stationary power generation [21].
- *Energy Storage:* Nanotechnology plays a crucial role in advancing energy storage systems, especially through the use of graphene-based supercapacitors and nanostructured electrode materials. These innovations lead to faster charge/discharge cycles, higher energy density, longer battery life, and greater thermal and mechanical stability in lithium-ion and next-generation batteries [25].

- *Air Pollution Control:* Nanoscale air filtration systems utilize materials such as titanium dioxide, carbon nanotubes, and electrospun nanofibers to effectively capture ultrafine particulate matter (PM2.5 and smaller), volatile organic compounds (VOCs), and harmful gases. These nanotech-enabled filters are being applied in both industrial emissions control and consumer-grade air purifiers to combat environmental and indoor air pollution [4].

20.5.4 Agriculture and the Food Industry

Agriculture and the food industry have already been revolutionized by nanotechnology through the better protection of crops, food safety and the preservation of remaining life time [5].

- *Nanofertilizers and Nanopesticides:* These nanotechnology-based agrochemicals provide controlled and sustained release of nutrients and active agents, reducing leaching into the soil and minimizing harmful environmental impacts. By improving the uptake efficiency and targeting specific pests or deficiencies, nanoformulations help increase crop yield while promoting sustainable farming practices [6].
- *Smart Sensors:* Nanosensors embedded in agricultural systems enable real-time monitoring of critical parameters such as soil nutrient levels, moisture content, and the presence of disease-causing pathogens. These sensors support precision agriculture by providing farmers with accurate data to optimize irrigation, fertilization, and pest management, ultimately enhancing productivity and resource efficiency [16].
- *Food Packing:* Nanocoatings and microbiocidal nanoparticles (e.g., silver, zinc oxide) are integrated into food contact materials to inhibit microbial growth, block UV radiation, and create moisture barriers. This nanotechnology-enhanced packaging helps extend shelf life, maintain food quality, and reduce foodborne contamination throughout the supply chain [17].
- *Nutrient Delivery:* Nanoencapsulation techniques enhance the bioavailability, stability, and targeted delivery of essential nutrients and bioactives in functional foods and dietary supplements. By improving solubility and protecting sensitive compounds from degradation during digestion, nanocarriers ensure more effective nutrient absorption and health benefits [18, 39].

20.6 ETHICAL ISSUES OF NANOTECHNOLOGY

Nanotechnology is transforming a wide range of fields, including medicine, electronics, agriculture, energy, and environmental protection. However, these advances are

FIGURE 20.5 Categories of ethical issues faced in nanotechnology.

accompanied by significant ethical challenges. Key concerns include potential health risks, environmental impacts, threats to privacy, issues of fairness and equity, regulatory gaps, human enhancement implications, and the responsibilities of researchers. These considerations highlight the need for careful oversight and responsible innovation.

Nanotechnology, a revolutionary field that encompasses medicine, electronics, environmental protection, agriculture, and energy, faces strong moral implications in tandem with bio-technologies. These challenges include health and safety risks, environmental consequences, privacy issues, equity, gaps in the regulatory system, the enhancement of humans, and the ethical duty of researchers. This section discusses these challenges in detail and provides an overview of the recent studies that point out the true costs to society.

20.6.1 Health and Safety Issues

The potential health and safety impacts of nanotechnology represent a critical ethical challenge. Due to their small size (1–100 nm), nanoparticles can penetrate biological barriers, such as cell membranes and the blood–brain barrier, leading to potential toxicity [20]. Research indicates that the toxicity of engineered nanomaterials depends on physicochemical properties like size, shape, surface area, and composition [20, 22]. For instance, gold and silver nanoparticles, widely used in cancer therapy, exhibit varying toxicity levels. Silver nanoparticles have shown higher cytotoxicity, surpassing asbestos at specific concentrations, while gold nanoparticles are generally less toxic [21].

- *Toxic and Genotoxic Effects:* Nanoparticles, especially those composed of heavy metals (e.g., silver, titanium dioxide) and carbon (e.g., carbon nanotubes), can cause oxidative stress, DNA adducts, and inflammatory responses in cells [20, 21]. A study of cerium oxide nanoparticles showed that they induced DNA damage and oxidative stress in human dermal fibroblasts, indicating genotoxic hazards. Smaller particles (e.g., ~50 nm in diameter) are taken up more effectively and are more cytotoxic than larger ones, and the shape (e.g., nanorods versus spherical particles) affects biological interactions [20]. These effects have caused concern regarding their safe application for medical and industrial purposes.
- *Work and Occupational Exposure:* Nanotechnology-based industry workers are also at risk of exposure through inhalation or dermal absorption of airborne NPs that may result in respiratory or dermal toxicity. These studies simulating occupational exposure to titanium dioxide nanoparticles reported that they have tendency to aggregate into larger ones, which enhances the respiratory effects [5]. A major impediment to risk assessment has been the lack of harmonized safety and toxicokinetic (i.e., absorption, distribution, metabolism, and excretion ADME) protocols. This highlights the importance of strong workplace safety standards that protect workers.
- *Long-Term Health Effects:* In particular, nanoparticle accumulation in organs such as the lungs, liver, or brain is a concern for chronic diseases as cancer and neurodegenerative diseases. MWCNT penetration and retention have been reported in lung tissue, causing fibrosis comparable to asbestos and increasing the risk of cancer [16]. The dearth of longitudinal human studies has, however, led to unclear understanding of these long-term effects, and therefore more research is required on safe nanotechnology applications [5, 16, 20, 21].

20.6.2 Environmental Implications

The environmental impacts of nanotechnology are two sided, providing remediation of pollution on the one hand and threatening ecosystems on the other hand. Nanoparticles can be deposited in soil, water, and air, posing a disruption to the ecological system [26].

- *Water and Soil Pollution:* Nanoparticles that are discharged from industrial applications or consumer products may contaminate water and soil, which can have negative impacts on aquatic organisms and drinking water quality. For instance, food packaging–derived silver and titanium dioxide nanoparticles may enter waste water and affect aquatic organisms [25]. Nanotechnology offers remediation technologies, including membrane bioreactors and antifouling membranes applied in water treatment, leading to savings of chemicals and lessening environmental harm [25, 26]. The ethical calculus of balancing these benefits with the risks of pollution proves key.
- *Persistence and Bioaccumulation:* Nanoparticles are usually slow to degrade, at least as compared to larger particles; therefore, they may persist in the environment for long periods of time. Carbon nanomaterials, such as graphene, have also been shown to bioaccumulate in organisms, bioaccumulating at the next trophic level and enhancing their toxicity. This resistance poses challenges for waste management, since conventional processes may not be effective for nanoscaled materials. Proper disposal and recycling methods are needed to guard against this risk [24].
- *Effects on Biodiversity:* NPs can affect favorable microorganisms, plants, and animals, leading to disrupted ecosystems and biodiversity. For example, soil microbial communities and nitrogen cycling and plant growth are disturbed by silver nanoparticles [18]. On the other hand, nanotechnology can also support environmental recovery

by helping to restore structures damaged by pollution. For example, in nano-enhanced bioremediation of oil spills, nanoadditives can accelerate the microbial degradation of pollutants, improving the effectiveness of ecosystem restoration efforts [27]. Supportive deployment has to balance the maximization of remediation benefits against the minimization of unintended ecological side effects [18, 19, 26–28].

20.6.3 Privacy and Surveillance Issues

Advances in nanotechnology have enabled the manufacture of advanced nanosensors and monitoring apparatus, which provide major advantages but simultaneously raise serious threats to individual privacy and enhance fears of surveillance [29]. The ability of such technologies to surreptitiously gather data undermines autonomy and consent and as a result demands strong ethical and regulatory guidelines.

- *Advancements in Nanoscale Surveillance Technologies:* Nanoscale devices like nanosensors can now be manufactured that can monitor environmental conditions or human activity with a previously unattainable level of accuracy [17, 18, 20]. The idea of "smart dust" micro-wireless sensors that monitor and transmit data hints at the possibility of omniscient surveillance. Used without permission, such technologies would be capable of following where people travel, eavesdropping on conversations and observing their behavior; effectively creating a surveillance society and the death of privacy. The moral implications of such power are far reaching and replete with concerns regarding privacy versus advancement. Regulatory incentives should ensure that these technologies are used in a transparent and opt-in manner [29].
- *Ethical Implications of Biometric Applications:* Biometric tools (like fingerprint and iris recognition) are also enhanced by nanotechnology for better performance and results [30, 31]. But, like all good things, it comes at a cost, one of which is privacy. Accumulating a large amount of highly sensitive biometric data calls for a very secure environment to avoid unauthorized access or abuse of biometric data. For example, if biometric databases are compromised, it could result in irreversible privacy compromise, because biometric identifiers are not revocable in the way traditional passwords are. From an ethical perspective, the goal of these systems should be to protect the privacy and autonomy of citizens through appropriate data management and informed consent so that individuals remain in ultimate control of their private data [30].
- *Challenges in Data Governance:* Nanosensors' capacity to aggregate real-time, detailed information from a range of health, location, or activity data represents additional ethical tensions [17, 18, 20]. Autonomy is violated when data is harvested without our permission. The commercialization of such knowledge, moreover, poses the threat of turning people into objects within data markets. Resolution of these issues will require clear data collection procedures, strong consent models, and the implementation of privacy-by-design principles in nanotechnologies. These actions are absolutely essential to preserve public confidence and ensure integrity [29].
- *Workplace Surveillance and Ethical Considerations:* In occupational settings, nanosensors integrated into wearable devices or implants could be used to monitor employees' productivity, health, or safety [17]. While such applications may enhance workplace efficiency and safety, they also raise significant concerns regarding worker autonomy and privacy. Ethical deployment of these technologies requires clear policies on data usage, voluntary participation, and protections against coercive monitoring. Without such safeguards, trust between employers and employees may be undermined, underscoring the need for strong ethical guidance on employment-related applications [17, 18].

20.6.4 Equity and Accessibility

Although the potential for nanotechnology to impact several areas is great, promises of new technologies run the risk of not being available to all and could lead to income inequalities. The high cost and expertise required to develop it can lead to it not being available to disadvantaged groups unless it is specifically promoted as being inclusive [32].

- *Economic Barriers and Global Disparities:* The largescale cost associated with the research and development of nanotechnology makes it often unaffordable for less developed countries or the less endowed sectors [24]. In medicine, nanoscale medicine, such as targeted drug delivery, could revolutionize treatment, but the cost may put it out of reach for poorer areas. In agriculture as well, nanotechnology may improve productivity in the form of novel fertilizers and pest-resistant crops, but small farmers in developing countries may not have access to the capital required to take advantage of these innovations. Such economic division helps to create inequality, as well as slowing down our ability to work together to tackle global challenges, such as famine and disease [32].
- *Intellectual Property and Access Restrictions:* Nanotechnological advances can also have a

negative impact, as patents based on nanotechnologies can limit access to specific technologies, thus constituting a monopoly over crucial technologies [33]. Nanomedicines or energy solutions which are patented can be too expensive for resource-poor countries to afford. This brings to bear a vicious circle of disadvantage and raises ethical issues of profit over human well-being. International approaches, including technology transfer arrangements or perhaps supported licensing, are critical to widen access and ensure that nanotechnology benefits the global community [24, 33].

- *Policy Interventions for Equitable Distribution:* Attaining fair access to nanotechnology requires proactive implementation of policies. Governments and international agencies must increase funds for research and development of devices for the underserved, like affordable, dozen-liter nanoscale water purification or diagnostic [32]. Training of expertise in developing countries can result in less reliance on individual experts from outside and create longer term sustainability [24]. Ethical approaches have to include the voices of marginalized communities so that their needs guide the direction of development in nanotechnology.
- *Global Health and Nanotechnology:* In the field of global health, nanotechnology has shown promise in addressing urgent needs, most notably in the context of the COVID-19 pandemic, taking the forms of rapid diagnostic assays and vaccines [32]. But unequal access to the technologies showed the need for fair distribution mechanisms. Responsible use of nanotechnology in global health must take into account socioeconomic, cultural, and legal issues to eliminate disparities in access to emerging technologies for the poorer citizens of rich nations and those in the developing world [24, 32].

20.6.5 Regulatory and Governance Challenges

The pace of advancement of nanotechnology far exceeds the development of policies to govern these new technologies, and there are important gaps in regulatory oversight and ethical governance [31]. This disparity represents significant risks to global health, to environmental governance, and to global equity and therefore requires a balanced and diversified regulatory approach to innovation and responsibility.

- *Absence of Standardized Safety Protocols:* One of the challenges in the regulation of nanotechnology is that there are no safety guidelines designed specifically for nanomaterials. In contrast to bulk materials, nanomaterials have their own physicochemical characteristics (e.g., size, shape, and surface reactivity), which may cause unanticipated interactions with biological and environmental systems. As a result, many nanoproducts, including those incorporated into food packaging, are being released to the market without having undergone extensive toxicity testing, which may potentially be a risk to public health. For example, silver and titanium dioxide nanoparticles used in antimicrobial coatings have been demonstrated to migrate into food under some circumstances, and concerns have been raised about the effect that this may have on human health in the long term [32]. The unavailability of test systems tailored to nanomaterials further increases these risks, as conventional safety procedures are not appropriate for nanoscale behavior. Establishing rigorous, soundly based protocols is the first step to prevent the unintended consequences overweighing the benefits of beneficial nanotechnology applications.
- *Risk Uncertainty:* The lack of knowledge about nanomaterials brings a high degree of uncertainty when there is a need to extrapolate in vitro data about their impact on the behavior of biological and environmental substances. As opposed to those with macroscopic scales, nanoscale materials may have quantum effects and high surface reactivity, making the risk evaluation more difficult. Existing regulative architectures relying on data of bulk materials are not suitable to represent these complexities. For instance, although promising, it is necessary to carefully assess the persistence and bioaccumulation of these NPs in ecosystems, as for nanoenhanced bioremediation of oil spills [34]. In this uncertainty, flexible regulatory policies are required that grow and learn with the science, incorporating an ever-broadening body of evidence over time. This means that regulators that lack this flexibility are more likely to make decisions on the basis of partial information that may jeopardize public safety and environmental sustainability.
- *International Cooperation:* In light of the international nature of nanotechnology, proper regulations will need international collaboration, especially in setting standards and ethical guidelines. Mismatches in regulations on a national basis are opening up gaps in safety standards that the unscrupulous can exploit and can act as blocks to access to the benefits of the technology for all. Coordinated efforts as exemplified by organizations like the International Organization for Standardization (ISO) and the Organization for Economic Cooperation and Development (OECD) are essential for establishing the harmonized framework between innovation and safety [22].

- Furthermore, global collaboration is essential to discuss transnational environmental issues of nanotechnology, since nanomaterials are able to transport through air, water, and soil to influence ecosystems beyond national boundaries [31]. By promoting mutual understanding of risks and benefits, international cooperation would enable nanotechnology to contribute to, rather than hinder, global advances.
- *Capacity Building and Public Engagement:* In addition to these knowledge gaps, a good governance approach must involve building up regulatory capacity in developing countries where the application of nanotechnology exceeds local skills [31]. Capacity building measures, such as education and technology transfer programs, can enable these countries to improve their safety standards [22]. Furthermore, public involvement is essential for the construction of trust, and for that to be guaranteed, the regulatory structures should also represent the values of society. Open communication with respect to the risks and benefits of nanotechnology, as proposed by international activities, may fill the chasm between scientific discovery and public comprehension, enabling decisions that are well informed [31].

20.6.6 Human Enhancement and Ethical Boundaries

The possibility provided by nanotechnology to enhance human abilities through interfaces between the machines that humanity developed and the machines nature provides (such as brains), as well as through nanoscale implants and performance-enhancing drugs, provides unprecedented opportunities and yet calls into question whether this represents an ethical transgression [28]. These developments challenge established concepts of human identity and require responsible oversight to avoid abuse and exploitation.

- *Cognitive Enhancement:* The possibility to improve cognitive performance, including learning, memory, intelligence, and decision-making ability, via nanomedicine may raise challenging ethical implications concerning justice and social equity [27]. The prevalence of cognitive enhancement will depend heavily on the cost of such enhancers and, consequently, on the degree of wealth inequality within a society. If only the wealthy can afford cognitive enhancers, resentment and social tension may arise among those who cannot. On the one hand, the transformative potential of the applications of nanotechnology in sustainable agriculture is clear; on the other hand, the very same technologies could be refocused on cognitive enhancement, with the result being an increase of the gap between those with resources to access these upgrades and those without [27]. Ethical paradigms should promote justice in access to these technologies, with the goal of ensuring that their cognitive enhancements are for the common good, not benefits just for the few who can afford them.
- *Physical and Genetic Modifications:* The power of nanomedicine to enable both physical and genetic alteration introduces a further ethical layer. The ability to manipulate genes or characteristics at the nano level leads to concerns about the potential impact and acceptance of such interventions. For instance, while it is reasonable to aspire for nanotechnology to contribute to evidence-based solutions for global environmental challenges, applying similar arguments to human enhancement could lead to unforeseen health risks or create societal pressures to conform to enhancement norms [35, 36]. These changes would be less likely to marginalize individuals who choose not to participate, as they would create a culture in which enhancement is viewed as a collective expectation rather than a purely personal choice. Ethical points to be considered also concern the individual's freedom in contrast to that of the society, taking carefully into account a responsible pursuit of modifications.
- *Informed Consent and Autonomy:* The concept of informed consent is key to the ethical application of human enhancement technologies. People need complete information about the risks, benefits, and ethical concerns of nano-enhancements in order to be able to decide autonomously. This calls for openness in behavior by researchers and health workers and guidance to safeguard individuals' rights. For example, it is important that the ethical scrutiny of nanotechnologies as a tool for environmental remediation be clear about the necessity of transparency and accountability, as these principles equally apply to human enhancement [33]. We need to put up a shield against coercion, especially in situations where economic or social pressures could force people to enhance. The ethical limits should respect human dignity, and they should prevent the selling of human features.
- *Societal Implications and Long-Term Risks:* The societal impact of nano-enhancements could be significant as well, affecting the work market, educational systems, and human relations [28]. For instance, the enhanced might crowd out others in competitive arenas, displace the unenhanced, and provoke new forms of discrimination [27]. Furthermore, the potential long-term risks of nanoscale interventions, including unforeseen health or psychological effects, are not well understood, and studies with long-term follow-up are required to inform ethical considerations [35].

Ethical vigilance should therefore be extended from the concern for consent to concern for broader societal issues at stake and ensure that nanotechnologies can increase human capacity without eroding social stability.

20.7 ETHICAL RESPONSIBILITIES IN RESEARCH AND INNOVATION

Given the rapid development of nanotechnology in a wide range of areas, including medicine, environmental management, and industry, a sound ethical structure is needed to facilitate the development of the field. For nanotechnology, responsible research and innovation (RRI) needs to become an integral part of it, meaning that ethical, legal, and societal aspects need to be integrated in the core of technological development in order to shape it in a way that is beneficial for society, respects public opinions, and deals with risks in a responsible way. This section explores the varied ethical obligations of researchers and industries, highlighting transparency, resource distribution, stakeholder inclusion, risk management, IP, and moral learning. Informed by recent studies, it highlights the relevance of such obligations to maintaining the trust of the public and delivering fair, sustainable results.

- *Fostering Public Awareness and Transparency:* Transparency is one of the foundations of ethical nanotechnology research, integral to building public confidence and promoting an informed public discourse. Governments and industry will need to invest in clear, accessible information on the benefits and risks of nanotechnology applications. For example, in the case of nanotechnology used in water purification, transparent and outreaching communication about its safety, performance, and environmental effects is a prerequisite for public acceptance and responsible deployment [6]. For instance, the National Nanotechnology Initiative (NNI) in the US and the European Union's Horizon 2020 framework are committed to public engagement and believe in sharing the full spectrum of potential impacts of nanotechnology (National Nanotechnology Initiative; European Union, 2020). Such efforts provide a bridge between scientific breakthroughs and public understanding, but some stakeholders warn that too much transparency could forestall innovation, since it could emphasize perceived risks.
- *Ethical Allocation of Research Resources:* The allocation of research funds according to plans is an ethical duty, since it is indicative of the priorities and mentality of a society. Funding should favor such investigations over technology-driven ones while remaining attuned to societal, environmental, and health impacts in the long term of the development and deployment of nanotechnology. Specifically, in the area of nanodrug delivery systems, funding should also be directed at research that evaluates safety, efficacy, and commercial feasibility for society-wide use [23]. The NNI's 2025 budget dedicates substantial resources for foundational research and applications-focused research, with a strong focus on understanding the broader impacts of nanotechnology (National Nanotechnology Initiative). This direction also allows resources to be used to reduce potential risks associated with nanoparticles, such as toxicity or environmental persistence, rather than being solely focused on commercialization. But questions remain on the tradeoff between money spent on innovative breakthroughs versus safety-focused research, and some believe the focus on ethics may slow the field's progress.
- *Responsible Research and Innovation:* The RRI framework offer a holistic approach on how to integrate ethical issues into developing nanotechnology. RRI focuses on anticipation, reflection, and action in relation to societal needs, ensuring that research is implemented in accordance with public values. RRI brings in ethical, legal, and social issues (ELSI) to encourage researchers to reflect on the broader context of their work and its consequences, for example, that nanotechnology may widen social disparities or adversely affect the environment. The European Environment Agency nanotechnology report emphasizes the need for early warnings and anticipatory ethical reflection in order to avert such external effects [1]. Embedding RRI principles in the field of nanotechnology will ensure that nanotechnology development is accomplished in a responsible manner, addressing societal needs while minimizing inherent risks.
- *Stakeholder Engagement and Inclusivity:* Involving a broad spectrum of stakeholders that includes the public, policy makers, and industry and civil society organizations is essential to ensure that nanotechnology develops according to social preferences and also addresses the public's concerns. Collaborative programs such as those conducted by the Centers for Nanotechnology in Society housed at Arizona State University and the University of California, Santa Barbara, foster discussion among consumers, ethicists, scientists, and policy-makers to focus on the ethical aspects of nanotechnology. Actions like these serve to highlight potential risks and benefits to policies that can foster more equitable and socially responsible outcomes. But inspiring harmonized action among a variety of stakeholders is challenging, and efforts will need to be well facilitated in order to be inclusive.

- *Risk Assessment and Management:* Appropriate risk assessment is an ethical imperative for nanotechnology, because nanomaterials have novel properties which could have unknown effects on health and environment. The safety of the particular engineered nanomaterials used has to be considered by researchers, and safety measures should be applied to avoid hazards such as toxic nanoparticles in drug delivery or industrial applications [23]. This lack of prevailing scientific knowledge regarding the risks related to the use of nanomaterials demonstrates the necessity for prudent decision-making and repetition research (workplace ethics). In addition, to develop harmonized safety regulations for the transboundary impacts of nanotechnology, international cooperation, as promoted by bodies such as the OECD, is also necessary [1]. Striking a balance between prudent risk management and innovation continues to be a contentious issue, with some claiming that tight regulation could hinder advancement.
- *Intellectual Property and Equitable Access:* Patenting in nanotechnology is an area where tension between the positive and negative aspects of patenting has been said to pose an ethical dilemma. There is therefore an ethical responsibility to strike a good balance between protecting IP and fostering equitable access and, in this respect, to ensure that IP rights should not be considered just a legal right but also a social right, especially for emerging countries. For example, nanoapplications in water treatment, healthcare (e.g., disease treatment), and others, need to be accessible to underserved populations to address international issues like clean water or public health [6, 23]. Barriers can be lessened through mechanisms such as technology transfer agreements and open-access publishing to ensure nanotechnology's benefits reach a broad audience (National Nanotechnology Initiative). It is imperative that there be ethical ways of managing IP to avoid global inequity. The development of an ethics-sensitized research community remains a critical need for responsible innovation in nanotechnology. Training and education programs must include the ELSA (ethical, legal, and social aspects) issues of nanotechnology, providing researchers with the ability to navigate complex moral dimensions. The Research Ethics Program of the University of California, San Diego, exemplifies this common sense, provides training in the responsible conduct of research, and works to sensitize scientists to ethical problems. Such initiatives help ensure our next generation of innovators value public welfare as well as scientific advancement. Furthermore, public educational campaigns can increase social awareness on nanotechnology, which is instrumental in shaping the human right aspects.
- *Societal Implications and Long-Term Vision:* The ethical duties of the nanotechnologist cannot only be defined as an obligation to consider the present but also include an obligation to strive for the future (for the well-being of our society). Academics need to think about how their work contributes to future technological futures, thinking about questions of ethics, sustainability, and human respect. For instance, the potential of nanotechnology to contribute to a range of societal solutions, such as sustainability through better water cleaning systems, needs to be weighed and measured against risks of unintended biological or ecological harm [6, 31]. By taking those responsibilities, the nanotechnology community can steer a course that maximizes human potential while preserving moral considerations. Since nanotechnology is a global enterprise, cooperation at the international level is required to build consensus on ethical guidelines that help to maximize benefits while minimizing risks [2, 31].

20.8 SUSTAINABILITY IN NANOTECHNOLOGY

Impact nanotechnology, a transformative enabling technology, presents an unprecedented opportunity to solve the challenges of global sustainability in renewable energy, clean water, and recycling. By controlling properties of materials at the nanoscale (1–100 nm) level, scientists have begun creating new and effective means for improving efficiency, decreasing environmental impact, and conserving resources. Yet with such advances come ethical and societal concerns, together with the associated likelihood of adverse environmental effects and the importance of clear regulatory oversight. This section analyzes the role of nanotechnology in clean energy, water, and waste management through discussion of recent research to highlight its contribution to sustainability, challenges associated with its use, and ethical issues.

20.8.1 Nanotechnology in Clean Energy

Nanotechnology is expected to transform clean energy production, storage, and utilization by improving the efficiency, cost, and environmental impact of energy-based systems. Capitalizing on the extraordinary features of nanomaterials (e.g., high surface area, controllable optical properties, and improved conductivity), a series of solutions were proposed to realize these fundamental challenges in alternative energy conversion and storage.

- *Solar Energy Innovations:* Nanotechnology has dramatically enhanced photovoltaics, making solar cells cheaper and more efficient. Quantum dots (QDs) are semiconductor nanoparticles with size-dependent optical properties that can be tuned to absorb light with a higher efficiency than conventional silicon-based solar cells. Perovskite solar cells using nanostructured materials have attained efficiencies similar to silicon-based cells, with the added benefits of flexibility and cost-effective solution processing [33]. These developments are said to have the potential to enable solar energy to become a main energy source, shielding us from the side effects of burning fossil fuel. However, the durability of perovskite materials and sustainability of manufacturing nanomaterials need to be considered at the same time [31].

- *Energy Storage Advancements:* Nanotechnology has contributed significantly in the energy storage sector; most importantly, it has led to the development of high-performance rechargeable batteries, essential for electric vehicles and the integration of renewable energy. Nanostructured electrodes, including graphene-based ones, can increase electrical conductance and facilitate ion migration, bringing greater energy to batteries and making them charge in a short time. For example, graphene-based nanocomposites have been demonstrated to improve battery capacity and cycle life and hold great promise of being useful in next-generation energy storage systems. Despite these advances, the energy-intensive production of graphene and other nanomaterials has raised concerns about their overall environmental benefits, and lifecycle assessments are needed to evaluate the sustainability of these materials [32, 33].
- *Fuel Cells and Energy Efficiency:* Nanotechnology enables fuel cells that convert hydrogen into electric power with high efficiency and zero emissions (except for water). The use of nanocatalysts such as Pt particles supported on carbon nanotubes [19, 24, 29, 37] allows higher fuel cell efficiency by providing larger available surfaces for reactions without requiring high-cost precious metals. In addition, nanotechnology helps to cut energy usage by, for example, enhancing the thermal insulation of buildings with applications such as nanocoatings and by improving wind turbine blades using lightweight nanocomposites that aid energy capture and will last longer. These advances demonstrate the potential of nanotechnologies as part of a transition to a low-carbon economy, as long as issues of scale and pricing can be overcome [19].
- *Challenges and Ethical Considerations:* Nanotechnology has its own challenges for clean energy system integration. Nanomaterials are often energy and resource intensive to produce, potentially negating any net environmental benefits. Also, at the end of their life cycle, nanomaterials that are potential toxic pollutants have to be disposed of, which requires efficient waste handling systems. Ethical issues include that clean energy technologies should be accessible to everyone, especially in the case of developing countries, in order to prevent an increase in global inequality. Further exploration and responsible innovation are required in order to face these problems and allow for the maximum achievement of nanotechnology's potential for sustainable energy [19, 29].

20.8.2 Water and Waste Management

Nanotechnology has provided dynamic approaches for water treatment and waste disposal, which are crucial environmental concepts (water shortage, pollution, and waste management). The highly reactive and selective properties of nanomaterials have enabled the demonstration of a range of technologies that have the potential to improve the efficiency and sustainability of water and waste treatment processes.

- *Advanced Water Purification:* These days, nanotechnology is being used to change the world of water treatment and filtration as we know it. Nanofiltration membranes which feature nanoscale pores filter out dissolved salts, heavy metals, and organic molecules to yield clean drinking water where existing resources prove insufficient. Nanomaterials such as titania (TiO_2) materials have been used to remove organic pollutants under UV light irradiation through the process of photocatalysis, providing an environmentally friendly alternative to chemical methods [6]. These treatments also improve water quality in both industrial and household settings and are particularly valuable in regions where access to clean water is limited [36]. But concern arises that the release of nanomaterials into aquatic environments might affect ecology; hence the necessity to strictly supervise and control this kind of material.
- *Wastewater Treatment and Contaminant Removal:* Nanotechnology can be involved in wastewater treatment through adsorption and catalytic degradation of pollutants. Particles, like iron oxide or carbon-based materials, can adsorb heavy metals and organic pollutants, and nanoscale catalysts can degrade persistent pollutants. For instance, nano-zero-valent iron was used for groundwater remediation of contaminated water by chlorinated solvents to be converted into less hazardous chemicals [31]. Nano-facilitated biosensors are advancing water management by enabling real-time monitoring of micropollutants and allowing timely responses to pollution events. These developments are increasing the efficacy of waste water treatment, although there is controversy over the toxicity of some nanoparticles, such as silver, to aquatic systems.

- *Nano-Enhanced Bioremediation:* Nanotechnology plays an important role in bioremediation, a method for the treatment of environmental pollution using microorganisms. Nanomodified bioremediation involves the use of engineered nanoparticles to facilitate microbial activity or act as nutrient vehicles to enhance degradation of pollutants, such as oil spills. For example, nanoparticles disperse droplets of oil, increasing the surface area and making the oil droplets available to bacteria that degrade oil, according to recent research. It is a sustainable method for environmental remediation, but the long-term ecological impact of nanoparticles on the environment warrants further research for safety [38].
- *Air Purification and Solid Waste Management:* Apart from water, nanotechnology finds application in air filtration and treatment of solid waste. Nanostructure materials such as carbon nanotubes are used as filters in high-efficiency particulate air (HEPA) to filter ultrafine particle and volatile organic compound to purify air. Nanotechnology has numerous applications in solid waste management by contributing to recycling through better material separation and degradation. One possibility is that in some applications nanomaterials could help polystyrene degrade or aid in making composting work better [33, 38]. However, nanomaterials are environmentally persistent and have the potential to accumulate in ecosystems. These concerns warrant stringent disposal measures.
- *Challenges and Ethical Considerations:* There are some challenges in using nanotechnology for water and waste treatment. The likely toxicity of nanomaterials like silver nanoparticles to aquatic organisms and useful microorganisms make it necessary to conduct thorough safety investigations. In addition, cost and the necessity for a dedicated infrastructure for nanotechnology-based systems can limit their applications, especially in developing areas. Ethical issues involve the equitable development of such technologies as a response to global water and waste issues without increasing disparities. Sensible innovation, supported with regulation, is important for finding a balance between gains from new technologies like nanotech and their possible risks [30].

20.8.3 Sustainable Manufacturing and Environmentally Friendly Materials

Nanotechnology is considered a revolution in the manufacturing processes through which new materials with excellent properties, low resource consumption, and a small environmental footprint can be produced. These developments make it easier to develop products in line with sustainable development objectives by enabling a reduction of waste and the propagation of ecologically preferable options across sectors [24].

- *Lightweight and High-Strength Materials:* Thanks to nanotechnology, the new microstructured materials are light and strong enough to perform at peak efficiency while demanding less material. Examples of these are carbon nanotubes and CNT-amplified composites, which are closely connected with the aerospace and automobile sector, in which they reduce vehicle weight and consequently fuel consumption and greenhouse gas emissions [18]. These materials have excellent hardness, so relatively small amounts may be used with sufficient durability. One study shows that CNT composites have tensile strengths that are 100 times higher than those of conventional materials, thus improving structural performance [18].
- However, despite requiring only small material inputs, the energy-intensive synthesis of CNTs raises doubts about their overall environmental friendliness, underscoring the need for further advances in green manufacturing technologies to support sustainable development [24].
- *Nano-Catalysts for Efficient Chemical Processes:* Nano-catalysts have a large surface area and are highly active, which results in more efficient industrial processes that can save energy and reduce wastage. For instance, compared with bulk catalysts, green hydrogen production using platinum nanoparticles as a catalyst achieves more efficient and more eco-friendly industrial applications in the chemical reactions [21]. They make it possible for reactions to take place at a lower temperature and under less pressure, and thus resource consumption and emissions are reduced. Studies that detail their applications in biofuel production, for example, prove extremely valuable as nanocatalysts not only increase the yield but also minimize the damage caused to the environment. However, the possible non-biodegradability of some nanocatalysts, due to their potentially toxic influence when released into the environment, demands lifecycle assessment for their safe application [34].
- *Self-Cleaning and Antimicrobial Surfaces:* Nanocoatings for self-cleaning and antimicrobial surfaces are examples of extraordinary solutions for environment-friendly maintenance and enable reduced use of environmentally hazardous chemical cleaning agents. For example, titanium dioxide (TiO_2) nanocoatings are photocatalytic (i.e. break down organic pollutants on their surface under the action of sunlight) and keep surfaces clean with negligible environmental impact. Silver nanoparticles, as a coating with antimicrobial features, have increased the lifetime of products (e.g.,

medical equipment, textiles) [22]. These developments help reduce dependence on environmentally harmful cleaners and serve sustainability goals. Nevertheless, the risks of nanoparticle introduction into an ecosystem emphasize the importance of strict safety evaluations [24].

- *Biodegradable Nanomaterials:* The development of biodegradable nanomaterials is a great leap forward towards environmentally friendly production. Polymeric nanomaterials like packaging can be designed to degrade naturally not only to suppress plastic waste but also to protect the environment [24]. According to studies, nanocomposite biodegradable plastic often possesses equal strength compared to conventional plastic and has better degradation efficiency in natural environments [18]. The shorts also contribute to the circular economy by reducing long-term waste. However, scaling the manufacture of biodegradable nanomaterials is a challenge because cost of production and performance reliability should be improved [24].
- *Challenges in Sustainable Manufacturing:* However, there are barriers that reduce the viability of sustainable nanomanufacturing, such as high production costs and energy-demanding synthesis strategies. The environmental impact of nanomaterial manufacture, in particular CNTs and metal-based NPs, could undo sustainability benefits unless it is mitigated by greener production processes [34]. Furthermore, due to the absence of standardized nanomaterial safety regulations, their large-scale usage in industries is challenging and demands strong policy making to ensure environmental and human health safety [24].

20.8.4 Environmental Remediation

Nanotechnology provides powerful solutions for environmental problems and helps to clean contaminated soil, water, and air more effectively. It is key to recover ecosystems and alleviate the impacts from industrial activities, although it also represents potential risks to be carefully managed [19].

- *Soil Remediation Using Nanoparticles:* Nanoparticles, such as nZVI or others, have proved highly efficient in the treatment of organic pollutants and heavy metals from contaminated soils, and after treatment, these soils can be used in agricultural or urban areas. nZVI nanoparticles degrade chlorinated hydrocarbons and immobilize hazardous metals such as lead and arsenic, providing convenient remediation at a lower cost as compared to conventional methods. It is reported that field experiments have shown that nZVI can accomplish up to 90% of pollutant removal, which greatly promotes soil restoration [28]. Nevertheless, concerns such as the potential alteration of soil microflora by nZVI and the persistence of environmental residues in ecosystems have raised important questions about long-term ecological safety, which require further investigation [19].
- *Air Pollution Abatement:* The quality of air can be purified using nanotechnology by the production of nanofilters and nanocatalysts that trap fine particles and toxic gases which are produced by industry. For example, carbon nanotube filters can capture ultrafine particles (PM2.5) effectively to reduce the health risk of air pollution. Similarly, nanocatalysts, including those composed of titanium dioxide, promote the decomposition of volatile organic compounds (VOCs) in industrial discharge, which is favorable to air quality [31]. These technologies help to ensure that industrial activities have less of an environmental impact. However, the possible release of nanoparticles from these filters might be risky and needs careful design to avoid secondary airborne pollution [19].
- *Oil Spill Remediation:* Bioremediation assisted by nanomaterials provides new strategies to combat an environmental menace: oil spills. Hydrophobic nanomaterials, for instance, oleophilic polymer–encapsulated silica nanoparticles can selectively adsorb oil from the surface of water, enabling faster and more efficient cleaning. They're also more effective than traditional dispersants, which would add other environmental problems. Studies have shown that nanoenhanced bioremediation can increase the microbiological degradation of hydrocarbons and may reduce the remediation timescale by half as compared to the conventional method [27]. However, the fate of these nanomaterials in the environment, including their bioaccumulation potential in marine systems, is an important and challenging issue which should be carefully researched [19].
- *Sensor Technologies for Pollution Monitoring:* Nanotechnology provides tools for making ultrasensitive nanosensors for monitoring toxins in the environment. These nanosensors can be designed to detect levels in the environment in real time. With nanosensors, individuals can wear protective clothing. Research focused on liposomes, nanoscale phospholipid-based vesicles is expanding rapidly. These can be used to deliver medication into the human body more accurately than previously thought possible. If someone suffers a head injury, magnetic resonance imaging (MRI) may be used to detect damage. These nanodroplets are micelles; they self-assemble and are used to deliver cancer treatment directly to the cells

within the body. Nanotechnology can help in the cure of cancer. For instance, graphene-derived nanosensors can sense a few parts per trillion levels of heavy metals and organic pollutants in water and air, which may serve as an early warning system before leading to environmental illness [19]. These sensing units contribute to proactive environmental stewardship, ensuring complete and accurate reporting for regulatory compliance and remediation designs. Utilization of these particles aids in sustainable operations by minimizing resource-demanding laboratory analysis, although the long-term stability and environmental impact of these systems need to be investigated [31].

- *Challenges in Environmental Remediation:* Despite the great benefits of nanotechnology for environmental remediation, challenges include the possible ecotoxicity and longevity of nanomaterials in the environment. For example, silver nanoparticles applied as antimicrobial agents may perturb aquatic microbial communities when released into a water environment [19]. Moreover, the magnitude of nanoremediation is restricted due to the high cost of production and laborious site-specific optimization [28]. Ethical concerns such as ensuring that nanomaterial-based remediation does not introduce new environmental risks highlight the importance of comprehensive lifecycle assessments, phased implementation approaches, and stringent regulatory oversight [19, 31].

20.8.5 Challenges in Sustainable Nanotechnology Implementation

The incorporation of nanotechnology into sustainable technologies faces multiple challenges in the environmental, economic, regulatory, and societal realms. These issues will need to be dealt with in an organized and rational manner to permit the exploitation of the benefits of nanotechnology while maintaining human health and environmental integrity [31].

- *Potential Environmental Risks:* The possible environmental risk and sustainability of these nanomaterials is a major question to be answered. Small-scale and highly reactive, nanomaterials could remain in ecosystems, affecting biodiversity sources. For example, the silver NPs in antimicrobial coatings can be released to water bodies, impairing aquatic organisms and microbial consortia that contribute to nutrient turnover [32]. Studies have shown that these substances are food web bioaccumulative, thus enhancing their environmental impact in the long term [19]. The unknown long-term environmental effects emphasize the importance of a thorough lifecycle assessment for nanomaterials in order to evaluate nanomaterial safety from production to disposal [31].
- *Energy-Intensive Manufacturing Processes:* Due to high energy consumption for the fabrication of nanomaterials, they are not deemed sustainable. For instance, the fabrication process of CNTs and graphene is high-temperature energy consuming, which may offset the environmental benefits of lightweight or energy storage applications [34]. Research shows that the carbon footprint of nanomaterial fabrication can be higher than that of analogous products, begging the question of their net sustainability contribution to [24]. Energy-efficient methods with minimal energy input are essential to help nanotechnologies comply with green principles; however, existing methodologies still pose a significant obstacle for sustainable penetration into large-volume processes [34].
- *Lack of Stringent Regulatory Frameworks:* The lack of uniform nomenclature in a harmonized global regulatory framework represents a significant obstacle to achieve sustainable use of nanomaterials. Current regulation frameworks, frequently focused on bulk materials, do not consider the extraordinary features of nanomaterials, which are more reactive and potentially toxic, and many nanoproducts like food packaging and medical products are being introduced into the marketplace without adequate safety tests, threatening human health and the environment [32]. Without agreed-upon standards for nanomaterial safety, exercising effective oversight is difficult, and there is a need to promote responsible use through nanomaterial-specific regulations [31]. International collaboration is necessary to define coherent recommendations between safety and innovation [22].
- *Societal Perception and Acceptability:* Public attitude is very important for the acceptance of nanotechnology in sustainable applications. Safety and ethical issues associated with nanomaterials can have an inverse relationship to lack of transparency, perception of risk, or past controversies around emerging technologies, inhibiting public acceptance [22]. For example, the application of nanotechnology into agriculture, such as nanofertilizer, may be limited if there is not proper communication about human or environmental risk [6]. Studies prioritize public engagement to reduce misunderstandings and develop trust; however, there is still a lack of activities to inform and involve stakeholders in many countries [31]. Overcoming societal distrust involves proactive communication that emphasizes the benefits of nanotechnology while being transparent about risks.
- *Scalability and Economic Viability:* Economical and technical limitations hinder the upscaling

of sustainable nanotechnology applications. Nanotechnology, for example, is improving water purification, giving rise to water systems with excellent performance, but it is prohibitively expensive for resource poor regions due to cost and infrastructure constraints [25]. Similarly, the industrialization of biodegradable nanomaterials is obstructed on the basis of the headway of production efficiency and performance uniformity [24]. Such financial restrictions disproportionately impact developing countries and widen global inequalities in the adoption of sustainable technologies [32]. These challenges have to be tackled with an investment in low-cost production techniques and supportive policies for market assimilation [34].

- *Future Directions and Recommendations:* To guide the responsible and sustainable development of nanotechnology, recommendations and strategies must ensure responsible innovation, environmental protection, and fairness. Such endeavors need cooperative efforts between scientists, policy makers, industries, and the public guiding them through this complex path of sustainable conversion [28].
- *Development of Biodegradable Nanomaterials:* In the future, to reduce environmental persistence and ecological toxicity, it will be necessary to develop biodegradable nanomaterials. Studies on polymeric nanomaterials, for example, as used in green packaging, also show some motivation towards designing materials that can naturally degrade with non-toxic products [6]. As an example, by using nanoenhanced biodegradable composites, it is now possible to replace traditional plastics and to do so without wasteful accumulation but with a nod towards performance [18]. Further investment in materials science research both to refine degradation rates and to enable commercial scale-up as knowledge gaps are addressed would allow biodegradable nanomaterials to become practical across multiple sectors [28].
- *Advancement of Green Nanomanufacturing:* Guided by principles of green chemistry, green nanomanufacturing provides a route to reduce the environmental impact of nanomaterial synthesis. For example, bioinspired synthesis and low-energy processes can reduce energy consumption and generate less waste [34]. For instance, plant extracts have become an alternative in the synthesis of metal nanoparticles due to their potential to reduce the use of dangerous reagents [6]. Upscaling of these approaches will require the synergistic collaboration of multiple disciplines in the development of cheap, high-yield processes that meet sustainability targets. Policy incentives such as subsidies can encourage the adoption of green manufacturing practices [34].
- *Strengthening Regulatory Frameworks:* Strong nanomaterial-specific regulations are necessary for safe and sustainable use. Such frameworks should include reproducible test procedures for assessing the toxicity, environmental safety, and lifecycle risk of nanomaterials [32]. International institutions, like the Organization for Economic Co-operation and Development, are seen as important to standardizing regulations to address the cross-border nature of nanotech's impacts [22]. Regulatory bodies will also need to phase in adaptive systems that keep abreast of technological developments so that oversight and regulation can keep up with innovation [31]. Public–private partnerships may usefully expedite the construction of these frameworks, ensuring a tradeoff between industry requirements and social safety.
- *Enhancing Public Engagement and Education:* Creating trust and acceptance among the general public will require detailed engagement and education programs. Open communication about both the benefits of nanotechnology such as its role in environmental remediation and its potential risks can help overcome public concerns and misunderstandings [31]. Through educational programs directed at schools, communities, and policymakers, societal awareness can be improved, leading to informed decision-making [22]. For example, open fora and citizen science projects, as strongly advised by the National Nanotechnology Initiative, are ways to allow for community engagement and help guide nanotechnology's path forward [28]. These initiatives are essential to provide a supportive infrastructure for sustainable nanotechnology.
- *Promoting Interdisciplinary and Global Collaboration:* Scientists, engineers, ethicists, and policymakers need to work together across disciplines in bringing nanotechnology research to bear on sustainable policy. Collaborative networks between research partners, for example, that funded by European Union's Horizon 2020 program, enable sharing of science and innovation [28]. International cooperation is also vital to ensure equal opportunity to experience the benefits of nanotechnology, especially for developing countries. Mechanisms such as technology transfer agreements and capacity-building initiatives can support the adoption of sustainable nanotechnology solutions in resource-poor settings [22]. These joint initiatives should emphasize inclusiveness and incorporate views from a variety of stakeholders in shaping the international nanotechnology agenda [31].
- *Lifecycle Assessments and Circular Economy Integration:* Applying LCAs is a fundamental suggestion to assess whether nanotechnology

products could be considered sustainable applications. LCAs can help to highlight potentially harmful environmental and health impacts from cradle to grave, thus aiding the design of safe nanomaterials [19]. Basing on this, we can see that the embedding of nanotechnology in the circular economy through recyclable/reusable nanomaterials is also a promising solution for sustainability processes [24]. As an example, recyclability-driven nanomaterials, which are easy to recycle, can minimize waste generation in fields such as electronics and packaging design [18]. Research and policy making should therefore aim to develop nanomaterial-tailored LCA methods, which would support their incorporation within sustainable economic processes [34].

20.8.6 Future Directions and Recommendations

To ensure that nanotechnology contributes effectively to global sustainability, future strategies must prioritize long-term resilience, responsible technological growth, and equitable global access. Achieving these goals will require coordination among scientific, regulatory, industrial, and community stakeholders who collectively influence the sustainable trajectory of emerging nanotechnologies [28].

- *Innovation Pathways for Next-Generation Sustainable Materials:* Looking ahead, research should focus on developing next-generation sustainable nanomaterials that combine high performance with minimal environmental impacts. Building on current efforts in biodegradable polymeric systems, advanced materials designed to degrade predictably and safely could replace persistent plastics in packaging, consumer products, and specialized industrial applications [6]. Nanoenhanced biodegradable composites, for example, offer a route to simultaneously reduce plastic waste and improve product durability [18]. Progress in this area will depend on greater investment in material design, refinement of degradation mechanisms, and scaling of synthesis methods so that biodegradable nanomaterials become economically competitive across diverse sectors [28].
- *Transitioning Nanomanufacturing Toward Low-Impact Production Systems:* Future nanomanufacturing must move beyond minimizing harm to proactively embedding sustainability principles into production processes. Bioinspired and low-energy synthetic techniques can significantly reduce resource use and prevent hazardous waste generation [34]. The emergence of plant-extract-based nanoparticle synthesis illustrates how alternative chemistries can lessen reliance on toxic reagents while maintaining functional performance [6]. Long-term implementation will require integrated research programs aimed at designing cost-effective, high-yield manufacturing strategies aligned with environmental targets, supported by regulatory incentives such as green-production subsidies [34].
- *Building Adaptive and Globally Coherent Governance Systems:* Sustainable development of nanotechnology will also depend on governance structures that evolve in step with scientific progress. Strengthened regulatory systems tailored to nanomaterials should incorporate harmonized testing standards, transparent risk-evaluation processes, and clear protocols for environmental and lifecycle assessments [32]. International organizations such as the OECD play a central role in facilitating cooperation on cross-border regulatory issues and aligning global practices [22]. Adaptive, science-responsive oversight frameworks can help maintain safety while avoiding regulatory stagnation that might impede technological innovation [31]. Collaborative public–private efforts will be essential in balancing industrial innovation with societal protection.
- *Expanding Societal Dialogue and Responsible Technology Culture:* Future progress requires a societal environment where nanotechnology is understood, questioned, and shaped through open dialogue. Public trust is strengthened when communication conveys both the benefits (e.g., environmental remediation applications) and the potential risks of nanotechnology in a transparent and accessible manner [31]. Expanding educational opportunities across schools, communities, and policymaking sectors can support informed public participation and reduce misconceptions [22]. Initiatives such as community fora and citizen science projects promoted by the National Nanotechnology Initiative provide valuable platforms for co-developing responsible technological pathways [28]. A culture of shared decision-making will help embed sustainability at the core of nanotechnology's societal acceptance.
- *Advancing Global Knowledge-Sharing Networks:* A key future direction lies in developing inclusive international networks that promote collaboration between scientists, engineers, ethicists, and policymakers. Multinational programs such as those supported under the EU's Horizon 2020 framework demonstrate the value of coordinated research and shared innovation [28]. For developing nations, equitable access to nanotechnology's benefits depends on mechanisms such as technology-transfer partnerships and capacity-building programs that support local adoption of green nanotechnology approaches [22]. Ensuring that diverse cultural, economic, and disciplinary perspectives are incorporated in decision-making

processes will help shape a fair and globally relevant nanotechnology agenda [31].

- *Embedding Lifecycle Thinking into Sustainable Technology Design:* Looking forward, sustainability assessment tools must be more deeply integrated into the design and implementation of nanotechnology. Comprehensive lifecycle assessments (LCAs) can identify environmental and health impacts from production through end-of-life and are essential for guiding the creation of safer next-generation nanomaterials [19]. Embedding nanotechnology into circular-economy models, such as through recyclable or reusable nanomaterials can provide additional leverage for reducing waste and improving resource efficiency [24]. Recyclability-driven materials designed for easier recovery in sectors like electronics and packaging illustrate promising advancements in this direction [18]. Continued research should prioritize developing nanomaterial-specific LCA methodologies and policies that enable their incorporation into sustainable economic systems [34].

20.9 CONCLUSION

Nanotechnology is filled with exciting potential as a disruptive technology for sustainability that can deliver entirely new solutions, such as clean energy from quantum dots and perovskite solar cells, water purification by graphene oxide, and resource-saving materials for aerospace. The unique characteristics of nanomaterials, such as penetration of biological barriers, are driving sustainability advantages, albeit with significant uncertainties. For example, their use in renewable energy could lessen the reliance on fossil fuels and mitigate climate change, and use in environmental clean-up such as soil contaminant decomposition by zero-valent iron nanoparticles holds the potential of cleaner environments.

But this promise is inevitably coupled with an intricate web of ethical issues that must be untangled for nanomaterials to be a force for good in a sustainable future. These challenges, public health risks, environmental impacts, privacy concerns, equity disparities, and regulatory voids, are not isolated barriers but interconnected factors of the trajectory the technology will take. Research suggests that nanoparticles, such as silver or carbon nanotubes, can cause oxidative stress, DNA damage, or inflammation that threatens workers breathing airborne particles or consumers exposed through products. Chronic disease, which occurs when toxic elements are deposited over long periods into organs such as the liver or lungs, could overburden healthcare systems and challenge the concept of sustainability should public health be affected. This duality extends to the environment, and the capabilities of nanotechnology to restore, as compared to disrupting, ecosystems. Also, particles released by manufacturing or discarding these nanoparticles may not ever leave; they can stick around in water bodies, harming life in the water and disrupting ecological systems while contaminating water that could be used for drinking. Whilst they contribute to resource efficiency, their long lifespans call into question the sustainability of their lifecycle. Similarly, health and security are enhanced by nanosensors, but such devices enable omnipresent surveillance, which erodes privacy, trust, social foundations for the cohabitation of people in societies. Social justice enters in as well: the high cost and private ownership of nanotechnology could worsen global inequality and prevent developing countries from harnessing its sustainable potential, and it would undermine shared efforts to battle climate change and scarcity of resources. These problems are exacerbated by regulatory gaps that have led to a disconnect between the pace of nanotechnology and the pace of safety oversight. The majority of nanoproducts get to the marketplace without being properly tested for toxicity, providing humans and ecosystems with unknown hazards. A coordinated global response lags, with governance remaining piecemeal when common standards are essential. But even human implant and enhancement applications, such as brain–machine interfaces, muddle the ethical water by delivering quality-of-life improvements but at the same time pushing towards social splits and invasions of autonomy when access is uneven or take-up forced. These are all reasons that together constitute the issue of whether nanotechnology can offer sustainability without undue baggage.

To do otherwise is to court disaster for both human health and ecological integrity, but to treat them wisely offers an opportunity for nanotechnology to enhance its contribution to sustainable development. But the path forward is to braid moral responsibility into innovation. Scientists need to concentrate on robust testing, lifecycle analysis, and biodegradable forms to ensure they do not inflict too much harm. We need consistent policies to keep people safe and working. This makes for transparency and trust. Through merging health, environmental, privacy, equity, and regulatory concerns in a single strategy, nanotechnology has the prospect to fulfill its promises of sustainability: optimizing resources, waste reduction, and promoting clean technologies, but without undermining well-being or justice. Scientists, governments, industry, and communities: all must come together to navigate the complexity; thus nanotechnology becomes a basis for a healthy, equitable, and sustainable world.

REFERENCES

1. Khan, I., Saeed, K., & Khan, I. (2019). Nanoparticles: Properties, applications and toxicities. *Arabian Journal of Chemistry, 12*(7), 908–931. https://doi.org/10.1016/j.arabjc.2017.05.011
2. Wilson, H. (2019). Gold and silver nanoparticles in cancer therapy. *Biomedical Nanotechnology Review, 12*(4), 310–325. https://doi.org/10.1080/21691401.2018.1449118
3. Lee, D., & Chen, Y. (2022). Quantum dots for next-generation displays. *Nanoelectronics Journal, 8*(1), 45–63. https://doi.org/10.1039/D0NH00556H

4. Peng, L.-M., Zhang, Z., & Wang, S. (2014). Carbon nanotube electronics: Recent advances. *Materials Today, 17*(9), 433–442. https://doi.org/10.1016/j.mattod.2014.07.008
5. Nishu, & Kumar, S. (2023). Smart and innovative nanotechnology applications for water purification. *Hybrid Advances, 3*. https://doi.org/10.1016/j.hybadv.2023.100044
6. Farajollahi, A. (2025). Nanotechnology in solar energy: From active systems to advanced solar cells. *Frontiers in Energy Research, 13*. https://doi.org/10.3389/fenrg.2025.1560718
7. *National Nanotechnology Initiative (NNI)* Official Budget Data, 2025. NNI Supplement to the President's 2025 Budget—National Nanotechnology Coordination Office (NNCO).
8. *Global Governmental Investment in Nanotechnologies* PMC, 2025. https://doi.org/10.2174/157341305774642957
9. *Government Nanotech Innovation: Policy Insights: How Government Initiatives Shape Nanotech Markets*, Faster Capital, 2025. Government Nanotech Innovation: Policy Insights: How Government Initiatives Shape Nanotech Markets - FasterCapital
10. *NNI Supplement to the President's 2024 Budget*, 2024. NNI Supplement to the President's 2024 Budget—National Nanotechnology Coordination Office (NNCO).
11. Ankita Rathore, G. Mahesh, Public perception of nanotechnology: A contrast between developed and developing countries. *Technology in Society*. https://doi.org/10.1016/j.techsoc.2021.101751.
12. *Worldwide Public Funding for Research and Development in Nanotechnology* AzoNano, 2025. https://www.azonano.com/article.aspx?ArticleID=1077
13. Khatoon, U. T., & Velidandi, A. (2025). An overview on the role of government initiatives in nanotechnology innovation for sustainable economic development and research progress. *Sustainability, 17*(3), 1250. https://doi.org/10.3390/su17031250
14. *Top 5 Countries Leading in Nanotechnology for Nanoscale Innovation* World Nano Foundation, 2025. https://www.worldnanofoundation.com/new-blog/2023/9/29/faqs-top-5-countries-leading-in-nanotechnology-for-nanoscale-innovation
15. Rambaran, T., & Schirhagl, R. (2022). Nanotechnology from lab to industry—a look at current trends. *Nanoscale Advances, 4*(18), 3664–3675. https://doi.org/10.1039/d2na00439a
16. Saleem, H., Zaidi, S. J., Ismail, A. F., & Goh, P. S. (2022). Advances of nanomaterials for air pollution remediation and their impacts on the environment. *Chemosphere, 287*(Pt 2), 132083. https://doi.org/10.1016/j.chemosphere.2021.132083
17. Zain, M., Ma, H., Ur Rahman, S., Nuruzzaman, M., Chaudhary, S., Azeem, I., Mehmood, F., Duan, A., & Sun, C. (2024). Nanotechnology in precision agriculture: Advancing towards sustainable crop production. *Plant Physiology and Biochemistry, 206*(Pt 2), 108244. https://doi.org/10.1016/j.plaphy.2023.108244
18. Mohammad, Z. H., & Ahmad, F. (2023). Nanocoating and its application as antimicrobials in the food industry: A review. *International Journal of Biological Macromolecules, 254*, 127906. https://doi.org/10.1016/j.ijbiomac.2023.127906
19. De Oliveira, J. L., Campos, E. V. R., Bakshi, M., Abhilash, P. C., & Fraceto, L. F. (2014). Application of nanotechnology for the encapsulation of botanical insecticides for sustainable agriculture: Prospects and promises. *Biotechnology Advances, 32*(8), 1550–1561. https://doi.org/10.1016/j.biotechadv.2014.10.010
20. Podila, R., & Brown, J. M. (2013). Toxicity of engineered nanomaterials: A physicochemical perspective. *Journal of Biochemical and Molecular Toxicology, 27*(1), 50–55. https://doi.org/10.1002/jbt.21442
21. Patra, J. K., Das, G., Fraceto, L. F., Campos, E. V. R., Rodriguez-Torres, M. del P., Acosta-Torres, L. S., Diaz-Torres, L. A., Grillo, R., Swamy, M. K., Sharma, S., Habtemariam, S., & Shin, H. S. (2018). Nano based drug delivery systems: Recent developments and future prospects. *Journal of Nanobiotechnology, 16*, 71. https://doi.org/10.1186/s12951-018-0392-8
22. Schulte, P. A., & Salamanca-Buentello, F. (2007). Ethical and scientific issues of nanotechnology in the workplace. *Environmental Health Perspectives, 115*(1), 5–12. https://10.1289/ehp.9456
23. Noury, H., Rahdar, L.F.R. Ferreira, Z. Jamalpoor (2025). AI-driven innovations in smart multifunctional nanocarriers for drug and gene delivery: A mini--review-. *Critical Reviews in Oncology/Hematology, 210*, 104701. https://doi.org/10.1016/j.critrevonc.2025.104701
24. Ajayan, P. M., Schadler, L. S., & Braun, P. V. (2003). *Nanocomposite Science and Technology*. Wiley-VCH. https://doi.org/10.1002/3527602127
25. Shannon, M. A., Bohn, P. W., Elimelech, M., Georgiadis, J. G., Mariñas, B. J., & Mayes, A. M. (2008). Science and technology for water purification in the coming decades. *Nature, 452*(7185), 301–310. https://doi.org/10.1038/nature06599
26. Manam, V. K. (2022). Nanotechnology in energy and environment. *Frontiers in Nanotechnology*, 1–11.
27. Khan, N., & Bano, A. (2024). Role of nanotechnology in sustainable agriculture and environmental management. In *Nanotechnology for Sustainable Agriculture, Food and Environment* (pp. 267–282). Springer. https://doi.org/10.1007/978-981-97-2104-7_14
28. Varghese, A., Mathew, S., Pinheiro, D., & Pai, K. R. (2023). Graphene-based nanocomposites for energy conversion and storage. In *Nanotechnology for Next-Generation Energy Storage: Trend and Applications* (pp. 65–103). Elsevier. https://doi.org/10.1016/B978-0-443-30065-3.00008-9
29. Rogers, J. A., Someya, T., & Huang, Y. (2010). Materials and mechanics for stretchable electronics. *Science, 327*(5973), 1603–1607. https://doi.org/10.1126/science.1182383
30. Khin, M. M., Nair, A. S., Babu, V. J., Murugan, R., & Ramakrishna, S. (2012). A review on nanomaterials for environmental remediation. *Energy & Environmental Science, 5*(8), 8075–8109. https://doi.org/10.1039/C2EE21818F
31. Copeland, B. R., Shapiro, J. S., & Taylor, M. S. (2022). Globalization and the environment. In G. Gopinath, E. Helpman, & K. Rogoff (Eds.), *Handbook of International Economics* (vol. 5, pp. 61–146). Elsevier. https://doi.org/10.1016/bs.hesint.2022.02.002
32. Duncan, T. V. (2011). Applications of nanotechnology in food packaging and food safety: Barrier materials, antimicrobials and sensors. *Journal of Colloid and Interface Science, 363*(1), 1–24. https://doi.org/10.1016/j.jcis.2011.07.017
33. Tratnyek, P. G., & Johnson, R. L. (2006). Nanotechnologies for environmental cleanup. *Nano Today, 1*(2), 44–48. https://doi.org/10.1016/S1748-0132(06)70048-2
34. Pete, A. J., Bharti, B., & Benton, M. G. (2021). Nano--enhanced bioremediation for oil spills: A review-. *ACS

ES&T Engineering, *1*(6), 928–946. https://doi.org/10.1021/acsestengg.0c00217

35. Mamalis, A. G., Ramsden, J. J., Holt, G. C., Vortselas, A. K., & Mamali, A. A. (2011). The effect of nanotechnology on mitigation and adaptation strategies in response to climate change. *Nanotechnology Perceptions*, *7*, 159–179.
36. Alsukaibi, A. K., Algarni, H., Bafail, A. O., Alzahrani, A., & Khormi, M. (2024). Nanotechnology in water treatment: A review on current status and future prospects. *Fuel*, *357*, 128–680. https://doi.org/10.1016/j.fuel.2023.128680
37. Bruce, P. G., Scrosati, B., & Tarascon, J.-M. (2008). Nanomaterials for rechargeable lithium batteries. *Angewandte Chemie International Edition*, *47*(16), 2930–2946. https://doi.org/10.1002/anie.200702505
38. Zhang, L., Qiu, Z., Bai, Y., & Huang, Y. (2023). Recent progress in nanotechnology for water treatment: A review. *Environmental Nanotechnology, Monitoring & Management*, *20*, 103–543. https://doi.org/10.1016/j.jwpe.2023.103543
39. De Oliveira, J. L., Campos, E. V. R., Bakshi, M., Abhilash, P. C., & Fraceto, L. F. (2020). Application of nanotechnology for the encapsulation of botanical insecticides for sustainable agriculture: Prospects and promises. *Biotechnology Advances*, *40*, 100–467. https://doi.org/10.1016/j.rineng.2022.100467

21 Nickel Oxide/Reduced Graphene Oxide Nanocomposites

Synthesis, Properties, and Applications

Dhruva Kumar, Joydeep Biswas, and Dheeraj Kumar

21.1 INTRODUCTION

In recent years, researchers worldwide have been interested in transition metal oxides/reduced graphene oxides due to their superior properties, which are required for several applications. Nickel oxide/reduced graphene oxide (NiO/rGO) is the forerunner among them. NiO is already used in the field for electrochemical applications due to its high thermal stability [1], large energy bandgap [2], and immense catalytic potential [3, 4].

On the other hand, rGO has unique properties owing to its two-dimensional carbon lattice, which gives it a large area [5], high electrical conductivity [6], and a high degree of mechanical robustness [7, 8].

To harness the best of both materials, researchers have combined them through various synthesizing techniques, and the final nanocomposite, NiO/rGO, is found to have superlative properties and can be used in multiple fields, like energy storage [9], catalysis [10], sensors [11], and environmental remediation [12]. Even though there are many advantages of NiO/rGO, the main issue identified in various studies is the difficulty of scaling up production [13] and keeping it stable [14]. This chapter provides a thorough review of the various synthesis methods, properties, applications, and challenges associated with NiO/rGO nanocomposites.

21.2 SYNTHESIS METHODS

Several methods are available for synthesizing NiO/rGO nanocomposites, but the synthesis method plays a crucial role in determining the properties of the resulting nanocomposite. So, the selection of synthesis methods should be based on the intended application of the nanocomposites.

21.2.1 Sol-Gel Method

The sol-gel technique is a popular method for synthesizing NiO/rGO nanocomposites, as the synthesized material exhibits uniformly sized particles [15] and allows for the controlled formation of shapes [16]. Initially, nickel salts such as nickel acetate or nickel nitrate are dissolved in a solvent to form a precursor solution. After mixing in graphene oxide (GO), a gel forms through hydrolysis, followed by polycondensation of the nickel solution. The gel is dried and then heated at high temperatures to produce the NiO/rGO composite [17]. One of the significant advantages of this method is that it helps control the size of the NiO particles, which are evenly spread across rGO sheets. However, the high-temperature treatment can lower rGO's conductivity as oxygen functionalities get removed [18].

21.2.2 Hydrothermal and Solvothermal Methods

The hydrothermal method involves heating and pressurizing an aqueous solution in a specially designed closed reaction vessel to produce a high-temperature, high-pressure reaction environment. It enables the creation of well-formed NiO nanoparticles uniformly distributed on the surface of rGO, making them ideal for preparing NiO/rGO nanocomposites. [19]. In hydrothermal methods, GO and nickel precursors are heated together in an autoclave, reducing GO into rGO and depositing NiO on top of it [14]. Similar principles apply to solvothermal procedures, which alter the size and shape of the NiO particles by using organic solvents rather than just water. Although these techniques offer advantages such as regulated forms and elevated yields, scaling up production can be challenging due to the use of high-pressure machinery [20, 21].

21.2.3 Electrodeposition

Applying a voltage to materials immersed in nickel ions + GO solutions creates an electric field that deposits NiO onto rGO through electrodeposition. This process, known as deposition, involves the reduction of nickel ions onto the surface of rGO while simultaneously reducing GO into rGO [22]. It provides precise control over the thickness of NiO layers, which is ideal for supercapacitors or batteries. One interesting feature of these composites is that they can be grown directly on conductive electrodes [23]. However, this procedure is so time consuming that scaling it up is challenging and requires a considerable amount of time [24].

21.2.4 Chemical Vapor Deposition

By dissolving nickel sources while carbon sources remain present, chemical vapor deposition (CVD) enables the growth of thin films of NiO on rGO. To ensure that NiO

DOI: 10.1201/9781003644439-21

forms smoothly on top of rGO with consistent consistency throughout, the experiment is performed inside a vacuum chamber at high temperatures [25]. However, complex CVD equipment is needed, making large-scale production challenging.

21.3 STRUCTURAL AND MORPHOLOGICAL PROPERTIES

To fully comprehend the performance of NiO/rGO nanocomposites in various applications, it is essential to examine their structure and morphology thoroughly. Researchers frequently employ a variety of instruments, such as Raman spectroscopy, X-ray diffraction (XRD), transmission electron microscopy (TEM), and scanning electron microscopy (SEM), which provide valuable insights.

21.3.1 Morphology and Surface Area

SEM and TEM are frequently used to investigate the surface features and particle distribution of NiO on rGO. SEM provides insights into the surface texture. Ideally, the NiO nanoparticles should be well dispersed across the rGO sheets to maximize surface area and interaction between the two components. Pore et al. found urchin-like structures in the NG-60, NG-120, and NG-180 samples, which vanished once GO was added to NiO. This was because rGO sheets coat the urchin-like NiO nanostructures. The surface of NiO nanostructures has an even distribution of rGO sheets. It is also evident that the rGO is closely compounded with NiO from the lack of urchin-like NiO spikes due to rGO wrapping. By acting as conducting networks, these rGO sheets enhance the efficacy of nonenzymatic glucose sensing and electrochemical supercapacitors. Additionally, when used in supercapacitors, rGO is an excellent Electrochemical Double-Layer Capacitor (EDLC) material that improves electrochemical performance [26]. Narwade et al. found that Ni/NiO NPs are well decorated on rGO sheets [27]. An et al. prepared NiO-rGO composite, and the SEM and TEM images of the composite showed that the thickness of the layered composite was less than 10 nm, and the layered rGO was well preserved, as shown in Figure 21.1a, b. This suggested that there were few rGO layers and that the rGO sheet did not clump together due to the uniform anchoring of NiO nanoparticles in it. Additionally, there was no significant NiO aggregation, and every NiO nanoparticle was securely attached to the rGO, suggesting that the composite's morphology was good. The NiO nanoparticles in Figure 21.1c are homogenous and range in size from ten to tens of nanometers. The size of the NiO crystal was approximately 5.5 ± 1.5 nm, and the crystal plane spacing of the NiO nanoparticles was approximately 0.24 nm, which corresponds to the (111) plane of the cubic NiO crystal, as shown in Figure 21.1d.

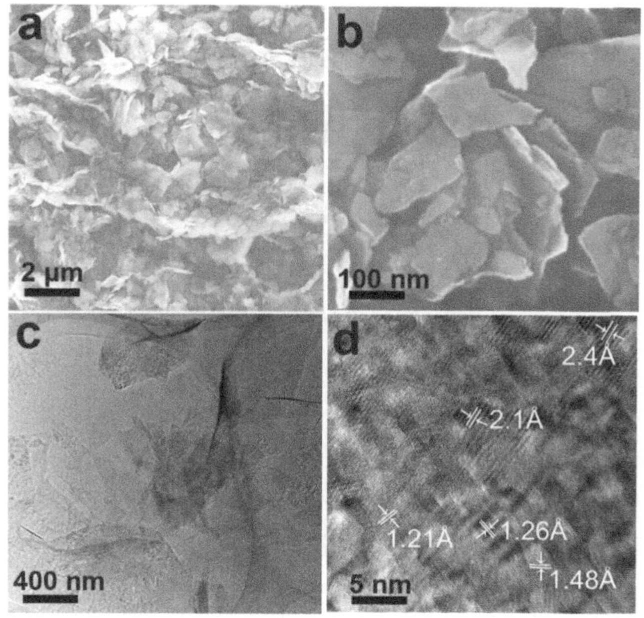

FIGURE 21.1 (a, b) NiO-rGO composite SEM image. (c, d) NiO-rGO composite TEM image [28].

21.3.2 Crystallographic Structure

XRD is used to determine the crystalline structure of the NiO/rGO nanocomposite. The typical XRD pattern of NiO shows peaks corresponding to its cubic phase, while rGO produces a broad peak around 25°, corresponding to the (002) plane. An et al. observed prominent diffraction peaks of the NiO–rGO composite at 37.2°, 43.3°, 62.8°, 75.4°, and 79.3°, corresponding to crystal plane diffractions of (111), (200), (220), (311), and (222), respectively, in Figure 21.2. The presence of these peaks confirms the formation of the NiO/rGO composite [28]. Xu et al. prepared NiO/rGO composite material and observed that the diffraction peak of the graphene oxide present at 10.2° disappeared. The diffraction peak of graphene appeared at 23.9°, indicating that the graphene oxide moiety was reduced to form rGO. In contrast, other diffraction peaks were the same as the NiO diffraction peak, which proves a successful synthesis of NiO/rGO composite [29]. The crystallite size of NiO can be estimated using the Scherrer equation, which is crucial for understanding the relationship between particle size and electrochemical performance [30].

21.3.3 Raman Spectroscopy

Raman spectroscopy is employed to observe the structural changes in GO during reduction and the interaction between NiO and rGO in the nanocomposite. GO typically exhibits two characteristic peaks: the D-band (~1350 cm^{-1}) and the G-band (~1590 cm^{-1}). Upon reduction to rGO, the ratio of the D-band to the G-band increases, indicating the removal of oxygen-containing groups and the restoration of sp^2 domains. In NiO/rGO nanocomposites, shifts

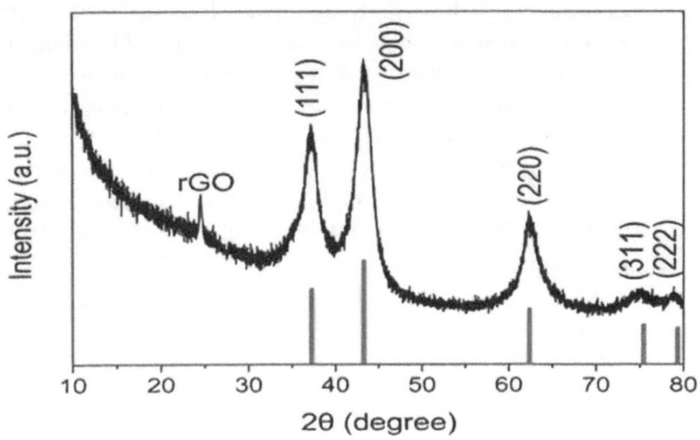

FIGURE 21.2 XRD spectra of the NiO-rGO nanocomposite [28].

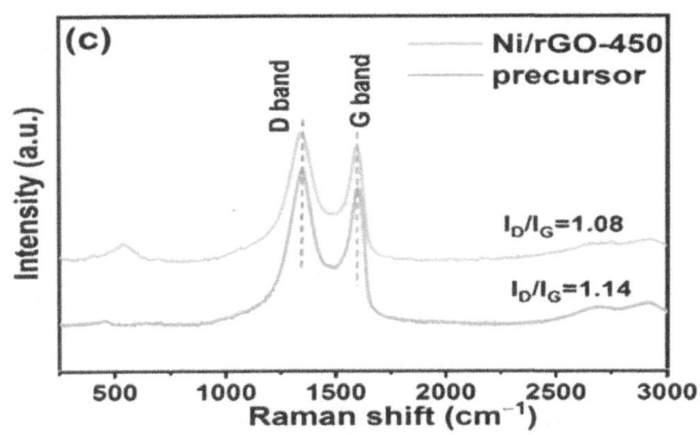

FIGURE 21.3 Raman spectra of precursor and Ni/rGO-450 [31].

in these bands can reveal insights into the level of interaction between NiO and rGO [31]. The peak observed at 531 cm^{-1} was attributed to the vibrational peaks of Ni–O [32] and signifies the successful anchoring of Ni NPs on the graphene sheets in the as-synthesized Ni/rGO catalysts, as shown in Figure 21.3.

21.4 ELECTROCHEMICAL PROPERTIES

The electrochemical properties of NiO/rGO nanocomposites are essential for their application in energy storage and conversion technologies. NiO's redox activity, combined with rGO's conductivity, results in improved charge storage capabilities, rapid charge/discharge cycles, and long-term stability [29].

21.4.1 Conductivity and Charge Storage

NiO's pseudocapacitive behavior, resulting from reversible redox reactions involving Ni^{2+}/Ni^{3+}, is key to its charge storage capacity [33]. However, NiO's relatively low electrical conductivity limits its performance. The addition of rGO addresses this limitation by creating a conductive network that facilitates efficient charge transport. Moreover, the large surface area of rGO provides additional sites for electrochemical reactions, enhancing overall performance [34].

21.4.2 Cyclic Voltammetry

Cyclic voltammetry (CV) is a widely used technique for investigating the electrochemical behavior of NiO/rGO nanocomposites. This method tracks the current response as the voltage is changed to set limits. We see clear redox in NiO/rGO nanocomposites. These peaks show the transition of Ni^{2+} to Ni^{3+}. When the area under the CV curve is larger, there's more capacitance. Additionally, the shape of the curve provides insights into how charge storage operates [17].

21.4.3 Galvanostatic Charge/Discharge

This technique involves charging and discharging NiO/rGO nanocomposites at a constant current to determine their specific capacitance. We can find the particular capacitance using the discharge curve. It's interesting to note that NiO/rGO nanocomposites often show high capacitance. This is because of NiO's pseudocapacitance combined with rGO's ability to hold an electrical double layer [35].

21.5 APPLICATIONS

The excellent properties of NiO/rGO nanocomposites make them a potent material for various applications, some of which are discussed in the following.

21.5.1 Energy Storage

NiO/rGO nanocomposite is an ideal material for lithium-ion batteries and supercapacitors due to its rapid charge/discharge rates [17], extended cycle life [36], and high specific capacitance [17]. The combination of rGO's remarkable conductivity and NiO's pseudocapacitance significantly enhances the performance of these devices. Additionally, their large surface area and excellent conductivity aid in the improved storage of lithium ions in batteries [34].

21.5.2 Sensors

The high surface area and excellent conductivity of NiO/rGO make them ideal for gas detectors with extreme sensitivity [37]. NiO reacts particularly well with gases such as ethanol, methane, and hydrogen. Simultaneously, rGO increases sensor sensitivity through improved charge transfer [38]. These fascinating nanocomposites have the potential to be utilized in biosensors for detecting biomolecules such as dopamine and glucose [39].

21.5.3 Photocatalysis

Water splitting and the degradation of contaminants are two applications where NiO/rGO nanocomposites have demonstrated considerable promise in photocatalysis [40]. Electron-hole pairs are created when light strikes NiO, but the efficiency of these pairs is reduced by their propensity to recombine. On the other hand, rGO enhances photocatalytic activity and reduces the likelihood of recombination, acting as an electron acceptor [41].

21.6 CHALLENGES AND FUTURE PROSPECTS

NiO/rGO nanocomposite nanomaterials possess numerous excellent properties; however, they also present several recurring challenges, including issues of stability and scalability. Also, their synthesis techniques are costly and cumbersome. Some more straightforward and scalable techniques will be necessary for their commercialization. Additionally, it is essential to note that extending the lifespan of these composites is a crucial goal that warrants further attention. Over time, effectiveness decreases as rGO becomes oxidized and NiO aggregation occurs.

21.7 CONCLUSION

NiO/rGO nanocomposites have considerable potential in the field of energy storage. These materials, used in devices such as supercapacitors and batteries, can be utilized for water filtration and the removal of pollutants. Due to their high sensitivity, they are considered prospective materials for gas sensors. NiO/rGO nanocomposites are highly promising, but the challenge of synthesizing them on a large scale while maintaining their quality remains a significant barrier to their widespread adoption. This chapter also analyzed synthesis methods, characteristics, and potential applications of NiO/rGO nanocomposites to identify the gaps that require further research in this field.

REFERENCES

1. J. Kacher, P. Elizaga, S.D. House, K. Hattar, M. Nowell, I.M. Robertson, Thermal stability of Ni/NiO multilayers, Mater. Sci. Eng. A. 568 (2013) 49–60 https://doi.org/10.1016/j.msea.2013.01.033.
2. H.K. Jung, S.J. Lee, D. Han, A.-R. Hong, H.S. Jang, S.H. Lee, J.H. Mun, H. Lee, S.H. Han, D. Yang, D.H. Kim, Au-incorporated NiO nanocomposite thin films as electrochromic electrodes for supercapacitors, Electrochim. Acta. 330 (2020) 135203 https://doi.org/10.1016/j.electacta.2019.135203.
3. T. Zahra, K.S. Ahmad, C. Zequine, R.K. Gupta, A. Guy Thomas, M. Azad Malik, Evaluation of electrochemical properties for water splitting by NiO nano-cubes synthesized using Olea ferruginea Royle, Sustain. Energy Technol. Assessments. 40 (2020) 100753 https://doi.org/10.1016/j.seta.2020.100753.
4. T. Munawar, A. Bashir, S. Sardar, M.S. Nadeem, F. Mukhtar, S. Manzoor, M.N. Ashiq, S.A. Khan, M. Koc, F. Iqbal, Electrochemical behavior of V/Ce co-doped carbon shell-coated NiO nanocomposite for alkaline OER and supercapacitor applications, J. Energy Storage. 76 (2024) 109556 https://doi.org/10.1016/j.est.2023.109556.
5. S. Tamang, S. Rai, R. Bhujel, N.K. Bhattacharyya, B.P. Swain, J. Biswas, A concise review on GO, rGO and metal oxide/rGO composites: Fabrication and their supercapacitor and catalytic applications, J. Alloys Compd. 947 (2023) 169588 https://doi.org/10.1016/j.jallcom.2023.169588.
6. P. Viprya, D. Kumar, S. Kowshik, Study of different properties of Graphene Oxide (GO) and reduced graphene oxide (rGO), in: RAiSE-2023, MDPI, Basel, Switzerland, 2023, p. 84 https://doi.org/10.3390/engproc2023059084.
7. K. Li, L. Han, X. Wang, F. Gao, J. Zhang, J. Cheng, MOF-derived CoNC@rGO/amine-rich@rGO/fluorinated-epoxy nanocomposites with EMI shielding, mechanical robustness, superamphiphobicity and long-term anticorrosion properties, Chem. Eng. J. 455 (2023) 140542 https://doi.org/10.1016/j.cej.2022.140542.
8. T. Taniguchi, L. Nurdiwijayanto, N. Sakai, K. Tsukagoshi, T. Sasaki, T. Tsugawa, M. Koinuma, K. Hatakeyama, S. Ida, Revisiting the two-dimensional structure and reduction process of graphene oxide with in-plane X-ray diffraction, Carbon N. Y. 202 (2023) 26–35 https://doi.org/10.1016/j.carbon.2022.11.009.
9. H. Torun, S.R. Tokgöz, M. Arslan Çarpan, M.H. Somay, A. Peksöz, Fabrication of a high-performance symmetrical supercapacitor based on electrochemically interconnected Cu:NiO@rGO:PANI/NF electrodes, J. Energy Storage 88 (2024) 111462 https://doi.org/10.1016/j.est.2024.111462.
10. S. Nagarani, J.-H. Chang, M. Yuvaraj, Mohanraj Kumar, Dipsikha Ganguly, S. Balachandran, Evaluation of a ZnO–NiO/rGO hybrid electrocatalyst for enhanced oxygen reduction reaction (ORR) applications, New J. Chem. 48 (2024) 10910–10924 https://doi.org/10.1039/D3NJ05305A.
11. S.M. Mali, S.S. Narwade, B.B. Mulik, R.V. Digraskar, R.R. Harale, B.R. Sathe, Enhanced electrochemical ethanol sensitivity on Ni/NiO-rGO hybrids nanostructures at room temperature, ChemistrySelect. 8 (2023) https://doi.org/10.1002/slct.202204328.
12. S.A. Ansari, N. Parveen, G.M. Alsulaim, A.A. Ansari, S.A. Alsharif, K.M. Alnahdi, H.A. Alali, V.R.M. Reddy, Emerging NiO–rGO nanohybrids for antibiotic pollutant degradation under visible-light irradiation, Surfaces and Interfaces. 40 (2023) 103078 https://doi.org/10.1016/j.surfin.2023.103078.
13. S. Seenivasan, S. Dhinesh, F. Maiz, M. Shkir, Electrochemical investigation of NiO@MnO2@rGO ternary nanocomposite based electrode material for high-performance supercapacitor applications, Ionics (Kiel). 29 (2023) 3641–3652 https://doi.org/10.1007/s11581-023-05110-y.
14. S. Yuan, M. Yang, C. Lu, X. Wang, Synthesis of a rGO/NiO composite with a hierarchical porous structure by self-assembly and its electrochemical performance as a supercapacitor electrode, New Carbon Mater. 35 (2020) 731–738 https://doi.org/10.1016/S1872-5805(20)60525-X.
15. M.M. Ullah, U. Draz, A. Tariq, R. Nasir, M. Irfan, S.M. Ramay, S. Atiq, Utilization of NiO-rGO nanoarchitectures-based composite electrodes for high-performance electrochemical applications, J. Electrochem. Soc. 171 (2024) 080523 https://doi.org/10.1149/1945-7111/ad6d01.
16. S. Rafiq, M. Aadil, M.F. Warsi, S. Yousaf, M.T. Alotaibi, S.M. El-Bahy, M. Shahid, NiO nanoparticles and their nanohybrid with flat rGO sheets: As an ideal electroactive material for hybrid capacitor applications, Ceram. Int. 48 (2022) 14596–14605 https://doi.org/10.1016/j.ceramint.2022.01.353.

17. Y. Zhu, W. Chu, N. Wang, T. Lin, W. Yang, J. Wen, X.S. Zhao, Self-assembled Ni/NiO/RGO heterostructures for high-performance supercapacitors, RSC Adv. 5 (2015) 77958–77964 https://doi.org/10.1039/C5RA14790E.
18. A. Ahmed, A. Singh, S.-J. Young, V. Gupta, M. Singh, S. Arya, Synthesis techniques and advances in sensing applications of reduced graphene oxide (rGO) composites: A review, Compos. Part A Appl. Sci. Manuf. 165 (2023) 107373 https://doi.org/10.1016/j.compositesa.2022.107373.
19. X. Li, L. Fan, X. Li, H. Shan, C. Chen, B. Yan, D. Xiong, D. Li, Enhanced anode performance of flower-like NiO/RGO nanocomposites for lithium-ion batteries, Mater. Chem. Phys. 217 (2018) 547–552 https://doi.org/10.1016/j.matchemphys.2018.06.050.
20. S. Srirattanapibul, P. Nakarungsee, C. Issro, I.-M. Tang, S. Thongmee, Performance of NiO intercalated rGO nanocomposites for NH3 sensing at room temperature, Mater. Sci. Semicond. Process. 137 (2022) 106221 https://doi.org/10.1016/j.mssp.2021.106221.
21. J. Zhao, Y. Tian, A. Liu, L. Song, Z. Zhao, The NiO electrode materials in electrochemical capacitor: A review, Mater. Sci. Semicond. Process. 96 (2019) 78–90 https://doi.org/10.1016/j.mssp.2019.02.024.
22. S. Liu, S. Shao, C. Xue, J. Li, Optimized 3D interconnected foam-like NiO@rGO composite as a binder-free electrode for high-performance asymmetric supercapacitor, Mater. Res. Express. 10 (2023) 105505 https://doi.org/10.1088/2053-1591/ad04ae.
23. M. Mirzaee, C. Dehghanian, Pulsed electrodeposition of reduced graphene oxide on Ni NiO foam electrode for high-performance supercapacitor, Int. J. Hydrogen Energy. 43 (2018) 12233–12250 https://doi.org/10.1016/j.ijhydene.2018.04.173.
24. H. Razmi, H. Shirdel, R. Mohammad-Rezaei, NiO nanoparticles electrodeposited on reduced GO–CuO nanocomposite bulk modified CCE as a sensitive glucose sensor, Micro Nano Lett. 12 (2017) 217–222 https://doi.org/10.1049/mnl.2016.0566.
25. H.H. El-Maghrabi, A.A. Nada, S. Roualdes, M.F. Bekheet, Design of Ni/NiO–TiO2/rGO nanocomposites on carbon cloth conductors via PECVD for electrocatalytic water splitting, Int. J. Hydrogen Energy. 45 (2020) 32000–32011 https://doi.org/10.1016/j.ijhydene.2020.08.259.
26. O.C. Pore, A.V. Fulari, V.G. Parale, H.H. Park, R.V. Shejwal, V.J. Fulari, G.M. Lohar, Facile hydrothermal synthesis of NiO/rGO nanocomposite electrodes for supercapacitor and nonenzymatic glucose biosensing application, J. Porous Mater. 29 (2022) 1991–2001 https://doi.org/10.1007/s10934-022-01313-2.
27. S.S. Narwade, S.M. Mali, R.V. Digraskar, V.S. Sapner, B.R. Sathe, Ni/NiO@rGO as an efficient bifunctional electrocatalyst for enhanced overall water splitting reactions, Int. J. Hydrogen Energy. 44 (2019) 27001–27009 https://doi.org/10.1016/j.ijhydene.2019.08.147.
28. Q. An, X. Zhao, S. Suo, Y. Bai, Novel Lithium-Ion capacitor based on a NiO-rGO composite, Materials (Basel). 14 (2021) 3586 https://doi.org/10.3390/ma14133586.
29. J. Xu, L. Wu, Y. Liu, J. Zhang, J. Liu, S. Shu, X. Kang, Q. Song, D. Liu, F. Huang, Y. Hu, NiO-rGO composite for supercapacitor electrode, Surfaces and Interfaces. 18 (2020) 100420 https://doi.org/10.1016/j.surfin.2019.100420.
30. Z.T. Khodair, M.A. Al-Jubbori, A.M. Shano, F.I. Sharrad, study of optical and structural properties of (NiO)1-x(CuO) x nanostructures thin films, Chem. Data Collect. 28 (2020) 100414 https://doi.org/10.1016/j.cdc.2020.100414.
31. Y. Wang, J. Qian, J. Li, J. Xing, L. Liu, Facile fabrication of Nickel supported on reduced graphene oxide composite for oxygen reduction reaction, Nanomaterials. 13 (2023) 3087 https://doi.org/10.3390/nano13243087.
32. S. Arunpandiyan, A. Raja, S. Bharathi, A. Arivarasan, Fabrication of ZnO/NiO:rGO coated Ni foam binder-free electrode via hydrothermal method for supercapacitor application, J. Alloys Compd. 883 (2021) 160791 https://doi.org/10.1016/j.jallcom.2021.160791.
33. W.G. Nunes, A.N. Miranda, B. Freitas, R. Vicentini, A.C. Oliveira, G. Doubek, R.G. Freitas, L.M. Da Silva, H. Zanin, Charge-storage mechanism of highly defective NiO nanostructures on carbon nanofibers in electrochemical supercapacitors, Nanoscale. 13 (2021) 9590–9605 https://doi.org/10.1039/D1NR00065A.
34. J.-S. Gao, T. Lian, Z. Liu, Y. He, Efficient NiO/RGO combination for high-cycling-stability supercapacitors by an alkaline hydrothermal strategy, J. Appl. Electrochem. 52 (2022) 1045–1052 https://doi.org/10.1007/s10800-022-01694-x.
35. R. Kumar, R. Thangappan, Electrode material based on reduced graphene oxide (rGO)/transition metal oxide composites for supercapacitor applications: A review, Emergent Mater. 5 (2022) 1881–1897 https://doi.org/10.1007/s42247-021-00339-7.
36. A. Liu, H. Zhang, G. Wang, J. Zhang, S. Zhang, Sandwich-like NiO/rGO nanoarchitectures for 4 V solid-state asymmetric-supercapacitors with high energy density, Electrochim. Acta. 283 (2018) 1401–1410 https://doi.org/10.1016/j.electacta.2018.07.099.
37. H. Ren, C. Gu, S.W. Joo, J. Zhao, Y. Sun, J. Huang, Effective hydrogen gas sensor based on NiO@rGO nanocomposite, Sensors Actuators B Chem. 266 (2018) 506–513 https://doi.org/10.1016/j.snb.2018.03.130.
38. J. Zhang, D. Zeng, S. Zhao, J. Wu, K. Xu, Q. Zhu, G. Zhang, C. Xie, Room temperature NO 2 sensing: What advantage does the rGO–NiO nanocomposite have over pristine NiO?, Phys. Chem. Chem. Phys. 17 (2015) 14903–14911 https://doi.org/10.1039/C5CP01987G.
39. X. Zhang, Z. Zhang, Q. Liao, S. Liu, Z. Kang, Y. Zhang, Nonenzymatic glucose sensor based on in situ reduction of Ni/NiO-graphene nanocomposite, Sensors. 16 (2016) 1791 https://doi.org/10.3390/s16111791.
40. P. Vivek, R. Sivakumar, E. Selva Esakki, S. Deivanayaki, Fabrication of NiO/RGO nanocomposite for enhancing photocatalytic performance through degradation of RhB, J. Phys. Chem. Solids. 176 (2023) 111255 https://doi.org/10.1016/j.jpcs.2023.111255.
41. S.A.H. Moradi, N. Ghobadi, S.M. Tabatabaeinejad, Facile and rapid preparation of progressive ZnO/NiO/rGO nanophotocatalyst and investigation its mechanism and reaction kinetics while decomposition of pharmaceuticals pollutant, Surfaces and Interfaces. 39 (2023) 102939 https://doi.org/10.1016/j.surfin.2023.102939.

22 CMOS Active Full Wave Rectifier for Wirelessly Powered Biomedical Implants

Sushma P S, Shrividya G, and Usha Desai

22.1 INTRODUCTION

Wearable and implantable biomedical devices such as cochlear implants, brain-machine interface systems for recording and stimulation are wirelessly powered. They are small and more widely used as they have wireless features. The energy harvesting circuits for these devices are placed on the skin and the AC signals captured inductively coupled to the rectifier circuit, which is on chip in the implanted device operating within the body either as prosthetic device or recording and transmission module or as stimulators. The frequency typically used in wirelessly powered biomedical devices is 13.56 MHz as this frequency provides good coupling with near field communication links. In addition, at this frequency, power attenuation characteristics due to skin impedance found to be good. Further, no extra data coil is required, and the sizes of the output filtering capacitor and the resonance of LC tank reduced. This allows the entire system to be integrated on-package or on-chip.

The design of rectifier circuits in inductive power links is one of most important tasks. The basic block diagram of wirelessly powered biomedical implant shown in Figure 22.1 consists of inductively coupled power receiver. Due to relative motion of the primary and secondary coils, the coupled AC input value of amplitude may vary by a few times; therefore, the rectifier designed for a wide input amplitude range.

The overdrive voltage requirement of the diode rectifier reduces its conversion efficiency. In active rectifiers, all four diodes driven off and on as switches completely to minimize the value of voltage drop. Amongst a number of choices for the rectifier, compared to the Schottky diode rectifier, CMOS active rectifiers have compatibility with other building blocks and low-cost.

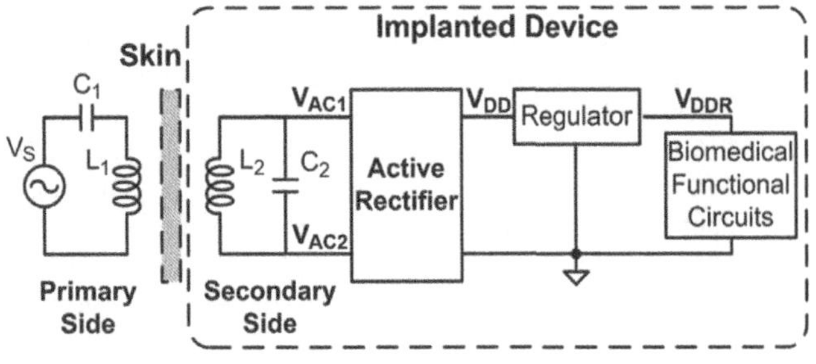

FIGURE 22.1 Basic block diagram of wirelessly powered biomedical implant [1]

CMOS full-wave active rectifier with low loss consists two symmetrically dynamically biased and matched active diodes with reverse-current control, each realized by voltage comparator driven nMOS switch [1]. Two cross-coupled pMOS transistors replace two diodes of the diode bridge rectifier. A high on/off current fraction can be achieved by a special assemble; the gates of these pMOSFETs are driven by a bigger voltage swing than the diode-connected pMOSFETs. However, to block the reverse current, the bottom two diodes still using the implementation of diode-connected nMOS transistors and the efficiency is not optimized. An unbalanced comparator controlled switches are used in an active rectifier designed for 0.125–1 MHz is found to have improved efficiency [2], however not suitable for biomedical implants. In full wave rectifier proposed by [3], unbalanced-biased comparators are used to reduce the voltage drop in the conducting path. Comparators with offset control and high speed [4], cross coupled latched comparators [5] are used to improve the power conversion efficiency, delay and reverse current. To reduce the threshold voltage of active diodes bootstrapped capacitors are used for selected MOS switches [6]. It achieves a low voltage-drop and considerable increase in the overall power efficiency. The on/off delay compensation is provided by PWM method in [7–9] to improve the efficiency of power conversion of a 2 MHz rectifier circuit. In low power designs, comparators are not preferred as they consume lot of power[10–12]. The current controlled delay lines, multiple pulsing methods are used [13, 14].

The switching mechanism used to turn on active diode determines the maximum power conversion efficiency achievable. When the comparator is used for switching the rectifying device, the offset voltage and the maximum operating frequency of the comparator affect the active diode switching voltage and frequency. Therefore, choice of circuit topology used for comparator plays an important role in the design of high efficiency active rectifier.

22.2 METHODOLOGY

In this work comparator, based active full wave rectifier is studied [15]. It is bridge rectifier where the top two diodes are realized by gate cross-coupled normally on pMOSFETs and the remaining two are realized using

DOI: 10.1201/9781003644439-22

nMOSFETs where switching is controlled by comparators. The differential pair of comparators reduce the zero cross over delay because of the forward voltage of the diode in the rectifier output with respect to the input signal. Figure 22.2 shows the schematic of the method used to realise the active rectifier.

The main building block of active rectifiers is active diode, which is having almost an ideal diode characteristic such as almost zero reverse saturation current and forward voltage drop as low as 20 mV. This active diode is controlled nMOSFET switch, which quickly turned on by the comparator. The inherent switching delay of the diodes is reduced using comparators and switched-offset biasing which do not allow the reverse current to flow. This improves the efficiency of power conversion.

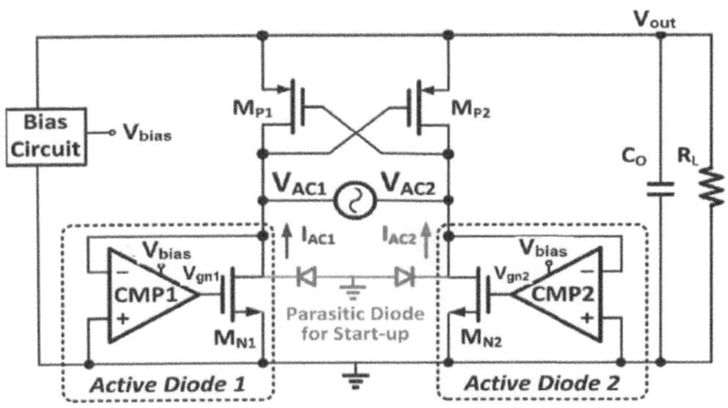

FIGURE 22.2 CMOS active full wave rectifier [1]

22.3 CMOS ACTIVE RECTIFIER

22.3.1 Schematic of Comparator

The Figure 22.3 shows transistor level circuit for comparators CMP1 and CMP2 of CMOS active rectifier depicted in Figure 22.2. It uses push pull comparator topology, which reduces its switching time, with input stage being a differential amplifier. The output of the comparator is connected to a buffer and then is used as gate signal for nMOSFET switch of the rectifier circuit. To prevent multiple pulsing, the NOR SR-latch is included.

The input stage of the comparator has two NMOS current mirrors, which turned on only when the input is negative. During positive cycles of the input PMOS current mirror is ON. This makes the comparator output to switch and hence generate gate turn on voltage to nMOSFET switch, only when input is present.

The transient analysis of this comparator is shown in Figure 22.4 and switching pulses are produced during the negative half cycles. The switching characteristics of the comparator is shown in Figure 22.5, that represents PEX extracted comparator simulation. The comparators have faster response and more instant current.

22.3.2 Schematic of Active Rectifier

The transistor level circuit of CMOS active rectifier is shown in Figure 22.6. Simulation of the PEX extracted schematic for a 1.2V sinusoidal signal at 13.56 MHz with a 300pF capacitor and 500 Ω load gave an output V_{out} as shown in Figure 22.7. Table 22.1 summarizes the

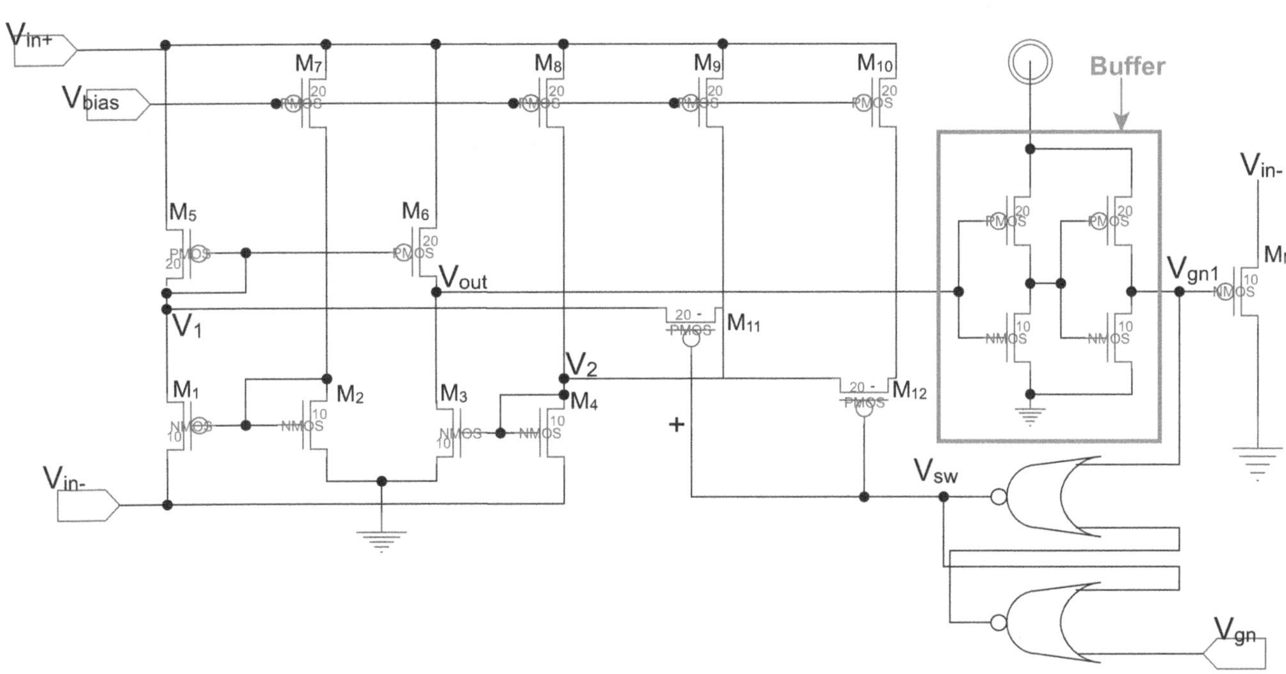

FIGURE 22.3 Transistor level circuit of comparator

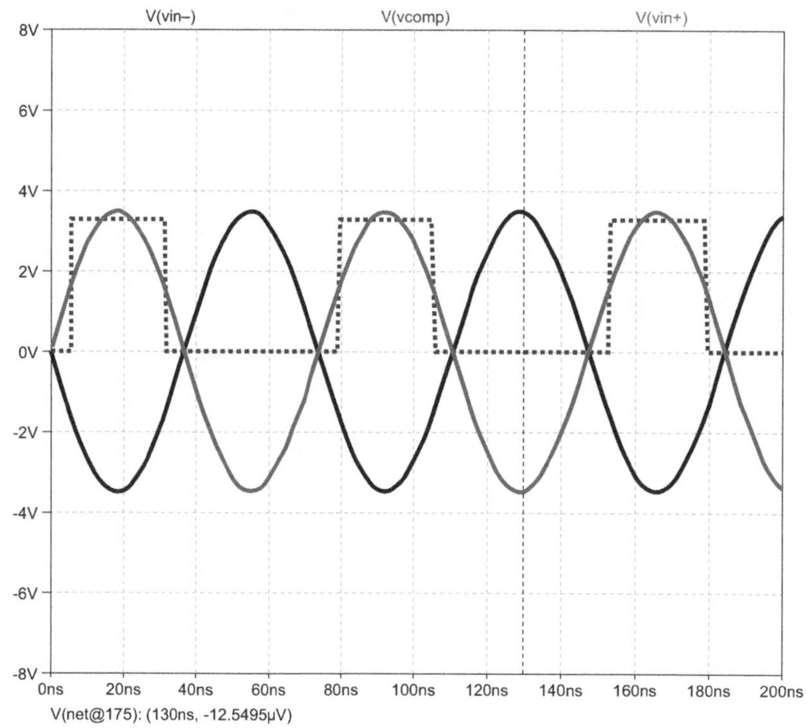

FIGURE 22.4 Transient response of comparator

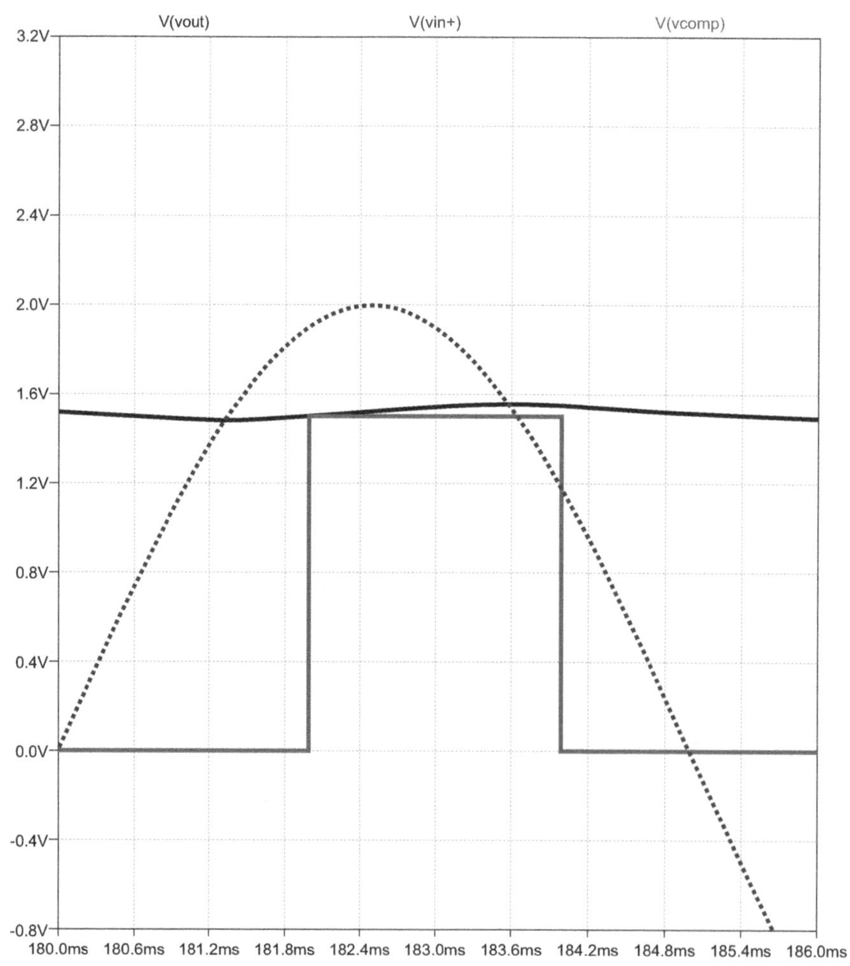

FIGURE 22.5 Switching characteristics of comparator

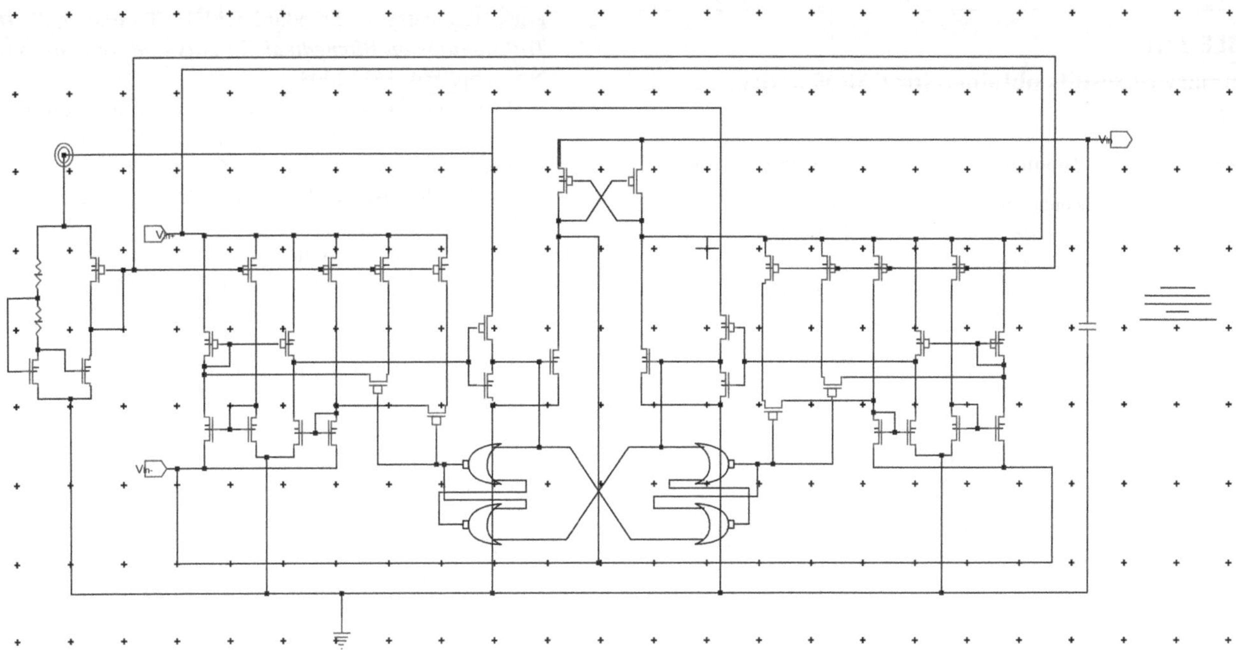

FIGURE 22.6 Schematic of CMOS Active full wave rectifier

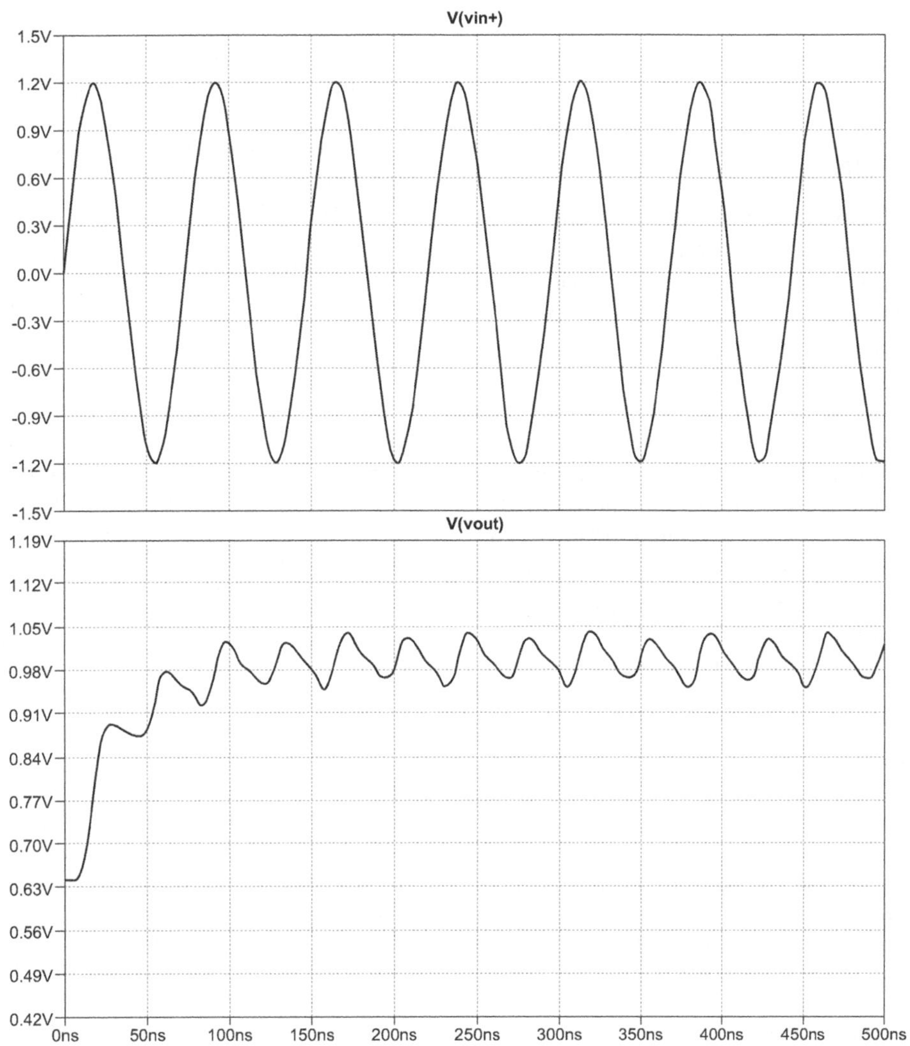

FIGURE 22.7 Input output waveforms of active rectifier with capacitor filter

TABLE 22.1
Summary of results obtained for CMOS active rectifier

Sl. No	Parameters	Values obtained
1	Technology	180 nm
2	Frequency	13.56 MHz
3	Input amplitude	1.2–3.3 V
4	Output voltage	0.95–2.94 V (R_L =500Ω)
5	Max. P_{DC}	20 mW
6	PCE	85–90%

performance parameters of this work. This active rectifier is useful for inductively coupled powering schemes. The expression for power conversion efficiency (PCE) of this rectifier is given by

$$PCE = \frac{P_{DC}}{P_{AC}} = \frac{\frac{V_{DD}^2}{R_L}}{\frac{1}{NT}\int_{t_0}^{t_0+NT} V_{AC}(t) I_{AC}(t) dt} \quad (1)$$

where, N is the number of cycles, T is the period of the input sinusoidal signal.

22.4 CONCLUSION

A CMOS active full-wave rectifier with offset-controlled high speed comparators with operating frequency of 13.56 MHz and 300 pF capacitive filter and load resistance of 500Ω in standard TSMC 0.18µm CMOS technology is studied here. The four diodes traditionally used in diode rectifier are replaced by two comparators controlled NFETs and a pair of cross-coupled PFETs. Using the active elements, the diode's overdrive voltage is removed, allowing much higher voltage and power conversion ratios. The inherent switching delay of the diodes is reduced using comparators and switched-offset biasing which do not allow the reverse current to flow. The higher conduction time for active diodes and the proper design of the digital logic eliminates the multiple pulsing problem associated with the dynamic offset sensing in the comparators. This rectifier is designed and implemented in a 0.18µm 6-Metal/2-Poly TSMC 3.3V process using Electric VLSI Design Tool.

REFERENCES

[1] Y. H. Lam, W. H. Ki and C. Y. Tsui, "Integrated Low-Loss CMOS Active Rectifier for Wirelessly Powered Devices", *IEEE Transactions on Circuits and Systems II: Express Briefs*, Vol. 53, No. 12, pp. 1378–1382, 2006.

[2] G. Bawa and M. Ghovanloo, "Active High Power Conversion Efficiency Rectifier with Built-In Dual-Mode Back Telemetry in Standard CMOS Technology", *IEEE Transactions on Biomedical Circuits and Systems*, Vol. 2, No. 3, pp. 184–192, 2008.

[3] S. Guo and H. Lee, "An Efficiency-Enhanced CMOS Rectifier with Unbalanced Biased Comparators for Transcutaneous-Powered High-Current Implants", *IEEE Journal of Solid-State Circuits*, Vol. 44, No. 6, pp. 1796–1804, 2009.

[4] H. M. Lee and M. Ghovanloo, "An Integrated Power-Efficient Active Rectifier with Offset-Controlled High Speed Comparators for Inductively Powered Applications", *IEEE Transactions on Circuits and Systems I: Regular Papers*, Vol. 58, No. 8, pp. 1749 –1760, 2011.

[5] H. K. Cha, W. T. Park and M. Je, "A CMOS Rectifier with a Cross Coupled Latched Comparator for Wireless Power Transfer in Biomedical Applications", *IEEE Transactions on Circuits and Systems II: Express Briefs*, Vol. 59, No. 7, pp. 409–413, 2012.

[6] S. S. Hashemi, M. Sawan and Y. Savaria, "A High-Efficiency Low Voltage CMOS Rectifier for Harvesting Energy in Implantable Devices", *IEEE Transactions on Biomedical Circuits and Systems*, Vol. 6, No. 4, pp. 326–335, 2012.

[7] Y. Lu and W. H. Ki, "Circuit Design of CMOS Rectifiers", in: *CMOS Integrated Circuit Design for Wireless Power Transfer. Analog Circuits and Signal Processing*. Springer, Singapore, 2018.

[8] K. Zheng, X. Liu, Q. Su, X. Wang and Y. Liu, "A 2MHz CMOS Active Rectifier with PWM Mode Adaptive On/Off Delay COMPENSATION for Wireless Power Transfer Systems", *2022 IEEE International Symposium on Circuits and Systems (ISCAS)*, Austin, TX, 2022, pp. 2448–2452.

[9] S. Shahsavari and M. Saberi, "A Power-Efficient CMOS Active Rectifier with Circuit Delay Compensation for Wireless Power Transfer Systems", *Circuits, Systems, and Signal Processing*, Vol. 38, pp. 947–966, 2019.

[10] Z. Xue, S. Fan, D. Li, L. Zhang, W. Gou and L. Geng, "A 13.56 MHz, 94.1% Peak Efficiency CMOS Active Rectifier with Adaptive Delay Time Control for Wireless Power Transmission Systems", *IEEE Journal of Solid-State Circuits*, Vol. 54, No. 6, pp. 1744–1754, 2019.

[11] K. S. Zheng, X. Liu, X. Wang, Q. Su and Y. Liu, "A 6.78 MHz CMOS Active Rectifier with Hybrid Mode Delay Compensation for Wireless Power Transfer Systems," *IEEE Access*, Vol. 10, pp. 46176–46186, 2022.

[12] A. Banerjee, T. K. Bhattacharyya and S. Nag, "High Efficiency CMOS Active Rectifier with Adaptive Delay Compensation", *Microelectronics Journal*, Vol. 112, 2021.

[13] Y. Zhang, J. Ma and X. Tang, "A CMOS Active Rectifier with Efficiency-Improving and Digitally Adaptive Delay Compensation for Wireless Power Transfer Systems", *Energies*, Vol. 14, p. 8089, 2021. https://doi.org/10.3390/en14238089

[14] C. Huang, T. Kawajiri and H. Ishikuro, "A Near-Optimum 13.56 MHz CMOS Active Rectifier with Circuit-Delay Real-Time Calibrations for High-Current Biomedical Implants", *IEEE Journal of Solid-State Circuits*, Vol. 51, No. 8, pp. 1797–1809, 2016.

[15] G. Bawa and M. Ghovanloo, "Analysis, Design, and Implementation of a High-Efficiency Full-Wave Rectifier in Standard CMOS Technology", *Analog Integrated Circuits and Signal Processing*, Vol. 60, pp. 71–81, 2009.

23 The Impact of Multiwalled Carbon Nanotubes on the Mechanical Properties of Glass Fiber–Reinforced Concrete

Ali Ebrahim, S. Kandasamy, and N. Pannirselvam

23.1 INTRODUCTION

Since ancient times, concrete has been a key material in the construction industry due to its outstanding durability, versatility, and high compressive strength.

Although concrete is widely used, the industry still grapples with certain inherent limitations, such as low tensile strength and poor ductility, which can lead to cracking and compromised long-term durability [1–3]. Several factors, including the water–cement ratio, aggregate size and type, admixtures, cement type and content, and curing conditions [4, 5], play a critical role in determining concrete quality, necessitating careful control during the casting process. When cement, water, and aggregate are mixed, they form a liquid slurry that can be molded into different shapes [6]. However, the concrete matrix contains a weak area called the interfacial transition zone (ITZ), which makes it susceptible to tensile forces [6]. To overcome this weakness, studies have demonstrated that incorporating fibers into concrete improves its mechanical and durability properties [7–9]. Various types of fibers, such as polyester, cellulose, natural, carbon, glass, steel, polypropylene, and polyvinyl alcohol, have been utilized in both traditional and reinforced concrete to enhance the static performance of structures, resulting in the advancement of fiber-reinforced concrete [3, 10, 11].

Conventional reinforcement techniques, such as the use of glass fibers, have significantly improved concrete's performance and durability [1]. These enhancements include increased compressive and flexural strength, better crack and impact resistance, and improved durability in aggressive environments [7, 12]. Glass fibers provide corrosion resistance, help reduce shrinkage cracking, and enhance toughness, making concrete loads that are more resilient to dynamic and sudden impacts. Additionally, they contribute to fire resistance and lower permeability, making concrete more resistant to water intrusion [13]. Glass fiber–reinforced concrete (GFRC) is highly adaptable, enabling a wide range of applications and intricate designs [14]. Although the initial costs may be higher, the long-term benefits, such as reduced maintenance and repair expenses, make it a cost-effective option [15]. Furthermore, incorporating glass fibers supports sustainability by recycling waste glass and extending the material's service life, reducing the need for replacements [16, 17].

At the same time, the rise of nanotechnology in civil engineering introduces groundbreaking materials and solutions, driving a transformative shift in the field. There are many types of nanomaterials, and each of them has an effect on construction materials, like carbon nanotubes (CNTs), titanium dioxide nanoparticles (TiO_2), and nanoclay [18]. Nanoparticles have a profound impact due to their distinct properties at the nanoscale, enhancing material performance, enabling self-healing features, and promoting the development of smart infrastructure. Nanotechnology plays a role in environmental remediation; supports the creation of lightweight, high-strength materials; and boosts energy efficiency [19, 20]. In geotechnical engineering, it aids in soil stabilization and contributes to stronger, more stable foundations. It also improves precision in construction processes and enables the miniaturization of components, while nanomaterials reinforce traditional materials [21]. Carbon nanotubes exhibit exceptional mechanical properties, including tensile strengths ranging from 65 to 93 GPa and a stiffness of around 1 TPa [22, 23]. Their distinctive graphene-based electrical structure allows them to display metallic, quasi-metallic, and semiconducting behaviors [22, 24, 25].

Their high aspect ratio and unique properties make carbon nanotubes ideal for use as reinforcing agents, significantly enhancing the mechanical and electrical properties of cementitious materials in structural applications [22, 26]. Multiwalled carbon nanotubes (MWCNTs) consist of multiple concentric layers of graphene cylinders. Primarily their length characterizes these tubular nanostructures, but in many applications, the aspect ratio—the ratio of length to diameter—is a critical factor. This ratio greatly influences MWCNTs' properties and determines their suitability for various applications. For MWCNTs, with diameters ranging from 7 to 100 nm, the aspect ratio can vary widely, from 50 to 4000 [27, 28]. Incorporating MWCNTs into concrete offers numerous benefits, including improved mechanical properties, enhanced electrical conductivity, increased durability, and a reduced carbon footprint [27, 29, 30]. Achieving uniform dispersion of MWCNTs in concrete is essential for maximizing their reinforcement potential, though their tendency to agglomerate presents a challenge. Addressing this issue often requires the use of specific dispersion techniques and additives [27, 31].

DOI: 10.1201/9781003644439-23

The aim of this research is to investigate the impact of different proportions of MWCNTs and glass fibers on the mechanical properties of the concrete matrix. Additionally, SEM analysis conducted to closely examine the microstructure of the samples, offering detailed insights into the distribution of MWCNTs and the alignment of fibers in various concrete mixtures. This allowed for an evaluation of how MWCNTs and varying glass fiber ratios effectively mitigate micro- and nanocracks within the concrete matrix.

23.2 EXPERIMENTAL DETAILS

23.2.1 Materials

This study utilized ordinary portland cement (OPC) of grade 53, in accordance with the specifications of Indian Standard IS: 269–2015 [32], with a specific gravity of 3.15. The fine aggregate (FA) used was river sand from zone II, conforming to IS: 383–2016 [33], while the coarse aggregate (CA) consisted of well-graded crushed limestone particles ranging in size from 5 to 20 mm. The superplasticizer (SP) employed in the study was Classic Superflo, a brown liquid based on naphthalene formaldehyde condensates produced by Primadon, which met the requirements of IS: 9103–2018 [34]. The glass fibers (GFs) used were chopped fibers, produced by Angle Enterprises in India. Figure 23.1. shows an overview of the glass fibers, and their properties are listed in Table 23.1.

FIGURE 23.1 Overview of glass fibers.

TABLE 23.1
Properties of Glass Fibers Used in This Research

property	Value
Length (mm)	12
Filament diameter (μm)	14
Aspect ratio of the filament	850
Tensile strength (MPa)	1600
Moisture (%)	0.9
Specific gravity	2.68
Density (kg/m3)	2600
Modulus of elasticity (GPa)	74

23.2.2 Proportions of Concrete Mixture

The detailed mix proportions are presented in Table 23.2. MWCNTs were incorporated based on the cement weight, while GFs were added relative to the total weight of the concrete matrix.

TABLE 23.2
Details of Concrete Mixes

Mix No	Mix ID	Cement (kg/m^3)	CA (kg/m^3)	FA (kg/m^3)	Water (kg/m^3)	SP (kg/m^3)	MWCNTs (g/m^3)	GF (kg/m^3)
Mix1	Reference concrete	340	1280	730	150	4.2	–	–
Mix2	0.05wt%MWCNT+0.5wt%GF	340	1280	730	150	4.5	170	12.50
Mix3	0.05wt%MWCNT+1.0wt%GF	340	1280	730	150	5.0	170	25.0
Mix4	0.05wt%MWCNT+1.5wt%GF	340	1280	730	150	5.2	170	37.50
Mix5	0.10wt%MWCNT+0.5wt%GF	340	1280	730	150	4.5	340	12.50
Mix6	0.10wt%MWCNT+1.0wt%GF	340	1280	730	150	5.0	340	25.0
Mix7	0.10wt%MWCNT+1.5wt%GF	340	1280	730	150	5.2	340	37.50
Mix8	0.15wt%MWCNT+0.5wt%GF	340	1280	730	150	4.5	510	12.50
Mix9	0.15wt%MWCNT+1.0wt%GF	340	1280	730	150	5.0	510	25.0
Mix10	0.15wt%MWCNT+1.5wt%GF	340	1280	730	150	5.2	510	37.50

CA: Coarse aggregates, FA: Fine aggregates, SP: Superplasticizer, MWCNTS: Multiwalled carbon nanotubes, GF: Glass fibers

TABLE 23.3
Physical Properties of MWCNTs

Properties	Value
Type	MWCNT CAS. NO. 7782-42-5
Purity	~99 %
Outer diameter (nm)	10–30
Inner diameter (nm)	5–10
Length (μm)	5
Surface area (m^2/g)	110–350
CNT content (%)	95–99
Bulk density (g/cm^3)	0.04
Chemical formula	C
Physical form	Fluffy, very light powder
Odor	Odorless
Color	Black

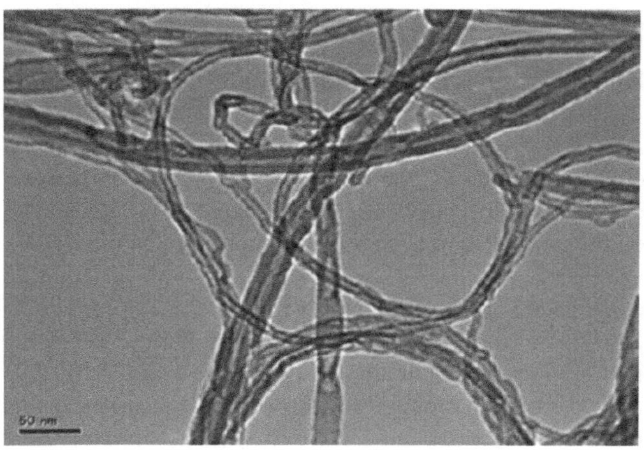

FIGURE 23.2 TEM image of MWCNTs [35].

23.2.3 Multiwalled Carbon Nanotubes

The MWCNTs used in this study are about five μm in length, with an outer diameter ranging between 10 and 30 nm, and produced by AdNano Technologies Pvt. Ltd in India. Additional information regarding their physical properties, as supplied by the manufacturer, is listed in Table 23.3. Moreover, a transmission electron microscope (TEM) image of the MWCNTs is displayed in Figure 23.2.

23.2.4 Dispersion of MWCNTs in Water

MWCNTs tend to agglomerate or clump together due to van der Waals forces and their high aspect ratio, which can lead to poor dispersion within the concrete matrix. This agglomeration can prevent the uniform distribution of MWCNTs, limiting their effectiveness in enhancing the concrete's properties [36]. When properly dispersed, MWCNTs can influence the cement hydration process by acting as nucleation sites for the formation of calcium-silicate-hydrate (C-S-H). Their high aspect ratio and large surface area can promote the early precipitation and growth of C-S-H crystals, potentially resulting in a denser microstructure and faster strength development [37, 38].

FIGURE 23.3 Dispersion of MWCNTs in water process.

The dispersion of MWCNTs in water was accomplished through a systematic procedure [39]. The process began with precise weighing of the required MWCNT quantity. The measured MWCNTs were then gradually added to water under constant stirring to minimize clumping. A VCX 130 sonicator (Newtown, CT, USA) was utilized to enhance the dispersion quality. During the sonication process, careful temperature monitoring was essential, as exceeding 60°C could potentially damage the MWCNT structure [40, 41]. The initial sonication phase lasted 30 minutes, after which temperature readings were taken. When temperature exceeded acceptable limits, adjustments were made to sonication either duration or power settings. In cases where optimal dispersion not reached after the initial phase, additional sonication performed in incremental periods of 5–10 minutes until achieving the desired dispersion level. The final assessment of MWCNT dispersion quality was conducted through visual examination. The complete systematic dispersion process is illustrated in Figure 23.3.

23.2.5 Specimen Preparation and Curing

All concrete mixtures were prepared using a consistent water–cement ratio of 0.42. The experimental design incorporated varying concentrations of MWCNTs (0.05%, 0.10%, and 0.15% by cement weight) combined with different proportions of glass fibers (0.5%, 1.00%, and 1.50% by total mix weight). To ensure compliance with IS: 456–2000 [42], workability was maintained within a slump range of 80–100 mm. The plasticizer dosage was fine-tuned based on preliminary trial mixes and slump test outcomes. Since fiber-reinforced concrete typically shows reduced workability compared to conventional concrete [8], plasticizer quantities were progressively increased in proportion to the glass fiber content to achieve the target workability levels.

The concrete mixing procedure was conducted using a pan mixer with a 55-liter capacity. The process began by blending the superplasticizer with water to ensure uniform distribution. Subsequently, the dry ingredients (cement, fine aggregates, and coarse aggregates) were combined and dry-mixed for 1 minute. Following this step, half of the designated water quantity was introduced, and mixing continued for a 3-minute duration. The final phase involved the gradual addition of glass fibers and the pre-dispersed MWCNT solution, accompanied by simultaneous water addition to prevent fiber agglomeration.

Moreover, for optimal performance, the fibers spread throughout the concrete mix uniformly. This creates a 3D web-like structure that offers several advantages, like preventing cracks from forming, bridging them, and reducing pressure at crack tips so that, to ensure a uniform distribution, continuous mixing was carried out for 5 minutes [7, 43]. The concrete was placed into molds and compacted using a needle vibrator. After 24 hours, the specimens were demolded and underwent water curing until the designated testing date.

23.2.6 Specimen Testing

To evaluate the compressive strength of each mix, 150 mm cubes of each mix were cast and tested following IS: 516–2018 [44]. A constant loading rate of 140 kg/cm²/min was applied to the cube specimens until they failed. The crushing load was automatically recorded throughout the testing process. For split tensile strength, 150-mm-diameter and 300-mm-height cylinders were cast and tested following IS: 5816–2018 [45]. Prism specimens of size 100 × 100 × 500 mm were used in the flexural tensile tests in accordance with IS: 516–2018 [44] guidelines. The concrete prisms underwent a flexural test using three-point loading to evaluate a particular strength parameter, with three specimens cast for each concrete mix.

23.2.7 SEM Observation

The SEM analysis was done in compliance with the ASTM C1723–2010 standard [46]. A Thermo Scientific Apreo S Scanning Electron Microscope was employed to examine the micromorphology of the concrete specimens. Following a 28-day curing period, the 100 × 100 × 100 mm^3 cube specimens were fractured to obtain samples. For SEM analysis, a small portion of the specimen, approximately 20 mm × 15 mm × 10 mm in size, was extracted. This section encompassed the entire concrete matrix, including aggregate particles, cement, fibers, and MWCNTs. After coating, the specimens were placed on the SEM scanning table.

23.3 RESULTS AND DISCUSSION

23.3.1 Compressive Strength

Figure 23.4 illustrates the compressive strength results of the various concrete mixes. The inclusion of MWCNTs and GFs significantly enhanced compressive strength compared to the control concrete. Specifically, incorporating 0.05 wt% MWCNTs with varying GF contents (0.5, 1.0, and 1.5 wt%) resulted in moderate improvements ranging from 4.9% to 19.0% at 28 days of curing. Increasing the MWCNT dosage to 0.10 wt% while maintaining the same GF percentages led to more substantial improvements, with compressive strength increases ranging from 9.7% to 28% after 28 days. However, further increasing MWCNT content to 0.15 wt% with 1.5 wt% GF resulted in a slight decrease (−0.6%) in compressive strength compared to the control. The curing period significantly influenced the compressive strength of all mixes, with a steady increase observed between 7 and 28 days. Notably, Mix 6 (0.10 wt% MWCNT + 1.0 wt% GF) demonstrated the highest improvement, reaching a strength of approximately 42.5 MPa at 28 days compared to the control concrete (32.1 MPa). These findings suggest that the optimal MWCNT content for maximizing the benefits of GF reinforcement may vary depending on the specific GF dosage employed.

23.3.2 Flexural Strength

Figure 23.5 summarizes the flexural strength results of the various concrete mixes containing MWCNTs and GFs. All mixes incorporating these nanomaterials exhibited enhanced flexural strength compared to the reference concrete at all curing ages. The influence of MWCNT content on flexural strength appears to be dependent on the GF dosage. While adding 0.05 wt% MWCNTs with different GF contents led to moderate improvements (2.1% to 5.9%) at 28 days, increasing the MWCNT dosage to 0.10 wt% resulted in more significant enhancements (13.9% to 20%). Incorporating 0.15 wt% MWCNTs with various GF dosages yielded moderate improvements (12.1% to 17.2%) compared to the control mix. Overall, higher MWCNT dosages generally led to increased flexural strength, especially at later curing ages. Mix 6 (0.10 wt% MWCNT + 1.0 wt% GF) demonstrated the most significant improvement, with a strength increase of approximately 4.62 MPa at 28 days compared to the reference concrete (3.52MPa). These findings suggest a possible synergistic effect between MWCNTs and GFs, likely attributed to their combined ability to bridge microcracks and enhance the tensile strength of the concrete matrix.

23.3.3 Splitting Tensile Strength

Figure 23.6 presents the splitting tensile strength results of the various concrete mixes incorporating MWCNTs and GFs. All of these mixes exhibited superior splitting tensile strength compared to the control mix. Adding 0.05 wt% MWCNTs with varying GF percentages (0.5, 1.0, and 1.5 wt%) led to moderate improvements (3.1% to 8.2%) at 28 days. However, increasing MWCNT dosage to 0.10 wt% with the same GF percentages resulted in more substantial enhancements (15.2% to 23.2%) at 28 days. Incorporating 0.15 wt% MWCNTs with various GF percentages also demonstrated moderate improvements (15.3% to 20.8%).

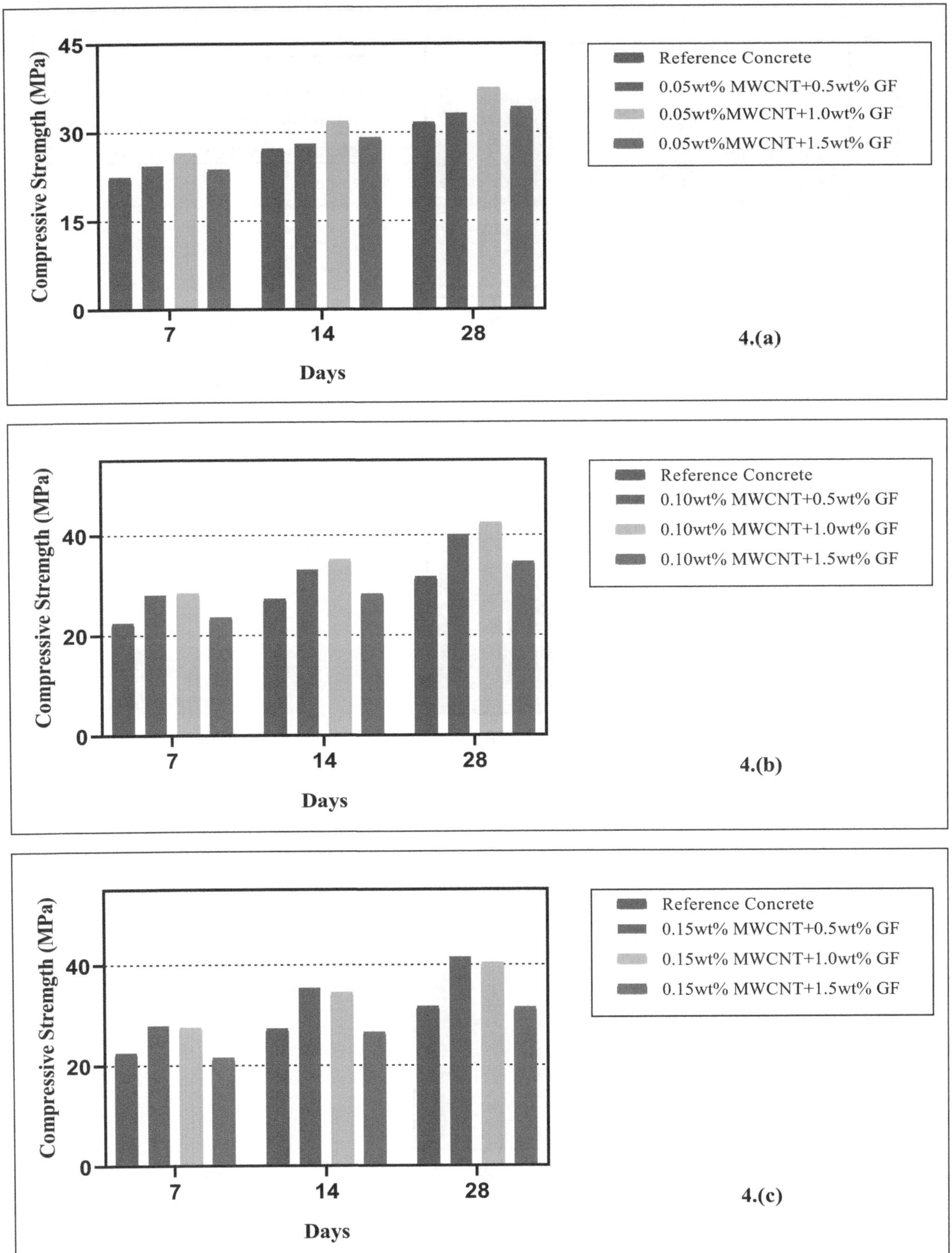

FIGURE 23.4 Compressive strength vs age.

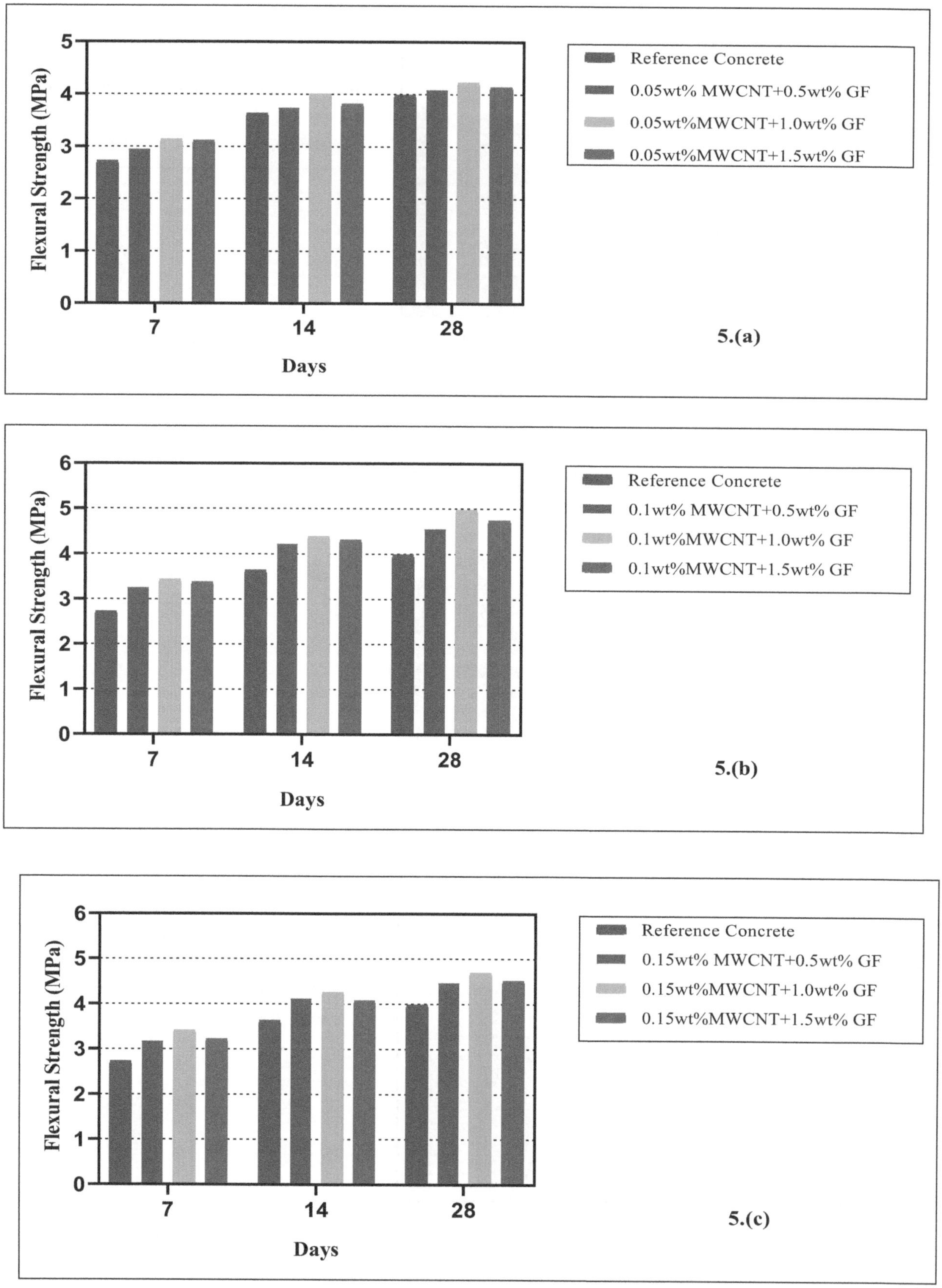

FIGURE 23.5 Flexural strength vs age.

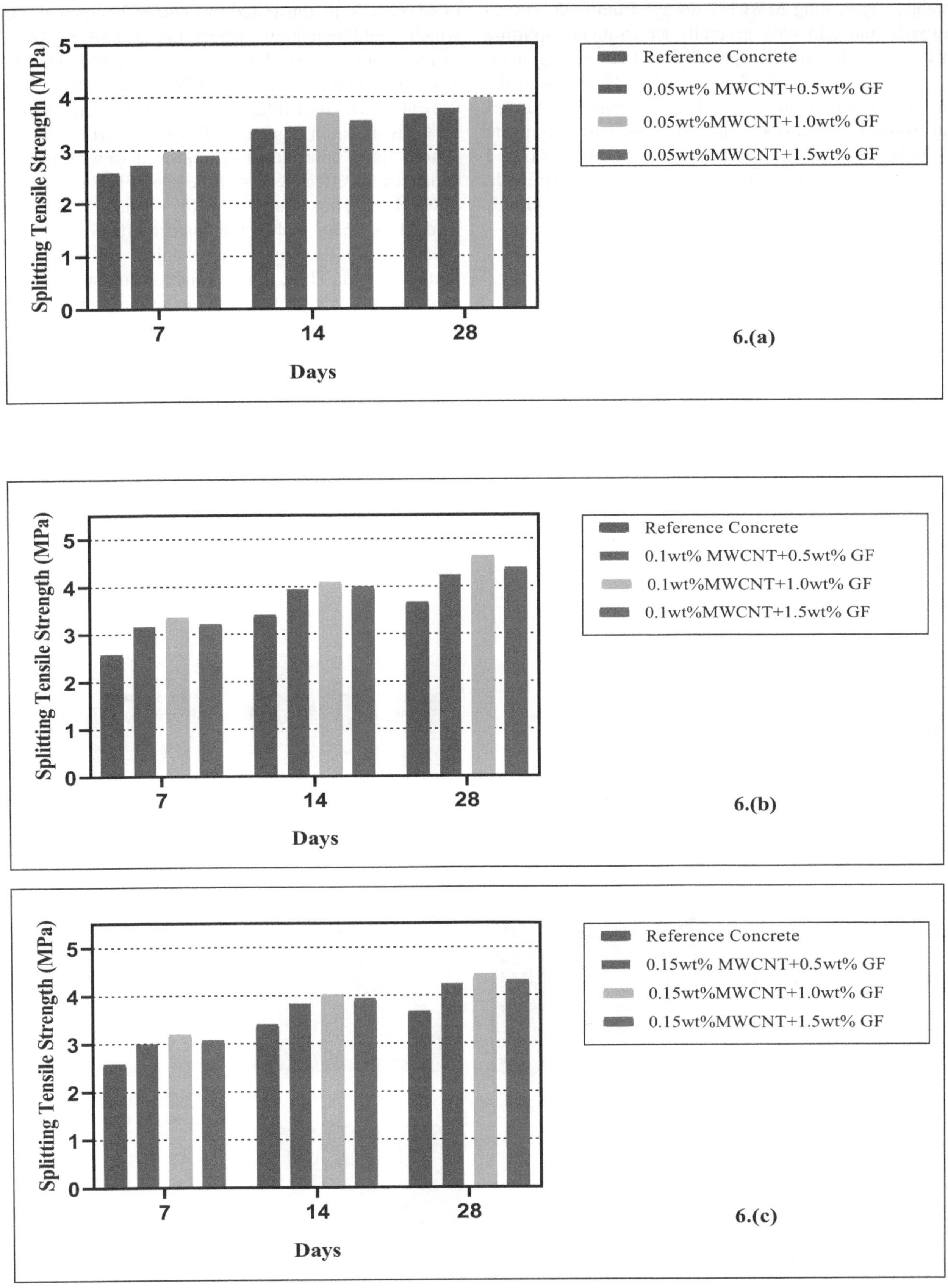

FIGURE 23.6 Splitting tensile strength vs age.

Notably, augmenting MWCNT dosage from 0.05 wt% to 0.10 wt% and 0.15 wt% generally led to higher splitting tensile strength values, especially at later curing ages. Mix 6 (0.10 wt% MWCNT + 1.0 wt% GF) exhibited the most significant enhancement, with a strength increase of approximately 4.46 MPa compared to the control concrete (3.2 MPa). The observed trends suggest that MWCNTs and GFs have a synergistic effect on splitting tensile strength, with MWCNTs potentially bridging microcracks and GFs enhancing the tensile strength of the concrete matrix.

23.3.4 Microstructural Analysis

Microstructural analysis revealed that glass fibers were generally well distributed within the concrete mixes, as depicted in Figures 23.7(c) and 23.9(b). Similarly, MWCNTs were observed evenly dispersed in all concrete mixes, as shown in Figures 23.7(a) and 23.8(a). The dispersion of MWCNTs is influenced by the percentage of glass fibers. At lower fiber contents (0.5 wt%), both MWCNTs and glass fibers demonstrated excellent distribution within the matrix, resulting in a dense and uniform microstructure. However, increasing the fiber dosage (1.0% to 1.5 wt%) led to a tendency for fiber agglomeration and non-uniform distribution of MWCNTs, as illustrated in Figures 23.8(c) and 23.9(c), which could potentially affect the overall performance of the composite. SEM images further highlighted favorable interactions between MWCNTs, glass fibers, and the cement matrix, particularly at lower fiber contents (0.5, 1.0 wt%), as shown in Figures 23.7(b) and 23.8(b). These interactions likely contributed to improved stress transfer and enhanced the effectiveness of the glass fibers.

SEM imaging analysis confirms that evenly distributed glass fibers within the matrix effectively facilitate crack bridging and arrest mechanisms as illustrated in Figure 23.9(a). These fibers function as barriers against crack propagation, thereby enhancing the composite's fracture resistance and durability. As shown in Figure 23.8(a), MWCNTs, characterized by their elongated structure and nanometric dimensions, contribute additional crack resistance and matrix reinforcement at the nanoscale, leading to improved mechanical properties, including enhanced toughness, strength, and durability. Simultaneously, glass fibers provide reinforcement at the microscale by withstanding tensile forces and enhancing the composite's overall mechanical performance. This dual-scale reinforcement system creates synergistic improvements in both mechanical and functional properties of the composite material.

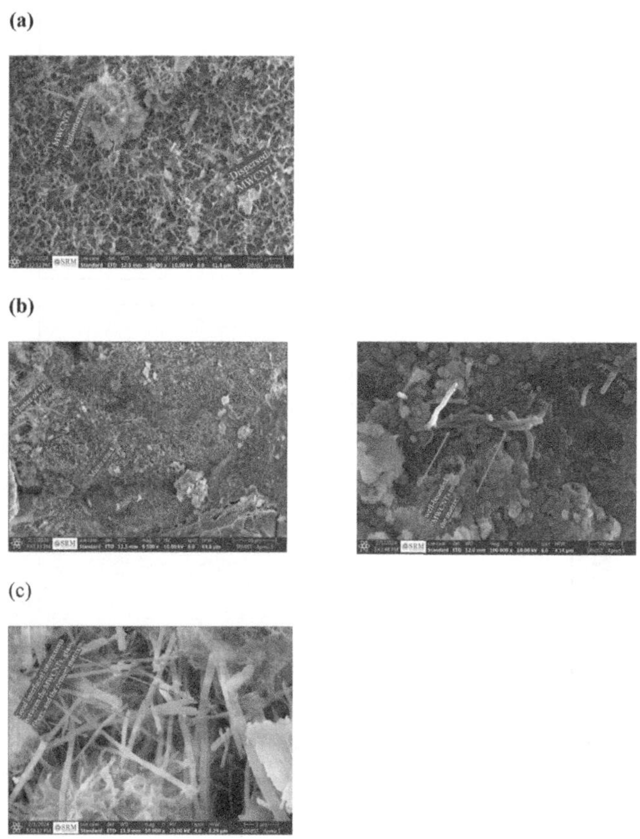

FIGURE 23.7 SEM images of concrete samples modified with 0.05wt% MWCNT and varying glass fiber content: (a) 0.5wt%, (b) 1.0wt%, and (c) 1.5wt%.

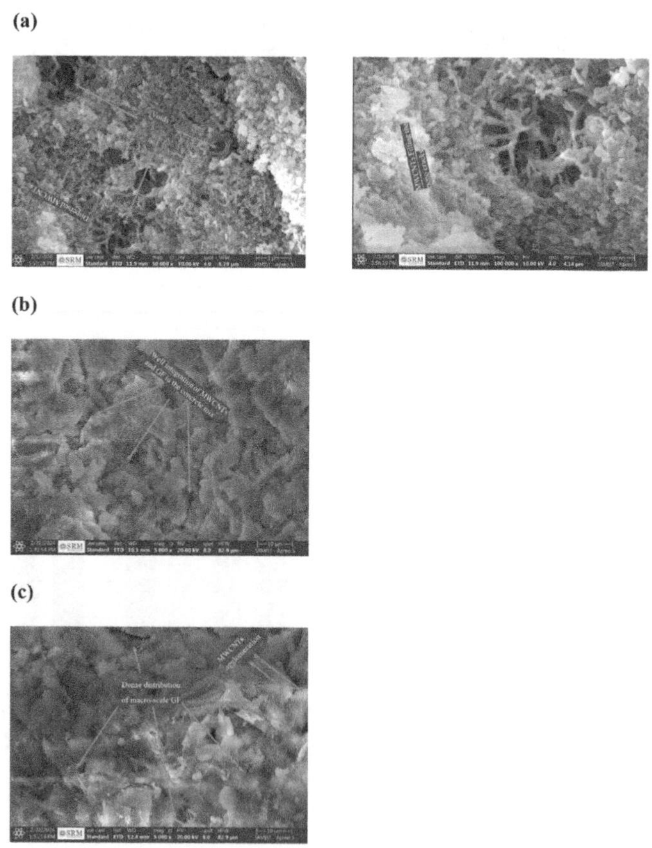

FIGURE 23.8 SEM images of concrete samples modified with 0.1wt% MWCNT and varying glass fiber content: (a) 0.5wt%, (b) 1.0wt%, and (c) 1.5wt%.

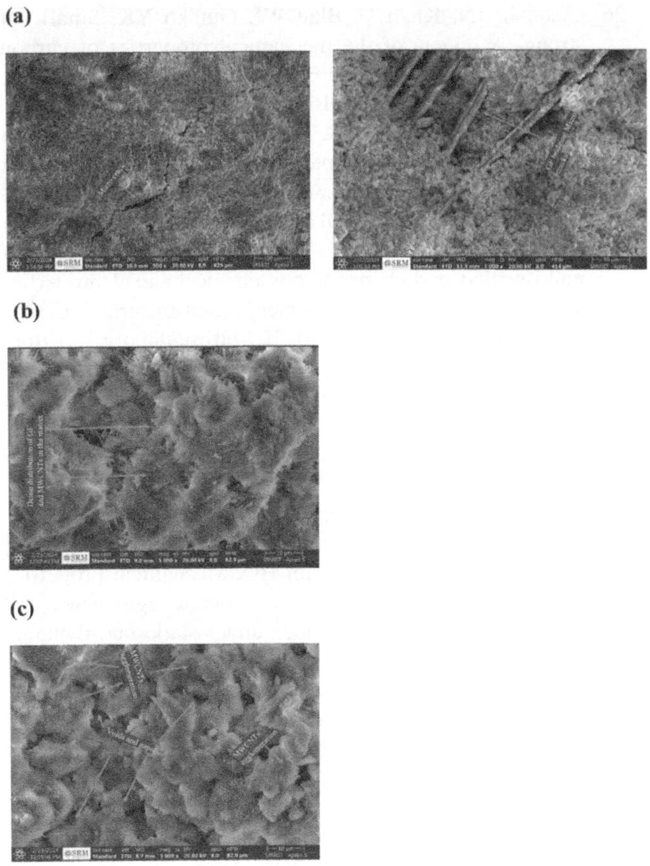

FIGURE 23.9 SEM images of concrete samples modified with 0.15wt% MWCNT and varying glass fiber content: (a) 0.5wt%, (b) 1.0wt%, and (c) 1.5wt%.

Microstructural examination reveals that at lower fiber concentrations (0.5 wt%), the matrix maintains density and uniformity with minimal void formation. However, as fiber content increases to 1.5 wt%, the material exhibits increased porosity and agglomeration tendencies.

23.4 CONCLUSIONS

The following conclusions were derived from the current experimental study and SEM analysis:

1. The inclusion of MWCNTs and GFs at various dosages significantly enhanced all measured mechanical properties. The most notable improvements occurred with a combination of 0.1 wt% MWCNTs and 1.0 wt% GFs, resulting in a 28% increase in compressive strength, a 20% increase in flexural strength, and a 23% increase in splitting tensile strength.
2. MWCNTs and GFs appear to be more effective against direct tensile forces than bending stresses. This is evident from the faster rate of increase in tensile strength compared to flexural strength.
3. The combination of MWCNTs and glass fibers provides synergistic reinforcement. MWCNTs offer nanoscale reinforcement, while glass fibers provide microscale reinforcement and crack bridging.
4. The percentage of glass fibers influences the distribution of MWCNTs within the concrete mix. At lower fiber contents (0.5%), both MWCNTs and glass fibers are well dispersed. However, increasing the fiber content (1.0% to 1.5%) can lead to agglomeration of MWCNTs and increased porosity.

REFERENCES

1. Ruben N, Venkatesh C, Durga CS, Chand MS. Comprehensive study on performance of glass fibers-based concrete. *Innov Infrastruct Solut*. 2021;6(2):112. https://doi.org/10.1007/s41062-021-00490-4.
2. Tassew ST, Lubell AS. Mechanical properties of glass fiber reinforced ceramic concrete. *Constr Build Mater*. 2014;51:215–224. https://doi.org/10.1016/j.conbuildmat.2013.10.046.
3. Sridhar R, Abbas M. Experimental study on hybrid fiber reinforced concrete with composite layers. *Mater Today Proc*. 2022;65:489–494. https://doi.org/10.1016/j.matpr.2022.03.032.
4. Ahmad S, Alghamdi SA. A study on effect of coarse aggregate type on concrete performance. *Arab J Sci Eng*. 2012;37:1777–1786. https://doi.org/10.1007/s13369-012-0282-6.
5. Ayanlere SA, Ajamu SO, Odeyemi SO, Ajayi OE, Kareem MA. Effects of water-cement ratio on bond strength of concrete. *Mater Today Proc*. 2023;86:134–139. https://doi.org/10.1016/j.matpr.2023.04.686.
6. Chand N, Jain D. Effect of sisal fibre orientation on electrical properties of sisal fibre reinforced epoxy composites. *Compos—A: Appl Sci Manuf*. 2005;36(5):594–602. https://doi.org/10.1016/j.compositesa.2004.08.002.
7. Xue G, Yilmaz E, Song W, Cao S. Mechanical, flexural and microstructural properties of cement-tailings matrix composites: Effects of fiber type and dosage. *Compos B Eng*. 2019;172:131–142. https://doi.org/10.1016/j.compositesb.2019.05.039.
8. Ali B, Qureshi LA. Influence of glass fibers on mechanical and durability performance of concrete with recycled aggregates. *Constr Build Mater*. 2019;228:116783. https://doi.org/10.1016/j.conbuildmat.2019.116783.
9. Algburi AH, Sheikh MN, Hadi MN. Mechanical properties of steel, glass, and hybrid fiber reinforced reactive powder concrete. *Front Struct Civ Eng*. 2019;13:998–1006. https://doi.org/10.1007/s11709-019-0533-7.
10. Das CS, Dey T, Dandapat R, Mukharjee BB, Kumar J. Performance evaluation of polypropylene fibre reinforced recycled aggregate concrete. *Constr Build Mater*. 2018;189:649–659. https://doi.org/10.1016/j.conbuildmat.2018.09.036.
11. Nehdi ML, Najjar MF, Soliman AM, Azabi TM. Novel steel fibre-reinforced preplaced aggregate concrete with superior mechanical performance. *Cem Concr Compos*. https:/doi.org/10.1016/j.cemconcomp.2017.07.002.
12. Yuan Z, Jia Y. Mechanical properties and microstructure of glass fiber and polypropylene fiber reinforced concrete: An

experimental study. *Constr Build Mater.* 2021;266:121048. https://doi.org/10.1016/j.conbuildmat.2020.121048.
13. Chen H, Wang P, Pan J, Lawi AS, Zhu Y. Effect of alkali-resistant glass fiber and silica fume on mechanical and shrinkage properties of cement-based mortars. *Constr Build Mater.* 2021;307:125054. https://doi.org/10.1016/j.conbuildmat.2021.125054.
14. Ahmad J, González-Lezcano RA, Majdi A, Ben Kahla N, Deifalla AF, El-Shorbagy MA. Glass fibers reinforced concrete: Overview on mechanical, durability and microstructure analysis. *J Mater.* 2022;15(15):5111. https://doi.org/10.3390/ma15155111.
15. Hussain I, Ali B, Akhtar T, Jameel MS, Raza SS. Comparison of mechanical properties of concrete and design thickness of pavement with different types of fiber-reinforcements (steel, glass, and polypropylene). *Case Stud Constr Mater.* 2020;13(e00429). https://doi.org/10.1016/j.cscm.2020.e00429.
16. Ali B, Qureshi LA, Khan SU. Flexural behavior of glass fiber-reinforced recycled aggregate concrete and its impact on the cost and carbon footprint of concrete pavement. *Constr Build Mater.* 2020;262:120820. https://doi.org/10.1016/j.conbuildmat.2020.120820.
17. Chen Y, He Q, Liang X, Jiang R, Li H. Experimental investigation on mechanical properties of glass fiber reinforced recycled aggregate concrete under uniaxial cyclic compression. *J Mater.* 2022;6:100164. https://doi.org/10.1016/j.clema.2022.100164.
18. Mohajerani A, Burnett L, Smith JV, Kurmus H, Milas J, Arulrajah A, Horpibulsuk S, Abdul Kadir A. Nanoparticles in construction materials and other applications, and implications of nanoparticle use. *J Mater.* 2019;12(19):3052. https://doi.org/10.3390/ma12193052.
19. Firoozi AA, Naji M, Dithinde M, Firoozi AA. A review: Influence of potential nanomaterials for civil engineering projects. *Iran J Sci Technol—Trans Civ Eng.* 2021;1–12. https://doi.org/10.1007/s40996-020-00474-x.
20. Onaizi AM, Huseien GF, Lim NH, Amran M, Samadi M. Effect of nanomaterials inclusion on sustainability of cement-based concretes: A comprehensive review. *Constr Build Mater.* 2021;306:124850. https://doi.org/10.1016/j.conbuildmat.2021.124850.
21. Ghasabkolaei N, Choobbasti AJ, Roshan N, Ghasemi SE. Geotechnical properties of the soils modified with nanomaterials: A comprehensive review. *Arch Civ Mech Eng.* 2017;17(3):639–650. https://doi.org/10.1016/j.acme.2017.01.010.
22. Konsta-Gdoutos MS, Danoglidis PA, Falara MG, Nitodas SF. Fresh and mechanical properties, and strain sensing of nanomodified cement mortars: The effects of MWCNT aspect ratio, density and functionalization. *Cem Concr Compos.* 2017;82:137–151. https://doi.org/10.1016/j.cemconcomp.2017.05.004.
23. Salvetat JP, Bonard JM, Thomson N, Kulik A, Forro L. Mechanical properties of carbon nanotubes. *Appl Phys A.* 1999;69:255–260. https://doi.org/10.1007/s003390050999.
24. Issi JP, Langer L, Heremans J, Olk CH. Electronic properties of carbon nanotubes: Experimental results. *Carbon.* 1995;33(7):941–948. https://doi.org/10.1016/0008-6223(95)00023-7.
25. Bandaru PR. Electrical properties and applications of carbon nanotube structures. *J Nanotechnol.* 2007;7(4–5):1239–1267. https://doi.org/10.1166/jnn.2007.307.
26. Coleman JN, Khan U, Blau WJ, Gun'ko YK. Small but strong: A review of the mechanical properties of carbon nanotube—polymer composites. *Carbon.* 2006;44(9):1624–1652. https://doi.org/10.1016/j.carbon.2006.02.038.
27. Ebrahim A, Kandasamy S. The effect of using multi-walled carbon nanotubes on the mechanical properties of concrete: A review. *Innov Infrastruct Solut.* 2023;8(9):251. https://doi.org/10.1007/s41062-023-01219-1.
28. Gao F, Tian W, Wang Z, Wang F. Effect of diameter of multi-walled carbon nanotubes on mechanical properties and microstructure of the cement-based materials. *Constr Build Mater.* 2020;260:120452. https://doi.org/10.1016/j.conbuildmat.2020.120452.
29. Mohsen MO, Alansari M, Taha R, Senouci A, Abutaqa A. Impact of CNTs' treatment, length and weight fraction on ordinary concrete mechanical properties. *Constr Build Mater.* 2020;264:120698. https://doi.org/10.1016/j.conbuildmat.2020.120698.
30. Allujami HM, Abdulkareem M, Jassam TM, Al-Mansob RA, Ibrahim A, Ng JL, Yam HC. Mechanical properties of concrete containing recycle concrete aggregates and multi-walled carbon nanotubes under static and dynamic stresses. *Case Stud Constr Mater.* 2022;17(e01651). https://doi.org/10.1016/j.cscm.2022.e01651.
31. Qissab MA, Abbas ST. Behaviour of reinforced concrete beams with multiwall carbon nanotubes under monotonic loading. *Eur J Environ Civ Eng.* 2018;22(9):1111–1130. https://doi.org/10.1080/19648189.2016.1232661.
32. IS 269–2015. Ordinary Portland Cement—Specification, Bureau of Indian Standards, New Delhi.
33. IS 383–2016. Specification for Coarse and Fine Aggregates for Concrete from Natural Sources for Concrete, Bureau of Indian Standards, New Delhi.
34. IS 9103–2018. Specification for Concrete Admixtures, Bureau of Indian Standards, New Delhi.
35. Alarifi, S, Ali, D. Mechanisms of multi-walled carbon nanotubes–induced oxidative stress and genotoxicity in mouse fibroblast cells. *Int J Toxicol.* 2015;34(3): 258–265. https://doi.org/10.1177/1091581815584799.
36. Sarvandani MM, Mahdikhani M, Aghabarati H, Fatmehsari MH. Effect of functionalized multi-walled carbon nanotubes on mechanical properties and durability of cement mortars. *J Build Eng.* 2021;41:102407. https://doi.org/10.1016/j.jobe.2021.102407.
37. Naqi A, Abbas N, Zahra N, Hussain A, Shabbir SQ. Effect of multi-walled carbon nanotubes (MWCNTs) on the strength development of cementitious materials. *J Mater Res Technol.* 2019;8(1):1203–1211. https://doi.org/10.1016/j.jmrt.2018.09.006.
38. Cerro-Prada E, Pacheco-Torres R, Varela F. Effect of multi-walled carbon nanotubes on strength and electrical properties of cement mortar. *J Mater.* 2020;14(1):79. https://doi.org/10.3390/ma14010079.
39. Assi L, Alsalman A, Bianco D, Ziehl P, El-Khatib J, Bayat M, Hussein FH. Multiwall carbon nanotubes (MWCNTs) dispersion & mechanical effects in OPC mortar & paste: A review. *J Build Eng.* 2021;43:102512. https://doi.org/10.1016/j.jobe.2021.102512.
40. Shchegolkov AV, Nachtane M, Stanishevskiy YM, Dodina EP, Rejepov DT, Vetcher AA. The effect of multi-walled carbon nanotubes on the heat-release properties of elastic nanocomposites. *J Compos Sci.* 2022;6(11):333. https://doi.org/10.3390/jcs6110333.

41. Jóźwiak B, Greer HF, Dzido G, Kolanowska A, Jędrysiak R, Dziadosz J, Dzida M, Boncel S. Effect of ultrasonication time on microstructure, thermal conductivity, and viscosity of ionanofluids with originally ultra-long multi-walled carbon nanotubes. *Ultrason Sonochemistry.* 2021;77:105681. https://doi.org/10.1016/j.ultsonch.2021.105681.
42. IS 456–2000. Code of Practice for Plain and Reinforced Concrete, Bureau of Indian Standards, New Delhi.
43. Kandasamy S, Sarathkumar T, Narasimhulu K, Kumar CM. Influence of permeable formwork on the performance of polypropylene fibre reinforced concrete. *J Build Pathol Rehabil.* 2023;8(1):34. https://doi.org/10.1007/s41024-023-00278-2.
44. IS 516–2018. Methods of Tests for Strength of Concrete, Bureau of Indian Standards, New Delhi.
45. IS 5816–2018. Method of Test for Splitting Tensile Strength of Concrete, Bureau of Indian Standards, New Delhi.
46. ASTM C1723–2010. Standard Guide for Examination of Hardened Concrete Using Scanning Electron Microscopy, Annual Book of ASTM Standard.

24 Nanotechnological Advancements in Medical Devices

P. R. Lighittha, M. Sundara Srivathsan, and G. Lakshmi Priya

24.1 INTRODUCTION TO NANOTECHNOLOGY IN HEALTHCARE INNOVATION

Nanotechnology has emerged as a transformative force in health care innovation, offering unprecedented opportunities for improved care, diagnosis, and treatment. This chapter explores the multifaceted impact of nanotechnology on medical device applications, including its role in enhancing diagnostics through medical imaging, revolutionising drug delivery for targeted therapies, enabling implantable sensors and real-time monitoring, and facilitating the fusion of diagnostics and therapeutics in the concept of theranostics. While nanotechnology promises a brighter future for health care, it also brings forth crucial considerations of nanotoxicology, biocompatibility, and regulatory frameworks to ensure safety and efficacy. This chapter navigates the promising landscape of nanotechnological advancements, reshaping health care with materials at the nanoscale.

24.2 NANOTECHNOLOGY'S IMPACT ON CARE, DIAGNOSIS, AND TREATMENT

Nanotechnology has garnered a lot of attention because it possesses immense potential and great importance in a plethora of applications [1]. With respect to the size of these particles, nanoparticles (NPs) range in size from 1 to 100 nm, but in medicine, they can range from 5 to 250 nm [2]. The size of an NP is an important factor, as its application in different fields relies on it. NPs and nanodevices can display novel features and functions due to their incredibly small sizes. The NPs have a very high surface area–to–volume ratio, which may be their principal benefit because of their small size. Although it may seem unimportant, this characteristic really increases their dependability and reproducibility [3]. To apply, different NPs in the imaging, diagnosis, and therapy of critical diseases, effective changes made to their size, shape, and surface features. These functionalised NPs can deliver medications to specific areas or tissues, enabling controlled-release therapy [4].

24.2.1 Nanosystems—Usage in the Health Care Sectors

Whatever the subject or application, nanotechnology focuses on a few basic nanostructures. NPs, carbon nanotubes (CNTs), dendrimers, nanoprobes, quantum dots (QDs), nanowires, and nanodiamonds are a few of the significant ones. NPs have distinctive properties, and their astoundingly small size enables them to pass through membranes and minuscule pores easily. The five main categories of NPs are metal, lipid, ceramic, polymeric, and semiconductor NPs. The following are the nanostructures and their respective applications [3, 5].

- CNTs—Used as vectors for the transfer of genes, for the delivery of fibrinogen and bovine protein to cells, and for the treatment of breast cancer, osteoporosis, and fractured bones.
- Dendrimers—Used for gene delivery, antibacterial medicines, anticancer medications, vaccine formulation improvement, and the treatment of ophthalmic illnesses, among other uses.
- Nano-diamonds—Used in imaging and therapy, in the early diagnosis of cancer, and in the treatment of brain and breast tumours. Also used to treat bone disorders via targeted medication delivery (bone regeneration).
- QDs—Applied for intracellular tracking, therapeutic drug delivery, real-time in vivo bio-imaging, and disease control.
- Nanofilms—Used in electrochemical devices as helpful biological, chemical, and nanomechanical sensors, for controlled medication release, and as nanopatches to close incisions following open surgery.
- Liposomes—Used for drug administration, they can hold both hydrophobic and hydrophilic medications and guard against enzymatic and chemical destruction.

24.2.1.1 Ocular Conditions

Due to complicated barriers such as the tear film, ocular surface epithelium, internal blood-aqueous and blood–retinal barriers, it is difficult to transfer medications to the eye effectively. Due to their small size and changeable surface qualities, NPs can deliver medications to the desired site without having any negative side effects [6]. NPs can lessen undesirable side effects by increasing bioavailability in anterior eye illnesses such as retinoblastoma and cataracts, which are frequently treated with eye drops (see Figure 24.1).

DOI: 10.1201/9781003644439-24

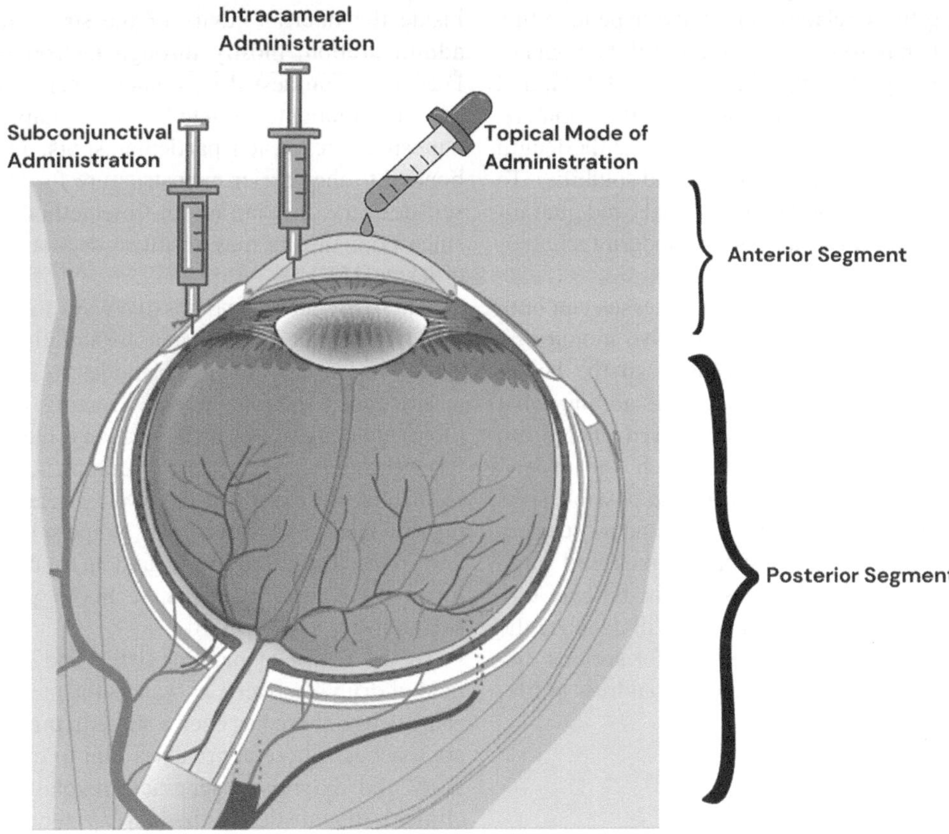

FIGURE 24.1 Modes of drug delivery in the human eye.

24.2.1.2 Brain Diseases

As it prevents medications from accessing the central nervous system and upholds brain homeostasis, the blood–brain barrier (BBB) is a critical barrier in the treatment of brain illnesses. However, medications can be effectively delivered to injured brain regions via NPs. The BBB is penetrated by NPs using both organic and inorganic materials as a core; appropriate inorganic materials include silica, molybdenum, cerium, iron, and gold. Because of their small size, high drug-loading capacity, and effective imaging capabilities, NPs have therapeutic efficacy [7]. NPs' shape and charge also affect how deeply they penetrate.

24.2.1.3 Oncology (Cancer)

Nanotubes, nanocantilevers, NP probes, and nanowire arrays can all be used to detect cancer early. Without using radioactive labelling, micro-cantilevers can detect single nucleotide polymorphisms in DNA target oligonucleotides [8]. Information about the tumour microenvironment and cancer markers is provided by probes with molecularly tailored recognition agents. Systems with nanoprobes can spot DNA alterations and DNA loss. Early stages of cancer can be detected using CNTs [9]. Growth factors, cytokines, and antibodies—which are cancer-targeting ligands—can facilitate vehicle uptake into target cells.

24.2.1.4 Nervous System Diseases

NPs have demonstrated promise in the treatment of spinal cord injuries, neurodegenerative diseases, and genetic disorders. They may also aid in the regeneration of injured brain cells and axons, which are frequently prevented from doing so by inhibitory molecules and age-related reductions in intrinsic regenerative potential. Nanotechnology is proving a highly successful tool for treating problems of the central nervous system, and researchers are looking at different methods to suppress these processes [10]. Traditional spinal cord injury treatments have come under fire for being ineffective; however, nano-assemblies have demonstrated improved neurologic deficit scores and quicker motor recovery. To reduce inflammation without causing harm, polymethyl methacrylate NPs have been utilised to target particular macrophage cell types [11].

24.2.2 Towards the Future

Although still in its infancy, nanotechnology has shown promise in addressing inefficiencies in the health care sector. Researchers should plan their studies to create novel solutions to overcome obstacles in order to hasten this process. Hope for the successful treatment of life-threatening and crippling illnesses has been raised by the progressive development of nanotechnology-based techniques. Nanotechnology does face some difficulties and restrictions, however, including those related to toxicity, dependability, reproducibility, and environmental effects [12]. Nanomaterials come in a variety of compositions, sizes, forms, and effects on living systems. The performance of these materials is controlled by several variables, including the length of exposure, coating, aggregation, charge, and solubility. To swiftly characterise synthesised nanomaterials and guarantee excellent yields, stability, and purity, modern technologies and methods are required.

The treatment and diagnosis of major diseases can both be considerably improved by effective in vivo monitoring devices. Before applying nanotechnologies to the health care sector, mechanisms that aid in understanding what happens to NPs after use must be addressed. There are many problems that need thorough research and experimentation to provide a solution, including the possible risks posed by nanomaterials. To create the safest and most effective therapy regimens, various nanomedicines and nano formulations targeting specific diseases must be carefully developed. In conclusion, nanotechnology will advance the development of more effective treatments for serious diseases and give researchers powerful tools to address bottlenecks in the healthcare sector.

24.3 NPS IN DRUG DELIVERY

Drug delivery systems for distributing therapeutic agents and active chemicals with natural sources to their targeted targets for treating various illnesses have made great strides in recent years. Although several drug delivery systems have shown success, there are still challenges to address, necessitating the development of advanced technologies for effective drug delivery. This has led to the exploration of nano-based drug delivery systems that hold promise for enhancing drug delivery mechanisms.

The field of medicine known as nanomedicine uses nanotechnology to treat and prevent disease utilising nanoscale components such as biocompatible NPs and nanorobots [13]. These materials have uses in the diagnosis, medication delivery, sensing, and actuation of living things. Low solubility medications frequently have problems with absorption, restricted membrane diffusion, and large doses for intravenous administration, leading to unfavourable side effects. However, drug delivery methods utilising nanotechnology can get over these restrictions. Due to its potential benefits, such as the capacity to alter solubility, drug release patterns, diffusivity, bioavailability, and immunogenicity, drug design at the nanoscale has drawn a lot of attention. This may result in the creation of delivery methods that are more practical, decreased toxicity, fewer adverse effects, enhanced drug dispersion, and lengthened drug life cycles. Engineered drug delivery systems sometimes involve self-assembly to produce clearly defined structures and are meant for either precise targeting or controlled release at specified places [13, 14].

Passive and self-delivery are the two main ways that nanostructures deliver pharmaceuticals. Drugs are sealed inside the internal cavity of the structure during passive administration, mostly through hydrophobic interactions. Due to the modest drug content enclosed in a hydrophobic environment, targeted release happens when these nanostructures reach particular spots. Drugs are directly bonded to the carrier nanostructure for effective release in self-delivery, in contrast. In this method, timing is crucial since poor timing may result in decreased bioactivity and efficacy [15].

There are active and passive ways to target medications using nanomaterials or nanoformulations. To bind molecules like antibodies and peptides to receptors expressed at the target region, active targeting entails connecting these molecules with drug delivery mechanisms. Passive targeting directs drug carrier complexes to the target site by affinity or binding in the blood-stream by taking advantage of factors including pH, temperature, molecular site, and shape. Targeting cancer and its treatment is the present focus of the majority of nanotechnology-mediated medication delivery systems [16].

In the realm of nanomedicine and the ever-advancing field of drug discovery, as well as the mechanisms for delivering drugs, there has been a surge in innovative therapeutic approaches. These approaches aim to enhance drug specificity and diagnostic accuracy, revolutionising traditional clinical diagnostic methods [17]. One notable focus is exploring new avenues for drug administration while ensuring precise targeting in specific areas of the body. This dual objective seeks to minimise drug toxicity and boost the availability of drugs within the body [18].

A promising aspect of the search for new therapeutic agents based on a deep understanding of biological targets in the context of drug design is emerging. The development of experimental methods for characterising and purifying proteins, peptides, and biological targets, as well as the quick development of computer sciences, are crucial factors in the advancement of this discipline [19]. The rational design of various compounds has also been the subject of numerous studies and reviews, which emphasises the need to research various drug release mechanisms. Additionally, natural products provide exciting and useful approaches to address difficulties in drug design, acting as a source of ideas for the discovery of medications with desirable physicochemical features. Further, the significance of drug delivery systems has significantly increased in recent years. These systems can be easily developed and excel in promoting the controlled release of active ingredients within the body [20, 21].

It is important to note that every drug delivery system has chemical, physical, and morphological properties. These properties frequently show an affinity for various drug polarities through chemical interactions (like covalent and hydrogen bonds) or physical interactions (like electrostatic and van der Waals forces). For instance, a study by Mattos et al. [22] showed how chemical interactions affected the release profile of biogenic silica NPs grafted with neem bark extract as opposed to biogenic silica NPs loaded with neem bark extract. These variables greatly affect how nanocarriers work with biological systems and control the rate at which active substances are released into the body [23].

Understanding drug delivery profiles also heavily depends on the nature of nanocarriers (organic, inorganic, or hybrid materials), as well as how pharmaceuticals are incorporated into them (e.g., core–shell systems, or matrix systems). Studies have looked into a variety of drug release methods from nanocarriers, including diffusion, solvent effects, chemical reactions, and stimulus-controlled release. Even though many nanocarriers have unique medication release profiles, ongoing research aims to increase the specificity of nanostructures for accurate targeting inside the body. These techniques additionally aim to reduce immunogenic reactions by coating or chemically altering the target material with chemicals, such as polymers, natural polysaccharides, antibodies, cell membranes, tunable surfactants, peptides, and others [24–31].

24.3.1 Chitosan in Drug Delivery

Chitosan is a biopolymeric material with adhesive properties. It is used in various drug delivery systems because it can adhere to tight junctions in different epithelial tissues like the mouth, intestines, nasal passages, eyes, and lungs. For example, researchers developed NPs containing chitosan to deliver antibiotics to the eye. These NPs improved the drug's lifespan in the eye and did not harm eye cells. They also found applications for oral drug delivery using chitosan-based NPs, although further refinements are needed:

- **Chitosan and Oral Drug Delivery**: In one study [32], NPs made from chitosan, alginate, and pectin were considered for drug delivery to the oral cavity. Alginate NPs showed the most stability in artificial saliva, while chitosan NPs were the most biocompatible, even though they were less stable. Zn2+'s employment as a crosslinking agent, however, might have had some cytotoxic effects. Each formulation has its pros and cons, and further improvements are necessary.
- **Intra-Nasal Drug Delivery**: Carboxymethyl chitosan NPs were used to release carbamazepine intra-nasally, allowing the drug to bypass the BBB. This resulted in higher drug concentrations in the brain, improving treatment efficacy and reducing systemic exposure [33].
- **Oral Drug Delivery for Colon Cancer**: Researchers studied the release profile of 5-fluorouracil (5-FU), a drug used in colon cancer treatment. They coated chitosan NPs with hyaluronic acid and administered them orally. These NPs ensured a high local concentration of 5-FU in the colon by preventing its release in the stomach and small intestine. This approach aimed to enhance the drug's effectiveness against colon cancer while reducing systemic side effects [34].

24.3.2 Liposomes in Drug Delivery

Liposomes are vesicles composed of phospholipids and steroids, typically ranging from 50 to 450 nm in size, and are widely used as drug delivery vehicles [35]. They are biocompatible and biodegradable, with a membrane structure like cell membranes, allowing for the incorporation of hydrophilic and hydrophobic drugs. Liposomes can improve drug stability, biodistribution, and bioavailability. There are four main types of liposomes: conventional, PEGylated, ligand-targeted, and theranostic. Drug loading in liposomes can be achieved through active and passive approaches. Liposomes face challenges such as interaction with the reticuloendothelial system, opsonisation, and immunogenicity. However, enhanced permeability and retention can be utilised to improve drug delivery efficiency. PEGylated liposomes have been developed to overcome some of the challenges associated with liposome drug delivery. As possible candidates for inexpensive drug and protein encapsulation, continuous flow methods have been created for the manufacture and functionalisation of liposomes [36, 37].

In drug delivery systems, cellulose and its derivatives play a part in changing drug solubility and managing drug release. Liposomes are versatile drug delivery vehicles with various types and loading methods, offering advantages in drug stability and distribution, although they face challenges related to the body's immune system. Researchers are continuously working on improving liposome synthesis and functionality for cost-effective drug encapsulation [38]. Other NPs utilised in drug delivery systems include alginate, xanthan gum, polymeric micelles, dendrimers, inorganic and organic NPs, nanocrystals, QDs, and protein and polysaccharides NPs [13].

24.4 NANOTECHNOLOGICAL APPLICATIONS IN THERANOSTICS

Over the past decade, we have witnessed remarkable strides in the field of treatments and diagnostics due to advances in medicine and pharmaceuticals. One exciting development is the emergence of theranostics, a therapeutic approach that combines both therapy and diagnostic strategies within a single nanostructure. This nanostructure, which often takes the shape of an NP, combines therapeutic (such as pharmaceutical) and diagnostic/imaging components.

Theranostic drug delivery systems are receiving more attention in the scientific community, especially for the treatment of conditions like cancer and inflammation. Research and development in nanotheranostics have been accelerated by recent advances in nanodelivery devices. The exact targeting of areas, the ability to react to variables like temperature, and the distribution of therapeutic combinations are just a few benefits that NPs can provide [44].

Different imaging and diagnostic agents are used in nanocarriers in theragnostic. These substances, which enable optical imaging, magnetic resonance imaging (MRI), nuclear imaging, and computed tomography (CT), include QDs, fluorescent dyes, and superparamagnetic iron oxides. This strategy offers numerous opportunities for fusing medicinal and imaging components for better disease management [39, 40, 44, 45].

Nanomaterials are incredibly versatile and valuable in the field of nano-medicine. We can categorise these materials based on their dimensions, like 0D, 1D, 2D, and 3D. It is fascinating that they can be shaped, stretched into wires, or flattened into sheets. QDs (0D), nanorods, nanowires (1D), nanosheets (2D), and nanocubes (3D) are a few examples. These NPs can be produced in a variety of ways, including bottom-up construction and top-down dismantling of bigger materials [46–53].

24.4.1 0D Nanomaterials in Theranostics

These semiconductor-based NPs glow and produce fluorescence when excited by photons, with the quantised energy levels determined by the particle's size [39]. CdTe, CdSe, InAs, and InP are a few QD types [40]. They stand out for their diminutive size, high photostability, and the power to customise their optical characteristics [41]. Protein-QD nanohybrid hybrid QDs are one type of hybrid QD that is used to increase the stability of QDs in biological fluids [42].

Another family of nanomaterials with optical characteristics, biocompatibility, and excellent photostability is carbon-based QDs, or C-dots. They are made from carbonaceous materials and are used for medication delivery and visualising biological components [43]. C-dots are carbon based, in contrast to QDs, which are made of metal, and they have slightly different photoluminescence characteristics.

24.4.2 1D Nanomaterials in Theranostics

These materials possess unique physicochemical properties due to their restricted electron confinement in two dimensions. Nanorods can be synthesised using various methods, including polyol and surfactant-assisted approaches. In their synthesis, common surfactants like CTAB and CTAC are employed [54]. For in vivo imaging applications, recent advances substitute phospholipids like DOPC and DSPE for CTAB [55]. These nanorods are stable in biological fluids and exhibit high biocompatibility. When exposed to near-infrared (NIR) light, they generate heat, which can be harnessed for killing cancer cells, opening up new possibilities in cancer nanomedicine. The principle behind this is the oscillation of valence electrons upon photon absorption, leading to heat generation.

Nanotubes are another valuable tool in theranostics. Common methods for their synthesis include CVD, laser ablation, and arc discharge, with CVD being a cost-effective and controllable approach. CNTs, particularly single-walled CNTs (SWCNTs), stand out for their high photothermal capabilities and NIR absorption. SWCNTs can selectively destroy cancer cells during aggregation, sparing normal tissues [56, 57].

Other nanotubes, like hybrid nanotubes (HNTs) coated with porphyrin-covered silica, find applications in tracking macrophages in vivo. Magnetic nanotubes, synthesised through hydrothermal methods and stabilised with polyvinylpyrrolidone (PVP), are used for hyperthermia in cancer theranostics [58]. These 1D nanomaterials offer passive targeting in cancer therapy, allowing materials to circulate for extended periods until they reach their intended targets. 1D nanomaterials hold great promise in the field of theranostics, with nanorods and nanotubes paving the way for innovative diagnostic and therapeutic applications.

24.4.3 2D Nanomaterials in Theranostics

In the world of theranostics, 2D nanomaterials like nanosheets and nano-plates are making waves. These materials have unique electron confinement properties, limiting electrons to just one direction. Take graphene, for instance. It is a 2D carbon structure that can exist as a single layer or in multiple layers [59]. The surface area grows as the number of stacked layers decreases. Sp^2 hybridisation, a high Young's modulus, and both thermal and electrical conductivity are among graphene's outstanding attributes. Both bottom-up (self-assembly or CVD) and top-down (exfoliation) approaches can be used to create it [60]. The biological effects of graphene on cellular absorption, interactions with cells, transport across the blood-brain barrier, and renal clearance show how versatile this material is.

One exciting development involves multifunctional NPs that serve both diagnostic and therapeutic purposes in disease monitoring. By modifying graphene surfaces, researchers enhance their reactivity. Acid treatment can make them more reactive, while hydrophobic interactions help encapsulate drugs with high loading capacity. These nanosheets are a promising tool in combating multidrug-resistant cancers [61].

Graphene oxide (GO) and reduced graphene oxide (rGO) are also stars in biomedicine, especially in theranostics. The key difference between GO and rGO lies in their surface oxygen content and conductivity. rGO, with its high conductivity, finds applications as an electrode material for supercapacitors. PEGylated-GO, for instance, is being used to load tumour-targeting peptides and anticancer drugs for targeted cancer therapy. The interaction between the anticancer drug doxorubicin (DOX) and GO, based on π–π interaction and weak hydrogen bonding, helps deliver DOX to cells effectively [62, 63]. Functional rGO is also employed for gene therapy due to its excellent photo-absorption properties, particularly in the NIR region. It boasts high photothermal capabilities compared to GO-based nanomaterials [64, 65]. Other nanomaterials like MXenes and gold nanoplates are also making strides in theranostics. What sets 2D nanomaterials apart is their exceptional photothermal efficiency, a feature that sets them apart in the world of therapeutics and diagnostics.

24.4.4 3D Nanomaterials in Theranostics

Within the realm of theranostics, 3D nanomaterials, such as nanocubes and nanostars, have emerged as focal points of research and application. These materials are distinguished by their inability to confine electrons in any specific direction, presenting unique properties and applications.

For example, gold nanorods and silver nanocubes are becoming more common in photothermal applications. The polyol technique with sulfur-mediated reduction in the presence of PVP is used to synthesise silver nanocubes. They function as models for the production of gold nanocages, which are excellent for photothermal applications. Utilising the surface plasmon resonance (SPR) concept, NIR light is absorbed by gold nanocages when the outermost electrons fluctuate [65, 66]. This results in the generation of heat, a capability exploited for effectively targeting and eradicating cancer cells in theranostics.

Another notable nanomaterial is cobalt–iron oxide (Co_xFe_3-xO_4) nanocubes, synthesised through a thermal decomposition process involving cobalt, iron, decanoic acid, and other components [67]. These magnetic nanocubes find application in magnetic hyperthermia, a technique employed in cancer therapy to convert electromagnetic energy into heat.

Challenges in cancer theranostics include effectively treating deep tissue. Lanthanum hexabromide nanocubes (LaB_6-NCs) have recently emerged as a promising solution. These nanocubes exhibit a broad range of optical absorption across biological windows, including the NIR I & II windows, outperforming conventional dyes and organic photosensitisers. Additionally, they facilitate photodynamic therapy (PDT), addressing the issue of hypoxia within tumours. By utilising intracellular water as a source for generating free radicals, LaB6-NCs effectively treat tumours without exacerbating hypoxia [68].

Three-dimensional nanomaterials hold a distinctive advantage in theranostics due to their capacity to accommodate substantial drug payloads and generate significant heat for hyperthermia applications. They represent a promising frontier in the arena of biomedical sciences, offering innovative solutions for both diagnosis and therapy.

Theranostics, which integrates diagnostic and therapeutic functions using NPs, stands as a pivotal frontier in nanomedicine. Nanotechnology, with its array of NP dimensions and properties, offers innovative solutions. From 0D QDs serving as an alternative to conventional dyes to 1D gold nanorods and CNTs employed for imaging and photothermal therapy and 2D graphene nanosheets with encapsulation capabilities to 3D nanocubes and nanostars boasting potent photoluminescence and photothermal effects, nanomaterials are revolutionising cancer theranostics, ushering in new horizons within the realm of biomedical sciences.

24.5 NANOBIOSENSORS AND APPLICATIONS

Nanobiosensors represent a remarkable integration of NPs into biosensor fabrication, revolutionising bioanalytical applications in fields like bioimaging, diagnostics, drug delivery, and therapy. Among nanomaterials, NPs take the spotlight due to their diverse bioanalytical functionalities.

In the realm of nanobiosensors, amperometric tools gauge enzyme-based reactions, while fluorescent QD devices excel at measuring binding efficiency and immunolabeling, utilising conjugated NPs for analyzing biomolecular interactions. NPs are a promising material for biosensors because they have built-in optical qualities and can be combined with fluorescent markers. Carbon-based nanostructures like CNTs, rGO, graphene, and graphene sheets are being used in more and more electro/chemiluminescent assays, fluorescent-based assays, and biological field-effect transistor assays. The quenching properties of graphene are widely used to identify it in fluorescent-based tests [69, 70].

QDs, whose size and shape determine certain fluorescence characteristics; have several uses in sensors and imaging. For accurate identification, functionalising QDs with polymers like polyethylene glycol and polysaccharides is crucial. In comparison to conventional dyes, QDs have a number of benefits, including better molar extinction and quantum coefficient, a wider absorption range, smaller emission spectra, and photobleaching resistance [71]. They are widely employed in field-effect transistors and sandwich assays. Beyond these nanostructures, various distinct nanobased biosensors are emerging, serving diverse purposes. Noteworthy applications include developing glucose sensors for diabetes management, identifying HIV/AIDS antigens, monitoring microbial bioburden in urinary tract infections, and conducting diagnostic tests for cancer. Consequently, nanobiosensors hold immense potential in tissue engineering for both diagnostic and therapeutic purposes.

24.5.1 Graphene-Based Biosensors

In the field of nanotechnology, graphene is referred to as a "wonder material" due to its exceptional physicochemical properties. Graphene is the perfect material for transducing because of its properties, which include excellent conductivity, low charge carrier resistance, and a superb surface-to-volume ratio. Notably, the one-atomic-layer–thick, 2D substance known as graphene offers special benefits. It is important to note that graphene has the lowest resistance in the world, outperforming copper by a wide margin with a conductivity of 10^6 S/m and a resistivity of only 10^{-6} ohm-cm. [72, 73].

While various methods exist for synthesising graphene, large-scale production techniques are still under development, limiting its widespread utilisation in various applications. GO, derived from graphene through oxidation, can be further reduced to produce rGO [74]. For the physicochemical synthesis of graphene and its derivatives, a variety of techniques are available, such as chemical oxidation of graphite, scotch tape exfoliation, chemical vapour deposition, and more [75, 76].

In the domain of biosensors, graphene has demonstrated its immense potential. For instance, scientists have developed a very sensitive graphene optical biosensor (GOB) that can identify live cancer cells in response to the chemotherapy drug paclitaxel when the medicine is being administered. This innovative optical approach detects minute refractive index (RI) changes, enabling label-free sensing and dynamic cell monitoring without electromagnetic interference or surface contamination. The GOB achieved remarkable sensitivity, detecting ultra-small RI changes with a rapid response [77]. Researchers have successfully developed graphene-based biosensors for COVID-19 spike protein detection, achieving impressive limits of detection (LODs) and rapid response times [78]. These biosensors represent valuable tools for real-time disease detection.

24.5.2 Gold NP Biosensors

Due to their unique colourimetric features, which change based on their shape, size, and state of aggregation in the presence of analytes, gold NPs (GNPs) have become a promising platform in biosensing [79]. In a liquid medium, GNPs change colour visibly, usually from red to blue, as they go from a monodispersed state to an aggregated state. GNPs, which have a diameter ranging from 1 to 100 nm with a high surface-to-volume ratio and considerable surface energy, allow for quick direct electron transport between different electrodes and electroactive species. As a result, GNPs are used in biosensing, particularly in the creation of optical, electrochemical, and piezoelectric biosensors.

Changes in light or photon output provide the foundation for sensing and producing the right signals in optical biosensors. Due to their optical sensing capability and surface plasmon resonance behaviour, GNPs are used in optical biosensing. The amplification of SPR signals by GNPs can lead to fluctuations in the dielectric constant and luminescence from the metal–liquid interface by enhancing the local electromagnetic field [80]. These unique qualities of GNPs have been extensively utilised in the creation of biosensors for applications relating to food safety, particularly in the detection of various allergens and pollutants.

Particularly noteworthy is the use of GNPs in the development of micro-cantilever-based biosensors for the detection of protein biomarkers. High detection precision is achieved by creating a microcavity at the cantilever's free end to aid in local biological reactions. A remarkable detection limit of about 21 pg/mL, which is 70 times smaller than that attained without GNP amplification, has been reached in these biosensors thanks to amplification of the GNPs [81]. The research team led by Yuan used GNPs to identify DNA linked to allergies to sesame, soybean, and peanuts using colourimetric readouts. Long double-stranded DNA from the GNP surface is released by the allergen gene during the detection process, causing them to assemble and turn blue when NaCl is added. DNA probes bind to the GNPs if the gene sample falls below the cutoff level, avoiding aggregation and retaining the red hue. With a low LOD of 0.5 nM, this method effectively identified allergy gene targets [82].

24.5.3 CNTs in Biosensors

A special type of carbon known as CNTs is renowned for its excellent mechanical, electrical, electrocatalytic, and thermal capabilities. They are simply rolled-up graphene layers, and the number of walls they have influences their characteristics [83]. CNTs have drawn a lot of attention in the realm of biosensors because of their exceptional qualities and distinctive geometry. SWCNTs, multiwalled CNTs (MWCNTs), and double-walled CNTs (DWCNTs) are the three primary varieties of CNTs. MWCNTs can have diameters up to 100 nm, while SWCNTs have the smallest diameters, which are typically between 1 and 2 nm. Between SWCNTs, DWCNTs, and MWCNTs, DWCNTs have fewer flaws than MWCNTs while still having many of the beneficial characteristics of SWCNTs [84].

24.5.4 QD-Based Biosensors

NPs with a diameter of less than 10 nm are referred to as QDs, also referred to as fluorescent semiconductor nanocrystals. Their exceptional optical, electrical, and size-dependent luminous characteristics have made them a key instrument in cutting-edge biosensing technology [85]. These characteristics have been used by researchers to create extremely sensitive biosensors. The performance of biosensors can be improved by using QDs because of their excellent charge carrier transport capabilities and high surface-to-volume ratio. Due to its low toxicity, high solubility, chemical stability, and adaptability, carbon QDs have emerged as a leading QD in sensing research. Metal–organic frameworks (MOFs) and covalent organic frameworks (COFs) are two different kinds of nanomaterials that have attracted a lot of interest outside of the context of biosensors.

COFs are organic molecule–derived crystalline solids with nano/micro-porous structures joined by covalent bonds. They have a completely conjugated structure, a huge surface area, and an atomic-level thickness, among other benefits. As a result, COFs have found use in a number of industries, including electrochemical devices, energy storage, sensors, and electrocatalysis. They are useful for biosensing applications due to their exceptional structural qualities, chemical and physical robustness, strong covalent bonding, and high surface area because of porosity. Due to its ability to store energy and possessing optoelectronic qualities, COFs have been utilised in gas detection, storage, catalysis, and biosensing. The interactions between aromatic organic molecules are facilitated by the COFs' low inter-layer distance, which increases their usefulness in biosensing applications. [74, 86, 87].

By comparison, MOFs are distinguished by their fascinating structures, created by assembling metal clusters and multifunctional organic linkers. High porosity, a sizeable surface area, and adaptable and tunable qualities are just a few of their remarkable qualities. MOFs have found use in a variety of industries, including biosensors, medication delivery, gas storage, biotechnology, biomedicine, and biocatalysis. MOFs have become more popular in biosensing, especially in electrochemical sensing, because of their superior insulating qualities. The uses of MOFs in biosensing have considerably increased as a result of the incorporation of molecules including enzymes, antibodies, and aptamers. This has made it possible to detect proteins and DNA in a variety of settings, including biomedicine, food, industry, and the environment [88, 89].

Biosensors have reached new heights thanks to nanotechnology, which has improved sensor qualities like the LOD, accuracy, and dependability. Modern nanobiosensors are becoming more and more valuable because they can quickly identify target analytes, accomplish single-molecule detection, and improve transduction outputs. The release of NPs into the environment and quantum phenomena that result in high sensitivity, random noise, and background signals are drawbacks of this technique. Some analytes may exhibit cross-sensitivity, nonlinearity, and unpredictability in response to nanobiosensor exposure. Additionally, mass production of promising nanomaterials like graphene remains a challenge that needs to address for the widespread adoption of nanotechnology in biosensors.

Machine learning (ML) and other cutting-edge technologies have been used to solve these problems. For data analysis in biosensing applications, ML methods such as support vector machines, random forests, artificial neural networks, convolutional neural networks, naive Bayes, and *k*-nearest neighbours are utilised. ML algorithms are excellent at identifying patterns and can assist nanobiosensors in deriving useful information from unstructured data. By removing unimportant information from sensor outputs, they can classify raw sensor data; lessen cross-sensitivity, and lower misclassification, which ultimately lowers detection limits.

Furthermore, the use of biosensors in biomarker detection has promising implications for personalised therapeutics, diagnostic tools, and medical devices. Surgically implanted biosensors can provide real-time monitoring of treatment outcomes, allowing for better assessment of drug efficacy before clinical testing. The use of biosensor chip technology to identify DNA changes in blood paves the way for the early diagnosis of diseases. Wearable biosensors, integrated into devices like smart bracelets, can non-invasively monitor various biological samples, enabling continuous health monitoring and early disease detection.

In the medical profession, biosensors have several uses that are advantageous to patients and physicians. These uses include illness prevention, clinical diagnosis, access to health information, and therapy evaluation. Their powers have been further increased by the recent use of nanomaterials in the production of biosensors. Biosensors are essential for classifying individuals for individualised treatment, and revolutionising healthcare procedures. The construction of stimuli-responsive materials in highly accurate biosensors using LC (liquid crystal) interface principles has opened up a wide range of therapeutic and diagnostic possibilities because of this paradigm shift. The success of this sector heavily depends on the choice of target molecules, immobilisation techniques, and enzymes. Utilising smart nanostructure-based techniques can enhance the performance of liquid crystal (LC) biosensors, including achieving low detection limits and high sensitivity. These sensors are cost-effective, easily fabricated, require minimal power, and deliver consistent and repeatable results across different batches. Overall, nano-enabled LC-based biosensors hold significant potential for the recognition of various biological substances, making them valuable tools [69].

But there are also technological difficulties, like prolonging the life of biosensors. Biosensors play a crucial role in various fields, including biotechnology, drug discovery, pathogen detection, gene therapy, and more. Their applications in health care and medicine are expanding rapidly, particularly in health management, patient surveillance, diagnosis, and disease detection.

24.6 REGULATORY CONSIDERATIONS FOR NANOTECHNOLOGY-BASED MEDICAL DEVICES

The regulatory landscape for nanotechnology-based medical products is continually evolving. Understanding the future regulatory trends, applications, and challenges is essential for the responsible development, evaluation, and approval of these innovative technologies.

- i Future Regulatory Trends for Nanotechnology-Based Medical Devices: The US Food and Drug Administration (FDA) currently employs a case-by-case approach for evaluating these products. However, challenges arise in adapting traditional regulatory frameworks to novel nanomedicine products. This section discusses the evolving regulatory landscape, adequacy of current regulations, and novel risks associated with nanotechnology.
- ii Challenges in the Regulatory Classification of Nanomaterials: Classifying nanomaterials for regulatory purposes presents several challenges, including assessing the impact of physicochemical properties, ensuring reliable safety data, and harmonising regulations across regions. These challenges highlight the need for regulatory agility and improved guidance [55].
- iii Challenges in the Safety Evaluation of Nanomedicines: Nanomedicines pose specific safety evaluation challenges due to their unique properties and interactions with biological systems. These challenges include the lagging safety evaluation methods, the particulate nature of nanomedicines, and the need for specific toxicological assessments.
- iv Challenges in Regulating Nanomedicines: Regulatory challenges in nanomedicines encompass classification

and categorisation, a lack of specific guidance, safety assessment, environmental affect manufacturing and quality control, and nanotoxicology. These challenges influence the development, evaluation, and regulation of nanomedicine products.

v Future Regulatory Trends and Applications in Nanomedicines: While not explicitly discussed, the document suggests that future regulatory trends may involve improving guidance and standardisation for nanomedicine products. It also hints at the potential applications in gene therapy, targeted drug delivery, and precision medicine, requiring further regulatory considerations and adaptations.

vi Future Regulatory Trends and Applications for Nanotechnology-Based In Vitro Diagnostics (nanoIVDs): Although not explicitly addressed, the document implies that regulatory trends for nanoIVDs may involve the development of specific evaluation guidelines and technical regulations for nanomaterials [54]. The increasing experience in regulating these innovative diagnostics is expected to lead to improved regulatory guidance.

The evolving regulatory landscape for nanotechnology-based medical products will require ongoing assessments of the regulatory framework, addressing novel risks, integrating scientific developments, and ensuring safety and efficacy. Challenges in classification, safety evaluation, and regulation must be addressed to facilitate the responsible development and use of nanomedicines (Tables 24.1 and 24.2, and Figure 24.2).

24.6.1 The Promising Future of Health Care with Nanotechnology

Through advanced medical devices, nanotechnology has ushered in a new era of health care innovation, revolutionising diagnostics, treatment, and patient care. With novel opportunities for the diagnosis and treatment of numerous illnesses, nanotechnology has been identified as a promising tool in the field of paediatric medicine [90]. From enhancing medical imaging for precise diagnostics to enabling targeted drug delivery and real-time monitoring through implantable sensors, nanotechnology has reshaped the landscape of healthcare [51]. The fusion of diagnostics and therapeutics in theranostics holds immense promise for personalised medicine [52].

However, to fully realise these advancements, it is essential to address challenges related to nanotoxicology, biocompatibility, and regulatory considerations to ensure the safety and efficacy of nanotechnology-based medical devices. As we embrace this transformative journey, the future of health care with nanotechnology shines brightly, offering boundless possibilities for improved patient outcomes and a healthier world [91].

TABLE 24.1
Comparative Analysis of Medical Device Regulations by Country

Country	Regulatory Authority	Key Requirements	Device Classifications
United States	FDA (Food and Drug Administration)	—**510(k) Clearance** or **Premarket Approval (PMA)** based on device risk level. —**Quality System Regulation (QSR)** compliance (21 CFR Part 820). —**Labelling and Reporting** of adverse events.	—**Class I**: Low risk (e.g., surgical instruments). —**Class II**: Moderate risk (special controls). —**Class III**: High risk (e.g., pacemakers).
European Union	EMA and Conformité Européenne (CE) Marking	—**CE Marking** under EU directives (MDR/IVDR). —**Clinical Evaluation** per MEDDEV. —**Notified Body** involvement. —**Unique Device Identification (UDI)** compliance.	—**Class I, IIa, IIb, III**: Based on risk (from low to high).
Canada	Health Canada	—**Medical Device License (MDL)** required. —**Quality System** compliance with **ISO 13485**.	—**Class I, II, III, IV**: Increasing risk levels, similar to EU.
Japan	Pharmaceuticals and Medical Devices Agency (PMDA)	—**Pre-market Certification**for Class II and III. —Compliance with **Pharmaceutical Affairs Law (PAL)**. —**Good Manufacturing Practice (GMP)** adherence.	—**Class I, II, III**: With subcategories based on risk.
Australia	Therapeutic Goods Administration (TGA)	—**Conformity Assessment** under Essential Principles. —Listing on the **Australian Register of Therapeutic Goods (ARTG)**. —**Post-market Surveillance** for adverse event monitoring.	—**Class I, IIa, IIb, III**: Risk-based classification similar to EU and Canada.
India	Central Drugs Standard Control Organization (CDSCO)	—**Registration and Licensing** required for manufacturers/ importers. —Compliance with **Medical Device Rules (MDR)**. —Conformance with **BIS Standards** for safety and quality.	—Devices classified based on risk: **Low**, **Moderate**, and **High** risk categories.

TABLE 24.2
Comparison of Nanotechnology-Based and Traditional Medical Devices—Regulatory Challenges

Regulatory Challenge	Nanotechnology-Based Medical Devices	Traditional Medical Devices
Unique Materials	Nanomaterials introduce novel materials with distinct properties, raising concerns about biocompatibility, toxicity, and long-term effects.	Use well-characterised materials with known biocompatibility profiles.
Size Considerations	Operate at the nanoscale, posing challenges in manufacturing, characterisation, and quality control.	Operate at larger scales, which are easier to handle and standardise.
Complex Interactions	May exhibit complex interactions at the cellular or molecular level, making safety and efficacy assessment more intricate.	Typically have simpler mechanisms of action, facilitating safety evaluations.
Lack of Regulations	The regulatory landscape is still evolving, with gaps in guidelines and standards.	Benefit from well-established regulatory frameworks with comprehensive guidance.
Uncertainty in Testing	Require adapted or novel testing methods due to unique properties, leading to uncertainty in evaluation.	Testing methods are well-defined and standardised.
Multidisciplinary Expertise	Evaluation demands specialised expertise in nanotechnology, which regulatory bodies may not fully possess.	Fall within the domain of expertise of existing regulatory agencies.
Global Harmonisation	Harmonising regulations is challenging due to diverse national approaches and lack of unified frameworks.	Global harmonisation is relatively more established and consistent.
Longer Approval Times	Regulatory approval can be prolonged due to the need for novel assessments and lack of precedent.	Approval pathways are well-defined, leading to shorter timelines.
Patient Safety Concerns	Long-term safety and potential side effects are less understood, raising concerns.	Safety risks are often better understood based on historical data and long-term usage.

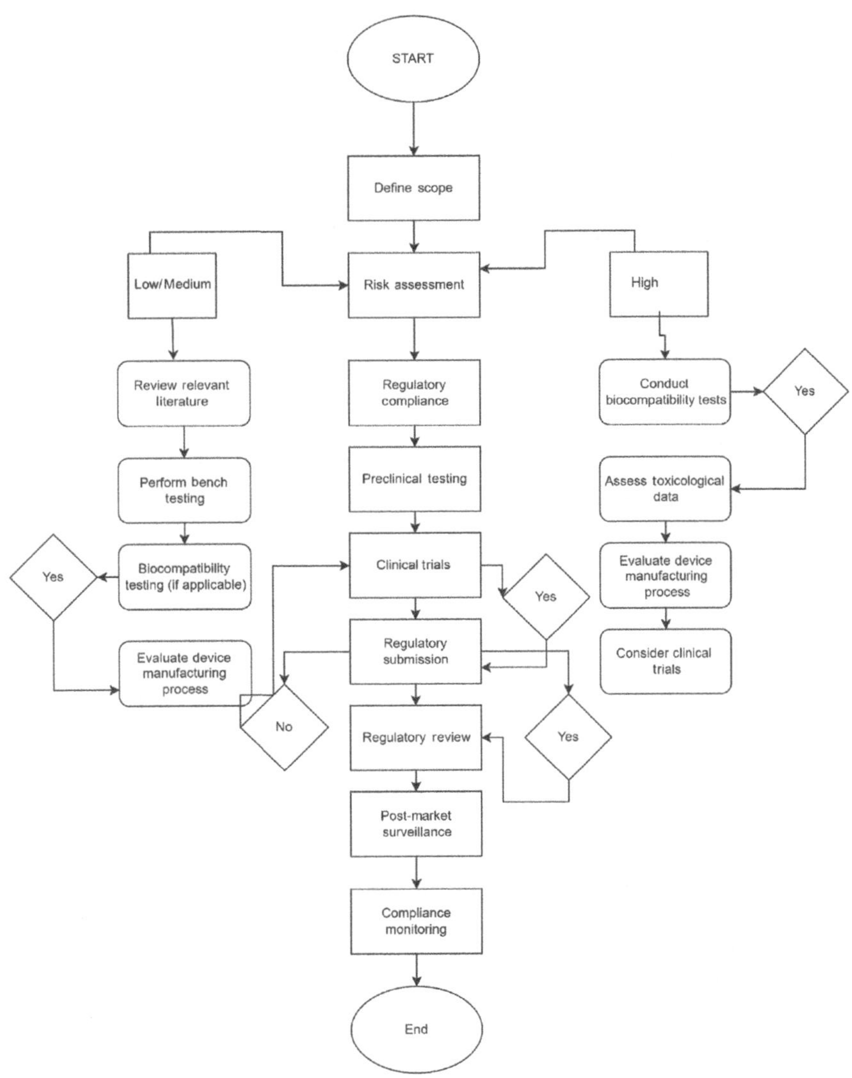

FIGURE 24.2 Flow chart for safety assessment and risk management.

REFERENCES

1. Hobson, D.W. The commercialization of medical nanotechnology for medical applications. In *Intracellular Delivery III* (pp. 405–449). Springer, Berlin/Heidelberg, Germany; 2016.
2. Sindhwani, S.; Chan, W.C. Nanotechnology for modern medicine: Next step towards clinical translation. *J. Intern. Med.* 2021, 290, 496–498.
3. Laouini, S.; Bouafia, A.; Tedjani, M. Catalytic activity for dye degradation and characterization of silver/silver oxide nanoparticles green synthesized by aqueous leaves extract of Phoenix Dactylifera L. *Res. Sq.* 2021. https://doi.org/10.21203/rs.3.rs-139856/v1
4. Agarwal, H.; Kumar, S.V.; Rajeshkumar, S. A review on green synthesis of zinc oxide nanoparticles—An eco-friendly approach. *Resour. Effic. Technol.* 2017, 3, 406–413.
5. Anjum, S.; Ishaque, S.; Fatima, H.; Farooq, W.; Hano, C.; Abbasi, B.H.; Anjum, I. Emerging applications of nanotechnology in healthcare systems: Grand challenges and perspectives. *Pharmaceuticals* 2021, 14, 707.
6. Zhou, H.-Y.; Hao, J.-L.; Wang, S.; Zheng, Y.; Zhang, W.-S. Nanoparticles in the ocular drug delivery. *Int. J. Ophthalmol.* 2013, 6, 390–396.
7. Niu, X.; Chen, J.; Gao, J. Nanocarriers as a powerful vehicle to overcome blood-brain barrier in treating neurodegenerative diseases: Focus on recent advances. *Asian J. Pharm. Sci.* 2019, 14, 480–496.
8. Ferrari, M. Cancer nanotechnology: Opportunities and challenges. *Nat. Rev. Cancer* 2005, 5, 161–171.
9. Liu, Y.; Zhu, F.; Dan, W.; Fu, Y.; Liu, S. Construction of carbon nanotube based nanoarchitectures for selective impedimetric detection of cancer cells in whole blood. *Analyst* 2014, 139, 5086–5092.
10. Liu, K.; Tedeschi, A.; Park, K.K.; He, Z. Neuronal intrinsic mechanisms of axon regeneration. *Annu. Rev. Neurosci.* 2011, 34, 131–152.
11. Klein, P.M.; Wagner, E. Bioreducible polycations as shuttles for therapeutic nucleic acid and protein transfection. *Antioxid. Redox Signal.* 2014, 21, 804–817.
12. Chavda, V.P.; Patel, M.; Patel, R.; Thakkar, K.; Patel, D.; Balar, P.C.; Patel, P.; Desai, U. Nanotechnology advancements in HMI. In *Human-Machine Inter-Face Technology Advancements and Applications* (pp. 111–142). CRC Press; 2024.
13. Patra, J.K. et al. Nano based drug delivery systems: Recent developments and future prospects. *J. Nanobiotechnol.* 2018, 16, 71.
14. Mirza, A.Z.; Siddiqui, F.A. Nanomedicine and drug delivery: A mini review. *Int. Nano Lett.* 2014, 4, 94.
15. Lu, H.; Wang, J.; Wang, T.; Zhong, J.; Bao, Y.; Hao, H. Recent progress on nano-structures for drug delivery applications. *J. Nanomater.* 2016, 2016, 20.
16. Kumari, A.; Kumar, V.; Yadav, S. Nanotechnology: A tool to enhance therapeutic values of natural plant products. *Trends Med. Res.* 2012, 7, 34–42.
17. Mignani, S.; El Kazzouli, S.; Bousmina, M.; Majoral, J.P. Expand classical drug administration ways by emerging routes using dendrimer drug delivery systems: A concise overview. *Adv. Drug Deliv. Rev.* 2013, 65, 1316–1330.
18. Lounnas, V.; Ritschel, T.; Kelder, J.; McGuire, R.; Bywater, R.P.; Foloppe, N. Current progress in structure-based rational drug design marks a new mindset in drug discovery. *Comput. Struc. Biotechnol. J.* 2013, 5, e201302011.
19. Mavromoustakos, T.; Durdagi, S.; Koukoulitsa, C.; Simcic, M.; Papadopoulos, M.; Hodoscek, M.; Golic Grdadolnik, S. Strategies in the rational drug design. *Curr. Med. Chem.* 2011, 18, 2517–2530.
20. Prachayasittikul, V.; Worachartcheewan, A.; Shoombuatong, W.; Songtawee, N.; Simeon, S.; Prachayasittikul, V.; Nantasenamat, C. Computeraided drug design of bioactive natural products. *Curr. Top. Med. Chem.* 2015, 15, 1780–1800.
21. Chen, G.; Roy, I.; Yang, C.; Prasad, P.N. Nanochemistry and nanomedicine for nanoparticle-based diagnostics and therapy. *Chem. Rev.* 2016, 116, 2826–2885.
22. Mattos, B.D.; Rojas, O.J.; Magalhaes, W.L.E. Biogenic silica nanoparticles loaded with neem bark extract as green, slow-release biocide. *J. Clean. Prod.* 2017, 142, 4206–4213.
23. Kinnear, C.; Moore, T.L.; Rodriguez-Lorenzo, L.; Rothen-Rutishauser, B.; PetriFink, A. Form follows function: Nanoparticle shape and its implications for nanomedicine. *Chem. Rev.* 2017, 117, 11476–11521.
24. Mattos, B.D.; Tardy, B.L.; Magalhaes, W.L.E.; Rojas, O.J. Controlled release for crop and wood protection: Recent progress toward sustainable and safe nanostructured biocidal systems. *J. Control. Release* 2017, 262, 139–150.
25. Siepmann, F.; Herrmann, S.; Winter, G.; Siepmann, J. A novel mathematical model quantifying drug release from lipid implants. *J. Control. Release* 2008, 128, 233–240.
26. Almalik, A.; Benabdelkamel, H.; Masood, A.; Alanazi, I.O.; Alradwan, I.; Majrashi, M.A.; Alfadda, A.A.; Alghamdi, W.M.; Alrabiah, H.; Tirelli, N.; Alhasan, A.H. Hyaluronic acid coated chitosan nanoparticles reduced the immunogenicity of the formed protein corona. *Sci. Rep.* 2017, 7, 10542.
27. Martens, T.F., Remaut, K., Deschout, H., Engbersen, J.F.J., Hennink, W.E., van Steenbergen, M.J., Demeester, J., De Smedt, S.C., Braeckmans, K. Coating nanocarriers with hyaluronic acid facilitates intravitreal drug delivery for retinal gene therapy. *J. Control. Release* 2015, 202, 83–92.
28. Kolhar, P.; Anselmo, A.C.; Gupta, V.; Pant, K.; Prabhakarpandian, B.; Ruoslahti, E.; Mitragotri, S. Using shape efects to target antibody-coated nanoparticles to lung and brain endothelium. *Proc. Natl. Acad. Sci. USA* 2013, 110, 10753–10758.
29. Gao, W.W.; Zhang, L.F. Coating nanoparticles with cell membranes for targeted drug delivery. *J. Drug Target.* 2015, 23, 619–626.
30. Muller, J.; Bauer, K.N.; Prozeller, D.; Simon, J.; Mailander, V.; Wurm, F.R.; Winzen, S.; Landfester, K. Coating nanoparticles with tunable surfactants facilitates control over the protein corona. *Biomaterials* 2017, 115, 1–8.
31. Gao, H.; Yang, Z.; Zhang, S.; Cao, S.; Shen, S.; Pang, Z.; Jiang, X. Ligand modifed nanoparticles increases cell uptake, alters endocytosis and elevates glioma distribution and internalization. *Sci. Rep.* 2013, 3, 2534.
32. Pistone, S.; Goycoolea, F.M., Young, A.; Smistad, G., Hiorth, M. Formulation of polysaccharide-based nanoparticles for local administration into the oral cavity. *Eur. J. Pharm. Sci.* 2017, 96, 381–389.
33. Liu, S.; Yang, S.; Ho, P.C. Intranasal administration of carbamazepine-loaded carboxymethyl chitosan nanoparticles for drug delivery to the brain. *Asian J. Pharm. Sci.* 2018, 13, 72–81.
34. Jain, A.; Jain, S.K. Optimization of chitosan nanoparticles for colon tumors using experimental design methodology. *Arti.f Cells Nanomed. Biotechnol.* 2016, 44, 1917–1926.

35. Bozzuto, G.; Molinari, A. Liposomes as nanomedical devices. *Int. J. Nanomedicine* 2015, 10, 975.
36. Sercombe, L.; Veerati, T.; Moheimani, F.; Wu, S.Y., Sood, A.K.; Hua, S. Advances and challenges of liposome assisted drug delivery. *Front. Pharmacol.* 2015, 6, 286.
37. Akbarzadeh, A.; Rezaei-Sadabady, R.; Davaran, S.; Joo, S.W.; Zarghami, N.; Hanifehpour, Y.; Samiei, M.; Kouhi, M.; Nejati-Koshki, K. Liposome: Classification, preparation, and applications. *Nanoscale Res. Lett.* 2013, 8, 102.
38. Sun, B.; Zhang, M.; Shen, J.; He, Z.; Fatehi, P.; Ni, Y. Applications of cellulosebased materials in sustained drug delivery systems. *Curr. Med. Chem.* 2017, 26, 2485–2501.
39. Bohn, B.J.; Simon, T.; Gramlich, M.; Richter, A.F.; Polavarapu, L.; Urban, A.S.; Feldmann, J. dephasing and quantum beating of excitons in methylammonium lead iodide perovskite nanoplatelets. *ACS Photonics* 2018, 5, 648–654.
40. Xie, R.; Chen, K.; Chen, X.; Peng, X. InAs/InP/ZnSe core/shell/shell quantum dots as near-infrared emitters: Bright, narrow-band, non-cadmium containing, and biocompatible. *Nano Res.* 2008, 1, 457–464.
41. Haleem, A.; Javaid, M.; Singh, R.P.; Rab, S.; Suman, R. Applications of nano-technology in medical field: A brief review. *Global Health Journal* 2023, 7(2), 70–77.
42. Anjum, S.; Ishaque, S.; Fatima, H.; Farooq, W.; Hano, C.; Abbasi, B.H.; Anjum, I. Emerging applications of nanotechnology in healthcare systems: Grand challenges and perspectives. *Pharmaceuticals (Basel)* 2021, 14(8), 707. https://doi.org/10.3390/ph14080707
43. Saha, M. Nanomedicine: Promising tiny machine for the healthcare in future: A review. *Oman Med. J.* 2009, 24(4), 242–247. https://doi.org/10.5001/omj.2009.50
44. Paramasivam, G.; Palem, V.V.; Sundaram, T.; Sundaram, V.; Kishore, S.C.; Bellucci, S. Nanomaterials: Synthesis and applications in theranostics. *Nanomaterials* 2021, 11, 3228. https://doi.org/10.3390/nano11123228
45. McHugh, K.J.; Jing, L.; Behrens, A.M.; Jayawardena, S.; Tang, W.; Gao, M.; Langer, R.; Jaklenec, A. Biocompatible semiconductor quantum dots as cancer imaging agents. *Adv. Mater.* 2018, 30, 1706356.
46. Tang, S.-Y.; Yang, C.-C.; Su, T.-Y.; Yang, T.-Y.; Wu, S.-C.; Hsu, Y.-C.; Chen, Y.-Z.; Lin, T.-N.; Shen, J.-L.; Lin, H.-N. Design of core–shell quantum dots– 3D WS2 nanowall hybrid nanostructures with high-performance bifunctional sensing applications. *ACS Nano* 2020, 14, 12668–12678.
47. Farshbaf, M.; Davaran, S.; Rahimi, F.; Annabi, N.; Salehi, R.; Akbarzadeh, A. Carbon quantum dots: Recent progresses on synthesis, surface modification and applications. *Artif. Cells Nanomed. Biotechnol.* 2018, 46, 1331–1348.
48. Wu, H.; Su, W.; Xu, H.; Zhang, Y.; Li, Y.; Li, X.; Fan, L. Applications of carbon dots on tumour theranostics. *View* 2021, 2, 20200061.
49. Lee, G.-J.; Shin, S.-I.; Kim, Y.-C.; Oh, S.-G. Preparation of silver nanorods through the control of temperature and pH of reaction medium. *Mater. Chem. Phys.* 2004, 84, 197–204.
50. Roach, L.; Booth, M.E.; Ingram, N.; Paterson, D.A.; Batchelor, D.V.; Moorcroft, S.C.; Bushby, R.J.; Critchley, K.; Coletta, P.L.; Evans, S.D. Evaluating phospholipid-functionalized gold nanorods for in vivo applications. *Small* 2021, 17, 2006797.
51. Rao, R.; Pint, C.L.; Islam, A.E.; Weatherup, R.S.; Hofmann, S.; Meshot, E.R.; Wu, F.; Zhou, C.; Dee, N.; Amama, P.B. Carbon nanotubes and related nano-materials: Critical advances and challenges for synthesis toward mainstream commercial applications. *ACS Nano* 2018, 12, 11756–11784.
52. Prajapati, S.K.; Malaiya, A.; Kesharwani, P.; Soni, D.; Jain, A. Biomedical applications and toxicities of carbon nanotubes. *Drug Chem. Toxicol.* 2020, 43, 1–16.
53. Hayashi, K.; Nakamura, M.; Miki, H.; Ozaki, S.; Abe, M.; Matsumoto, T.; Ishimura, K. Near-infrared fluorescent silica/porphyrin hybrid nanorings for in vivo cancer imaging. *Adv. Funct. Mater.* 2012, 22, 3539–3546.
54. Chen, K.; Han, H.; Tuguntaev, R.G.; Wang, P.; Guo, W.; Huang, J.; Gong, X.; Liang, X.-J. Applications and regulatory of nanotechnology-based innovative in vitro diagnostics. *View* 2020, 2(2), 20200091.
55. Foulkes, R.; Man, E.; Thind, J.; Yeung, S.; Joy, A.; Hoskins, C. The regulation of nanomaterials and nanomedicines for clinical application: Current and future perspectives. *Biomater. Sci.* 2020, 8, 4653–4664.
56. Paradise, J. Regulating nanomedicine at the food and drug administration. *AMA J. Ethics* 2019, 21(4), E347–E355.
57. van der Zee, M.; de Vries, C.; Masa, M. et al. Regulatory aspects of a nano-material for imaging therapeutic cells. *Drug Deliv. Transl. Res.* 2023, 13, 2693–2703.
58. De Jong, W.H.; Geertsma, R.E.; Borchard, G. Regulatory safety evaluation of nanomedical products: Key issues to refine. *Drug Deliv. Transl. Res.* 2022, 12(9), 2042–2047. https://doi.org/10.1007/s13346-022-01208-4
59. Lee, X.J.; Lim, H.N.; Gowthaman, N.; Rahman, M.B.A.; Abdullah, C.A.C.; Muthoosamy, K. In-situ surface functionalization of superparamagnetic reduced graphene oxide–Fe3O4 nanocomposite via Ganoderma lucidum extract for targeted cancer therapy application. *Appl. Surf. Sci.* 2020, 512, 145738.
60. Darrigues, E.; Nima, Z.A.; Griffin, R.J.; Anderson, J.M.; Biris, A.S.; Rodriguez, A. 3D cultures for modeling nanomaterial-based photothermal therapy. *Nanoscale Horiz.* 2020, 5, 400–430.
61. Knights, O.B.; McLaughlan, J.R. Gold nanorods for light-based lung cancer theranostics. *Int. J. Mol. Sci.* 2018, 19, 3318.
62. Lee, H.; Choi, M.; Lim, J.; Jo, M.; Han, J.-Y.; Kim, T.M.; Cho, Y. Magnetic nanowire networks for dual-isolation and detection of tumor-associated circulating biomarkers. *Theranostics* 2018, 8, 505.
63. Su, Y.; Wei, X.; Peng, F.; Zhong, Y.; Lu, Y.; Su, S.; Xu, T.; Lee, S.-T.; He, Y. Gold nanoparticles-decorated silicon nanowires as highly efficient near-infrared hyperthermia agents for cancer cells destruction. *Nano Lett.* 2012, 12, 1845–1850.
64. Wang, Y.-S.; Shao, D.; Zhang, L.; Zhang, X.-L.; Li, J.; Feng, J.; Xia, H.; Huo, Q.-S.; Dong, W.-F.; Sun, H.-B. Gold nanorods-silica Janus nanoparticles for theranostics. *Appl. Phys. Lett.* 2015, 106, 173705.
65. Mahmoud, M.; El-Sayed, M. Metallic double shell hollow nanocages: The challenges of their synthetic techniques. *Langmuir* 2012, 28, 4051–4059.
66. De Sio, L.; Placido, T.; Comparelli, R.; Curri, M.L.; Striccoli, M.; Tabiryan, N.; Bunning, T.J. Next-generation thermo-plasmonictechnologies and plasmonic nanoparticles in optoelectronics. *Prog. Quantum Electron.* 2015, 41, 23–70.
67. Yang, W.; Yu, Y.; Wang, L.; Yang, C.; Li, H. Controlled synthesis and assembly into anisotropic arrays of magnetic

cobalt-substituted magnetite nanocubes. *Nanoscale* 2015, 7, 2877–2882.
68. Kuthala, N.; Vankayala, R.; Chiang, C.S.; Hwang, K.C. Unprecedented theranostic LaB6 nanocubes-mediated NIR-IIb photodynamic therapy to conquer hypoxia-induced chemoresistance. *Adv. Funct. Mater.* 2020, 30, 2002940.
69. Ramesh, M.; Janani, R.; Deepa, C.; Rajeshkumar, L. Nanotechnology-enabled biosensors: A review of fundamentals, design principles, materials, and applications. *Biosensors* 2023, 13, 40. https://doi.org/10.3390/bios13010040
70. Tran, M.V.; Le, P.M.L. Nanoflake manganese oxide and nickel-manganese oxide synthesized by electrode position for electrochemical capacitor. *J. Nanomater.* 2015, 2015, 230.
71. Kargozar, S.; Hoseini, S.J.; Milan, P.B.; Hooshmand, S.; Kim, H.W.; Mozafari, M. Quantum dots: A review from concept to clinic. *Biotechnol. J.* 2020, 15, 2000117.
72. Sengupta, J.; Hussain, C.M. Graphene-based field-effect transistor biosensors for the rapid detection and analysis of viruses: A perspective in view of COVID-19. *Carbon Trends* 2021, 2, 100011.
73. Afsahi, S.; Lerner, M.B.; Goldstein, J.M.; Lee, J.; Tang, X.; Bagarozzi, D.A.; Pan, D.; Locascio, L.; Walker, A.; Barron, F.; et al. Novel graphene-based biosensor for early detection of Zika virus infection. *Biosens. Bioelectron.* 2018, 100, 85–88.
74. Singh, S.; Arshid, N.; Cinti, S. Electrochemical nano biosensors for the detection of extracellular vesicles exosomes: From the benchtop to everywhere? *Biosens. Bioelectron.* 2022, 216, 114635.
75. Hong, Y.; Wang, Z.; Jin, X. Sulfuric acid intercalated graphite oxide for graphene preparation. *Sci. Rep.* 2013, 3, 3439.
76. Coleman, J.N. Liquid exfoliation of defect-free graphene. *Acc. Chem. Res.* 2013, 46, 14–22.
77. Wang, Y.; Zhang, S.; Xu, T.; Zhang, T.; Mo, Y.; Liu, J.; Yan, L.; Xing, F. Ultrasensitive and ultra-fast detection of whole unlabeled living cancer cell responses to paclitaxel with a graphene-based biosensor. *Sens. Actuators B Chem.* 2018, 263, 417–425.
78. Seo, G.; Lee, G.; Kim, M.J.; Baek, S.H.; Choi, M.; Ku, K.B.; Lee, C.S.; Jun, S.; Park, D.; Kim, H.G.; et al. Rapid detection of COVID-19 causative virus (SARS-CoV-2) in human nasopharyngeal swab specimens using field-effect transistor-based biosensor. *ACS Nano* 2020, 14, 5135–5142.
79. Aldewachi, H.; Chalati, T.; Woodroofe, M.N.; Bricklebank, N.; Sharrack, B.; Gardiner, P. Gold nanoparticle-based colorimetric biosensors. *Nanoscale* 2017, 10, 18–33.
80. Xu, G.; Li, H.; Ma, X.; Jia, X.; Dong, J.; Qian, W. A cuttlebone-derived matrix substrate for hydrogen peroxide/glucose detection. *Biosens. Bioelectron.* 2009, 25, 362.
81. Teles, F.R.R.; Fonseca, L.P. Trends in DNA biosensors. *Talanta* 2008, 77, 606–623.
82. Yuan, D.; Fang, X.; Liu, Y.; Kong, J.; Chen, Q. A hybridization chain reaction coupled with gold nanoparticles for allergen gene detection in peanut, soybean and sesame DNAs. *Analyst* 2019, 144, 3886–3891.
83. Iijima, S. Helical microtubules of graphitic carbon. *Nature* 1991, 354, 56–58.
84. Simon, J.; Flahaut, E.; Golzio, M. Overview of carbon nanotubes for biomedical applications. *Materials* 2019, 12, 624.
85. Farzin, M.A.; Abdoos, H. A critical review on quantum dots: From synthesis toward applications in electrochemical biosensors for determination of disease-related biomolecules. *Talanta* 2020, 224, 121828.
86. Yola, M.L.; Atar, N. Amperometric galectin-3 immunosensor-based gold nanoparticle-functionalized graphitic carbon nitride nanosheets and core–shell Ti-MOF@ COFs composites. *Nanoscale* 2020, 12, 19824–19832.
87. Boyacıoğlu, H.; Yola, B.B.; Karaman, C.; Karaman, O.; Atar, N.; Yola, M.L. A novel electrochemical kidney injury molecule-1 (KIM1) immunosensor based covalent organic frameworks-gold nanoparticles composite and porous NiCo2S4@ CeO2 microspheres: The monitoring of acute kidney injury. *Appl. Surf. Sci.* 2022, 578, 152093.
88. Carrasco, S. Metal-organic frameworks for the development of biosensors: A current overview. *Biosensors* 2018, 8, 92.
89. Nangare, S.N.; Sangale, P.M.; Patil, A.G.; Boddu, S.H.; Deshmukh, P.K.; Jadhav, N.R.; Tade, R.S.; Patil, D.R.; Pandey, A.; Mutalik, S.; et al. Surface architectured metal organic frameworks-based biosensor for ultrasensitive detection of uric acid: Recent advancement and future perspectives. *Microchem. J.* 2021, 169, 106567.
90. Omidian, H.; Mfoafo, K. Exploring the potential of nanotechnology in pediatric healthcare: Advances, challenges, and future directions. *Pharmaceutics* 2023, 15, 1583.
91. Dobrovolskaia, M.A.; McNeil, S.E. Understanding the correlation between in vitro and in vivo immunotoxicity tests for nanomedicines. *J. Control. Release* 2013, 172(2), 456–466. https://doi.org/10.1016/j.jconrel.2013.05.025

25 Sustainable Nanomaterials and Green Synthesis

Biogenic Synthesis of Barium Sulphate Nanoparticles Using Azadirachta indica *Leaf Extract and Assessment of Its Antibacterial Potential*

Muktha H, Ishika Jain, Jayashree D R, and Dileep Kumar M G

25.1 INTRODUCTION

All that exists must be measured, and without a common system, it would be anarchy. The internationally accepted metric system, or International System of Units, allows us to measure sizes correctly. The microscopic realm, which was hidden from us, became intriguing with the realization of microorganisms. Then, with the advent of nanotechnology, contemporary science is progressing in a completely new and interesting direction (1).

From the Latin *nanus*, "dwarf," and the Greek for "very small," the prefix "nano" denotes a factor of 10^{-9} in the SI system. Nanotechnology involves matter at the nanoscale—1 to 100 nanometers—where atoms and molecules exhibit special properties (2). Nanotechnology investigates and controls small structures, resulting in novel applications, particularly in medicine where conventional approaches are inadequate. Contrary to the belief that it concentrates on tiny objects, nanotechnology also deals with bigger materials with nanoscale characteristics. During the past few years, nanoscience research has quickly spread across different industries and product groups, revolutionizing material development (1).

Nanotechnology is the process of making new materials at the atomic, molecular, and macromolecular levels. Solid drug carriers with a diameter of less than 100 nm that are either biodegradable or not are known as pharmaceutical nanoparticles (3). The two types of nanoparticles are nanospheres, where the drug is evenly dispersed, and nanocapsules, where the drug is enclosed in a polymeric shell (4).

The goal of nanoparticle formation is to achieve targeted drug delivery by precisely controlling the size of the particles, their surface characteristics, and the rate at which the active pharmaceutical ingredient is released (5). These characteristics are necessary to guarantee that the medication is delivered to the site of action at the appropriate rate and dosage (6). Higher stability and controlled delivery are two advantages of polymeric nanoparticles in particular (7, 8). They are therefore ideal for efficient therapeutic delivery since they also help to preserve the integrity of medications and proteins (5, 9).

As drug delivery vehicles, nanoparticles offer a number of advantages. Their high drug loading capacity, which allows drugs to be loaded without chemical modification, is one of their main advantages. By using magnetic guidance or binding specific ligands to the surface of the nanoparticle, targeted drug delivery can be achieved (10, 11). Liposomes and polymer-based nanoparticles pose little risk because they are typically biodegradable and do not accumulate in the body (12). More of the medication concentrates at the intended location because of their small particle size, which enables intravasation into small capillaries (10, 13). It is simple to administer nanoparticles via a variety of methods (14, 15). In order to achieve passive or active targeting for maximum therapeutic activity and minimal side effects, they can also readily alter their surface dimensions and characteristics after parenteral delivery (16). For optimal performance and controlled delivery, even the rate of drug release and nanoparticle degradation can be managed with the right matrix material selection (17–19).

Although they are useful, nanoparticles also have some drawbacks such as low drug loading and quick burst release because they are small (20). Prior to clinical or commercial application, issues like physical property changes leading to particle agglomeration need to be overcome (21). They are difficult to process in liquid or solid form because of their small size and large surface area and become very reactive in biological environments (22, 23).

25.2 TOXICITY

These fine particles can penetrate the body through the skin, the lungs, or the gastrointestinal tract and become lodged in one or more organs, where they can cause profound adverse

DOI: 10.1201/9781003644439-25

biological effects by modifying the physiochemical characteristics of tissue (24, 25). Non-biodegradable particles, when employed for drug delivery, can deposit in the delivery area and cause long-term inflammatory reactions (26). Particulate matter inhalation causing lung and cardiovascular disease is the predominant mode of most nanoparticulate toxicities (25, 27).

25.3 TYPES OF NANOPARTICLES

Polymeric nanoparticles are colloidal delivery systems that are made of semi-synthetic or synthetic polymers. The matrix of the nanoparticle holds the drug back, dissolves it, encloses it, or binds to it. Depending on how they are prepared, they can produce nanospheres or nanocapsules. In nanospheres, the drug is dispersed homogeneously throughout the matrix, but in nanocapsules, the drug is sealed within a cavity by a layer of polymer (28). Typical polymer-based systems include chitosan, chitosan-polylactic acid, polylactic-co-glycolic acid, and polycaprolactone (29). Also, solid lipid nanoparticles (SLNs) have been proposed as a new colloidal drug delivery system for intravenous administration, comprising solid lipid particles dispersed in aqueous or surfactant solution (30).

25.4 CLASSIFICATION OF NANOPARTICLES

There are different methods of classifying nanomaterials. Nanoparticles can be categorized according to their dimensions—one, two, and three (31).

One-dimensional nanoparticles have been utilized for decades in electronics, chemistry, and engineering. One-dimensional systems, including thin films or built surfaces, are also common (32). Thin film (dimension 1–100 nm) or monolayer fabrication is common in solar cell or catalysis devices. Thin films are used in a wide variety of technological applications, including optical devices, chemical and biological sensors, fiberoptic devices, and information storage devices (33).

Carbon nanotubes (CNTs) are nanoparticles whose two dimensions are created by rolling a graphite sheet into a cylinder of hexagonal carbon atoms, at about 1 nm across and nanometers in length. There exist two kinds: single-walled (SWCNTs) and multiwalled (MWCNTs). Their mechanical, electrical, and physical properties of note make these alloys unique. CNTs can be either metallic or semiconducting according to their atomic arrangement. Their behavior matches superconductors, as current densities reach a billion amperes per square meter. They are nearly 60 times stronger than steel while providing outstanding molecular absorption plus chemical stability (34).

Fullerenes like C60 are three-dimensional ball-shaped molecules of 28 and more than 100 carbon atoms. C60 is like a soccer ball with bonded carbon hexagons and pentagons that form a hollow sphere. Fullerenes have unusual physical properties, including the capacity for high compression strength and recovery. Because they don't interact, fullerenes have excellent potential as lubricants. Their remarkable electrical properties render them ideal for use in nanoelectronics, such as data storage and solar cells. Their tubular structure enables them to transport other molecules, presenting potential medical applications (35).

Because of their numerous surface functional groups that improve drug targeting, dendrimers are highly developed, nanomolecular polymers (10–100 nm) with a highly controlled architecture that are widely used in drug delivery and medical imaging. Modern dendrimers can be made to include therapeutic or diagnostic agents in their cores thanks to extensive structural and functional research. They can encapsulate a variety of compounds and are essential building blocks in the creation of larger organic and inorganic nanostructures, including metallic forms and nanotubes. They are helpful in biomedical applications because they are biocompatible with organic substances like DNA. Dendrimers are used in high-throughput drug screening as well as formulations for pro-drug, anti-inflammatory, antiviral, antibacterial, and anticancer therapies (36). However, there are worries about possible toxicity because of their positive surface charge, which can be harmful to cell membranes (37).

With a droplet of free electrons, quantum dots (QDs) are tiny semiconductor nanocrystals that typically have a diameter of 2 to 10 nm. They can be made by electrochemistry or colloidal methods using semiconductors like indium phosphide (InP), indium arsenide (InAs), cadmium selenide (CdSe), and cadmium telluride (CdTe). A quantum dot's electrons can range in size from one to thousands, and it is possible to precisely control their shape, size, and quantity (38). QDs have potential uses in optoelectronic devices, quantum computing, and information storage. They can be made to be metals, semiconductors, magnetic materials, or insulators. Because of their large surface area and capacity to hold therapeutic agents, they are used in DNA screening and also have potential for in vivo imaging, tissue engineering, and drug delivery (39).

Proteins, polysaccharides, and synthetic polymers are examples of natural substances that are used to create nanoparticles. Particle size, drug stability and solubility, surface charge, biodegradability, biocompatibility, toxicity, drug release pattern, and antigenicity must all be considered when selecting an inert matrix material (40).

25.5 SYNTHESIS OF NANOPARTICLES

Nanoparticle production technologies are categorized as various approaches which include the bottom-up approach, chemical reaction method, top-down method, and combination method, as depicted in Figure 25.1. Bottom-up procedures entail precipitating nanoparticles from solutions of molecules. On the contrary, the top-down technique consists of agglomeration of large particles into nanoparticles. The combination method blends both the bottom-up and top-down approaches for the creation of nanoparticles. Finally, the chemical reaction technique directly generates

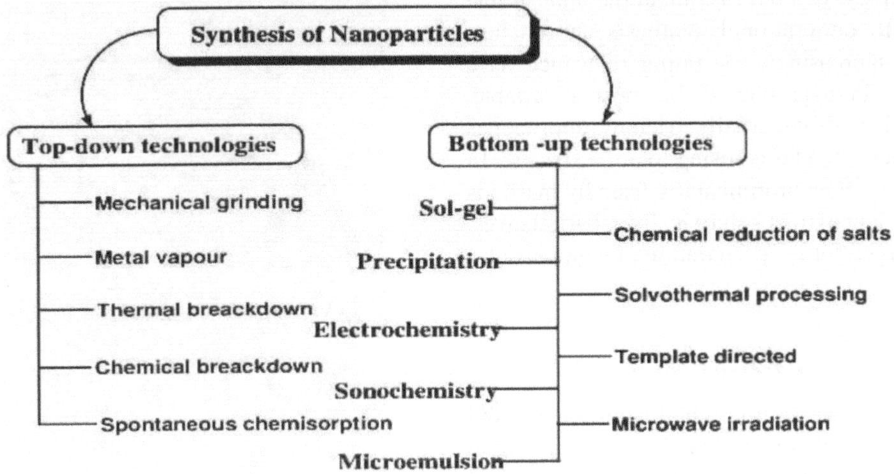

FIGURE 25.1 Synthesis of nanoparticles.

nanoparticles after a phase of chemical reaction without depending on larger particle agglomeration (41). These all have other merits based on what characteristics or functions are desired for the resulting nanoparticles (31).

Chemical processes, including polymerizations, can produce nanoparticles but are generally not applied to pure API drug nanoparticles. These processes are crucial industrially for the manufacture of pharmaceutical coatings such as latex dispersions (5). Chemical processes can form polymeric nanoparticles with an API in a matrix, although drug loading is typically less than 100%, which differentiates them from conventional particle size reduction methods (42).

In bottom-up approaches, drug molecules dissolved in solution start to develop larger structures due to the alteration in the system of solution. In traditional precipitation, a poorly soluble API is dissolved in a solvent for an organic compound that is miscible with water. Upon mixing with an aqueous phase, precipitation takes place (43). This technique is also referred to as the "solvent/antisolvent" technique and was invented in the 1980s by Sucker and others. Ever since, research institutes have optimized traditional precipitation methods and brought about more sophisticated technologies known as particle engineering. Popular among these, evaporative precipitation into aqueous solution (EPAS) involves dissolving the API in a solvent organic that is water immiscible. This is later sprayed into water that has been heated, triggering immediate evaporation of the solvent and resulting in drug nanoparticles. Other methods, including spray-freezing into liquid (SFL) and ultrarapid freezing (URF), have been created (44).

Top-down technology, which reduces large active pharmaceutical ingredient particles to tiny nanoparticles, is an alternative to the bottom-up approach. The most widely used and economically viable particle size reduction method is now this one. One of the well-known methods used in this process is wet ball milling (WBM). It involves using a milling chamber filled with beads such as polystyrene, zirconium dioxide, or selenium nitride, along with a stabilizer or surfactant solution and micronized API. The drug is ground into nanoparticles by the shear forces produced by the bead movement. In order to produce the required size, the suspension is continuously passed through the mill in a circulation mode during bulk production. The drug particles are then separated from the beads in a filter or gap (45). This process is already employed by five commercial products today, with scores more in the pipeline.

High-pressure homogenization (HPH) is another important top-down method. This method, which was initially created for aqueous processes at room temperature, pumps a coarse suspension through a narrow homogenization gap where collisions, cavitation, and shear cause the particle size to decrease. Use in non-aqueous systems is made possible by advancements. High-pressure fluid streams colliding in jet stream homogenizers also produce nanoparticles. However, this method only produces one commercial product and requires 50–100 cycles (46). The combinative approach, a more modern production technique, blends top-down and bottom-up size-reduction procedures (47).

25.6 GREEN SYNTHESIS OF NANOPARTICLES

Green synthesis has proved a reliable, sustainable, and eco-friendly process for the production of different nanomaterials, such as metal oxides, hybrids, and biomimetic compounds, in materials science. Metallic nanoparticles have fascinated scientists for a long time and are presently extensively applied in engineering and healthcare disciplines. Their synthesis is a critical sector in nanotechnology, thanks to their characteristic nanoscale dimensions (1–200 nm), shape-sensitive properties, and multifaceted usage in areas of medicine, catalysis, production of biofuels, electronics, and biology. In a bid to combat the undesirable impacts linked with conventional synthesis approaches, green synthesis is increasingly assuming relevance as a harmless substitute. Perhaps one of the most affordable, uncomplicated, and environmentally friendly approaches is metal nanoparticle synthesis using plant extracts. In recent years, a variety of environmentally friendly methods involving aqueous plant extract solutions from bark, leaves, and roots have emerged for the preparation of nanoparticles (48, 49).

25.7 *AZADIRACHTA INDICA*

Medicinal plants possess antimicrobial substances, and many of them show the potential to fight bacterial infection. Neem (*Azadirachta indica*) is widely used as a medicinal plant throughout Southeast Asia and is commonly found in the tropical and subtropical parts of the world (50). Because of its powerful antibacterial activity, it is an important part of Ayurvedic, Unani, and homeopathic medicine. Neem has more than 140 bioactive compounds, of which the most significant in its leaves are azadirachtin, nimbin, and nimbidine. Different parts of the neem tree, including its leaves, bark, flowers, roots, seeds, and fruits, are utilized for treating skin-related diseases, gums, and teeth. Neem leaves are also the most effective among these for treating skin diseases and wound healing, particularly chickenpox and smallpox (51). The leaf extract has been proved to possess antibacterial activity against bacteria such as *Staphylococcus* spp., *Streptococcus* spp., *Pseudomonas* spp., and *E. coli.* Experiments have demonstrated its capacity to boost humoral and cell-mediated immunity in immunosuppressed birds. Chicken diseases like fowl cholera, typhoid, and colibacillosis caused by causative agents like *Pasteurella multocida, Salmonella pullorum, Salmonella gallinarum*, and *Escherichia coli* are prevalent in Bangladesh and cause significant harm to the poultry industry. In contrast to antibiotics, which cause multidrug resistance, neem extract provides a natural option without the accompanying resistance (52).

The purpose of this study was to test the antibacterial action of neem (*Azadirachta indica*) leaf extract against multidrug-resistant (MDR) poultry pathogens like *P. multocida, S. pullorum, E. gallinarum,* and *E. coli.* The investigation also delved into the synthesis of barium sulfate nanoparticles ($BaSO_4$ NPs) using neem leaf extract. Neem is a quick-growing tree belonging to the family Meliaceae and is widely available in India. It contains quercetin and β-sitosterol, polyphenolic flavonoids with known antimicrobial properties. Neem leaves have an extensive array of therapeutic activities such as antibacterial, antifungal, anti-inflammatory, antioxidant, and antimalarial activity. Because it is antibacterial, neem extract was used as a reducing agent of biological origin in the synthesis of $BaSO_4$ NPs. It helps to stabilize nanoparticles and reduce barium and sodium salts to $BaSO_4$ NPs, amplifying neem's antimicrobial activity (53).

FIGURE 25.2 *Azadirachta indica* (neem leaves).

25.8 ANTIBACTERIAL ACTIVITY

Escherichia coli (*E. coli)* is a Gram-negative, rod-shaped, facultative anaerobic bacterium that inhabits the lower intestines of warm-blooded animals. Although the majority of *E. coli* strains are harmless and serve a beneficial function in the gut by producing vitamin K2 for clotting blood and preventing disease-causing bacterial colonization, certain serotypes (e.g., EPEC and ETEC) have pathogenic potential and are recognized to cause foodborne illness, sometimes leading to product recall (54). *E. coli* is found in feces and will quickly proliferate under aerobic conditions in fresh feces for 3 days before declining in number over time. Because of its wide scientific significance, *E. coli* is heavily researched as a model organism for microbiology and biotechnology. It is highly preferred for use in research owing to its simple cultivation, high rate of growth, and cheap maintenance in laboratories (55).

Pseudomonas aeruginosa is a Gram-negative, rod-shaped, encapsulated bacterium that is strictly aerobic but can also grow anaerobically in the presence of nitrate. This bacterium has the ability to infect plants and animals, including humans. It is found extensively in nature and is a clinically important, multidrug-resistant pathogen. *P. aeruginosa* has complex antibiotic resistance mechanisms and is most often associated with severe hospital-acquired infections like ventilator-associated pneumonia and septic conditions. It is an opportunistic pathogen which mainly infects patients with a pre-existing disease such as cystic fibrosis, extensive burns, or compromised immune systems but can also infect healthy patients (56). Its resistance to a wide array of antibiotics has made it difficult to treat and thus the organism of choice in antibacterial studies (57).

Staphylococcus aureus is a Gram-positive, round bacterium belonging to the Firmicutes phylum. It is a common inhabitant of the human microbiota, specifically on the skin and in the upper respiratory tract. As a facultative anaerobe, it can exist without oxygen and exhibits catalase and nitrate-reducing activity. Though *S. aureus* generally lives harmlessly within the body, it can become an opportunistic pathogen under some circumstances. It is a frequent cause of food poisoning, sinus infections, and a range of skin and respiratory infections such as abscesses. Pathogenic isolates produce virulence factors such as powerful toxins and proteins that inactivate antibodies, leading to diseases ranging from mild skin infection to serious disease such as pneumonia, meningitis, toxic shock syndrome, and sepsis (58).

Enterococcus faecalis is a Gram-positive commensal bacterium normally residing in the gastrointestinal tract of the human. Like other species of *Enterococcus*, *E. faecalis* inhabits healthy humans and may be used as a probiotic. However, it is also a highly resilient pathogen with the potential to cause dangerous infections, especially within hospitals, where it displays innate antibiotic resistance. It is also often seen in reinfected root canal-treated teeth and presents approximately nine times more than with primary infections. Though largely non-pathogenic in healthy individuals, *E. faecalis* is implicated in disease processes such as endocarditis, sepsis, and urinary tract infections. It is also known to be tyrosine decarboxylase-producing, and L-dopa may be decarboxylated by the enzyme, with implications in treating Parkinson's disease (59).

25.9 PHARMACEUTICAL APPLICATION OF BARIUM SULFATE NANOPARTICLES

Because of their unique mechanical, optical, and chemical properties, barium sulfate nanoparticles ($BaSO_4$ NPs) are being used increasingly as fillers in coatings, such as car finishes. $BaSO_4$ NPs have also been used in a variety of applications in diagnostic imaging and orthopedic surgery. Pellethane, a polyurethane elastomer, exhibits antibacterial activity in vitro when mixed with $BaSO_4$ NPs. Yet aerosolized $BaSO_4$ NP inhalation can happen at different phases, such as production, processing, transportation, and waste disposal (60). Chronic exposure to high concentrations of micron-sized $BaSO_4$ particles, especially among miners, has been associated with pneumoconiosis, namely baritosis, which presents potential occupational health issues (61).

25.10 MATERIALS AND METHODS

Materials: *Azadirachta indica,* Ciprofloxacin (powder), Bacterium strains—*Escherichia coli* (ATCC 25922), *Pseudomonas aeruginosa* (ATCC 27853), *Staphylococcus aureus* (ATCC 25923), and *Enterococcus faecalis* (ATCC 29212).

25.11 SYNTHESIS OF BARIUM SULFATE NANOPARTICLES

25.11.1 Preparation of $BaSO_4$ Nanoparticles from Fresh Plant Extract

Washed fresh neem leaves (500 g) were ground with 100 ml of water into fine paste. The paste was first filtered through cheesecloth, and Whatman's paper was used to obtain clear extract (pH 4–5). Forty ml of the extract was agitated and 80 ml of 0.05 M $BaCl_2$ added dropwise and then 80 ml of 0.05 M Na_2SO_4 to form a white $BaSO_4$ precipitate. The reaction mixture was incubated at 37°C for 24 hours to allow growth of the nanoparticles. The precipitate was centrifuged at 6000 rpm for 10 min and decanted, washed thrice with ethanol, and air-dried to obtain $BaSO_4$.

25.11.2 Preparation of $BaSO_4$ Nanoparticles From Dry Plant Extract

Fresh green neem leaves (500 g) were properly washed with double distilled water to eliminate surface pollutants and then dried in the shade at room temperature for 7 days. The dry leaves were finely powdered, and 60 g was blended with 300 ml of a solution of water and ethanol, followed by reflux at 30°C for 60 minutes. The extract, after cooling, was filtered on Whatman filter paper and set aside to cool for 30 minutes. Forty ml of the filtrate was stirred while adding 80 ml of 0.05 M $BaCl_2$ dropwise while sonication was in progress, then 80 ml of 0.05 M Na_2SO_4. The resulting $BaSO_4$ precipitate was incubated at 37°C for 24 hours, centrifuged (6000 rpm, 10 min), washed with ethanol, and air-dried, giving 0.40 g of $BaSO_4$.

25.12 CHARACTERIZATION OF $BASO_4$ NANOPARTICLES

25.12.1 Fourier Transform Infrared Spectroscopy

Fourier transform infrared (FTIR) spectrometry is one of the most common instrumental analysis methods using infrared, which identifies molecules in terms of their characteristic spectrum, much like a fingerprint. It is done by transmitting infrared radiation through a specimen and observing which wavelengths are absorbed by the sample. This technique is preferred for its speed, non-destructive nature, and sensitivity. With the application of an interferometer and Fourier transformation, it transforms interferograms into IR spectra. FTIR is widely used in disciplines such as polymer science, pharmacy, petrochemicals, and food science to investigate vibrations of molecular bonds.

25.12.2 X-Ray Diffraction

X-ray diffraction (XRD) is a popular analytical method to identify the crystalline phases and determine unit cell parameters of materials. The sample is ground into fine powder, mixed, and then analyzed to uncover its general composition. Having evolved from Max von Laue's discovery in 1912 that crystals behave as three-dimensional gratings for diffraction of X-rays, XRD employs the principle of Bragg's law ($n = 2d \sin\theta$) relating to constructive interference between monochromatic X-rays and planes of the crystal. The equipment consists of an X-ray tube, sample, holder, and detector. X-rays are produced by bombarding a target with electrons, and monochromatic beams are produced by using filters. XRD facilitates the determination of atomic spacing, crystal structures, and phase identification by comparing diffraction patterns with standard reference databases

25.13 EVALUATION OF ANTIBACTERIAL ACTIVITY OF BASO₄ NANOPARTICLES

The antibacterial potential of $BaSO_4$ NPs, which were prepared with *Azadirachta indica* (neem), against Gram-positive bacteria (*Staphylococcus aureus, Enterococcus faecalis*), Gram-negative bacteria (*Escherichia coli, Pseudomonas aeruginosa*), and clinical isolates was investigated. Minimum inhibitory concentration (MIC) was assessed with a 96-well microtitre plate assay, where 150 µL CaMHB broth, different concentrations of NPs, 50 µL bacterial inoculum (~1.0×10^7 CFU/mL) were added and attained ~5.0×10^5 CFU/mL. Broth, culture, and vehicle controls also were added. Bacterial growth was quantitated by turbidity and OD at 600 nm after 24 hours of incubation at 37°C.

Test compounds: $BaSO_4$ NPs (DBS, WBS)
Positive control: Ciprofloxacin (powder)
Bacterium (strain): Escherichia coli (ATCC 25922), *Pseudomonas aeruginosa* (ATCC 27853), *Staphylococcus aureus* (ATCC 25923), and *Enterococcus faecalis* (ATCC 29212)

TABLE 25.1
Concentration of Test Compounds and Antibiotics for MIC Assay

Sl. No.	Compound Name	Assay Concentration
1.	DBS	0.06–64 (µg/ml)
2.	WBS	0.06–64 (µg/ml)
3.	Ciprofloxacin	0.01–16 (µg/ml)

25.13.1 Inoculum Preparation

One day prior to the experiment, *Escherichia coli* (ATCC 25922), *Pseudomonas aeruginosa* (ATCC 27853), *Staphylococcus aureus* (ATCC 25923), and *Enterococcus faecalis* (ATCC 29212) were picked from isolated colonies and inoculated into sterile CaMHB tubes. The overnight incubation was carried out at 37°C to achieve the exponential phase. After incubation, Gram's staining was carried out for identification, and the optical density (OD) was standardized to 1.0×10^7 prior to use in the assay.

25.13.2 Preparation of Master Stock Solution of Test Compounds and Ciprofloxacin

Stock of DBS, WBS, and ciprofloxacin in a concentration of 3200 µg/mL was made by dissolving 3.2 mg of each one into 1 mL of DMSO/DMA and water for WBS and DBS, respectively, to make the MS solution. The MS (3.2 mg/mL) was then serially diluted twofold to make WS solutions. The concentration of assay solvent was 2%, with a final volume of 200 µL.

25.13.3 MIC Assay

The MIC assay was performed in a 96-well microplate, to which 150 µL of CaMHB broth with varying concentrations of test compounds was added per well. Subsequently, 50 µL (~1.0×10^7 CFU/mL) was inoculated to achieve ~5.0×10^5 CFU/mL. Controls were Broth Control (BC), Culture Control (CC), and Vehicle Control (VC, with 2% solvent). The plates were incubated at 37°C for 24 hours, then checked for turbidity, with optical density at 600 nm measured. The assay was run in duplicate.

25.13.4 Data Analysis

MIC was designated as the lowest concentration of test compound to inhibit bacterial growth (absence of turbidity or the OD at 600 nm compared with no growth control).

25.14 RESULTS AND DISCUSSION

25.14.1 Synthesis of Barium Sulfate Nanoparticles

25.14.1.1 Preparation of $BaSO_4$ Nanoparticles from Fresh Plant Extract

After the fresh leaves were ground using distilled water, a fine neem leaf paste was obtained. Upon adding $BaCl_2$ and Na_2SO_4 at an interval of 5 minutes, the leaf extract under magnetic stirring shifted from dark green to pale green, yellow, and whitish yellow, reflecting completion of the co-precipitation reaction as observed in Figure 25.3. The yield of the dried $BaSO_4$ nanoparticles was 0.46 g.

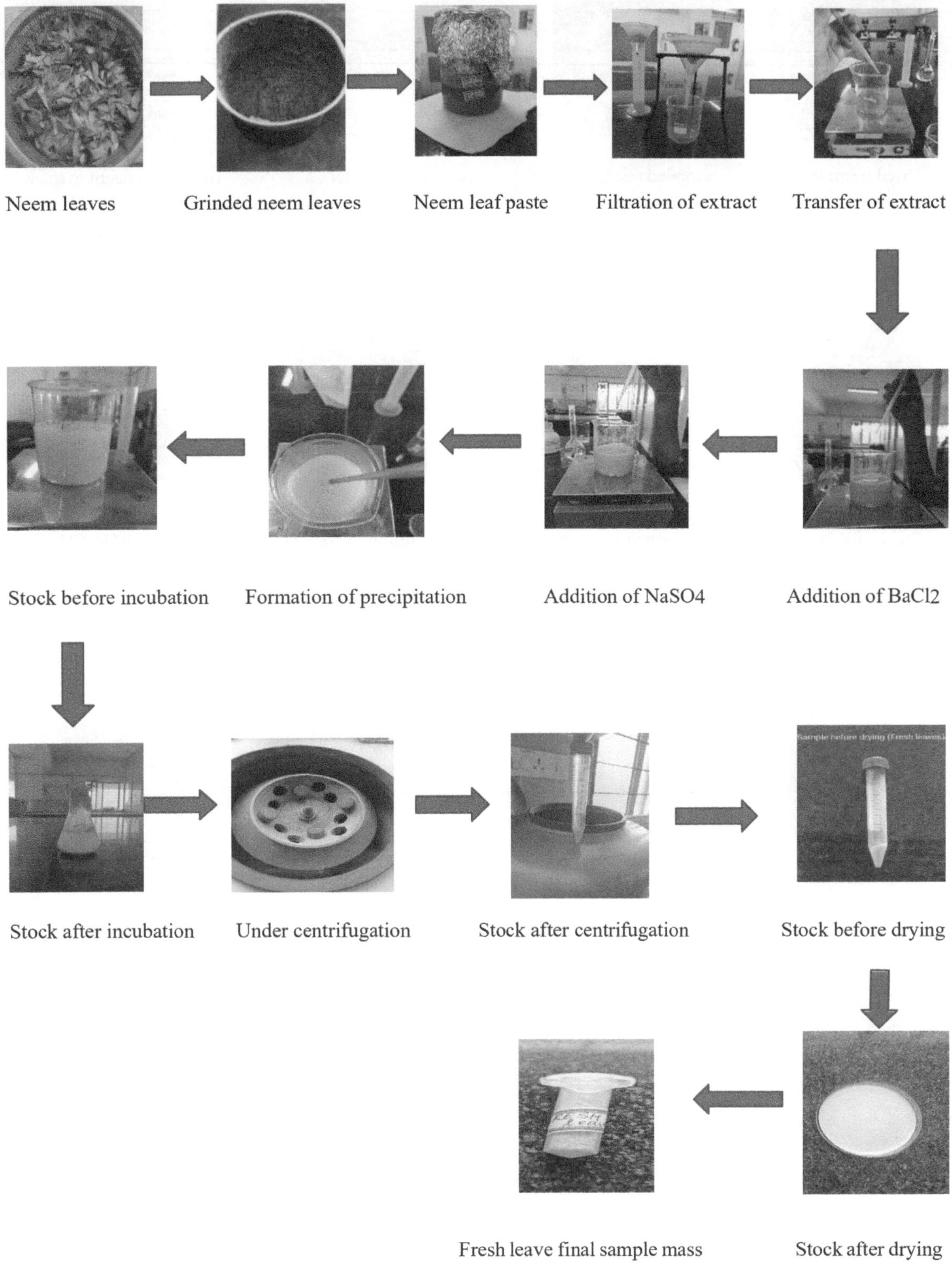

FIGURE 25.3 Extraction of fresh neem leaf–mediated barium sulfate nanoparticle synthesis.

25.14.1.2 Preparation of $BaSO_4$ Nanoparticles from Dry Plant Extract

Once the dried leaves were ground, a fine neem leaf powder was obtained. After the addition of $BaCl_2$ and Na_2SO_4 with a 5-minute interval, magnetic stirring changed the color of the leaf extract from dark green to pale green, yellow, and whitish yellow, showing that the co-precipitation reaction was complete, as observed in Figure 25.4. The yield of dried $BaSO_4$ nanoparticles was 0.40 g.

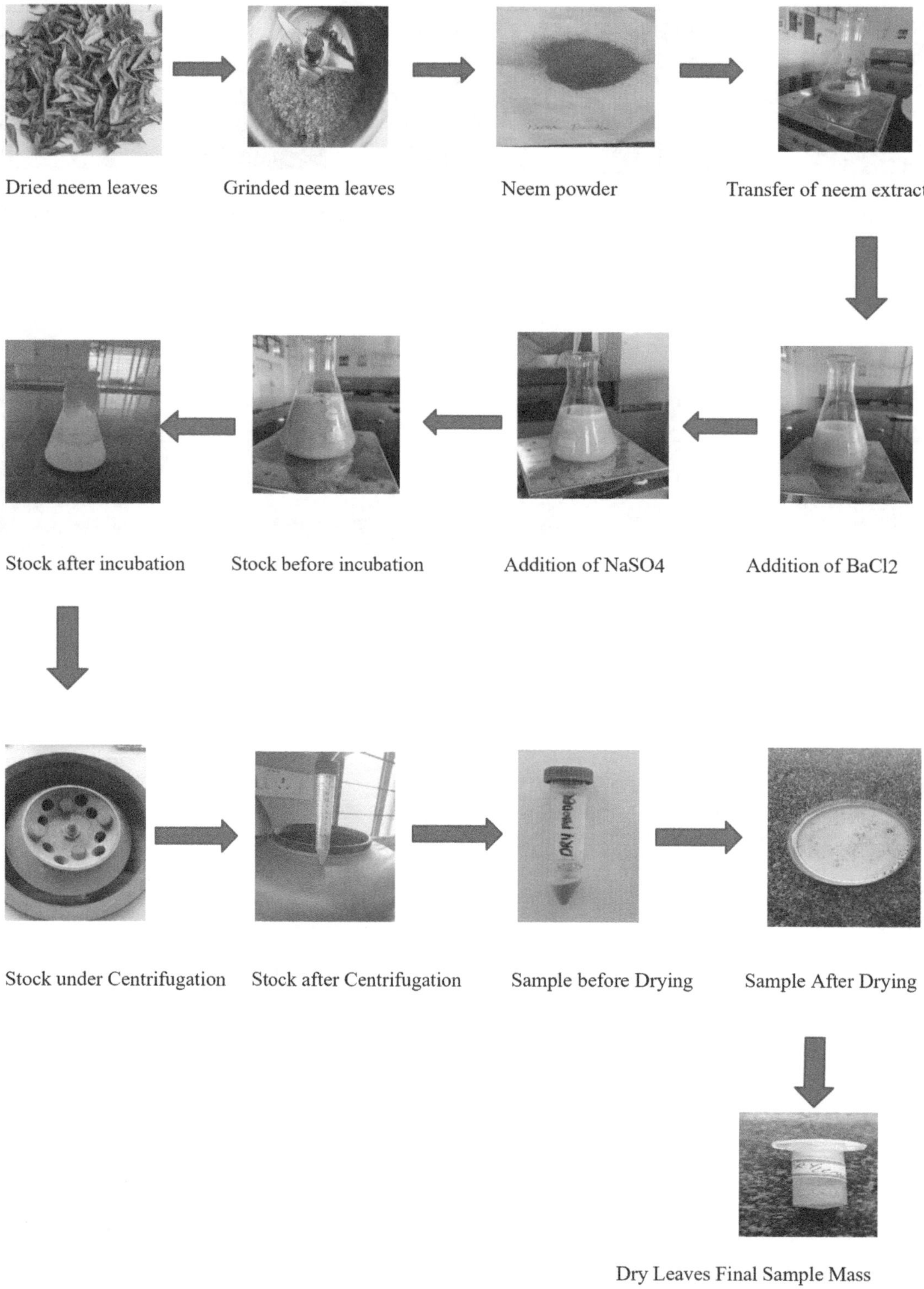

FIGURE 25.4 Extraction of dry plant extract of neem leaf–mediated barium sulfate nanoparticle synthesis.

25.15 CHARACTERIZATION OF $BASO_4$ NANOPARTICLES

25.15.1 FT-IR Analysis

The FTIR spectrum of $BaSO_4$ prepared by the co-precipitation process with *Azadirachta indica* dry leaf extract (Figure 25.5a) exhibits a peak at 605 cm^{-1}, which represents SO_4^{2-} out-of-plane bending, whereas peaks at 1131, 1125, and 984 cm^{-1} represent symmetric vibrational stretching of SO_4^{2-}. The sulfur-oxygen (S-O) stretching region (1131–1072 cm^{-1}) verifies the successful synthesis of $BaSO_4$. Likewise, the FTIR spectrum of $BaSO_4$ prepared from fresh

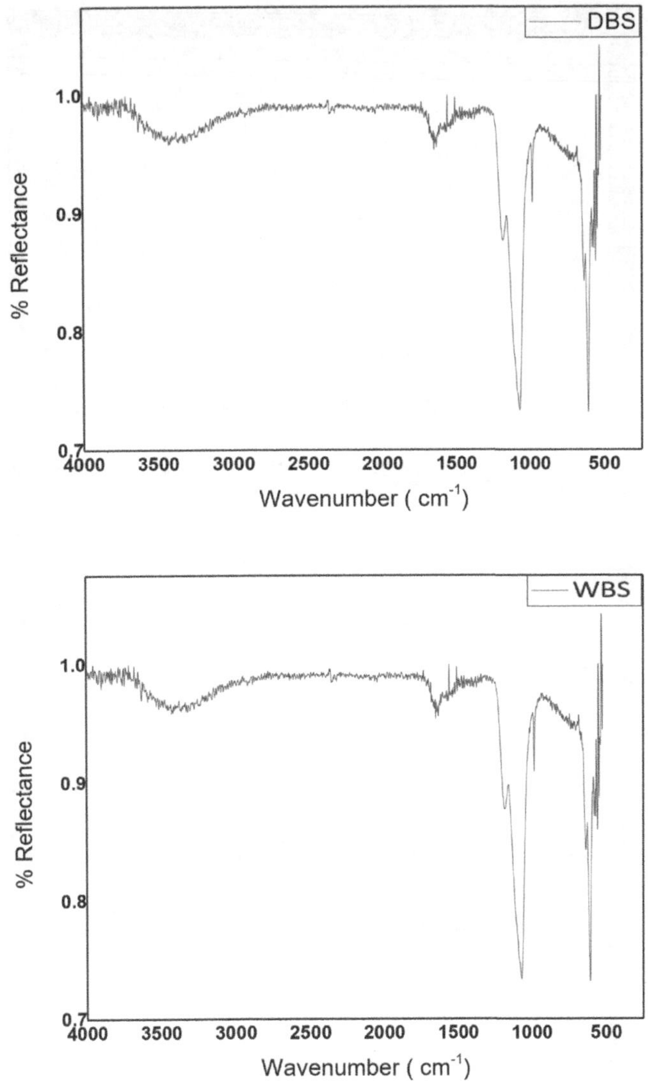

FIGURE 25.5 FTIR analysis results of (a) dry plant extract of neem leaves and (b) fresh plant extract of neem leaf–mediated barium sulfate nanoparticle synthesis.

leaf extract (Figure 5b) indicates peaks at 1132, 1127, and 983 cm^{-1} for SO_4^{2-} stretching and 606 cm^{-1} for out-of-plane bending.

25.15.2 X-Ray Diffraction

The microstructural properties of cryo-ground $BaSO_4$ nanoparticles that were prepared from *Azadirachta indica* leaf extracts were analyzed through X-ray diffraction patterns. The presence of well-defined, sharp peaks in the XRD spectra suggests a high level of crystallinity in the prepared $BaSO_4$ nanoparticles. Also, no extra or unwanted peaks indicates the high purity of the sample. The diffraction pattern is that of the orthorhombic phase of $BaSO_4$. The crystallite size was calculated from the Debye–Scherrer formula:

$$D = (k\lambda/\beta \cos \theta)$$

where k is the Scherrer constant, λ is the wavelength of the X-ray beam used (1.54, 184 Å), β is the full width at half maximum (FWHM) of the peak, and θ is the Bragg angle.

For $BaSO_4$ prepared from dry leaf extract, significant diffraction peaks existed at 24°, 27°, 30°, 33°, 34°, and 44°, whereas for fresh leaf extract, peaks occurred at 22°, 24°, 27°, 28°, 29°, and 44° (Figure 25.6a and b, respectively).

25.15.3 Evaluation of Antibacterial Activity of $BaSO_4$ Nanoparticles

The MIC of DBS was reported to be >64 μg/ml for *Escherichia coli, Pseudomonas aeruginosa, Staphylococcus aureus*, and *Enterococcus faecalis.* MIC value of the standard reference drug (ciprofloxacin) is presented in Table 25.2.

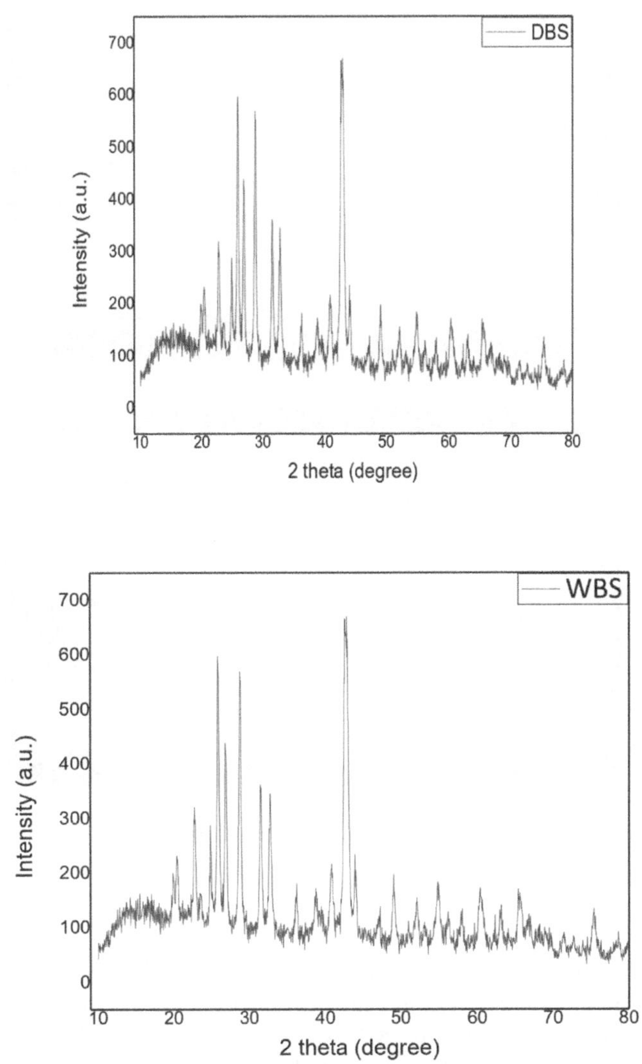

FIGURE 25.6 XRD analysis results of (a) dry plant extract of neem leaf–mediated barium sulfate nanoparticle synthesis and (b) fresh plant extract of neem leaf–mediated barium sulfate nanoparticle synthesis.

TABLE 25.2
Results of DBS, WBS and Ciprofloxacin

Organism Name	MIC µg/ml DBS	MIC µg/ml WBS	MIC µg/ml Ciprofloxacin
Escherichia coli (ATCC 25922)	>64	>64	0.03
Pseudomonas aeruginosa (ATCC 27853)	>64	>64	0.125
Staphylococcus aureus (ATCC 25923)	>64	>64	0.25
Enterococcus faecalis (ATCC 29212)	>64	>64	0.5

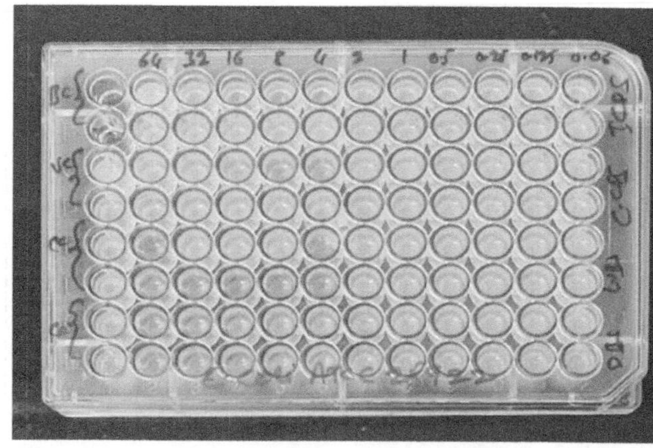

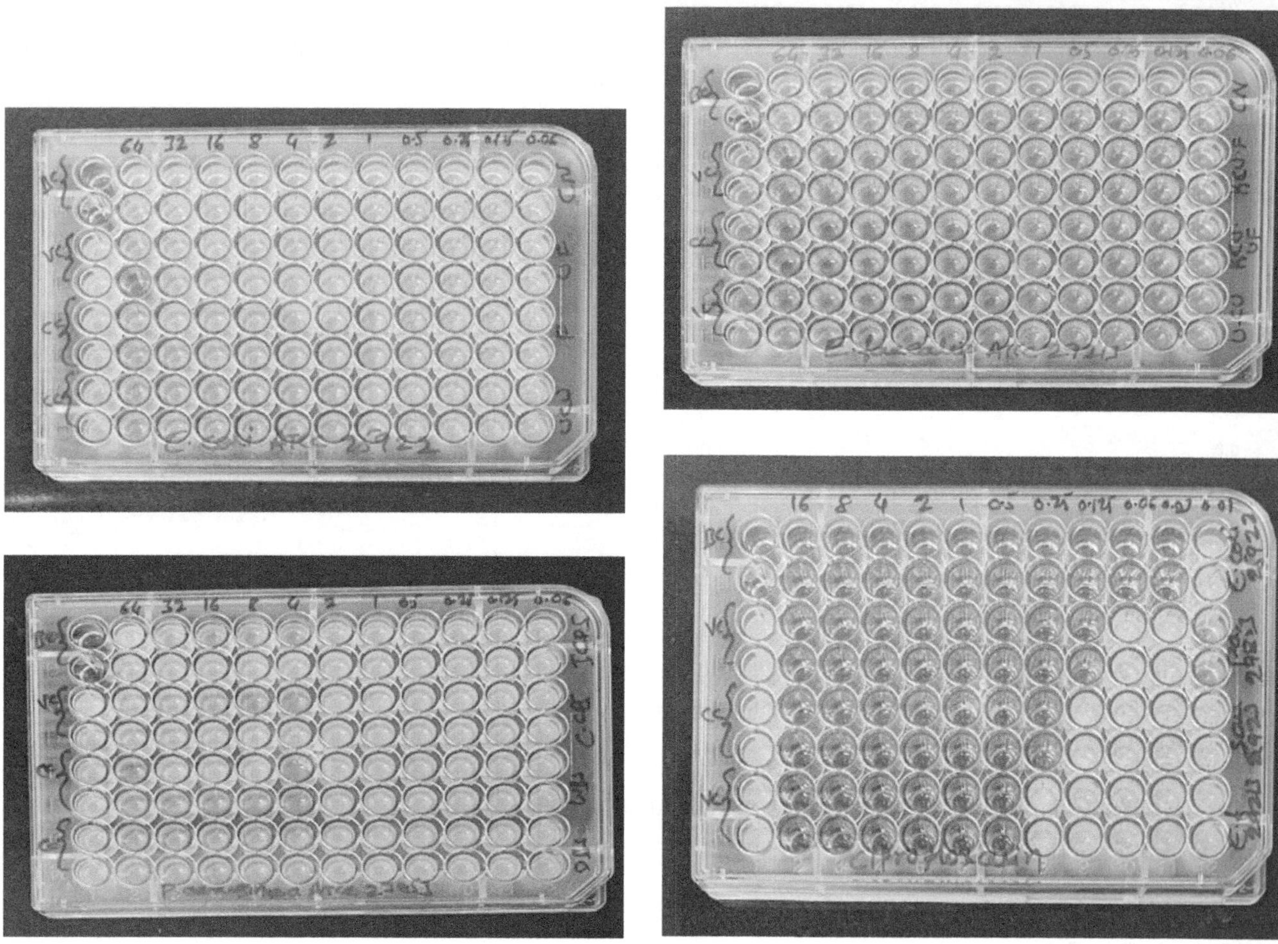

FIGURE 25.7 Minimum inhibitory concentration of the compound against all bacterial strains (*E. coli, P. aeruginosa, S. aureus, E. faecalis,* ciprofloxacin).

The MIC of WBS was >64 µg/ml for *Escherichia coli*, *Pseudomonas aeruginosa*, *Staphylococcus aureus*, and *Enterococcus faecalis*. MIC value of the standard reference drug (ciprofloxacin) is presented in Table 25.2.

The MIC of ciprofloxacin against *Escherichia coli* (ATCC 25922), *Pseudomonas aeruginosa* (ATCC 27853), *Staphylococcus aureus* (ATCC 25923), and *Enterococcus faecalis* (ATCC 29212) was 0.03, 0.125, 0.25, and 0.5 µg/ml, respectively, as shown in Table 25.2 and Figure 25.7.

25.16 CONCLUSION

In the present chapter, an innovative, quick, energy-saving, and green method is proposed to synthesize size-controlled stable $BaSO_4$ nanoparticles by adopting *Azadirachta indica* (neem) as a reducing/capping agent in two forms, namely dry leaf powder and fresh leaf extract. The procedure was carried out without the presence of surfactants or additives, and therefore, the methodology is green and sustainable. The formed $BaSO_4$ nanoparticles were of extremely high purity, orthorhombic, crystalline in nature, and thermally stable, with crystallite dimensions of 20–45 nm. The neem-derived $BaSO_4$ nanoparticles showed little antibacterial effect against four microorganisms—*Escherichia coli, Pseudomonas aeruginosa, Staphylococcus aureus,* and *Enterococcus faecalis*—with their minimum inhibitory concentration values more than 64 µg/mL, signifying low antibacterial activity. $BaSO_4$ nanoparticles have attracted great attention because of their extensive industrial applications, such as in oil drilling, electronics, TV screens, ceramics, vehicle filters, paints, and medicine. Many techniques have been used to regulate the size and morphology of $BaSO_4$ nanoparticles, including organic modification, membrane separation, and direct precipitation. Techniques mentioned include the application of polycarboxylate modifiers, non-ionic microemulsions, impinging stream reactors, and sodium hexametaphosphate as a stabilizer. Recently, $BaSO_4$ nanoparticles have also been synthesized using chemical precipitation without the employment of polymer stabilizers.

REFERENCES

1. Chithrakumar, T., Thangamani, M, and Pon Miruthu Vasini, S. (2021). Exploring nanotechnology for diagnostic, therapy and medicine. *IOP Conference Series: Materials Science and Engineering*, 1091(1), 012062. https://doi.org/10.1088/1757-899x/1091/1/012062
2. Whitesides, George M. (2004). Nanoscience, nanotechnology, and chemistry. *Small*, 1(2), 172–179. https://doi.org/10.1002/smll.200400130
3. Shah, Parag (2006). Use of nanotechnologies for drug delivery. *MRS Bulletin*, 31(11), 894–899. https://doi.org/10.1557/mrs2006.209
4. Patra, Jayanta Kumar et al. (2018). Nano based drug delivery systems: Recent developments and future prospects. *Journal of Nanobiotechnology*, 16(1). https://doi.org/10.1186/s12951-018-0392-8
5. Castro, Karine Cappuccio de, Costa, Josiel Martins, and Campos, Maria Gabriela Nogueira (2020). Drug-loaded polymeric nanoparticles: A review. *International Journal of Polymeric Materials and Polymeric Biomaterials*, 71(1), 1–13. https://doi.org/10.1080/00914037.2020.1798436
6. Wen, Hong, Jung, Huijeong, and Li, Xuhong (2015). Drug delivery approaches in addressing clinical pharmacology-related issues: Opportunities and challenges. *The AAPS Journal*, 17(6), 1327–1340. https://doi.org/10.1208/s12248-015-9814-9
7. Charelli, Letícia Emiliano et al. (2022). Polymeric nanoparticles as therapeutic agents against coronavirus disease. *Journal of Nanoparticle Research*, 24(1). https://doi.org/10.1007/s11051-022-05396-5
8. Adhikari, Chandan (2021). Polymer nanoparticles-preparations, applications and future insights: A concise review. *Polymer-Plastics Technology and Materials*, 1–29. https://doi.org/10.1080/25740881.2021.1939715
9. Banik, Brittany L., Fattahi, Pouria, and Brown, Justin L. (2015). Polymeric nanoparticles: The future of nanomedicine. *WIREs Nanomedicine and Nanobiotechnology*, 8(2), 271–299. https://doi.org/10.1002/wnan.1364
10. Dadwal, Ankita, Baldi, Ashish, and Kumar Narang, Raj (2018). Nanoparticles as carriers for drug delivery in cancer. *Artificial Cells, Nanomedicine, and Biotechnology*, 46(sup2), 295–305. https://doi.org/10.1080/21691401.2018.1457039
11. Liu, Yun et al. (2020). Development of high-drug-loading nanoparticles. *ChemPlusChem*, 85(9), 2143–2157. https://doi.org/10.1002/cplu.202000496
12. Buse, Joshua, and El-Aneed, Anas (2010). Properties, engineering and applications of lipid-based nanoparticle drug-delivery systems: Current research and advances. *Nanomedicine*, 5(8), 1237–1260. https://doi.org/10.2217/nnm.10.107
13. Chenthamara, Dhrisya et al. (2019). Therapeutic efficacy of nanoparticles and routes of administration. *Biomaterials Research*, 23(1). https://doi.org/10.1186/s40824-019-0166-x
14. Wang, Lu et al. (2021). Administration route governs the therapeutic efficacy, biodistribution and macrophage targeting of anti-inflammatory nanoparticles in the lung. *Journal of Nanobiotechnology*, 19(1). https://doi.org/10.1186/s12951-021-00803-w
15. Chandrakala, V., Aruna, Valmiki, and Angajala, Gangadhara (2022). Review on metal nanoparticles as nanocarriers: Current challenges and perspectives in drug delivery systems. *Emergent Materials*, 5(6), 1593–1615. https://doi.org/10.1007/s42247-021-00335-x
16. Chenthamara, Dhrisya et al. (2019). Therapeutic efficacy of nanoparticles and routes of administration. *Biomaterials Research*, 23(1). https://doi.org/10.1186/s40824-019-0166-x
17. Yang, Fang et al. (2015). Controlled drug release and hydrolysis mechanism of polymer–magnetic nanoparticle composite. *ACS Applied Materials & Interfaces*, 7(18), 9410–9419. https://doi.org/10.1021/acsami.5b02210
18. Pitukmanorom, Pemakorn, Yong, Tseh-Hwan, and Ying, Jackie Y. (2008). Tunable release of proteins with polymer–inorganic nanocomposite microspheres. *Advanced Materials*, 20(18), 3504–3509. https://doi.org/10.1002/adma.200800930
19. Beck-Broichsitter, Moritz et al. (2010). Novel 'nano in nano' composites for sustained drug delivery: Biodegradable nanoparticles encapsulated into nanofiber non-wovens. *Macromolecular Bioscience*, 10(12), 1527–1535. https://doi.org/10.1002/mabi.201000100
20. Sultana, Shaheen et al. (2020). Stability issues and approaches to stabilised nanoparticles based drug delivery system. *Journal of Drug Targeting*, 28(5), 468–486. https://doi.org/10.1080/1061186x.2020.1722137
21. Ashraf, Muhammad Aqeel et al. (2018). Effects of size and aggregation/agglomeration of nanoparticles on the interfacial/interphase properties and tensile strength of polymer nanocomposites. *Nanoscale Research Letters*, 13(1). https://doi.org/10.1186/s11671-018-2624-0
22. Buzea, C., Pacheco, I. I., and Robbie, K. (2007). Nanomaterials and nanoparticles: Sources and toxicity. *Biointerphases*, 2(4), MR17–MR71. https://doi.org/10.1116/1.2815690
23. Schrand, A. M., Rahman, M. F., Hussain, S. M., Schlager, J. J., Smith, D. A., and Syed, A. F. (2010). Metal-based nanoparticles and their toxicity assessment. *Wiley Interdisciplinary*

Reviews. Nanomedicine and Nanobiotechnology, 2(5), 544–568. https://doi.org/10.1002/wnan.103

24. Srivastava, Varsha, Gusain, Deepak, and Sharma, Yogesh Chandra (2015). Critical review on the toxicity of some widely used engineered nanoparticles. *Industrial & Engineering Chemistry Research*, 54(24), 6209–6233. https://doi.org/10.1021/acs.iecr.5b01610
25. Bakand, Shahnaz, Hayes, Amanda, and Dechsakulthorn, Finance (2012). Nanoparticles: A review of particle toxicology following inhalation exposure. *Inhalation Toxicology*, 24(2), 125–135. https://doi.org/10.3109/08958378.2010.642021
26. Kohane, Daniel S. et al. (2006). Biodegradable polymeric microspheres and nanospheres for drug delivery in the peritoneum. *Journal of Biomedical Materials Research Part A*, 77A(2), 351–361. https://doi.org/10.1002/jbm.a.30654
27. Miller, Mark R. et al. (2017). Inhaled nanoparticles accumulate at sites of vascular disease. *ACS Nano*, 11(5), 4542–4552. https://doi.org/10.1021/acsnano.6b08551
28. Frank, Luiza A. et al. (2015). Improving drug biological effects by encapsulation into polymeric nanocapsules. *WIREs Nanomedicine and Nanobiotechnology*, 7(5), 623–639. https://doi.org/10.1002/wnan.1334
29. Murugesan, Selvakumar, and Scheibel, Thomas (2021). Chitosan-based nanocomposites for medical applications. *Journal of Polymer Science*, 59(15), 1610–1642. https://doi.org/10.1002/pol.20210251
30. Paliwal, Rishi et al. (2020). Solid lipid nanoparticles: A review on recent perspectives and patents. *Expert Opinion on Therapeutic Patents*, 30(3), 179–194. https://doi.org/10.1080/13543776.2020.1720649
31. Joudeh, Nadeem, and Linke, Dirk (2022). Nanoparticle classification, physicochemical properties, characterization, and applications: A comprehensive review for biologists. *Journal of Nanobiotechnology*, 20(1). https://doi.org/10.1186/s12951-022-01477-8
32. Xia, Y. et al. (2003). One-dimensional nanostructures: Synthesis, characterization, and applications. *Advanced Materials*, 15(5), 353–389. https://doi.org/10.1002/adma.200390087
33. Hora, Jitesh et al. (2017). Inorganic thin film deposition and application on organic polymer substrates. *Advanced Engineering Materials*, 20(5). https://doi.org/10.1002/adem.201700868
34. Sinnott, Susan B., and Andrews, Rodney (2001). Carbon nanotubes: Synthesis, properties, and applications. *Critical Reviews in Solid State and Materials Sciences*, 26(3), 145–249. https://doi.org/10.1080/20014091104189
35. Bhushan, Bharat et al. (1993). Fullerene (C60) films for solid lubrication. *Tribology Transactions*, 36(4), 573–580. https://doi.org/10.1080/10402009308983197
36. Wu, Lin-ping et al. (2015). Dendrimers in medicine: Therapeutic concepts and pharmaceutical challenges. *Bioconjugate Chemistry*, 26(7), 1198–1211. https://doi.org/10.1021/acs.bioconjchem.5b00031
37. Abbasi, Elham et al. (2014). Dendrimers: Synthesis, applications, and properties. *Nanoscale Research Letters*, 9(1). https://doi.org/10.1186/1556-276x-9-247
38. Valizadeh, Alireza et al. (2012). Quantum dots: Synthesis, bioapplications, and toxicity. *Nanoscale Research Letters*, 7(1). https://doi.org/10.1186/1556-276x-7-480
39. Kagan, Cherie R. et al. (2020). Colloidal quantum dots as platforms for quantum information science. *Chemical Reviews*, 121(5), 3186–3233. https://doi.org/10.1021/acs.chemrev.0c00831
40. Mahapatro, Anil, and Singh, Dinesh K (2011). Biodegradable nanoparticles are excellent vehicle for site directed in-vivo delivery of drugs and vaccines. *Journal of Nanobiotechnology*, 9(1), 55. https://doi.org/10.1186/1477-3155-9-55
41. Ijaz, Irfan et al. (2020). Detail review on chemical, physical and green synthesis, classification, characterizations and applications of nanoparticles. *Green Chemistry Letters and Reviews*, 13(3), 223–245. https://doi.org/10.1080/17518253.2020.1802517
42. Quintanar-Guerrero, David et al. (1998). Preparation techniques and mechanisms of formation of biodegradable nanoparticles from preformed polymers. *Drug Development and Industrial Pharmacy*, 24(12), 1113–1128. https://doi.org/10.3109/03639049809108571
43. Ahmadi Tehrani, Ali et al. (2019). Formation of nanosuspensions in bottom-up approach: Theories and optimization. *DARU Journal of Pharmaceutical Sciences*, 27(1), 451–473. https://doi.org/10.1007/s40199-018-00235-2
44. Bajaj, Amrita et al. (2012). Nanocrystallization by evaporative antisolvent technique for solubility and bioavailability enhancement of telmisartan. *AAPS PharmSciTech*, 13(4), 1331–1340. https://doi.org/10.1208/s12249-012-9860-x
45. Leleux, Jardin, and Williams, Robert O. (2013). Recent advancements in mechanical reduction methods: Particulate systems. *Drug Development and Industrial Pharmacy*, 40(3), 289–300. https://doi.org/10.3109/03639045.2013.828217
46. Souto, Eliana B. et al. (2019). Key production parameters for the development of solid lipid nanoparticles by high shear homogenization. *Pharmaceutical Development and Technology*, 24(9), 1181–1185. https://doi.org/10.1080/10837450.2019.1647235
47. Jenning, V., Lippacher, A., and Gohla, S. H. (2002). Medium scale production of solid lipid nanoparticles (SLN) by high pressure homogenization. *Journal of Microencapsulation*, 19(1), 1–10. https://doi.org/10.1080/713817583
48. Salem, Salem S., and Fouda, Amr (2020). Green synthesis of metallic nanoparticles and their prospective biotechnological applications: An overview. *Biological Trace Element Research*, 199(1), 344–370. https://doi.org/10.1007/s12011-020-02138-3
49. Gour, Aman, and Jain, Narendra Kumar (2019). Advances in green synthesis of nanoparticles. *Artificial Cells, Nanomedicine, and Biotechnology*, 47(1), 844–851. https://doi.org/10.1080/21691401.2019.1577878
50. Kharwar, Ravindra N. et al. (2020). Harnessing the phytotherapeutic treasure troves of the ancient medicinal plant azadirachta indica (Neem) and associated endophytic microorganisms. *Planta Medica*, 86(13/14), 906–940. https://doi.org/10.1055/a-1107-9370
51. Brahmachari, Goutam (2004). Neem—an omnipotent plant: A retrospection. *ChemBioChem*, 5(4), 408–421. https://doi.org/10.1002/cbic.200300749
52. Sarkar, Subendu, Singh, Rajender Pal, and Bhattacharya, Gorachand (2021). Exploring the role of Azadirachta indica (neem) and its active compounds in the regulation of biological pathways: An update on molecular approach. *3 Biotech*, 11(4). https://doi.org/10.1007/s13205-021-02745-4
53. Ulaeto, Sarah B. et al. (2019). Biogenic ag nanoparticles from neem extract: Their structural evaluation and antimicrobial effects against *Pseudomonas nitroreducens*

and *Aspergillus unguis* (NII 08123). *ACS Biomaterials Science & Engineering*, 6(1), 235–245. https://doi.org/10.1021/acsbiomaterials.9b01257
54. Wang, G., Zhao, T., and Doyle, M. P. (1996). Fate of enterohemorrhagic Escherichia coli O157:H7 in bovine feces. *Applied and Environmental Microbiology*, 62(7), 2567–2570. https://doi.org/10.1128/aem.62.7.2567-2570.1996
55. Martinson, Jonathan N. V., and Walk, Seth T. (2020). *Escherichia coli* residency in the gut of healthy human adults. *EcoSal Plus*, 9(1). https://doi.org/10.1128/ecosalplus.esp-0003-2020
56. El Zowalaty, Mohamed E. et al. (2015). *Pseudomonas Aeruginosa*: Arsenal of resistance mechanisms, decades of changing resistance profiles, and future antimicrobial therapies. *Future Microbiology*, 10(10), 1683–1706. https://doi.org/10.2217/fmb.15.48
57. Strateva, Tanya, and Yordanov, Daniel (2009). Pseudomonas aeruginosa—a phenomenon of bacterial resistance. *Journal of Medical Microbiology*, 58(9), 1133–1148. https://doi.org/10.1099/jmm.0.009142-0
58. Tong, Steven Y. C. et al. (2015). Staphylococcus aureus infections: Epidemiology, Pathophysiology, clinical manifestations, and management. *Clinical Microbiology Reviews*, 28(3), 603–661. https://doi.org/10.1128/cmr.00134-14
59. Kayaoglu, Güven, and Ørstavik, Dag (2004). Virulence factors of *Enterococcus faecalis:* Relationship to endodontic disease. *Critical Reviews in Oral Biology & Medicine*, 15(5), 308–320. https://doi.org/10.1177/154411130401500506
60. Schwotzer, Daniela et al. (2017). Effects from a 90-day inhalation toxicity study with cerium oxide and barium sulfate nanoparticles in rats. *Particle and Fibre Toxicology*, 14(1). https://doi.org/10.1186/s12989-017-0204-6
61. Konduru, Nagarjun et al. (2014). Biokinetics and effects of barium sulfate nanoparticles. *Particle and Fibre Toxicology*, 11(1). https://doi.org/10.1186/s12989-014-0055-3

26 Emerging Trends and Applications of Green Nanotechnology in Healthcare

Chaitanya Lakshmi S R, Kuthambakam Baby Rani, Bhimavaram Keerthi, Lokesh Ravi, Asha P Johnson, Rajasekaran Subbarayan, Rupendra Shrestha, and Jeevan Kumar M

26.1 INTRODUCTION

Nanotechnology has emerged as a transformative field with significant applications in medicine, engineering, diagnosis, and biotechnology. It is inherently interdisciplinary (Wagner et al. 2006). Green nanotechnology focuses on synthesizing nanomaterials and nanodevices utilizing biologically and ecologically acceptable materials. This area combines green chemistry with biological synthesis, minimizing the impact of toxic chemicals in conventional methods of synthesis (Iravani, 2011). This method involves the use of natural sources, including plant extracts, algae, microorganisms and biopolymers. While this method avoids toxic substances and harmful byproducts, it also minimizes the biological and environmental damage associated with the nanoparticles (Ahmed et al. 2016). This chapter primarily discusses the applications of plant-based nanoparticle synthesis, their modes of action, their advantages and the limitations faced in their development and use (Kumar et al., 2026).

26.2 TYPES OF GREEN NANOPARTICLES

Green synthesis uses biological sources such as plant extracts and microbes to make nanoparticles in an environmentally sustainable manner (Iravani, 2011; Mittal et al., 2013). There are different types of green nanoparticles, such as gold nanoparticles (AuNPs), silver nanoparticles (AgNPs), zinc oxide nanoparticles (ZnO-NPs) and others. These nanoparticles are used in many biomedical applications for their specific characteristics and biocompatibility (I. Khan et al., 2019).

AuNPs show good anticancer, antibacterial and antioxidant activities. Studies have shown that AuNPs prepared from plant extracts shows anticancer activity against various cell lines (Dasgupta et al., 2017). AgNPs show antibacterial activity against microorganisms like *Staphylococcus aureus* and *Escherichia coli* (Ahmed et al., 2016). ZnO-NPs show significant antioxidant, antibacterial and anticancer activities. These nanoparticles are used as drug delivery carriers and therapeutic agents in treating cancer (I. Khan et al., 2019). Copper oxide nanoparticles (CuO-NPs) show antibacterial and anticancer properties. They are used in medical equipment as antibacterial agents and in wound healing (Mittal et al., 2013).

Iron oxide nanoparticles (IONPs) are involved in various applications like drug delivery, bioimaging agents, hyperthermic agents, cell labeling and gene transport. These nanoparticles reduce toxicity and enhance biocompatibility (Dasgupta et al., 2017). Titanium dioxide nanoparticles (TiO_2-NPs) show anticancer, antibacterial, and antioxidant activities. They are used as drug delivery agents and in photodynamic treatment (Iravani, 2011). Selenium nanoparticles show antioxidant and anticancer activity. They increase immune responses and act as a therapeutic agent for cancer treatment (I. Khan et al., 2019). Cerium oxide nanoparticles are used in photocatalytic dye degradation and show various biological activities. The redox properties which occur in these nanoparticles are useful for neuroprotective applications (Mittal et al., 2013). Stannic oxide nanoparticles are used in photocatalytic, antioxidant, biosensing and antibacterial applications (Ahmed et al., 2016). Silver chloride nanoparticles are used in environmental and medical applications such as antibacterial and wound healing activities (Iravani, 2011).

26.3 ANTIMICROBIAL PROPERTIES

Green nanoparticles (GNPs) synthesized from plant extracts demonstrate antimicrobial properties, which are essential for the maintenance of diabetes. These nanoparticles offer therapeutic benefits because of their biocompatibility and reduced cytotoxicity when compared to chemically synthesized compounds (Dasgupta et al., 2017; Mittal et al., 2013). The mechanism of antimicrobial activity in diabetes patients has been illustrated, where diabetes impairs the immune system, hence making the patient susceptible to infections. GNPs can reduce risk by disrupting microbial cell membranes. In this process, the GNPs attach to the bacterial cell wall, causing structural damage that ultimately leads to cell death. Additionally, the generation of reactive oxygen species (ROS) by these nanoparticles induces oxidative stress in microbial cells, resulting in apoptosis (Costa et al., 2020).

DOI: 10.1201/9781003644439-26

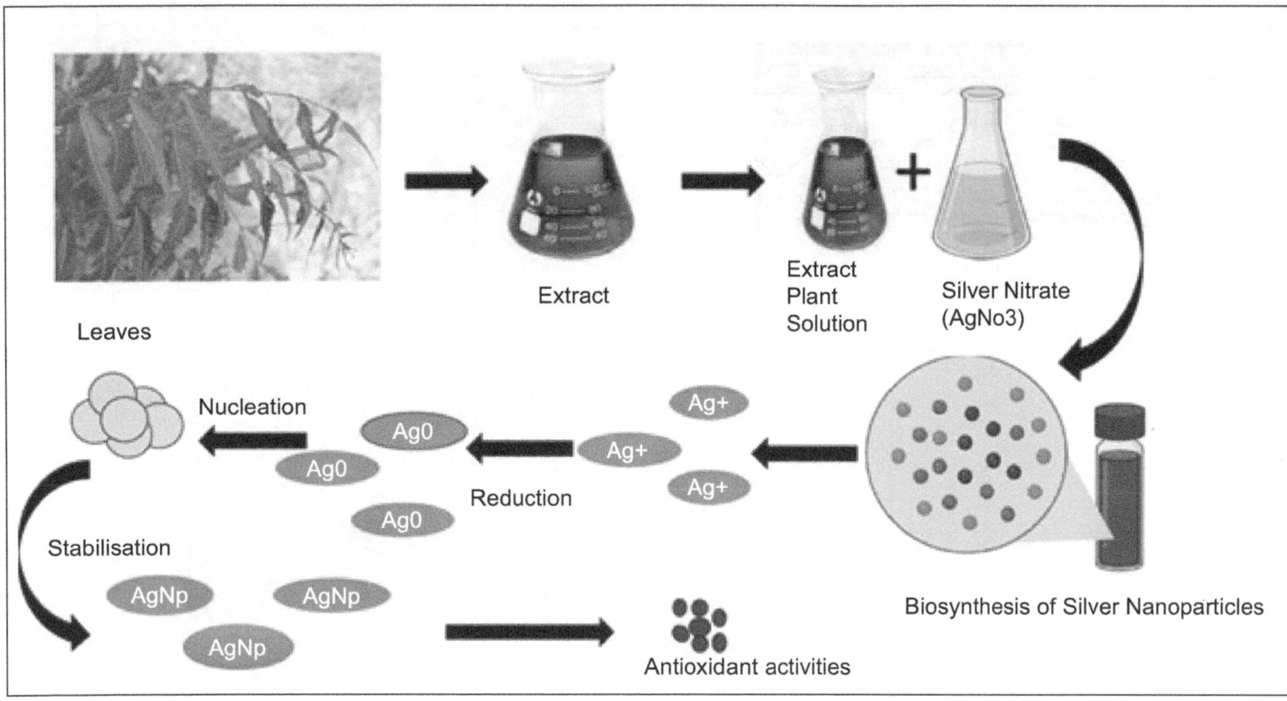

FIGURE 26.1 Biosynthesis of nanoparticles—antiviral properties.

TABLE 26.1
Various Plants Utilized to Synthesize GNPs with Antimicrobial and Antidiabetic Properties

Plant Name	Nanoparticle Synthesized	Biomedical Applications	Sources
Azadirachta indica	Silver nanoparticles	Exhibits antimicrobial activity against pathogens and demonstrates antidiabetic properties by modulating liver function and antioxidant enzyme activities.	(Rehman et al., 2023)
Allium sativum	Silver nanoparticles	Shows significant antidiabetic activity by reducing hyperglycemia.	(Andleeb et al. 2020)
Fagonia cretica	Silver nanoparticles	Demonstrates antimicrobial properties and potential applications in treating diabetes and other diseases.	(Kumar et al., 2023)

26.4 ANTIVIRAL PROPERTIES

Green-synthesized nanoparticles demonstrate strong antiviral activity by causing oxidative stress that damages viral structures and by preventing the attachment and entry of viruses into host cells. These mechanisms work together to restrict viral replication and enhance the effectiveness of host immune responses (Galdiero et al., 2011). Green-synthesized AgNPs are used in coatings for medical devices and surfaces to inhibit contamination and transmission of viruses. AuNPs and SeNPs act as carriers for antiviral medications, improving their bioavailability and stability and providing specific targeted drug delivery to infected cells. In addition to their optical properties, green nanoparticles help in the creation of biosensors for the fast and highly precise identification of viral particles.

26.5 ANTICANCER PROPERTIES

Green nanoparticles made using plant extracts show promising anticancer activity because of their biocompatibility and effectivity. Synthesized metal nanoparticles from plants, such as gold, silver, zinc, and copper, act as reducing and stabilizing agents. Toxicity capacity has been reduced using this method. These green nanoparticles exhibit certain lethal effects on specific cancer cell lines, HeLa, HL-60 and MCF-7, by regulating oxidative stress, inducing cell death and interfering in cell function. For instance, AgNPs extracted from *Saccharina japonica* show toxic activity against cervical cancer cells, and AuNPs are stable and compatible for cancer therapy (Thakur & Verma, 2021).

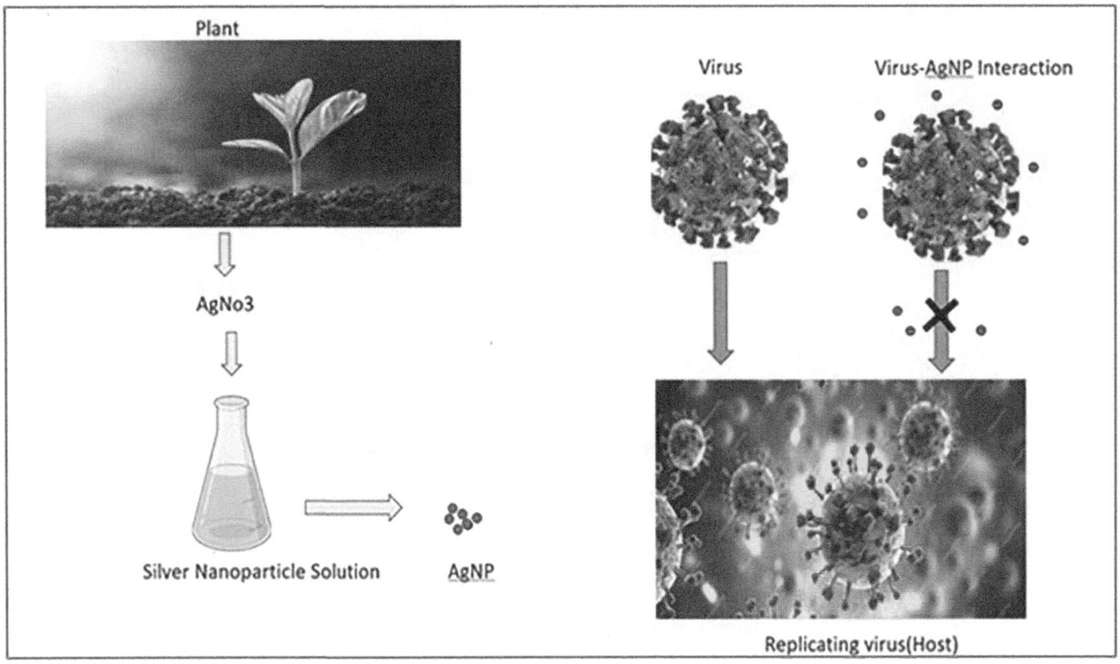

FIGURE 26.2 Green synthesis of nanoparticles with antiviral properties.

TABLE 26.2
Different Types of Green Nanoparticles with Antiviral Properties

Type of Green Nanoparticle	Mechanism of Action	Target Virus	Biomedical Application	Source
Silver Nanoparticles	Direct virucidal effect, inhibition of viral entry, immune modulation	HIV, HSV, SARS-CoV-2, influenza	Antiviral coatings, drug delivery, wound healing	(Lara et al., 2010)
Gold Nanoparticles	Inhibition of viral replication, blocking viral attachment	HIV, Hepatitis C, influenza	Antiviral therapy, biosensors	(Bhatti et al., 2023)
Zinc Oxide Nanoparticles	Blocking viral attachment, immune stimulation	HSV, influenza, COVID-19	HSV, influenza, COVID-19	(Gupta et al., 2022)
Copper Nanoparticles	Disrupting viral genome replication, oxidative stress induction	SARS-CoV-2, influenza, norovirus	Antiviral surface coatings, drug carriers	Frontiers in Bioengineering
Iron Oxide Nanoparticles	Inhibition of viral replication, magnetic targeting	Hepatitis B, influenza	Targeted drug delivery, biosensors	(Kumar et al., 2019)
Silicon-Based Nanoparticles	Preventing viral attachment, disrupting envelope integrity	HIV, dengue, H1N1	Diagnostic tools, drug delivery systems	(Parfenyuk et al., 2025)

26.5.1 Anticancer Mechanisms of Plant-Based Nanoparticles

Plant-based nanoparticles show strong anticancer effects through several mechanisms. One key process is the change in mitochondrial membrane potential, which triggers the release of pro-apoptotic proteins, especially cytochrome C, into the cytoplasm. This process ultimately causes programmed cell death. Many nanoparticles alter the function of apoptosis-related genes. This involves tumor suppressor gene activation such as P53. Through upregulating these pathways, they can induce apoptosis in specific cells. Furthermore, green NPs can interact with membrane proteins and lipids in cancer cells. This interaction disturbs the electrochemical balance and eventually leads to cellular disintegration (Das et al., 2023).

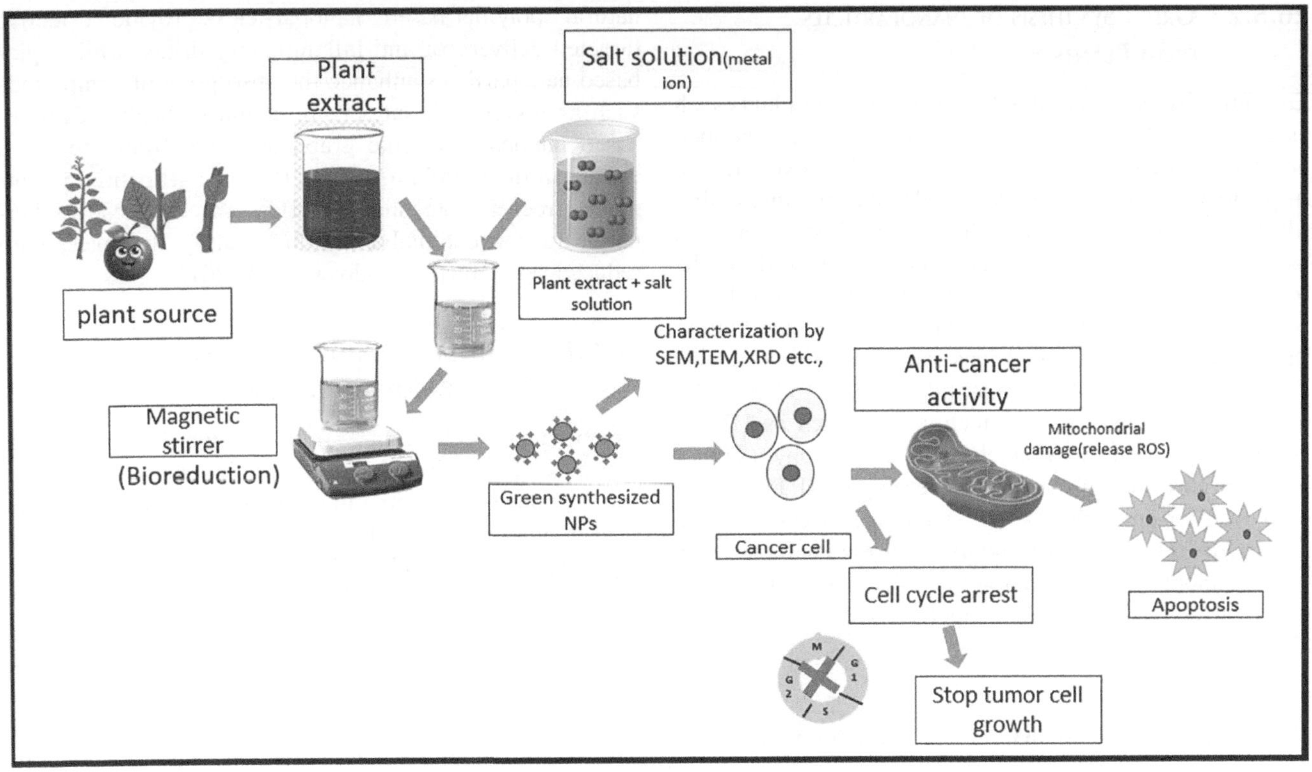

FIGURE 26.3 Plant-mediated synthesis of nanoparticles for anti-cancer applications.

TABLE 26.3
Green Plant Extract Nanoparticles with Anticancer Activity

Plants	Type of Green Nanoparticle	Mechanism of Action	Biomedical Application	Source
Scutellaria barbata	Gold nanoparticles	Induce oxidative stress, apoptosis, and protein/gene modulation, making them a promising anti-cancer agent for pancreatic cancer	Anticancer activity in pancreatic cancer cell (PANC-1)	(L. Wang et al., 2019)
Dendropanax morbifera	D-AgNPs and D-AuNPs	Cytotoxicity in a human keratinocyte cell line and in A549 human lung cancer cell line	Human lung cancer (anticancer)	(C. Wang et al., 2016)
Artemisia turcomanica	Silver nanoparticles	Apoptosis induction on gastric cancer cell line (AGS)	Anti-cancer effect against gastric cancer cell lines	(Mousavi et al., 2025)
Psoralea corylifolia	Iron oxide nanoparticles	Involve reactive oxygen species generation, apoptosis induction and caspase activation	Anti-cancer effect against breast, lung, liver, prostate, cervical, colorectal and leukemia cells	(Nagajyothi et al., 2017)
Ocimum species	Silver nanoparticles	Induction of apoptosis via ROS generation	Anticancer against breast cancer (MCF-7)	(Alex et al., 2024)
Tridax procumbens	AgNPs	Disruption of mitochondrial membrane potential	Anticancer against cervical cancer (HeLa)	(Pungle et al., 2022)
Diospyros kaki L.	AgNPs	Induction of cytotoxic effects through DNA damage	Anticancer against lung cancer(A549)	(Keskin et al., 2023)
Couroupita guianensis	Gold nanoparticles	Inhibition of cancer cell proliferation	Anticancer against breast cancer (MCF-7)	(Geetha et al., 2013)
Holigarna arnottiana	Silver nanoparticles	Disrupt cancer cell growth and induce apoptosis	Anticancer against colon cancer (HT-29)	(Ajaykumar et al., 2023)

26.5.2 Green Synthesis of Nanoparticles from Plants

To synthesize plant-based NPs, various parts of plants such as leaves, bark, stems and roots are obtained, chopped and then boiled in distilled water. The plant extract, rich in phytochemicals such as flavonoids and proteins, is first filtered and then combined with a metal salt solution, such as $HAuCl_2$. The mixture is left undisturbed to allow the reaction to occur. The phytochemicals serve dual roles as reducing agents and natural capping agents, facilitating the transformation of metal ions into nanoparticles (NPs) like gold and silver. The formation of NPs is indicated by a color change, and various techniques—such as UV-vis spectroscopy, transmission electron microscopy (TEM), and Fourier transform infrared spectroscopy (FTIR)—are employed for analysis. The NP synthesis process can take anywhere from minutes to hours, with the characteristics of the nanoparticles, including their size and shape, influenced by factors such as pH, temperature, and reaction time (Bharadwaj et al., 2021).

26.6 ANTI-INFLAMMATORY PROPERTIES

Green nanotechnology uses plants in an eco-friendly way to synthesize nanoparticles without harmful chemicals. These biologically synthesized NPs alleviate inflammation by controlling oxidative damage and immune response. Metallic NPs such as those made from silver, gold and zinc oxide exhibit anti-inflammatory activity. For instance, AuNPs from turmeric are used to treat arthritis, whereas aloe vera–derived ZnO NPs are used for wound healing. Additionally, natural polymer-based nanoparticles provide effective targeted delivery of anti-inflammatory drugs, while lipid-based nanoparticles enhance the absorption of compounds such as resveratrol, supporting cognitive health. Carbon-based nanoparticles like graphene oxide from green tea extract mitigate inflammation effects in arthritis. In summary, green nanotechnology offers an eco-friendly alternative to conventional anti-inflammatory therapies, with reduced side effects (Moldovan et al., 2017).

26.6.1 Anti-Inflammatory Mechanisms of Plant-Based Nanoparticles

Green nanoparticles reduce inflammation through various processes. They inhibit harmful proteins such as cytokines, which induce pain and inflammation. Specific elements, such as gold nanoparticles, act as antioxidants by reducing harmful chemicals called reactive oxygen species, which can damage cells. Some nanoparticles have anti-inflammatory effects; for example, selenium NPs inhibit the NF-κB pathway, which controls inflammation, while silver NPs mitigate COX-2 enzyme levels, hence diminishing pain-inducing chemicals, and also reduce VEGF, which regulates blood flow and edema. Gold NPs control cellular signals that govern inflammation and stop excessive production of mucus. These natural nanoparticles help in reducing inflammation more than conventional pharmaceuticals (Agarwal et al., 2019). To illustrate, David et al. documented the anti-inflammatory activity of silver NPs with the use of European black elderberry fruit extract; the anti-inflammatory effect was demonstrated by diminishing cytokine production (David et al., 2014).

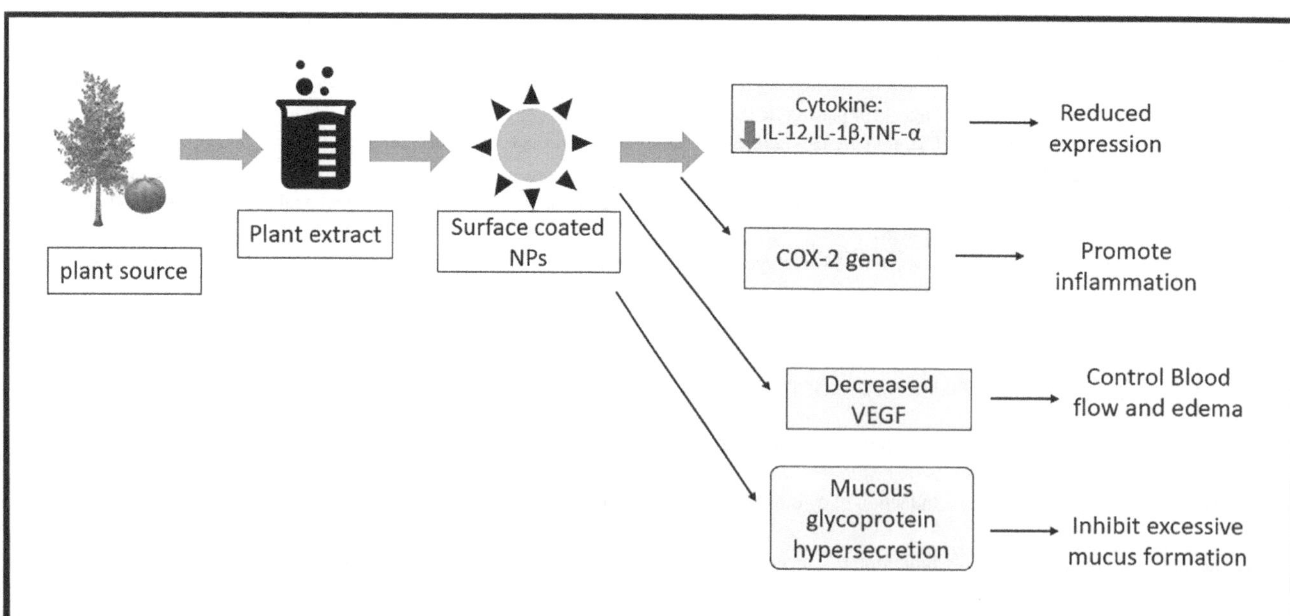

FIGURE 26.4 Plant-mediated synthesis of nanoparticles for anti-inflammatory applications.

TABLE 26.4
Green-synthesized nanoparticles with plant extracts are utilized in anti-inflammatory treatment

Plants	Type of Green Nanoparticle	Mechanism of Action	Biomedical Application	Source
Pelargonium odoratissimum (L.)	ZnO nanoparticles	Involves membrane stabilization, preventing lysosomal enzyme release and oxidative damage	Anti-inflammatory activity to inhibit hypotonicity-induced RBC membrane lysis	(Abdelbaky et al., 2022)
Astragalus tribuloides Delile	Silver nanoparticles	Inhibits lysosomal component release of neutrophils at the inflammation location	Anti-inflammatory effect to inhibit protein denaturation	(Sharifi-Rad et al., 2020)
Viburnum opulus L.	Silver nanoparticles	Involves antioxidant effects, apoptosis regulation and modulation of inflammatory mediators	Downregulating pro-inflammatory cytokines (TNF-α, IL-6), reducing oxidative stress and inhibiting neutrophil infiltration	(Moldovan et al., 2017)
Polycladia crinita	Selenium	Exhibits anti-inflammatory activity by modulating inflammatory mediators	Inhibiting COX-2, IL-1β and PGE2 while reducing oxidative stress	(Almurshedi et al., 2023)
Cystoseira tamariscifolia	Gold nanoparticle	Inhibits COX, LOX and NF-κB, reducing pro-inflammatory mediators	Helping mitigate oxidative stress and inflammation	(Machado et al., 2019)
Polygala tenuifolia	Zinc oxide nanoparticles	Suppresses both mRNA and protein expressions of iNOS, COX-2, IL-1β, IL-6 and TNF-α	Regulating the expressions of various genes encoding pro-inflammatory cytokines	(Nagajyothi et al., 2015)
Curcuma pseudomontana	Gold nanoparticles	Involves membrane stabilization by inhibiting lysosomal enzyme release, similar to the action of standard anti-inflammatory drugs like diclofenac	HRBC membrane stabilization, which prevents hemolysis caused by hypotonic stress	(Muniyappan et al., 2021)

26.7 ANTIOXIDANT PROPERTIES

Silver nanoparticles made from natural extracts have the potential to donate electrons and neutralize harmful free radicals. The antioxidant properties are confirmed by 2,2-diphenyl-1-picrylhydrazyl (DPPH), Cupric Ion Reducing Antioxidant Capacity (CUPRAC), Silver Nanoparticle Antioxidant Capacity (SNPAC) tests (Scampicchio et al., 2006). These nanoparticles are established and maintained by natural components of plant extracts like flavonoids, polyphenols and proteins. When these chemicals adhere to the surface of nanoparticles, antioxidant activity is increased.

AgNPs also carry a potent negative charge, which prevents them from clumping together. Therefore, they maintain stability and are active for a longer period. Their antioxidant properties depend on their size, shape and surface charge, which differ according to the plant extract used in synthesis (Flieger et al., 2021).

26.7.1 Antioxidant Mechanisms of Plant-Based Nanoparticles

Green-synthesized nanoparticles are involved in several ways to fight against oxidative stress. They can neutralize free radicals either by donating or accepting electrons, which alters reactive oxygen species and decreases oxidative stress. The bioactive substances on the surface of nanoparticles also help by acting as hydrogen or electron donors to improve antioxidant effects. The compounds contain the hydroxyl group (–OH); the number present also will play a role in radical scavenging. Antioxidant activity also depends on the concentration of nanoparticles used and can be evaluated by assays such as DPPH and ABTS. Also, natural reducing agents enable nanoparticle production and act as stabilizes, improving their efficacy in free radical scavenging (Bedlovičová et al., 2020). For instance, in bacterial cells, ROS formation and potent inhibition of the activity of catalase (CAT), a main enzyme helpful in the removal of hydrogen peroxide and superoxide dismutase in *E. coli* cells, is achieved by SeNP synthesis, which has antioxidant activity (Grudniak et al., 2025).

26.8 DIAGNOSTIC IMPLEMENTATION

Green nanotechnology in diagnostics uses nanoparticles for early disease detection, DNA sequencing through nanopores, single-virus identification, and imaging. Quantum dots are utilized for imaging techniques, while polymeric nanospheres provide accurate imaging. The advancements improve disease screening, enhance imaging contrast and allow real-time monitoring of biomolecules in cells (Yezdani et al., 2018).

Green nanotechnology in agriculture involves using nano-biosensors for rapid identification of plant diseases. In the current era of pesticide resistance and residual toxicity, incorporating nanoparticles has become a novel approach to disease management. Metal-based nanoparticles, particularly silver, zinc, and copper, are used to control diseases due to their antibacterial and antifungal properties (Vinay et al., 2021).

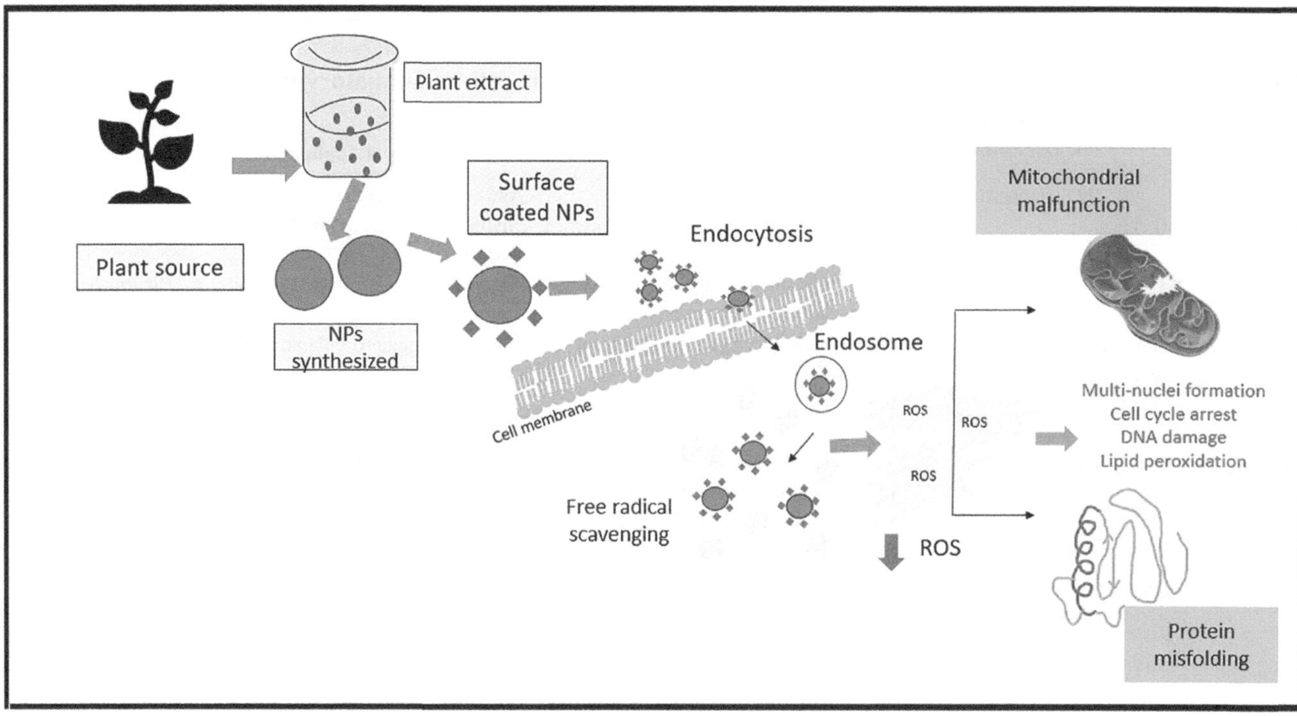

FIGURE 26.5 Plant-mediated synthesis of nanoparticles for antioxidant applications.

TABLE 26.5
Green Plant Extract with Nanoparticles in Application with Antioxidant Activity

Plants	Type of Green Nanoparticle	Mechanism of Action	Biomedical Application	Source
Hibiscus rosa-sinensis	Copper	Scavenging free radicals, such as hydrogen peroxide, thereby mitigating oxidative stress.	It shows antimicrobial properties against pathogens like *E. coli* and *Bacillus subtilis* and potential anti-lung cancer activity.	(Ramasamy & Selvam, 2015)
Dioscorea bulbifera	Copper	Scavenging free radicals (DPPH, nitric oxide, superoxide) due to ascorbic acid content.	Includes antidiabetic effects by inhibiting α-amylase and α-glucosidase, helping to manage type 2 diabetes.	(Piyush More et al., 2015)
Couroupita guianensis	Gold	Neutralizing free radicals through electron donation and stabilizing reactive oxygen species.	Fruits protect cells from oxidative stress-related damage and are used to treat colds and stomach aches, while leaves have been used to cure skin-related diseases.	(Kumar et al., 2016)
Rhus coriaria	Gold	DPPH radical scavenging effect, reduction in oxidative stress.	Anti-esophageal cancer activity, organ protection (hepatic, renal and gastric).	(Liu et al., 2020)
Hippophae rhamnoides	Silver	Act as antioxidants by scavenging oxygen-based free radicals.	Reduce infections in burn treatment and arthroplasty, bacteria colonization on prostheses, urinary catheters, vascular grafts, dental materials, human skin irritation, human immunodeficiency virus (HIV), and SARS-COVID-19.	(Moges & Goud, 2022)

In diagnostic applications, green nanotechnology produces biocompatible gold nanoparticles synthesized from natural components. These nanoparticles are used in various imaging modalities, including X-ray contrast agents, single-photon emission computed tomography (SPECT), positron-emission tomography (PET), optical coherence tomography (OCT), fluorescence imaging, magnetic resonance imaging (MRI), surface-enhanced Raman spectroscopy (SERS), and ultrasound imaging. They not only enhance imaging sensitivity and specificity but also reduce toxicity (Thipe et al. 2022).

Green-synthesized nanoparticles have shown promising applications in drug delivery, cancer therapy, antimicrobial treatments and tissue regeneration due to the nanoparticles' unique properties like high surface area and bioactive compounds, improved drug delivery and reduced side effects. Moreover, platinum nanoparticles (PNPs) fight against antimicrobial resistance and support wound healing. Plant-derived nanoparticles involve early identification of disease and pathogens via targeted imaging utilizing biofunctionalized NPs and biosensors (Karnwal et al., 2024).

AuNPs synthesized from green tea leaves were employed for the detection of CD4, a cancer biomarker, via ball resonator optical fibers. The integration of ball resonator optical fibers and nanoparticles improves both efficacy and sensitivity. Bekmurzayeva et al. documented the application of a fiberoptic ball resonator for the detection of cancer biomarkers. In accordance with these concepts, Ashikbayeva et al. integrated green-synthesized gold nanoparticles (2023).

The application of phytonanotherapy in managing diabetes mellitus, a multi-organ problem, exhibits a synergistic impact by mixing plant and metal nanoparticles, hence surpassing synthetic monotherapy (Anand et al., 2017).

According to Medina-Cruz et al., osteogenic diagnosis is possible through plant-based nanoparticles, as it is a nontoxic diagnostic platform and biologically acceptable. This process reduces harm to the environment and improves image quality, facilitating early disease detection (Medina-Cruz et al. 2020). Moreover, selenium nanoparticles, which are green synthesized, can be used to identify seizures in epilepsy conditions precisely. Selenium NPs enhance biosensing and imaging quality and facilitate early detection and effective monitoring of epilepsy (Othman et al., 2023).

Utilizing gold NPs synthesized from *Silybum marianum* for the diagnosis of gallstones, which is a non-invasive method of detection, AuNPs enhance imaging and promote early identification of cholesterol-fed animals, which improves accuracy and reduce toxicity (Chandran, 2020). Similarly, AuNPs and ZnO nanorods in combination with plant-based enzymes from moss extract have improved the detection of phenylalanine in the phenylketonuria (PKU) disease condition. The modification of paper-based electrodes provides a precise identification of PKU through biosensing applications (Rahimi-Mohseni et al., 2021).

In the case of cardiovascular tissue engineering, green-synthesized nanoparticles were used to produce nanofeatures which crosslink nanomaterials into cardiac patches and grafts, hence enhancing cell adhesion and differentiation, as well as improving scaffold properties (Benko et al., 2022). They are further used for myocardial regeneration after myocardial infraction (MI), also known as a heart attack (Tang et al., 2024).

26.9 GREEN NANOPARTICLES IN DRUG DELIVERY

Green nanoparticles are used in drug delivery applications for their distinct physiochemical properties and environmentally sustainable synthetic processes. The nano size and high surface area-to-volume ratio of green nanoparticles improved drug delivery, hence enhancing bioavailability, controlled drug release and site-specific drug delivery. Many studies have shown the capability of green-synthesized nanoparticles in the delivery of antidiabetic and antitumor drugs, antibiotics and other therapeutic molecules to reduce unwanted effects (Zafar et al., 2020).

The main advantage of using these nanoparticles is their capacity to encapsulate therapeutic molecules to protect them from premature denaturation. Encapsulation of nanoparticles will improve the site-specific targeting, stability and enhanced bioavailability of therapeutic molecules (Garg et al., 2021). In IV drug administration, nanoparticles enhance bioavailability and minimize off-target specific accumulation (Jahangirian et al., 2017). The biocompatible nature and effective clearance of green-synthesized nanoparticles from the body decrease toxicity and improve the treatment effectively. Studies have shown that green nanoparticles facilitate controlled and prolonged release in the body (Garg et al., 2021). Usage of these green nanoparticles for drug delivery is an advantage because of their small dimensions, drug encapsulation, high affinity to the targeted receptor and improved stability properties (Garg et al., 2021; Jahangirian et al., 2017; Velusamy et al., 2016). The site-specific delivery of nanoparticles in anticancer activity is promising, as the nanoparticles deliver the therapeutic molecules to the target cancer tissue while not affecting healthy tissues. Green nanoparticles are used to enhance the bioavailability and solubility of weak polar therapeutic molecules for their efficacy. The unique properties and less toxic effects make them more suitable than conventional drug delivery systems (Garg et al., 2021).

Conventional drug delivery systems have limitations like low bioavailability, low solubility, high dosages, low selectivity and stability issues of the molecules. Second-generation drug delivery systems (1980–2010) focused on the existing limitations and conquering them by introducing nanotechnology, and third-generation drug delivery systems (post-2010) focused on green synthesis of nanoparticles for improving biocompatibility, drug delivery, environmental sustainability and reducing the usage of hazardous chemicals, which pose a risk to humans (Kanwar et al., 2019).

Green-synthesized metal nanoparticles are used in the biomedical field for their unique characteristics, such as catalytic activity, optical properties and antibacterial abilities. Silver and gold nanoparticles are synthesized by using green chemistry principles by using reducing agents like chitosan, starch, tannic acid and other natural polymers (Velusamy et al., 2016). The synthesis of metal nanoparticles is done by using biological sources like plants, algae and microtubes through NADPH-dependent reductase enzymes which convert metal ions into stable nanoparticles. These nanoparticles interact with various biological molecules, leading to cellular permeability, oxidative stress, gene expression and protein synthesis (Sharma et al., 2022).

26.10 TOXICITY OF GREEN NANOPARTICLES

Green nanoparticles have shown hazardous effects, which are determined by their dimensions, concentration and

TABLE 26.6
Different Types of Green Nanoparticles Used in Drug Delivery

Source	Types of Green Nanoparticles	Biomedical Applications	Reference
Cyanobacteria	Carbon quantum dots (C-QDs)	DOX-loaded C-QDs enhanced anticancer activity against MCF-7 and HepG2 cells	(Lee et al., 2014)
Aloe barbadensis leaf extract	Green-synthesized zinc oxide nanoparticles	ZnO-NPs are used for cancer drug delivery	(Batool et al., 2021)
Rhubarb Extract	Silver nanoparticles	Ag-NPs is used in pharmaceutical formulations for antimicrobial properties	(Palem et al., 2018)
Phoenix dactylifera pollen extract	Gold and silver nanoparticles	Au-NPs and Ag-NPs are used in cytotoxicity against MCF-7 breast cancer cells	(Banu et al., 2018)
Aloe barbadensis leaf extract	PEGylated zinc oxide nanoparticles (PEG-ZnO-NPs)	PEG-ZnO-NPs are used as a pH-responsive targeted drug which releases in the tumor microenvironment	(Batool et al., 2021)
Fusarium oxysporum	Gold-silver alloy nanoparticles	Au-AgNPs are used in extracellular synthesis via nitrate-dependent enzyme and shuttle quinone reduction	(Senapati et al., 2005)
Poa annua extract	Silver nanoparticles	Ag-NPs are evaluated for biocompatible anti-cancer therapy	(Gul et al., 2021)

surface characteristics. Their nano size makes them simple cellular entry as small particles can enter easily, leading to circulation within the bloodstream, which accumulates in the organs, resulting in hazardous consequences. Smaller nanoparticles (1–10 nm) have enhanced cytotoxicity due to high cellular uptake, whereas bigger nanoparticles (50–100 nm) accumulate in the tissues over time. Positively charged nanoparticles have high interaction with cell membranes, resulting in increased cytotoxicity, whereas PEG-coated nanoparticles have shown the ability to change immunological responses.

The toxicity of green nanoparticles is caused mainly by oxidative stress, as green nanoparticles produce reactive oxygen species, leading to DNA damage, lipid peroxidation and altered antioxidant enzyme levels. These actions can cause cellular death, mitochondrial malfunction and reduced cell viability. Furthermore, green nanoparticles have shown the capacity to modify gene expression, generate chromosomal abnormalities and cause DNA strands to break, which results in genotoxicity. Inflammatory responses are characterized by the release of pro-inflammatory cytokines such as interleukin-6 (IL-6), interleukin-1β (IL-1β) and tumor necrosis factor (TNF-α), causing persistent inflammation and organ-specific damage.

Inhalation of green nanoparticles like AgNPs is risky, as these nanoparticles can infiltrate the respiratory system, resulting in oxidative stress, lipid peroxidation and cytotoxic effects in airway epithelial cells. Long-term exposure to these nanoparticles may lead to damage to lung tissues or respiratory irritation, and in extreme situations, tumor formation can be observed.

Organ-specific toxicity is a major issue for the toxicity of green nanoparticles, as these nanoparticles concentrate in the kidneys, spleen, liver and brain, causing renal impairment, hepatic injury and neurotoxicity. Specific nanoparticles can cross the blood–brain barrier (BBB) and accumulate in neurons, causing gene expression to change, resulting in neurodegenerative consequences. Nanoparticles affect sperm viability and motility in the reproductive system, and some studies have shown that estrogen is built up in granulosa cells, resulting in hormonal imbalances. Skin toxicity can occur based on a particular nanoparticle type, as nanoparticles cause an increase in apoptosis and oxidative stress in keratinocyte cell lines. These toxic effects of nanoparticles in the organs are raising concerns over their usage for a long period of time, particularly in biomedical applications (Garg et al., 2021).

Silver and gold nanoparticles exhibit different degrees of toxicity from other metal nanoparticles. AgNPs emit harmful silver ions (Ag^+), which increase oxidative stress, induce DNA damage and impair mitochondrial function, whereas AuNPs show reduced toxicity compared to AgNPs. The toxicity of AgNPs works based on a Trojan horse mechanism where the cells ingest nanoparticles and release silver ions, thereby interrupting cellular functions. Smaller AgNPs (<20 nm) exhibit greater toxicity, causing enhanced cellular absorption and reactivity. ZnO nanoparticles have shown the capacity to generate oxidative stress, apoptosis and inflammation; hence, these nanoparticles are used as anti-cancer medicines. However, this also raises concerns over their toxicity in biological systems (Amuthavalli et al., 2021; de Souza et al., 2019).

Despite the therapeutic effects of green nanoparticles in drug delivery, their clinical applications have been limited due to toxicity issues. Certain nanodrug delivery technologies have shown adverse effects on cell membranes, impairing normal cellular activities and raising safety concerns about prolonged usage. Green-synthesized AgNPs are used in antibacterial activity, but they continue to have toxicity hazards. These nanoparticles, after releasing into the environment, experience aggregation, oxidation and adsorption, which affects their bioavailability and toxicity. Oxidation releases harmful silver ions, and aggregation causes bioavailability modification, occasionally reducing uptake of nanoparticles. The adsorption of natural organic matter (NOM) can either stabilize AgNPs or extend the exposure of hazards. (de Souza et al., 2019; Uppal et al., 2018).

The extensive usage of green nanoparticles has resulted in accumulation in the environment, causing contamination, which is potentially harmful for living beings on continuous exposure. Their non-selective antimicrobial characteristics cause hazards to eukaryotic cells, typically in the range of <50 nm. In the environment, green nanoparticles are present in air and water, resulting in bioaccumulation and ecological modifications such as adjustments in pH and salinity. Each green nanoparticle's toxicity varies based on composition, coating and solubility. For example, green-synthesized AgNPs and CuNPs exhibit more toxicity than TiO_2NPs. Cytotoxic effects can occur based on the dosage, as elevated doses (>50 mg/mL) can cause reduced cell viability. In plants, green nanoparticles can either enhance or delay seed germination and root development in agricultural applications, which raises concerns about toxicity issues at elevated concentrations (Rana et al., 2020).

Bioavailable silver ions are recognized as principal factors posing harmful effects in zebrafish embryos. In vitro studies have shown that AgNP exposure results in mitochondrial dysfunction and membrane leakage, causing cytotoxicity in murine germline stem cells. Furthermore, AgNPs have been documented to increase harmful responses by influencing cytokine expression and proliferation by the generation of peripheral blood mononuclear cells (PBMCs). In vivo studies on rats have shown that AgNP exposure causes cytotoxicity, with histological findings indicating bile duct hyperplasia, pigmentation and fibrosis resulting from excessive AgNP exposure (Jabeen et al., 2021).

26.11 LIMITATIONS

Green nanoparticles have generated interest in biomedical research for their environmentally good manufacturing, biocompatibility and future advancements in drug transport, treatment of cancer and antimicrobial therapies. Despite their benefits, several issues have raised concerns about utilizing these nanoparticles broadly in medicine (Iravani, 2011).

One of the limitations of using green nanoparticles is the absence of sustainable large-scale production, as some plants used as sources to create nanoparticles are not readily available. *Lithodora hispidula*, used for PdNP production, is restricted to Cyrenaica in Libya. Other plants such as *Euphorbia* and *Sapium sebiferum* are used in the production of PdNP and are predominantly located in subtropical locations, hence limiting their availability for wide applications (Sharma et al., 2022). For the synthesis of green nanoparticles, finding raw materials is an issue, as some plants can only be harvested in certain growth seasons. For example, the synthesis of AgNP requires cotton leaves, which can be harvested during the flowering season, or *Sargassum fusiforme*, which can be accessible only in a particular region. Some raw materials are classified as secondary products and need additional processing, hence making the processing more complex and expensive. The coffee powder used in CuO-NP synthesis requires further refining, which rises questions about economic viability (Ying et al., 2022).

In addition to the raw materials, processing for the synthesis of green nanoparticles also has challenges: it consumes high energy, relying on industrial chemical reagents and prolonged durations of reactions. Preserving plant extracts for later usage is also a challenge. For example, *Azadirachta indica* leaf extract and dried grass extracts should be stored in low-temperature refrigeration at 4°C. This raises operational expenses and energy usage; hence, we need to choose room-temperature synthesis as a better option for overcoming this obstacle. The extraction procedure is time consuming, making it unsuitable for large-scale production. Furthermore, limited information about the biosynthesis pathway makes it difficult to identify accurate chemical reactions. For the scalability of green synthesis, the lack of established production rules for mass balance and stoichiometric ratios makes it a challenge. Nanoparticles synthesized from various plants differ in their characteristics, resulting in different particle sizes. This situation complicates the maintenance of nanoparticle properties during large-scale manufacturing, hence limiting dependence on green synthesis. The yield and conversion rate of green nanoparticles are frequently low, limiting their usage in industrial applications (Ying et al., 2022).

Toxicity and biocompatibility are issues associated with the usage of green nanoparticles even if green synthesis minimizes using hazardous chemicals. As discussed regarding toxicity, green nanoparticles may provoke DNA damage, oxidative stress and death in mammalian cells, resulting in inflammatory and immunogenic reactions (Siddiqi et al., 2018). Inhalation of these nanoparticles can result in lung damage and serious illness. The usage of polyvinyl alcohol as a stabilizing agent in drug delivery is raising concerns. Nanoparticles induce alveolar inflammation, cytotoxicity and autonomic dysfunction, hence directly impacting vascular and cardiac functions (Parveen et al., 2016).

When we synthesize green nanoparticles, we use plant extracts and others as a source. When plant extracts are used, their phytochemical content makes it difficult to understand the redox reactions in nanoparticles, which in turn affects the size, shape, and crystal structure of silver nanoparticles (AgNPs). These phytochemicals are either upregulated or downregulated based on intrinsic or extrinsic factors influencing the repeatability, homogeneity, scalability and polydispersity of AgNPs (Velidandi et al., 2020).

Further research is needed to enhance the physicochemical properties of green nanoparticles for specific applications. A key to better control over nanoparticle size and shape is optimizing reactants like microbes, plant extracts and fermentation media and factors like pH, temperature and stirring speed. We need to understand the plant and microbial components involved in green synthesis.

Major issues like the absence of standard regulatory frameworks, limited toxicological data and characterization methods hinders approval and usage in biomedical fields (Iravani, 2011). Quality control of these nanoparticles is difficult, as they vary in production conditions, affecting commercial viability (Ahmed et al., 2016). To fully utilize these nanoparticles in the medical field, advancements in synthesis, safety testing and large-scale production are required.

26.12 FUTURE PROSPECTS

The potential of using green nanotechnology in biomedical applications is good and efficient but needs improvements soon. In drug delivery, green nanoparticles are anticipated to enhance treatment by selectively targeting drugs and reducing toxic effects. In cancer therapy, plant-derived AgNPs and AuNPs are significant for their capacity to selectively target and destroy cancer cells by inducing oxidative stress and causing DNA damage. Antimicrobial applications are used widely, as green nanoparticles are used for wound dressings, medical device coatings and implants to reduce severe infections, particularly infections raised by antibiotic-resistant bacteria.

In anti-inflammatory therapies, plant-derived nanoparticles may provide new methods to reduce inflammation and oxidative stress, thus being beneficial for diseases like arthritis and wound healing. Green nanotechnology is anticipated to enhance biosensors and imaging modalities like MRI and PET scans for early and precise diagnosis in detecting diseases. These diagnosis methods should be less invasive and more efficient. In tissue engineering, the usage of green nanoparticles in making cardiovascular tissue is beneficial and can increase the scaffold materials used in heart tissue repair, getting better results for patients suffering from cardiac problems.

Although these developments are encouraging, issues need to be addressed. Guaranteeing the safety of using nanoparticles for the body and preventing their accumulation in organs over time is a concern which needs to be addressed. Mass production is a significant challenge, as we need to identify the methods by which these nanoparticles are produced to preserve their distinct qualities. Regulatory approvals and clinical trials will influence the rate of these technologies to achieve widespread availability. Current research is focusing on providing safer, more sustainable and highly effective treatment, diagnostics and disease prevention options of green nanotechnology, keeping in mind the obstacles mentioned before.

26.13 CONCLUSION

Green nanotechnology has evolved as a new and sustainable method in biomedical applications, providing sustainable solutions to different healthcare issues. This chapter emphasizes different applications of green-synthesized nanoparticles in drug delivery, tissue engineering, diagnostics, wound healing and antimicrobial therapy. Green nanotechnology provides a successful alternative to traditional nanoparticle production and processes by reducing toxicity, improving biocompatibility and encouraging environmentally sustainable solutions. The chapter also shows how plant extracts and other natural sources might be used for therapeutic purposes in nanoparticles while reducing side effects. Continued research, especially involving clinical validation and integrating AI/ML tools, is important to optimize nanoparticle design, ensuring safe clinical use and predicting biological responses. As advancements are occurring, green nanotechnology plays an important role in shaping a safer, more sustainable and innovative future for the healthcare sector.

REFERENCES

Abdelbaky, A. S., Abd El-Mageed, T. A., Babalghith, A. O., Selim, S., & Mohamed, A. M. H. A. (2022). Green synthesis and characterization of ZnO nanoparticles using Pelargonium odoratissimum (L.) aqueous leaf extract and their antioxidant, antibacterial and anti-inflammatory activities. *Antioxidants*, *11*(8), 1444. https://doi.org/10.3390/antiox11081444

Agarwal, H., Nakara, A., & Shanmugam, V. K. (2019). Anti-inflammatory mechanism of various metal and metal oxide nanoparticles synthesized using plant extracts: A review. *Biomedicine & Pharmacotherapy*, *109*, 2561–2572. https://doi.org/10.1016/j.biopha.2018.11.116

Ajaykumar, A. P., Sabira, O., Binitha, V. S., Varma, S. R., Mathew, A., Jayaraj, K. N., Janish, P. A., Zeena, K. V., Sheena, P., Venugopal, V., Palakkapparambil, P., & Aswathi. (2023). Bio-fabricated silver nanoparticles from the leaf extract of the poisonous plant, holigarna arnottiana: Assessment of antimicrobial, antimitotic, anticancer, and radical-scavenging properties. *Pharmaceutics*, *15*(10), 2468. https://doi.org/10.3390/pharmaceutics15102468

Alex, A. M., Subburaman, S., Chauhan, S., Ahuja, V., Abdi, G., & Tarighat, M. A. (2024). Green synthesis of silver nanoparticle prepared with Ocimum species and assessment of anticancer potential. *Scientific Reports*, *14*(1). https://doi.org/10.1038/s41598-024-61946-y

Almurshedi, A. S., El-Masry, T. A., Selim, H., El-Sheekh, M. M., Makhlof, M. E. M., Aldosari, B. N., Alfagih, I. M., AlQuadeib, B. T., Almarshidy, S. S., & El-Bouseary, M. M. (2023). New investigation of anti-inflammatory activity of Polycladia crinita and biosynthesized selenium nanoparticles: Isolation and characterization. *Microbial Cell Factories*, *22*(1), 173. https://doi.org/10.1186/s12934-023-02168-1

Amuthavalli, P., Hwang, J. S., Dahms, H. U., Wang, L., Anitha, J., Vasanthakumaran, M., et al. (2021). Zinc oxide nanoparticles using plant Lawsonia inermis and their mosquitocidal, antimicrobial, anticancer applications showing moderate side effects. *Sci Rep.*, *11*(1), 8837.

Andleeb, S., Tariq, F., Muneer, A., Nazir, T., Shahid, B., Latif, Z., et al. (2020). In vitro bactericidal, antidiabetic, cytotoxic, anticoagulant, and hemolytic effect of green-synthesized silver nanoparticles using Allium sativum clove extract incubated at various temperatures. *Green Processing and Synthesis*, *9*(1), 538–53.

Anand, K., Tiloke, C., Naidoo, P., & Chuturgoon, A. A. (2017). Phytonanotherapy for management of diabetes using green synthesis nanoparticles. *Journal of Photochemistry and Photobiology B: Biology*, *173*, 626–639. https://doi.org/10.1016/j.jphotobiol.2017.06.028

Ashikbayeva, Z., Bekmurzayeva, A., Myrkhiyeva, Z., Assylbekova, N., Atabaev, T. Sh., & Tosi, D. (2023). Green-synthesized gold nanoparticle-based optical fiber ball resonator biosensor for cancer biomarker detection. *Optics & Laser Technology*, *161*, 109136. https://doi.org/10.1016/j.optlastec.2023.109136

Banu, H., Renuka, N., Faheem, S. M., Ismail, R., Singh, V., Saadatmand, Z., et al. (2018). Gold and Silver Nanoparticles Biomimetically Synthesized Using Date Palm Pollen Extract-Induce Apoptosis and Regulate p53 and Bcl-2 Expression in Human Breast Adenocarcinoma Cells. *Biol Trace Elem Res [Internet]*. *186*(1), 122–34. Available from: https://doi.org/10.1007/s12011-018-1287-0

Batool, M., Khurshid, S., Daoush, W. M., Siddique, S. A., Nadeem, T. (2021). Green synthesis and biomedical applications of ZnO nanoparticles: Role of pegylated-zno nanoparticles as doxorubicin drug carrier against MDA-MB-231(TNBC) Cells Line. *Crystals (Basel)*, *11*(4).

Bedlovičová, Z., Strapáč, I., Baláž, M., & Salayová, A. (2020). A brief overview on antioxidant activity determination of silver nanoparticles. *Molecules*, *25*(14), 3191. https://doi.org/10.3390/molecules25143191

Benko, A., Truong, L. B., Medina-Cruz, D., Mostafavi, E., Cholula-Díaz, J. L., & Webster, T. J. (2022). Green nanotechnology in cardiovascular tissue engineering. In *Tissue Engineering* (pp. 237–281). Elsevier. https://doi.org/10.1016/B978-0-12-824064-9.00012-5

Bharadwaj, K. K., Rabha, B., Pati, S., Sarkar, T., Choudhury, B. K., Barman, A., Bhattacharjya, D., Srivastava, A., Baishya, D., Edinur, H. A., Abdul Kari, Z., & Mohd Noor, N. H. (2021). Green synthesis of gold nanoparticles using plant extracts as beneficial prospect for cancer theranostics. *Molecules*, *26*(21), 6389. https://doi.org/10.3390/molecules26216389

Chandran, M., and Kosala, D. (2019). Antilithiasic and hypolipidaemic effect of AuNPs of Silybum marianum on lithogenic diet fed mice gallbladder", *International Journal of Emerging Technologies and Innovative Research*, *6*(6), 345–360.

Cristina Silva Costa D. (2020). Additions to non-activated alkenes: Recent advances. *Arabian Journal of Chemistry*, *13*(1), 799–834.

Das, C. G. A., Kumar, V. G., Dhas, T. S., Karthick, V., & Kumar, C. M. V. (2023). Nanomaterials in anticancer applications and their mechanism of action—A review. *Nanomedicine: Nanotechnology, Biology and Medicine*, *47*, 102613. https://doi.org/10.1016/j.nano.2022.102613

David, L., Moldovan, B., Vulcu, A., Olenic, L., Perde-Schrepler, M., Fischer-Fodor, E., Florea, A., Crisan, M., Chiorean, I., Clichici, S., & Filip, G. A. (2014). Green synthesis, characterization and anti-inflammatory activity of silver nanoparticles using European black elderberry fruits extract. *Colloids and Surfaces B: Biointerfaces*, *122*, 767–777. https://doi.org/10.1016/j.colsurfb.2014.08.018

Flieger, J., Franus, W., Panek, R., Szymańska-Chargot, M., Flieger, W., Flieger, M., & Kołodziej, P. (2021). Green synthesis of silver nanoparticles using natural extracts with proven antioxidant activity. *Molecules*, *26*(16), 4986. https://doi.org/10.3390/molecules26164986

Galdiero, S., Falanga, A., Vitiello, M., Cantisani, M., Marra, V., Galdiero, M. (2011). Silver Nanoparticles as Potential Antiviral Agents. *Molecules*, *16*(10), 8894–918.

Garg, R., Rani, P., Garg, R., Eddy, N. O. (2021). Study on potential applications and toxicity analysis of green synthesized nanoparticles. *Turk J Chem.*, *45*(6), 1690–706.

Geetha, R., Ashokkumar, T., Tamilselvan, S., Govindaraju, K., Sadiq, M., & Singaravelu, G. (2013). Green synthesis of gold nanoparticles and their anticancer activity. *Cancer Nanotechnology*, *4*(4–5), 91–98. https://doi.org/10.1007/s12645-013-0040-9

Grudniak, A., Folcik, J., Szmytke, J., & Sentkowska, A. (2025). Mechanism of antioxidant activity of selenium nanoparticles obtained by green and chemical synthesis. *International Journal of Nanomedicine*, *20*, 2797–2811. https://doi.org/10.2147/IJN.S507712

Gul, A. R., Shaheen, F., Rafique, R., Bal. J., Waseem, S., Park, T. J. (2021). Grass-mediated biogenic synthesis of silver nanoparticles and their drug delivery evaluation: A biocompatible anti-cancer therapy. Chemical Engineering Journal [Internet]. *407*, 127202. Available from: https://www.sciencedirect.com/science/article/pii/S1385894720333295

Iravani S. (2011). Green synthesis of metal nanoparticles using plants. *Green Chemistry*. *13*(10), 2638.

Jabeen, S., Qureshi, R., Munazir, M., Maqsood, M., Munir, M., Shah, S. S. H., et al. (2021). Application of green synthesized silver nanoparticles in cancer treatment—a critical review. *Mater Res Express*, *8*(9), 092001.

Jahangirian, H., Lemraski, E. G., Webster, T. J., Rafiee-Moghaddam, R., Abdollahi, Y. (2017). A review of drug delivery systems based on nanotechnology and green chemistry: green nanomedicine. Int J Nanomedicine [Internet]. *12*(null), 2957–78. Available from: https://www.tandfonline.com/doi/abs/10.2147/IJN.S127683

Jorge de Souza, T. A., Rosa Souza, L. R., Franchi, L. P. (2019). Silver nanoparticles: An integrated view of green synthesis methods, transformation in the environment, and toxicity. *Ecotoxicol Environ Saf.*, *171*, 691–700.

Kanwar, R., Rathee, J., Salunke, D. B., Mehta, S. K. (2019). Green Nanotechnology-Driven Drug Delivery Assemblies. *ACS Omega [Internet]*, *4*(5), 8804–15. Available from: https://doi.org/10.1021/acsomega.9b00304

Karnwal, A., Jassim, A. Y., Mohammed, A. A., Sharma, V., Al-Tawaha, A. R. M. S., & Sivanesan, I. (2024). Nanotechnology for healthcare: Plant-derived nanoparticles in disease treatment and regenerative medicine. *Pharmaceuticals*, *17*(12), 1711. https://doi.org/10.3390/ph17121711

Keskin, C., Ölçekçi, A., Baran, A., Baran, M. F., Eftekhari, A., Omarova, S., Khalilov, R., Aliyev, E., Sufianov, A., Beilerli, A., & Gareev, I. (2023). Green synthesis of silver nanoparticles mediated Diospyros kaki L. (Persimmon): Determination of chemical composition and evaluation of their antimicrobials and anticancer activities. *Frontiers in Chemistry*, *11*. https://doi.org/10.3389/fchem.2023.1187808

Sathishkumar G., Pradeep K. Jha, Vignesh V., Rajkuberan C., Jeyaraj M., Selvakumar M., Rakhi Jha, Sivaramakrishnan S. (2016). Cannonball fruit (Couroupita guianensis, Aubl.) extract mediated synthesis of gold nanoparticles and evaluation of its antioxidant activity. *J Mol Liq*. 215, 229–36.

Kumar, S., Choudhir, G., Sharma, S., Vimala, Y., Joshi, N., Tiwari, A., et al. (2023). Computational probing of Nigella sativa bioactive metabolites against chickungunya nsP2 cysteine protease. *J King Saud Univ Sci.*, *35*(4), 102651.

Kumar, C., Singh, A. K., Tomar, A., Amami, R., Ghazouani, H. (2026). Artificial intelligence in sustainable and smart agrochemicals. In: *Artificial Intelligence in Chemical Engineering*. Elsevier, p. 443–70.

Lee, H. U., Park, S. Y., Park, E.S., Son, B., Lee, S. C., Lee, J. W., et al. (2014). Photoluminescent carbon nanotags from harmful cyanobacteria for drug delivery and imaging in cancer cells. *Sci Rep [Internet]*, *4*(1), 4665. Available from: https://doi.org/10.1038/srep04665

Liu, J., Zangeneh, A., Zangeneh, M. M., & Guo, B. (2020). Antioxidant, cytotoxicity, anti-human esophageal squamous cell carcinoma, anti-human Caucasian esophageal carcinoma, anti-adenocarcinoma of the gastroesophageal junction, and anti-distal esophageal adenocarcinoma properties of gold nanoparticles green synthesized by *Rhus coriaria* L. fruit aqueous extract. *Journal of Experimental Nanoscience*, *15*(1), 202–216. https://doi.org/10.1080/17458080.2020.1766675

Machado, S., Gonçalves, A., González-Ballesteros, N., Magalhães, L., Sárria, M., Rodríguez-Argüelles, M. C., & Gomes, A. C. (2019). Comparing toxicity and bioactivity of gold nanoparticles synthetized using two brown algae Cystoseira sp. *Frontiers in Marine Science*, *6*. https://doi.org/10.3389/conf.fmars.2019.08.00155

Moges, A., & Goud, V. V. (2022). Optimization, characterization, and evaluation of antioxidant and antibacterial activities of silver nanoparticles synthesized from Hippophae salicifolia D. Don. *Inorganic Chemistry Communications*, *146*, 110086. https://doi.org/10.1016/j.inoche.2022.110086

Moldovan, B., David, L., Vulcu, A., Olenic, L., Perde-Schrepler, M., Fischer-Fodor, E., Baldea, I., Clichici, S., & Filip, G. A. (2017). In vitro and in vivo anti-inflammatory properties of green synthesized silver nanoparticles using Viburnum opulus L. fruits extract. *Materials Science and Engineering C*, *79*, 720–727. https://doi.org/10.1016/j.msec.2017.05.122

Mousavi, B., Tafvizi, F., & Zaker Bostanabad, S. (2025). Green synthesis of silver nanoparticles using *Artemisia turcomanica* leaf extract and the study of anti-cancer effect and apoptosis induction on gastric cancer cell line (AGS). *Artificial Cells, Nanomedicine, and Biotechnology*, *46*(supl), 499–510. https://doi.org/10.1080/21691401.2018.1430697

Muniyappan, N., Pandeeswaran, M., & Amalraj, A. (2021). Green synthesis of gold nanoparticles using Curcuma pseudomontana isolated curcumin: Its characterization, antimicrobial, antioxidant and anti-inflammatory activities. *Environmental Chemistry and Ecotoxicology*, *3*, 117–124. https://doi.org/10.1016/j.enceco.2021.01.002

Nagajyothi, P. C., Cha, S. J., Yang, I. J., Sreekanth, T. V. M., Kim, K. J., & Shin, H. M. (2015). Antioxidant and anti-inflammatory activities of zinc oxide nanoparticles synthesized using Polygala tenuifolia root extract. *Journal of Photochemistry and Photobiology B: Biology*, *146*, 10–17. https://doi.org/10.1016/j.jphotobiol.2015.02.008

Nagajyothi, P. C., Pandurangan, M., Kim, D. H., Sreekanth, T. V. M., & Shim, J. (2017). Green synthesis of iron oxide nanoparticles and their catalytic and in vitro anticancer activities. *Journal of Cluster Science*, *28*(1), 245–257. https://doi.org/10.1007/s10876-016-1082-z

Othman, M. S., Obeidat, S. T., Aleid, G. M., Al-Bagawi, A. H., Fareid, M. A., Hameed, R. A., Mohamed, K. M., Abdelfattah, M. S., Fehaid, A., Hussein, M. M., Aboelnaga, S. M. H., & Abdel Moneim, A. E. (2023). Green synthetized selenium nanoparticles using syzygium aromaticum (Clove) extract reduce pentylenetetrazol-induced epilepsy and associated cortical damage in rats. *Applied Sciences*, *13*(2), 1050. https://doi.org/10.3390/app13021050

Palem, R. R., Ganesh, S. D., Kronekova, Z., Slávikavá, M., Saha, N., Saha, P. (2018). Green synthesis of silver nanoparticles and biopolymer nanocomposites: a comparative study on physico-chemical, antimicrobial and anticancer activity. *Bulletin of Materials Science [Internet]*., *41*(2), 55. Available from: https://doi.org/10.1007/s12034-018-1567-5

Parveen, K., Banse, V., Ledwani, L. (2016). Green synthesis of nanoparticles: *Their advantages and disadvantages*. In 2016. p. 020048.

Piyush More, S. G., Soham Jagtap, R. N., & Chippalkatti, R. (2015). Antidiabetic and antioxidant properties of copper nanoparticles synthesized by medicinal plant dioscorea bulbifera. *Journal of Nanomedicine & Nanotechnology*, *s6*. https://doi.org/10.4172/2157-7439.s6-007

Pungle, R., Nile, S. H., Makwana, N., Singh, R., Singh, R. P., & Kharat, A. S. (2022). Green synthesis of silver nanoparticles using the tridax procumbens plant extract and screening of its antimicrobial and anticancer activities. *Oxidative Medicine and Cellular Longevity*, *2022*, 1–14. https://doi.org/10.1155/2022/9671594

Rahimi-Mohseni, M., Raoof, J. B., Aghajanzadeh, T. A., & Ojani, R. (2021). Phenylketonuria monitoring in human blood serum by mosses extract/ZnO@Au nanoarrays-loaded filter paper as a novel electrochemical biosensor. *Microchemical Journal*, *160*, 105739. https://doi.org/10.1016/j.microc.2020.105739

Ramasamy, S., & Selvam, M. (2015). Green synthesis of copper nanoparticles from Hibicus rosasinensis and their antimicrobial, antioxidant activities. *Article in Research Journal of Pharmaceutical Biological and Chemical Sciences*. https://www.researchgate.net/publication/281992972

Rana, A., Yadav, K., Jagadevan, S. (2020). A comprehensive review on green synthesis of nature-inspired metal nanoparticles: Mechanism, application and toxicity. *J Clean Prod.*, *272*, 122880.

Rehman, G., Umar, M., Shah, N., Hamayun, M., Ali, A., Khan, W., et al. (2023). Green Synthesis and Characterization of Silver Nanoparticles Using Azadirachta indica Seeds Extract: In Vitro and In Vivo Evaluation of Anti-Diabetic Activity. *Pharmaceuticals*, *16*(12), 1677.

Scampicchio, M., Wang, J., Blasco, A. J., Sanchez Arribas, A., Mannino, S., & Escarpa, A. (2006). Nanoparticle-based assays of antioxidant activity. *Analytical Chemistry*, *78*(6), 2060–2063. https://doi.org/10.1021/ac052007a

Senapati, S., Ahmad, A., Khan, M. I., Sastry, M., Kumar, R. (2005). Extracellular Biosynthesis of Bimetallic Au–Ag Alloy Nanoparticles. *Small.*, *1*(5), 517–20.

Sharifi-Rad, M., Pohl, P., Epifano, F., & Álvarez-Suarez, J. M. (2020). Green synthesis of silver nanoparticles using astragalus tribuloides delile. Root extract: Characterization, antioxidant, antibacterial, and anti-inflammatory activities. *Nanomaterials*, *10*(12), 2383. https://doi.org/10.3390/nano10122383

Sharma, P., Navneet, Kaushal, A. (2022). Green nanoparticle formation toward wound healing, and its application in drug delivery approaches. *European Journal of Medicinal Chemistry Reports [Internet]*., *6*, 100088. Available from: https://www.sciencedirect.com/science/article/pii/S2772417422000607

Siddiqi, K.S., Husen, A., Rao, R. A. K. (2018). A review on biosynthesis of silver nanoparticles and their biocidal properties. *J Nanobiotechnology.*, *16*(1), 14.

Tang, C., Zhou, K., Wu, D., & Zhu, H. (2024). Nanoparticles as a novel platform for cardiovascular disease diagnosis and therapy. *International Journal of Nanomedicine*, *19*, 8831–8846. https://doi.org/10.2147/IJN.S474888

Thakur, P. K., & Verma, V. (2021). Retracted article: A review on green synthesis, characterization and anticancer application of metallic nanoparticles. *Applied Biochemistry and Biotechnology*, *193*(7), 2357–2378. https://doi.org/10.1007/s12010-021-03598-6

Thipe, V. C., Karikachery, A. R., Çakılkaya, P., Farooq, U., Genedy, H. H., Kaeokhamloed, N., Phan, D.-H., Rezwan, R., Tezcan, G., Roger, E., & Katti, K. V. (2022). Green nanotechnology—An innovative pathway towards biocompatible and medically relevant gold nanoparticles. *Journal of Drug Delivery Science and Technology*, *70*, 103256. https://doi.org/10.1016/j.jddst.2022.103256

Uppal, S., Italiya, K. S., Chitkara, D., Mittal, A. (2018). Nanoparticulate-based drug delivery systems for small molecule anti-diabetic drugs: An emerging paradigm for effective therapy. *Acta Biomater.*, *81*, 20–42.

Velidandi, A., Dahariya, S., Pabbathi, N. P. P., Kalivarathan, D., Baadhe, R. R. (2020). A review on synthesis, applications, toxicity, risk assessment and limitations of plant extracts synthesized silver nanoparticles. *NanoWorld J.*, *6*(3), 35–60.

Velusamy, P., Kumar, G. V., Jeyanthi, V., Das, J., Pachaiappan, R. (2016). Bio-Inspired Green Nanoparticles: Synthesis, Mechanism, and Antibacterial Application. *Toxicol Res [Internet]*. *32*(2), 95–102. Available from: https://doi.org/10.5487/TR.2016.32.2.095

Vinay, J. U., Iliger, K. S., & Chikkannaswamy. (2021). Green nanotechnology in agriculture: Plant disease diagnosis to management. In *Innovative Approaches in Diagnosis and Management of Crop Diseases* (1st ed., p. 29). Apple Academic Press.

Wang, C., Mathiyalagan, R., Kim, Y. J., Castro-Aceituno, V., Singh, P., Ahn, S., Wang, D., & Yang, D. C. (2016). Rapid green synthesis of silver and gold nanoparticles using Dendropanax morbifera leaf extract and their anticancer activities. *International Journal of Nanomedicine*, *11*, 3691–3701. https://doi.org/10.2147/IJN.S97181

Wang, L., Xu, J., Yan, Y., Liu, H., Karunakaran, T., & Li, F. (2019). Green synthesis of gold nanoparticles from *Scutellaria barbata* and its anticancer activity in pancreatic cancer cell (PANC-1). *Artificial Cells, Nanomedicine, and Biotechnology*, *47*(1), 1617–1627. https://doi.org/10.1080/21691401.2019.1594862

Wagner, V., Dullaart, A., Bock, A.K., Zweck, A. (2006). The emerging nanomedicine landscape. *Nat Biotechnol.* Oct;24(10):1211–7.

Yezdani, U., Gayoor Khan, M., Kushwah, N., Verma, A., Khan, F., & Khan, M. G. (2018). Application of nanotechnology in diagnosis and treatment of various diseases and its future advances in medicine. *World Journal of Pharmacy and Pharmaceutical Sciences*, *7*(11), 1611–1633. https://doi.org/10.20959/wjpps201818–12703

Ying, S., Guan, Z., Ofoegbu, P. C., Clubb, P., Rico, C., He, F., et al. (2022). Green synthesis of nanoparticles: Current developments and limitations. *Environ Technol Innov.*, *26*, 102336.

Zafar, N., Madni, A., Khalid, A., Khan, T., Kousar, R., Naz, S. S., et al. (2020). Pharmaceutical and Biomedical Applications of Green Synthesized Metal and Metal Oxide Nanoparticles. *Curr Pharm Des.*, *26*(45), 5844–65.

27 An Integrated Approach to Fabricate, Visualize, and Analyze Nanostructured Images Using EBL, DIP, and ToF-SIMS

Mbadiwe S. Benyeogor, Olusegun I. Lawal, Nnaemeka M. Mbachu, Kolawole I. Agbaogun, and Kabiru Alani Olaiya

27.1 INTRODUCTION

Nanotechnology is the art, science, and engineering of controlling and structuring matter on an atomic, molecular, and supramolecular scale (dimensions between approximately 1 and 100 nm) for scientific, biomedical, and industrial applications and has been inspired by the scientific exploration of materials, solid states, complex organic compounds, and biomolecules. The practice of nanotechnology draws strongly on technical know-how from the semiconductor industry (microelectronics, microsystem technology, and thin-film technology), microscopy, pharmacy, biotechnology, and the printing press (i.e. lithographic printing process). For this reason, certain apparatus in these fields find application in a clean room (i.e. confined laboratory or facility where nanotechnology is practiced).

The very idea of miniaturizing physical systems (e.g., electronic circuits, sensors, machines, etc.) dates back to 1949 when the German engineer Werner Jacobi at Siemens AG filed a patent for an integrated semiconductor amplifier (AG 1949). In 1959, the first integrated circuit (IC) was invented by Robert Noyce of Fairchild Semiconductor using the planar process technology. The emergence of nanotechnology, as is known today, can be traced back to a famous Christmas lecture in 1959 by the theoretical physicist Prof. Richard Feynman, during which he envisioned nanotechnology and its potential applications by emphasizing the possibility of making tiny machines (Junk and Riess 2006). The decade that followed quickly saw the development of the first computer microprocessor and the microelectromechanical system (MEMS) by microfabrication and surface micromachining, respectively (Dragoman and Dragoman 2005). However, the dream of Richard Feynman was still far fetched until the year 1981, when the scanning tunneling microscope (STM) was invented by Gerd Binnig and Heinrich Rohrer at IBM Zurich, with which it is now possible to visualize and manipulate individual atoms and also align them into various nanostructures of choice (Quate 2009). Quantitatively, the nano-scale is 1000 times smaller than the micro-scale. But in practice, there are few credible distinctions between nanotechnology and microtechnology, as both have almost the same objective. Therefore, several techniques of microelectronics and microtechnology have become the technical basics of nanofabrication.

Fundamentally, nanotechnology is more about making small and functional material entities such as nanochips, nanosensors, nanomachines, nanobots, pharmaceuticals, biomarkers, and lab-on-chip by a process known as nanofabrication, which is the subject of the present chapter. Hence, the aim of the present chapter is to practicalize some of the basic steps that are crucial during the creation of nanostructures, in particular those involving the use of wafers (e.g. a silicon substrate) as the structural base for nanostructures. We nanoengraved an image on a silicon chip by electron beam lithography (EBL) and analyzed it microscopically, computationally, and spectrometrically. The last two involve digital image processing (DIP) and time-of-flight secondary ion mass spectrometry (ToF-SIMS). In essence, this work aims to highlight the advantages of combining EBL and ToF-SIMS as a viable approach to nanotechnology-based research and product development, as well as the applications of DIP to analyze micrographs of nanostructures.

27.2 BACKGROUND

Photolithography has been the conventional means of transferring and realizing structural patterns onto a chip. However, its resolution is limited by the wavelength of light, which is about 250 nm in the UV region of the electromagnetic spectrum. For highly miniaturized nanostructures, the EBL is more applicable, since the electron has a wavelength shorter than that of light photons (Groves 2014).

27.2.1 The EBL System

The EBL process presents a robust way to create structures on the nanoscale with high precision and resolution. This tool was developed from the modification of the scanning electron microscope (SEM) (Groves 2014). The EBL has the same working principle as photolithography, but unlike the latter, which uses UV light, the EBL uses an electron beam for structuring patterns on substrates. Moreover, following the synthesis of polymethylmethacrylate (PMMA)—an

DOI: 10.1201/9781003644439-27

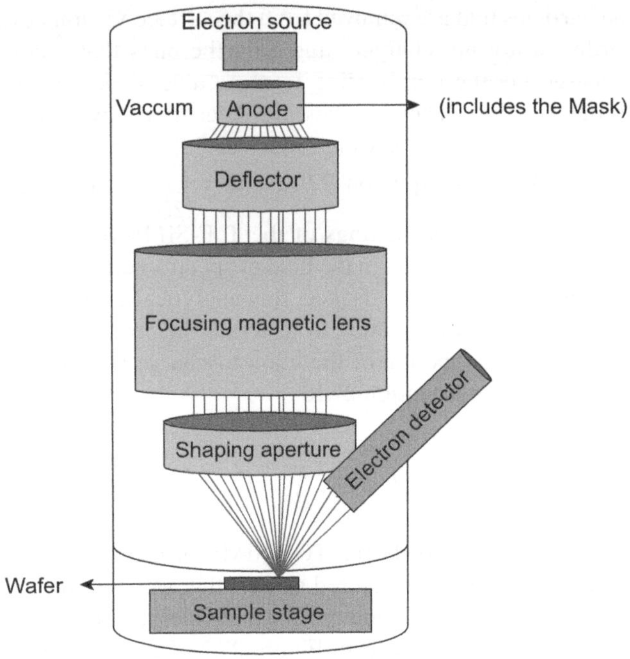

FIGURE 27.1 Diagram of a typical EBL system.

electron-resistant material, which is the EBL equivalent of the photoresist in conventional microtechnology—the EBL has opened a new paradigm for further miniaturization of devices, structures, and patterns for various applications. A typical EBL system, as described in Figure 27.1, consists of a chamber, an electron gun, a column containing all the electron optics (or magnetic lenses) for focusing and directing the electron beam onto the target substrate or some portion of it, and an electron detector with which the process can be microscopically visualized and monitored. To effectively operate this system, the following parameters are considered.

27.2.1.1 Writing Field

This is the largest surface area of the substrate that is exposable to the e-beam, while its mounting stage is stationary. Typically, this ranges from a few tens of μm to nm.

27.2.1.2 Stitching

This is the division of fields that are larger than the maximum writing field of an EBL system into smaller writing fields, which are then stitched together by software-assisted stage movement to recreate the large field or area pattern. Precise stage movement is essential to minimize the stitching error, and this can be achieved by using the laser interferometer-controlled stage. In addition, electron beam deflection must be adjusted to match the stage movement, which is referred to as "writing field alignment."

27.2.1.3 Exposure Dosage

This is a measure of the intensity of e-beam energy incident per unit area A of the substrate. At constant e-beam acceleration voltage, the exposure dosage D is a function of the current I of energy powering the e-beam source, as expressed in Equation 27.1.

$$D = \frac{I}{A} \times t \tag{27.1}$$

where t is the minimum time to expose the given area A for the given dose D, which is measured in μCcm^{-2}. However, the effect of electron scattering by the resist and substrate leads to an undesired effect in the regions adjacent to those exposed to the e-beam. This effect is called the proximity effect, which in turn could affect the quality of the fabricated structure (Hanapiah et al. 2008). To minimize the proximity effect, we employed both experimental and computational means to find the optimal exposure dosage for nano-engraving an image onto our sample.

Overall, the EBL is a key to realizing patterns as nanostructures, opening up possibilities for fabricating artistic patterns on a chip.

27.2.2 ToF-SIMS System

As shown in Figure 27.2, the ToF-SIMS system is mass spectrometry technology for analyzing the chemistry of materials, thin films, organic membranes, and nanostructures in vacuum (Fearn 2015). It is a surface-sensitive analytical method that uses a pulsed primary ion beam to bombard and remove atoms/molecules from the outermost surface of a sample. The particles are removed from atomic monolayers on the surface. These particles are then charged (i.e., secondary ions) as they are accelerated into an electrified flight tube, and their mass is determined by measuring the exact time at which they reach the detector (i.e. time-of-flight). To mathematically explain how this time of flight is calculated, let the potential energy E_p of a charged particle

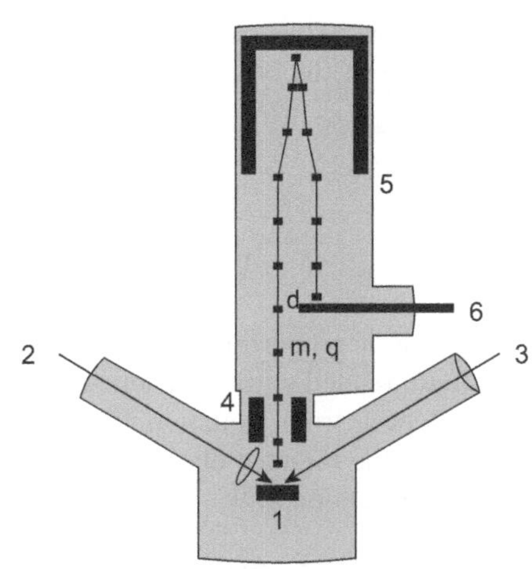

FIGURE 27.2 Internal structure of ToF-SIMS instrument[a].

[a]1: Coolable target holder, 2: primary ion gun (Ar^+), 3: ion gun capable of generating O^{2+} or Cs^+ for depth profiling, 4: secondary metal ion optics (magnetic lens), 5: flight-tube/reflectron, and 6: detector.

in an electric field be related to the quantity of charge q on it by Equation 27.2:

$$E_p = qU \tag{27.2}$$

Here, U is the electric field potential difference or voltage. Upon accelerating the charged particle up the flight tube, E_p is converted to kinetic energy E_k:

$$E_k = \frac{1}{2}mv^2 \tag{27.3}$$

so that:

$$qU = \frac{1}{2}mv^2 \tag{27.4}$$

The final velocity of the accelerated charged particle is uniform since it ascends through a field-free flight tube. Therefore, this uniform velocity can be calculated as the length d of the flight tube divided by the time of flight t:

$$v = \frac{d}{t} \tag{27.5}$$

By substituting Equation 27.5 into Equation 27.4, we have:

$$qU = \frac{1}{2}m\left(\frac{d}{t}\right)^2 \tag{27.6}$$

By making t in Equation 27.6 the subject of the formula, the time of flight of the charged particle is derived as:

$$t = \frac{d}{\sqrt{2U}} \times \sqrt{\frac{m}{q}} \tag{27.7}$$

As factor $\frac{d}{\sqrt{2U}}$ contains the intrinsic parameter of the ToF-SIMS instrument, which in principle would remain constant irrespective of the types of ions being analyzed, we can take $k = \frac{d}{\sqrt{2U}}$ so that:

$$t = k \times \sqrt{\frac{m}{q}} \tag{27.8}$$

In this way, k is a proportional constant representing parameters that are related to the instrument's settings and programming. Hence, from Equation 27.7, the mass-to-charge ratio m/q of the targeted chemical element or molecule is derived in Equation 27.9:

$$m/q = \frac{2U}{d^2} \times t^2 \tag{27.9}$$

The removed particles are then charged (i.e., secondary ions) and accelerated into the electrified flight tube. Most of the particles that are removed from the surface via ion bombardment are not charged, and only the ones that receive a charge via the bombarding ions are able to be detected in the analyzer. Their masses are determined by measuring the exact time t at which they reach the detector (i.e., time-of-flight in Equation 27.7), and $\frac{d}{\sqrt{2U}}$ contains the intrinsic parameters/settings of the ToF-SIMS instrument. Notably, it is a tradition in the field of spectrometry to represent the quantity q as z. Hence, this analytical cycle can be repeated at high frequency to generate the complete mass spectrum as a function of the mass-to-charge ratio m/z, as re-expressed in Equation 27.10:

$$m/z = \frac{2U}{d^2} \times t^2 \tag{27.10}$$

In the present work, the TOF-SIMS instrument is used to analyze the nanostructured sample that we fabricated by EBL, as described in Section 27.3. The internal structure of this instrument is shown in Figure 27.2. The value of t in essence corresponds physically to the value of m/z for each charged element or molecule in flight, and it is measured by the detector, which transmits the data to the supporting computer-controlled instrumentation of the ToF-SIMS apparatus. Here, look-up tables are programmed into the system to map the streaming time logs t to the values of m/z, the intensity of which provides quantitative information about the presence of the corresponding chemical species on the analyte.

27.2.2.1 Dynamics of the ToF-SIMS Process

Any atomic or molecular species x that can be ionized and transported can be analyzed by mass spectroscopy (Fearn 2015). Therefore, ionization occurs near the emission of particles from the surface, and the secondary ion current I_s^x is expressed in Equation 27.11:

$$I_s^x = I_p C_x S \gamma F \tag{27.11}$$

where I_p is the primary ion beam current, C_x the concentration of x, S the Sputter ion yield of x, γ the ionization efficiency, and F the transmittance of x (Note: I_s^x also means the secondary ion counts of species x, and it varies directly as C_x). As shown in Figure 27.3, the primary ions penetrate the surfaces, and energy is lost through a collision cascade. The primary ion is implanted into the solid, and the secondary particles (i.e., ions and neutrals) exit the surface with low energy. This is accompanied by electron (and sometimes x-ray) emission. The escape depth of sputtered species is about a few Å (i.e., ≈ 50 nm). Another important microphysical interaction that is associated with ToF-SIMS is the "matrix effect," which is the influence of the sample's surface chemistry on the ionization process—the formation of secondary ions (Seah and Shard 2018). Consequently, the secondary ions depend on the surrounding chemical profile, especially the material analyte. Thus, the result of any secondary ion mass

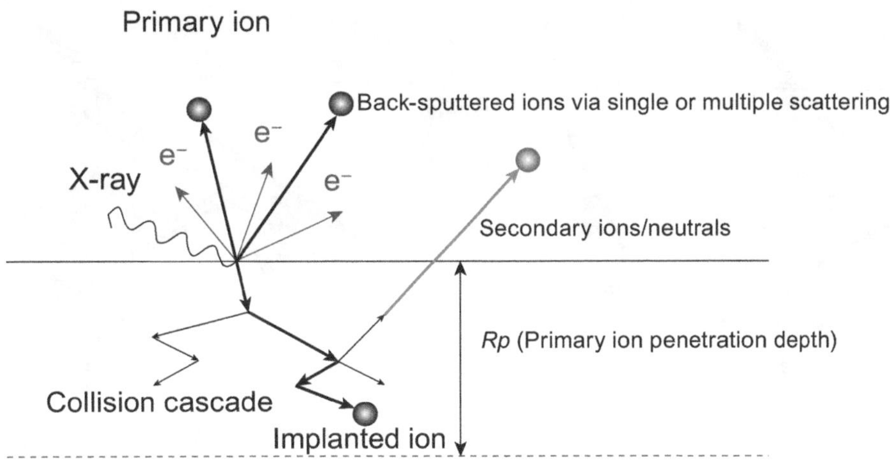

FIGURE 27.3 Generation of secondary ions.

spectrometry such as ToF-SIMS could be affected. For this reason, ToF-SIMS is not very suitable for chemical quantitative analysis (Priebe et al. 2020).

27.2.2.2 Operational Modes of ToF-SIMS

Four operational modes are available for the application of ToF-SIMS: surface imaging (ion images), surface spectroscopy (mass spectra), depth profiling, and 3D analysis. The first two modes involve static ToF-SIMS (which uses less reactive primary ions like Ar^+ or Bi^{3+}), while the last two modes involve dynamic ToF-SIMS (which uses corrosive reagents O^{2+} or Cs^+). Details of our analysis using the ToF-SIMS instrument are discussed in Section 27.4.

27.3 ELECTRON BEAM LITHOGRAPHY METHOD

As EBL is the practice of scanning a focused beam of electrons to draw custom shapes on a surface covered with a film of electron-sensitive resist. Just like in optical lithography, the electron beam changes the solubility of the resist material to enable selective removal of either the exposed or non-exposed regions of the resist by developing it in a solvent, thereby creating miniaturized structures in the resist that can subsequently be transferred onto the substrate material by methods such as etching (McCord and Rooks 2000). According to Equation 27.1, the quality of the nanostructure that is realized is a function of the dosage of the e-beam that is exposed to the target substrate. The main advantage of EBL is that it can be used to draw custom patterns (i.e., direct-write) with sub-10-nm resolution. In this experiment, we practiced how to engrave pictures, text, and QR codes on a chip using the EBL system. For effective EBL, the pictures and other encryption to be structured must be formatted in black-and-white coloration. The picture selected for experimenting with this fabrication method, as sourced from Benyeogor et al. (2022), is shown in Figure 27.4(a). This picture is formatted into the suitable computer-aided design (CAD) formats, shown in Figure 27.4(b), for mask production, using the KLayout CAD software, which was then realized as an engraved chromium-coated glass using a laser writer. The chromium engravement on the glass is an inscription of the image in Figure 27.4 for the EBL fabrication process. To fabricate the pictorial pattern on a chip, a silicon substrate is cleaned by standard clean room procedure and baked at 180°C for 2 min, after which it is spin-coated with the polymethyl methacrylate (PMMA) 4.5 resist at 1700 rpm.

The PMMA is a polymeric material that is suitable for both imaging and nonimaging nanofabrication. It is commonly used as a high-resolution positive resist for direct write, offering extremely high resolution, ease of handling, and excellent film characteristics. It is also insensitive to electromagnetic radiation with wavelengths longer than about 300 nm (Carbaugh et al. 2015). For this reason, we selected it as the resist material for our EBL experiment. After spin-coating the substrate with PMMA, it is once again baked at 180°C for 2 min. The sample is then exposed to an electron beam during the EBL process using various electron beam doses from 600 to 1800 μCcm^{-2} at a current of 50 nA. The exposed sample is then developed in a methyl isobutyl ketone-isopropyl alcohol (MIBK:IPA) solution mixture of 1:3 for 4 min. The development reaction was stopped by rinsing with IPA, after which the developed sample was sputtered with 100 nm film of aluminum for 1 min using the sputtering machine. In the lift-off stage, we submerged the sample in a beaker filled with acetone, sealed it tight with paraffin to prevent the vaporization of the acetone (a volatile compound), and left it for about 12 h to completely lift off. After this time, we sonicated the sample for 2 min to totally remove the resist material. Finally, we rinsed it with water and then dried it with the liquid nitrogen jet. The resulting nanostructure, shown in Figure 27.5, is analyzed by optical/electron microscopy. In addition, the chemical profile of the structure's surface and depth is analyzed by ToF-SIMS. These are discussed as follows.

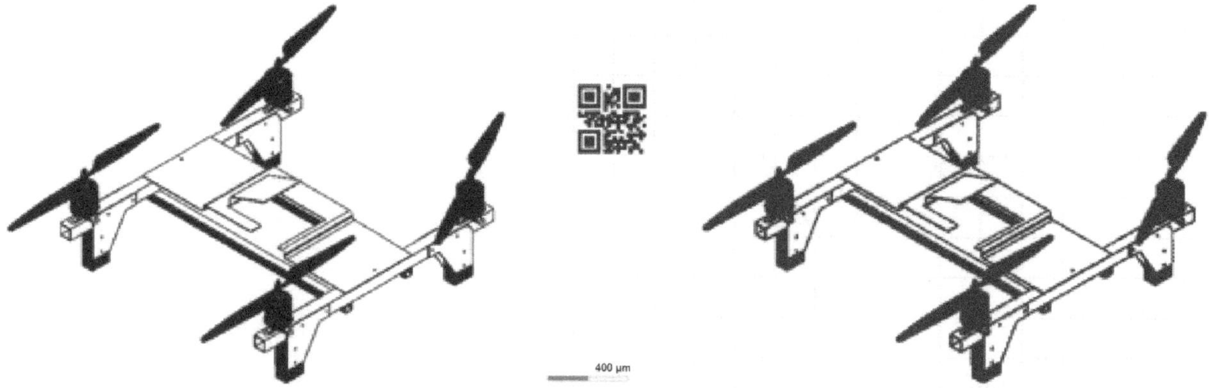

FIGURE 27.4 Adopted image for EBL experiment.

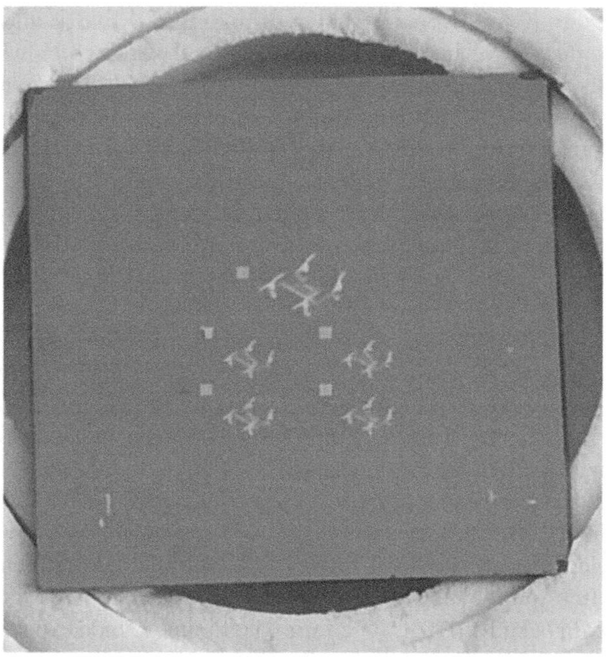

FIGURE 27.5 On-chip nanostructured images fabricated by EBL.

27.4 ANALYSIS OF NANOSTRUCTURES

To evaluate the quality of the EBL-enabled directly written nanostructured end product (shown in Figure 27.5) of the fabrication process described in Section 27.3, two characterization techniques are employed: micrographic analysis of the samples and ToF-SIMS analyses of the sample's chemical composition. These are discussed as follows.

27.4.1 Micrographical Analysis and Visualization

Here, the resulting image structures on the nanochip are micrographically and computationally examined, as hinted earlier, using light microscopy, with which we compared these EBL-fabricated structures of different exposure dosages. Some of the captured micrographs are shown in Figure 27.6.

27.4.1.1 Image Quality Analysis with Python

To determine the optimal exposure dosage that was used in the lithography of the images in Figure 27.6, we applied an image quality assessment (IQA) algorithm based on the Python-encrypted blind/referenceless imaging spatial quality evaluator (BRISQUE) library that extracts natural scene statistics (NSS) and feature vectors of the surface images to predict their image quality scores (Chow and Rajagopal 2017; Mittal et al. 2012; Saad et al. 2012). The lower the score, the better the quality of an image, and vice versa. An example of the global Python program that calls and implements this algorithm is given in Listing 1, in which the name micrograph.png is ascribed to the image file to be analyzed, and it is stored in the same folder as the image processing program or called by specifying the path to it in the computer memory. Similarly, another image file can be processed by simply specifying its file name or/and path in Listing 1. With regard to the current study, the results of this computation are plotted as bars in Figure 27.7. It is therefrom deduced that the pattern that was structured under an exposure dose of 1000 μCcm^{-2} has the optimal quality in terms of resolution and lithographic precision.

LISTING 1

Example of BRISQUE-Based Python Program for Image Quality Assessment of Captured Micrographs

```
# importing useful libraries
import imquality. brisque as brisque
import PIL. Image
import matplotlib. pyplot as plt
# image path
path = ' micrograph. png '
# reading image
img = PIL. Image. open(path)
plt. imshow(img)
plt. axis (' off')
# quality assessment of above image
print ("image-quality { }". format (brisque. score (img)))
```

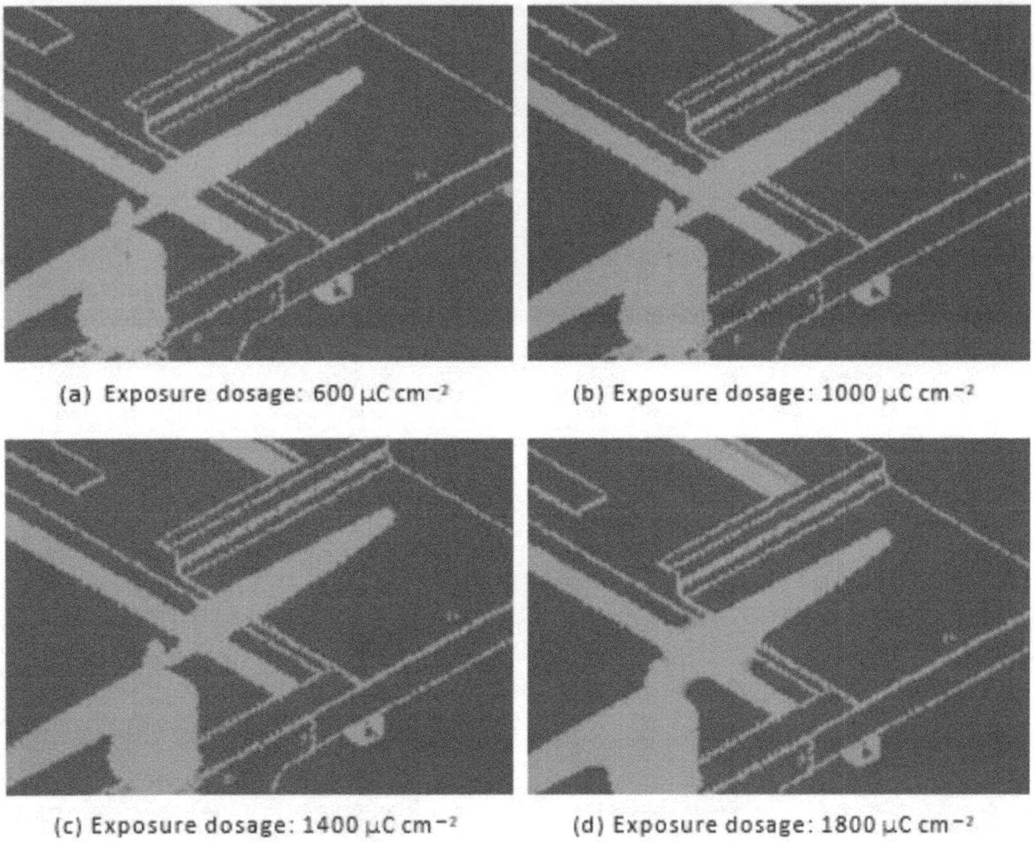

FIGURE 27.6 Optical micrographs of EBL-fabricated sample at different exposure dosages.

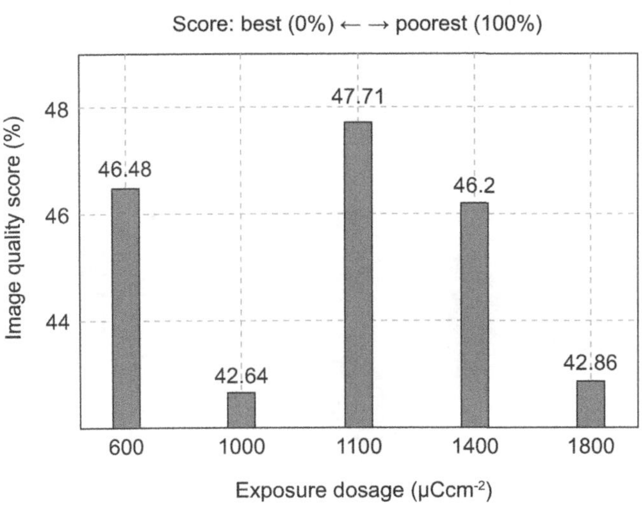

FIGURE 27.7 Bar chart of image quality score vs exposure dosage.

This reveals several pieces of information about the precision, stability, and resolution of the EBL process. Therefore, these micrographs (in Figure 27.6) show a close resemblance to the source images in Figure 27.4. In addition to the proximity effect, other factors such as the acoustical stability of the EBL machine and the presence of ambient contaminants can affect the quality of the fabricated nanostructures. As a precaution, in addition to regular clean-room protocols, the EBL machine was already installed in airtight and air-conditioned confinement. This is specifically meant to avoid the introduction of dust and other unwanted elements or chemical compounds into the samples.

27.4.1.2 Visualization of Structural Defects and Outlines

To further validate the image quality assessment and identify defects in the fabricated nanostructures, we employed the Python Imaging Library (PIL) to create color contours of the micrographs, as depicted in Figure 27.8. These contours visually represent the spatial variation of the deposited materials. Based on textural homogeneity, the micrograph corresponding to the exposure dose of 1000 μCcm^{-2} stands out by exhibiting the highest quality. By applying Algorithm 1, which is based on the Gaussian blur filter (Wang et al. 2014) and Canny edge detector (Ma et al. 2012), we also generated the high-resolution visuals in Figure 27.9 to clearly show the defects as well as the structural outlines or edges. Here, the resulting imagery highlights the neatness of the fabricated nanostructures, which underscores the precision of the EBL process. To verify the chemical cleanliness of the sample, ToF-SIMS techniques are employed, as described in the next analysis.

ALGORITHM 1

Edge Detection Algorithm (Note: *i, j, m,* and *n* are computational parameters)

```
image ← FETCH[micrograph.png]
function EDGE_DETECTION(image)
                                        ▷ Apply Gaussian blur to reduce noise
    blurred ← GAUSSIANBLUR
      (image, (i, j), k)
                                        ▷ Apply Canny edge detector
    edges ← CANNY(blurred, m, n)
    return edges
end function
```

27.4.2 ToF-SIMS Analysis

To analyze the surface chemistry of the nanoengraved pictures displayed in Section 27.3, the ToF-SIMS method described in Section 27.2.2 is applied. This involves the analysis of surface ion images, ion spectra, and depth profile (which in turn includes the 3D ion models) of the resulting nanostructured surface using the ToF-SIMS instrument depicted in Figure 27.10. These characterization procedures are consecutively used to analyze the EBL-fabricated nanochip onto which our image is engraved.

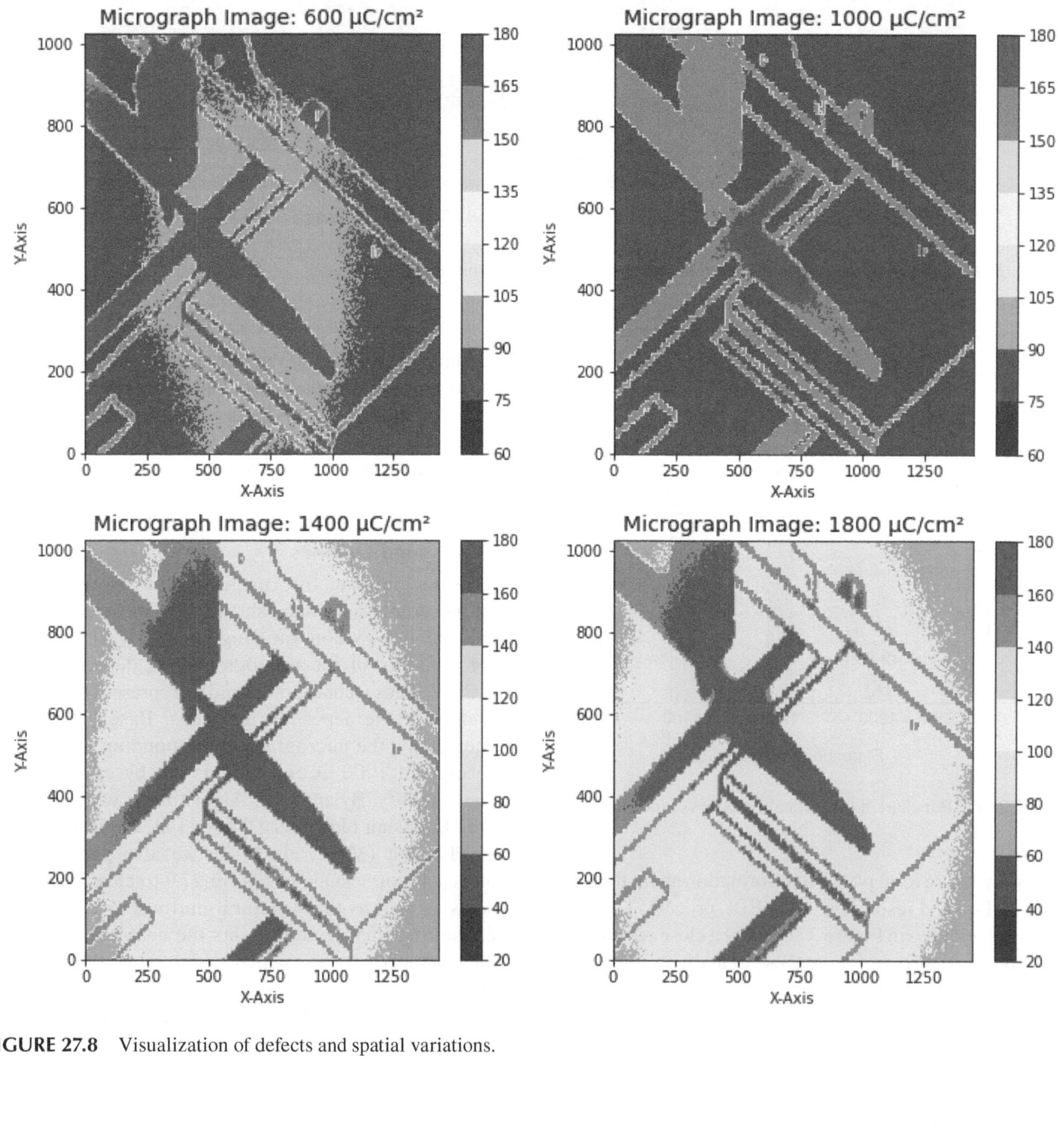

FIGURE 27.8 Visualization of defects and spatial variations.

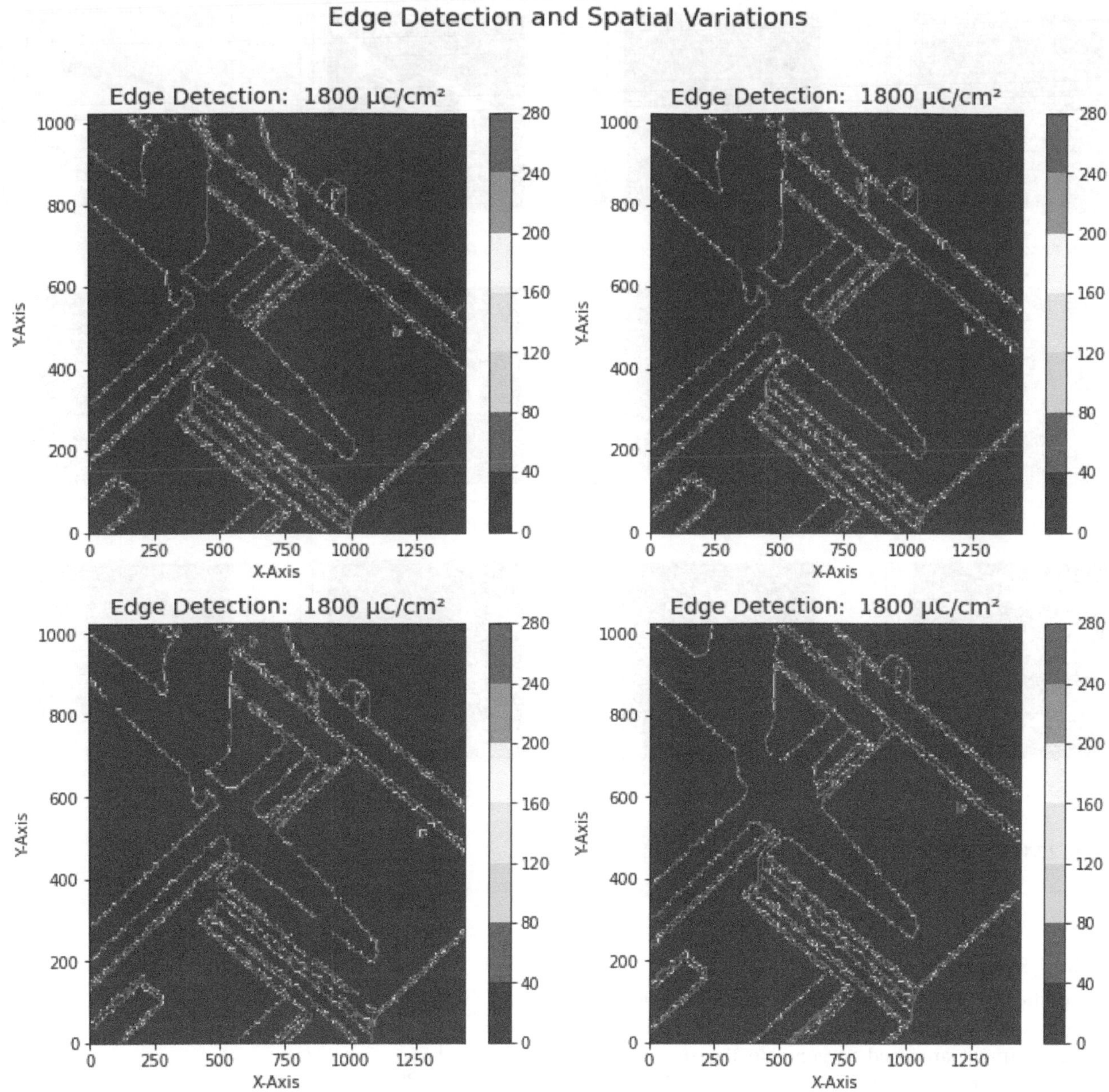

FIGURE 27.9 Visualization of defects and outlines.

FIGURE 27.10 Adopted ToF-SIMS instrument.

Courtesy: **IonToF GmBH, with copyright permission.**

27.4.2.1 Analysis of Surface Ion Images

To just identify the chemical elements and molecules present on the surface of the sample, the sputtered secondary ions are selectively detected and used to generate 2D images of the nanostructure on the sample. Roughly, the clarity and resolution of the generated pixels are functions of the concentration of the chemical element or molecule used in generating the image. Based on the ion images in Figure 27.11, the clearest images of the nanostructures are generated with Al^{+3} and Si^{+4}, which are the predominant elements in the samples, and thus form the negative image color of each other. At a glance, this imagery also reveals the presence of trace ions and radicals such as Na^{+}, K^{+}, Mg^{2+}, Ca^{+2}, Cr^{+2}, $C_2H_30_2^+$, $C_4H_30_2^+$, and $C_4H_30^+$.

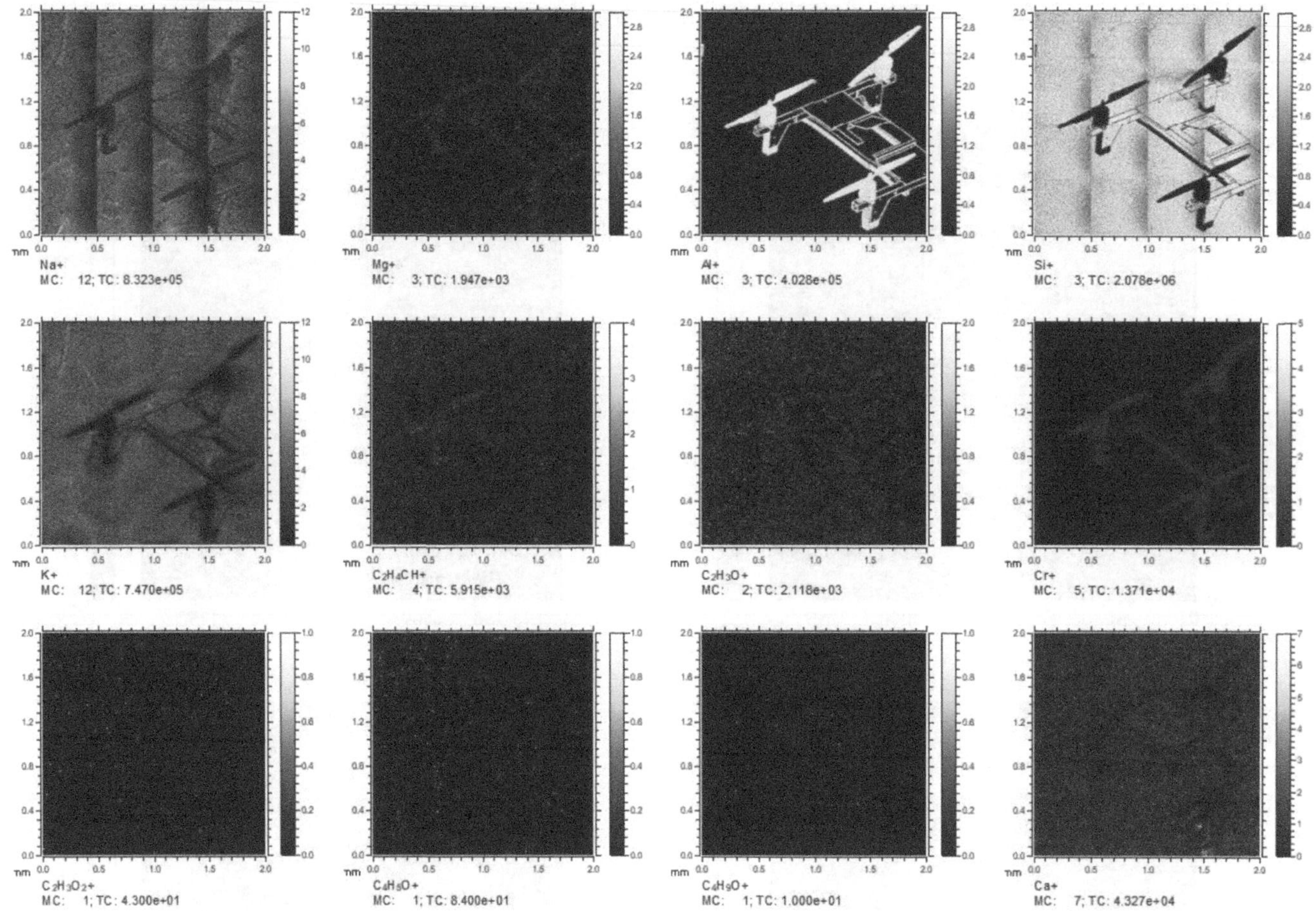

FIGURE 27.11 Surface ion imagery.

27.4.2.2 Surface Ion Mass Spectrometry

To quantitatively target and identify chemical elements and compounds present in the sample, data from the ToF-SIMS experiment are used to generate the spectral plot of secondary ions and radicals that are ejected from its nanostructured surface, and the result of this is presented in Figure 27.12. Here, bismuth cations are used as the primary ions, which enable nondestructive sampling of the nanostructured surface. In spite of the associated detriments due to the matrix effect, this analysis can still provide information about the concentration of each chemical species on the nanostructured surface. This also identified most of the ions and radicals that were already identified by means of ion imaging.

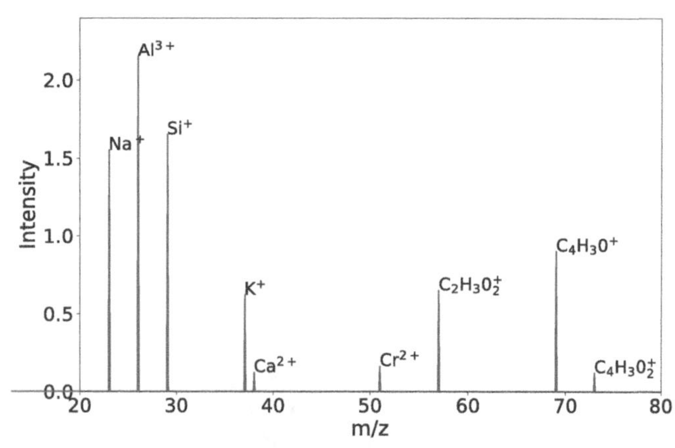

FIGURE 27.12 Plot of surface ion mass spectra.

27.4.2.3 Analysis of Depth Profile

Using the ToF-SIMS instrument, depth profiling of the EBL-fabricated sample is performed. This is to gain some understanding of how the chemical composition of the sample varies with depth into its nano-strata and how well defined the structures of its elemental interfaces are. Here, an additional ion source (i.e., the sputter ion gun) is engaged. This dynamically sped up the sputtering rate of the secondary ion to expose further strata or layers of atoms and molecules of the nanostructure, thereby enabling the measurement of mass spectra as a function of

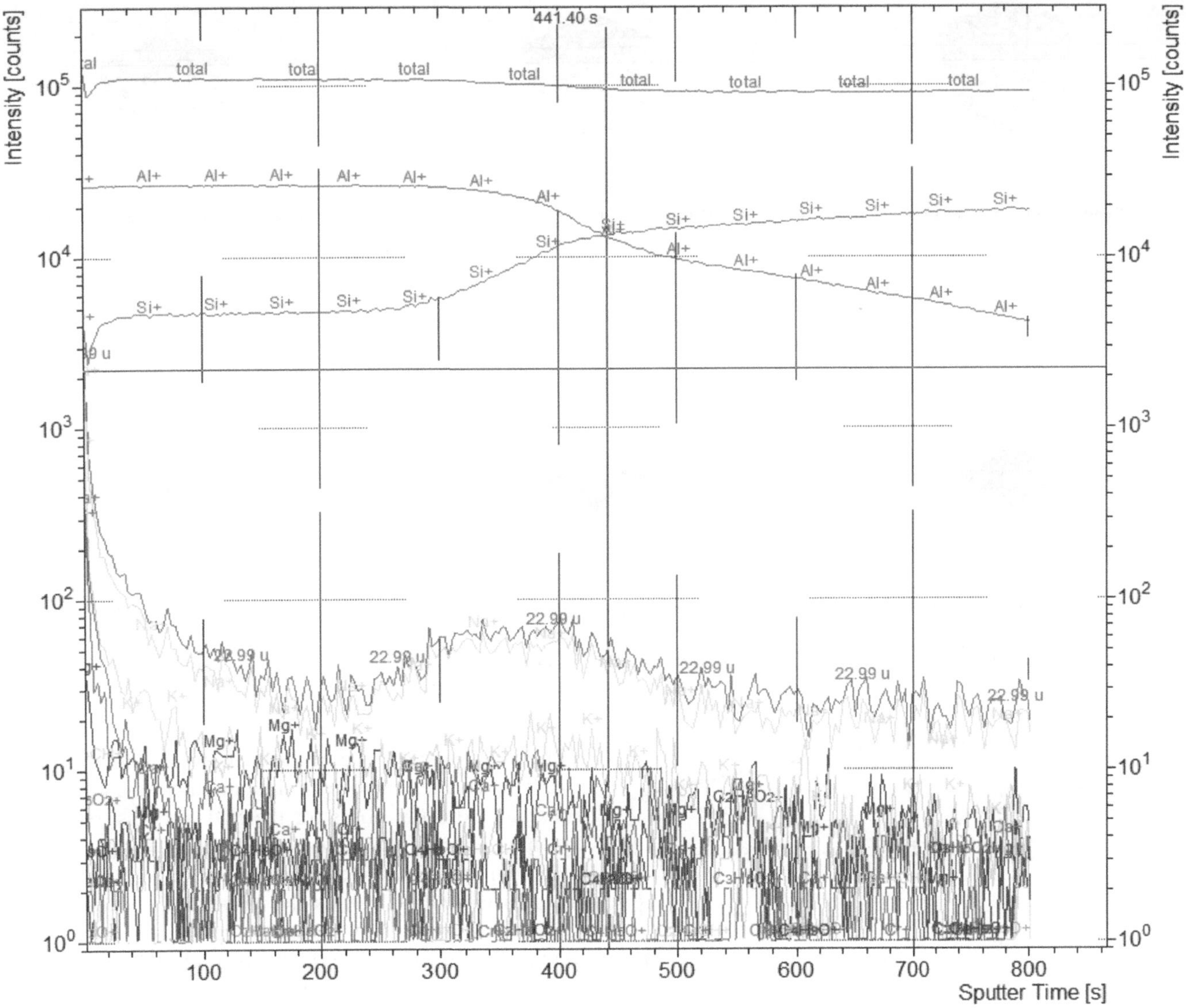

FIGURE 27.13 Depth profiling of samples.

depth. In the present experiment, oxygen ions are used as the additional primary ion that is produced by the sputter ion gun. This amplifies the destructive effect of secondary ion sputtering.

The resulting mass spectra across the depth of our sample are shown in Figure 27.13. This also reveals the presence of the aforementioned trace molecules and metal atoms at the surface of the nanostructure, particularly aluminum and silicon (i.e., the deposited material and substrate, respectively). By further analysis of deeper chemical profiles of the sample, a widespread layer of aluminum is observed, which continuously diminishes as the underlying silicon substrate atoms become more detectable. The 3D modeling of this evolution is captured in Figure 27.14. In addition, Figure 27.15 depicts the distribution of sodium (a significant contaminant) at several layers of the nanostructure. These sodium atoms (and other trace metals) may have been intrinsically present as impurities in the deposited aluminum or the substrate. Possible sources of organic contaminants include accidental handling of the samples by bare hand during the micrographic examination or cleaning with acetone (an organic solvent). The ToF-SIMS analysis of the sample reveals that it is relatively clean, thus complementing the micrographical and DIP-based characterization and quality assessment of the nanostructured materials.

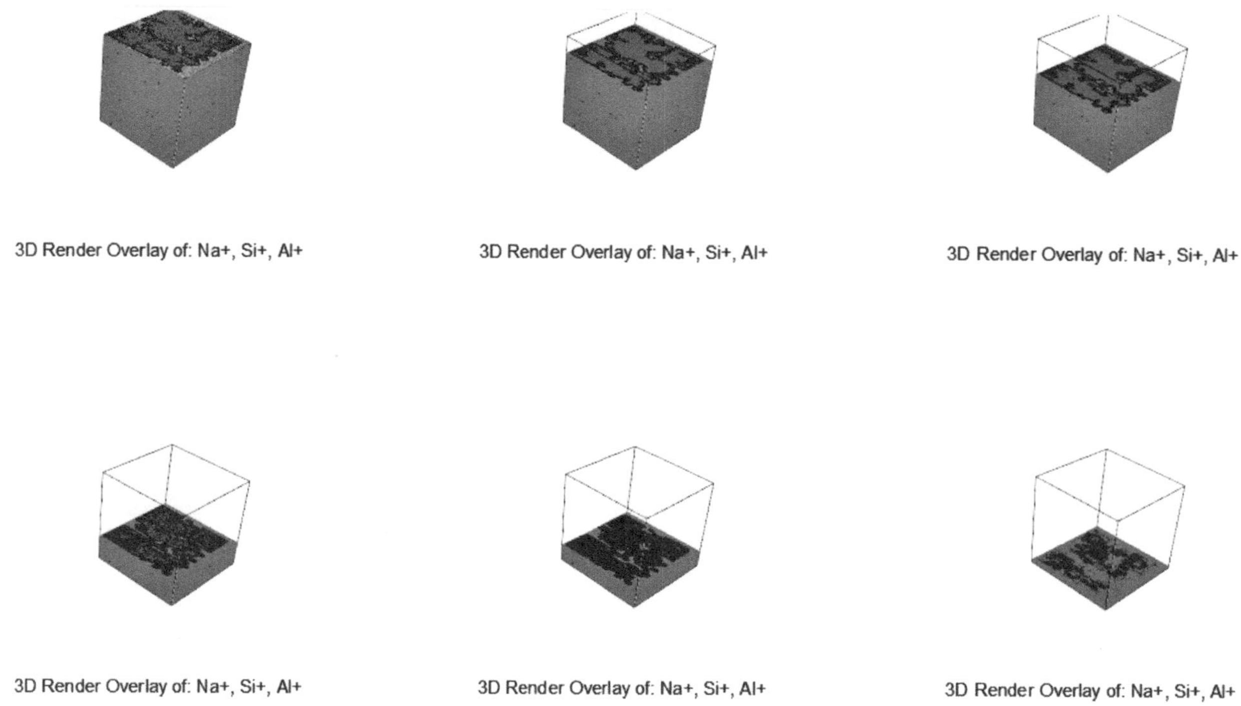

FIGURE 27.14 3D model of the depth profile of Al-atoms (red) and Si-atoms (blue).

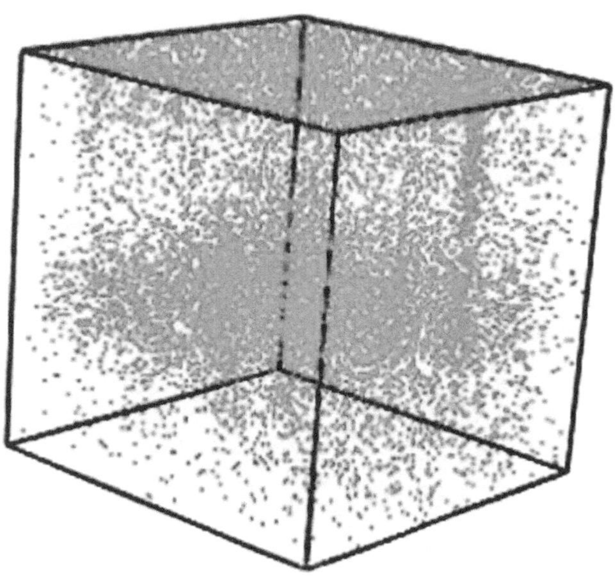

FIGURE 27.15 3D render overlay of Na^+ (green).

27.5 CONCLUSION

Nanotechnology as an attempt to re-engineer and control the microphysical world is opening new frontiers in research and industry, the potential of which has yet to be fully exploited. This work is a treatise on our glimpses into the tiny world of wonders. Hence, practical metrics on spin-coating and the use of the EBL system are exemplified. This enabled us to optimize the thickness of sample films and exposure doses for the EBL process, based on which we engraved an image onto a chip. This gave a full understanding of the technical rudiments and physics of nanofabrication, as well as the associated chemical processes. Various instruments and computational techniques were used to characterize the nanostructured image. This gave insight into why nanodevices could fail, which may arise from inappropriate configurations of fabrication equipment, such as the spin-coater, the EBL machine, and the sputtering system. In terms of contribution to knowledge, we have demonstrated the applicability and robustness of visual computing as a way to assess the quality of micrographs and images of nanostructures as opposed to mere visual inspection by a human investigator, which is difficult to quantify and prone to poor judgment as a result.

Through the EBL experiment, we fulfilled Feynman's dream of writing information onto a chip, but on a lesser scale, by engraving a picture as a nanochip. The results suggest that our exposure parameter (of 1000 μCcm^{-2}) and the adopted analytical methods (i.e., microscopy, BRISQUE-based image quality assessment, Gaussian blur filter, Canny edge detection algorithm, and ToF-SIMS) are optimal and reliable for the demonstrated nanotechnology-based imaging and segmentation. The ToF-SIMS analysis provided insight into the material variation and chemical nature of the nanostructured surface. Our work is an innovative blend of art, science, and technology that may be used to inspire young researchers, instructors, and students about the field of nanotechnology. Overall, the utilization of EBL, DIP, and ToF-SIMS highlights the interdisciplinary nature of our approach.

ACKNOWLEDGMENTS

The samples were prepared and characterized at the Center for Nanotechnology, Münster, Germany.

REFERENCES

AG, W. J. (1949). *Halbleiterverstärker Priority Filing on 14 April 1949, Published on 15 May 1952*. DE 833366 W. Jacobi/ SIEMENS AG.

Benyeogor, M. S., Olakanmi, O. O., Nnoli, K. P., Odeyemi, K. O., and Gratton, E. J. (2022). Supplementary material: Optimization of low-weight cargo UAV with real-time controller by CAD design, FEM simulation and dynamic modeling. https://doi.org/10.21227/7gkv-3h45. IEEE Dataport.

Carbaugh, D. J., Wright, J. T., Parthiban, R., and Rahman, F. (2015). Photolithography with polymethyl methacrylate (PMMA). *Semiconductor Science and Technology*, 31(2):025010.

Chow, L. S., and Rajagopal, H. (2017). Modified-BRISQUE as no reference image quality assessment for structural MR images. *Magnetic Resonance Imaging*, 43:74–87.

Dragoman, M., and Dragoman, D. (2005). Microelectromechanical systems. In Bassani, F., Liedl, G. L., and Wyder, P., editors, *Encyclopedia of Condensed Matter Physics*, pages 415–423. Elsevier, Oxford.

Fearn, S. (2015). Introduction. In *An Introduction to Time-of-Flight Secondary Ion Mass Spectrometry (ToF-SIMS) and Its Application to Materials Science, 2053–2571*, pages 1–1 to 1–6. Morgan Claypool Publishers.

Groves, T. (2014). 3—electron beam lithography. In Feldman, M., editor, *Nanolithography*, pages 80–115. Woodhead Publishing.

Hanapiah, M. M. A., Sin, Y. K., and Ibrahim, K. (2008). The effect of dose exposure in electron beam lithography. *AIP Conference Proceedings*, 1017(1):99–103.

Junk, A., and Riess, F. (2006). From an idea to a vision: There's plenty of room at the bottom. *American Journal of Physics*, 74(9):825–830.

Ma, X., Li, B., Zhang, Y., and Yan, M. (2012). The canny edge detection and its improvement. In Lei, J., Wang, F. L., Deng, H., and Miao, D., editors, *Artificial Intelligence and Computational Intelligence*, pages 50–58, Berlin, Heidelberg. Springer Berlin Heidelberg.

McCord, M., & Rooks, M. (1997). Electron beam lithography. In P. Rai-Choudhury (Ed.), *Handbook of microlithography, micromachining, and microfabrication* Vol. 1. Microlithography (Chap. 2). SPIE Press. https://doi.org/10.1117/3.2265070

Mittal, A., Moorthy, A. K., and Bovik, A. C. (2012). No-reference image quality assessment in the spatial domain. *IEEE Transactions on Image Processing*, 21(12):4695–4708.

Priebe, A., Xie, T., Bürki, G., Pethö, L., and Michler, J. (2020). The matrix effect in tof-sims analysis of two-element inorganic thin films. *Journal of Analytical Atomic Spectrometry*, 35:1156–1166.

Quate, C. F. (10 Dec. 2009). *Scanning Tunneling Microscope*. https://www.britannica.com/technology/scassnning-tunneling-microscope. Accessed 29 August 2021.

Saad, M. A., Bovik, A. C., and Charrier, C. (2012). Blind image quality assessment: A natural scene statistics approach in the DCT domain. *IEEE Transactions on Image Processing*, 21(8):3339–3352.

Seah, M., and Shard, A. (2018). The matrix effect in secondary ion mass spectrometry. *Applied Surface Science*, 439:605–611.

Wang, M., Zheng, S., Li, X., and Qin, X. (2014). A new image denoising method based on Gaussian filter. In *2014 International Conference on Information Science, Electronics and Electrical Engineering, IEEE*, volume 1, pages 163–167. DOI: 10.1109/ InfoSEEE.2014.6948089.

Index

For Product Safety Concerns and Information please contact our
EU representative GPSR@taylorandfrancis.com Taylor & Francis
Verlag GmbH, Kaufingerstraße 24, 80331 München, Germany